DISCRETE MATHEMATICS

Books by the Same Author

- *Operations Research: Theory and Applications (4nd Edition)*
- *Operations Research: Problems and Solution (3nd Edition)*
- *Linear Programming: Theory and Applications*
- *Quantitative Techniques for Managerial Decisions*
- *Quantitative Methods*
- *Management of Systems*

DISCRETE MATHEMATICS

FIFTH EDITION

Dr J K Sharma
Professor, Amity Business School
Amity University, Noida

TRIUNITY™
PRESS
(An Imprint of Laxmi Publications Pvt. Ltd.)
An ISO 9001:2015 Company
BENGALURU • CHENNAI • CGUWAHATI • HYDERABAD
KOCHI • KOLKATA • LUCKNOW • RANCHI
NEW DELHI

DISCRETE MATHEMATICS

Printed and bound in India

First Published : 2003 (*Reprinted 2 times*), *Second Edition* : 2005 (*Reprinted 5 times*)
Third Edition : 2011, *Reprinted* : 2013, *Fourth Edition* : 2015, *Reprinted* : 2018,
Fifth Edition : 2023; *Edition* : 2026

ISBN : 978-93-5138-143-3

Branches

Bengaluru	080-26 75 69 30
Chennai	044-24 34 47 26
Guwahati	0361-254 36 69
Hyderabad	040-27 55 53 83
Kochi	0484-405 13 03
Kolkata	033-40 04 77 79
Lucknow	0522-430 36 13
Ranchi	0651-224 24 64

PUBLISHED IN INDIA BY

TRIUNITY PRESS
(An Imprint of Laxmi Publications Pvt. Ltd.)

An ISO 9001:2015 Company
113, GOLDEN HOUSE, GURUDWARA ROAD, DARYAGANJ,
NEW DELHI - 110002, INDIA

Corporate Office : B-33, Sector-5, Noida,
Gautambuddha Nagar (U.P.) - 201 301, INDIA
Telephone : +91-120-44 46 900
www.laxmipublications.com info@laxmipublications.com

C—0000/026/02
Printed at : Ajit Printing Press, Delhi.

Preface to the Fifth Edition

This edition of *Discrete Mathematics* is an outcome of continuous feedback received from its ``readers—students, faculty members, and researchers. The first edition of this book was written with following objectives in mind which remain unchanged even in this edition:

- To demonstrate readers that this branch of mathematics is simple to understand, relevant in use and interesting to learn.
- To explain the concepts and theories of Discrete Mathematics in a simple and lucid way so that readers could memorise them easily and be able to use them in practice.
- To provide a book that has a comprehensive coverage for students of MSc(Maths), MTech, MCA, BTech/BE and BCA students and other similar programmes.

Most chapters have been revised and updated with new text material to respond to its readers feedback. A number of solved examples have been included to make text material more illustrative. Particular care has been taken to ensure the correctness of equations, formulae and solved problems.

I express my gratitude to all my colleagues from Management Schools, Engineering Institutes and Universities in India for adopting the book; for making suggestions for improvement in the text material and for extending their support and encouragement

While writing this book I have benefited immensely by referring too many books and publications. I express my gratitude to all such authors, publishers; many of them are listed in the references. If any such source has been left out inadvertently, I seek their pardon.

I would like to thank the editorial and production team of the publishers for the efficient and throughly professional way in which the whole project was managed. My wife and children also deserve sincere thanks for the unflagging support and encouragement they gave me while I worked on this book.

All efforts have made to make this text both pedagogically sound and error free. However, I retain the responsibility of any kind of errors in the book. Suggestions to improve contents of the book are always welcome and will be appreciated and acknowledged.

J K Sharma

Preface to the First Edition

Discrete Mathematics, the study of finite mathematical systems, is a hybrid subject. The topics covered in this book have been chosen keeping in view the knowledge required to understand the functioning of the digital computer because many of its properties can be understood and interpreted within the frame work of finite mathematical systems. This book provides reasonably rigorous and compact treatment of major mathematical techniques available for study in theoretical computer science. An attempt has been made to cover elementary to advance concepts in each chapter to take care of the needs of students who may earlier not have knowledge of these topics.

Each chapter contains sufficiently large number of solved examples and illustrations to explain definitions, principles, theorems and other descriptive material. This is followed by sets of *Self Practice Problems* with *Hints and Answers* to stimulate further learning and interest.

The book is designed to be self-contained and comprises of 13 chapters. It can be used by students of BSc (Maths), BCA/BIS/BIT and BE/BTech as an introduction to the fundamental concepts of discrete mathematics and by students of MSc, MCA, MTech for the development of more advanced mathematical concepts required in computer science applications.

—**J K Sharma**

List of Symbols

a, b, c, x, y, z	elements of a set
$\in$	belongs to
$\notin$	do not belong to
I	set of all integers
I_+	set of all positive integers
N	set of real numbers
Q	set of rational numbers
Φ	empty (null) set
$\sim$ or $\equiv$	equivalent sets
$\subseteq$	subset or is contained in
P(A)	power set or set of all sets
$A \Rightarrow B$	A implies B
$A \Leftrightarrow B$	A is equivalent to B
U	universal set
$A \cup B$	union of sets A and B
$A \cap B$	intersection of sets A and B
$A - B$	A difference B
$\overline{A}$ or A^c	complement of set A
$A \oplus B$	Symmetric difference of sets A and B
$n(A)$ or $\|A\|$	Cardinality of set A
$A \times B$	cartesian product of A and B
$\prod_{i=1}^{n} A_i$	cartesian product of n sets
xRy	x is related to y
M_R	adjacency or relation matrix
R^{-1}	inverse relation
I_A	identity relation
A/R	partition of set A determined by equivalence relation R on A
$[a]$ or A_a	equivalence class of a
$f: A \to B$	a function f from the set A to set B
$f(x)$	image of x under f
f^{-1}	f inverse
$f^{-1}(x)$	inverse image of x
gof	composition of function f with function g
$\lceil x \rceil$	ceiling function, the smallest integer $\geq x$
$\lfloor x \rfloor$	floor function, the largest integer $\leq x$
f_A	characteristic function of a set A
$a \equiv b \pmod{m}$	a congruent to b modulo m
GCD (a, b)	greatest common divisor of a and b
LCM (a, b)	least common divisor of a and b
$n!$	n-factorial
^nP_r	number of permutations of n objects taken r at a time
^nC_r	number of combinations of n objects taken r at a time
P (A)	probability of the event A
$\overline{A}$	complement of an event A
$P(A \cup B)$	probability of occurrence of event A or B

$P(A \cap B)$	probability of occurrence of event A and B
$P(A/B)$	conditional probability of event A given that event B has occurred
Δ	forward or descending difference operator
E	shift operator
$G(a\ z)$	generating function of sequence of numbers a
$G(V\ E)$	graph with vertices V and edges E
deg (V)	degree of vertex V
indeg (V)	indegree of the vertex
outdeg (V)	outdegree of the vertex
$d(u\ v)$	distance between vertices u and v
diam (b)	diameter the maximum distance between two vertices
$E = \Phi$	null graph
K_n	complete graph with n vertices
K_{mn}	bipartite graph
M_{ij}	incidence matrix
c (G)	chromatic number of G
$*$	a binary operation
$a * b$	image of the pair (a b)
S/R	quotient semigroup of a semigroup S
(G $*$)	a group under the operation $*$
$+_m$	addition modulo m
$\times_p$	multiplication modulo p
$(R + \cdot)$	a ring under the operation + and ·
$(D + *)$	an integral domain under the operator + and $*$
$(F + \cdot)$	a field under the operation + and ·
glb	greatest lower bound
lub	least upper bound
$a \wedge b$	'a meet b' least upper bound of a and b
$a \vee b$	'a join b' greatest lower bound of a and b
$(L \wedge \vee)$	a lattice L under the operation $\wedge$ and $\vee$
D_n	set of positive integers that are divisors of n
$\bar{x}$	complement of an element x
$\sim p$	not p (negation operation)
$p \wedge q$	p and q (conjunction operation)
$p \vee q$	p or q (disjunction operation
$p \Rightarrow q$	p implies q a conditional operation
$p \Leftrightarrow q$	p is equivalent to q a biconditional operation
$p \uparrow q$	negation after ANDing statements p and q
$p \downarrow q$	negation after ORing statements p and q
$p \oplus q$	XORing of statements p and q
$\forall$	for all or every (universal quantifier)
$\exists$	there exist for some (existential quantifier)
$[A + \cdot -]$	a mathematical system with operation + · and –
1	tautology
0	contradiction
$[A \wedge \vee \sim]$	a Boolean algebra with operation $\vee$ (and) $\wedge$ (or) and $\sim$ (NOT)

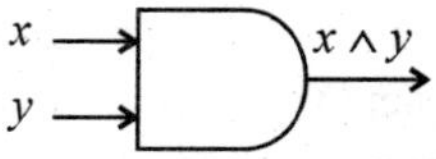

AND gate, 304

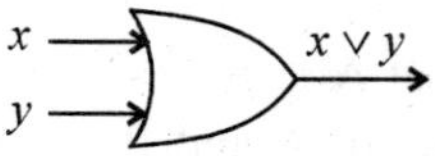

OR gate, 304

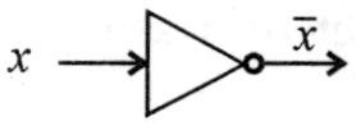

NOT (inverter) gate, 305

Contents

Chapter 9 Graph Theory 258–312

Chapter 10 Trees 313–345

Chapter 1

Set Theory

PREVIEW

This chapter introduces some of the basic tools of discrete mathematics – beginning with sets, subsets, and their operations, next it deals with algebra of sets and duality along with computer representation of sets, index/indexed sets, misets, partition of sets, etc. Concepts of fuzzy sets along with their operations have also been introduced. This gives readers the background needed to being with exploration of mathematical structures.

LEARNING OBJECTIVES

After studying this chapter, you should be able to

- use the notation of set theory
- find the union, intersection, difference, complement, and cartesian product of sets.
- identify binary and unary operations on sets
- prove set identities
- apply the multiplication principle and addition principle to solve counting problems.
- represent sets in computer language.
- distinguish between fuzzy sets and normal sets along with their operations.

1.1 INTRODUCTION

This chapter introduces the notations and terminology of set theory which was originated in the 1895 by the German mathematician G. Cantor. In our daily lives, we often use phrases or words such as a *bunch* of keys, a *pack* of cards, a *class* of students, a *team* of players, etc. The words bunch, pack, class and team all denote a collection of several discrete objects. Also, the dictionary meaning of *set*, is *a group* or *a collection* of distinct, definite and distinguishable objects selected by means of some rules or description.

In this chapter, we shall elaborate and discuss the importance of this word in the application of mathematics, computer science and engineering to real-life problems. Venn diagram representation of sets can be applied to understand relationship between set theory and logical arguments in Boolean algebra. Moreover, this concept simplify, clarify and unify many other mathematical concepts.

1.2 SET AND ITS ELEMENTS

Definition A set is any well-defined collection or class of distinct objects.

By a *well-defined* collection, we mean that there exists a rule with the help of which we should be able to say whether any given object or entity belongs to or not to the collection under consideration. The following are some of the examples of the set:

(*i*) The even positive integers not greater than 8
(*ii*) Rivers in India
(*iii*) The planets: Sun, Moon and Earth
(*iv*) The set of all points in a plane
(*v*) Students who speak either Hindi or English
(*vi*) Countries in the world
(*vii*) The vowels in English alphabet
(*viii*) Students in a college who specialise in computer science.

1.2.1 Elements of a Set

The objects in a set are called its *elements* or *members*. The elements in the set must be *distinct* and *distinguishable*. By 'distinct', we mean that no element is repeated and by 'distinguishable', we mean that given any object, it is either in the set or not in the set.

Generally, capital letters A, B, C, X, Y, Z, etc., are used to denote sets and its elements by lower case letters a, b, c, x, y, z, etc.

The symbol $\in$ (*epsilon* – a Greek alphabet) is used to indicate, *belongs to* (a member or elements of). If x is an element of a set A, then symbolically we write $x \in$ A, which is read as 'x is an element of the set A' or 'x belongs to the set A'.

The symbol $\notin$ is used to indicate, *does not belong to*. Thus, if x is not an element of the set A, then we write $x \notin$ A, which is read as 'x is not an element of the set A'.

1.2.2 Standard Sets and Symbols

The choice of the name for a set is much like the choice of an identifier name in programming. There are few sets of numbers that we will use frequently in the text and so we use standard symbols for them, such as:

N = {1, 2, 3, 4, . . .}, the set of natural numbers.

I = {. . ., – 3, – 2, – 1, 0, 1, 2, 3, . . .}, the set of integers.

I_+ = {1, 2, 3, . . .}, the set of positive integers.

$$Q = \left\{x : x = \frac{p}{q}, \text{ where } p \text{ and } q \text{ are integers and } q \neq 0\right\}, \text{ the set of rational numbers.}$$

Q_+ = the set of positive rational numbers.

R = the set of real numbers.

$C = \{x : x = a + ib;\ a, b \in R,\ i = \sqrt{-1}\}$, the set of complex numbers.

P = the set of prime numbers.

Remark

1. Before we go further into details, let us make a clear difference between a 'collection' and a 'set'. Every collection is not a set. For a collection to be a set, it must be *well-defined.* For a collection to be well-defined it should be known whether a given object does or does not belong to that collection. For example, consider the following collection:

   A = the collection of any five positive integers.

   B = the collection of first five even positive integers.

   The collection A is not well-defined because the positive integer 4 (say) may or may not belong to A. On the other hand, collection B is well-defined because the members of the collection are 2, 4, 6, 8, 10 and, therefore, the collection B is a set.
2. Collection of qualitative attributes in certain objects such as honest persons, rich persons, beautiful women, good players, etc., do not form sets.
3. $a, b \in A$ is an abbreviated way of saying $a \in A$ and $b \in A$. Commas in mathematics are read as 'and'.

Illustrations

1. V = set of vowels in the English alphabet. Then $e \in V$, but $x \notin V$.
2. P = set of the Presidents of India. Then Morarji Desai $\notin$ P, Neelam Sanjiva Reddy $\in$ P.

1.3 SET DESCRIPTION

The way of describing a set and its elements vary depending on the specific set. The most common methods of describing the elements of a set are as follows.

1.3.1 Roster Method

In this method, elements of a set are described by enumerating (listing or writing) them exhaustively within braces. For example, the elements of the set V of all the vowels in the English alphabet can be represented as:

$$V = \{a, e, i, o, u\}$$

Sometimes, it is not possible to list all the elements, but after knowing a few elements we can understand as to what the other elements are. For example,

N = {1, 2, 3, . . .}, the set of all natural numbers.

A = {a, b, c, . . .}, the set of letters of the English alphabet.

S = {1, 9, 25, 49, . . .}, the set of squares of positive odd integers.

Roster method is also known as *Tabular* or *Enumeration Method.*

Remark No element in the set should be repeated. For example, we never write {3, 5, 5}, but write {3, 5}. Since a set is a collection of objects in a well-defined manner, the order in which its elements are listed does not make any difference. The three sets {2, 4, 3}, {4, 2, 3} and {3, 2, 4} are different listings of the same set. The elements distinguish the sets but not the order in which these are written.

1.3.2 Set Builder Method

In this method, the elements of a set are represented by stating a property (qualitative, quantitative or both) that uniquely characterise them. To understand this method, let us consider the sets:

(*a*) A = {2, 4, 6, 8, . . .} (*b*) B = {a, e, i, o, u}

If we try to search out some property among the elements of sets A and B, we find that all elements of the set A are even positive integers; all elements of the set B are the vowels of the English alphabet. Thus, we can use the letter 'x' to represent an arbitary element of the set together with the property of x. Therefore, sets A and B may be represented as:

A = {2, 4, 6, 8, . . .} = {x : x is an even positive integer}

B = {a, e, i, o, u} = {x : x is a vowel in English alphabet}

where the colon (:) is read as: *such that* or *where* and is used interchangeably with a vertical line. The symbol {x : . . .} is read 'the set of all x such that . . . .' The property follows ':' help us determine the elements of the set that is being described.

Set builder method is also known as *set selection method.*

1.3.3 Cardinal Number (Dimension or Order) of a Set

The number of distinct elements in a set A is also called the cardinal number of the set and is denoted as n(A) or O(A) or | A |.

(a) If n(A) = n(B), A and B are said to be equivalent sets.

(b) If A = {3, 5, 7, 9, 11}, then the cardinal number of this set is 5 because there are only five elements in this set and in written as: n(A) = | A | = 5.

(c) {0} and {ϕ} are equivalent sets each with cardinality equal to one.

1.4 TYPES OF SETS

Finite and infinite sets A set is *finite* if it contains finite numbers of different elements. For example,

(*i*) The set of months in a year.

(*ii*) The set of rivers in India.

(*iii*) The set of students in a class.

(*iv*) The set of vowels in English alphabets.

A set which contains infinite number of elements is known as *infinite* set. For example,

N = {1, 2, 3, 4, . . . }, the set of natural numbers.

I = {. . ., – 3, – 2, – 1, 0, 1, 2, 3, . . .}, the set of integers.

A = $\{x : x$ is the set of points in the Euclidean planes$\}$.

Singleton set A set which contains only one element is called a *singleton* (or *unit*) set. For example,

A = $\{x : 4 < x < 6, x$ is an integer$\}$

N = {5}

I = set of all those negative integers whose square is 9,

= $\{x : x^2 = 9$ and x is a negative integer$\}$

Null set A set which contains no element is called an *empty (null* or *void)* set. This set is denoted by a Greek letter Φ (read as *phi*) or { }. For example,

(*i*) Φ = set of all those x which are not equal to x itself

= $\{x : x \neq x\}$

(*ii*) A = $\{x : x^2 + 4 = 0, x$ is real$\}$

(*iii*) Set of all integers whose square is 5

= $\{x : x^2 = 5$ and x is an integer$\}$

We note that $|\phi| = 0$ but $\{0\} \neq \phi$. Also $\phi \neq \{0\}$ because $\{\phi\}$ is a set with one element, namely, the null set. A set which is not a null set, i.e., which has at least one element is called *non-empty* set.

Equality of sets Two sets A and B are said to be equal if every element of A is an element of B and also every element of B is an element of A. The equality of two sets A and B is denoted by A = B. Symbolically,

$$A = B \text{ if and only if } x \in A \Leftrightarrow x \in B$$

This is known as *Axiom of Extension* or *Identity*. For example, if

(*a*) A = {4, 3, 2, 1} and B = {1, 3, 2, 4}

then A = B, because both have same and equal number of elements. Also, if

(*b*) A = {2, 3}, B = {3, 2} and C = $\{x : x^2 - 5x + 6 = 0\}$

then A = B = C, because each element which belongs to any one of the sets also belongs to the other two sets.

(*c*) Let A = set of all integers whose square is 25.

B = set of all roots of the equation, $x^2 - 25 = 0$

C = {– 5, + 5}

Then A = B = C.

Equivalent sets If the elements of one set can be put into one-to-one correspondence with the elements of another set, then the two sets are called *equivalent* sets.

In other words, two sets A and B are said to be equivalent sets if and only if there exists a one-to-one correspondence between their elements. By one-to-one correspondence we mean that for each element in A, there exists and match with one element in B and vice versa.

The symbol, ~ or ≡ is used to denote equivalent sets.

Illustrations

1. Let A = $\{a, b, c, d\}$ and B = {1, 2, 3, 4} be two sets. Clearly the set A is not equal to the set B. However, the elements of A can be put into one-to-one correspondence with those of B, therefore, we write, A ≡ B.
2. A = $\{x : x$ is a letter in the word BOAT}
   B = $\{x : x$ is a letter in the word CART}
   Thus A ≡ B.

Subset Let A and B be two non-empty sets. The set A is a *subset* of B (or A is contained in B) if and only if every element of A is an element of B. In other words, the set A is a subset of B if $x \in A \Rightarrow x \in B$. Symbolically, this relationship is written as:

$$A \subseteq B \text{ if } x \in A \Rightarrow x \in B$$

which is read as 'A is a subset of B' or 'A is contained in B'. If A ⊆ B, then B is called the *superset* of A and we write B ⊇ A which is read as 'B is a superset of A' or 'B contains A'.

If the set A is not a subset of the set B, i.e., if at least one element of A does not belong to B and we write, A ⊈ B. In other words, if $x \in A \Leftrightarrow x \notin B$ which is read as 'A is not a subset of B.'

Illustrations Suppose

1. A = {1, 3, 5}, B = {5, 4, 3, 2, 1}. Since each element of A is also the element of B, A ⊆ B.
2. The set of rational numbers is a subset of the set of real numbers. Finally, the set of natural numbers is a subset of the set of integers. Symbolically, N ⊆ I ⊆ Q ⊆ R.
3. A = {3, 5, 6}, B = {5, 3, 6}. Since each element of A belongs to B and vice versa, therefore, A = B. Thus, it follows that *every set is a subset of itself.*
4. The set, A = {1, 3, 5} is not the subset of the set, B = {1, 3, 6, 7}, since 5 ∈ A but 5 ∉ B.
5. Null set Φ is a subset of every set.

Properties of Subsets

1. If the set A is a subset of the set B, then the set B is called *superset* of the set A.
2. If the set A is a subset of the set B and the set B is a subset of the set A, then the sets A and B are said to be equal, i.e., A ⊆ B and B ⊆ A ⇒ A = B.
3. If the set A is a subset of the set B and the set B is a subset of C, then A is a subset of C, i.e., A ⊆ B and B ⊆ C ⇒ A ⊆ C.

Proper subset The set A is a proper subset of the set B or A is properly contained in B, if and only if

(*i*) every element of the set A is also an element of set B, i.e., A ⊆ B, and

(*ii*) there is at least one element in set B that is not in set A, i.e., A ≠ B.

If A is a proper subset of B, then we write, A ⊂ B. When we write A ⊂ B, then both possibilities (*i*) A is a proper subset of B, and (*ii*) A = B are included in it. In order to show that A ⊂ B we have to show that (*i*) A ⊆ B, and (*ii*) there exists at least one element of B that is not in A. Symbolically,

$$A \subset B \Leftrightarrow A \subseteq B \quad \text{and} \quad x \in B \quad \text{but} \quad x \notin A$$

and when A, B are finite

$$A \subseteq B \Rightarrow |A| \leq |B| \quad \text{and} \quad A \subset B \Rightarrow |A| < |B|.$$

Illustrations

1. Set of natural numbers, set of integers and set of rational numbers are the proper subsets of the set of real numbers.
2. Each set is an improper subset of itself, i.e., $A \subseteq A$, where A is any set.
3. Null set Φ, is unique and is the subset of every set.

Family of sets If the elements of a set are set themselves, then such a set is called the *family of sets*. The word *collection* and *class* are also used for a *set of sets*.

For example, if $A = \{a, b\}$, then the set $\{\Phi, \{a\}, \{b\}, \{a, b\}\}$ is the family of sets whose elements are subsets of the set A.

Remark It should be noted that $\in$ connects an element and a set while $\subseteq$ and $\subset$ connect two sets, i.e., if $A = \{a, b, c\}$ then $a \in A$, $\{a\} \subset A$ are correct statements, while $a \subset A$, $\{a\} \in A$ are wrong statements.

Power set If S is any set, then the family of all the subsets of S is called the power set of S and is denoted by P(S), i.e., $P(S) = \{A : A \subset S\}$. Obviously, Φ and S are both members of P(S).

Illustrations

1. Let $A = \{a\}$, then $P(A) = \{\Phi, \{a\}\}$
2. Let $A = \Phi$, then $P(A) = \{\Phi\}$
3. If $A = \{a, b\}$, then $P(A) = \{\Phi, \{a\}, \{b\}, \{a, b\}\}$

Remark If A is a finite set of n elements, then the total number of subsets of A are 2^n. In other words, the number of elements in P(A) is 2^n.

Number of element in power set If the number of elements in any set is n, then the number of elements in its power set is:

$$^nC_n + {}^nC_{n-1} + {}^nC_{n-2} + \dots + {}^nC_1 + 1 = 2^n.$$

Universal set In applications of set theory all the sets under discussion are assumed to be the subset of the fixed large set, called the *universal set* or *universal of discourse*. This set is usually denoted by letter U or E. Obviously, the set U is the superset of every set.

Illustrations

1. All the people in the world constitute the universal set in any study of human population.
2. A set of integers may be considered as a universal set for a set of odd or even integers
3. A set of CAs in India may be considered as a universal set for a set of fellows of CAs.
4. A set of all points in the plane form a universal set.

1.5 VENN-EULER DIAGRAMS

A device known as *Venn-Euler diagram* or simply *Venn diagram* is a pictorial representation of sets. In Venn diagrams, a universal set U is represented by the interior of a rectangle and each subset of U is represented by the circle inside the rectangle.

If the sets A and B are equal, then the same circle represents both A and B. If the sets A and B are disjoint, i.e., they have no elements in common, then circles representing A and B are drawn in such a way that they have no common area as shown in Fig. 1.1(*b*). However, if few elements are common both in A and B, then sets A and B are in general represented as in Fig. 1.1(*c*).

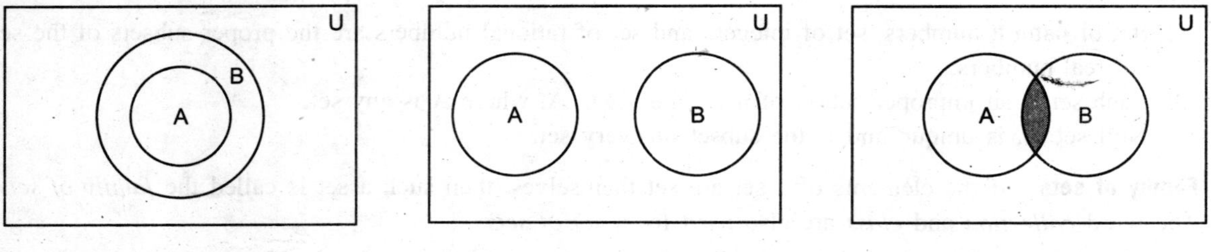

(a) $A \subseteq B$ (b) A and B are disjoint (c) $A \cap B$

Fig. 1.1 Venn Diagrams

Venn diagrams are used to illustrate the set relations such as the subset and the operations with sets. But these cannot be used to prove any statement(s) regarding sets.

Example 1 Which of the following sets are different? (a) Φ; (b) $\{0\}$; (c) $\{\Phi\}$.

Solution Each set is different from other, because

(a) Φ is a set containing no element, i.e., null set.

(b) $\{0\}$ is a set containing the single element zero, i.e., singleton set.

(c) $\{\Phi\}$ is a set containing one element, i.e., empty set Φ, or it is a set of set.

Example 2 Write the following set in tabular form.

$$A = \{a : a \in N, a < 10, a \text{ is odd}\}$$

Solution As per description of the nature of elements of the set A, it can be written as, $A = \{1, 3, 5, 7, 9\}$. The elements of A are the odd natural numbers less than 10.

Example 3 Are the following sets equal?

$$A = \{x^2 : x^2 - 3x = -2\}, \quad B = \{2, 1\} \text{ and } C = \{1, 2, 2, 1\}$$

Solution $A = \{x : x^2 - 3x = -2\} = \{x : x^2 - 3x + 2 = 0\}$

$= \{x : (x - 2)(x - 1) = 0\} = \{2, 1\}$

also $C = \{1, 2, 2, 1\} = \{1, 2\}$

Since elements of A, B and C are same and equal in number, $A = B = C$.

Example 4 Let $A = \{x : x = 5n + 2, n \in N\}$ and $B = \{x : x = 5n - 3, n \in N\}$. Find which of the following statements is correct and why?

(a) $A \subset B$; (b) $B \subset A$; (c) $A = B$; (d) None of these.

Solution Putting $n = 1, 2, 3, \ldots$, we list different elements of the sets as:

$$A = \{7, 12, 17, 22, \ldots\} \text{ and } B = \{2, 7, 12, 17, \ldots\}$$

Hence, $A \subset B$.

Self-practice Problems I

1. Define a set and give examples to illustrate the difference between a collection and a set. What are different ways to specify a set? Give examples.
2. Write down the following sets:
   (*i*) A = $\{x : x^2 = 4\}$ (*ii*) B $\{x : x^2 = 9, x - 3 = 5\}$
   (*iii*) C = $\{x : x^2 + 1 = 0, x$ is a complex number$\}$ (*iv*) D = $\{x : x^2 + 1 = 0; x$ is a real number$\}$
3. Prove that for all sets A, B and C if A ⊂ B, B ⊂ C, and C ⊂ A, then B = C.
4. Which of the following statements is true?
   (*i*) $\{a, b, c\} = \{c, b, a\}$. (*ii*) Every subset of an infinite set is infinite.
   (*iii*) The set {1, 2, 3, 4} has in all 32 subsets.
5. Which of the following sets are equal?
   (*i*) A = {1, 2, 2, 3} (*ii*) B = $\{x : x^2 - 2x + 1 = 0\}$ (*iii*) C = {1, 2, 3}
   (*iv*) D = $\{x : x^3 - 6x^2 + 11x - 6 = 0\}$ (*v*) E = {3, 2, 1}
6. Find out which of the following sets are null, singleton?
   (*i*) A = $\{x : x^2 = 11, x$ is an even integer$\}$ (*ii*) B = {0}
   (*iii*) C = $\{x : x + 6 = 6\}$ (*iv*) D = $\{x : x^2 = 7, 3x = 5\}$
7. Let A = $\{x : x^2 - 3x = -2\}$, B = {2, 1}, and C = {2, 3, 4}. Find x if
   (*i*) $x \in$ A, $x \notin$ B (*ii*) $x \in$ A, $x \notin$ C
8. List the elements of the following sets:
   (*i*) A = $\{x : x \in N, 5 < x < 14\}$ (*ii*) B = $\{x : x \in N, x$ is odd, $x < 21\}$
   (*iii*) C = $\{x : x \in N, 6 + x = 3\}$
9. Show that A = {2, 3, 5, 7} is not a subset of B = $\{x : x \in N, x$ is odd$\}$.
10. Show that A = {3, 5, 7, 8} is a proper subset of B = {1, 2, 3, . . ., 8, 9}.
11. Determine the power sets of the following sets: (*i*) $\{a\}$, (*ii*) $\{\{a\}\}$, (*iii*) {Φ, {Φ}}.

Hints and Answers

1. See the text
2. A = {2, – 2}, B = Φ, C = $\{i, -i\}$, D = Φ
4. (*i*) true (*ii*) false (*iii*) false
5. A = C = D = E
6. A = Φ, B and C are singleton, D = Φ
7. (*i*) $x = 0$, (*ii*) $x = 1$.
8. A = {6, 7, 8, . . ., 12, 13}
9. Since 2 ∈ A but 2 ∉ B because B consists of odd numbers, thefefore, A is not subset of B.
10. Each element A is also an element of B so A ⊆ B. On the other, 1, 2 ∈ B but 1, 2 ∉ A. Hence, A ≠ B. Therefore, A is the proper subset of B.

1.6 SET OPERATIONS AND LAWS OF SET THEORY

You are familiar with the mathematical operations such as addition, subtraction, multiplication and division on numbers. In this section we shall discuss certain operations on sets so as to yield new sets with the given sets. These operations play an important role in many applications of discrete mathematics.

1.6.1 Union of Sets

Let A and B be two non-empty sets. The union of A and B is the set of all elements which are either in A or in B or in both A and B. The union of A and B is denoted by $A \cup B$. It is usually read as *A union B*. Union is also known as *join* or *logical sum* of A and B. Symbolically,

$$A \cup B = \{x : x \in A \text{ or } x \in B\}$$

Here the word 'or' is used in the sense of representing 'and/or'. Figure 1.2 is a Venn diagram in which $A \cup B$ is shaded.

Illustrations

1. If $A = \{1, 2, 3, 4\}$ and $B = \{2, 6, 8\}$, then $A \cup B = \{1, 2, 3, 4, 6, 8\}$.
2. If $A = \{x : x \in I, 0 < x < 7\}$ and $B = \{y : y \in I, 8 < x < 12\}$, then $A \cup B = \{z : z \in I, 0 < z < 7; 8 < z < 12\}$ $= \{1, 2, 3, 4, 5, 6, 9, 10, 11\}$.
3. If A = set of positive real numbers, and B = set of negative real numbers, then $A \cup B$ = set of all real numbers except zero.
4. If A = set of all natural numbers, and B = set of all even natural numbers, then $A \cup B = A$, i.e., set of all natural numbers.

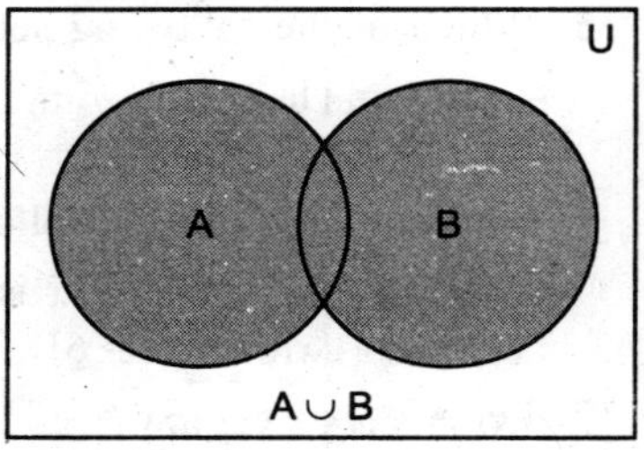

Fig. 1.2 Union of Sets A and B (Shaded Area)

Properties of Union of Sets

1. The union of sets is commutative If A and B are any two sets, then $A \cup B = B \cup A$.

Proof For proving $A \cup B = B \cup A$, we shall prove $A \cup B \subseteq B \cup A$ and $B \cup A \subseteq A \cup B$.

Let x be an arbitary element of $A \cup B$. Then

$$x \in A \cup B \Rightarrow x \in A \text{ or } x \in B \quad \text{[by definition of union]}$$
$$\Rightarrow x \in B \text{ or } x \in A$$
$$\Rightarrow x \in B \cup A$$
$$\therefore \quad A \cup B \subseteq B \cup A \quad ...(i)$$

Again, let $y \in B \cup A$. Then

$$y \in B \cup A \Rightarrow y \in B \text{ or } y \in A$$
$$\Rightarrow y \in A \text{ or } y \in B$$
$$\Rightarrow y \in A \cup B$$
$$\therefore \quad B \cup A \subseteq A \cup B \quad ...(ii)$$

From (*i*) and (*ii*), we have $A \cup B = B \cup A$.

Remark The results (*i*) and (*ii*) can be proved simultaneously in compact form also by using both ways implication sign $\Leftrightarrow$ (read as 'implies and implied by') as shown below.

$$x \in A \cup B \Leftrightarrow x \in A \text{ or } x \in B$$

$\Leftrightarrow x \in B$ or $x \in A$

$\Leftrightarrow x \in B \cup A$

proving thereby $A \cup B = B \cup A$.

We shall adopt this shorter and easier way to prove other results of above type.

2. The union of sets is associative If A, B and C are any three sets, then

$$A \cup (B \cup C) = (A \cup B) \cup C$$

Proof Let x be an arbitrary element of set $A \cup (B \cup C)$. Then

$x \in A \cup (B \cup C) \Leftrightarrow x \in A$ or $x \in (B \cup C)$

$\Leftrightarrow x \in A$ or ($x \in B$ or $x \in C$)

$\Leftrightarrow$ ($x \in A$ or $x \in B$) or $x \in C$

$\Leftrightarrow x \in (A \cup B)$ or $x \in C$

$\Leftrightarrow x \in (A \cup B) \cup C$

$\therefore \quad A \cup (B \cup C) = (A \cup B) \cup C.$

3. The union of sets is idempotent If A is any set, then $A \cup A = A$.

Proof Let x be an arbitrary element of the set $A \cup A$. Then

$x \in A \cup A \Leftrightarrow x \in A$ or $x \in A$

$\Leftrightarrow x \in A$

$\therefore \quad A \cup A = A.$

4. If A and B are any sets, then $A \subseteq A \cup B$ and $B \subseteq A \cup B$.

Proof Let x be an arbitrary element of the set A. Then

$x \in A \quad \Leftrightarrow x \in A$ or $x \in B$

$\Leftrightarrow x \in A \cup B$

$\therefore \quad A \subseteq A \cup B$. Similarly, it can be proved that $B \subseteq A \cup B$.

5. If A is any set, then $A \cup \Phi = A$, where Φ is the null set.

Proof Let x be an arbitrary element of the set $A \cup \Phi$. Then

$x \in A \cup \Phi \quad \Leftrightarrow x \in A$ or $x \in \Phi$

$\Leftrightarrow x \in A$ [by definition of null set, $x \notin \Phi$]

$\therefore \quad A \cup \Phi = A.$

6. If A is any subset of the universal set U, then $A \cup U = U$.

Proof We know that every set is a subset of the universal set. Consequently, $A \cup U \subseteq U$. Also by Property 4, $U \subseteq A \cup U$. Thus, $U = A \cup U$.

7. If $A \subseteq B$, then $A \cup B = B$ and if $B \subseteq A$ then $A \cup B = A$.

Proof From Property 4, we have

$$B \subseteq A \cup B \qquad \text{... }(i)$$

Now let $\quad x \in A \cup B \Rightarrow x \in A$ or $x \in B$

$\Rightarrow x \in B$, as $A \subseteq B$.

$$\therefore \quad A \cup B \subseteq B \qquad \text{...}(ii)$$

From (*i*) and (*ii*), we have $A \cup B = B$.

Similarly, it can be proved that if $B \subseteq A$, then $A \cup B = A$.

1.6.2 Intersection of Sets

Let A and B be two non-empty sets. The intersection of A and B is the set of all elements which are in both A and B. Intersection of A and B is denoted by A ∩ B. It is usually read as *A intersection B*. Symbolically,

$$A \cap B = \{x : x \in A \text{ and } x \in B\}$$

To find the intersection of two sets means finding elements common to A and B. Figure 1.3 is a Venn diagram in which A ∩ B is shaded.

Illustrations

1. If $A = \{1, 2, 3, 4, 5, 6, 7\}$, and $B = \{3, 5, 7, 9, 11\}$, then
   $A \cap B = \{3, 5, 7\}$.
2. If $A = \{x : x \in I, 0 < x < 5\}$ and $B = \{y : y \in I, 4 \le y < 7\}$, then
   $A \cap B = \{z : z \in I, 4 \le z < 5\} = \{4\}$.
3. If A = set of all natural numbers, and B = set of all odd natural numbers, then
   A ∩ B = B, i.e., set of all odd natural numbers.

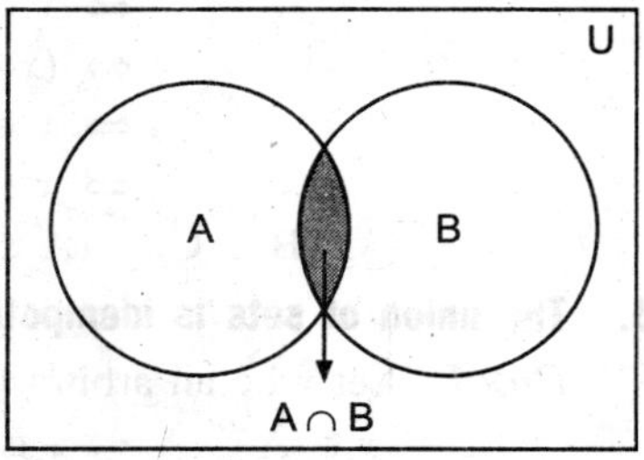

Fig. 1.3 Intersection of Sets A and B

Remarks For any two sets A and B, we have
(*i*) $A \cap B \subseteq A \cup B$, (*ii*) $A \cap B \subseteq A$, (*iii*) $A \subseteq A \cup B$

Properties of Intersection of Sets

1. The intersection of sets is commutative If A and B are any two sets, then $A \cap B = B \cap A$.

Proof Let $x \in A \cap B \Leftrightarrow x \in A \text{ and } x \in B$
$\Leftrightarrow x \in B \text{ and } x \in A.$
$\Leftrightarrow x \in B \cap A.$

Hence, $A \cap B = B \cap A$.

2. The intersection of sets is associative If A, B and C are any three sets, then

$$A \cap (B \cap C) = (A \cap B) \cap C$$

Proof Let x be an arbitrary element of $A \cap (B \cap C)$. Then
$x \in A \cap (B \cap C) \Leftrightarrow x \in A \text{ and } x \in (B \cap C)$
$\Leftrightarrow x \in A \text{ and } (x \in B \text{ and } x \in C)$
$\Leftrightarrow (x \in A \text{ and } x \in B) \text{ and } x \in C$
$\Leftrightarrow x \in (A \cap B) \cap C.$

Hence, $(A \cap B) \cap C = A \cap (B \cap C)$.

3. The intersection of sets is idempotent If A is any set, then $A \cap A = A$.

Proof Let x be an arbitrary element of the set $A \cap A$. Then
$x \in A \cap A \Leftrightarrow x \in A \text{ and } x \in A$
$\Leftrightarrow x \in A.$

Hence, $A \cap A = A$.

4. If A and B are any two sets, then $A \cap B \subseteq A$ and $A \cap B \subseteq B$.

Proof Let $x \in A \cap B$. Then

$$x \in A \cap B \Leftrightarrow x \in A \text{ and } x \in B$$
$$\Leftrightarrow x \in A \text{ as } A \cap B \subseteq A$$

$\therefore$ $A \cap B \subseteq A$. Similarly, it can be proved that $A \cap B \subseteq B$.

5. If A is any set, then $A \cap \Phi = \Phi$, where Φ is the null set.

Proof We know that the empty set is the subset of every set. Consequently,

$$\Phi \subseteq A \cap \Phi.$$

Again, by Property 4, $A \cap \Phi \subseteq \Phi$. Thus,

$$\Phi \subseteq A \cap \Phi \text{ and } A \cap \Phi \subseteq \Phi \Leftrightarrow A \cap \Phi = \Phi$$

6. If A is any subset of the universal set U, then $A \cap U = A$.

Proof As in Property 4, we can show that

$$A \cap U \subseteq A \quad \text{... } (i)$$

Let $x \in A \Rightarrow x \in A$ and $x \in U$ [since $A \subseteq U$]

$$\Rightarrow x \in A \cap U.$$

$\therefore$ $A \subseteq A \cap U$ ... (*ii*)

From (*i*) and (*ii*), we have $A = A \cap U$.

Example 5 Let $A = \{\varphi, b\}$, construct the following sets.

(a) $A - \varphi$ (b) $\{\varphi\} - A$ (c) $A \cup P(A)$ (d) $A \cap P(A)$ (d) $\varphi - A$

Solution Given that $A = \{\varphi, b\}$, therefore,

(a) $A - \varphi = \{\varphi, b\} - \varphi = \{\varphi, b\} = A$

(b) $\{\varphi\} - A = \{\varphi\} - \{\varphi, b\}$ set having elements of φ and not having elements φ and $b = \varphi$

(c) $A \cup P(A) = \{\varphi, b\} \cup \{\varphi, \{\varphi\}, \{b\}, \{\varphi, b\}\} = \{\varphi, b, \{\varphi\}, \{b\}, \{\varphi, b\}\} = P(A)$

(d) $A \cap P(A) = \{\varphi, b\} \cap \{\varphi, \{\varphi\}, \{b\}, \{\varphi, b\}\} = \{\varphi, b) = A$

(e) $\varphi - A = \varphi$

1.6.3 Disjoint Sets

Let A and B be two non-empty sets. These two sets are said to be disjoint (or mutually exclusive) if they have no common elements, i.e., their intersection is a null set. Symbolically,

$$A \cap B = \Phi$$

Figure 1.4 is a Venn diagram in which disjoint sets A and B are shown.

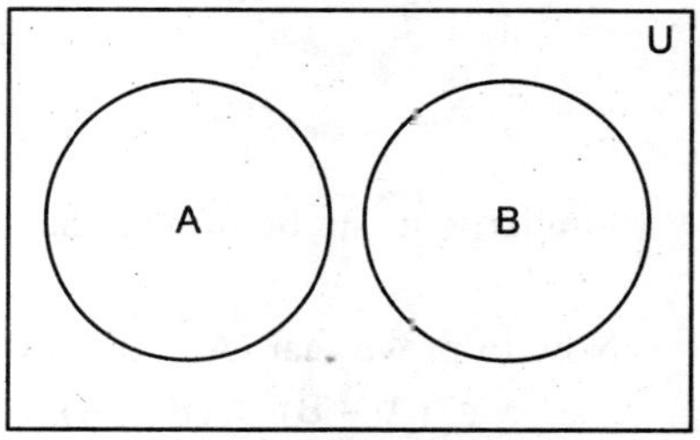

Fig. 1.4 Disjoint Sets A and B

Illustrations

1. If $A = \{1, 2, 3, 4\}$ and $B = \{5, 6, 7\}$, then $A \cap B = \Phi$.
2. If A = set of all positive integers, and B = set of negative integers, then $A \cap B = \Phi$, i.e., there is no number which is positive and negative both.

1.6.4 Difference of Two Sets

If A and B are any two sets, then the difference of A and B denoted by A – B or A\B is the set of elements which belong to A but do not belong to B. The difference of A and B is read as *A difference B* or simply *A minus B*. It is also denoted by A ~ B. Symbolically,

$A - B = \{x : x \in A \text{ and } x \notin B\}$ and similarly

$B - A = \{x : x \in B \text{ and } x \notin A\}$

Figure 1.5 is a Venn diagram in which A – B is shaded.

Illustration If $A = \{0, 2, 4, 9\}$ and $B = \{0, 3, 6, 8, 9\}$, then

$A - B = \{2, 4\}$ and $B - A = \{3, 6, 8\}$.

Obviously, $A - B \neq B - A$.

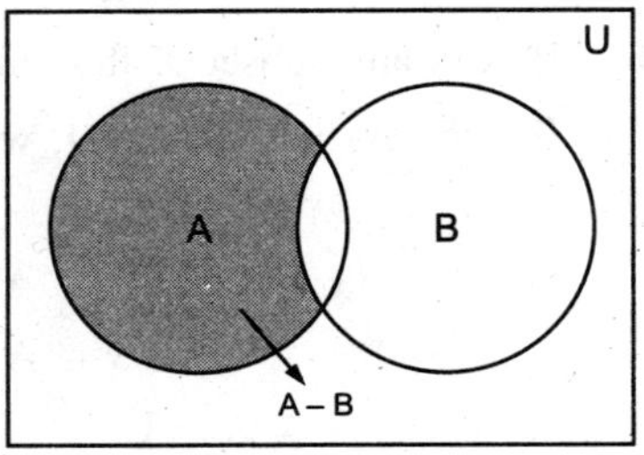

Fig. 1.5 Difference of Sets A and B

Properties of Difference of Two Sets

1. $A - A = \Phi$ 2. $A - \Phi = A$
3. $A - B \subseteq A$
4. $A - B$, $A \cap B$ and $B - A$ are mutually disjoint.
5. $(A - B) \cap B = \Phi$ 6. $(A - B) \cup A = A$.

Proof 1. Let $x \in A - A \Rightarrow x \in A$ and $x \notin A$. But there is no element satisfying both these conditions. Hence, no element belongs to A – A, i.e., $A - A = \Phi$.

2. Let $x \in A - \Phi \Rightarrow x \in A$ and $x \notin \Phi$. This simply means that $x \in A$, since there is no element belonging to Φ. Conversely, $x \in A \Leftrightarrow x \in A - \Phi$. That is, $A - \Phi = A$.

3. Let $x \in A - B \Leftrightarrow x \in A$ and $x \notin B$

$\Leftrightarrow x \in A$

$\therefore \quad A - B \subseteq A$.

4. To prove $(A - B) \cap (A \cap B) = \Phi$, let

$x \in (A - B) \cap (A \cap B) \Leftrightarrow x \in (A - B)$ and $x \in (A \cap B)$

$\Leftrightarrow (x \in A$ and $x \notin B)$ and $(x \in A$ and $x \in B)$

$\Leftrightarrow x \in A$ and $x \in \Phi$,

[since there is no element satisfying both $x \in B$ and $x \notin B$]

$\Leftrightarrow x \in \Phi$

Similarly, it can be proved that

$(B - A) \cap (A \cap B) = \Phi$

Now to prove that $(A - B) \cap (B - A) = \Phi$, let

$x \in (A - B) \cap (B - A) \Leftrightarrow x \in (A - B)$ and $x \in (B - A)$

$\Leftrightarrow (x \in A$ and $x \notin B)$ and $(x \in B$ and $x \notin A)$

$\Leftrightarrow x \in \Phi$

since there is no element satifying both ($x \in A$ and $x \notin A$) or ($x \in B$ and $x \notin B$).

From the above discussion it is clear that (A – B), (B – A) and ($A \cap B$) are disjoint sets.

5. To prove $(A - B) \cap B = \Phi$, let

$$x \in (A - B) \cap B \Leftrightarrow x \in (A - B) \text{ and } x \in B.$$
$$\Leftrightarrow (x \in A \text{ and } x \notin B) \text{ and } x \in B.$$
$$\Leftrightarrow x \in A \text{ and } x \in \Phi$$
$$\Leftrightarrow x \in \Phi$$

since there is no element satisfying both $x \in B$ and $x \notin B$.

6. To prove $(A - B) \cup A = A$, let

$$x \in (A - B) \cup A \Leftrightarrow x \in (A - B) \quad \text{or} \quad x \in A$$
$$\Leftrightarrow (x \in A \text{ and } x \notin B) \quad \text{or} \quad x \in A$$
$$\Leftrightarrow x \in A \text{ and } x \notin B$$
$$\Leftrightarrow x \in A$$

Hence, $(A - B) \cup A = A$.

1.6.5 Complement of a Set

Let A be any set. The complement of A is the set of elements that belong to the universal set but do not belong to A. Thus, if U is the universal set, the complement of A is the set U – A and is denoted by A', A^c, $\overline{A}$ or $-A$. Symbolically.

$$A^c = U - A = \{x : x \in U \text{ and } x \notin A\} = \{x : x \notin A\}$$

Figure 1.6 is the Venn diagram in which A^c is shaded.

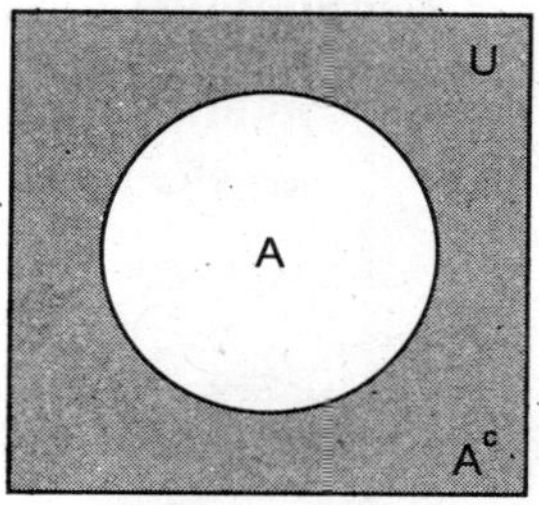

Fig. 1.6 Complement of Set A (Shaded Area)

Illustrations

1. If $N = \{1, 2, 3, 4, \ldots\}$ is the universal set and let $A = \{1, 3, 5, 7, \ldots\}$, then

$$A^c = N - A = \{2, 4, 6, 8, \ldots\}$$

2. If U = R, then Q^c = the set of irrational numbers
3. $U^c = \Phi$ and $\Phi^c = U$

Properties of Complement of a Set

1. $A \cup A^c = U$
2. $A \cap A^c = \Phi$
3. $U^c = \Phi$
4. $\Phi^c = U$
5. $(A^c)^c = A$
6. $(A - B) = A \cap B^c$
7. If $A \subseteq B$, then $A \cup (B - A) = B$.
8. $A \cap A^c = \phi$
9. $(A \cup B)^c = A^c \cap B^c$
10. $(A \cap B)^c = A^c \cup B^c$.

Proof 1. Since every set is a subset of the universal set, we have

$$A \cup A^c \subseteq U \quad \ldots(i)$$

Now let $x \in U \Rightarrow x \in A$ or $x \in A^c$

$$\Rightarrow x \in A \cup A^c$$

$\therefore \quad U \subseteq A \cup A^c. \quad \ldots(ii)$

From (*i*) and (*ii*), we have $A \cup A^c = U$.

2. Let $x \in A \cap A^c \Rightarrow x \in A$ and $x \in A^c$

$$\Rightarrow x \in \Phi$$

since there is no element x satisfying both $x \in A$ and $x \in A^c$. Therefore,

$$A \cap A^c = \Phi.$$

3. Let $x \in U^c \Rightarrow x \notin U$. Since there is no such element which does not belong to the universal set, therefore, $U^c = \Phi$.

4. Let $\Phi^c = \{x : x \in U \text{ and } x \notin \Phi\} = \{x : x \in U\} = U$

which is a universal truth. Hence, Φ^c is the set consisting of all possible elements in the universal set. Hence, $\Phi^c = U$.

5. To prove $(A^c)^c = A$, let $x \in (A^c)^c \Leftrightarrow x \notin A^c \Leftrightarrow x \in A$

Hence, $A = (A^c)^c$.

6. Let $x \in (A - B) \Leftrightarrow x \in A$ and $x \notin B$

$\Leftrightarrow x \in A$ and $x \in B^c \Leftrightarrow x \in A \cap B^c$.

Hence, $A - B = A \cap B^c$.

7. Given $A \subseteq B$. Then to prove $A \cup (B - A) = B$, let

$x \in A \cup (B - A) \Leftrightarrow x \in A$ or $x \in (B - A)$

$\Leftrightarrow (x \in A$ or $x \in B)$ and $(x \in A$ or $x \in A^c)$

$\Leftrightarrow x \in A \cup B$ and $x \in U$ [since $A \cup A^c = U$]

$\Leftrightarrow x \in A \cup B$

$\Leftrightarrow x \in B$ [since $A \subseteq A \cup B = B$]

Hence, $A \cup (B - A) = B$.

Example 6 Simplify the expression $[\{(A \cup B) \cap C\}^c \cup B^c]^c$.

Solution $[\{(A \cup B) \cap C\}^c \cup B^c]^c$

$= [\{A \cup B) \cap C\}^c]^c \cap (B^c)^c$ due to DeMorgan's law

$= \{(A \cup B) \cap C\} \cap B$ due to law of double complement

$= (A \cup B) \cap (C \cap B\}$ due to associative law of intersection

$= (A \cup B) \cap (B \cap C) = [(A \cup B) \cap B] \cap C$ due to associative law of intersection

$= B \cap C$ due to absorption law

Example 7 Express $(A - B)^c$ in terms of $\cup$ (union) and c (complement).

Solution Since $A - B = \{x : x \in A \text{ and } x \notin B\}$

$= A \cap \overline{B}$

Thus $(A - B)^c = (A \cap B^c)^c$

$= A^c \cup (B^c)^c$ due to DeMorgan's law

$= A^c \cup B$

1.6.6 Distributive Laws

1. Union is distributive over intersection If A, B and C are any three sets, then

$$A \cup (B \cap C) = (A \cup B) \cap (A \cup C)$$

Proof Let x be an arbitrary element of $A \cup (B \cap C)$. Then

$$\begin{aligned} x \in A \cup (B \cap C) &\Leftrightarrow x \in A \text{ or } x \in (B \cap C) \\ &\Leftrightarrow x \in A \text{ or } (x \in B \text{ and } x \in C) \\ &\Leftrightarrow (x \in A \text{ or } x \in B) \text{ and } (x \in A \text{ or } x \in C) \\ &\Leftrightarrow x \in (A \cup B) \text{ and } x \in (A \cup C) \\ &\Leftrightarrow x \in (A \cup B) \cap (A \cup C) \end{aligned}$$

Hence, $A \cup (B \cap C) = (A \cup B) \cap (A \cup C)$.

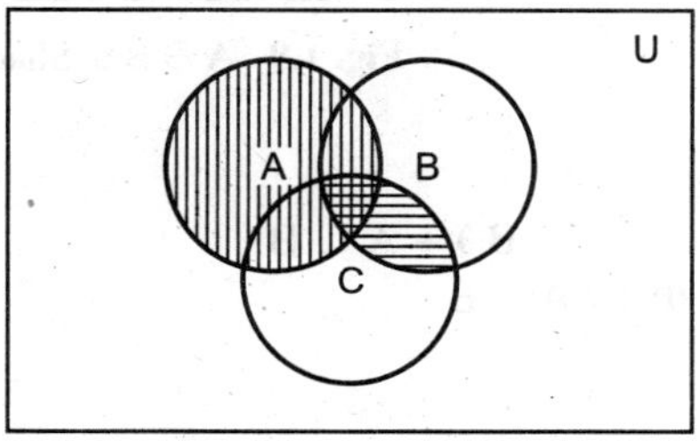

(*a*) $A \cup (B \cap C)$
Shaded Area

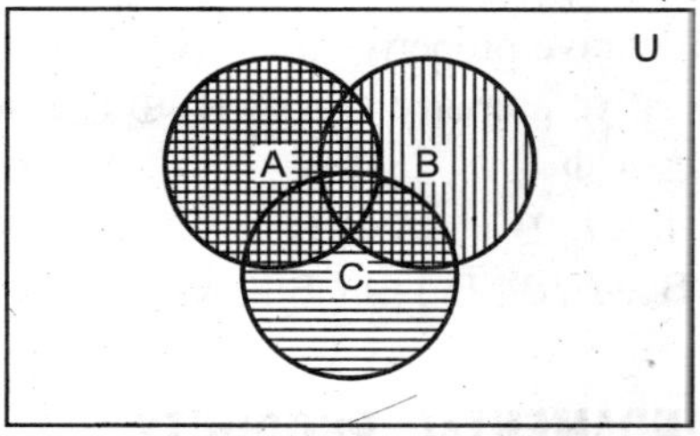

(*b*) $(A \cup B) \cap (A \cup C)$
Cross Hatche Area

Fig. 1.7 Union Distributes over Intersection

2. Intersection is distributive over union If A, B and C are any three sets, then

$$A \cap (B \cup C) = (A \cap B) \cup (A \cap C)$$

Proof Let x be an arbitrary element of $A \cap (B \cup C)$. Then

$$\begin{aligned} x \in A \cap (B \cup C) &\Leftrightarrow x \in A \text{ and } x \in (B \cup C) \\ &\Leftrightarrow x \in A \text{ and } (x \in B \text{ or } x \in C) \\ &\Leftrightarrow (x \in A \text{ and } x \in B) \text{ or } (x \in A \text{ and } x \in C) \\ &\Leftrightarrow x \in (A \cap B) \text{ or } x \in (A \cap C) \\ &\Leftrightarrow x \in (A \cap B) \cup (A \cap C) \end{aligned}$$

Hence $A \cap (B \cup C) = (A \cap B) \cup (A \cap C)$.

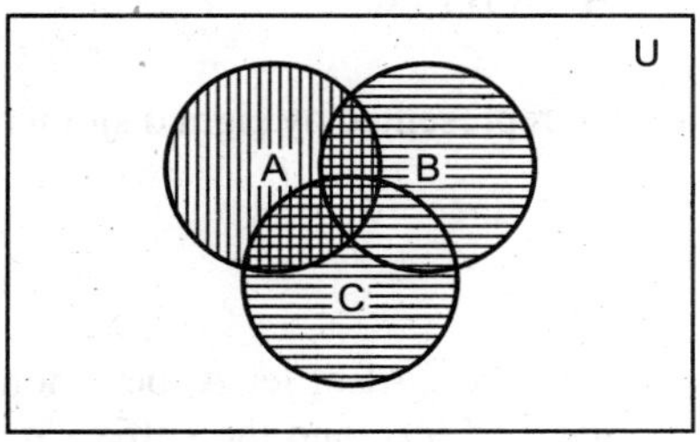

(*a*) $A \cap (B \cup C)$
Cross Hatche Area

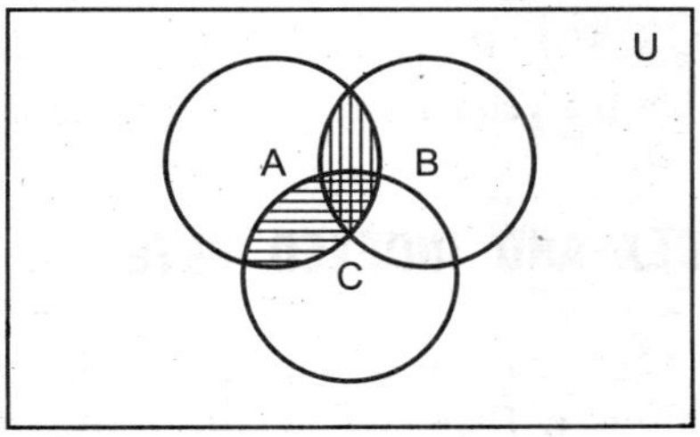

(*b*) $(A \cap B) \cup (A \cap C)$
Shaded Area

Fig. 1.8 Intersection Distributes over Union

1.6.7 Symmetric Difference of Sets

Let A and B be two non-empty sets. Then the *symmetric difference* of A and B, denoted by $A \oplus B$ or $A \Delta B$, is the set of elements which either belong to A or B but not to both. It may be noted that

$$A \oplus B = (A \cup B) - (B \cap A)$$
$$= (A - B) \cup (B - A)$$
$$= \{x : x \in A \text{ or } x \in B\}$$

For example, let $A = \{1, 2, 3\}$, $B = \{3, 4, 5\}$. Then

$$A \oplus B = (A \cup B) - (B \cap A)$$
$$= \{1, 2, 3, 4, 5\} - \{3\} = \{1, 2, 4, 5\}$$

Figure 1.9 is a Venn diagram in which $A \oplus B$ is shaded.

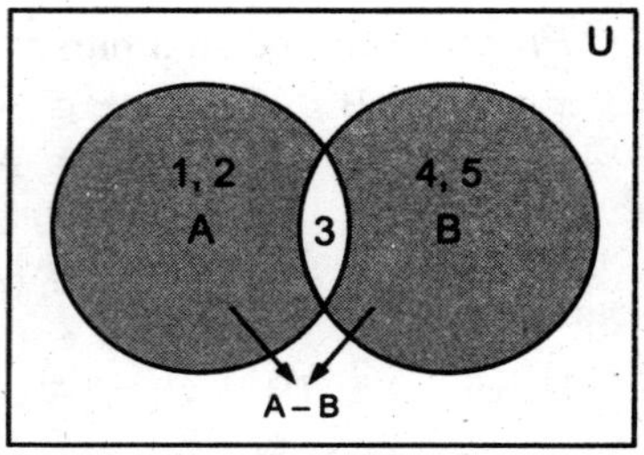

Fig. 1.9 $A \oplus B$ is Shaded

Properties of Symmetric Difference

1. Commutative property $A \oplus B = B \oplus A$
2. Associative property $(A \oplus B) \oplus C = A \oplus (B \oplus C)$
3. $A \oplus A = \Phi$
4. $A \oplus (A \cap B) = \Phi$
5. $A \oplus (A \cap B) = A \cup B$
6. $(A \oplus B) \cup (A \cap B) = A \cup B$
7. $A \oplus B = (A - B) \cup (B - A)$
8. $A \oplus B = \Phi \Leftrightarrow A = B.$

1.7 FUNDAMENTAL PRODUCTS

The fundamental product of n distinct sets $A_1, A_2, \ldots, A_n$ is a set defined as:

$$A_1^* \cap A_2^* \cap \ldots \cap A_n^*$$

where A_i^* $(i = 1, 2, \ldots, n)$ is either A_i or A_i^c. Few properties of such products are as under:

(*a*) The total such fundamental products are 2^n.

(*b*) Any two fundamental products are disjoint.

(*c*) The union of all the fundamental products is the universal set, U.

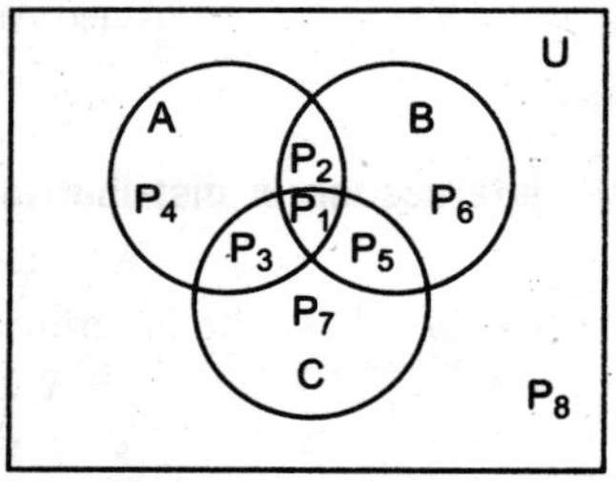

Fig. 1.10 Fundamental Products

Illustration Consider the following fundamental products of those distinct sets A, B and C, i.e., $2^3 = 8$.

$P_1 = A \cap B \cap C$ $\quad P_2 = A \cap B \cap C^c$ $\quad P_3 = A \cap B^c \cap C$

$P_4 = A \cap B^c \cap C^c$ $\quad P_5 = A^c \cap B \cap C$ $\quad P_6 = A^c \cap B \cap C^c$

$P_7 = A^c \cap B^c \cap C$ $\quad P_8 = A^c \cap B^c \cap C^c$

Figure 1.10 is the Venn diagram showing these eight products which are represented by eight disjoint regions.

1.8 INDEX AND INDEXED SETS

Let $I = \{1, 2, \ldots, n\}$ be a non-empty set, called *index set* or *set of indices*. Also, let A_i be a non-empty set for each $i \in$ 1. Then non-empty sets $A_1, A_2, \ldots, A_n$ are called *indexed sets*, and the suffix $i \in I$ of A_i is called an index. The collection $A = \{A_i : i \in I\}$ or $\{A_i\}$ of sets is said to be *indexed family of sets* or *indexed classes of sets*.

If $A = \{A_n : n \in N\}$, where $A_n = \{x : x \in N, x \text{ is a multiple of } n\}$, n is natural number, then A is the indexed family of sets A_n. Thus

$$A_1 = \{1, 2, 3, \ldots\}, A_2 = \{2, 4, 6, \ldots\}, A_3 = \{3, 6, 9, \ldots\}, \ldots, A_n = \{n, 2n, 3n, \ldots\}$$

Under these conditions, the union and intersection of these sets are defined as:

$$A_1 \cup A_2 \cup \ldots \cup A_n = \bigcup_{i \in I} A_i = \{x : x \in A_i \text{ for at least one } i \in I\}$$

that is, the union consists of those elements which belong to at least one of the sets, and $A_1 \cap A_2 \cap \ldots A_n = \bigcap_{i \in I} A_i = \{x : x \in A_i \text{ for every } i \in I\}$ i.e., the intersection consists of those elements which belong to all the sets.

An indexed family of sets $\{A_i\}$ is said to be disjoint when no element is shared by all sets of the family, i.e., $\bigcap_{i \in I} A_i = \phi$. Similarly, the indexed family of sets is said to be pairwise disjoint when any two distinct sets share a common element, i.e., $A_i \cap A_j = \phi$, for $i \neq j$. For example, the sets $\{1, 2\}$, $\{3, 6\}$, $\{4, 7\}$ are pairwise disjoint, while the sets $\{1, 2\}$, $\{3, 6\}$, $\{6, 7\}$ are disjoint but not pairwise disjoint because element 6 is common in last two sets.

Illustrations

1. Let $I = \{3, 4, 5, 6\}$ and for each $i \in I$ let $A_i = \{1, 2, 3, \ldots, n\}$ be a subset of some universal set $U = I_+$ (set of positive integers). Then

$$\bigcup_{i \in I} A_i = \bigcup_{i=1}^{6} = \{1, 2, 3, 4, 5, 6\} = A_6$$

$$\bigcap_{i \in I} A_i = \{1, 2, 3\} = A_3$$

2. Let $U = R$ and $I = R^+$. If for each $r \in R^+$, $A_r = \{-r, r\}$, then

$$\bigcup_{r \in I} A_r = R \quad \text{and} \quad \bigcap_{r \in I} A_r = \{0\}$$

3. If the index set I is the set infinite positive integers I_+, then we can write

$$\bigcup_{i \in I_+} A_i = \bigcup_{i=1}^{\infty} A_i \quad \text{and} \quad \bigcap_{i \in I_+} = \bigcap_{i=1}^{\infty} A_i$$

4. Let $U = R$ and $I = I_+$ (set of positive integers). For each $n \in I_+$, let $A_n = \{-2n, 3n\}$. Then

(*a*) $A_3 = \{-6, 9\}$ (*b*) $\bigcap_{n=1}^{7} A_n = A_7$ (*c*) $\bigcup_{n \in I_+}^{\infty} A_n = I_+$

Example 8 Let I be an index set and $\{A_i : i \in I\}$ be a collection of subsets of some universal set U that is indexed by I. Then show that

(*a*) $\left\{\bigcup_{i \in I} A_i\right\}^c = \bigcap_{i \in I} A_i^c$ and (*b*) $\left\{\bigcap_{i \in I} A_i\right\}^c = \bigcap_{i \in I} A_i^c$

Solution (*a*) Let $x \in \left\{\bigcup_{i \in I} A_i\right\}^c$. Then $x \notin \bigcup_{i \in I} A_i$ and so $x \notin A_i$ for all $i \in I$. This further implies that $x \in A_i^c$ for all $i \in I$ and hence $x \in \bigcap_{i \in I} A_i^c$.

(*b*) Left as an exercise for the reader.

Example 9 Let $A_n = \{x : x \text{ is divisible by } n\}$ when $n \in N$. Find $A_3 \cap A_5$ and $A_3 \cup A_5$.

Solution We have

$$A_3 \cap A_5 = \{x : x \text{ is divisible by both 3 and 5}\}$$
$$= \{x : x \text{ is divisible by 15}\}$$
$$= \{0, 15, 30, 45, \ldots\}$$

and

$$A_3 \cup A_5 = \{x : x \text{ is divisible by 3 or 5}\}$$
$$= \{0, 3, 5, 6, 9, 10, 12, 15, \ldots\}$$

Example 10 If $A = [0, i]$ where $i \in I$, the set of integers, then find (*a*) $A_1 \cup A_2$, (*b*) $A_3 \cap A_4$ and (*c*) $\bigcup_{i=5}^{20} A_i$.

Solution (*a*) $A_1 \cup A_2$ = set of all real numbers in the interval [0, 1] and [0, 2]

$$= [0, 2] = A_2$$

(*b*) $A_3 \cap A_4$ = set of all real numbers lies in both intervals [0, 3] and [0, 4]

$$= [0, 3] = A_3$$

(*c*)
$$\bigcup_{i=1}^{10} A_i = A_1 \cup A_2 \cup \ldots \cup A_{10} = [0, 1] \cup [0, 2] \cup \ldots \cup [0, 10]$$
$$= [0, 10] = A_{10}$$

1.9 PARTITIONS OF SETS

A bigger set can be partitioned into smaller non-overlapping, non-empty subsets in order to make the task easy for analysing each part separately. For example, students in a class may be partitioned into small groups on the basis of certain criteria so as to keep track of their progress and regular internal assessment.

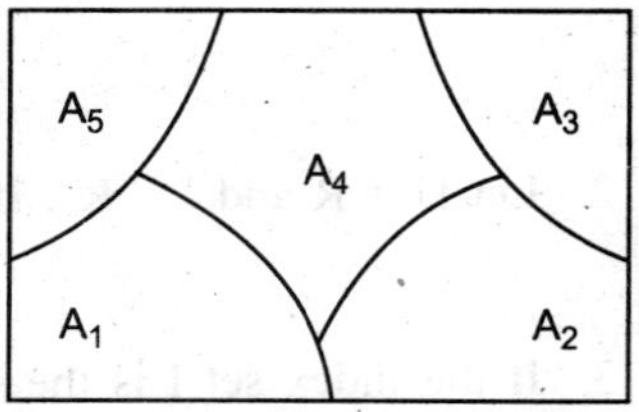

Fig. 1.11 Partitions of Sets

Let A be a non-empty set. The partition of A is any set of non-empty, non-overlapping subsets (cells) $A_1, A_2, \ldots, A_n$ such that

(*i*) $A = A_1 \cup A_2 \cup \ldots \cup A_n$.

(*ii*) The subsets A_i are mutually disjoint, i.e., $A_i \cap A_j = \Phi$ for $i \neq j$.

Figure 1.11 is the Venn diagram showing partition of the set A into five subsets (or cells), A_1 A_2, A_3, A_4 and A_5.

Illustrations

1. Let $A = \{a, b, c\}$. Then $A_1 = \{a\}$, $A_2 = \{b, c\}$ are the partitions of A.
2. The set of integers may be partitioned as: $\{\{n\} : n \in I\}$, and $\{\{n\} : n \in I, n < 0\}, \{0\}, \{n : n \in I, n > 0\}\}$. The set of subsets $\{\{n : n \in I, n > 0\}, \{n : n \in I, n < 0\}\}$ is not a partition because the two have a non-empty intersection.

Example 11 Given that $A \oplus B = A \oplus C$, is it necessary that $B = C$? Justify your answer.

Solution Let B = C. Then $A \oplus B = (A \cup B) - (A \cap B)$ is the set that contains elements which are either in A or B but not in both. But, if $A \oplus B = A \oplus C$, then

$$(A \cup B) - (A \cap B) = (A \cup C) - (A \cap C) \Rightarrow A \cup B = A \cup C \text{ and } A \cap B = A \cap C$$
$$\Rightarrow B = C.$$

Example 12 Find all the partition of $X = \{a, b, c, d\}$. [*UPTU, BTech., 2006*]

Solution Each partition of X contains either 1, 2, 3 or 4 distinct sets as shown below:

1. $[\{a, b, c, d\}]$
2. (*i*) $[\{a\}, \{b, c, d\}], [\{b\}, \{a, c, d\}], [\{c\}, \{a, b, d\}]$
   (*ii*) $[\{d\}, \{a, b, c\}], [\{a, b\}, \{c, d\}], [\{b, c\}, \{a, d\}], [\{a, c\}, \{b, d\}]$
3. $[\{a\}, \{b\}, \{c, d\}], [\{a\}, \{c\}, \{b, d\}], [\{a\}, \{b\}, \{b, c\}]$
   $[\{b\}, \{c\}, \{a, d\}], [\{b\}, \{d\}, \{a, c\}], [\{c\}, \{d\}, \{a, b\}]$
4. $[\{a\}, \{b\}, \{c\}, \{d\}]$

Example 13 Let $A = \{a, b, c, d, e, f, g\}$ and suppose A_1, A_2, A_3, A_4 and A_5 are the subsets of A such that $A_1 = \{a, c, e, f, g\}$, $A_2 = \{a, b, c, d\}$, $A_3 = \{a, c, e, g\}$, $A_4 = \{b, d\}$, $A_5 = \{f\}$.

Here $\{A_1, A_2\}$ is not a partition because $A_1 \cap A_2 \neq \varnothing$, similarly $\{A_1, A_5\}$ is also not a partition because $A_1 \cap A_5 \neq \varnothing$, $(f \in A_1 f \in A_5)$. The collection of sets, $\{A_3, A_4, A_5\}$ is a partition because it satisfies condition (*a*), i.e., each element of A belongs to anyone of A_3, A_4, A_5.

It also satisfies condition (*b*), i.e., $A_3 \cap A_4 = A_3 \cap A_5 = A_4 \cap A_5 = \varnothing$.

1.10 MINSETS

Let $\{B_1, B_2, \ldots, B_n\}$ be a set of subsets of a set A. Then the set of the form $C_1 \cap C_2 \cap C_3 \ldots \cap C_n$, where each C_i may be either B_i or its complement B_i^c is called a *minset* or *minterm*, generated by B_is $(i = 1, 2, \ldots, n)$

From the definition of minset it is clear that each of minset (or minterm) is formed by taking the intersection of two sets and number of minsets, in general, are 2^n, where n is number of subsets of a set A. The set of non-empty minsets generated by $B_1, B_2, \ldots, B_n$ is a partition of set A.

For example, let $B_1 = \{1, 4, 6\}$ and $B_2 = \{1, 3, 4\}$ be two subsets of the set $A = \{1, 2, 3, 4, 5, 6\}$. In order to have a partition of A without repeatition of elements, we may describe B_1 and B_2 as follows:

$$C_1 = B_1 \cap B_2^c = \{1, 4, 6\} \cap \{2, 5, 6\} = \{6\}$$
$$C_2 = B_1 \cap B_2 = \{1, 4, 6\} \cap \{1, 3, 4\} = \{1, 4\}$$
$$C_3 = B_1^c \cap B_2 = \{2, 3, 5\} \cap \{1, 3, 4\} = \{3\}$$
$$C_4 = B_1^c \cap B_2^c = \{2, 3, 5\} \cap \{2, 5, 6\} = \{2, 5\}$$

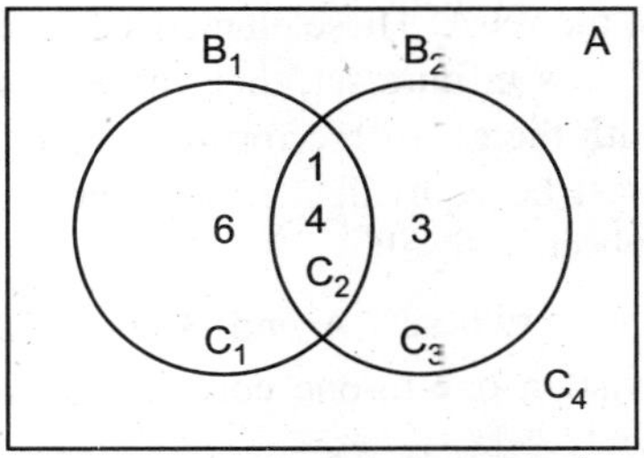

Fig. 1.12 Minsets of Set A

As shown in Fig. 1.12 since none of the sets C_1, C_2, C_3 and C_4 have elements in common, therefore, the set generated by B_1 and B_2 is the partition of A.

1.11 COUNTABLE AND UNCOUNTABLE SETS

The size or cardinality of a finite set is defined to be the number of elements in that set. In this section, we shall discuss this concept when the given set is infinite. To develop this concept we need to understand the concept of 'mapping' (see next chapter for detail discussion).

Definition Let A and B be two non-empty sets. If a *rule* or *correspondence* denoted by f is such that it relates (or associates) with each element $x \in$ A, to a unique (one and only one) element $y \in$ B, then such a rule is said to be *one-to-one*. Also, if with each element $y \in$ B there exists exactly one element $x \in$ A, then this rule is said to be *one-to-one* and *onto*. Symbolically, the function f expressed as:

$$f: \mathrm{A} \to \mathrm{B} \text{ such that } y = f(x), \text{ for } x \in \mathrm{A} \text{ and } y \in \mathrm{B}$$

describes one-to-one correspondence if f is both one-to-one and onto.

Illustrations

1. Let A be a set of positive integers and B = $\{2x : x \in \mathrm{I}_+\}$ = {2, 4, 6, . . .}. The function $f: \mathrm{A} \to \mathrm{B}$ defined by $y = f(x) = 2x$, is a one-to-one correspondence. Moreover,

   (*i*) for $x_1, x_2 \in$ A, we have $f(x_1) = f(x_2) \Rightarrow 2x_1 = 2x_2 \Rightarrow x_1 = x_2$, so f is one-to-one.

   (*ii*) But if $y \in$ B, then $y = 2x$ for a unique element $x \in$ A and $f(x) = 2x = y$. This makes f an onto function.

   Two non-empty sets A and B are said to be equivalent (written as A ~ B) if there exists a one-to-one correspondence between their elements.

2. From Illustration 1, we see that I_+ = {1, 2, 3, 4, 5, 6, . . .} has the same number of elements as $2\mathrm{I}_+$ = {2, 4, 6, . . .}, even though it appears that $2\mathrm{I}_+$ has less number of elements than I_+ because $2\mathrm{I}_+ \subseteq \mathrm{I}_+$. If we define a function as:

   $$g : \mathrm{B} \to \mathrm{A} \text{ such that } g(2x) = x, \text{ where } \mathrm{B} = 2\mathrm{I}_+ \text{ and } \mathrm{A} = \mathrm{I}_+, \text{ then}$$

   (*i*) $g(2x_1) = g(2x_2) \Rightarrow x_1 = x_2$, so g is one-to-one.

   (*ii*) for each $x \in$ A, we have $2x \in$ B with $g(2x) = x$, so g is also onto function.

   Consequently, g is a one-to-one correspondence and hence B ~ A.

Finite set Any non-empty set A is called a finite set if either it is an empty set, i.e., A = ϕ or it is equivalent to a set of the form {1, 2, 3, . . ., n} for some positive integer n.

If a set is not finite, then it is called *infinite set*. The set N of natural numbers, the set I of all integers, the set Q of rational numbers, etc. are few examples of infinite sets.

From above definition of a finite set A, it appears that there is a one-to-one correspondence f: {1, 2, 3, . . ., n} $\to$ A for some positive integer n. The function f gives a listing g (1), g (2), . . ., g (n) of the elements of the set A. These elements can be counted as first, second, . . . and so on upto nth (last) element. However, if A is infinite set, then there exists no positive integer n for which there is a one-to-one correspondence with the set of the form {1, 2, 3, . . ., n}. The main difference between the two sets is that an infinite set must be equivalent to a proper subset of itself, whereas a finite set cannot be equivalent to any proper subset of itself.

If two sets A and B are infinite sets, then obviously both of these have same cardinality, i.e., there exists a one-to-one correspondence between A and B.

Countably infinite set An infinite set A is said to be countably infinite (or denumerable) set if it is equivalent to the set N of natural numbers.

Countable set A set A is called countable when it is either (*i*) empty, i.e., A = ϕ, or (*ii*) finite or countably infinite.

Since $2\mathrm{I}_+ \sim \mathrm{I}_+$, $3\mathrm{I}_+ \sim \mathrm{I}_+$, it follows that I_+, $2\mathrm{I}_+$ and $3\mathrm{I}_+$ are all countable sets. The function $f: \mathrm{I}_+ \to k\mathrm{I}_+$ such that $f(x) = kx$ for all $k \in$ I and $x \in \mathrm{I}_+$, is a one-to-one correspondence, so $k\mathrm{I}_+$ is a countable set. Consequently $k\mathrm{I}_+$ has same cardinality as I_+.

Example 14 Show that the set I of all integers is countable.

Solution Since the set I of all integers is an infinite set, it must be countably infinite. To prove that I is countably infinite, we need to prove that there exist an one-to-one correspondence between I and the set N of natural numbers. For this, consider the function f : N $\to$ I defined as:

$$f(n) = \begin{cases} \dfrac{n}{2} & \text{, when } n \text{ is even positive integer} \\ \dfrac{1-n}{2} & \text{, when } n \text{ is odd positive integer} \end{cases}$$

Suppose n_1, $n_2 \in$ N with $f(n_1) = f(n_2)$. Then

(*i*) If n_1 and n_2 are both even, then $f(n_1) = f(n_2) \Rightarrow n_1/2 = n_2/2 \Rightarrow n_1 = n_2$

(*ii*) If n_1 and n_2 are both odd, then $f(n_1) = f(n_2) \Rightarrow (1 - n_1)/2 = (1 - n_2)/2 \Rightarrow 1 - n_1 = 1 - n_2 \Rightarrow n_1 = n_2$

(*iii*) If n_1 is even and n_2 is odd, then $f(n_1) = f(n_2) \Rightarrow n_1/2 = (1 - n_2)/2 \Rightarrow n_1 = 1 - n_2$
$\Rightarrow n_1 - 1 = - n_2$ with $n_1 - 1 \geq 1$ and $- n_2 < 0$. This case is not possible.

Consequently, the function f is one-to-one. Moreover, for all $n \in$ I,

(*i*) if $n = 0$, then $f(1) = 0$

(*ii*) if $n > 0$, then $2n \in$ N and $f(2n) = 2n/2 = n$,

(*iii*) if $n < 0$, then $1 - 2n \in$ N and $f(1 - 2n) = [1 - (1 - 2n)]/2 = n$. Consequently the function f is also an onto function. Hence I is countable.

Example 15 Show that the set [0, 1] of all real numbers is not a countable set.

Solution Suppose the set [0, 1] is countable. Then all the real numbers $0 < x \leq 1$ can be written as a sequence of distinct terms, say $x_1, x_2, x_3, \ldots$. Representing real numbers $x_1, x_2, x_3, \ldots$ in (0, 1] by a non-terminating decimal expansion as:

$$x_1 = 0 \cdot a_{11}\ a_{12}\ a_{13}\ a_{14} \ldots$$
$$x_2 = 0 \cdot a_{21}\ a_{22}\ a_{23}\ a_{24} \ldots$$
$$\vdots$$
$$x_n = 0 \cdot a_{n1}\ a_{n2}\ a_{n3}\ a_{n4} \ldots$$

where each a_{ij} (for all i, j) is one of the numbers of the set {0, 1, 2, 3, . . . , 9}. For example, if $x_1 = 0 \cdot 572316 \ldots$ then $a_{11} = 5$, $a_{12} = 7$, $a_{13} = 2$, $a_{14} = 3$, and so on.

Let us consider a new real number $x = 0 \cdot b_1\ b_2\ b_3 \ldots$ for all $i \in$ N, the set of natural numbers or positive integers, where

$$b_i = \begin{cases} 1 & \text{if } a_{ii} = 9 \\ 9 - a_{ii} & \text{if } a_{ii} = 0, 1, 2, \ldots, 8 \end{cases}$$

for all i. We choose an expansion which ends with 9 such as 1/2 = 0.50000 . . . = 0.49999. Such an expansion ensure a unique representation of all numbers. Then obviously the real number $x = 0{\cdot}b_1b_2b_3 \ldots$ belong to the interval (0, 1), which does not have trailing 0's. But for every $i \in$ N, the number x differ from any of the numbers $x_1, x_2, x_3, \ldots$, i.e., $x \notin \{x_1, x_2, x_3, \ldots\}$ because it differs from the first number in the first digit, the second number in the second digit, and so on. Thus a real number x which belongs to the interval $0 < x \leq 1$, but is not in the list, contradicts the assumption that $(0, 1] = \{x_1, x_2, x_3, \ldots\}$, i.e., the set is countably infinite. Hence, the set of real numbers in the interval (0, 1] is uncountable.

Theorem 1.1 Any infinite set has a countable (denumerable) set.

Proof Let S be a non-empty infinite countable set. Since any non-empty countable set can be written as a sequence of distinct elements, therefore, for any natural number n we can write n distinct elements $a_1, a_2, \ldots, a_n$ of S.

Let the set A be the subset of S. Now let $b \in$ A. Since $A \subseteq S = \{s_1, s_2, s_3, \ldots\}$, we can write $b = s_m$ for any $m \in$ N. If $m = a_1$, then $s_m = a$. But if $m \neq a_1$, then there is a smallest element $r \in$ N such that $a_{r-1} < m \leq a_r$ (because $a_1 < a_2 < a_3 \ldots < a_n$). From the definition we know that $a_r = \min \{t : t \in N, x > a_{r-1}$ and $s_t \in A\}$. Also, as $m > a_{r-1}$ and $s_m \in$ A, we have $a_r \leq m$. Since $a_r \leq m$ and $m \leq a_r \Rightarrow a_r = m$, therefore, $s_m = b$.

Theorem 1.2 The set $Q \cap [0, 1]$ is countable.

Proof Since any element belongs to the set $Q \cap [0, 1]$ has to be expressed in the form p/q, where p, $q \in I_+$, the set of positive integers, and no common divisor other than 1. Let us define a function as:

$$f : Q \cap [0, 1] \rightarrow I_+ \times I_+, \text{ such that } f(p/q = (p, q)$$

Now for any two elements p/q, $u/v \in Q \cap [0, 1]$, there exists elements $f(p/q) = f(u/v) \Rightarrow (p, q) = u/v \Rightarrow p = u$ and $q = v$. This further implies that the function f is one-to-one function. Consequently $Q \cap [0, 1] \sim$ A, a subset of the countable set $I_+ \times I_+$. Hence, the set $Q \cap [0, 1]$ is countable.

Consequently A is subset of S and is equivalent to N.

Corollary The set of all real numbers is an uncountable set. Since the subset [0, 1] of the set R of all real numbers is uncountable, therefore, by Theorem 1, the set R is uncountable.

Theorem 1.3 The set Q of all rational numbers is countable.

Proof From Theorem 1.2, we recall that the set $A = Q \cap [0, 1]$ is countable. Considering for each non-zero integer n, $A_n = Q \cap [n, n + 1]$. Define a function as:

$$f : A_n \rightarrow A_0 \text{ such that } f(x) = x - n.$$

Then $f(x_1) = f(x_2) \Rightarrow x_1 - n = x_2 - n \Rightarrow x_1 = x_2$, so f is one-to-one function. Consequently $A_n \subseteq A_0$. From Theorem 1.2 since any subset of infinite countable set is countable, therefore, A_n is countable. Moreover, for all, distinct elements $m, n \in$ I, $A_m \cap A_n = \phi$. Since the set of all rational numbers in [0, 1] is countable, the elements of the set $\{A_0, A_1, A_{-1}, A_2, A_{-2}, \ldots\}$ are countable sets and disjoint. Hence, the set of rational numbers is countable.

Theorem 1.4 The union of any countable collection of countable sets is countable.

Proof Let $A = \{A_1, A_2, A_3, \ldots\}$ be any collection of countable sets. Construct the countable collection $B = \{B_1, B_2, B_3, \ldots\}$ of sets such that $A_1 = B_1$ and $B_n = A_n - \bigcup_{i=1}^{n-1} A_i$ for $n \geq 2$. Then B is a countable disjoint collection. Since each set in the collection of countable disjoint sets is countable, therefore, their union is also countable, and hence $\bigcup_{i=1}^{\infty} A_i = \bigcup_{i=1}^{\infty} B_i$.

1.12 ALGEBRA OF SETS AND DUALITY

Let S be an equation in the algebra of sets as discussed earlier in this chapter, involving sets and the operations such as, $\cap$, $\cup$, U, Φ, c, etc. The dual of S, denoted S^d, is obtained from S by replacing

(*i*) ϕ and U in S by U and ϕ, respectively, and

(*ii*) $\cap$ and $\cup$ in S by $\cup$ and $\cap$, respectively.

Few of the important results on algebra of sets and the *principle of duality* are discussed below:

1. Idempotent laws

(*a*) $A \cap A = A$ (*b*) $A \cup A = A$

2. Identity laws

(*a*) $A \cup \Phi = A$ (*b*) $A \cap U = A$

(*c*) $A \cup U = U$ (*d*) $A \cap U = \Phi$

3. Inverse laws

(*a*) $A \cup A^c = U$ (*b*) $A \cap A^c = \Phi$

(*c*) $U^c = \Phi$ (*d*) $\Phi^c = U$

4. Domination laws

(*a*) $A \cup U = U$ (*b*) $A \cap \phi = \Phi$

5. Absorption laws

(*a*) $A \cup (A \cap B) = A$ (*b*) $A \cap (A \cup B) = A$

6. Demorgan's laws

(*a*) $(A \cup B)^c = A^c \cap B^c$ (*b*) $(A \cap B)^c = A^c \cup B^c$

7. (*a*) Complement of the union of two sets is the intersection of their complements (see Fig. 1.13).

$$(A \cup B)^c = A^c \cap B^c.$$

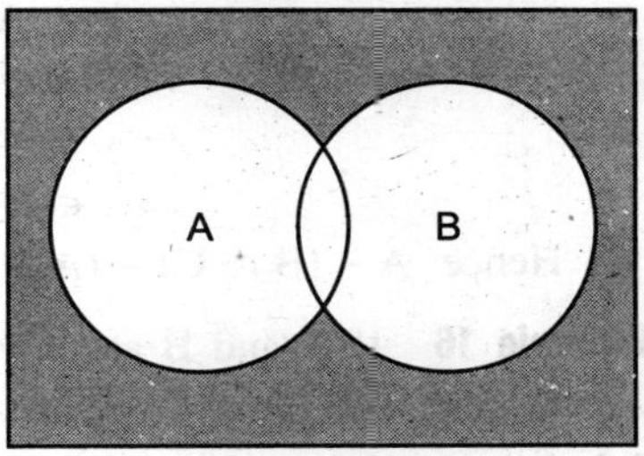

Fig. 1.13 $A^c \cap B^c$ (Shaded Region)

Proof Let x be any element of $(A \cup B)$. Then

$x \in (A \cup B)^c \Leftrightarrow x \in U$ and $x \notin (A \cup B)$

$\Leftrightarrow x \in U$ and $(x \notin A$ or $x \notin B)$

$\Leftrightarrow (x \in U$ but $x \notin A)$ and $(x \in U$ but $x \notin B)$

$\Leftrightarrow x \in A^c$ and $x \in B^c$

$\Leftrightarrow x \in A^c \cap B^c$

Hence, $(A \cup B)^c = A^c \cap B^c$.

(*b*) Complement of the intersection of two sets is the union of their complements, i.e.,

$$(A \cap B)^c = A^c \cup B^c$$

Proof Let x be any element of $(A \cap B)^c$. Then

$x \in (A \cap B)^c \Leftrightarrow x \notin (A \cap B)$

$\Leftrightarrow x \notin A$ or $x \notin B$

$\Leftrightarrow x \in A^c$ or $x \in B^c$

$\Leftrightarrow x \in A^c \cup B^c$

Hence, $(A \cap B)^c = A^c \cup B^c$.

8. (*a*) $A - (B \cup C) = (A - B) \cap (A - C)$

Proof Let x be any arbitrary element of $A - (B \cup C)$. Then

$x \in A - (B \cup C) \Leftrightarrow x \in A$ and $x \notin (B \cup C)$

$\Leftrightarrow x \in A$ and $(x \notin B$ or $x \notin C)$

$\Leftrightarrow (x \in A$ but $x \notin B)$ and $(x \in A$ but $x \notin C)$

$\Leftrightarrow x \in (A - B)$ and $x \in (A - C)$

$\Leftrightarrow x \in (A - B) \cap (A - C)$

Hence, A – (B ∪ C) = (A – B) ∩ (A – C). [see Fig. 1.14]

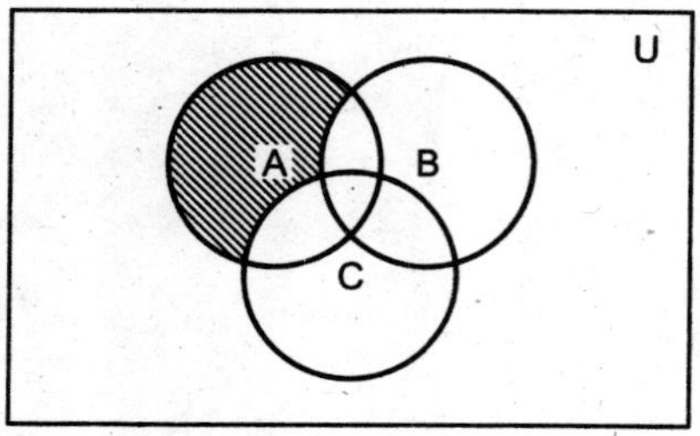

(*a*) A – (B ∪ C)
Shaded Area

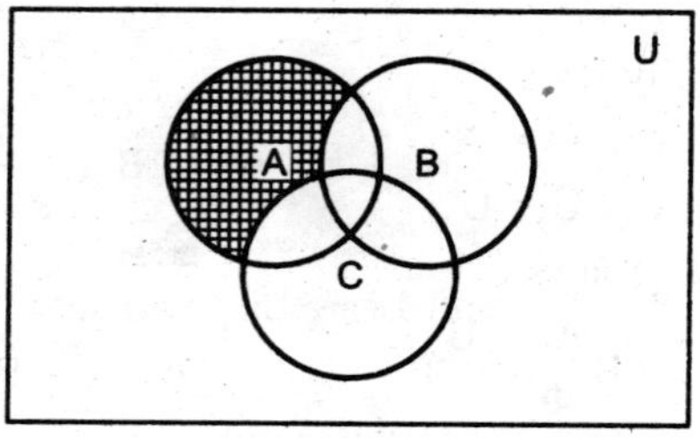

(*b*) (A – B) ∩ (A – C)
Cross Hatche Area

Fig. 1.14 A – (B ∪ C) = (A – B) ∩ (A – C)

(*b*) A – (B ∩ C) = (A – B) ∪ (A – C)

Proof Let x be an arbitrary element of A – (B ∩ C). Then

$x \in A - (B \cap C) \Leftrightarrow x \in A$ and $x \notin (B \cap C)$
$\Leftrightarrow x \in A$ and ($x \notin B$ and $x \notin C$)
$\Leftrightarrow$ ($x \in A$ and $x \notin B$) or ($x \in A$ and $x \notin C$)
$\Leftrightarrow x \in (A - B)$ or $x \in (A - C)$
$\Leftrightarrow x \in (A - B) \cup (A - C)$

Hence A – (B ∩ C) = (A – B) ∪ (A – C).

Example 16 If A and B are any two sets, then

(*a*) A ∩ (B – A) = Φ (*b*) (A – B) ∩ B = Φ.

Solution (*a*) To prove A ∩ (B – A) = Φ, let

$x \in A \cap (B - A) \Leftrightarrow x \in A$ and $x \in (B - A)$
$\Leftrightarrow x \in A$ and ($x \in B$ and $x \notin A$)
$\Leftrightarrow$ ($x \in A$ and $x \notin A$) and $x \in B$
$\Leftrightarrow x \in \Phi$ and $x \in B$
$\Leftrightarrow x \in \Phi$

Hence, A ∩ (B – A) = Φ.

(*b*) Let $x \in (A - B) \cap B \Leftrightarrow x \in (A - B)$ and $x \in B$.
$\Leftrightarrow$ ($x \in A$ and $x \notin B$) and $x \in B$
$\Leftrightarrow x \in A$ and ($x \notin B$ and $x \in B$)
$\Leftrightarrow x \in A$ and $x \in \Phi$
$\Leftrightarrow x \in \Phi$

Hence, (A – B) ∩ B = Φ.

Example 17 If A and B are any two sets, then A ∪ B = (A – B) ∪ B.

Solution To prove the result, let

$x \in A \cup B \Leftrightarrow x \in A$ or $x \in B$
$\Leftrightarrow x \in B$ or $x \in A$

$\Leftrightarrow (x \in B \text{ or } x \in A) \text{ and } (x \in B \text{ or } x \notin B)$

[since $x \in B$ or $x \notin B$ is always a true statement]

$\Leftrightarrow x \in B \text{ or } (x \in A \text{ and } x \notin B)$

$\Leftrightarrow x \in B \text{ or } x \in (A - B)$

$\Leftrightarrow x \in (A - B) \text{ or } x \in B$

$\Leftrightarrow x \in (A - B) \cup B$

Hence, $A \cup B = (A - B) \cup B$.

Example 18 If A and B are any two sets, then prove that

(*a*) $A - B = A \cap B^c$ (*b*) $A - B = B^c - A^c$.

Solution (*a*) To prove this result, let

$x \in A - B \Leftrightarrow x \in A \text{ and } x \notin B.$

$\Leftrightarrow x \in A \text{ and } x \in B^c$

$\Leftrightarrow x \in A \cap B^c$

Hence, $A - B = A \cap B^c$.

(*b*) Let $x \in A - B \Leftrightarrow x \in A \text{ and } x \notin B$

$\Leftrightarrow x \notin A^c \text{ and } x \in B^c$

$\Leftrightarrow x \in B^c \text{ and } x \notin A^c$

$\Leftrightarrow x \in B^c - A^c$

Hence, $A - B = B^c - A^c$.

Example 19 If A and B are any two sets, then prove that

(*a*) $A - B = A \Leftrightarrow A \cap B = \Phi$ (*b*) $A - B = \Phi \Leftrightarrow A \subseteq B$.

Solution (*a*) Suppose that $A - B = A$. Then we have to show that $A \cap B = \Phi$. For this, let

$x \in A \cap B \Leftrightarrow x \in A \text{ and } x \in B$

$\Leftrightarrow x \in (A - B) \text{ and } x \in B$ [since $A = A - B$]

$\Leftrightarrow (x \in A \text{ and } x \notin B) \text{ and } x \in B$

$\Leftrightarrow x \in A \text{ and } (x \in B \text{ and } x \notin B)$

$\Leftrightarrow x \in A \text{ and } x \in \Phi$

$\Leftrightarrow x \in \Phi$

Thus $A - B = A \Rightarrow A \cap B = \Phi$. ...(*i*)

Conversely, suppose that $A \cap B = \Phi$. Then to show that $A - B = A$, let

$x \in A \Rightarrow x \in A \text{ and } (x \in B \text{ or } x \notin B)$

$\Rightarrow (x \in A \text{ and } x \in B) \text{ or } (x \in A \text{ and } x \notin B)$

$\Rightarrow x \in (A \cap B) \text{ or } x \in (A - B)$

$\Rightarrow x \in \Phi \text{ or } x \in (A - B)$ [since $A \cap B = \Phi$]

$\Rightarrow x \in A - B$ [since $x \notin \Phi$]

Thus $A \cap B = \Phi \Rightarrow A - B = A$. ...(*ii*)

From (*i*) and (*ii*), it is concluded that $A - B = A \Leftrightarrow A \cap B = \Phi$

(*b*) $A - B = \Phi \Rightarrow A \cap B^c = \Phi$

$\Rightarrow$ A and B^c are disjoint

$\Rightarrow A \subseteq B$ ...(*i*)

Since A and B^c are disjoint, therefore, for an arbitrary chosen $x \in A \Rightarrow x \notin B^c$ or $x \in B$.

Conversely, assume that $A \subseteq B$. Then to show that $A - B = \Phi$, let

$$A - B = \{x : x \in A \text{ and } x \notin B\}$$
$$\subseteq \{x : x \in B \text{ and } x \notin B\} \qquad [\text{since } A \subseteq B]$$

Thus $A \subseteq B \Rightarrow A - B = \Phi$ ...(*ii*)

Hence, from (*i*) and (*ii*), it is concluded that $A - B = \Phi \Leftrightarrow A \subseteq B$.

Example 20 If A and B are any two sets, then prove that

$$A \cup B = A \cap B \Leftrightarrow A = B$$

Solution Suppose $A \cup B = A \cap B$. Then to show that $A = B$, let

$x \in A \Leftrightarrow x \in A$ or $x \in B$.

$\Leftrightarrow x \in A \cup B$

$\Leftrightarrow x \in A \cap B$ [since $A \cup B = A \cap B$]

$\Leftrightarrow x \in A$ and $x \in B$

$\Leftrightarrow x \in B$

Thus $A \cup B = A \cap B \Rightarrow A = B$ ...(*i*)

Conversely, suppose that $A = B$. Then to show that $A \cup B = A \cap B$, let

$x \in A \cup B \Leftrightarrow x \in A$ or $x \in B$

$\Leftrightarrow x \in A$ or $x \in A$ [since $A = B$]

$\Leftrightarrow x \in A$

$\Leftrightarrow x \in A$ and $x \in A$

$\Leftrightarrow x \in A$ and $x \in B$ [since $A = B$]

$\Leftrightarrow x \in A \cap B$.

Thus $A = B \Rightarrow A \cup B = A \cap B$ ....(*ii*)

Hence, from (*i*) and (*ii*), we get

$$A \cup B = A \cap B \Leftrightarrow A = B.$$

Example 21 Prove that, $A - (A - B) = A \cap B$.

Solution To prove this result, let

$x \in A - (A - B) \Leftrightarrow x \in A$ and $x \notin (A - B)$

$\Leftrightarrow x \in A$ and $x \notin (A \cap B^c)$ [since $A - B = A \cap B^c$]

$\Leftrightarrow x \in A$ and $(x \notin A$ and $x \notin B^c)$

$\Leftrightarrow x \in A$ and $(x \notin A$ and $x \in B)$

$\Leftrightarrow (x \in A$ and $x \notin A)$ and $(x \in A$ and $x \in B)$

$\Leftrightarrow x \in \Phi$ and $x \in (A \cap B)$

$\Leftrightarrow x \in A \cap B$ [since $x \notin \Phi$]

Hence, $A - (A - B) = A \cap B$.

Example 22 Prove that powerset $(A \cap B)$ is equal to the powerset (A) $\cap$ powerset (B)

Solution Let C be a non-empty set and $C \in$ powerset $(A \cap B)$. Then ...(*i*)

$C \subseteq A \cap B$ i.e., $C \subseteq A$ and $C \subseteq B$

or $x \in$ Powerset A and $x \in$ Powerset B

or $x \in$ Powerset A $\cap$ Powerset B ...(*ii*)

From (i) and (ii)

$$\text{Powerset }(A \cap B) \subseteq \text{Powerset }(A) \cap \text{Powerset }(B) \quad ...(iii)$$

Similarly, it can also be proved that

$$\text{Powerset }(A) \cap \text{Powerset }(B) \subseteq \text{Powerset }(A \cap B) \quad ...(iv)$$

From (iii) and (iv), it is concluded that

$$\text{Powerset }(A \cap B) = \text{Powerset }(A) \cap \text{Powerset }(B)$$

Example 23 Show that for any two sets A and B: A -- (A $\cap$ B) = A – B

Solution Let $x \in A - (A \cap B)$. This implies that

$x \in$ A and ($x \notin A \cap B$) = $x \in$ A and ($x \notin$ A or $x \notin$ B)

$\Rightarrow$ ($x \in$ A and $x \notin$ A) or ($x \in$ A and $x \notin$ B)

$\Rightarrow$ ($x \in \phi$) or [$x \in A - B$] $\Rightarrow x \in \phi \cap (A - B)$, i.e., $x \in A - B$

Hence $A - (A \cap B) \subseteq A - B$ ...(*i*)

Again, let $x \in A - B$. This implies that

$x \in$ A and $x \notin$ B or $x \in$ A and $x \notin A \cap B$, (since $x \notin$ B)

or $x \in A - (A \cap B)$ ...(*iii*)

Hence $(A - B) \subseteq A - (A \cap B)$

From (*i*) and (*ii*), we have $(A - B) = A - (A \cap B)$

Example 24 Prove that $A \oplus B$ or $(A - B) \cup (B - A) = (A \cup B) - (A \cap B)$.

Solution To prove this result, let

$x \in (A - B) \cup (B - A)$ $\Leftrightarrow x \in (A - B)$ or $x \in (B - A)$

$\Leftrightarrow$ ($x \in$ A and $x \notin$ B) or ($x \in$ B and $x \notin$ A)

$\Leftrightarrow$ [($x \in$ A and $x \notin$ B) or $x \in$ B] and [($x \in$ A and $x \notin$ B) or $x \notin$ A]

$\Leftrightarrow$ [$x \in$ B or ($x \in$ A and $x \notin$ B)] and [$x \notin$ A or ($x \in$ A and $x \notin$ B)]

$\Leftrightarrow$ [($x \in$ B or $x \in$ A) and ($x \in$ B or $x \notin$ B)]
and [($x \notin$ A or $x \in$ A) and ($x \notin$ A or $x \notin$ B)]

$\Leftrightarrow x \in (A \cup B)$ and $x \notin (A \cap B)$

$\Leftrightarrow x \in (A \cup B) - (A \cap B)]$

Hence, $(A - B) \cup (B - A) = (A \cup B) - (A \cap B)$.

Example 25 Prove that $(A - C) \cap (B - C) = (A \cap B) - C$

Solution To prove this result, let

$x \in (A - C) \cap (B - C)$ $\Leftrightarrow x \in (A - C)$ and $x \in (B - C)$

$\Leftrightarrow$ ($x \in$ A and $x \notin$ C) and ($x \in$ B and $x \notin$ C)

$\Leftrightarrow$ ($x \in$ A and $x \in$ B) and $x \notin$ C

$\Leftrightarrow x \in A \cap B$ and $x \notin$ C

$\Leftrightarrow x \in (A \cap B) - C$

Hence, $(A - C) \cap (B - C) = (A \cap B) - C$.

Example 26 Prove that $A \cap (B - C) = (A \cap B) - (A \cap C)$.

Solution To prove this result, let

$$\begin{aligned} x \in A \cap (B - C) &\Leftrightarrow x \in A \text{ and } x \in (B - C) \\ &\Leftrightarrow x \in A \text{ and } (x \in B \text{ and } x \notin C) \\ &\Leftrightarrow (x \in A \text{ and } x \in B) \text{ and } x \notin C \\ &\Leftrightarrow x \in (A \cap B) \text{ and } x \notin C \\ &\Leftrightarrow x \in (A \cap B) \text{ and } x \notin (A \cap C) \qquad [\text{since } x \notin C \Rightarrow x \notin (A \cap C)] \\ &\Leftrightarrow x \in (A \cap B) - (A \cap C) \end{aligned}$$

Hence, $(A \cap B) - (A \cap C) = A \cap (B - C)$.

Example 27 Show that $A \cup (B - C) \neq (A \cup B) - (A \cup C)$.

Solution To prove this result, let

$$\begin{aligned} x \in A \cup (B - C) &\Leftrightarrow x \in A \text{ or } x \in (B - C) \\ &\Leftrightarrow x \in A \text{ or } (x \in B \text{ and } x \notin C) \\ &\Leftrightarrow (x \in A \text{ or } x \in B) \text{ and } (x \in A \text{ or } x \notin C) \\ &\Leftrightarrow (x \in A \text{ or } x \in B) \text{ and } (x \in A \text{ or } x \in C^c) \\ &\Leftrightarrow x \in (A \cup B) \cap (A \cup C^c) \end{aligned}$$

Hence, $A \cup (B - C) \neq (A \cup B) - (A \cup C)$.

Example 28 There are two sets $S_1 = \{a, b, c, 2\}$ and $S_2 = \{1, 2, 3\}$. Perform the operations of union and intersection on these sets. Also find $S_1 - S_2$ and $S_2 - S_1$.

Solution (*a*) $S_1 \cup S_2 = \{a, b, c, 2, 1, 2, 3\} = \{a, b, c, 1, 2, 3\}$

(*b*) $S_1 \cap S_2 = \{2\}$

(*c*) $S_1 - S_2$ = set of those elements of S_1 which are not in $S_2 = \{a, b, c\}$

(*d*) $S_2 - S_1$ = set of those elements of S_2 which are not in $S_1 = \{1, 3\}$

Example 29 If $A = \{3, 4, 2\}$, $B = \{3, 4, 5, 6\}$ and $C = \{2, 4, 6, 8\}$, verify by actual writing the sets that

$$A \cap (B - C) = (A \cap B) - (A \cap C)$$

Solution Given that $A = \{3, 4, 2\}$, $B = \{3, 4, 5, 6\}$ and $C = \{2, 4, 6, 8\}$. Therefore,

$$B - C = \{3, 4, 5, 6\} - \{2, 4, 6, 8\} = \{3, 5\}$$

Also $A \cap (B - C) = \{3, 4, 2\} \cap \{3, 5\} = \{3\}$ ...(*i*)

Again $A \cap B = \{3, 4, 2\} \cap \{3, 4, 5, 6\} = \{3, 4\}$

$A \cap C = \{3, 4, 2\} \cap \{2, 4, 6, 8\} = \{4, 2\}$

$(A \cap B) - (A \cap C) = \{3, 4\} - \{4, 2\} = \{3\}$ ...(*ii*)

From (*i*) and (*ii*), we get

$$A \cap (B - C) = (A \cap B) - (A \cap C)$$

Example 30 If $A = \{1, 3, 5\}$, $B = \{2, 4, 6, 8\}$, $C = \{2, 5, 10\}$ and $U = \{1, 2, 3, 4, 5, 6, 7, 8, 9, 10\}$, verify by actual writing the sets that

(*a*) $(A \cap B)^c = A^c \cup B^c$ (*b*) $A \cap (B \cup C) = (A \cap B) \cup (A \cap C)$.

Solution (*a*) $A \cap B = \{1, 3, 5\} \cap \{2, 4, 6, 8\} = \Phi$. Therefore,

$(A \cap B)^c = U - (A \cap B) = U - \Phi = U$ ...(*i*)

Also $A^c = U - A = \{1, 2, 3, 4, 5, 6, 7, 8, 9, 10\} - \{1, 3, 5\} = \{2, 4, 6, 7, 8, 9, 10\}$

and $B^c = U - B = \{1, 2, 3, 4, 5, 6, 7, 8, 9, 10\} - \{2, 4, 6, 8\} = \{1, 3, 5, 7, 9, 10\}$

$$A^c \cup B^c = \{2, 4, 6, 7, 8, 9, 10\} \cup \{1, 3, 5, 7, 9, 10\}$$
$$= \{1, 2, 3, 4, 5, 6, 7, 8, 9, 10\} = U \quad ...(ii)$$

Thus, from (*i*) and (*ii*), we get

$$(A \cap B)^c = A^c \cup B^c$$

(*b*) Left as an exercise for the reader.

Example 31 Let $U = \{x : x \in N, 1 \le x \le 12\}$ be the universal set and

$$A = \{1, 9, 10\}, B = \{3, 4, 6, 11, 12\} \text{ and } C = \{2, 5, 6\}$$

are subsets of U. Find the sets

(*a*) $(A \cup B) \cap (A \cup C)$ (*b*) $A \cup (B \cap C)$ (*c*) $(A \cup B \cup C)^c$.

Solution Given that, $U = \{x : x \in N, 1 \le x \le 12\} = \{1, 2, 3, 4, 5, 6, 7, 8, 9, 10, 11, 12\}$

(*a*) $A \cup B = \{1, 9, 10\} \cup \{3, 4, 6, 11, 12\} = \{1, 3, 4, 6, 9, 10, 11, 12\}$

and $A \cup C = \{1, 9, 10\} \cup \{2, 5, 6\} = \{1, 2, 5, 6, 9, 10\}$

Thus $(A \cup B) \cap (A \cup C) = \{1, 3, 4, 6, 9, 10, 11, 12\} \cap \{1, 2, 5, 6, 9, 10\} = \{1, 6, 9, 10\}$

(*b*) $B \cap C = \{3, 4, 6, 11, 12\} \cap \{2, 5, 6\} = \{6\}$. Thus

$$A \cup (B \cap C) = \{1, 9, 10\} \cup \{6\} = \{1, 6, 9, 10\}$$

(*c*) $(A \cup B \cup C) = \{1, 9, 10) \cup \{3, 4, 11, 12\} \cup \{2, 5, 6\} = \{1, 2, 3, 4, 5, 6, 9, 10, 11, 12\}$

Thus $(A \cup B \cup C)^c = U - (A \cup B \cup C) = \{7, 8\}$.

Example 32 If $X = Y \cup Z$, show that $X - Y = Z \cap Y^c$.

Solution Writing X – Y as follows:

$$X - Y = X \cap Y^c = (Y \cup Z) \cap Y^c \quad [\text{since } X = Y \cup Z]$$
$$= (Y \cap Y^c) \cup (Z \cap Y^c) = \Phi \cup (Z \cap Y^c) = Z \cap Y^c.$$

1.13 COMPUTER REPRESENTATION OF SETS

Bit string method can be used to represent sets in the computer by storing their elements in an ordered manner. Since set operations like union, intersection, difference, etc. of sets take large amount of time in searching their elements, therefore, an arbitrary ordering of the elements of the universal set to store the elements is commonly used to represent sets.

Suppose an universal set $U = \{x_1, x_2, \ldots, x_n\}$ has n elements. Then its subsets can be represented with a bit string of length n. A *bit string* is a string over the alphabet {0, 1}. If the set A is a subset of U, then it is represented by bit string where ith bit of the string is 1 when $x_i \in A$ and 0 when $x_i \notin A$. This rule permit us to represent a universal set as an array of length n in the computer. Assignment of either 0 or 1 to each location A [k] of the array specifies a unique subset of U.

Illustration Let U = {1, 2, 3, 4, 5, 6} be a universal set and A = {1, 3}, B = {3, 5, 6} be its subsets. The bit string of the set A is of length 6 and has value 1 when $x = 1$ or 3, otherwise has value 0. It is, therefore, represented as: 101000. Similarly, the bit string of the set B has value 1 when $x = 3$ or 5 or 5, otherwise has value 0. It is represented as: 001011.

It is easy to find complement, union, intersection of sets using bit strings method to represent sets as follows:

1. The *complement* of a set from the bit string can be obtained by changing each 1 to 0 and vice versa.
2. The bit string for the *union* is the bitwise OR (a Boolean operation) of the bit strings for two sets.
3. The bit string for the intersection is the bitwise AND (a Boolean operation) of the bit strings for two sets.

Example 33 If U = {1, 2, 3, 4, 5, 6} A = {1, 2, 3, 4} and B = {3, 4, 5, 6}. Find the bit string for the sets A and B and use them to find union, intersection of A and B. Also find A^c and B^c.

Solution The bit string of sets A and B is of length 6 and its representation for the set A is 111100, and for the set B is 001111.

The bit string for the union of sets A and B is given by

$$111100 \cup 001111 = 111111$$

which corresponds to the universal set, U = {1, 2, 3, 4, 5, 6}

The bit string for the intersection of sets A and B is given by

$$111100 \cap 001111 = 001100$$

which corresponds to the set {3, 4}.

The complement of the set A is 000011 which corresponds to the set {5, 6}. Similarly, the complement of the set B is 110000 which corresponds to the set {1, 2}.

Example 34 If U = {1, 2, 3, 4, 5, 6, 7}, then find the set specified by each of the following bit strings

(*a*) 1010100 (*b*) 0101010 (*c*) 0011001 (*d*) 1110001

Solution (*a*) 1010100 = {1, 3, 5} (*b*) 0101010 = {2, 4, 6]

(*c*) 0011001 = {3, 4, 7} (*d*) 1110001 = {1, 2, 3, 7}

Example 35 What is the bit string corresponding to the difference of two sets? Use it to find A – B, where A = {2, 3, 5} and B = {3, 5}.

Solution The bit in the *i*th position of the bit string of the difference of two sets is 1 when *i*th bit of the first is 1 and *i*th bit of the second string is 0, and 0 otherwise. Difference of two sets is: A – B = {2} = 01000.

Self-practice Problems II

12. Let U = {0, 1, 2, 3, 4, 5, 6, 7}, A = {0, 2, 4, 6}, B = {1, 3, 5, 7} and C = {0, 3, 6}. Find

(*i*) $A \cup B$ (*ii*) $B \cap C$ (*iii*) B^c (*iv*) $A - B$ (*v*) $B \oplus C$

13. Let A = {1, 2, 3}, B = {3, 4, 5}. Find A – B.

14. If A, B and C are three sets, then prove that

(*i*) $A \cup (B \cup C) = (A \cup B) \cup C$ (*ii*) $A \cup (B \cap C) = (A \cup B) \cap (A \cup C)$

(*iii*) $A - (B \cap C) = (A - B) \cup (A - C)$ (*iv*) $A \cap (B - C) = (A \cap B) - (A \cap C)$

15. For A = {*a*, *b*, *c*, *d*}, B = {*b*, *e*}, C = {*a*, *e*, *f*}. Verify that $A \cup (B \cap C) = (A \cup B) \cap (A \cup C)$.

16. If A and B are any two sets, then prove that

(*i*) $A \subseteq (A \cup B)$ and $B \subseteq (A \cup B)$ (*ii*) $(A \cap B) \subseteq A$ and $(A \cap B) \subseteq B$

(*iii*) $(A \cap B) \subseteq A \subseteq (A \cup B)$
(*iv*) $(A - B) \subseteq A$
(*v*) $A \cap (B - A) = \Phi$
(*vi*) $(A \cup B)^c = A^c \cap B^c$
(*vii*) $(A \cap B)^c = A^c \cup B^c$
(*viii*) $A \cup B = (A - B) \cup B$
(*ix*) $A - B = A - B^c = B^c - A$
(*x*) $(A - B) \cup (B - A) = (A \cup B) - (A \cap B)$
(*xi*) $A - (A - B) = A \cap B$
(*xii*) $(A \cup B) \cap A = A$
(*xiii*) $(A - B) - C = A - (B \cup C)$
(*xiv*) $(A - B) - C = (A - C) - B$
(*xv*) $(A - B) - C = (A - C) - (B - C)$

17. Prove that the empty set is the subset of every set.

18. Show with the help of Venn diagrams.

(*i*) $(A^c)^c = A$
(*ii*) $(A \cap B)^c = A^c \cup B^c$.

19. Prove the following:

(*i*) $A \cap (A^c \cup B) = A \cap B$
(*ii*) $A^c - B^c = B - A$
(*iii*) $B - A^c = B \cap A$
(*iv*) $A \cup (A^c \cap B) = A \cup B$
(*v*) $A^c - B^c = (A \cap B)^c$
(*vi*) $A \cap B = \Phi \Rightarrow A \cup B^c = B$.

20. If A and B are subsets of a set X, then prove that $A \subseteq B \Leftrightarrow X - B \subseteq X - A$.

21. Prove the following:

(*i*) $A \subseteq B \Leftrightarrow A \cup B = B$
(*ii*) $A \subseteq B \Leftrightarrow A \cap B = A$
(*iii*) $C \subseteq A$ and $C \subseteq B \Leftrightarrow C \subseteq A \cap B$
(*iv*) $C \subseteq A$ and $B \subseteq C \Leftrightarrow A \cup B \subseteq C$.

22. Let $U = \{x : |x| < 5, x \in 1\}$ be universal set,

$A = \{x : x^3 + x^2 - 2x = 0\}$ and $B = \{x : x \text{ is even integer}\}$

Find $A - B$ and $B - A$.

23. If $U = \{x : x < 12, x \in N\}$, $A = \{x : x = 1, 2, 3, 4\}$, $B = \{x : x < 11 \text{ and } x \text{ is even natural number}\}$; and $C = \{x : x \in N, 2 < x < 7\}$. Find (*i*) A^c (*ii*) B^c (*iii*) $(B - C)^c$.

24. If $A = \{1, 2, 3, 4\}$, $B = \{2, 3, 4, 5\}$, $C = \{1, 3, 4, 5, 6, 7\}$, verify that,

$A \cap (B \cup C) = (A \cap C) \cup (A \cap C)$

25. Let $A = \{1, 2, 3, 4, \ldots, 8, 9\}$, $B = \{2, 4, 6, 8\}$, $C = \{1, 3, 5, 7, 9\}$, $D = \{3, 4, 5\}$ and $E = \{3, 5\}$. Then which set can equal X if we are given the following information?

(*i*) X and B are disjoint
(*ii*) $X \subseteq A$ but $x \not\subseteq C$
(*iii*) $X \subseteq D$ but $X \not\subseteq B$
(*iv*) $X \subseteq C$ but $x \not\subseteq A$

26. List the sets A, B and C, given that $A \cup B = \{p, q, r, s\}$, $A \cup C = \{q, r, s, t\}$, $A \cap B = \{q, r\}$ and $A \cap C = \{q, s\}$.

27. Let A, B, C be subsets of U. Given that $A \cap B = A \cap C$ and $A^c \cap B = A^c \cap C$. Is it necessary that $B = C$? Justify your answer.

28. Comment on the sets A and B, if

(*i*) $A \cap B = A$ (*ii*) $A \cup B = A$ (*iii*) $A \oplus B = A$ (*iv*) $A \cap B = A \cup B$.

29. Prove that $A \cap B = A \cap C$ without $B = C$.

30. Let A and B be any two sets. Prove that

(*i*) A is the disjoint union of $A - B$ and $A \cap B$.

(*ii*) $A \cup B$ is the disjoint union of $A - B$, $A \cap B$, and $B - A$.

31. If $A - B = A \cap B^c$, then define the union $A \cup B$ in terms of the operations of intersection and complement.

32. Write the dual of each set equation

(*i*) $A \cup B = (B^c \cap A^c)^c$ (*ii*) $A = A \cup (A \cup B)$

(*iii*) $A = (B^c \cap A) \cup (A \cap B)$ (*iv*) $(A \cup B \cup C)^c = (A \cup C)^c \cap (A \cup B)^c$

33. (*a*) Let N = {1, 2, 3, . . . } be the set of natural numbers, and

$A = \{x : x \leq B\}$, $B = \{x : 6 \leq x \leq 11\}$, $C = \{1, 3, 5, 7, 9, 11\}$, $D = \{2, 3, 5, 7, 8\}$

Find (*i*) $A \oplus B$ (*ii*) $B \oplus C$ (*iii*) $A \cap (B \oplus C)$ (*iv*) $(A \cap B) \oplus (A \cap D)$

(*b*) Let $A = \{1, 2, 3, 4\}$, $B = \{1, 5\}$ and $C = \{1, 5\}$. If $A \oplus B = A \oplus C$, then prove that $B = C$.

(*c*) Determine the set (*i*) $\Phi \oplus \{a, \Phi, \{\Phi\}\}$, (*ii*) $\{\Phi\} \oplus \{a, \Phi, \{\Phi\}\}$.

34. Let A, B, C be three non-empty sets. Prove the following properties of the symmetric difference

(*i*) Associative law : $A \oplus (B \oplus C) = (A \oplus B) \oplus C$

(*ii*) Commutative law : $A \oplus B = B \oplus A$

(*iii*) Cancellation law : $A \oplus B = A \oplus C \Rightarrow B = C$

(*iv*) Distributive law : $A \cap (B \oplus C) = (A \cap B) \oplus (A \cap C)$

35. (*i*) Partition the set A = {1, 2, 3, . . ., 10} into the minsets generated by the sets $B_1 = \{1, 7, 8\}$, $B_2 = \{1, 6, 9, 10\}$ and $B_3 = \{1, 9, 10\}$.

(*ii*) How many different subsets of the set A may be generated using B_1, B_2 and B_3 with standard set operations?

(*iii*) Do these exits subsets C_1, C_2 and C_3 that will generate every subset of A?

(*iv*) How many elements of the power set of A can be generated by B_1, B_2 and B_3?

36. Let A = {1, 2, 3, 4, 5, 6}. Determine whether or not each of the following is a partition of A.

(*i*) $P_1 = \{1, 2, 3\}, \{1, 4, 5, 6\}$ (*ii*) $P_2 = \{1, 2\}, \{3, 5, 6\}$

(*iii*) $P_3 = \{1, 3, 5\}, \{2, 4\}, \{6\}$ (*iv*) $P_4 = \{1, 3, 5\}, \{2, 4, 6\}$

37. Determine whether or not each of the following is a partition of the set, I, of positive integers:

(*i*) $\{n : n > 5\}, \{n : n < 5\}$ (*ii*) $\{n : n > 5\}, \{0\}, \{1, 2, 3, 4, 5\}$

(*iii*) $\{n : n^2 > 11\}, \{n : n^2 < 11\}$

38. For each $n \in N$, let $A_n = [n, 2n]$. Determine the following:

(*a*) $A_2 \cap A_6$ (*b*) $A_5 \cap A_8$ (*c*) $\bigcup_{n=4}^{8} A_n$

39. For each $n \in N$, let $A_n = [-2n, 3n]$. Determine the following:

(*a*) A_3 (*b*) $A_3 - A_4$ (*c*) $\bigcup_{n=1}^{7} A_n$ (*d*) $\bigcup_{n \in N} A_n$

40. Let $A_n = \{x : x \text{ is a multiple of } n \in N\} = \{n, 2n, 3n, \ldots\}$. Determine the following

(*a*) $A_2 \cap A_5$ (*b*) $A_4 \cap A_6$

41. Let A_n (0, 1/*n*), for $n \in N$. Find (*a*) $A_3 \cup A_5$, (*b*) $A_4 \cap A_{10}$ and (*c*) $\bigcap_{i \in N} A_i$.

42. (*a*) If A, B are uncountable, then prove that $A \cap B$ is countable set.

(*b*) If A, B are uncountable, then prove that $A - B$ is countable set.

(*c*) Let A, B be sets with A uncountable. If $A \subseteq B$, prove that B is uncountable.

43. (*a*) If A is uncountable set and B is countable set, prove that A – B is uncountable set.
(*b*) Show that the set of all bit strings is countable.
(*c*) Show that the set of integers that are multiples of 5 are countable.

Hints and Answers

12. (*i*) {0, 1, 2, 3, 4, 5, 6, 7} (*ii*) {3} (*iii*) {0, 2, 4, 6} (*iv*) {0, 2, 4, 6} (*v*) {0, 1, 5, 6, 7}

13. {1, 2, 4, 5}

16. (*xi*) $A - (A - B) = A - (A \cap B^c) = A \cap (A \cap B^c)^c = A \cap (A^c \cup B) = (A \cap A^c) \cup (A \cap B)$
$= \Phi \cup (A \cap B) = A \cap B.$

23. $A^c = \{5, 6, 7, 8, 9, 10, 11\}$; $B^c = \{1, 3, 5, 7, 9, 11\}$; $(B - C)^c = \{1, 3, 4, 5, 6, 7, 8, 9, 11\}$.

24. $A \cap (B \cup C) = \{1, 2, 3, 4\}$.

25. (*i*) X = C, E; (*ii*) X = A, B and D (*iii*) X = D, E (*iv*) X = none.

26. $A = \{q, r, s\}$; $B = \{p, q, r\}$; $C = \{q, s, t\}$.

29. Let A = {2, 3}, B = {3, 4}, and C = {3, 5}. Then $A \cap B = \{3\}$, $A \cap C = \{3\}$. Thus $A \cap B = A \cap C$. But $B \neq C$.

31. $A \cup B = (A^c \cap B^c)^c$

32. (*i*) $A \cap B = (B^c \cup A^c)^c$ (*ii*) $A = A \cup (A \cup B)$
(*iii*) $A = (B^c \cup A) \cap (A \cup B)$ (*iv*) $(A \cap B \cap C)^c = (A \cap C)^c \cup (A \cap B)^c$

33. (*a*) (*i*) {1, 2, 3, 4, 5, 9, 10 11} (*ii*) {1, 3, 5, 6, 10}
(*iii*) {1, 3, 5, 6} (*iv*) {3, 5, 6, 8}
(*c*) (*i*) $\{a, \Phi, \{\Phi\}\}$ (*ii*) $\{a, \{\Phi\}\}$.

35. (*i*) {1}, {2, 3, 4, 5}, {6}, {7, 8}, {9, 10} (*iv*) 2^5

36. (*i*) No (*ii*) No (*iii*) Yes (*iv*) Yes

37. (*i*) No (*ii*) No (*iii*) Yes

38. (*a*) ϕ (*b*) {8, 10} (c) {8}

39. (*a*) {– 6, 9} (*b*) ϕ (*c*) A_7 (*d*) R

40. (*a*) A_{10} (*b*) A_{12}

41. (*a*) A_3 (*b*) A_{10} (*c*) ϕ

1.14 THE INCLUSION AND EXCLUSION PRINCIPLE

The number of elements in finite sets such as $A \cup B$, $A \cap B$, $A \oplus B$, etc., are obtained by adding as well as excluding certain terms. This method of finding the number of elements in a finite set (also called *cardinal number*) is known as *inclusion-exclusion principle*. The *cardinal number* of a set A is denoted by *n*(A). For example, if A = {1, 2, 3, 4}, then *n* (A) = 4 as A contains only four elements.

Some simple but important results on the cardinality of sets are

1. $n(A \cup B) \le n(A) + n(B)$ when $A \cap B = \phi$
2. $n(A \cap B) \le \text{Min } \{n(A), n(B)\}$
3. $n(A \oplus B) = n(A) + n(B) - 2n\ (A \cap B)$
4. $n(A - B) \ge n(A) - n(B)$
5. $n(A \cup B) = n(A) + n(B) - n(A \cap B)$

   It is because when elements of A and B are added, the elements of $A \cap B$ had been counted twice, since $A \cap B$ is the subset of both A and B. But if sets A and B are disjoint, i.e., $A \cap B = \Phi$, then

$$n(A \cup B) = n(A) + n(B).$$

6.(*a*) $n(A \cup B \cup C) = n(A) + n(B) + n(C) - n(A \cap B) - n(B \cap C) - n(A \cap C) + n(A \cap B \cap C)$

   (*b*) If the sets A, B and C are mutually disjoint, then

$$n(A \cup B \cup C) = n(A) + n(B) + n(C)$$

7. $n(A^c) = n(U) - n(A)$

8.(*a*) $n(A) = n[(A \cap B^c) \cup (A \cap B)] = n[(A - B) \cup (A \cap B)] = n(A - B) + n(A \cap B)$

   (*b*) $n(B) = n(B - A) + n(A \cap B)$.

9. $n(A \cup B) = n(A - B) + n(A \cap B) + n(B - A)$.

Example 36 In a group of 200 people, each of whom is at least accountant or management consultant or sales manager, it was found that 80 are accountants, 110 are management consultants and 130 are sales managers, 25 are accountants as well as sales managers, 70 are management consultants as well as sales managers, 10 are accountants, management consultants as well as sales managers. Find the number of those people who are accountants as well as management consultants but not sales managers.

Solution Suppose A, M and S denote the set of accountants, management consultants and sales managers, respectively. Then it is given that

$$n(A) = 80,\ n(M) = 110,\ n(S) = 130,\ n(A \cap S) = 25$$
$$n(M \cap S) = 70 \text{ and } n(A \cap M \cap S) = 10$$

Now we should find the number of those people who are accountants as well as management consultants but not sales managers, i.e., $n\ (A \cap M \cap S^c)$.

$$n(A \cup M \cup S) = n(A) + n(M) + n(S) - n(A \cap M) - n(M \cap S) - n(A \cap S) + n(A \cap M \cap S)$$

or $$200 = 80 + 110 + 130 - n(A \cap M) - 70 - 25 + 10$$

$$200 = 330 - 95 - n\ (A \cap M)$$

or $$n(A \cap M) = 330 - 200 - 95 = 35$$

Now $$n(A \cap M) = n(A \cap M \cap S^c) + n(A \cap M \cap S)$$

$$35 = n(A \cap M \cap S^c) + 10$$

or $$n(A \cap M \cap S^c) = 35 - 10 = 25$$

Example 37 A TV survey shows that 60 per cent people see programme A, 50 per cent see programme B, 50 per cent see programme C, 30 per cent see programme A and B, 20 per cent see programme B and C, 30 per cent see programme A and C, and 10 per cent do not see any programme. Find

(*a*) what per cent see programme A, B and C?

(*b*) what per cent see exactly two programmes?

(*c*) what per cent see programme A only?

Solution Suppose X, Y and Z denote the set of people who see programme A, B and C, respectively. Then it is given that

$$n(X) = 60,\ n(Y) = 50,\ n(Z) = 50,\ n(X \cap Y) = 30,\ n(Y \cap Z) = 20,\ n(X \cap Z) = 30$$
$$n[(X \cup Y \cup Z)^c] = 10$$

(*a*) Let $n(X \cup Y \cup Z) + n[(X \cup Y \cup Z)^c] = 100$. Then

$$n(X \cup Y \cup Z) = 100 - n[(X \cup Y \cup Z)^c] = 100 - 10 = 90$$

But $$n(X \cup Y \cup Z) = n(X) + n(Y) + n(Z) - n(X \cap Y) - n(Y \cap Z) - n(X \cap Z) + n(X \cap Y \cap Z)$$

$$90 = 60 + 50 + 50 - 30 - 20 - 30 + n(X \cap Y \cap Z)$$

or $$n(X \cap Y \cap Z) = 90 - 80 = 10$$

Thus, 10 per cent people see programme A, B and C.

(*b*) Since the set of people who see programme A and B but not C, i.e., $X \cap Y \cap Z^c$ and the set of people who see all the programmes A, B and C, i.e., $X \cap Y \cap Z$ are disjoint sets, therefore,

$$n(X \cap Y \cap Z^c) + n(X \cap Y \cap Z) = n(X \cap Y)$$

or $$n(X \cap Y \cap Z^c) = n(X \cap Y) - n(X \cap Y \cap Z) = 30 - 10 = 20,$$

Similarly $$n(X \cap Y^c \cap Z) + n(X \cap Y \cap Z) = n(X \cap Z)$$

$$n(Z \cap Y^c \cap Z) = n(X \cap Z) - n(X \cap Y \cap Z) = 30 - 10 = 20,$$

and $$n(X^c \cap Y \cap Z) + n(X \cap Y \cap Z) = n(Y \cap Z)$$

$$n(X^c \cap Y \cap Z) = n(Y \cap Z) - n(X \cap Y \cap Z) = 20 - 10 = 10.$$

Thus, the percentage of people who see exactly two programmes = 20 + 20 + 10 = 50.

(*c*) Since $$n(X \cap Y^c \cap Z^c) = n(X) - n(X \cap Y) - n(X \cap Z) + n(X \cap Y \cap Z)$$

$$= 60 - 30 - 30 + 10 = 10$$

Thus, the percentage of people who see programme A only is 10.

Example 38 In a pollution study of 1,500 Indian rivers, the following data were reported: 520 were polluted by sulphur compounds, 335 were polluted by phosphates, 425 were polluted by crude oil, 100 were polluted by both crude oil and sulphur compounds, 180 were polluted by sulphur compounds and phosphates, 150 were polluted by both phosphates and crude oil and 28 were polluted by sulphur compounds, phosphates and crude oil. How many of the rivers were polluted by at least one of the three impurities? How many rivers were not polluted by exactly one of the three impurities? How many of the rivers were not polluted?

Solution Suppose S, P and C denote the set of rivers polluted by sulphur compounds, phosphates and crude oil, respectively. Then it is given that

$$n(S) = 520, \quad n(P) = 335, \quad n(C) = 425, \quad n(C \cap S) = 100,$$

$$n(S \cap P) = 180, \quad n(P \cap C) = 150, \quad n(S \cap P \cap C) = 28.$$

(*a*) Number of rivers polluted by at least one of the three impurities

$$n(S \cup P \cup C) = n(S) + n(P) + n(C) - n(S \cap P) - n(P \cap C) - n(C \cap S) + n(S \cap P \cap C)$$

$$= 520 + 335 + 425 - 100 - 180 - 150 + 28 = 878$$

(*b*) Number of rivers polluted by exactly sulphur compounds

$$n(S \cap P^c \cap C^c) = n[S \cap (P \cup C)^c] = n(S) - n(S \cap P) - n(S \cap C) + n(S \cap P \cap C)$$

$$= 520 - 180 - 100 + 28 = 268$$

(*c*) Number of rivers polluted by exactly phosphate

$$n(P \cap S^c \cap C^c) = n[P \cap (S \cup C)^c] = n(P) - n(P \cap S) - n(P \cap C) + n(P \cap S \cap C)$$

$$= 335 - 180 - 150 + 28 = 33$$

(*d*) Number of rivers polluted by exactly crude oil

$$n(C \cap P^c \cap S^c) = n[C \cap (P^c \cup S^c)] = n(C) - n(C \cap P) - n(C \cap S) + n(C \cap S \cap P)$$

$$= 425 - 150 - 100 + 28 = 203.$$

Thus, the number of rivers which were polluted by exactly one of the three impurities $= 268 + 33 + 203 = 504$ and so the number of rivers which were not polluted at all $= 1500 - 872 = 622$.

Example 39 Of a group of 20 persons, 10 are interested in music, 7 are interested in photography, and 4 like swimming; further more 4 are interested in both music and photography, 3 are interested in both music and swimming, 2 are interested in both photography and swimming and one is interested in music, photography and swimming. How many are interested in photography but not in music and swimming?

Solution Supppose M, P and S denote the set of persons interested in music, photography and swimming, respectively. Then it is given that

$$n(M) = 10, \quad n(P) = 7, \quad n(S) = 4, \quad n(M \cap P) = 4, \quad n(M \cap S) = 3$$

$$n(P \cap S) = 2 \text{ and } n(M \cap P \cap S) = 1.$$

Number of persons interested in photography but not in music and swimming is given by

$$n(P \cap M^c \cap S^c) = n[P \cap (M \cup S)^c] = n(P) - n(P \cap M) - n(P \cap S) + n(P \cap M \cap S)$$

$$= 7 - 4 - 2 + 1 = 2$$

Example 40 Out of 450 students in a school, 193 students read Science and 200 students read Commerce, 80 students read neither. Find out how many read both.

Solution Suppose A and B denote the set of students who read Science and Commerce, respectively. It is given that

$$n(A) = 193, \quad n(B) = 200, \quad n(U) = 450 \quad \text{and} \quad n(A^c \cap B^c) = 80.$$

Now we should find the number of those students who read Science as well as Commerce, i.e., $n(A \cap B)$.

Since $A^c \cap B^c = (A \cup B)^c$, therefore, $n(A \cup B)^c = 80$. But

$$n(A \cup B)^c = n(U) - n(A \cup B)$$

$$80 = 450 - n(A \cup B) \text{ or } n(A \cup B) = 450 - 80 = 370$$

Further, we know that $n(A \cup B) = n(A) + n(B) - n(A \cap B)$

or $\quad 370 = 193 + 200 - n(A \cap B)$ or $n(A \cap B) = 23$

Thus, 23 students read both science and commerce.

Example 41 In a survey concerning the smoking habits of consumers, it was found that 55 per cent smoke cigarette A, 50 per cent smoke B, 42 per cent smoke C, 28 per cent smoke A and B, 20 per cent smoke A and C, 12 per cent smoke B and C and 10 per cent smoke all the three cigarettes.

(*a*) What percentage do not smoke?

(*b*) What percentage somke exactly two brands of cigarettes?

Solution Let A, B, C be the sets of persons who smoke brand A, B and C, respectively. It is given that

$$n(A) = 55; n(B) = 50; n(C) = 42; n(A \cap B) = 28; n(A \cap C) = 20; n(B \cap C) = 12; \text{ and}$$
$$n(A \cap B \cap C) = 10.$$

(*a*) Now (A ∪ B ∪ C) is the set of all person who smoke either A or B or C; or any two brands or all the brands and $(A \cup B \cup C)^c$ is the set of persons who do not smoke. Therefore

$$n(A \cup B \cup C) = n(A) + n(B) + n(C) - n(A \cap B) - (B \cap C) - n(A \cap C) + n(A \cap B \cap C)$$

$$= 55 + 50 + 42 - 28 - 12 - 20 + 10 = 97$$

But, $n(A \cup B \cup C)^c = 100 - n(A \cup B \cup C) = 100 - 97 = 3$

Hence, 3 per cent persons do not smoke.

(*b*) $A \cap B \cap C^c$ is the set of persons who smoke A and B, but not C. $(A \cap B \cap C)$ is the set of persons who smoke all the 3 brands A, B and C. Therefore,

$$n(A \cap B) = n(A \cap B \cap C^c) + n(A \cap B \cap C)$$

$$28 = n(A \cap B \cap C^c) + 10 \text{ or } n(A \cap B \cap C^c) = 18.$$

Similarly, $n(A \cap B^c \cap C) = 10$ and $n(A^c \cap B \cap C) = 2$.

Hence, total required number = 18 + 10 + 2 = 30.

Example 42 In a class of 25 students, 12 have taken economics, 8 have taken economics but not political science. Find the number of students who have taken economics and political science and those who have taken politics but not economics.

Solution Suppose A and B denote the set of students who take economics and political science, respectively. It is given that

$$n(A) = 12, n(A \cup B) = 25, \text{ and } n(A \cap B^c) = 8.$$

Now we have to find the number of students who have taken economics and political science, i.e., $n(A \cap B)$ and the number of students who have taken political science but not economics, i.e , $n(B \cap A^c)$. Now

$$n(A) = n(A \cap B^c) + n(A \cap B) \qquad [\text{since } A = (A \cap B^c) \cup (A \cap B)]$$

or $12 = 8 + n(A \cap B)$ or $n(A \cap B) = 12 - 8 = 4$

Also $n(A \cup B) = n(A) + n(B) - n(A \cap B)$

$$25 = 12 + n(B) - 4 \text{ or } n(B) = 17.$$

Again, $n(B) = n(A \cap B) + n(B \cap A^c)$. Therefore,

$$17 = 4 + n(B \cap A^c) \text{ or } n(B \cap A^c) = 17 - 4 = 13.$$

Example 43 Determine the number of integers between 1 and 250 that are divisible by any of the integers 2, 3, 5 and 7.

Solution Suppose A_1, A_2, A_3 and A_4 denote the set of integers between 1 and 250 that are divisible by the integers 2, 3, 5 and 7, respectively. It is given that

$$n(A_1) = 250/2 = 125;\ n(A_2) = 250/3 = 83;\ n(A_3) = 250/4 = 50, \text{ and } n(A_4) = 250/7 = 35$$

Thus, we have

$$n(A_1 \cap A_2) = 250/2 \times 3 = 41 \qquad n(A_1 \cap A_3) = 250/2 \times 5 = 25$$
$$n(A_1 \cap A_4) = 250/2 \times 7 = 17 \qquad n(A_2 \cap A_3) = 250/3 \times 5 = 16$$
$$n(A_2 \cap A_4) = 250/3 \times 7 = 11 \qquad n(A_3 \cap A_4) = 250/5 \times 7 = 7$$

and $n(A_1 \cap A_2 \cap A_3) = 250/2 \times 3 \times 5 = 8$ $\quad n(A_1 \cap A_2 \cap A_4) = 250/2 \times 3 \times 7 = 5$

$n(A_1 \cap A_3 \cap A_4) = 250/2 \times 5 \times 7 = 3$ $\quad n(A_2 \cap A_3 \cap A_4) = 250/3 \times 5 \times 7 = 2$

$n(A_1 \cap A_2 \cap A_3 \cap A_4) = 250/2 \times 3 \times 5 \times 7 = 1.$

Hence, $n(A_1 \cup A_2 \cup A_3 \cup A_4) = n(A_1) + n\ (A_2) + n\ (A_3) + n\ (A_4) - n\ (A_1 \cap A_2)$

$- n(A_1 \cap A_3) - n\ (A_1 \cap A_4) - n(A_2 \cap A_3) - n(A_2 \cap A_4)$

$- n(A_3 \cap A_4) + \ldots - n(A_1 \cap A_2 \cap A_3 \cap A_4)$

$= 125 + 83 + 50 + 35 - 41 - 25 - 17 - 16 - 11 - 7 + 8 + 5$

$+ 3 + 2 - 1 = 193.$

Example 44 Among the first 500 positive integers,

(*a*) determine the integers which are neither divisible by 5, 7 nor 9.

(*b*) determine the integers which are divisible by 5 but not by 7 and 9.

Solution Suppose A, B and C are the set of integers divisible by 5, 7 and 9, respectively. Then

$$n(A) = 500/5 = 100;\ n(B) = [500/7] = 71, \text{ and } n(C) = [500/9] = 55$$

where [ ] indicates largest positive integer value

Also $n(A \cap B) = [500/(5 \times 7)] = 14,\ n(A \cap C) = [500/(5 \times 9)] = 11$

$n(B \cap C) = [500/(7 \times 9)] = 7$

$n(A \cap B \cap C) = [500/(5 \times 7 \times 9)] = 1$

(*a*) The number of integers divisible by 5, 7 and 9 are

$$n(A \cup B \cup C) = n(A) + n(B) + n(C) - n(A \cap B) - n(B \cap C) - n(C \cap A) + n(A \cap B \cap C)$$
$$= 100 + 71 + 55 - 14 - 7 - 11 + 1 = 195$$

Thus, the number of integers which are neither divisible by 5, 7 nor 9 are: 500 – 195 = 305.

(*b*) The number of integers which are divisible by 5, 7 and 9 are only one. Therefore, integers divisible by 5 and 7 but not by all three of them will be 14 – 1 = 13. Similarly, integers divisible by 5 and 9 but not by all three of them will be 11 – 1 = 10.

Hence, integers divisible by 5 but not 7 and 9 are: 100 – 13 – 10 – 1 = 76.

Example 45 Determine the cardinalities of the sets

(*a*) $A = \{n^7 : n \text{ is a positive integer}\}$ (*b*) $B = \{n^{109} : n \text{ is a positive integer}\}$ (*c*) $A \cup B$.

Solution (*a*) The cardinality of the set A is infinity because its elements, i.e., number of positive integers which it contains are infinite.

(*b*) The cardinality of the set is also infinite because of the same reason as in (*i*).

(*c*) Since both A and B are infinite sets, therefore the cardinality of their union is also infinite.

1.15 FUZZY SETS

The basic concept of the fuzzy set forms a generalisation of the *classical* or *crisp set*. The crisp set is defined in such a way as to classify the individuals in some given universe into two groups: (*i*) members which belong to the set, and (*ii*) non-members which do not belong to the set. A sharp distinction exists between the members and non-members of the class or category represented by the crisp set. However, many of the collections or categories such as the classes of tall people, animals, expensive cars, beautiful women, numbers much greater than 1, etc. do not constitute sets in the usual mathematical sense of these terms. Yet, the fact remains that such imprecisely defined 'classes' or sets play an important role in human thinking, particularly in the domain of pattern recognition, communication of information and abstraction.

The concept of fuzzy set provides a convenient point of departure for the construction of a conceptual framework which parallels in many respect the framework used in the case of ordinary sets, but is more general than the latter and potentially, may prove to have a much wider scope of applicability, particularly in the field of pattern classification and information processing. Essentially, such a framework provides a natural way of dealing with problems in which the source of imprecision in the absence of sharply defined criteria of class membership rather than the presence of random variables.

Let C be a crisp set defined on the universe U. Then for any point (object) of U, either $x \in$ C or $x \notin$ C. However, if a fuzzy set F is in U, then it is not necessary that either $x \in$ F or $x \not\in$ F.

In fuzzy set theory, the characteristic (membership) function $f_F(x)$ associates each point $x \in$ U a real number in the interval [0, 1] instead of two-element set {0, 1}. The value of $f_F(x)$ at x represents the 'grade of membership' of x in F. This grade corresponds to the degree to which that individual element is similar or campatible with the concept represented by the fuzzy set. The grade of membership to an element is given by a subjectively defined membership function. Thus, the nearer of $f_F(x)$ to 1, the higher the grade of membership of x in F, whereas closer the value of $f_F(x)$ to 0, the lower the grade of membership in the fuzzy set.

Definition Let U be an universal set. A fuzzy set (class) F in U is characterised by a membership (characteristic) function $f_F(x)$ as:

$$f_F(x) : U \rightarrow [0, 1]$$

This implies that every element $x \in$ U has a membership degree $f_F(x) \in [0, 1]$. Thus, the fuzzy set F can also be defined by the set of triples

$$F = \{x, f_F(x) : x \in U\}$$

Thus a fuzzy set can be defined as the set of pairs consisting of each possible individual in the universe of discourse and its grade of membership.

Illustrations

1. Let R be the real line and let A be a fuzzy set of numbers which are much greater than 1, then one can give a precise but subjective, characterisation of A by specifying $f_A(x)$ as a function on R. The representative values of such a function might be: $f_A(0) = 0$, $f_A(1) = 0$, $f_A(5) = 0.1$, $f_A(10) = 0.2$, $f_A(100) = 0.95$, $f_A(500) = 1$.
2. A realtor wants to classify the house he offers to his clients. One indicator of comfort of these houses is the number of bedrooms in it. Let U = {1, 2, . . . , 10} be the set of available types of houses described by x = number of bedrooms in a house. Then the fuzzy set of comfortable type of house for a 4-person family may be described as:

$$A = \{(1, 0.2), (2, 0.5), (3, 0.8), (5, 0.7), (6, 0.3)\}$$

3. A = real numbers close to 10 = $\left\{(x, f_A(x)) : f_A(x) = \dfrac{1}{1+(x-10)^2}\right\}$

If U = $\{x_1, x_2, \ldots, x_n\}$, then a fuzzy set A is also written as:

$$A = \left\{\frac{f_A(x_1)}{x_1}, \frac{f_A(x_2)}{x_2}, \frac{f_A(x_3)}{x_3}, \ldots, \frac{f_A(x_n)}{x_n}\right\}$$

Alternatively, the fuzzy set is also written as:

$$A = \{(x_1, f_A(x_1)), (x_2, f_A(x_2)), \ldots (x_n, f_A(x_n))\}$$

For example, the real numbers 5, 2, 0 with membership grades 0.2, 0.07 and 1 can be written as

$$A = \{(5, 0.2), (2, 0.07), (0, 1)\} = \left\{\frac{0.2}{5}, \frac{0.07}{2}, \frac{1}{0}\right\}$$

1.15.1 Some Useful Definitions

Containment Let A and B be two fuzzy sets in U with membership function f_A and f_B respectively. If the membership grade of each element in A is less than or equal to the membership grade is B, then A is said to be contained in B (equivalently, A is subset of B or A is smaller than or equal to B). Thus if

$$f_A(x) \leq f_B(x)$$

for every $x \in$ U, then A $\subseteq$ B. In other words,

$$A \subseteq B \Leftrightarrow f_A(x) \leq f_B(x).$$

Illustration If U = {20, 30, 40, 50} and two fuzzy sets A and B are defined as

$$A = \left\{\frac{0.1}{20}, \frac{0.2}{30}, \frac{0.4}{40}, \frac{0.5}{50}\right\}; \; B = \left\{\frac{0.8}{20}, \frac{1}{30}, \frac{1}{40}, \frac{1}{50}\right\}$$

then set A is the subset of B because for each element in U, $f_A(x) \leq f_B(x)$.

Normal Fuzzy Set A fuzzy set A of U is called a normal fuzzy set if and only if max $f_A(x) = 1$ for all $x \in$ U, otherwise A is abnormal.

For example, let U = {2, 3, 4, 5} and $A = \left\{\frac{0.8}{2}, \frac{1}{3}, \frac{0.3}{4}, \frac{0.2}{5}\right\}$. Since Max $f_A(x) = 1$, A is normal fuzzy set.

Support of Fuzzy Set Let A be the fuzzy set in the universal set U. The support of A is the crisp set which contains all the elements of U that have a non-zero membership grade in A, and is denoted as:

$$\text{Supp}(A) = \{x : f_A(x) > 0, x \in U\}.$$

α-Level set or α-cut A crisp set of elements that belong to the fuzzy set A at least to the degree α, is called the α-level set and is written as,

$$A_\alpha = \{x \in U : f_A(x) \geq \alpha\}$$

If α-level set has the condition

$$A'_\alpha = \{x \in U : f_A(x) > \alpha\}$$

then $f_A(x) > \alpha$ is called *strong α-level set.* This is also called strong α-cut.

A fuzzy set is convex if all α-level sets are convex.

Illustration Suppose U = {1, 2, 3, 4, 5, 6}. If a fuzzy set A of U is given by

$$A = \{(1, 0.2), (2, 0.5), (3, 0.7), (4, 1), (5, 0.8), (6, 0.3)\}$$

then all possible α-level sets are

$A_{0.2} = \{1, 2, 3, 4, 5, 6\}$ $A_{0.5} = \{2, 3, 4, 5\}$
$A_{0.8} = \{3, 4\}$ $A_{0.7} = \{3, 4, 5\}$
$A_1 = \{4\}$ $A_{0.3} = \{2, 3, 4, 5, 6\}$

Obviously $A_0 = U$, and $A_\alpha = \Phi$ for all $\alpha > 1$. The strong α-level set for $\alpha = 0.8$ is $A'_{0.8} = \{4\}$

Cardinality of Fuzzy Set The cardinality of a finite fuzzy set is defined as:

$$|A| = \sum_{x \in U} f_A(x)$$

The *relative cardinality* of the fuzzy set A is defined as: $\|A\| = \frac{|A|}{|U|}$.

Illustration For the fuzzy set 'comfortable type of house for a 4-person family' from previous illustration, the cardinality is given by $|A| = 0.2 + 0.5 + 0.8 + 1 + 0.7 + 0.3 = 3.5$. Also its relative cardinality is $\|A\| = 3.5/10 = 0.35$.

1.15.2 Operations on Fuzzy Sets

The operations in the fuzzy set theory are the complement, union and intersection. These operations with fuzzy sets are defined in terms of the respective membership functions.

Empty Set A fuzzy set is empty if and only if its membership function is identical to zero on universal set U.

Equal Set Two fuzzy sets are equal, written as A = B, if and only if $f_A(x)$ for all $x \in U$.

Instead of writing $f_A(x) = f_B(x)$, we may also write simply $f_A = f_B$.

Complement of Fuzzy Set The membership function of the complement of a fuzzy set A is denoted by $\overline{A}$ and is defined by

$$f_{\overline{A}}(x) = 1 - f_A(x) \text{ for all } x \in U.$$

For example, if an element in the fuzzy set A has a membership grade of 0.8, then its membership grade in the complement of A will be 0.2.

Union of Fuzzy Sets The union of two fuzzy sets A and B with respective membership function $f_A(x)$ and $f_B(x)$ is a fuzzy set C, written as $C = A \cup B$, whose membership function is related to those of A and B by

$$f_C(x) = \text{Max}\ \{f_A(x), f_B(x)\},\ x \in U.$$

In abbreviated form, we write $f_C = f_A \vee f_B$.

A more intuitively appealing way of defining the union is the following: The union of A and B is the smallest fuzzy set containing both A and B. More precisely, if D is any fuzzy set which contains both A and B, then it also contains the union of A and B.

To show that this definition is equivalent to earlier definition, we note, first that fuzzy set C contains both A and B, since

$$\text{Max}\ [f_A, f_B] \geq f_A \text{ and } \text{Max}\ [f_A, f_B] \geq f_B$$

Furthermore, if D is any fuzzy set containing both A and B, then

$$f_D \geq f_A \text{ and } f_D \geq f_B \text{ and hence } f_D \geq \text{Max}\ [f_A, f_B] = f_C$$

which implies that $C \subseteq D$.

Illustrations

1. Let $A = \{1, 0.2), (2, 0.5), (3, 0.8), (5, 0.7), (6, 0.3)$

$B = \{(3, 0.2), (4, 0.4), (5, 0.6) (6, 0.8), (7, 1), (8, 1)\}$

Then $A \cup B = \{(1, 0.2), (2, 0.5), (3, 0.8), (4, 0.4), (5, 0.7), (6, 0.8), (7, 1), (8, 1)\}$

2. If $U = \{25, 26, 27, 28\}$, $A = \frac{0.2}{25} + \frac{0.5}{26} + \frac{0.8}{27} + \frac{0.7}{28}$ and $B = \frac{1}{25} + \frac{0.8}{26} + \frac{0.5}{27} + \frac{0.2}{28}$, then union of A and B is given by

$$A \cup B = \text{Max}\left[\frac{(0.2, 1)}{25}\right] + \text{Max}\left[\frac{(0.5, 0.8)}{26}\right] + \text{Max}\left[\frac{(0.8, 0.5)}{27}\right] + \text{Max}\left[\frac{(0.7, 0.2)}{28}\right]$$

$$= \frac{1}{25} + \frac{0.8}{26} + \frac{0.8}{27} + \frac{0.7}{28}$$

$$= \{(25, 1), (26, 0.8), (27, 0.8), (28, 0.7)\}$$

Intersection of Fuzzy Sets The intersection of two fuzzy sets A and B with respective membership functions $f_A(x)$ and $f_B(x)$ is a fuzzy set C, written as $C = A \cap B$, whose membership function is related to those of A and B by

$$f_C(x) = \text{Min}\,[f_A(x), f_B(x)], \text{ for all } x \in U.$$

In abbreviated form we write, $f_C = f_A \wedge f_B$.

As in the case of the union, it is easy to show that the intersection of A and B is the largest fuzzy set which contained in both A and B.

Illustration Let $A = \{(1, 0.2), (2, 0.5), (3, 0.8), (5, 0.7), (6, 0.3)\}$

$B = \{(3, 0.2), (4, 0.4), (5, 0.6), (6, 0.8), (7, 1), (8, 1)\}$

Then

$$A \cap B = \text{Min}\left[\frac{(0, 0.2)}{1}\right] + \text{Min}\left[\frac{(0, 0.5)}{2}\right] + \text{Min}\left[\frac{(0.8, 0.2)}{3}\right]$$

$$+ \text{Min}\left[\frac{(0, 0.4)}{4}\right] + \text{Min}\left[\frac{(0.7, 0.6}{5}\right] + \text{Min}\left[\frac{(0.3, 0.8)}{6}\right]$$

$$+ \text{Min}\left[\frac{(0, 1)}{7}\right] + \text{Min}\left[\frac{(0, 1)}{8}\right]$$

$$= \frac{0}{1} + \frac{0}{2} + \frac{0.2}{3} + \frac{0}{4} + \frac{0.6}{5} + \frac{0.3}{6} + \frac{0}{7} + \frac{0}{8}$$

$$= \{(3, 0.2), (5, 0.6), (6, 0.3)\}$$

The intersection and union of two fuzzy sets are illustrated in Fig. 1.15. The membership function of the union is comprised of curve segments 1 and 2; that of the intersection is comprised of segments 3 and 4 (bold lines).

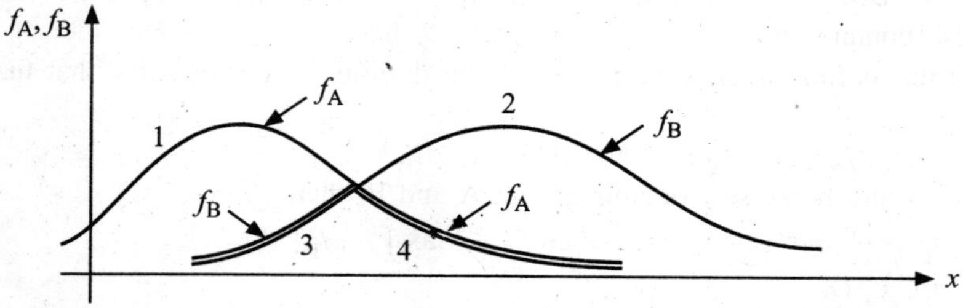

Fig. 1.15 Union and Intersection of Fuzzy Sets

Self-practice Problems III

44. A market research team interviews 160 students, asking each of them whether he reads any or all of the magazine: A – India Today; B – Business India; C – Today. The team returns the fol.owing data:

Category	*Number*	*Category*	*Number*
A	60	AC	20
B	50	AB	30
C	50	ABC	10
AC	30		

Are the returns consistent?

45. Two types of defects in shape and in weight were located in the production of an item. In a sample of 100 items, 20 were defective due to shape and 24 due to weights. Ten items had both types of defects. How many of them had exactly one defect?

46. In a certain examination 75 per cent passed in group A and 82 per cent in group B and 15 per cent failed in both the groups. Find the percentage of students passed in the examination.

47. In a recent survey of 400 students in a school, it was found that 100 students read magazine A and 150 read magazine B, 75 students read both the magazines. Find how many read either magazine.

48. In a survey of 100 students in a music school, the number of students learning one or more instruments: Tabla; Veena; Guitar were interviewed. The data is given below:

Category	*Number*	*Category*	*Number*
T	25	TV	10
V	32	TG	12
G	43	VG	4
TGV	3		

(*a*) How many students were learning none of these three musical instruments?

(*b*) How many students were learning only the guitar?

49. A survery reports that 80 per cent of the population read magazine M and 55 per cent are males. Find the least possible percentage of males reading M and females reading M.

50. In a town 60 per cent read magazine A, 25 per cent do not read magazine A but read magazine B. Calculate the percentage of those who do not read any of the magazine. Also find the highest and lowest possible figures of those who read magazine B.

51. In a survey of 1000 customers, the number of people that buy the various grades of coffee seeds were found to be as follows:

Grade	*Number*	*Grade*	*Number*
A	180	C but not B	80
C	480	None of A, B, C	240
A but not B	230		
A but not C	80		

(*a*) How many buy B grade coffee seeds?
(*b*) How many buy C grade, if and only if they do not buy B grade?
(*c*) How many buy C and B grades but not the A grade?

52. In a survey of 100 families, the numbers that read the most recent issues of various magazines were found to be as follows:

Magazine	*Number*	*Magazine*	*Number*
A	28	AB	8
B	30	AC	10
C	42	BC	5
		ABC	3

(*a*) How many read none of the three magazine?
(*b*) How many read magazine C only?
(*c*) How many read B, if and only if they read C?

53. A factory inspector examined the defects in hardness, finish and dimension of an item. After examining 100 items he gave the following report: All three defects – 5, defect in hardness and finish – 10, defect in dimension and finish – 8, defect in dimension and hardness – 20, defect in finish – 30, in hardness – 23, in dimension – 50. The inspector was fined. Why?

54. A sample of 586 families is given below according to their annual income and the number of vehicles owned by them.

Vehicles owned	*Less than Rs 2,500*	*Rs 2,500 to Rs 4,999*	*Rs 5,000 to Rs 7,499*	*Rs 7,500 and more*
Two or more	8	87	40	47
One	77	154	59	23
None	31	25	10	25

(*a*) Find the number of families owing one vehicle and income less than Rs 2,500.
(*b*) Find the number of families owing two or more vehicles or with a income of Rs 5,000 to Rs 7,499.

55. In a group of 40 persons, 13 are musicians, 18 are poets, 4 are musicians as well as poets, 5 are poets as well as dramatists, 3 are dramatists as well as musicians and 1 person having all the characteristics. How many of them are dramastists? How many of them are only musicians?

56. In a class of 42 students, each play at least one of the three games: Cricket, Hockey and Football. It is found that 14 play Cricket, 20 play Hockey and 24 play Football, 3 play both Cricket and Football, 2 play both Hockey and Football and none play all the three games. Find the number of students who play Cricket but not Hockey.

57. In a city, three daily newspapers A, B and C are published. 42 per cent of the people in that city read A, 56 per cent read B, 60 per cent read C, 24 per cent read A and B, 34 per cent read B and C, 32 per cent read C and A, 8 per cent do not read any of the three newspapers. Using the algebra of sets, find the percentage of persons who read all the three papers.

Hints and Answers

44. Conditions for consistency:

(*a*) $n(T) < (E)$; $T \subseteq E$ and (*b*) $n(A \cup B \cup C) \leq 160$

Given $ABC \subseteq AB$; $ABC \subseteq AC$: $ABC \subseteq BC$; $AB \subseteq A$; $AB \subseteq B$. Prove that

(*i*) $n(ABC) < n(AB)$ (*ii*) $n(ABC) < n(AC)$ (*iii*) $n(ABC) < n(BC)$
(*iv*) $n(AB) < n(A)$ (*v*) $n(AB) < n(B)$; $n(AB) < n(C)$, and (*vi*) $n(A \cup B \cup C) < 160$.

45. Given $n(A) = 20$; $n(B) = 24$, $n(A \cap B) = 10$

$A \oplus B = \{x : x \in (A \cap B^c) \text{ or } x \in (A^c \cap B)\} = (A \cap B^c) \cup (A^c \cap B) = (A - B) \cup (B - A) = 24$

46. 72% **47.** 22.5 **48.** (*a*) 23 (*b*) 30 **49.** 35% and 25%

50. (*a*)15% (*b*) 85% (*c*) 25% **51.** (*a*) 100 (*b*) 400 (*c*) 50 **52.** (*a*) 20 (*b*) 30 (*c*) 5

53. The number of items with dimension defect alone is 2.

54. (*a*) 8 (*b*) 251

56. $n(C) = n(C \cap H) + n(C - H)$ or $14 = 11 + n(C - H)$ or $n(C - H) = 3$

57. $n(A \cap B \cap C) = 25\%$.

Chapter 2

Number Theory

PREVIEW

This chapter introduces some basic properties of integers, congruence relation, congruence (or residue) classes, and congruence equations. Applications of congruence hashing function, random number and cryptography have been discussed.

LEARNING OBJECTIVES

After studying this chapter, you should be able to

- understand basic properties of intergers, greatest common divisor, and prime factors,
- understand whether any numbers is also prime number,
- understand properties of congruence relations, residue classes and arithmetic of residue classes,
- know the areas of application of congruences.
- develop congruence equations, and the criteria for existence of solutions.

2.1 INTRODUCTION

Number theory helps us to understand properties of integers, specially the properties of positive integers. Integer numbers are used in solving problems associated with the transmissions, coding and manipulation of numerical data. The development of electronic computers has enabled us to perform number-theoretic calculations much easily on personal computers or programmable calculators. In this chapter, we shall discuss some of the basic properties of integers particularly divisibility and congruence relation with their applications.

2.2 BASIC PROPERTIES OF INTEGERS

Let us denote the set of natural numbers (also called positive integers) N and the set of integers I as:

$$N = \{1, 2, 3, \ldots\}, \quad \text{and} \quad I = \{\ldots -3, -2, -1, 0, 1, 2, 3, \ldots\}$$

The following simple rules (without proof) concerning the addition and multiplication of integers are given below:

(*a*) Associative law for multiplication and addition

$$(a + b) + c = a + (b + c) \text{ and } (ab)c = a(bc), \text{ for all } a, b, c \in I$$

(*b*) Commutative law for multiplication and addition

$$a + b = b + a \text{ and } ab = ba, \text{ for all } a, b, c \in I$$

(*c*) Distributive law

$$a(b + c) = ab + bc, \text{ for all } a, b, c, \in I$$

(*d*) Additive identity 0 and multiplicative identity 1

$$a + 0 = 0 + a = a \text{ and } a.1 = 1.a = a$$

(*e*) Additive inverse $-a$ for any integer a

$$a + (-a) = (-a) + a = 0.$$

2.3 PROPERTIES OF INTEGERS

1. Divisor A non-zero integer a is said to be a *divisor* or *factor* of an integer b if there exists an integer q such that $b = aq$.

If 'a is a divisor of b', then we write $a \mid b$. (read as 'a is a divisor of b'). If a is a divisor of b, then we also say that 'b is *divisible* by a' or 'a is a *factor* of b' or 'b is a *multiple* of a'. Few examples of division are as under:

(*a*) $2 \mid 8$, since $8 = 2 \times 4$

(*b*) $-4 \mid 16$, since $16 = (-4) \times (-4)$

(*c*) $a \mid 0$ for all $a \in I$ and $a \neq 0$, because $0 = a \times 0$.

Note that division is not everywhere defined in the set of integers. For example, 3 is not the divisor of 4.

Remark If a divides b, then $-a$ also divides b because $b = aq$, then this implies that $b = (-a)(-q)$. Hence, it is enough to consider only positive divisors of an integer.

2. Proper and Improper Divisors Since $a \times 1 = (-a) \times (-1) = a$ for every $a \in I$, therefore, for every integer a, ± 1 and $\pm a$ are divisors of a. These are called *improper divisors* of a. If a has any divisor other than these, then these are called *proper divisors* of a.

A non-zero integer p is called a *prime* provided it is neither 1 nor -1 and its only divisors are ± 1 and $\pm p$. For example,

(*a*) Integers 2 and -11 are primes, whereas $6 = 2 \times 3$ and $-36 = 3 \times -12$ are not primes.

(*b*) The first few positive primes are: 2, 3, 5, 7, 11, 13, 17, 19, etc.

3. Composite Integer If an integer a can be written as $a = bq$ where b and q are integers such that $|b| > 1$ and $|q| > 1$, then a is called a composite integer.

If integer a is composite and $a = bq$, then $|b| < |a|$ and $|q| < |a|$. Since

$$a = bq \Rightarrow |a| = |bq| = |b|\ |q|, \text{ because } |b| > 1$$

therefore, $$|b|\ |q| > |q|, \text{ i.e., } |a| > |q| \text{ or } |q| < |a|.$$

For every integer $a \neq 0, \pm 1$ is either prime or composite. If a is a positive integer, then a is not a prime, i.e., a is the composite integer if and only if there exist positive integers a_1 and a_2 such that $a = a_1 . a_2$, where $1 < a_1 < a$ and $1 < a_2 < a$.

4. Associate Integers Two non-zero integers a and b are said to be associated provided 'a is a divisor of b' $(a \mid b)$ and 'b is the divisor of a' $(b \mid a)$.

Theorem 2.1 Let a, b, c be any integer numbers.

(*a*) If $a \mid b$, then for every integer q, $a \mid bq$

(*b*) If $a \mid b$ and $a \mid c$, then for every integers p and q, $a \mid (bp + cq)$

(*c*) If $a \mid b$ and $b \neq 0$, then $|a| \leq |b|$

(*d*) If $a \mid b$ and $b \mid c$, then $a \mid c$

(*e*) If $a \mid b$ and $a \mid c$, then $a \mid (b + c)$ and $a \mid (b - c)$.

Proof (*a*) We have $a \mid b \Rightarrow b = ak$ for any $k \in$ I. For any integer x, we can write $bx = (ak)x = a(kx)$. This implies that $a \mid bx$.

(*b*) We have $a \mid b \Rightarrow b = ax$ for any $x \in$ I and $a \mid c \Rightarrow c = ay$ for any $y \in$ I. Now $bp + cq = (ax)p + (ay)q = a(xp + yq)$. This implies that $a \mid (bp + cq)$.

(*c*) We have $a \mid b \Rightarrow b = ax$ for any $x \in$ I. Since $b \neq 0$, therefore $x \neq 0$ and consequently $|x| \geq 1$. Now

$$b = ax \Rightarrow |b| = |ax| = |a|\ |x|$$

Also $$|x| \geq 1 \Rightarrow |a|\ |x| \geq |a|, \text{ i.e., } |b| \geq |a|$$

(*d*) We have $a \mid b \Rightarrow b = ax_1$ and $b \mid c \Rightarrow c = bx_2$. Thus $(ax_1)x_2 = a(x_1x_2)$. This implies that $a \mid c$.

(*e*) We have $a \mid b \Rightarrow b = ax_1$ and $a \mid c \Rightarrow c = ax_2$ for any $x_1, x_2 \in$ I. Thus $b + c = (x_1 + x_2)\ a$, which implies $a \mid (b + c)$.

Proceed in the same manner to prove $a \mid (b - c)$.

2.3.1 Division Theorem (or Algorithm)

This is an important theorem in number theory. This theorem provides the basis for arithmetic process of long division.

Theorem 2.2 If a and b are any two integers, with $b > 0$, $a \neq 0$, then there exist unique integers q and r such that $a = bq + r$, $0 \leq r < |b|$.

Proof If b divides a, the result follows with $r = 0$, so we consider a case where b does not divide a. Consider the set A = $\{a - bx : x \in \text{I}\}$. If $b < 0$, then

$$b|a| \leq -|a| \leq a \Rightarrow a - b\ |a| \geq 0$$

Since $a - b|a| \in$ A as $a \in$ I; therefore when $b < 0$, A contains at least one non-negative integer, i.e., $a - b|a|$.

If $b > 0$ or $b \geq 1$, then $b\ (-|a|) \leq -|a| \leq a$. This implies $a - b(-|a|) \geq 0$. Thus $a - b(-|a|) \in$ A as $-|a| \in$ I. This means if $b > 0$, A contains at least one non-negative integer, i.e., $a - b(-|a|)$.

Let $r = a - bq$ be the smallest non-negative integer belonging to A. Let us claim that $r < |b|$. To prove this claim we should show that $r - |b| \in$ A, whether $b > 0$ or $b < 0$.

If $b > 0$ then

$$r - |b| = r - b = a - bq - b = a - (q + 1)\, b \in \text{A, since } (q + 1) \in \text{I}.$$

If $b < 0$, then

$$r - |b| = r - (-b) = a - bq + b = a - (q - 1)\, b \in \text{A, since } (q - 1) \in \text{I}.$$

Since $b \neq 0$, $r - |b| < r$. When $r \geq |b|$, we have $r - |b| \geq 0$, non-negative integer belonging to A, and $r - |b| < r$. This contradict the choice that r is the smallest integer in the set A. Hence, we must have $r < |b|$. This proves the existence of two integers q and r such that

$$r = a - bq \text{ or } a = bq + r, 0 \leq r < |b|$$

To prove the uniqueness of q and r, suppose they are not unique. Then it follows that there exists another pair q' and r' such that

$$a = bq + r, \quad 0 \leq r < b$$

and

$$a = bq' + r', \quad 0 \leq r' < b$$

Equating right hand sides, we have

$$bq + r = bq' + r' \text{ or } b(q' - q) = r - r'$$

This implies that b is a divisor of $r - r'$, which is possible only when $r' - r = 0$, i.e., $r' = r$. Putting r' and r in $bq' + r = bq + r$, we get $bq' = bq$, i.e., $q' = q$ (cancellation law, as $b \neq 0$). Hence q and r are unique.

Remark The numbers q and r in the division theorem are called *quotient* and *remainder*, respectively.

Illustrations

1. If $a = 16$, $b = 5$, then $18 = 3 \times 5 + 1$; $0 \leq 1 < |5|$
2. If $a = 17$, $b = -3$, then $17 = (-3) \times (-5) + 2$; $0 \leq 2 < |-3|$
3. If $a = -24$, $b = 7$, then $-24 = -4 \times 7 + 4$; $0 \leq 4 < |7|$
4. Let $a, b \in$ I. (*i*) If $a = bq$ for $q \in \text{I}_+$, then $-a = (-q)b$. Thus when $-a(< 0)$ is divided by $b(> 0)$, the quotient is $-q(< 0)$ and remainder is 0. (*ii*) If $a = bq + r$, for $q \in$ I and $0 < r < b$, then $-a = (-q)b - r = (-q)b - b + b - r = (-q - 1)\, b + (b - r)$. Thus when $-a(< 0)$ is divided by $b\,(> 0)$, the quotient is $(-q - 1) < 0$ and remainder is $b - r$, where $0 < b - r < b$.

Theorem 2.3 Prove that there exists an infinite number of primes.

Proof Let $p_1, p_2, \ldots, p_m$ be the list of finite number of positive primes arranged in order of their magnitude. Since every integer $n > 1$ can be written as a product of prime, therefore consider the integer $n = p_1 p_2 \ldots p_m$. Also consider an integer $n + 1$.

As no one of primes is a divisor of $n + 1$, it is possible to conclude that either $n + 1$ is a prime greater than p_m or has a prime greater than p_m as factor. This is not possible and so n is divisible by some other prime. This contradicts the assumption that $p_1 p_2 \ldots p_m$ are the only primes. Hence, there exists an infinite number of primes.

2.4 GREATEST COMMON DIVISOR

Let a and b be two positive integers. A positive integer d is said to be the *common divisor* of both a and b if d divides both a and b, i.e., $d \mid a$ and $d \mid b$.

If the common divisor, d of a and b is the largest among all other common divisors and also a multiple of every other common divisors, then d is said to be the *greatest common divisor* (gcd) of a and b.

In other words, the common divisor d is said to be the greatest common divisor of a and b, written as $d = \gcd(a, b)$ if and only if d has the following two properties:

(*i*) d divides both a and b, i.e., $d \mid a$ and $d \mid b$.

(*ii*) If c is any other common divisor of a and b, i.e., $c \mid a$ and $c \mid b$, then c is also the divisor of d, i.e., $c \mid d$.

(*iii*) $d = xa + yb$ for integers x and y (both may not be positive).

The greatest common divisor is also called the *highest common factor* (hcf).

Illustration $\pm 1, \pm 2, \pm 3, \pm 4, \pm 6$ and ± 12 are divisors of 12; and $\pm 1, \pm 2, \pm 4, \pm 5, \pm 20$ are divisors of 20. The common divisors of 12 and 20 are $\pm 1, \pm 2, \pm 4$. Hence, the greatest common divisor of 12 and 20 is 4. Thus $4 = \gcd(12, 20)$.

Similarly, $\gcd(12, -18) = 6$, $\gcd(14, 49) = 7$, $\gcd(15, 29) = 1$.

Relatively Prime Integer Two integers a and b are said to be relatively prime or *coprime* if their greatest common divisor is 1, i.e., $\gcd(a, b) = 1$. For example,

(*i*) $\gcd(4, -9) = 1$, (*ii*) $\gcd(21, 64) = 1$ (*iii*) $\gcd(-28, 45) = 1$, (*iv*) $\gcd(2, 7) = 1$.

When $\gcd(a, b) = 1$, it is also expressed by saying that a and b are *coprime* or *a is prime to b*. Accordingly if a and b are relatively prime, then there exist integers x and y such that $ax + by = 1$. Conversely if $ax + by = 1$, then a and b are relatively prime.

If $\gcd(a_1, a_2, \ldots, a_n) = 1$, then $a_1, a_2, \ldots, a_n$ are relatively prime. These numbers are relatively prime in pairs in case $\gcd(a_i, a_j) = 1$, for $i, j = 1, 2, \ldots, n$.

Illustration The integers 21, 85, 92, 143 are relatively prime in pairs because

$\gcd(21, 85) = 1$	$\gcd(21, 85) = 1$	$\gcd(21, 143) = 1$
$\gcd(85, 143) = 1$	$\gcd(85, 92) = 1$	$\gcd(92, 143) = 1$

2.4.2 Basic Properties of the Greatest Common Divisor

1. Since greatest common divisor should be positive, therefore $\gcd(a, b) = \gcd(a, -b) = \gcd(-a, b) = \gcd(-a, -b)$. This implies that $\gcd(a, b) = \gcd(|a|, |b|)$. For example, $\gcd(60, 24) = \gcd(60, -24) = 12$.
2. For any integer, $\gcd(a, a + 1) = 1$. It is due to the fact that any common divisor of a and $a + 1$ must divide their difference $(a + 1) - a = 1$.
3. If $a_1, a_2, \ldots, a_n$ are all relatively prime to b, then their product $a_1 . a_2 . \ldots . a_n$ is also prime to b.

4.(*a*) If $\gcd(a, b) = 1$, and a and b both divide c, then ab divides c.

(*b*) If c divides ab and $\gcd(a, b) = 1$, then c divides b.

5.(*a*) If $x > 0$ is any integer, then $\gcd(ax, bx) = x \gcd(a, b)$.

(*b*) If $d = \gcd(a, b)$, then $\gcd(a/d, b/d) = 1$.

(*c*) $x \gcd(a, b) = \gcd(a, b + ax)$, for any integer x.

(*d*) If $\gcd(a, b) = 1$, $\gcd(a, c) = 1$, then $\gcd(a, bc) = 1$.

Theorem 2.4 Prove that the greatest common divisor is unique.

Proof If possible, let d_1 and d_2 be two gcd of a and b. Then by definition of gcd, we have $d_1 \mid d_2$ and $d_2 \mid d_1$. This implies that d_1 and d_2 are associates and therefore $d_2 = \pm d_1$. But d_2 is positive, so must have $d_1 = d_2$. This proves the theorem.

2.5 EUCLIDEAN ALGORITHM

The algorithm developed by Euclid is an efficient method for finding the gcd of two integers based on the quotient and remainder technique. In other words, this theorem consists of repeatly applying the division algorithm to obtain gcd of two integers.

Lemma If $a = bq + r$, then $\gcd(a, b) = \gcd(b, r)$.

Proof Since $\gcd(a, b) = \gcd(|a|, |b|)$, therefore we can assume that both a and b are positive integers. By division algorithm we have $a = bq + r$, $0 \le r < b$.

If $r = 0$, then $a = bq$ and consequently b is a divisor of a. Thus b is the gcd of c and b, i.e., $b = \gcd(a, b)$. But if $r \ne 0$, then we can show that $\gcd(a, b) = \gcd(b, r)$ as follows.

Let $d = \gcd(a, b)$ and $d' = \gcd(b, r)$. It follows that $d \mid a$ and $d \mid b$. Thus d divides $r = a - bq$, i.e., $d \mid r$. This shows that d is a common divisor of b and r. As $d' = \gcd(b, r)$, therefore d divides d', i.e., $d \mid d'$. Similarly $d' = \gcd(b, r)$ implies d' divides b and d' divides r. This d' divides $a = bq + r$, i.e., $d' \mid a$. This shows that d' is a common divisor of a and b. Since $d = \gcd(a, b)$, therefore d' must divide d, i.e., $d' \mid d$.

Now $d \mid d'$ and $d' \mid d \Rightarrow d' = \pm d$

$\Rightarrow d' = d$ since both are positive.

Applying the division algorithm to b and r we get $b = rq_1 + r_1$; $0 \le r_1 < r$. Thus if $r_1 = 0$, then $\gcd(b, r) = r$. But if $r_1 \ne 0$, then $\gcd(a, b) = \gcd(b, r) = \gcd(r, r_1)$.

Euclidean Theorem 2.5 Let $a > b > 0$ (otherwise interchange a and b). Then a divides b to obtain remainder r_1, i.e., $a = bk_1 + r_1$; k_1 and r_1 are positive integers, and $0 \le r_1 < b$.

The procedure of successive division is as follows:

- divide b by r_1 so that $b = k_2r_1 + r_2$ $\quad 0 \le r_2 < r_1$.
- divide r_1 by r_2 so that $r_1 = k_3r_2 + r_3$ $\quad 0 \le r_3 < r_2$
- divide r_2 by r_3 so that $r_2 = k_4r_3 + r_4$ $\quad 0 \le r_4 < r_3$

⋮

- divide r_{n-2} by r_{n-1} so that $r_{n-2} = k_n\, r_{n-1} + r_n$ $\quad 0 \le r_n < r_{n-1}$
- divide r_{n-1} by r_n so that $r_{n-1} = k_{n+1}\, r_n + r_{n+1}$ $\quad 0 \le r_{n+1} + r_n$

Since $a > b > r_1 > r_2 \ldots$, eventually we obtain a remainder $r_{n+1} = 0$. It follows from the lemma:

$$\gcd(a, b) = \gcd(b, r_1) = \gcd(r_1, r_2) = \ldots = \gcd(r_{n-1}, r_n) = \gcd(r_n, 0) = r_n.$$

Theorem 2.6 Every pair of integers a and b, not both zero, has a unique $d = \gcd(a, b)$, which can be expressed in the form $d = ax + by$ for any integers x and y.

Proof The Euclidean theorem can also be used to express $d = \gcd(a, b)$ in the form $d = ax + by$, for all $x, y \in \mathrm{I}$ as follows:

$$\begin{aligned} r_n &= r_{n-2}\,(-k_n)r_{n-1} \\ &= r_{n-2} + [r_{n-3} + (-k_{n-1})r_{n-2}]\,(-k_n) \\ &= r_{n-3}\,(-k_n) + r_{n-2}\,(1 + k_{n-1}\,k_n) \end{aligned}$$

Continuing the back substituting for $r_{n-2}, r_{n-3} \ldots$ and so on until we reach $r_n = ax + by$ for any $x, y \in \mathrm{I}$.

Remark If $\gcd(a, b) = 1$, then $ax + by = 1$, but converse is not true. Hence, the necessary and sufficient condition for $\gcd(a, b) = 1$ is $ax + by = 1$ for some integers x and y.

Example 1 Find the gcd of (*a*) 45, 34; (*b*) 77, 128; and (*c*) 258, 60.

Solution (a) Using the Euclidean algorithm, we

divide $a = 45$ by $b = 34$ so that $45 = 1 \times 34 + 11(r_1)$
divide $b = 34$ by $r_1 = 11$ so that $34 = 3 \times 11 + 1(r_2)$
divide $r_1 = 11$ by $r_2 = 1$ so that $11 = 11 \times 1 + 0(r_3)$
Thus $\gcd(45, 34) = 1(= r_2)$.

(*b*) Using Euclidean algorithm, we

divide $a = 128$ by $b = 77$ so that $128 = 1 \times 77 + 51(r_1)$
divide $b = 77$ by $r_1 = 51$ so that $77 = 1 \times 51 + 26(r_2)$
divide $r_1 = 51$ by $r_2 = 26$ so that $51 = 1 \times 26 + 25(r_3)$
divide $r_2 = 26$ by $r_3 = 25$ so that $26 = 1 \times 25 + 1(r_4)$
divide $r_3 = 25$ by $r_4 = 1$ so that $25 = 25 \times 1 + 0(r_5)$
Thus $\gcd(128, 77) = 1(= r_4)$.

(*c*) Using Euclidean algorithm, we

divide $a = 258$ by $b = 60$ so that $258 = 4 \times 60 + 18(r_1)$
divide $b = 60$ by $r_1 = 18$ so that $60 = 3 \times 18 + 6(r_2)$
divide $r_1 = 18$ by $r_2 = 6$ so that $18 = 3 \times 6 + 0(r_3)$
Thus $\gcd(258, 60) = 6(= r_2)$.

Example 2 Find gcd of 26 and 118 and express it in the form $26x + 118y$.

Solution Applying division algorithm repeatedly, we have

$$118 = 4 \times 26 + 14 \quad \text{...}(i)$$
$$26 = 1 \times 14 + 12 \quad \text{...}(ii)$$
$$14 = 1 \times 12 + 2 \quad \text{...}(iii)$$
$$12 = 6 \times 2 + 0 \quad \text{...}(iv)$$

Since the last non-zero remainder is zero, $\gcd(118, 26) = 2$. Now we find x and y such that

$$2 = 26x + 118y$$

To find x and y we begin with the last non-zero remainder

$$\begin{aligned} 2 &= 14 - 1 \times 12 \\ &= 14 - 1 \times (26 - 14) = 2 \times 14 - 26, && \text{from Eq. } (ii) \\ &= 2\{118 - 4 \times 26\} - 26, && \text{from Eq. } (i) \\ &= 2 \times 118 - 9 \times 26 \end{aligned}$$

Thus $\gcd(26, 118) = 2$ provided $2 = 26x + 118y$, where $x = -9$ and $y = 2$.

Example 3 Find gcd of 427 and 616 and express it in the form $427x + 616y$.

Solution Applying division algorithm repeatedly, we have

$$616 = 1 \times 427 + 189$$
$$427 = 2 \times 189 + 49$$

$$189 = 3 \times 49 + 42$$
$$49 = 1 \times 42 + 7$$
$$42 = 6 \times 7 = 0$$

Since the last non-zero remainder is 7, gcd(427, 616) = 7.

To find integers x and y such that $7 = ax + by$, we begin with the last non-zero remainder as follows:

$$\begin{aligned} 7 &= 49 - 1 \times 42 = 49 - 1 \times (189 - 3 \times 49) \\ &= 4 \times 49 - 189 = 4 \times (427 - 2 \times 189) - 189 \\ &= 4 \times 427 - 9 \times 189 = 4 \times 427 - 9 \times (616 - 1 \times 427) \\ &= 13 \times 427 - 9 \times 616 \end{aligned}$$

Thus gcd(427, 616) = 7, provided $7 = 427x + 616y$, where $x = 13$ and $y = -9$.

2.5.1 Basic Properties of Prime Factors

Property 1 If two integers a and b are relatively prime, i.e., $\gcd(a, b) = 1$ and a divides bc, then a divides c.

Proof Since integers a and b are relatively prime, there exist integers x and y such that

$$ax + by = 1 \Rightarrow acx + bcy = c \text{ (multiplying with } c)$$

Given that a divides bc, i.e., $a \mid bc$. Therefore a divides bcy. Also a divides acx. Thus, it follows that a divides c, i.e., a divides $(acx + bcy)$ because $acx + bcy = c$.

Note that, if a and b are not prime, this property may not hold.

Property 2 If p is a prime integer and divides ab, where a and b are integers, then either p divides a or p divides b.

Proof Suppose p is not a divisor of a. Then let $\gcd(p, a) = s$, implies that $s \mid a$ and $s \mid b$. Since p is a prime integer, the only positive divisors of p are 1 and p, so either $s = 1$ or $s = p$. But s cannot be equal to p because p is not a divisor of a. This means we must have $s = 1$, so $\gcd(p, a) = 1$. This implies that p and a are relatively prime. Hence, by property 1, we have $p \mid ab$ implies p divides b.

Property 3 If p, a prime number, divides the product $a_1 . a_2 . \ldots a_n$ of certain integers, then p must divide at least one of the integers $a_1, a_2, \ldots, a_n$.

Proof Given that $p \mid (a_1 . a_2 \ldots a_n)$. If $p \mid a_1$, then we need not to prove further. But if p is not divisor of a_1, then p is relatively prime to a_1, so

$$p \mid a_1(a_2 . a_3 \ldots a_n) \text{ implies } p \mid (a_2 . a_3 . a_4 \ldots a_n).$$

If $p \mid a_2$, then proof ends. But if p is not a divisor of a_2, then p is relatively prime to a_2, so

$$p \mid a_2(a_3 . a_4 . a_5 \ldots a_n) \text{ implies } p \mid (a_3 . a_4 . a_5 \ldots a_n)$$

Repeating this process, till at least $p \mid a_n$, implying that p does not divide any of the integers $a_1, a_2, \ldots, a_n$.

Example 4 If p is a positive prime and a is any integer, then prove that either $\gcd(a, p) = 1$ or $\gcd(a, p) = p$.

Solution If p is a divisor of a, then we have $\gcd(a, p) = p$, so solution ends. So, let p is not a divisor of p. Consider $\gcd(a, p) = s$, implies that $s \mid p$ and $s \mid a$. Since p is a prime, therefore only positive divisors of p are 1 and p, so either $s = 1$ or $s = p$. As s cannot be equal to p because p is a divisor of a. Hence, we must have $s = 1$, i.e., $\gcd(a, p) = 1$.

Example 5 If gcd(a, c) = 1, gcd(b, c) = 1, then prove that gcd(ab, c) = 1.

Solution Given that gcd(a, c) = 1, i.e., a is relatively prime to c and gcd(b, c) = 1, i.e., b is relatively prime to c.

To prove that gcd(ab, c) = 1, i.e., ab is relatively prime to c, let us consider that ab is not relatively prime to c, so gcd(ab, c) = s, where $s > 1$. The integer s can be expressed as the product of a finite number of prime integers. Let the prime integer p be a factor of s, so that $p \mid s$. As $s \mid ab$ because gcd(ab, c) = s, $p \mid s$ and $s \mid ab$ implies that $p \mid ab$. Since p is the prime integer, $p \mid ab$ implies either $p \mid a$ or $p \mid b$.

Suppose $p \mid a$. Since gcd(ab, c) = s, therefore s divides c. But p divides s, so $p \mid s$ and $s \mid c$, implies that $p \mid c$. Thus $p \mid a$ and $p \mid c$ implies that a is not relatively prime to c. Which contradicts above assumption.

Also, suppose $p \mid b$. Then $p \mid c$ and $p \mid b$ implies that b is not relatively prime to c. Which, also contradicts above assumption.

Hence, ab must be relatively prime to c, i.e., gcd(ab, c) = 1.

Example 6 If gcd(a, b) = 1 and b divides ap, then prove that b divides p.

Solution Given gcd(a, b) = 1 and $b \mid ap$. Thus there exists integers x, y and z such that

$$ax + by = 1 \text{ and } ap = bz$$

or
$$p = pax + pby = (bz)x + pby = b(zx + py)$$

This implies that $b \mid p$.

Example 7 If gcd(a, b) = 1, then prove that gcd($a + b$, $a - b$) = 1 or 2.

Solution Let d = gcd($a + b$, $a - b$) implies that $a + b = k_1 d$ and $a - b = k_2 d$ for some integer k_1 and k_2. Adding these equations, we get $2a = (k_1 + k_2)d$. This implies that d divides $2a$.

Similarly, $2b = (k_1 - k_2)d$ prove that d divides $2b$. These two findings imply that

$$d \leq \gcd(2a, 2b) = 2\gcd(a, b) \leq 2$$

Hence gcd($a + b$, $a - b$) = 1 or 2.

Example 8 If a and b are two integers such that gcd(a, 4) = 2 and gcd(b, 4) = 2, then prove that gcd($a + b$, 4) = 4.

Solution We have gcd(a, 4) = 2, implies 2 divides a, i.e., $2 \mid a$ but 4 does not divide a. Let $a = 2x$ for some integer x so that gcd(2, x) = 1. Similarly, let $b = 2y$ for some integer y, so that gcd(2, y) = 1. This means both x and y are odd integers, so $x + y$ is an even integer. Let $x + y = 2n$ for some integer n.

Now $a + b = 2(x + y) = 4n$. Hence gcd($a + b$, 4) = gcd($4n$, 4) = 4.

2.6 LEAST COMMON MULTIPLE

Let $a_1, a_2, \ldots, a_n$ be positive integers, and k be a common multiple of $a_1, a_2, \ldots, a_n$ so that for each a_i, $k = a_i b_i$ for some integer b_i. If c is any unique smallest integer among infinitely many common multiples, then this number is called the least common multiple (lcm) of $a_1, a_2, \ldots, a_n$ and is written as $c = \text{lcm}(a_1, a_2, \ldots, a_n)$.

Theorem 2.7 Let $m = p_1^{a_1} \times p_2^{a_2} \times \ldots \times p_k^{a_k}$ and $n = p_1^{b_1} \times p_2^{b_2} \times \ldots \times p_k^{b_k}$. Then $\gcd(m, n) = p_1^{\min(a_1, b_1)} \times p_2^{\min(a_2, b_2)} \times \ldots \times p_k^{(a_k, b_k)} = \prod p_i^{\min(a_i, b_i)}$, where min($a$, b) represents the minimum of the two numbers a and b.

Proof To prove the result, we need to prove that the integer on the right hand side divides a and b, and no larger integer does. Let $d = \prod p^{\min(a,\, b)}$.

(*a*) Since $\min(a, b) \le a$, therefore $\prod p^{\min(a,\, b)}$ divides $\prod p^a = m$. Similarly, $\prod p^{\min(a,\, b)}$ divides $\prod p^b = n$. Hence, d divides both m and n.

(*b*) Let an integer n divides both m and n. It follows that n is the product of primes within $p_1, p_2, \ldots, p_k$. Let

$$c = p_1^{h_1} \times p_2^{h_2} \times \ldots \times p_k^{h_k} = \prod p^h$$

This implies that $\prod p^h$ divides both $\prod p^a$ and $\prod p^b$, so $h \le a$ and $h \le b$. This means $h \le \min(a, b)$, and therefore $\prod p^h$ divides $\prod p^{\min(a,\, b)}$. Hence, n divides d.

Theorem 2.8 Let $m = p_1^{a_1} \times p_2^{a_2} \times \ldots \times p_k^{a_k} = \prod p_i^{a_i}$ and $n = p_1^{b_1} \times p_2^{b_2} \times \ldots \times p_k^{b_k} = \prod p_i^{b_i}$. Then $\text{lcm}(m, n) = \prod p_i^{\max(a_i, b_i)}$.

Proof The proof of this theorem is on the same lines of Theorem 2.6. Let $d = \prod p^{\max(a,\, b)}$.

(*a*) Since $\max(a, b) \ge a$ and $\max(a, b) \ge b$, therefore $\prod p^{\max(a,\, b)}$ divides $\prod p^a = m$ and $\prod p^{\max(a,\, b)}$. Divides $\prod p^b = n$. Hence, d divides both m and n.

(*b*) Let $\prod p^a$ and $\prod p^b$ both divide an integer n, so $n = n_1 \prod p^{\max(a,\, b)}$, where n_1 is some integer ≥ 1. This means $n = n_1 d$ and hence d divides n.

Theorem 2.9 If a and b are two positive integers, then $\gcd(a, b) \times \text{lcm}(a, b) = ab$.

Proof Let $p_1, p_2, \ldots, p_k$ be all prime factors of either a or b. Then we can write

$$a = p_1^{a_1} \times p_2^{a_2} \times \ldots \times p_k^{a_k} \text{ and } b = p_1^{b_1} \times p_2^{b_2} \times \ldots \times p_k^{b_k}$$

where some of a_i's and b_i's may be zero. Thus

$$\gcd(a, b) = p_1^{\min(a_1,\, b_1)}\, p_2^{\min(a_2, b_2)} \ldots \times p_1^{\min(a_k,\, b_k)}$$

and $\text{lcm}(a, b) = p_1^{\max(a_1,\, b_1)} \times p_2^{\max(a_2,\, b_2)} \times \ldots \times p_k^{\max(a_k,\, b_k)}$

$$\text{Hence, } \gcd(a, b) \times \text{lcm}(a, b) = p_1^{a_1+b_1}\, p_2^{a_2+b_2} \ldots p_k^{a_k+b_k}$$

$$= \left(p_1^{a_1} p_2^{a_2} \ldots p_k^{a_k}\right).\left(p_1^{b_1} p_2^{b_2} \ldots p_k^{b_k}\right) = ab$$

Example 9 Find lcm of 340 and 304.

Solution Factoring 340 and 304, we obtain

$$a = 340 = 2^2 \times 5^1 \times 17^1 \text{ and } b = 304 = 2^4 \times 19^1$$

Thus all the prime factors of either a or b are $p_1 = 2$, $p_2 = 5$, $p_3 = 17$ and $p_4 = 19$. Then we can write

$$340 = 2^2 \times 5^1 \times 17^1 \times 19^0 \text{ and } 304 = 2^4 \times 5^0 \times 19^1$$

Then it follows that

$$\gcd((340, 304) = 2^{\min(2,\, 4)} \times 5^{\min(1,\, 0)} \times 17^{\min(1,\, 0)} \times 19^{\min(0,\, 1)}$$

$$= 2^2 \times 5^0 \times 17^0 \times 19^0 = 4$$

and

$$\text{lcm}(340, 304) = 2^{\max(2,\, 4)} \times 5^{\max(1,\, 0)} \times 17^{\max(1,\, 0)} \times 19^{\max(0,\, 1)}$$

$$= 2^4 \times 5^1 \times 17^1 \times 19^1 = 25{,}840$$

Thus gcd(340, 304) × lcm(340, 304) = 4 × 25,840

= 10,03, 360 = 340 × 304

Remark An association of integer 1 is called a unit, so the only units are 1 and –1.

Theorem *(Fundamental Theorem of Arithmetic)* **2.10** Every integer $n > 1$ can be expressed as a unit times a product of positive integers. This representation is unique except for the order of the prime factors. That is, $n > 1$ can be written uniquely as $p_1 . p_2 . \dots . p_n$, where $p_1 < p_2 < \dots < p_n$ are distinct primes that divide n.

Proof If n is a negative integer, then we can write $n = (-1)t$, where t is a positive integer. Let us take n as a positive integer. To prove this theorem, we need to prove that integer n can be expressed as the product of positive primes.

Let $2 \leq k < n$ for all integers k. If n is prime, then proof of the theorem is over. Suppose n is not a prime number, then n has the smallest divisor say p_1 which is prime because the smallest divisor (other than 1) of a composite number is a prime. Let us write $n = ab$, where $1 < a < n$ and $1 < b < n$. By induction process, b and c can be expressed as products of positive primes. Let

$$a = p_1 \cdot p_2 \cdot \dots \cdot p_r \text{ and } b = q_1 \cdot q_2 \cdot \dots \cdot q_s$$

where p_i's and q_i's are positive primes. Then

$$n = ab = (p_1 \cdot p_2 \cdot \dots \cdot p_r)(q_1 \cdot q_2 \cdot \dots \cdot q_s)$$

is the product of positive primes. Thus, in general, we may say that every positive integer greater than 1 can be expressed as a product of positive primes.

To prove factorisation is unique, let another factorisation of n as a product of positive primes be

$$n = p_1' \cdot p_2' \cdot \dots \cdot p_t'$$

Then $$p_1' \cdot p_2' \cdots p_t' = (p_1 \cdot p_2 \cdot \dots \cdot p_r)(q_1 \cdot q_2 \cdot \dots \cdot q_s), \text{ where } r < s.$$

Since p_1' is a prime, it must divide at least one of $p_1, p_2, \dots, p_r, q_1, q_2, \dots, q_s$. The product of integers is commutative, therefore we can assume that p_1' divides p_1, i.e., $p_1' | p_1$ and $p_1' = p_1$ because both of these are positive primes. Thus, we have

$$p_1 p_2' p_3' \cdots p_t' = p_1 p_2 \cdots p_r q_1 q_2 \cdots q_s$$

or $$p_2' q_3' \cdots p_t' = p_2 p_3 \cdots p_r q_1 q_2 \cdots q_s \text{ (due to cancellation law)}$$

Continuing this process, we are left with $1 = q_1 q_2 \dots, q_s$ since $r < s$ which is not possible. Hence, r cannot be less than s. Similarly, r cannot be greater than s. Hence $r = s$.

Remarks

1. The primes in the factorisation of n need not be distinct. Then n can be expressed uniquely in the form $n = p_1^{a_1} p_2^{a_2} \dots p_k^{a_k}$, where the a_i are positive and $p_1 < p_2 < p_3 \dots < p_k$. This is called the *canonical factorisation* of n.
2. The unique expression for the integer $n(\geq 2)$ as a product of primes is called the *prime factorisation* or *decomposition* of n.

2.7 TESTING FOR PRIME NUMBER

During any computation, usually a computer needs to decide whether a number is prime or composite. Such testing is important in the application of number theory to crypography. If a number n has only few decimal digits, then it is easy to show that the given number is prime by divisions upto square root. If the given number n is composite, then it must have a prime divisor $\sqrt{n}$.

Theorem 2.11 If $n > 1$ is a composite integer, then there exists a prime number p such that p divides n (i.e., n has a prime divisor) $p \leq \sqrt{n}$.

Proof If n is a composite integer, then there exist some integers a and b such that $n = ab$ where $2 \leq a$ and $b < n$. If $n = ab$, it implies that $n < a^2$ for $a < b$ or $a < \sqrt{n}$.

Now due to Theorem 2.9 (fundamental theorem of arithmetic), the integer a is either the prime or has prime divisor p. Hence, p divides a implies p divides n and $p \leq \sqrt{n}$.

Remark Theorem 2.10 makes it easy to check primality of an integer by simply verifying that it is not divisible by any prime number less than or equal to its square root.

For larger numbers, a test based on *Fermat's little* theorem states that if n is prime and it does not divide a number s, then $s^{n-1} = 1 \pmod{n}$. However, the converse of this statement is not true. In other words, to know that a number n is not prime, we need to compute only $s^{n-1} \pmod{n}$.

Example 10 Show that 29 is a prime number.

Solution Let $n = 29$, since $5 < \sqrt{29} < 6$, then 2 and 3 are the only prime numbers less than or equal to 5. But 29 is not divisible by 2 and 3. Thus 29 is a prime number.

Example 11 Show that $\sqrt{5}$ is not a rational number, i.e., $\sqrt{5} = p/q$, where p and $q \neq 0$ are integers.

Solution If possible, let $\sqrt{5}$ be a rational number and therefore we can write $\sqrt{5} = p/q$, where p and q are positive integer primes to each other. Now $\sqrt{5} = p/q$ or $5 = p^2/q^2$ or $p^2 = 5q^2$. This shows that 5 divides p^2, i.e., 5 is a factor of p, i.e., $5 \mid p$. Since 5 is a prime number, consider $p = 5m$. Then

$$p^2 = 5q^2 \text{ implies } 25m^2 = 5q^2 \text{ or } 5m^2 = q^2$$

This shows that 5 divides q^2, i.e., 5 is a factor of q, i.e., $5 \mid q$.

Thus, p and q have a common factor 5. This contradicts the assumption that p and q are primes to each other, i.e., $\gcd(p, q) = 1$. Hence, $\sqrt{5}$ is not a rational number.

Example 12 Find the integers x and y such that $37x + 249y = 1$

Solution We have

$$\begin{aligned}
249 &= 6 \times 37 + 27 \\
37 &= 1 \times 27 + 10 \\
27 &= 2 \times 10 + 7 \\
10 &= 1 \times 7 + 3 \\
7 &= 2 \times 3 + 1 \\
3 &= 1 \times 3 + 0
\end{aligned}$$

which shows that gcd(249, 37) = 1. Thus

$$\begin{aligned}1 &= 7 - 2 \times 3 = 7 - 2(10 - 1 \times 7) = 3 \times 7 - 2 \times 10\\ &= 3(27 - 2 \times 10) - 2 \times 10 = 3 \times 27 - 8 \times 10\\ &= 3 \times 27 - 8(37 - 1 \times 27) = 11 \times 27 - 8 \times 37\\ &= 11(249 - 6 \times 37) - 8 \times 37 = -74 \times 37 + 11 \times 249\end{aligned}$$

Example 13 For any $n \in I_+$, prove that the integers $8n + 3$ and $5n + 2$ are relatively primes.

Solution If $n = 1$, then gcd$(8n + 3, 5n + 2)$ = gcd(11, 7) = 1.

If $n \geq 2$, then we have $8n + 3 > 5n + 2$, so we may write

$$\begin{aligned}8n + 3 &= 1(5n + 2) + (3n + 1) &&, 0 < 3n + 1 < 5n + 2\\ 5n + 2 &= 1(3n + 1) + (2n + 1) &&, 0 < 2n + 1 < 3n + 1\\ 3n + 1 &= 1(2n + 1) + n &&, 0 < n < 2n + 1\\ 2n + 1 &= 2(n) + 1 &&, 0 < 1 < n\\ n &= n(1) + 0\end{aligned}$$

Since the last non-zero remainder is 1, gcd$(8n + 3, 5n + 2) = 1$ for all $n \geq 1$. But we may also notice that

$$(8n + 3)(-5) + (5n + 2)(8) = -15 + 16 = 1$$

As 1 is expressed as a linear combination of $8n + 3$ and $5n + 2$ and no smaller positive integer can have this property, it follows that the gcd of $8n + 3$ and $5n + 2$ is 1, for any positive integer n.

Self-practice Problems I

1. For each pair of integers a and b, find integers q and r such that $a = bq + r$ and $0 \leq r < |b|$
   (*a*) $a = 258, b = 12$ (*b*) $a = 573, b = -16$ (*c*) $a = -381, b = 14$
   (*d*) $a = -433, b = -17$ (*e*) $a = 4461, b = 16$ (*f*) $a = -262, b = 3$
2. Use the Euclidean algorithm to find the greatest common divisor of each pair of integers
   (*a*) $a = 143, b = 227$ (*b*) $a = 8316, b = 10,920$
   (*c*) $a = 37, b = 249$ (*d*) $a = 414, b = 662$
   (*e*) $a = 595, b = 252$ (*f*) $a = 2406, b = 654$
3. Find the gcd of following integer numbers and express the result as a linear combination of these integers
   (*a*) 231, 1820 (*b*) 1369, 2597 (*c*) 2689, 4001
4. Find the integers x and y such that
   (*a*) $6x + 10y = 104$ (*b*) $256x + 160y = 32$
   (*c*) $128x + 58y = 1$ (*d*) $155x - 135y = 5$
5. If d = gcd(a, b) for any integers a and b, then prove that gcd$(a/d, b/d) = 1$.
6. For each pair of integers a and b, find d = gcd(a, b) and express $d = ax + by$.
   (*a*) $a = 48$ and $b = 356$ (*b*) $a = 2310$ and $b = 168$
   (*c*) $a = 165$ and $b = 1287$ (*d*) $a = 195$ and $b = 968$

7. If $a = 5880$ and $b = 8316$, then

(*a*) express a and b as product of primes.

(*b*) find gcd(a, b) and lcm(a, b).

8. Prove that

(*a*) gcd(a, b) = gcd(a, $b + ka$), for some integer k

(*b*) gcd(na, nb) = n gcd(a, b).

9. Let d = gcd(a, b) for some $a, b, d \in I_+$. Then prove that d^2 divides ab.

10. If $d = a + bc$ for some $a, b, c \in I_+$, then prove that gcd(b, d) = gcd(a, b).

11. If gcd(a, b) = 1, then prove that lcm(a, b) = ab.

12. If gcd(a, b) = 1, then prove that gcd(a^2, b^2) = 1.

13. If gcd(a, b) = 1 with $a \mid c$ and $b \mid c$, then prove that $ab \mid c$.

14. If gcd(a, b) = 1 and $a \mid bc$ for some $a, b, c \in I_+$, then prove that $a \mid c$.

15. For any integer n, prove that gcd(n, $n + 2$) = 1 or 2, and gcd(n, $n + 3$) = 1 or 3.

16. If $a = 630$, gcd(a, b) = 105 and lcm(a, b) = 242, 550, then find the value of b.

17. For each positive integer n, find gcd(n, $n + 1$) and lcm(n, $n + 1$).

18. If gcd(a, b) = 1, then prove the following results

(*a*) gcd($2a + b$, $a + 2b$) = 1 or 3

(*b*) gcd($a + b$, $a^2 + b^2$) = 1 or 2

(*c*) gcd($a + b$, $a^2 - ab + b^2$) = 1 or 3.

19. If $a = bq + r$, then prove that gcd(a, b) = gcd(b, r).

20. (*a*) If $a > 2$ and $k > 1$, then prove that $a^k - 1$ is composite.

(*b*) If $n > 0$ and $2^n - 1$ is prime, then prove that n is prime.

21. Prove

(*a*) $3^n + 1$ is divisible by 2 when n is an even integer.

(*b*) $3^n + 1$ is divisible by 2^2 when n is an odd integer.

22. Prove that $\sqrt{p}$ is not a rational number for any prime p.

23. If $2^n - 1$ is a prime number for any positive integer n, then prove that $2^{n-1}(2^n - 1)$ is a perfect integer.

24. Prove each of the following statements

(*a*) The product of any three consecutive integers is divisible by 6.

(*b*) The product of any four consecutive integers is divisible by 24.

Hints and Answers

1. (*a*) $q = 21, r = 6$ (*b*) $q = -35, r = 13$ (*c*) $q = -28, r = 11$ (*d*) $q = 26, r = 9$ (*e*) $q = 278, r = 13$ (f) $q = -88, r = 2$

2. (*a*) 9 (*b*) 84 (*c*) 3 (*d*) 2 (*e*) 7 (*f*) 6

3. (*a*) gcd(1820, 231) = 7 = 1820(8) + 231(– 63)
(*b*) gcd(2597, 1369) = 1 = 2597(534) + 1369(– 1013)
(*c*) gcd(4001, 2689) = 1 = 4001(– 1117) + 2689(1662)

4. (*a*) $x = 14$ and $y = 2$ (*b*) $x = 2$ and $y = -3$ (*c*) $x = 41$ and $y = -9$ (*d*) $x = 7$ and $y = 8$

5. $\gcd(a, b) = d \Rightarrow d = ax + by$, for some $x, y \in I$
$\gcd(a, b) = d \Rightarrow a/d$ and $b/d \in I$
$1 = (a/d)x + (b/d)y \Rightarrow \gcd(a/d, b/d) = 1$.

6. (*a*) $d = 4 = 5(356) - 37(48)$ (*b*) $d = 33 = 8(165) - 1(1287)$
(*c*) $d = 42 = 14(168) - 1(2310)$ (*d*) $d = 1 = 139(195) - 28(968)$

7. (*a*) $a = 2^4.3.5.7^2$ and $b = 2^2.3^3.7.11$
(*b*) $\gcd(a, b) = 2^2.3.7$ and $\text{lcm}(a, b) = 2^4.3^3.5.7^2 = 1164240$

14. $\gcd(a, b) = 1 \Rightarrow ax + by = 1$. Then $acx + bcy = ca \mid acx$, $a \mid bcy$ because $a \mid bc \Rightarrow a \mid c$.

16. $b = 40, 425$

17. $\gcd(n, n + 1) = 1$; $\text{lcm}(n, n + 1) = n(n + 1)$

2.8 CONGRUENCE RELATION

Let m be a positive integer. Then an integer a is said to be congruent to another integer b modulo m provided m divides $a - b$. Symbolically, we write

$$a \equiv b \text{ (modulo } m) \text{ or } a \equiv b \text{ (mod } m)$$

It is read as 'a is congruent to b modulo m'. The integer m is called the *modulus,* and b is called a *residue* of a(mod m).

The negation of $a \equiv b$ (mod m) is written as $a \not\equiv b$ (mod m) and in such a case integers a and b are said to be *incongruent* modulo m.

From this definition of congruence, it may be concluded that $a \equiv b$(mod m) provided $a - b = km$ or $a = b + km$ for some integer k.

Illustrations

1. $89 \equiv 25$ (mod 4), since $89 - 25 = 64$ is divisible by 4. Consequently 25 is the residue of 89(mod 4) and 4 is the modulus of the congruence.
2. $153 \equiv -7$ (mod 8), since $153 - (-7) = 160$ is divisible 8. Thus, – 7 is the residue of 153(mod 8) and 8 is the modulus of the congruence.
3. $24 \not\equiv 3$(mod 5), since $24 - 3 = 21$ is not divisible 5. Thus, 24 and 3 are incongruent modulo 5.

Example 14 Find the least positive integers modulo 5 to which 19, 288 and – 28 are congruent.

Solution From given integer numbers, we find that

$$19 = 5 \times 3 + 4 \Rightarrow 19 \equiv 4 \text{(mod 5)}$$

$$288 = 5 \times 57 + 3 \Rightarrow 288 \equiv 3 \text{(mod 5)}$$

$$-28 = 5 \times -6 + 2 \Rightarrow -28 \equiv 2 \text{(mod 5)}$$

2.8.1 Properties of Congruences

Theorem 2.12 The relation 'congruence modulo m' is an equivalence relation in the set of integers. In other words, the relation is

(*i*) **Reflexive** For any integer a, we have $a \equiv a \pmod{m}$

(*ii*) **Symmetric** If $a \equiv b \pmod{m}$, then $b \equiv a \pmod{m}$

(*iii*) **Transitive** If $a \equiv b \pmod{m}$ and $b \equiv c \pmod{m}$, then $a \equiv c \pmod{m}$.

Proof (*i*) Let a be any integer. Then $a - a = 0$ is divisible by any fixed positive integer m. Thus, $a \equiv a \pmod{m}$ for all $a \in I$.

(*ii*) Given $a \equiv b \pmod{m}$, for any $a, b \in I$. This implies that $a - b$ is divisible by m. Since $b - a$ is also divisible by m, therefore we may write $b \equiv a$(mod). Hence,

$$a \equiv b \pmod{m} \Rightarrow b \equiv a \pmod{m}$$

(*iii*) Given $a \equiv b \pmod{m}$ and $b \equiv c \pmod{m}$. This means that both $a - b$ and $b - c$ are divisible by m. So the sum $(a - b) + (b - c) = a - c$ is also divisible by m. Consequently, we may write $a \equiv c \pmod{m}$. Hence,

$$a \equiv b \pmod{m} \text{ and } b \equiv c \pmod{m} \Rightarrow a \equiv c \pmod{m}.$$

Remark If a is any integer and m is a positive integer, then there exist integers q and r with $0 \leq r < m$, such that

$$\begin{aligned} a = mq + r &\Rightarrow mq = a - r \\ &\Rightarrow m \text{ divides } (a - r) \\ &\Rightarrow a \equiv r \pmod{m} \end{aligned}$$

Consequently, any integer a is congruent modulo m to a unique integer in the set $\{0, 1, 2, \ldots, m - 1\}$. The uniqueness is due to the fact that m cannot divide the difference of two such integers.

Theorem 2.13 Two integers a and b are congruent modulo m if and only if these have the same remainder when divided by m.

Proof Suppose a and b have two distinct remainders r_1 and r_2, respectively when divided by m. Then for some integers q_1 and q_2, we have

$$a = q_1 m + r_1,\ 0 \leq r_1 < m$$

and

$$b = q_2 m + r_2,\ 0 \leq r_2 < m$$

From these equations, we have

$$a - b = (q_1 - q_2)m + (r_1 - r_2) \Rightarrow r_1 - r_2 = (a - b) + m(q_1 - q_2)$$

Since $a \equiv b \pmod{m}$ implies m divides $a - b$. Also m divides $m(q_1 - q_2)$, therefore m divides $(a - b) + m(q_1 - q_2)$. Hence,

$$\begin{aligned} m \mid (r_1 - r_2) &\Rightarrow r_1 - r_2 = 0 \text{ because } 0 \leq (r_1 - r_2) < m \\ &\Rightarrow r_1 = r_2 \end{aligned}$$

Conversely, suppose that $r_1 = r_2$. Then to prove that $a \equiv b \pmod{m}$, consider

$$\begin{aligned} a - b &= m(q_1 - q_2) + (r_1 - r_2) \\ &= m(q_1 - q_2) \text{ for } r_1 = r_2 \end{aligned}$$

This implies that m divides $a - b$. Hence, $a \equiv b \pmod{m}$.

2.8.2 Congruence Arithmetic

Theorem 2.14 If $a \equiv b(\text{mod } m)$ and c is any integer, then

(*a*) $mc + a = b(\text{mod } m)$ (*b*) $a + c \equiv b + c(\text{mod } m)$

(*c*) $a - c \equiv b - c(\text{mod } m)$ (*d*) $ac \equiv bc(\text{mod } m)$

Proof (*a*) Let $a \equiv b(\text{mod } m)$. Then it implies that m divides $a - b$. But m divides mc for any integer c. Thus

$$m \mid \{mc + (a - b)\} \Rightarrow m \mid (mc + a) - b$$
$$\Rightarrow mc + a \equiv b(\text{mod } m)$$

Conversely, let $mc + a \equiv b(\text{mod } m)$. Then

$$m \mid (mc + a) - b \Rightarrow m \mid \{(mc + a) - b\} - mc, \text{ because } m \mid mc$$
$$\Rightarrow m \mid (a - b)$$
$$\Rightarrow a \equiv b(\text{mod } m).$$

(*b*) Given that

$$a \equiv b(\text{mod } m) \Rightarrow m \text{ divides } (a - b)$$
$$\Rightarrow m \text{ divides } (a + c) - (b + c), \text{ for all } c \in \text{I}$$
$$\Rightarrow a + c \equiv b + c(\text{mod } m)$$

(*c*) Same as part (*b*)

(*d*) Given that

$$a \equiv b(\text{mod } m) \Rightarrow m \text{ divides } (a - b)$$
$$\Rightarrow m \text{ divides } c(a - b) \text{ for all } c \in \text{I}$$
$$\Rightarrow m \text{ divides } ca - cb$$
$$\Rightarrow ca \equiv cb(\text{mod } m).$$

Theorem 2.15 If a, b, c and d are integers such that $a \equiv b(\text{mod } m)$ and $c \equiv d(\text{mod } m)$, then prove that

(*a*) $a + c \equiv b + d(\text{mod } m)$ (*b*) $a - c \equiv b - d(\text{mod } m)$

(*c*) $ac \equiv bd(\text{mod } m)$ (*d*) $a^k = b^k(\text{mod } m)$ for any positive integer k.

Proof (*a*) Given that

$$a \equiv b(\text{mod } m) \Rightarrow m \mid (a - b)$$
$$c \equiv d(\text{mod } m) \Rightarrow m \mid (c - d)$$

These two equations implies that the sum $(a - b) + (c - d)$ is also divisible by m. That is

$$m \mid (a - b) + (c - d) \Rightarrow m \mid (a + c) - (b + d)$$
$$\Rightarrow a + c \equiv b + d(\text{mod } m)$$

(*b*) Same as part (*a*)

(*c*) Since $m \mid (a - b)$ and $m \mid (c - d) \Rightarrow m \mid c(a - b)$ and $m \mid c(b - d)$

$$\Rightarrow m \mid c(a - b) + c(b - d)$$
$$\Rightarrow m \mid ca - bd$$
$$\Rightarrow ac \equiv bd(\text{mod } m)$$

(*d*) For $k = 1$, the statement is true, i.e., $a \equiv b(\text{mod})$. By induction theorem let the statement is true for some fixed integer k_1, i.e.,

$$a^k \equiv b^k(\text{mod } m)$$

Using the statement in part (*c*), we have

$$a\, a^k = b\, b^k (\text{mod } m) \text{ or } a^{k+1} \equiv b^{k+1} (\text{mod } m)$$

This implies that the statement is true for $k + 1$ also.

Remark Suppose $a_1 \equiv b_1(\text{mod } m)$, $a_2 \equiv b_2(\text{mod } m)$, . . ., $a_n \equiv b_n(\text{mod } m)$ are known. Then following results are also hold

$$(a)\ \sum_{i=1}^{n} a_i \equiv \sum_{i=1}^{n} b_i \ (\text{mod } m), \text{ and } (b)\ \prod a_i \equiv \prod b_i \ (\text{mod } m)$$

Theorem *(Cancellation law)* **2.16** Suppose $ab \equiv ac(\text{mod } m)$ and $d = \gcd(a, m)$. Then $b \equiv c(\text{mod } m/d)$.

Proof Given that m divides $ab - ac = a(b - c)$. Thus, there exists an integer x such that $a(b - c) = mx$. Dividing by d, we have

$$(a/d)\,(b - c) = (m/d)x$$

This implies that (m/d) divides $(a/d)\,(b - c)$. Since m/d and a/d are relatively primes, m/d divides $b - c$. Hence, this yields $b \equiv c(\text{mod } m/d)$, as required.

Theorem 2.17 Suppose $ab \equiv ac(\text{mod } m)$ and $\gcd(a, m) = 1$. Then $b \equiv c(\text{mod } m)$.

Proof The proof is same as of Theorem 2.15 when $d = 1$.

Illustrations Observe that $89 \equiv 25(\text{mod } 4)$ and $25 = 1(\text{mod } 4)$. Then

(*i*) $89 + 25 \equiv 25 + 1(\text{mod } 4)$ or $114 \equiv 26(\text{mod } 4)$

(*ii*) $89 \times 25 \equiv 25 \times 1(\text{mod } 4)$ or $2225 \equiv 25(\text{mod } 4)$

Remark The ordinary cancellation law such as $ab = ac \Rightarrow b = c$ for $a \neq 0$ does not hold for congruence. For example,

$$3.1 \equiv 3.5(\text{mod } 6) \text{ but } 1 \neq 5(\text{mod } 6)$$

That is, we cannot cancel 3 even though $3 \not\equiv 0(\text{mod } 6)$.

2.9 RESIDUE OR CONGRUENCE CLASSES

Since congruence modulo m is an equivalence relation in the set I of integers, it partition I into disjoint equivalence classes, called *residue classes modulo m* or *congruence classes modulo m* and is denoted by I_m or [I].

Hence, a residue consists of all those integers with the same remainder when divided by m. Thus, there are m such residue classes and each residue class contains exactly one of the integers in the set of possible remainders $\{0, 1, 2, . . ., m - 1\}$.

A set of m integers $\{a_1, a_2, . . ., a_n\}$ is said to be a complete residue system modulo m if each a_1 comes from a distinct residue class. Thus, the integers from 0 to $m - 1$ form a complete residue system modulo m.

If x is an integer, then the residue class (modulo m) containing x (i.e., those integers which are congruent to x) is given by $[x] = \{a : a \in \text{I and } a \equiv x(\text{mod } m)\} = \{x + mr : r \in \text{I}\}$ where $m \geq 2$. Accordingly, the residue classes can be denoted by [0], [1], [2], . . ., $[m - 1]$.

2.9.1 Properties of Residue Classes

The basic properties of the residue classes modulo m are

$[0] = \{\ldots -2m, -m, 0, m, 2m, \ldots\} = \{0 + mr : r \in I\}$

$[1] = \{\ldots -2m + 1, -m + 1, m + 1, 2m + 1, \ldots] = \{1 + mr : r \in I\}$

$[2] = \{\ldots -2m + 2, -m + 2, 2, m + 2, 2m + 2, \ldots\} = \{2 + mr : r \in I\}$

$\vdots$

$[m-1] = \{\ldots -m-1, -1, m-1, 2m-1, 3m-1, \ldots\} = \{(m-1) + mr : r \in I\}$

By using division algorithm, for all $n \in I$, we can write $n = qm + r$ where $0 \leq r < m$, so $n \in [r]$ or $[n] = [r]$. Often we use the notation I_n to denote [0], [1], [2], . . ., $[m-1]$.

The basic properties of the residue classes modulo m are:

1. If a and b are elements of the same residue class $[s]$, then $a \equiv b(\text{mod } m)$
2. If $[s]$ and $[t]$ are two distinct residue classes with

$$a \in [s] \text{ and } b \in [t], \text{ then } a \not\equiv b(\text{mod } m)$$

From these properties, it is obvious that [0] = [4] = [8] and so on. Similarly, [1] = [5] = [9] = [13] and so on.

Remarks

1. The set I_m of all residue classes of integers modulo m contains exactly m distinct elements. For example, the set $I_4 = \{[0], [1], [2], [3]\}$, contains 4 distinct elements. Of course, $[3] \in I_4 = [3] \in I_m$ if and only if $m = 4$.
2. The residue class [0] is called the *zero residue class*. Now $[a] = [0]$ if and only if m/a. The set of integers mod m will have $m - 1$ distinct non-zero residue classes.

   (*a*) If a and b are elements of the same residue class, then $a \equiv b(\text{mod } m)$

   (*b*) If $[x]$ and $[y]$ are two distinct residue classes with $a \in [x]$ and $b \in [y]$, then $a \not\equiv b(\text{mod } m)$.

2.9.2 Arithmetic of Residue Classes

Addition and multiplication are defined for an residue classes modulo m as follows:

$$[a] + [b] = [a + b] \text{ and } [a].\ [b] = [a]\ [b] = [a\ b]$$

Here the plus sign '+' on the left-hand side stands for addition of residue classes and '+' on the left hand side stands for addition of integers. For example, if $m = 7$, then [2] + [6] = [2 + 6] = [8] = [1], and [2] [6] = [12] = [5]. The residue modulo $m = 4$, i.e., $I_4 = \{0, 1, 2, 3\}$ are expressed in table form as shown below:

+	0	1	2	3
0	0	1	2	3
1	1	2	3	0
2	2	3	0	1
3	3	0	1	2

×	0	1	2	3
0	0	0	0	0
1	0	1	2	3
2	0	2	0	2
3	0	3	2	1

There are only a finite number of residue classes modulo m. Thus one can easily write down explicitly their addition and multiplication tables when m is small. Above tables show the addition and multiplication for the residue classes modulo $m = 4$. For example, $[a] + [b] = [r]$, where r is the remainder when $a + b$ or $a \times b$ is divided by m. Thus, if $m = 4$, we have different classes as shown above.

Example 15 Show that for all $n \in N$, $3^{2n+1} \equiv 3.2^n \pmod 7$. Obtain a similar result for 2^{n+2} and deduce that 7 divides $3^{2n+1} + 2^{n+2}$ for all n.

Solution Since $3^2 \equiv 2 \pmod 7$, and hence $3^{2n} = 2^n \pmod 7$

Now $3^{2n+1} = 3.3^{2n} = 3.2^n \pmod 7$.

Also $2^1 \equiv 2 \pmod 7$ and hence $2^n \equiv 2^n \pmod 7$. Now

$$2^{n+2} = 2^2.2^n = 4.2^n \pmod 7$$

Thus $3^{2n+1} + 2^{n+2} \equiv 7.\ 2^n \pmod 7 \equiv 0 \pmod 7$, which proves the result.

2.10 CONGRUENCE EQUATIONS

A polynomial congruence equation of degree n is of the form

$$f(x) = a_n x^n + a_{n-1} x^{n-1} + \ldots + a_1 x + a_0 \equiv 0 \pmod m$$

where $a_n \not\equiv 0 \pmod m$.

Any integer u is the solution of congruence equation $f(x) \equiv 0 \pmod m$, provided $x \equiv u \pmod m$. Thus, the number of solutions of the congruence equation is defined to be the number of incongruent solutions, or equivalently the number of solutions in the set $\{0, 1, 2, \ldots, m - 1\}$.

Illustrations Consider the equations

1. $x^2 + x + 1 \equiv 0 \pmod 4$. This equation has no solution because none of the values of $x = 0, 1, 2$ or 3 satisfy the equation.
2. $x^2 + 3 \equiv \pmod 6$. This equation has only one solution because from number of possible solutions $x = 0, 1, 2, \ldots, 5$, only $x = 3$ satisfy the equation. Thus, the general solution consists of the integers $3 + 6k$, for all $k \in I$.
3. $x^3 - 1 \equiv \pmod 8$. From number of solutions $x = 0, 1, 2, 3, \ldots, 8$, only four values $x = 1, 3, 5$ and 7 satisfy this equation. Thus, there are four solutions of this equation. Consequently, we may conclude that a congruence equation of degree n can have more than n solutions.

Remarks

1. The coefficients of a congruence equation can always be reduced modulo m. This resulted an equation with the same solutions. For example, the following equations

$$18x^2 + 21x + 16 \equiv 0 \pmod 5 \simeq 3x^2 + x + 1 \equiv 0 \pmod 5$$

are equivalent as the coefficients are congruent modulo $m = 5$. Usually, coefficients are chosen between 0 and $m - 1$ or between $-m/2$ and $m/2$.

2. Since we desire to have solution of congruence equation among residue classes modulo m rather than among the integers, therefore the congruence equation may be viewed as an equation over I_m (integer modulo m) rather than an equation over I (set of integers). This implies that a congruence equation can have solution(s) in I_m.

2.10.1 Linear Congruence Equation

An expression of the form

$$ax \equiv b(\text{mod } m),\ a \not\equiv 0\ (\text{mod } m)$$

where a, b and m are fixed integers with $m > 0$ is called a linear congruence equation modulo m.

Remark If x_1 is a solution of $ax \equiv b(\text{mod } m)$, then any other integer $x_2 \equiv x_1(\text{mod } m)$ is also a solution of the equation.

We know that if x_0 is an element of the residue class $[x_1]$ modulo m, then $x_0 \equiv x_1(\text{mod } m)$. Thus if x_1 is a solution of the equation $ax \equiv b(\text{mod } m)$, then each element of $[x_1]$ will also be the solution of the given equation. Hence, the congruence class $[x_1]$ is also called *congruence class of solutions* of the congruence equation $ax \equiv b(\text{mod } m)$.

For example, consider the equation: $6x \equiv 2(\text{mod } 4)$. If we put $x = 3$, we find that $6 \times 3(\text{mod } 4)$ is true. Hence, $x = 3$ is solution of the given equation. However, each element of the residue class

$$[3] = [\ldots, -13, -9, -5, -1, 3, 7, 11, 15, \ldots]$$

is also the solution of the equation. The least positive integer belongs to [3] is $x_0 = 3$.

2.10.2 Criteria for the Existence of Solutions

Theorem 2.18 If a and m are relatively prime numbers, then the equation $ax \equiv 1(\text{mod } m)$ has a unique solution, otherwise it has no solution.

Proof Let x_0 be a solution of the equation. Then m divides $ax_0 - 1$, consequently there exists an integer y_0 such that $my_0 = ax_0 - 1$. Thus using Euclidean algorithm we have $ax_0 + (-m)y_0 = 1$. This implies that a and m are relatively prime.

Conversely, if a and m are relatively prime, then there exist integers x_0 and y_0 such that $ax_0 + my_0 = 1$. From this it follows that $ax_0 \equiv 1(\text{mod } m)$, i.e., x_0 is a solution to the linear congruence equation.

To prove that the solution x_0 is an unique solution modulo m, suppose x_1 is another solution. Then $ax_0 = ax_1 \equiv 1\ (\text{mod } m)$. Since a and m are relatively prime, it follows that $x \equiv x_1(\text{mod } m)$.

Remark Suppose x_1 and x_2 are two solutions of the linear congruence equation $ax \equiv b(\text{mod } m)$. If $x_1 \not\equiv x_2(\text{mod } m)$, then x_1 and x_2 are called *incongruent solutions* of $ax \equiv b(\text{mod } m)$.

Illustration Consider the congruence equation $6x \equiv 2(\text{mod } 4)$. Since $x_1 = 1$ and $x_2 = 3$ are its solutions. But $1 \not\equiv 3(\text{mod } 4)$, consequently 1 and 3 are incongruent solutions of the equation. On the other hand, $x_0 = 5$ is also a solution of the equation, because $5 \equiv 1(\text{mod } 4)$. Thus, 5 and 1 are congruent solutions of the equation.

Theorem 2.19 The congruence $ax \equiv b(\text{mod } m)$ has a solution if and only if $\gcd(a, m) = d$ divides b and has exactly d incongruent solutions (mod m).

Proof Let $\gcd(a, m) = d$. Suppose congruence $ax \equiv b(\text{mod } m)$ has a solution $x = x_1$. Then

$$ax_1 \equiv b(\text{mod } m) \Rightarrow m \mid (ax_1 - b)$$

$$\Rightarrow ax_1 - b = mk \text{ for some integer } k$$

$$\Rightarrow b = ax_1 - mk$$

Now $d \mid a$ and $d \mid m$. Let $a = a_1 d$ and $m = m_1 d$ where a_1 and m_1 are some integers. Then $b = a_1 dx_1 - m_1 dk = d(a_1 x_1 - m_1 k)$. This implies that $d \mid b$. This shows that if a solution exists, then it is necessary that $d \mid b$.

Conversely, suppose d divides b, i.e., $d \mid b$. Then $b = b_1 d$ where b_1 is some integer. Since $d = \gcd(a, m)$, by Euclidean algorithm there exist integers u and v such that

$$\begin{aligned} d = ua + vm &\Rightarrow db_1 = uab_1 + vmb_1 \\ &\Rightarrow b = a(ub_1) + m(vb_1) \\ &\Rightarrow a(ub_1) - b = -(vb_1)m \\ &\Rightarrow m \mid \{a(ub_1 - b)\} \\ &\Rightarrow a(ub_1) \equiv b(\text{mod } m) \end{aligned}$$

This implies that $x = ub_1$ is a solution of $ax \equiv b(\text{mod } m)$. Thus, if d divides b, then $ax \equiv b(\text{mod } m)$ has a solution.

Illustrations

1. Consider the congruence $222x \equiv 12(\text{mod } 18)$. Here $\gcd(222, 18) = 6$ and 6 divides 12. Hence, there are 2 incongruent solutions of the given equation.
2. Consider the congruence $207x \equiv 6(\text{mod } 18)$. Here $\gcd(207, 18) = 9$. Since 9 does not divide 6, therefore the given equation has no solution.

Theorem 2.20 The congruence $ax \equiv b(\text{mod } m)$ has a unique incongruent solution modulo m when $\gcd(a, m) = 1$.

Proof Let $\gcd(a, m) = 1$. Then there exist integers s and t such that

$$\begin{aligned} 1 = as + mt &\Rightarrow b = bas + bmt \\ &\Rightarrow a(bs) - b = -(bt)m \\ &\Rightarrow m \mid \{a(bs) - b\} \\ &\Rightarrow a(bs) \equiv b(\text{mod } m) \end{aligned}$$

This implies that $x = bs$ is a solution of the given equation, and hence shows the existence of the solution.

To show that the congruence $ax \equiv b(\text{mod } m)$ has a unique incongruent solution, suppose x_1 and x_2 are two of its solutions. Thus, we may write

$$ax_1 \equiv b(\text{mod } m) \text{ and } ax_2 \equiv b(\text{mod } m)$$

This implies that $ax_1 = ax_2(\text{mod } m) \Rightarrow m \mid a(x_1 - x_2)$ or $m \mid (x_1 - x_2)$ because $\gcd(a, m) = 1$ means $x_1 \equiv x_2(\text{mod } m)$. This shows that any two solutions of the given equation are congruent to each other modulo m. Hence, the given equation has a unique incongruent solution modulo m.

Theorem 2.21 If a and m are relatively primes, then $ax \equiv 1(\text{mod } m)$ has a unique solution, otherwise it has no solution.

Proof Let s be the solution of the given congruence. Then m divides $as - 1$ and hence, there exists t such that $mt = as - 1$ or $as + mt = 1$, where a and m are relatively primes.

Conversely, if a and m are relatively primes, then there exist s and t satisfying the equation $as + mt = 1$ in which s is a solution of $ax \equiv 1(\text{mod } m)$.

To prove that the solution s is unique, let s_1 be another solution. Then $as \equiv 1 \equiv as_1(\text{mod } m)$. Since a and m are relatively primes, the cancellation law holds, and therefore $s \equiv s_1(\text{mod } m)$. This proves the theorem.

Example 16 Solve the following linear congruence equations:

(*a*) $259x \equiv 5 \pmod{11}$ (*b*) $11x \equiv 2 \pmod{317}$ (*c*) $52x \equiv 28 \pmod{20}$

Solution (*a*) Applying the Euclidean algorithm, we find gcd(259, 11) = 1. Hence, the congruence $259x \equiv 5 \pmod{11}$ has a unique incongruent solution.

Solving the given equation by applying division algorithm as follows:

$$259 = 23 \times 11 + 6 \text{ (because } 0 \leq 6 < 11) = 253 + 11$$

Now $253x \equiv 0 \pmod{11}$. From given equation and this equation, we have

$$259x - 253x \equiv 5 - 0 \pmod{11}$$

$$6x \equiv 5 \pmod{11} \text{ or } 5 \equiv 60 \pmod{11}$$

because 60 is the least positive integer divisible by 6 and also congruent to 5 modulo 11. Now

$$6x \equiv 5 \pmod{11} \Rightarrow 6x \equiv 60 \pmod{11} = 6 \times 11 \pmod{11}$$

or $$x \equiv 10 \pmod{11} \text{ because } \gcd(6, 11) = 1.$$

This shows that $x = 10$ is a solution. Since 10 is the least positive integer in the residue class $[10] \in I_{11}$, $x = 10$ is the required solution.

(*b*) Applying the Euclidean algorithm, we find that gcd(11, 317) = 1. Hence, the congruence $11x \equiv 2 \pmod{317}$ has a unique solution.

Since modulus is large, we shall solve the given equation by first finding the solution of $11x \equiv 2$(mod 317). The solution is obtained by finding integers m and n such that $1 = 11m + 317n$. We may write

$$317 = 11 \times 28 + 9$$
$$11 = 9 \times 1 + 2$$
$$9 = 2 \times 4 + 1$$
$$2 = 1 \times 2 + 0$$

Now $1 = 9 - 2 \times 4 = 9 - (11 - 9 \times 1).4 = 5 \times 9 - 4 \times 11 = 5(317 - 11 \times 28) - 4 \times 11 = 5 \times 317 + (-144)11$.

Thus $2 = 10 \times 317 - 288 \times 11$.

The linear congruence equation can now be written as:

$$\begin{aligned} 11x &\equiv 2 \pmod{317} \\ &\equiv 10 \times 317 + (-288)11 \pmod{317} \\ &\equiv (-288)11 \pmod{317} \end{aligned}$$

This shows that $x = -288$ is a solution to the given equation which is written in modular form as: $x \equiv -288 \pmod{317}$.

Since $-288 = -317 + 29$, therefore $x = 29$ is the least positive integer in the residue class $[-288] \in I_{317}$ and hence it is the required solution.

(*c*) First we find gcd(52, 20) = 4. Since 4 divides 28, therefore $52x \equiv 28 \pmod{20}$ has 4 incongruent solutions. Dividing the given equation by 4 throughout, we get $13x \equiv 7 \pmod{5}$. Here 13 and 7 are relatively prime and therefore the given equation has a unique solution modulo 5.

Also $13x \equiv 7 \pmod{5}$ is equivalent to $3x \equiv 7 \pmod{5}$ because $10x \equiv 0 \pmod{5}$. Now $3x \equiv 2$ mod(5), since $7 \equiv 2 \pmod{5}$

$$\Rightarrow 3x \equiv 12 \pmod{5}, \text{ since } 0 \equiv 10 \pmod{5}$$

$$\Rightarrow x = 4 \pmod{5}, \text{ since } \gcd(3, 5) = 1.$$

This shows that $x = 4$ is the solution of the given equation $13x \equiv 7 \pmod 5$.

Since 4 is the least positive integer in the residue class $[4] \in I_5$ and also satisfying the given linear congruence, $x = 4$ is the required solution.

Other incongruent solutions of $52x \equiv 28 \pmod{20}$ are: $x = 4, 4 + 1 \times 5, 4 + 2 \times 5, 4 + 3 \times 5$, i.e., $x = 4, 9, 14, 19$.

2.10.3 Simultaneous Linear Congruence

The simultaneous linear congruence arises when two or more congruences of first degree in one unknown variable have relatively prime moduli. For example, following simultaneous linear congruences

$$x \equiv 0 \pmod 5$$
$$x \equiv 2 \pmod 7$$

arise when it is required to find a number which when divide 5 and 7 leaves remainder 0 and 2, respectively.

Theorem *(Chinese Remainder Theorem)* **2.22** Let $m_1, m_2, \ldots, m_k$ be pairwise relatively prime positive integers. Then the system of congruences

$$x \equiv r_1 \pmod{m_1};\quad x \equiv r_2 \pmod{m_2} \ldots, x \equiv r_k \pmod{m_k},$$

where $r_i (i = 1, 2, \ldots, k)$ are given integers, has a unique solution modulo $M = m_1 m_2 \ldots m_k$.

Proof Let $M = m_1 m_2 \ldots m_k$ and $M_1 = \dfrac{M}{m_1}, M_2 = \dfrac{M}{m_2}, \ldots, M_k = \dfrac{M}{m_k}$. It is due to the reason that m_i are pairwise prime. It also follows that $\gcd(M_i, m_i) = 1$.

If $s_1, s_2, \ldots, s_k$, respectively are solutions of the congruence equations

$$M_1 x \equiv 1 \pmod{m_1},\ M_2 x \equiv 1 \pmod{m_2},\ \ldots,\ M_k x \equiv 1 \pmod{m_k}$$

then $x = M_1 s_1 r_1 + M_2 s_2 r_2 + \ldots + M_k s_k r_k$ is the solution of the system of congruences

Remark If moduli m_i are not pairwise relatively prime, then the solution of the simultaneous system of congruences is no longer guaranteed. However, if the solution exist, it is unique modulo $[m_1, m_2, \ldots, m_k]$, not modulo m.

Example 17 Solve the following simultaneous congruences

$$x \equiv 3 \pmod 5 \text{ and } x = 5 \pmod 7$$

Solution We are given $m_1 = 5$, $m_2 = 7$, $r_1 = 3$ and $r_2 = 5$. Since 5 and 7 are relatively primes, then there exists an unique solution. Calculating

$$M = 5 \times 7 = 35 \text{ and } M_1 = M/m_1 = 35/5 = 7,\ M_2 = M/m_2 = 35/7 = 5.$$

Thus, we need to find solutions to the equations

$$7x \equiv 1 \pmod 5 \text{ or } 2x \equiv 1 \pmod 5$$
$$5x \equiv 1 \pmod 7 \text{ or } 5x \equiv 1 \pmod 7$$

Thus, the solutions of these equations are $s_1 = 3$ and $s_2 = 3$. Hence, the required solution to the system of congruences is

$$x = M_1 s_1 r_1 + M_2 s_2 r_2 = 7 \times 3 \times 3 + 5 \times 3 \times 5 \pmod{35}$$
$$= 138 \pmod{35} = 33 \pmod{35}$$

Example 18 Solve the following simultaneous congruences

$$x \equiv 2 \pmod 3,\ x \equiv 4 \pmod 7 \text{ and } x \equiv 6 \pmod{10}.$$

Solution Given $m_1 = 3$, $m_2 = 7$, $m_3 = 10$, $r_1 = 2$, $r_2 = 4$, $r_3 = 6$. Since 3 and 7 are relatively primes, there exists a unique solution modulo $M = 3 \times 7 \times 10 = 210$.

We know that $M_1 = M/m_1 = 210/3 = 70$, $M_2 = M/m_2 = 210/7 = 30$ and $M_3 = M/m_3 = 210/10 = 21$. Thus we need to find solutions to the equations

$$70x \equiv 1 \pmod 3;\ 30x \equiv 1 \pmod 7 \text{ and } 21x \equiv 1 \pmod{10}$$

Reducing 70 modulo 3, 30 modulo 7 and 21 modulo 10, we get

$$x \equiv 1 \pmod 3,\ 2x \equiv 1 \pmod 7 \text{ and } x \equiv 1 \pmod{10}$$

The solutions of these equations are $s_1 = 1$, $s_2 = 4$ and $s_3 = 1$.

The required solution of the given system of congruences is obtained by substituting values in the formula:

$$\begin{aligned} x &= M_1 s_1 r_1 + M_2 s_2 r_2 + M_3 s_3 r_3 \\ &= 70 \times 1 \times 2 + 30 \times 4 \times 4 + 21 \times 1 \times 6 = 746 \pmod{210} \\ &= 116 \pmod{210} \end{aligned}$$

Theorem 2.23 The system $x \equiv b_1 \pmod{m_1}$ and $x \equiv b_2 \pmod{m_2}$ is solvable if and only if $b_1 - b_2$ is divisible by $\gcd(m_1, m_2)$, i.e., $b_1 = b_2 \bmod (m_1, m_2)$.

Proof Let $x = s$ be a solution of a given system. Then

$$s = b_1 \pmod{m_1} \text{ and } s = b_2 \pmod{m_2}$$

This implies that $b_1 = b_2(\text{mod } (m_1, m_2))$, which is the necessary condition.

Conversely, if $b_1 = b_2(\text{mod}(m_1, m_2))$, then the congruence $m_1 t = b_2 - b_1 \pmod{m_2}$ has solution $s \equiv d \pmod{m_2}$. Hence, we get

$$b_1 + m_1 d \equiv b_1 \pmod{m_1} \text{ and } b_1 + m_1 d \pmod{m_2}$$

Thus $x = b_1 + m_1 d$ is a solution of the given system, which is the sufficient condition.

Remark A system of simultaneous k congruences is solvable if and only if $\gcd(m_i, m_j)$ divides $b_i - b_j$ for every pairs of integers b_i and b_j in $s = \{b_1, b_2, \ldots, b_r\}$.

2.11 APPLICATION OF CONGRUENCES

Congruences are used in the design of hash function, generator of random numbers and in the area of cryptography.

2.11.1 Hashing Function

Each record in the computer memory is assigned its unique memory location or address so that it can be inserted, deleted or modified without affecting any other records in the memory. For example, a college keeps a record on each student, containing information on the student's enrolment number, name, address, class, subjects, etc. Such informations are called *fields*. While searching for a particular student's record, we use student's enrolment number as the *key* to record because it uniquely identifies that record. As a result, we develop a *hashing* (*or scattering*) function from the set of keys to the set of addresses in the *hash table*.

The hash table uses an algorithm (performing some arithmetic manipulation of key values) to determine the index (location) to store or retrive a data value. The idea is to produce a unique index to identify the value's storage location in a potentially large but sparsely populated array. There are two parts of data structure:

(*i*) A hash table consisting of a fixed array of m pointers to the table entries.

(*ii*) Table entries organised into m separate linked lists, called *buckets* (some buckets may be empty). Each record in the symbol table appears on exactly one of these lists.

Suppose a hash function h maps the key value k (a positive integer also called *identifier*) to the value $h(k)$. The value $h(k)$ is used as an *address* of k in the hash table. The address or location of k is obtained by computing some arithmetic function h of k and is referred to as the hash or home address of k.

(*a*) ***Division Method*** In this method a hash function is obtained by using modulo operator. The identifier k is divided by some number m and the remainder is used as the hash address for k. The hash function is

$$h(k) = k(\text{mod } m)$$

This function gives bucket addresses in the range 0 to $m - 1$, so that the hash table is at least of size m.

The success of this method depends upon the proper choice of the divisor. The range generated with a given divisor m is 0 to $m - 1$. The divisor must be greater than the maximum number of records in the file. This difficulty can be avoided by making m a prime number. It has been shown that an integer which has no prime factor less than 20 is considered as a good choice for divisor.

A *collision* is said to occur when two distinct key values are mapped to the same storage location. Collision is handled in a number of ways. The colliding records may be assigned to the next free location or may be assigned to an overflow area.

Example 19 Let the maximum number of records in a file be 100 and the key values be of 4-digit numbers. If $m = 101$, find the address of key values: 1676, 1682, 2987, 5875.

Solution The mapping keys to addresses is shown below:

Key Values	*Address*
1676	$h(1676) = 1676(\text{mod } 101) = 60$
1682	$h(1682) = 1682(\text{mod } 101) = 66$
2985	$h(1985) = 1985(\text{mod } 101) = 66$
5875	$h(5875) = 5875(\text{mod } 101) = 17$

Remark The collision in the memory address 66 may be avoided by using next memory location 67 which is free.

(*b*) ***Folding Method*** In this method the identifier k is partitioned into several parts, all but possibly the last being of the same length. These parts are added together to obtain the hash address of k, i.e., $h(k)$.

There are two ways in which identifier k can be partitioned: (*i*) In this method, all but the last partition are shifted so that the least significant bit of each lines up with the corresponding bit of the last partition. The different partitions are now added together to get $h(k)$ and (*ii*) *Folding at the boundaries:* In this method, the identifier k is folded at the partition boundaries, and digits falling into the same position are added together to obtain $h(k)$. This is equivalent to reversing every other partition and then adding.

Example 20 Let $k = 12320324111220$. Partition it into three decimal digits long.

Solution The partitions are $P_1 = 123$, $P_2 = 203$, $P_3 = 241$, $P_4 = 112$, $P_5 = 20$. Using the shift folding, we get

$$h(x) = \sum_{i=1}^{5} P_i = 123 + 203 + 241 + 112 + 20 = 699$$

In case folding at the boundaries is used, we first reverse P_2 and P_4 to obtain 302 and 211 respectively. Now adding five partitions, we get

$$h(x) = 123 + 302 + 241 + 211 + 20 = 897$$

2.11.2 Random Numbers

Random numbers play an important role in computer simulation. Several arithmetic methods of generating random numbers have been suggested, studied, and used in computer simulation. Deterministically generated numbers which appears to be random are called *pseudo-random* numbers. A fast and the most commonly used method (generator) based on some recurrence relation is called *congruential generator.*

Linear Congruential Generators This method consists of computing

$$x_{i+1} = ax_i + b \text{ (mod } m)$$

where x_i is the *i*th pseudo-random number; *a* and *b* are positive integers called the *multiplier* and the *increment*, respectively; and modulo *m* means that the number $ax_i + b$ is divided by *m* repeatedly till the remainder is less than *m*.

The iterative procedure of getting the next pseudo-random number x_{i+1} is as follows:

(*i*) Compute the value of $ax_i + b$

(*ii*) Divide $ax_i + b$ by m

(*iii*) Determine the largest integer I, so that $I \geq (ax_i + b)/m$

(*iv*) Compute a remainder R, so that $R = (ax_i + b) - mI$ because $I + R = (ax_i + b)/m$.

(*v*) The set the remainder R equal to the next number x_{i+1}.

To illustrate, suppose we start with an initial value $x_0 = 17$, called the *seed,* and parameters $m = 100$, $a = 47$ and $b = 20$ then the successive random number generated are:

i	$i + 1$	$ax_i + b$	$(ax_i + b)/m$	I	R	x_{i+1}
0	1	47(17) + 20	819/100 = 8.19	8	19	19
1	2	47(19) + 20	913/100 = 9.13	9	13	13
2	3	47(13) + 20	631/100 = 6.31	6	31	31

Since $m = 100$, it may be noted that all remainders fall between 0 and 100. Placing a decimal in front of these integers, the sequence of random numbers between 0 and 1 is as follows:

$$x_0 = 0.17, x_1 = 0.19, x_2 = 0.13, x_4 = 0.31, \ldots$$

The specific choice of x_0, *a*, *b* and *m* is important in determining the quality (statistical) of the generator as well as its cycle length (before the generated sequence repeats itself).

Multiplicative Congruential Generators Given the parameters x_0, *a*, *b* and *m*, a pseudo-random number can be generated from the formula:

$$x_{i+1} = x_i a \text{ (mod } m)$$

To illustrate, let $x_0 = 17$, $a = 47$ and $m = 100$. Then the successive random numbers generated are:

$$x_1 = 17 \times 47 \pmod{100} = 99$$
$$x_2 = 99 \times 47 \pmod{100} = 53$$
$$x_3 = 53 \times 47 \pmod{100} = 91$$
$$x_4 = 91 \times 47 \pmod{100} = 77$$

and so on, all between 0 and 100.

Since modulus arithmetic each of the numbers generated the equation $x_{i+1} = x_i a \pmod{m}$ must be one of the integers 0, 1, 2, . . ., $m - 1$. Thus eventually the series will repeat itself. The period of this generator will never be greater than m (but it can be shorter than m if a and x_0 are not chosen properly).

2.11.3 Cryptography

Cryptography is the study of methods of transmitting highly confidential information between geographically dispersed sites so that only intended recipient can read it. In cryptography, the important terms used are

- *Ciphers* are the codes in which messages are transmitted.
- *Plain text* is the message to be transmitted.
- *Cipher text* is the message in secret form.
- *Encryption* or *enciphering* is the process of converting messages in secret form.
- *Deciphering* or *decryption* is the process of determining the message from its secret form.

One method of transmitting a message is to substitute a different character of alphabet for each character in the message by ignoring the space between words and punctuation.

If P denotes the set of all possible plain text messages and C the set of all possible cipher text messages, then the mapping $P \to C$ is called *enciphering function* and the mapping $C \to P$ is called *deciphering function*. A mapping of the form $P \xrightarrow{f} C \xrightarrow{f^{-1}} P$ is called *cryptosystem*.

The cryptosystem (also called RSA cipher) developed by Rivest Shamir and Adleman (1978) provides one way method to secure communication. This method enciphers a message M (first coded into numeric form) using a public key (a, n) to obtain a cipher text C as follows:

$$f(M) = C = M^e \pmod{n}$$

where n is the product of two very large prime numbers p and q and the exponent e is coprime to $(p - 1)(q - 1)$.

The security of the cipher depends on the difficulty in factorising the number n into its parts p and q. The deciphering operation is similar, except that the exponent is different: $M = C^d \pmod{n}$.

Example 21 In the RSA system, take $p = 79$, $q = 89$ and $d = 119$. Determine the public key (a, n).

Solution We know $n = pq = 79 \times 89 = 7031$ and $(p - 1)(q - 1) = 78 \times 88 = 6864$. Since p and q are coprimes, there exist two integers a and b such that $119a + 6864 = 1$.

Applying Euclidean algorithm, we obtain $a = 2711$ and $b = -47$. Hence, the public key is (2711, 7031).

Self-practice Problems II

25. Show that if a, b, p are integers and p is either 1 or a positive prime, then $(a+b)^p \equiv a^p + b^p \pmod{p}$.

26. Find the least positive incongruent solutions of

(*a*) $235x \equiv 54 \pmod 7$
(*b*) $13x \equiv 9 \pmod{25}$
(*c*) $35x \equiv 14 \pmod{21}$
(*d*) $222x \equiv 12 \pmod{18}$
(*e*) $6x \equiv 1 \pmod{33}$
(*f*) $7x \equiv 1 \pmod 9$
(*g*) $3x \equiv 5 \pmod 8$
(*h*) $33x \equiv 38 \pmod{280}$

27. Prove that if $a, b, c, n \in$ I with $a, n > 0$ and $b \equiv c \pmod n$, then $ab \equiv ac \pmod{an}$.

28. Show that for all $n \in$ N, $3^{2n+1} \equiv 3.2^n \pmod 7$. Obtain a similar result for 2^{n+2} and deduce that $7 \mid (3^{2n+1} + 2^{n+2})$ on all n.

29. Show that if a, b, k and m are integers such that $k \geq 1$, $m \geq 2$ and $a \equiv b \pmod m$, then $a^k = b^k \pmod m$ whenever k is a positive integer.

30. Solve the following simultaneous equations

(*a*) $x \equiv 2 \pmod 3$; $x \equiv 4 \pmod 7$
(*b*) $x \equiv 3 \pmod 9$; $2x \equiv 1 \pmod{13}$
(*c*) $x \equiv 2 \pmod 7$; $x \equiv 7 \pmod{16}$

31. Solve the following system of congruences

(*a*) $x \equiv 3 \pmod 5$, $x \equiv 4 \pmod 7$, $x \equiv 6 \pmod 9$
(*b*) $x \equiv 2 \pmod 3$, $x \equiv 3 \pmod 5$, $x \equiv 4 \pmod{11}$.

32. Using the algorithm: $x_{i+1} = ax_i + b \pmod m$, find the random digit with

(*a*) $x_0 = 20$, $a = 3$, $b = 0$ and $m = 100$. Show that the sequence repeat itself with a cycle of four digits.

(*b*) $x_0 = 5$, $a = 5$, $b = 0$ and $m = 17$. Show that the sequence begins to repeat after 16 random digits.

Hints and Answers

25. Using binomial theorem, we get

$$(a+b)^p = a^p + pa^{p-1}b + \frac{p(p-1)}{2!}a^{p-2}b^2 + \ldots + pab^{p-1} + b^p$$

or

$$(a+b)^p - (a^p + b^p) = p\left[a^{p-1}b + \frac{(p-1)}{2!}a^{p-2}b^2 + \ldots + ab^{p-1}\right]$$

$$\Rightarrow p \text{ divides } \{(a+b)^p - (a^p + b^p)\}$$

$$\Rightarrow (a+b)^p \equiv (a^p + b^p) \pmod p$$

26. (*a*) $x = 3$, (*b*) $x = 18$, (*c*) $x = 1, 4, 7, 10, 13, 19$, (*d*) $x = 2, 5, 8, 11, 14, 17$, (*e*) $\gcd(6, 33) = 1 \Rightarrow$ no solution

(*f*) $\gcd(7, 9) = 1 \Rightarrow x = 4$ is unique solution, (*g*) $x = 7$ unique solution, (*h*) $x = 86 + 280k$, $k \in$ I.

27. Since $a \equiv b(\text{mod } n) \Rightarrow a = b + kn$ for some $k \in$ I. Also $m \mid n \Rightarrow n = sm$ for some $s \in$ I. Consequently, $a = b + kn = b + (sk)m$ and $a \equiv b(\text{mod } m)$.

30. (*a*) $x = 104(\text{mod } 21)$; (*b*) $x = 111(\text{mod } 117)$; (*c*) $x = 23(\text{mod } 112)$

31. (*a*) $x = 158$; (*b*) $x = 123$

Chapter 3

Relations and Digraphs

PREVIEW

The knowledge of binary relation may be used to formalized a relationship between people, numbers, sets, and many other entities. This chapter introduces the concepts and important results of cartesian product of sets, and binary relations along with algebraic methods of representing various objects. Important examples such as path in relations, equivalence relations and composition of relations have been discussed. Finally several useful types of algebraic manipulations including Hasse diagram have been introduced that may be performed on binary relations.

LEARNING OBJECTIVES

After studying this chapter, you should be able to

- identify ordered pairs related by a binary relation,
- identify the domian and range of a binary relation,
- test a binary relation for equivalence relation,
- represent a binary relation on finite sets for directed graph, arrow diagram and as a matrix,
- recognize partial orderings and construct a Hasse diagram for them,
- recognize an equivalence relation on a set and describe how it partitions the set into equivalence classes.

3.1 INTRODUCTION

The relations such as, *less than*, *perpendicular to*, *not equal to*, *subset of*, etc. are often used in mathematics and computer science problems concerning discrete objects. These relations help establishing a relation between pair of objects taken in a particular order. A relation between two sets can be defined by listing their elements as an *ordered pairs*. In this chapter, we will discuss relations defined on sets and various ways of representing finite relations along with their properties.

An ordered pair consists of two elements, say a and b in which one of them is designated as the first element and the other as the second element. An ordered pair is usually denoted by (a, b).

The element a is called the *first coordinate* (or first member) and the element b is called the *second coordinate* (or second member) of the ordered pair (a, b). There can be ordered pairs which have the same first and the second elements such as (1, 1), (a, a), (2, 2), etc.

Two ordered pairs (a, b) and (c, d) are said to be equal if and only if $a = c$ and $b = d$. In other words

$$(a, b) = (c, d) \Leftrightarrow a = c \text{ and } b = d.$$

Thus, the ordered pairs (2, 3) and (2, 3) are equal while the ordered pairs (2, 3) and (3, 2) are different. The reader must note the difference between the set {2, 3} and the ordered pair (2, 3). We have $\{2, 3\} = \{3, 2\}$ but $(2, 3) \neq (3, 2)$.

In coordinate geometry the points in the cartesian plane can be represented by an ordered pair (x, y) of real numbers, where the first member x is called the *abscissa* and the second member y is called *ordinate*.

3.2 Cartesian Product of Sets

If A and B are any two non-empty finite sets, then the set of all distinct ordered pairs whose first member (or coordinate) belongs to A and second member (or coordinate) belongs to B is called the *cartesian product of A and B* (in that order) and is denoted by A × B (read as *A cross B*). Symbolically,

$$A \times B = \{(a, b) : a \in A, b \in B\}$$

Remarks

1. In the product B × A of sets B and A, the first member (or coordinate) is of B and second is of A.
2. Elements in product sets A × B and B × A may be same but $A \times B \neq B \times A$ unless A = B. It follows that the cartesian product of two sets, in general, is not commutative.
3. If set A has m elements and the set B has n elements, then the product set A × B has mn elements.
4. If either A or B is infinite and the other is empty, then $A \times B = \Phi$.
5. If either A or B is infinite and the other is not empty, then A × B is infinite. If A and B are finite sets, then $n(A \times B) = n(A) \cdot n(B)$.

Illustration Let R be the set of real numbers. Then $R \times R (= R^2)$ denotes the cartesian plane or Euclidean 2-space, since the elements of this product set, viz., $\{(x, y) : x$ and y are real numbers$\}$ are the cartesian coordinate of a point in a plane.

Example 1 If A = {1, 2, 3} and B = {2, 3}, prove that $A \times B \neq B \times A$. Also find $n(A \times B)$.

Solution The product,

$$A \times B = \{(1, 2), (1, 3), (2, 2), (2, 3), (3, 2), (3, 3)\}$$

$$B \times A = \{(2, 1), (2, 2), (2, 3), (3, 1), (3, 2), (3, 3)\}$$

Since elements of A × B and B × A are not same, therefore $A \times B \neq B \times A$. Now

$$n(A \times B) = n(A) \times n(B) = 3 \times 2 = 6.$$

3.2.1 The Cartesian Product of *n* Sets

We may generalise the definition of cartesian product or cross product, to more than two sets. Let $A_1, A_2, \ldots, A_n$ be n-given sets. The set of ordered n-tuples $(a_1, a_2, \ldots, a_n)$, where $a_1 \in A_1, a_2 \in A_2, \ldots, a_n \in A_n$ is called the cartesian product of $A_1, A_2, \ldots, A_n$ and is denoted by $A_1 \times A_2 \times \ldots \times A_n$. In short, it is also denoted as $\prod_{i=1}^{n} A_i$.

This cartesian product of n sets is the set of all ordered n-tuples $(a_1, a_2, \ldots, a_n)$, for all $a_i \in A_i$ $(i = 1, 2, \ldots, n)$. In other words,

$$\prod_{i=1}^{n} A_i = \{(a_1, a_2, \ldots, a_n) : a_i \in A_i \text{, for all } i\}.$$

In particular,

$$\prod_{i=1}^{2} A_i = A_1 \times A_2 = \{(a_1, a_2) : a_1 \in A_1, a_2 \in A_2\}$$

$$\prod_{i=1}^{3} A_i = A_1 \times A_2 \times A_3 = \{(a_1, a_2, a_3) : a_1 \in A_1, a_2 \in A_2, a_3 \in A_3\}$$

The concept of the cartesian product of n sets can also be extended to the product of infinite sequence of sets.

Remarks

1. If R is the set of real numbers and $A_i = R$ for all i, then $\prod_{i=1}^{n} A_i = R$ is the set of all points in Euclidean n-space.
2. If $A_1 = A_2 = \ldots = A_n = A$, then $\prod_{i=1}^{n} A_i = A_n = \{(a_1, a_2, \ldots, a_n) : a_i \in A, \text{ for all } i\}$.

3.2.2 Important Results on Cartesian Product

1. If A, B and C are sets and $A \subseteq B$, then prove that $(A \times C) \subseteq (B \times C)$.

   Proof Let (a, c) be an arbitrary element of $A \times C$. Then

   $(a, c) \in A \times C \Rightarrow a \in A$ and $c \in C$

   $\Rightarrow a \in B$ and $c \in C$ [since $A \subseteq B$]

   $\Rightarrow (a, c) \in B \times C$

   Thus, $A \times C \subseteq B \times C$.

2. If $A \subseteq B$ and $C \subseteq D$, then prove that $(A \times C) \subseteq (B \times D)$.

   Proof Let (a, c) be an arbitrary element of $A \times C$. Then

   $(a, c) \in A \times C \Rightarrow a \in A$ and $c \in C$

   $\Rightarrow a \in B$ and $c \in D$ [since $A \subseteq B$ and $C \subseteq D$]

   $\Rightarrow (a, c) \in B \times D$

   Thus $A \times C \subseteq B \times D$.

3. If A, B, C are sets, prove that $A \times (B \cap C) = (A \times B) \cap (A \times C)$.

Proof Let (x, y) be an arbitrary element of $A \times (B \cap C)$. Then

$$
\begin{aligned}
(x, y) \in A \times (B \cap C) &\Leftrightarrow x \in A \text{ and } y \in (B \cap C)\\
&\Leftrightarrow x \in A \text{ and } (y \in B \text{ and } y \in C)\\
&\Leftrightarrow (x \in A \text{ and } y \in B) \text{ and } (x \in A \text{ and } y \in C)\\
&\Leftrightarrow (x, y) \in (A \times B) \text{ and } (x, y) \in (A \times C)\\
&\Leftrightarrow (x, y) \in (A \times B) \cap (A \times C)
\end{aligned}
$$

Thus $A \times (B \cap C) = (A \times A) \cap (A \times C)$.

4. If $A \subseteq B$, show that $A \times A = (A \times B) \cap (B \times A)$.

Proof Let (x, y) be an arbitrary element of the set $A \times A$. Then

$$
\begin{aligned}
(x, y) \in A \times A &\Leftrightarrow x \in A \text{ and } y \in A.\\
&\Leftrightarrow (x \in A \text{ and } y \in A) \text{ and } (x \in A \text{ and } y \in A)\\
&\Leftrightarrow (x \in A \text{ and } y \in B) \text{ and } (x \in B \text{ and } y \in A) && [\text{since } A \subseteq B]\\
&\Leftrightarrow (x, y) \in A \times B \text{ and } (x, y) \in B \times A\\
&\Leftrightarrow (x, y) \in (A \times B) \cap (B \times A)
\end{aligned}
$$

Hence, if $A \subseteq B$, $A \times A = (A \times B) \cap (B \times A)$.

5. Prove that $A \times (B \cup C) = (A \times B) \cup (A \times C)$.

Proof Let (x, y) be an arbitrary element of $A \times (B \cup C)$. Then

$$
\begin{aligned}
(x, y) \in A \times (B \cup C) &\Leftrightarrow x \in A \text{ and } y \in (B \cup C).\\
&\Leftrightarrow x \in A \text{ and } (y \in B \text{ or } y \in C)\\
&\Leftrightarrow (x \in A \text{ and } y \in B) \text{ or } (x \in A \text{ and } y \in C)\\
&\Leftrightarrow (x, y) \in A \times B \text{ or } (x, y) \in A \times C.\\
&\Leftrightarrow (x, y) \in (A \times B) \cup (A \times C).
\end{aligned}
$$

Hence, $A \times (B \cup C) = (A \times B) \cup (A \times C)$.

6. Prove that $(A \times B) \cap (S \times T) = (A \cap S) \times (B \cap T)$.

Proof Let (x, y) be an arbitrary element of $(A \times B) \cap (S \times T)$. Then

$$
\begin{aligned}
(x, y) \in (A \times B) \cap (S \times T) &\Leftrightarrow (x, y) \in A \times B \text{ and } (x, y) \in S \times T\\
&\Leftrightarrow (x \in A \text{ and } y \in B) \text{ and } (x \in S \text{ and } y \in T)\\
&\Leftrightarrow (x \in A \text{ and } x \in S) \text{ and } (y \in B \text{ and } y \in T)\\
&\Leftrightarrow x \in (A \cap S) \text{ and } y \in (B \cap T)\\
&\Leftrightarrow (x, y) \in (A \cap S) \times (B \cap T)
\end{aligned}
$$

Hence, $(A \times B) \cap (S \times T) = (A \cap S) \times (B \cap T)$.

7. For any three sets A, B and C, prove that $A \times (B - C) = (A \times B) - (A \times C)$.

Proof Let (x, y) be any arbitrary element of $A \times (B - C)$. Then

$$
\begin{aligned}
(x, y) \in A \times (B - C) &\Leftrightarrow x \in A \text{ and } y \in (B - C)\\
&\Leftrightarrow x \in A \text{ and } (y \in B \text{ and } y \notin C)\\
&\Leftrightarrow (x \in A \text{ and } y \in B) \text{ and } (x \in A \text{ and } y \notin C)\\
&\Leftrightarrow (x, y) \in (A \times B) \text{ and } (x, y) \notin (A \times C)\\
&\Leftrightarrow (x, y) \in (A \times B) - (A \times C)
\end{aligned}
$$

Hence, $A \times (B - C) = (A \times B) - (A \times C)$.

8. For a non-empty sets A and B, prove that $A \times B = B \times A \Leftrightarrow A = B$.

Proof Let A = B. Then $A \times B = A \times A$ and $B \times A = A \times A$. This shows that $A \times B = B \times A$.

Conversely, let $A \times B = B \times A$. Then, let

$$x \in A \Rightarrow (x, y) \in A \times B;\ y \in B$$
$$\Rightarrow (x, y) \in B \times A \qquad [\text{since } A \times B = B \times A]$$
$$\Rightarrow x \in B$$

This means, $A \subseteq B$. Similarly $B \subseteq A$. Consequently, A = B.

9. If A and B are two non-empty sets having n elements in common, then prove that $A \times B$ and $B \times A$ have n^2 elements in common.

Proof Let $C = A \cap B$. Then assume that $C \times C = (A \times B) \cap (B \times A)$

Also, $(x, y) \in C \times C \Leftrightarrow x \in C$ and $y \in C$

$$\Leftrightarrow x \in A \cap B \text{ and } y \in A \cap B \qquad [\text{since } C = A \cap B]$$
$$\Leftrightarrow (x \in A \text{ and } x \in B) \text{ and } (y \in A \text{ and } y \in B)$$
$$\Leftrightarrow (x \in A \text{ and } y \in B) \text{ and } (x \in B \text{ and } y \in A)$$
$$\Leftrightarrow (x, y) \in A \times B \text{ and } (x, y) \in B \times A$$
$$\Leftrightarrow (x, y) \in (A \times B) \cap (B \times A)$$

Thus $C \times C = (A \times B) \cap (B \times A)$.

Since $C \times C$ has n^2 elements, therefore $(A \times B) \cap (B \times A)$ also has n^2 elements. This implies that both $(A \times B)$ and $(B \times A)$ have n^2 elements in common.

Example 2 Let A = {1, 4}, B = {2, 3}, C = {3, 5}. Prove that $A \times B \neq B \times A$. Also find, $(A \times B) \cap (A \times C)$.

Solution $A \times B = \{(1, 2), (1, 3), (4, 2), (4, 3)\}$

$B \times A = \{(2, 1), (2, 4), (3, 1), (3, 4)\}$

Since in both the cases the elements are different, therefore $A \times B \neq B \times A$.

Now $A \times C = \{(1, 3), (1, 5), (4, 3), (4, 5)\}$.

Then $(A \times B) \cap (A \times C) = \{(1, 3), (4, 3)\}$.

Example 3 Let $A = \{a, b\}$, $B = \{p, q\}$ and $C = \{q, r\}$. Find

(*a*) $A \times (B \cup C)$ (*b*) $(A \times B) \cup (A \times C)$ (*c*) $A \times (B \cap C)$ (*d*) $(A \times B) \cap (A \times C)$.

Solution (*a*) First we compute $B \cup C$, i.e. $B \cup C = \{p, q, r\}$. Then

$$A \times (B \cup C) = \{a, b\} \times \{p, q, r\} = \{(a, p), (a, q), (a, r), (b, p), (b, q), (b, r)\}$$

(*b*) $A \times B = \{(a, p), (a, q), (b, p), (b, q)\}$

$A \times C = \{a, q), (a, r), (b, q), (b, r)\}$.

Then $(A \times B) \cup (A \times C) = \{(a, p), (a, q), (b, p)\ (b, q), (a, r), (b, r)\}$.

(*c*) $B \cap C = \{q\}$, therefore $A \times (B \cap C) = \{a, b\} \times \{q\} = \{(a, q), (b, q)\}$.

(*d*) $(A \times B) \cap (A \times C) = \{(a, q), (b, q)\}$.

Example 4 If A = {1, 2, 3}, B = {3, 4}, and C = {4, 5, 6}, find $A \times (B \cup C)$, $A \times (B \cap C)$ and $(A \times B) \cap (B \times C)$.

Solution Since $B \cup C = \{3, 4, 5, 6\}$ and $B \cap C = \{4\}$, therefore

$$A \times (B \cup C) = \{1, 2, 3\} \times \{3, 4, 5, 6\}$$
$$= \{(1,3), (1,4), (1,5), (1,6), (2,3), (2,4), (2,5), (2,6), (3,3), (3,4), (3,5), (3,6)\}$$

$$A \times (B \cap C) = \{1, 2, 3\} \times \{4\} = \{(1, 4), (2, 4), (3, 4)\}$$
$$A \times B = \{(1, 3), (1, 4), (2, 3), (2, 4), (3, 3), (3, 4)\}$$
$$B \times C = \{(3, 4), (3, 5), (3, 6), (4, 4), (4, 5), (4, 6)\}$$

Self-practice Problems I

1. If A = {0, 1} and B = {0, – 1} are two sets, find A × B and B × A. Hence, show that $A \times B \neq B \times A$.
   (*b*) Let A = {1, 4}, and B = {2, 3}. Prove that $A \times B \neq B \times A$.
2. Prove that if A, B and C are sets and $A \subseteq B$, then $A \times C \subseteq B \times C$.
3. If $A \subseteq B$, then show that $A \times A = (A \times B) \cap (B \times A)$.
4. If A, B and C are any three sets, then prove that $A \times (B \cup C) = (A \times B) \cup (A \times C)$.
5. Find the numbers x and y if the following ordered pairs are equal.
   (*a*) $(x + 1, y + 5) = (0, 5)$ (*b*) $(2x - 1, -5) = (x, y + 1)$.
6. If A = {1, 2, 3}, B = {2, 3, 4}, S = {1, 3, 4} and T = {2, 4, 5}, then verify that
   $$(A \times B) \cap (S \times T) = (A \cap S) \times (B \cap T).$$
7. If R is the set of real numbers, what is represented by R^2 and R^3?
8. Prove that $(A - B) \times C = (A \times C) - (B \times C)$.
9. Prove that
   (*a*) $(A \cap B) \times C = (A \times C) \cap (B \times C)$ (*b*) $(A \times B) \cap (C \times D) = (A \cap C) \times (B \cap D)$.
10. If A = {1, 4}, B = {4, 5}, and C {5, 7}, then find
    (*a*) $(A \times B) \cup (B \times C)$ (*b*) $(A \times B) \cap (B \times C)$.
11. A and B are two sets having two elements in common. If n (A) = 5 and n (B) = 3, then find n (A × B) and n [(A × B) ∩ (B × A)].

Hints and Answers

1. A × B = {(0, 0), (0, – 1), (1, 0), (1, – 1)}
   B × A = {(0, 0), (0, 1), (– 1, 0), (– 1, 1)}.
5. (*a*) $x = -1, y = 0$, (*b*) $x = 1, y = -6$.
10. A × B = {(1, 4), (1, 5), (4, 4), (4, 5)}; B × C = {(4, 5), (4, 7), (5, 5), (5, 7)}
    (A × B) ∪ (B × C) = {(1, 4), (1, 5), (4, 4), (4, 5), (4, 7), (5, 5), (5, 7)}
    (A × B) ∩ (B × C) = {(4, 5)}.
11. $n(A \times B) = n(A) \times n(B) = 5 \times 3 = 15$. Since 2 elements are common in A × B and B × A, therefore $n[(A \times B) \cap (B \times A)] = 2^2 = 4$ elements in common.

3.3 BINARY RELATIONS

The concept of ordered pairs can also be used in formulating the mathematical theory of relations in set language. For sets A, B, ⊆ U, any subset of A × B is called a *relation* from A to B. Any subset of A × A is called a *binary relation* on A.

Illustrations

1. The relation of marriage between man and woman is a binary relation, the relation of parenthood, on the other hand, is not a binary relation. It is ternary relation because we often say that x and y are parents of z. Thus, a relation of this kind involved three objects at a time. In the example of marriage, we see that this binary relation involves certain ordered pair (x, y) in which x is a man, y is a woman and x is married to y.
2. Let U = {1, 2, . . ., 7}, A = {2, 3, 4} and B = {4, 5}. Then following are some of the relations from A to B

   (*a*) Φ (*b*) {(2, 4)} (*c*) {(2, 4), (2, 5)}

   (*d*) {(2, 4), (3, 4), (4, 4)} (*e*) {(2, 4), (3, 4), (4, 5)}

Definition Let A and B be two non-empty sets. The binary relation, R from set A to B is defined to be a subset of A × B. Symbolically,

$$R : A \to B \text{ if and only if } R \subseteq A \times B \text{ and } (a, b) \in R$$

where $a \in$ A and $b \in$ B.

If this relation holds, then we say that a is related to b by R and we also write a R b. If a is not related to b by R, we write $a \not R\ b$.

Remark If A = B, then R is said to be the relation on A. In general, for finite sets with |A| = m and |B| = n, the set A × B will contain mn elements and so power set of (A × B) will have 2^{mn} elements. Hence, the total number of relations from A to B are 2^{mn} including the empty relation and the relation A × B itself. For any set A ⊆ U, A × Φ = Φ or Φ × A = Φ, because if $(a, b) \in$ A × Φ, then $a \in$ A and $b \in$ Φ which is impossible.

Illustrations

1. Let I be the set of all integers. The statement 'x is less than y, for $x, y \in$ I' determines a relation in I. If we denote this relation by R, then we may describe the set R in the set builder form as:

   $$R = \{(x, y) : x, y \in I, x < y\}$$

   Suppose R is a relation and (x, y) is an element of the relation R and $(x, y) \in$ R, then such relation is also expressed as: x R y which is read as, *x is R-related to y* and $(x, y) \notin$ R is read as *x is not R-related to y*.
2. Let I be the set of all integers and let R = $\{(x, y) : x, y \in I, x < y\}$.

   Since 2 is less than 3, therefore the ordered pair (2, 3) ∈ R and we write 2 R 3. On the other hand, 4 is not less than 3. Therefore, the ordered pair (4, 3) ∉ R.
3. Let R be the subset of N × N, where R = $\{(x, y) : y = 7x\}$. This relation R on N can also be defined recursively as:

   (*a*) (0, 0) ∈ R, and

   (*b*) If $(m, n) \in$ R then $(m + 1, n + 7) \in$ R.
4. Let A = $\{a, b\}$, A^2 = A × A = $\{aa, ab, ba, bb\}$ is the set of all words of length 2.

Example 5 Let $A \times B = \{(1, 2), (1, 4), (2, 2), (2, 4)$ and $R = \{(1, 2), (2, 4), (2, 1)\}$. State whether R is a relation from A to B or not. Give reasons.

Solution Since $R = \{(1, 2), (2, 4), (2, 1)\}$ is not the subset of $A \times B$ because $(2, 1) \notin (A \times B)$, therefore R is not a relation from A to B.

3.3.1 Binary Relation Defined in a Set

A binary relation is said to be defined in a set A, if for any ordered pair (x, y) of elements of A, it is meaningful to say whether x has or not the relation R to y. The relation R is defined to the set of all such ordered pair (x, y), where $x \, R \, y$ (i.e. x is related to y). Thus, $(x, y) \in R \Leftrightarrow x \, R \, y$.

Remarks

1. Since only binary relations are used here, therefore, the word *binary* to describe relations is omitted.
2. Since binary relations are sets of ordered pair, the notations of intersection, union, symmetric, difference, etc., follow directly from that of sets.

3.3.2 Domain and Range of a Relation

If R is the relation from a set A to another set B, then the set of elements in A that are related to some element or elements in B is called the domain of R and set B is called codomain of R. In other words, domain of R, a subset of A, is the set of all first elements in the ordered pairs which belongs to R. Symbolically,

$$D = \{x : (x, y) \in R, x \in A \text{ for some } y \in B\}$$

The range, say E of the relation R, is the set of elements in B that appear as second element in the ordered pair which belongs to R, i.e. all elements in B that are related to some element in A. Symbolically,

$$E = \{y : (x, y) \in R \text{ for some } x \in A\}$$

Illustrations

1. Let $A = \{1, 2, 3, 4\}$ and $B = \{a, b, c\}$. Then every subset of $A \times B$ is a relation from A to B. So if $R = \{(2, a), (4, a), (4, c)\}$, then the domain of R is the set $\{2, 4\}$ and the range of R is the set $\{a, c\}$.
2. Let I be the set of all integers. Then 'x is less than y for all $x, y \in I$' is a relation R in I and is described as:
$$R = \{(x, y) : x \in I, y \in I, x < y\}$$
3. For $A = \{1, 2, 3\}$ and $B = \{a, b, c, d\}$, $f = \{(1, a), (2, b), (3, b)\}$ is a function, and consequently a relation R from A to B. $R_1 = \{(1, a), (2, b)\}$ and $R_2 = \{(1, a), (2, a), (2, b), (3, d)\}$ are relations, but not functions from A to B.

3.4 SET OPERATIONS ON RELATIONS

Since all binary relations are set of ordered pairs, therefore set operations can also be carried on such sets. Let R and S be two relation, then few relations are defined as:

1. Intersection of R and S: $x(R \cap S)y = x \, R \, y \wedge x \, S \, y$
2. Union of R and S: $x(R \cup S)y = x \, R \, y \vee x \, S \, y$
3. Difference of R and S: $x(R - S)y = x \, R \, y \wedge x \, \not{S} \, y$
4. Complement of R: $x(R')y = x \, R \, y$

Example 6 If A = {2, 3, 5}, B = {6, 8, 10}, C = {2, 3} and D = {8, 10} are four non-empty sets. Suppose the relation R from A to B is defined as: R = {(2, 6), (2, 8), (3, 10)} and the relation S from C to D is defined as: S = {(2, 8), (3, 10)}. Find $R \cup S$, $R \cap S$, $R - S$ and R′.

Solution $R \cup S = \{(2, 6), (2, 8), (3, 10)\}$

$R \cap S = \{(2, 8), (3, 10)\}$

$R - S = \{(2, 6)\}$

R′ = set of all ordered pairs in A × B that are not in R.

$= \{(2, 10), (3, 6), (3, 8), (5, 6), (5, 8), (5, 10)\}$.

3.5 TYPES OF RELATIONS

Some special types of relations on a set that are useful for the applications of computer science are as follows:

Inverse relation Let R be any relation from a set A to a set B. The inverse relation, denoted by R^{-1}, is a relation from B to A and is denoted by

$$R^{-1} = \{(y, x) : x \in A, y \in B, (x, y) \in R\}.$$

In other words, the inverse relation is obtained by reversing each of the ordered pairs belong to R. Thus,

$$(x, y) \in R \Leftrightarrow (y, x) \in R^{-1} \text{ or } x\,R\,y \Leftrightarrow y R^{-1} x$$

Evidently, the range of R is the domain of R^{-1} and vice versa. If A = B, then R and R^{-1} are both relations on A.

Illustrations

1. Let R be the set of relations on the set R of real numbers defined by x is less than y, where $x, y \in$ R. Then R^{-1} is the relation defined by 'x is greater than y, where $x, y \in$ R'.
2. Let A = {2, 4, 6}, B = {a, b}. Then R = {(2, a), (2, b), (4, a)} is a relation from A to B. The inverse relation of R is

$$R^{-1} = \{(a, 2), (a, 4), (b, 2)\}.$$

Note that Dom (R) = Ran (R^{-1}) = {2, 4} and Ran (R) = Dom (R^{-1}) = {a, b}. Observe that if R is any relation, then $(R^{-1})^{-1} = R$. If R is a relation on any set A, then R^{-1} is also a relation on A.

Identity relation Let A be any set. Then the relation R in a set A denoted by I_A is said to be *identity relation* (or diagonal relation) if

$$I_A = \{(x, y) : x \in A, y \in A, x = y\}$$

In other words, the identity relation I_A in a set A is the set of all ordered pairs (x, y) of A × B for which $x = y$. The domain and range of I_A are both A.

Illustration Let A = {a, b, c, d}. Then the identity relation in A is

$$I_A = \{(a, a), (b, b), (c, c), (d, d)\}.$$

Universal relation A relation in a set A is said to be universal relation provided R = A × A.

For example, if A = {2, 3}, then R = A × A = {(2, 2), (2, 3), (3, 2), (3, 3)} is a universal relation.

Void relation A relation R in a set is said to be a void relation provided R is a null set, i.e. R = Φ.

For example, if A = {2, 3, 5} and the relation R is defined as: aRb if and only if a divides b, then R = Φ and hence A × A is a void relation.

3.5.1 Properties of Relations

Reflexive relation A relation R on a set A is reflexive if and only if each element in A is related to itself, i.e. a R a, for all $a \in$ A. In other words, for all $a \in$ A, $(a, a) \in$ R.

The necessary and sufficient conditions that a relation R on A be reflexive is that $I_A \subseteq R$. A relation R on a set A is not reflexive, if there is at least one element, say $b \in$ A such that $b \notin$ R.

Illustrations

1. Let A = {1, 2, 3, 4}. Then the relation

$$R_1 = \{(1, 1), (2, 4), (3, 3), (4, 1), (4, 4)\}$$

on A is not reflexive because 2 ∈ A but (2, 2) ∉ R_1. Whereas the relation

$$R_2 = \{(1, 1), (2, 2), (3, 3), (4, 4)\}$$

is a reflexive relation on A.

2. The relation in N defined by '*x is less than y*' is not reflexive, since '*x*' is not less than '*x*' for any natural number *x*. But the relation defined by *x is less than or equal to y* is reflexive.

Symmetric relation A relation R on a set A is said to be *symmetric* if any only if for all, a, $b \in$ R

$$a \text{ R } b \Rightarrow b \text{ R } a, \text{ i.e. } (a, b) \in R \Rightarrow (b, a) \in R$$

The necessary and sufficient conditions for a relation R to be symmetric is $R = R^{-1}$.

Illustrations

1. The relation in N defined by '*x is a divisor of y*' is not symmetric, since 2 divides 4 but 4 does not divide 2. But, the relation in N defined by $x + y = 8$ is symmetric because if $a + b = 8$, then $b + a = 8$.
2. Let A = {1, 2, 3} and R = {(1, 2), (2, 3), (2, 1), (3, 3)}. Then

$$R^{-1} = \{(2, 1), (3, 2), (1, 2), (3, 3)\}.$$

Since (2, 3) ∈ R but (2, 3) ∉ R^{-1}, therefore $R \neq R^{-1}$. Hence, R is not symmetric.

Anti-symmetric relation Let R be a relation on a set A. Then R is an anti-symmetric if and only if

$$a \text{ R } b \text{ and } b \text{ R } a \Rightarrow a = b, \text{ for } a, b \in A, \text{ i.e. } (a, b) \in R; (b, a) \in R \Rightarrow a = b.$$

Thus, R is an anti-symmetric if we never have both a R b and b R a except when $a = b$.

It is evident that a relation R on a set A is an anti-symmetric if $R \cap R^{-1} \subseteq I_A$, where I_A denotes the identity relation on A.

Illustrations

1. The relation in N defined by '*x divides y*' is anti-symmetric because *x* divides *y* and *y* divides *x* implies $x = y$.

2. Let $A = \{1, 2, 3, 4\}$ and $R = \{(1, 1), (2, 3), (3, 2)\}$

Then $R^{-1} = \{(1, 1), (3, 2), (2, 3)\}$ and $R \cap R^{-1} = \{(1, 1), (2, 3), (3, 2)\}$

But $R \cap R^{-1} \subseteq I_A$, where $I_A = \{(1, 1), (2, 2), (3, 3), (4, 4)\}$

Hence, R is not anti-symmetric.

Transitive relation A relation R on a set A is transitive if and only if for all $a, b, c \in R$

$a \text{ R } b$ and $b \text{ R } c \Rightarrow a \text{ R } c$, i.e. $(a, b), (b, c) \in R \Rightarrow (a, c) \in R$.

Illustrations

1. The relation in N defined by *x is less than y* is transitive relation because $x < y$ and $y < z \Rightarrow x < z$.
2. Let N be the set of all natural numbers. Let a relation R on N defined by *x is a divisor of y* where $x, y \in N$. Then
   (*a*) R is reflexive, since every natural number is a divisor of itself.
   (*b*) R is an anti-symmetric, since x divides y and y divides $x \Rightarrow x = y$.
   (*c*) R is transitive, since x divides y and y divides $z \Rightarrow x$ divides z.

Example 7 Given a finite set A with n elements. Find how many relations are on A and how many of these are reflexive and symmetric.

Solution Given a finite set A with n elements, we have n^2 ordered pairs in the set $A \times A$, i.e. $|A \times A| = n^2$. If a relation R on A is reflexive, then each of n ordered pairs $\{(a_i, a_i) : 1 \leq i \leq n\}$ must be in R. However, other $n(n - 1)$ ordered pairs in $A \times A$ of the form (a_i, a_j) where $i \neq j (1 \leq i, j \leq n)$ may not be in R. So by the rule of product there are $2^{n(n-1)}$ reflexive relations on A.

To prove symmetric relation on $A = \{a_1, a_2, \ldots, a_n\}$, we write $A \times A$ as $A_1 \cup A_2$, where $A_1 = \{(a_i, a_i) : i \leq i \leq n\}$ and $A_2 = \{(a_i, a_j) : i \neq j$ and $1 \leq i, j \leq n\}$, so that every order pair in $A \times A$ is in exactly one of A_1, A_2. The set A_2 contains $n(n - 1)/2$ subsets of the form $\{(a_i, a_j), (a_j, a_i)\}$ where $1 \leq i < j \leq n$. Thus for a symmetric relation R on A, for each ordered pairs $n(n - 1)/2$ taken from A_2 we can choose same number of ordered pairs from A_1. So by the rule of product there are $2^n . 2^{n(n-1)/2} = 2^{n(n+1)/2}$ symmetric relations on A.

For those relations on A that are both reflexive and symmetric we have only one choice for each ordered pair in A_1. Thus there are $2^{n(n-1)/2}$ relations on A that are both reflexive and symmetric.

Example 8 Let $A = \{1, 2, 3, 4\}$. Given a relation of R on A which is

(*a*) neither symmetric nor antisymmetric.
(*b*) anti-symmetric and reflexive but not transitive.
(*c*) transitive and reflexive but not anti-symmetric.

[*UPTU., BTech., 2006*]

Solution

(*a*) Since $A = \{1, 2, 3, 4\}$, the relation $R = \{(1, 1), (1, 2), (1, 3), (3, 1), (3, 3), (4, 4)\}$ is *not symmetric*, as $(1, 2) \in R$ but $(2, 1) \notin R$. Also, it is not *anti-symmetric*, as there exists a pair of distinct elements $1, 3 \in A$ such that $(1, 3) \in R$ and $(3, 1) \in R$.

(*b*) The relation, $R = \{(1, 1), (1, 3), (2, 2), (3, 3), (3, 4), (4, 4)\}$ is reflexive as $a\text{R}a$ for all $a \in A$. Also it is anti-symmetric as no pair of distinct elements $x, y \in A$ exist such that $(x, y) \in R$ and $(y, x) \in R$. However, the relation R is not transitive as 1R3, 3R4, i.e. $(1, 3) \in R$, $(3, 4) \in R$ but $(1, 4)$ does not belong to R.

(*c*) The relation R = {(1, 1), (1, 2), (2, 1), (2, 2), (3, 3), (3, 4), (4, 3), (4, 4)} is reflexive, as $a\text{R}a$ for all $a \in$ A. Also it is transitive, as $a\text{R}b$ and $b\text{R}c \Rightarrow a\text{R}c$ for $a, b, c \in$ A. However, the relation is not anti-symmetric as there exists a pair of distinct elements 1, 2, ∈ A such that (1, 2) ∈ R and (2, 1) ∈ R.

Example 9 Let R be a binary relation defined as: R = {$(a, b) \in$ R : $a - b \leq 3$), determine whether R is reflexive, symmetric, anti-symmetric and transitive. [*UPTU., BTech., 2007*]

Solution

(*i*) The relation R = {$(a, b) \in$ R : $a - b \leq 3$} is reflexive as there exist $a \in$ R, $a - a = 0 \leq 3$. It is not symmetric as $a, b \in$ R $\Rightarrow a - b \leq 3$ then $b - a \geq 3$.

(*ii*) The relation is anti-symmetric as $(a, b) \in$ R $\Rightarrow a - b \leq 3$ and $b - a \geq 3$ but $a - b \leq 3$ and $b - a \leq 3$ is possible only if $a = b$. Thus R is anti-symmetric.

(*iii*) The relation is not transitive because if $(a, b) \in$ R and $(b, c) \in$ R other $a - b \leq 3$ and $b - c \leq 3$. Adding these inequalities, we get $a - c \leq 6$. Thus R is not transitive.

Example 10 Let A = {1, 2, ..., 7} and R = {$(x, y) : x - y$ is divisible by 3}. Show that R is an equivalence relation. [*UPTU., BTech 2007*]

Solution Given that A = {1, 2, 3, 4, 5, 6, 7} and R = {$(x, y) : x - y$ is divisible by 3}

(*i*) *Reflexive:* There exists an element $x \in$ A such that $x - x$ is divisible by 3. This shown that $(x, x) \in$ A for all $x \in A$. Hence, R is reflexive.

(*ii*) *Symmetric:* Let $x, y \in$ A and $(x, y) \in$ R. This means, $x - y$ is divisible by 3 i.e., $x - y = 3m_1$, where m_1 is any integer. This implies that $y - x = 3m_2$, where m_2 is also an integer. Since $y - x$ is divisible by 3, R is symmetric.

(*iii*) *Transitive:* Let $x, y, z \in$ A. Also let $x - y = 3m_1$ and $y - z = 3m_2$, where m_1 and m_2 are integers.

Adding these two equations, we have, $x - z = 3(m_1 + m_2)$ where $m_1 + m_2$ is also an integer. So $x - z$ is also divisible by 3 or $(x, z) \in$ R. Thus R is transitive.

Hence, R is an equivalence relation.

3.6 PARTIAL ORDER RELATIONS

A relation, R on a set A is called a *partial order* or *partial ordering relation* if and only if following three conditions are satisfied:

1. R is reflexive, i.e. a R a, for all $a \in$ A.
2. R is anti-symmetric, i.e. a R b, b R a, if and only if $a = b$, for all $a, b \in$ A.
3. R is transitive, i.e. a R b, b R c implies that a R c, for all $a, b, c \in$ A.

The set A on which a partial order relation, R is defined is called a *partially ordered set* or simply a *poset*, and it is denoted by (A, R).

The relation '=' (equal to) defined on a set A is said to be *total order relation* provided for all $a, b \in$ A, we have $a = b$. This implies that $b = a$. In other words, a R b and b R a, and hence the relation '=' is also a symmetric relation.

Example 11 Show that the relation '$\subseteq$' (subset) defined on the power set P(A) of the set A is a partial order relation.

Solution To prove the relation, R a partial order relation, we prove that the following three properties hold.

(*a*) *Reflexive* For any element $A_1 \in P(A)$, we have $A_1 \subseteq A_1$, which is true because of the definition of subset. Hence, R is reflexive.

(*b*) *Anti-symmetric* Let $A_1, A_2 \in P(A)$ be any two different subsets of set A, such that $A_1 \subseteq A_2$. But $A_2 \subseteq A_1$ is not possible unless $A_1 = A_2$. Hence R is anti-symmetric.

(c) *Transitive* Let $A_1, A_2, A_3 \in P(A)$ be any three different subsets of the set A such that $A_1 \subseteq A_2$ and $A_2 \subseteq A_3$. It obviously follows that $A_1 \subseteq A_3$. Hence, R is transitive.

Since all three properties of partial order relation are satisfied, therefore the given relation is partial order relation.

Example 12 Show that the relation $\leq$ (less than or equal to) defined on the set of positive integers, I_+ is a partial order relation.

Solution To prove the relation, R a partial order relation, we prove that the following three properties hold.

(*a*) *Reflexive* For any element $x \in I_+$, we have $x \leq x$, which is true for all $x \in I_+$. Hence, R is reflexive.

(*b*) *Anti-symmetric* Let $x_1, x_2 \in I_+$ be any two elements, such that $x_1 \leq x_2$. But $x_2 \leq x_1$ is not possible unless $x_1 = x_2$. Hence, R is anti-symmetric.

(c) *Transitive* Let $x_1, x_2, x_3 \in I_+$ be any three elements, such that $x_1 \leq x_2$, and $x_2 \leq x_3$. It follows that $x_1 \leq x_3$. Hence, R is transitive.

Since all the three properties of partial order relation are satisfied, therefore the given relation is partial order relation.

Remarks

1. If R is a partial order relation on a set A, then its inverse R^{-1} is also a partial order relation on A. Thus, the poset (A, R^{-1}) is called the *dual* of the poset (A, R). For example, the poset $(R, \leq)$ is the dual of the poset $(R, \geq)$ on the set of real numbers. It is also true for the set of integers.
2. Let $(A, \leq)$ be a poset. Then elements $a, b \in A$ are said to be *comparable* if $a \leq b$ or $b \leq a$.
3. If every pair of elements in a poset, A is comparable, then it is said to be *linearly ordered set*, and partial order relation is called a *linear order relation.* For example, the relation $(I_+, \leq)$ is a partial as well as linear ordered on the set of positive integers. But the relation (I_+, divides) is not the linear order relation because elements $2, 5 \in I_+$ are not comparable.
4. If $(A, \leq)$ and $(B, \leq)$ are posets, then $(A \times B, \leq)$ is also a poset with partial order relation $\leq$ defined by $(a, b) \leq (a', b')$ for all $a \leq a'$ in A and $b \leq b'$ in B.

3.7 EQUIVALENCE RELATION

A relation R on a set A is called an equivalence relation (denoted by ~) if and only if following three conditions hold.

1. R is reflexive.
2. R is symmetric.
3. R is transitive.

Illustration

Let A = $\{a, b, c\}$ be any set of numbers. The relation R on the set A is defined by *is equal to* is an equivalence relation because it is reflexive since every element in A is equal to itself. It is symmetric because $a = b \Rightarrow b = a$. Also if $a = b$ and $b = c$, then $a = c$ so that R is transitive as well.

Example 13 Give an example of a relation that is reflexive but neither symmetric nor transitive.

Solution Let A = {1, 2, 3} and R is defined as:

$$R = \{(1, 1), (1, 2), (2, 3), (2, 2), (3, 3)\}.$$

Then R is reflexive since $(a, a) \in R$ for all $a \in A$. It is not symmetric since $(1, 2) \in R$ but $(2, 1) \notin R$. It is also not transitive since $(1, 2) \in R$ and $(2, 3) \in R$ but $(1, 3) \notin R$.

Example 14 A relation R on the set of integers, I is defined as: a R b if and only if $(a - b)$ is divisible by m(a positive integer). Show that R is an equivalence relation.

Solution (*a*) R is reflexive for each $a \in I$. Also $(a - b)$ is divisible by m and hence $b - a = -(a - b)$ is also divisible by m. Thus, $a \text{ R } b \Rightarrow b \text{ R } a$. Therefore, R is symmetric.

(*b*) For $a, b, c \in I$, we have a R b and $b \text{ R } c \Rightarrow a - b$ and $b - c$ are both divisible by m. This implies that

$$(a - b) + (b - c) \text{ is divisible by } m \text{ or } (a - c) \text{ is divisible by } m \text{ or } a \text{ R } c.$$

Therefore, R is transitive.

Since R is reflexive, symmetric and transitive, therefore R is an equivalence relation.

Example 15 Let N be the set of all natural numbers. The relation R on the set N × N of ordered pairs of natural numbers is defined as: (a, b) R (c, d) if and only if $ad = bc$. Prove that R is an equivalence relation.

[*UPTU, BTech., 2007*]

Solution (*a*) R is reflexive, since for each $(a, b) \in N \times N$, we have $ab = ba$, i.e. (a, b) R (b, a).

(*b*) Let (a, b) R (c, d). The $ad = bc \Rightarrow cb = da$. Thus (c, d) R (a, b), and R is symmetric.

(*c*) Let (a, b) R (c, d) and (c, d) R (e, f). Then $ad = bc$ and $cf = de$. Thus $(ad)(cf) = (bc)(de)$. Dividing on both sides by dc, we get

$$af = be, \text{ i.e. } (a, b) \text{ R } (e, f)$$

Therefore, R is transitive.

Since R is reflexive, symmetric and transitive, therefore R is an equivalence relation.

Example 16 The relation R on the set S of all real numbers is defined as: a R b if and only if $1 + ab > 0$. Show that this relation is reflexive and symmetric but not transitive.

Solution (*a*) Let a be any real number. Then $1 + aa = 1 + a^2 > 0$, Since $a^2 \geq 0$. Therefore a R a for all $a \in S$. Thus R is reflexive.

(*b*) Let $a, b \in S$. Then

$$a \text{ R } b \Rightarrow 1 + ab > 0$$

$$\Rightarrow 1 + ba > 0 \qquad \text{[since } ab = ba \text{, for real numbers]}$$

$$\Rightarrow b \text{ R } a$$

Thus, R is symmetric.

(*c*) Consider three real numbers as: 1, – 1/2 and – 4. Now we have

$$1 + 1(-1/2) = (1/2) > 0$$

Therefore, 1R (– 1/2).

Further $1 + (-1/2)(-4) = 3 > 0$. Therefore (– 1/2) R (– 4). But $1 + 1(-4) = -3$ (< 0). Therefore, 1 is not related to – 4. Thus, 1 R (– 1/2), (– 1/2) R (– 4) is not implying that 1 R (– 4). Hence, R is not transitive.

Example 17 If R and S are equivalence relations on a set A, prove that $R \cap S$ is an equivalence relations in A.

Solution Since R and S are two equivalence relations on a set A, therefore $R \subseteq A \times A$ and $S \subseteq A \times A$. Thus, $R \cap S \subseteq A \times A$ implies that $R \cap S$ is also a relation in A.

(*i*) Since R and S are reflexive, therefore $(a, a) \in R$ and $(a, a) \in S$ for all $a \in A$. Thus for all $a \in A$, $(a, a) \in R \cap S$. Hence, $R \cap S$ is reflexive.

(*ii*) Let $(a, b) \in R \cap S$. But

$$(a, b) \in R \cap S \Rightarrow (a, b) \in R \text{ and } (a, b) \in S \qquad \text{[since R and S are symmetric]}$$

Therefore $R \cap S$ is symmetric.

(*iii*) Let $(a, b) \in R \cap S$ and $(b, c) \in R \cap S$.

$$(a, b) \in R \cap S \text{ and } (b, c) \in R \cap S \Rightarrow [(a, b) \in R \text{ and } (a, b) \in S] \text{ and } [(b, c) \in R \text{ and } (b, c) \in S]$$
$$\Rightarrow [(a, b) \in R \text{ and } (b, c) \in R) \text{ and } [(a, b) \in S \text{ and } (b, c) \in S]$$
$$\Rightarrow (a, c) \in R \text{ and } (a, c) \in S \qquad \text{[since R and S are transitive]}$$
$$\Rightarrow (a, c) \in R \cap S.$$

Therefore $R \cap S$ is transitive.

Since $R \cap S$ is reflexive, symmetric and transitive, therefore $R \cap S$ is an equivalence relation.

Example 18 If R and S are relations from A to B, prove that

(*a*) $R^{-1} \subseteq S^{-1}$ when $R \subseteq S$

(*b*) $(R \cap S)^{-1} = R^{-1} \cap S^{-1}$

(*c*) $(R \cup S)^{-1} = R^{-1} \cup S^{-1}$

Solution (*a*) Since $R \subseteq S$, then for an element $(a, b) \in R^{-1}$ implies that $(b, a) \in R$ and $(b, a) \in S$. Also $(b, a) \in S^{-1}$ implies $(a, b) \in S$. Thus $R^{-1} \subseteq S^{-1}$.

(*b*) Let $(a, b) \in (R \cap S)^{-1} \Leftrightarrow (b, a) \in R \cap S$

$$\Leftrightarrow (b, a) \in R \text{ and } (b, a) \in S$$
$$\Leftrightarrow (a, b) \in R^{-1} \text{ and } (a, b) \in S^{-1}$$
$$\Leftrightarrow (a, b) \in R^{-1} \cap S^{-1}$$

Hence, $(R \cap S)^{-1} = R^{-1} \cap S^{-1}$

(*c*) Let $(a, b) \in (R \cup S)^{-1} \Leftrightarrow (b, a) \in R \cup S$

$$\Leftrightarrow (b, a) \in R \text{ or } (b, a) \in S$$
$$\Leftrightarrow (a, b) \in R^{-1} \text{ or } (a, b) \in S^{-1}$$
$$\Leftrightarrow (a, b) \in R^{-1} \cup S^{-1}$$

Hence, $(R \cup S)^{-1} = R^{-1} \cup S^{-1}$.

Example 19 If R is a relation on a set A, then prove that

(*a*) when R reflexive, R^{-1} is also reflexive.

(*b*) R is symmetric if and only if $R = R^{-1}$.

(*c*) R is anti-symmetric if and only if $R \cap R^{-1} \subseteq I_A$.

Solution (*a*) If R is reflexive, then any element (a, a) belongs to R also belongs to R^{-1} for all $a \in A$. This shows that R^{-1} is reflexive.

(*b*) Suppose R is symmetric. Then any element $(a, b) \in R^{-1} \Leftrightarrow (b, a) \in R$, and hence $(a, b) \in R$ because R is symmetric. This shows that $R^{-1} = R$.

Conversely, suppose $R^{-1} = R$. Then $(a, b) \in R \Leftrightarrow (a, b) \in R^{-1}$ and hence $(a, b) \in R$. This shows that R is symmetric. Hence, the set R is symmetric if and only if $R = R^{-1}$.

(*c*) Suppose R is anti-symmetric. Then any element

$(a, b) \in R \cap R^{-1} \Leftrightarrow (a, b) \in R$ and $(a, b) \in R^{-1}$

$\Leftrightarrow (a, b) \in R$ and $(b, a) \in R$ and hence, $a = b$ because R is anti-symmetric.

Hence, any element belonging to $R \cap R^{-1}$ is of the form (a, a) for all $a \in A$. This implies that $R \cap R^{-1} \subseteq I_A$.

Conversely, let $R \cap R^{-1} \subseteq I_A$, and let $(a, a) \in A \times A$ so that $(a, b) \in R$ and $(b, a) \in R$, or $(a, b) \in R^{-1}$. Thus $(a, b) \in R \cap R^{-1}$. Since $R \cap R^{-1} \subseteq I_A$, it implies that $b = a$, and hence, R is anti-symmetric.

Example 20 If R is an equivalence relations in a set A, then prove that R^{-1} is also an equivalence relation.

[*UPTU, MCA, 2003, 2006*]

Solution Since R is an equivalence relation in a set A, it is reflexive, symmetric and transitive.

(*a*) For all $a \in A$, an element $(a, a) \in R \Rightarrow (a, a) \in R^{-1}$. Thus R^{-1} is reflexive.

(*b*) For any element $(a, b) \in R^{-1}$ $\Rightarrow (b, a) \in R$ for all $a, b \in A$

$\Rightarrow (a, b) \in R$, as R is symmetric

$\Rightarrow (b, a) \in R^{-1}$

Thus R^{-1} is symmetric.

(*c*) For any elements $(a, b), (b, c) \in R^{-1} \Rightarrow (b, a), (c, b) \in R$, for all $a, b, c \in A$

$\Rightarrow (c, b), (b, a) \in R$

$\Rightarrow (c, a) \in R$, as R is transitive

$\Rightarrow (a, c) \in R^{-1}$

Thus R^{-1} is transitive.

Since relation R^{-1} has been proved reflexive, symmetric and transitive, it is an equivalence relation.

3.8 EQUIVALENCE CLASSES (OR SETS)

Let R be an equivalence relation on a set A, and let $a \in A$. For each $a \in A$, let $[a]$ be the set of all elements of A which are related to a under R. Then $[a]$ or A_a is an *equivalence class* of a in A with respect to R. Symbolically, an equivalence class is defined as:

$$A_a = \{x : x \in A \text{ and } (x, a) \in R, \text{ i.e. } x \, R \, a\}.$$

3.8.1 Quotient Set

The collection of all equivalence classes of elements of A under the equivalence relation R on a set A is defined as: A/R = {[a] : $a \in$ A}.

Illustrations

1. Let R be an equivalence relation defined by x *is congruent to* y on a set A of all triangles in a plane, where $x, y \in$ A. For any element $z \in$ A, the equivalence class, A_z is the set of all triangles of A congruent to the triangle z.

2. Let A = {1, 2, 3, 4} and R = {(1, 1), (2, 2), (2, 3), (3, 2), (3, 3), (4, 4)}. Then R is an equivalence relation on A. This equivalence relation has following equivalence classes:

$$[1] = \{1\}; [2] = \{2, 3\} = [3]; [4] = \{4\}$$

Example 21 Let I be the set of all integers and

$$R = \{(a, b) : a, b \in I \text{ and } a - b \text{ is divisible by } 3\}$$

is an equivalence relation. Find equivalence classes.

Solution Given that I = {. . ., – 3, – 2, – 1, 0, 1, 2, 3, . . .} is the set of integers. Also given that x R y if $x - y$ is divisible by 3. That is,

$$x \text{ R} y \text{ if } (x - y)/3 = m \text{ (say)}, m \in I$$

or

$$x \text{ R} y \text{ if } x - y = 3m \text{ or } x = 3m + y$$

This implies that if x is divided by 3, then remainder is y and x is equivalent to this remainder, where $0 \leq y \leq 2$. This means x will be equivalent to any one—0, 1, 2. Now,

$$\begin{aligned}\text{Equivalence class } [0] &= \{x : x \in I \text{ and } x \text{ R } 0\}\\ &= \{x : x - 0 = 3m;\ m \in I\}\\ &= \{\ldots -9, -6, -3, 9, 3, 6 \ldots\}\\ \text{Equivalence class } [1] &= \{x : x \in I \text{ and } x \text{ R } 1\}\\ &= \{x : x - 1 = 3m \text{ or } x = 3m + 1;\ m \in I\}\\ &= \{\ldots, -8, -5, -2, 1, 4, 7, 10, \ldots\}\\ \text{Equivalence class } [2] &= \{x : x \in I \text{ and } x \text{ R } 2\}\\ &= \{x : x - 2 = 3m \text{ or } x = 3m + 2;\ m \in I\}\\ &= \{-7, -4, -1, 2, 5, 8, \ldots\}\end{aligned}$$

and so on. It may also be verified that

$$[0] = [3] = [6] = \ \ldots = [-3] = [-6] = \ldots$$
$$[1] = [4] = [7] = \ \ldots = [-2] = [-5] = \ldots$$
$$[2] = [5] = [8] = \ \ldots = [-1] = [-4] = \ldots$$

Hence, the total number of distinct equivalence classes are [0], [1] and [3]. Obviously,

(*i*) I = [0] ∪ [1] ∪ [2]

(*ii*) Two equivalence classes are either disjoint or identical.

(*iii*) The elements of the same class are equivalent with respect to the given relation R.

Example 22 Let R be an equivalence relation on a non-empty set A. Let a and b be arbitrary elements in A. Then prove that

(*i*) $a \in [a]$, i.e. $[a]$ is non-empty

(*ii*) $b \in [a] \Leftrightarrow [b] = [a]$

(*iii*) $[a] = [b]$ if and only $(a, b) \in R$, i.e. $a \text{ R } b$.

(*iv*) equivalence classes determined by a and b are either disjoint or identical, i.e. either $[a] = [b]$ or $[a] \cap [b] = \Phi$.

Solution (*i*) Given that R is reflexive, therefore we have $(a, a) \in R$ for all $a \in A$. But

$$[a] = \{x : x \in A \text{ and } x \text{ R } a\}$$

Thus $$a \text{ R } a \Rightarrow a \in [a], \text{ for all } a \in A.$$

(*ii*) Let $b \in [a] \Rightarrow b \text{ R } a$. To prove $[b] = [a]$, let x be an arbitrary element of $[b]$. Then

$$x \in [b] \Leftrightarrow x \text{ R } b \text{ and } b \text{ R } a$$

$$\Leftrightarrow x \text{ R } a \quad \text{[since R is transitive]}$$

$$\Leftrightarrow x \in [a]$$

Since $x \in [b]$ and $x \in [a]$, therefore $[b] \subseteq [a]$.

Again, let x be an arbitrary element of $[a]$. Then

$$x \in [a] \Leftrightarrow x \text{ R } a$$

$$\Leftrightarrow x \text{ R } a \text{ and } a \text{ R } b$$

$$\Leftrightarrow x \text{ R } b \quad \text{[since R is symmetric]}$$

$$\Leftrightarrow x \in [b]$$

Since $x \in [a]$ and $x \in [b]$, therefore $[a] \subseteq [b]$.

Hence, from the proof given above, we get $[a] = [b]$.

(*iii*) Suppose $[a] = [b]$. Since R is reflexive, therefore $a \text{ R } a$.

$$a \text{ R } a \Rightarrow a \in [a]$$

$$\Rightarrow a \in [b] \quad \text{[since } [a] = [b]\text{]}$$

$$\Rightarrow a \text{ R } b$$

Thus $[a] = [b] \Rightarrow a \text{ R } b$.

Again, let $a \text{ R } b$. Then to prove $[a] = [b]$, let $x \in [a] \Rightarrow x \text{ R } a$, but given that $a \text{ R } b$. Therefore

$$x \text{ R } a \text{ and } a \text{ R } b \Leftrightarrow x \text{ R } b \quad \text{[R is transitive]}$$

$$\Leftrightarrow x \in [b]$$

This shows that $[a] = [b]$.

Since $[a] = [b] \Rightarrow a \text{ R } b$ and $a \text{ R } b \Rightarrow [a] = [b]$, therefore $[a] = [b]$ if and only if $a \text{ R } b$.

(*iv*) If for any two classes $[a]$ and $[b]$, we have $[a] \cap [b] = \Phi$, then the result is proved. Now let $[a] \cap [b] \neq \Phi$. Then to prove $[a] = [b]$, let

$$x \in [a] \cap [b] \Rightarrow x \in [a] \text{ and } x \in [b]$$

$$\Rightarrow x \text{ R } a \text{ and } x \text{ R } b$$

$$\Rightarrow a \text{ R } x \text{ and } x \text{ R } b \qquad \text{[R is symmetric]}$$

$$\Rightarrow a \text{ R } b \qquad \text{[R is transitive]}$$

$$\Rightarrow [a] = [b] \qquad \text{[using (iii)]}$$

Thus $[a] \cap [b] \neq \Phi \Rightarrow [a] = [b]$. But if $[a] \neq [b] \Rightarrow [a] \cap [b] = \Phi$.

Hence, either $[a] \cap [b] = \Phi$ or $[a] = [b]$.

Important result An equivalence relation R in a non-empty set A determines a partition of A and conversely, a partition of A defines an equivalence relation in A.

Illustration Let I be the set of all integers. Consider a relation

$$R = \{(a, b) : (a - b) \text{ is divisible by } 5\}.$$

The relation R is an equivalence relation on I. This relation partitions the set I into following five mutually disjoint equivalence classes. The equivalence classes determined by 0, 1, 2, 3, 4 are the sets consisting of all integers related to 0, 1, 2, 3 and 4, respectively. Obviously

$$[0] = \{a : a \in \text{I and } a \text{ R } 0\}$$

$$= \{a : a - 0 = 5m;\ m \in \text{I}\}$$

$$= \{\ldots -10, -5, 0, 5, 10, \ldots\}$$

Similarly

$$[1] = \{\ldots -9, -4, 1, 6, 11, \ldots\}$$

$$[2] = \{\ldots -8, -3, 2, 7, 12, \ldots\}$$

$$[3] = \{\ldots -7, -2, 3, 8, 13, \ldots\}$$

$$[4] = \{\ldots -5, -1, 4, 9, 14, \ldots\}.$$

Example 23 Let R be an equivalence relation on a set A = {1, 2, 3, 4} defined by partition, P = {{1}, {2}, {3}, {4}}. Determine the elements of equivalence relation and also find the equivalence classes of R.

Solution The elements of equivalence relation defined by partition P is

$$R = \{(1, 1), (2, 2), (3, 3), (4, 4)\}$$

The equivalence classes of R are

$$[1] = \{1\} \qquad [2] = \{2\}$$

$$[3] = \{3\} \qquad [4] = \{4\}.$$

3.9 REPRESENTATION OF RELATIONS ON FINITE SETS

Few of the ways in which a relation on finite sets can be represented are as follows:

3.9.1 Relation as an Arrow Diagram and Table

A relation can also be represented in a tabular or graphical form. This helps us understanding, at glance, a clear idea of the situation under consideration. For example,

1. A flow chart helps developing a computer program for solving the problem.
2. The graph of the relation defined by the equation, $x^2 + y^2 = 9$ helps in determining the nature of the function. The graph of this equation is a circle having its centre at (0, 0) and radius 3 as shown in Fig. 3.1.

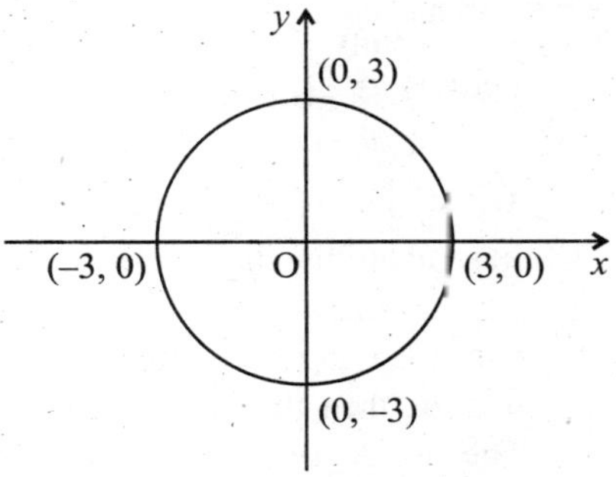

Fig. 3.1 Graph of $x^2 + y^2 = 9$

Illustration Let A = {*a*, *b*, *c*} be a set of students of IT and B = {*x*, *y*, *z*, *w*} be a set of companies that came for campus interviews for selection of students for jobs. We might have relation R_1 from A to B to describe that the companies had interviews with the students and relation R_2 from A to B to describe that the jobs were offered to the students by the companies. The elements of both the relations may be written as:

$$R = \{(x, a), (y, c), (z, a), (z, c), (w, b)\}$$

	a	*b*	*c*
x	1	0	0
y	0	0	1
z	1	0	1
w	0	1	0

Fig. 3.2 (*a*)

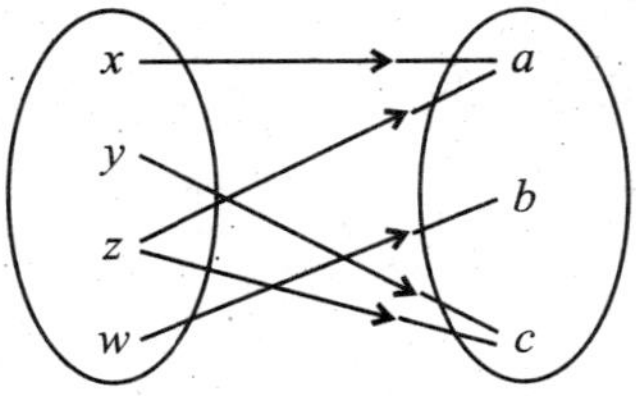

Fig. 3.2 (*b*)

Figures 3.2 (*a*) and (*b*) describe the relations in two ways. Here 1 or 0 in each position of the matrix represents whether elements are related or not related in two sets A and B, respectively.

3.9.2 Relation as a Directed Graph

Let R be a relation from a finite set A to itself. Draw a small circle for each element of A and label the circle with the corresponding element of A. These circles are called the *vertices* or *nodes* of the graph. Draw an *arrow* (also called *edge*) from vertex a_i to vertex b_j, if and only if a_i R b_j. This type of graph of a relation R is called a *directed graph* or *digraph of* R.

A vertex of a graph is also called a *node* (*point* or *junction*). An edge of a graph is also called an *arc* (*line* or *branch*). The edges in the graph of R correspond exactly to the pair of elements in R and the vertices correspond exactly to the elements of the set A. For practical purposes, we may refer to the pair of R as edges and the elements of A as vertices.

Definition Let A be a non-empty set. A directed graph (or diagraph) G of A is made up of the elements of A, called *vertices* or *nodes* of A, and a subset E of A × A that contains the directed *edges* or *arcs* of G. The set A is called *vertex set* of G and E is called the *edge set.* We write G = (A, E) to denote the graph.

If $a, b \in$ A and $(a, b) \in$ E, then there is an edge from *a* to *b*. Vertex *a* is called the *origin* or *source* edge with *b* the *terminus* or *terminating* vertex. If $a \neq b$, then $(a, b) \neq (b, a)$ and an edge of the form (a, a) is called a *loop* at *a*.

Illustrations

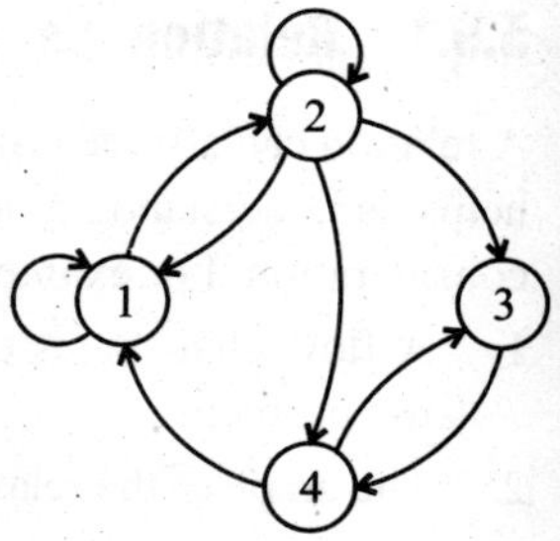

Fig. 3.3

1. Let $A = \{1, 2, 3, 4\}$ and

$$R = \{(1, 1), (1, 2), (2, 1), (2, 2), (2, 3), (2, 4), (3, 4), (4, 1), (4, 3)\}$$

The diagraph of R is as shown in Fig. 3.3.

In terms of the digraph of R, the *indegree* of a vertex is the number of edges terminating at the vertex, and the *outdegree* of a vertex is the number of edges leaving the vertex. In other words, if R is a relation on a set A and $a \in A$, then the indegree of a is the number $b \in A$ such that $(b, a) \in R$. The out-degree of a is the number $b \in A$ such that $(a, b) \in R$.

2. Consider the digraph (Fig. 3.3). Vertex 1 has indegree 3 and outdegree 2. Similarly, vertex 4 has indegree 2 and outdegree 2.

The directed graph and the matrix representing a relation R can also be used to determine whether the relation is an equivalence relation, i.e. reflexive, symmetric and transitive.

1.(*a*) **Reflexive** The relation R is reflexive if and only if there is a *loop at every vertex* of the directed graph so that ordered pairs of the form (a, a) occur in the relation.

(*b*) If all $m_{ij} = 1$ for $i = j$ in the relation matrix, then such a relation is called reflexive. That is, all elements in the main diagonal of the matrix are 1.

Illustration Following are the examples of reflexive relation

$$\begin{array}{c} \\ 1 \\ 2 \\ 3 \end{array}\begin{array}{c} \begin{array}{ccc} 1 & 2 & 3 \end{array} \\ \begin{bmatrix} 1 & 1 & 1 \\ 0 & 1 & 1 \\ 0 & 0 & 1 \end{bmatrix} \end{array}$$

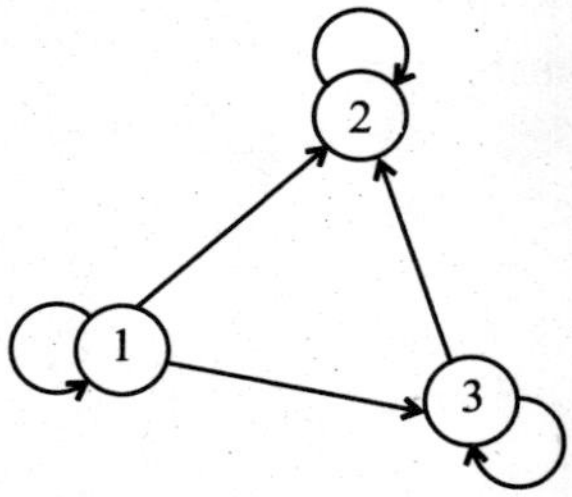

2.(*a*) **Symmetric** The relation R is symmetric if and only if for every edge between distinct vertices in its directed graph, there exists an edge in the opposite direction so that (b, a) is in the relation whenever (a, b) is in the relation.

A relation is anti-symmetric when no two distinct nodes in the directed graph have an edge between them in both directions.

(*b*) If $m_{ij} = m_{ji}$ for all i and j in the relation matrix, then such a relation is called *symmetric*. That is, relation matrix is symmetric with respect to the main diagonal. However, if $m_{ij} = 1$ necessitates $m_{ij} = 0$ in the relation matrix, then such a relation is called *anti-symmetric*.

Illustration The following are the examples of symmetric relation

$$\begin{array}{c} \\ 1 \\ 2 \\ 3 \end{array}\begin{array}{c} \begin{array}{ccc} 1 & 2 & 3 \end{array} \\ \begin{bmatrix} 1 & 0 & 1 \\ 0 & 1 & 0 \\ 1 & 0 & 0 \end{bmatrix} \end{array} \qquad \begin{array}{c} \\ 1 \\ 2 \\ 3 \end{array}\begin{array}{c} \begin{array}{ccc} 1 & 2 & 3 \end{array} \\ \begin{bmatrix} 0 & 0 & 1 \\ 0 & 1 & 0 \\ 0 & 1 & 0 \end{bmatrix} \end{array}$$

Symmetric Anti-symmetric

3.(*a*) **Transitive** The relation R is transitive if and only if there exists directed edge from a vertex, a to b and from vertex b to c, then there is also a directed edge from vertex a to c.

(*b*) If elements in the relation matrix are such that $m_{ij} = 1$ and $m_{jk} = 1$, then $m_{ik} = 1$ and we call a relation R as transitive relation. In other words, the relation R is transitive when $M_R^2 = M_R \cdot M_R$ has unit (i.e. 1) element in the position (i, k). That is, in particular, we must have $M_R^2 = M_R$. However, in this case converse is not true.

Illustration Following is the example of transitive relation

$$M_R = \begin{array}{c} \\ 1 \\ 2 \\ 3 \end{array}\begin{array}{c} \begin{array}{ccc} 1 & 2 & 3 \end{array} \\ \begin{bmatrix} 1 & 0 & 1 \\ 0 & 1 & 0 \\ 1 & 0 & 1 \end{bmatrix} \end{array}$$

3.9.3 Relation as a Matrix

Consider a relation R from a finite set, $A = \{a_1, a_2, \ldots, a_m\}$ to $B = \{b_1, b_2, \ldots, b_n\}$ containing m and n elements, respectively. We define *relation matrix* $M_R = [m_{ij}]$ of order $m \times n$, for R whose elements are given by

$$m_{ij} = \begin{cases} 1 & \text{if } a_i \text{ R } b_j \text{ or } (a_i, b_j) \in R \\ 0 & \text{otherwise} \end{cases}$$

The matrix M_R is also called the *adjacency matrix* (*Boolean matrix* or *relation matrix*).

Illustration Let $A = \{a_1, a_2, a_3\}$ and $B = \{b_1, b_2, b_3, b_4\}$ be two finite sets. Also, let the relation defined between them is

$$R = \{(a_1, b_1), (a_1, b_4), (a_2, b_2), (a_2, b_3), (a_2, b_3), (a_3, b_1), (a_3, b_2)\}.$$

The matrix of the given relation is shown below:

$$M_R = \begin{array}{c} \\ a_1 \\ a_2 \\ a_3 \end{array}\begin{array}{c} \begin{array}{cccc} b_1 & b_2 & b_3 & b_4 \end{array} \\ \begin{bmatrix} 1 & 0 & 0 & 1 \\ 0 & 1 & 1 & 0 \\ 1 & 1 & 0 & 0 \end{bmatrix} \end{array}$$

Example 24 Find the relations determined by the following digraphs.

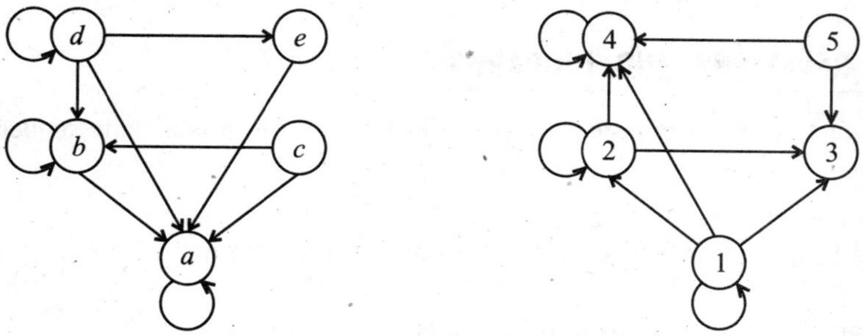

Solution The relations shown in digraphs are as follows:

(*i*) R = {(*a*, *a*), (*b*, *b*), (*d*, *d*), (*b*, *a*), (*d*, *a*), (*e*, *a*), (*c*, *a*), (*c*, *b*), (*d*, *b*), (*d*, *e*)}

(*ii*) R = {(1, 1), (2, 2), (4, 4), (1, 2), (1, 3), (1, 4), (2, 3), (2, 4), (3, 4), (5, 3), (5, 4)}

Example 25 Let R be a relation on the set A = {1, 2, 3, 4} defined by

$$R = \{(1, 1), (1, 2), (1, 3), (1, 4), (2, 2), (2, 4), (3, 3), (3, 4), (4, 4)\}$$

Construct the matrix and digraph of R.

Solution The matrix of the given relation and its digraph are shown in Figs 3.4(*a*) and (*b*), respectively.

	1	2	3	4
1	1	1	1	1
2	0	1	0	1
3	0	0	1	1
4	0	0	0	1

Fig. 3.4(*a*) Matrix of Relation R

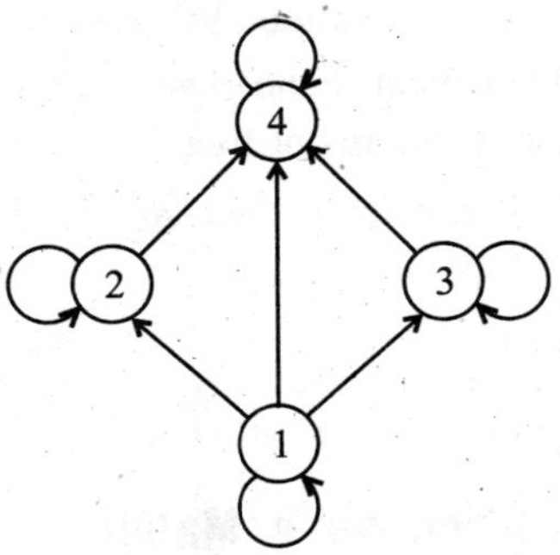

Fig. 3.4(*b*) Digraph of Relation R

Example 26 Let A = $\{a, b, c\}$ be a non-empty set, and R be the relation on A that has the matrix

$$M_R = \begin{bmatrix} 1 & 0 & 1 \\ 0 & 1 & 0 \\ 0 & 1 & 1 \end{bmatrix}$$

Construct the digraph of R, and list indegrees and outdegrees of all vertices.

Solution The digraph of R is shown in Fig. 3.5. The following table gives indegrees and outdegrees of all vertices.

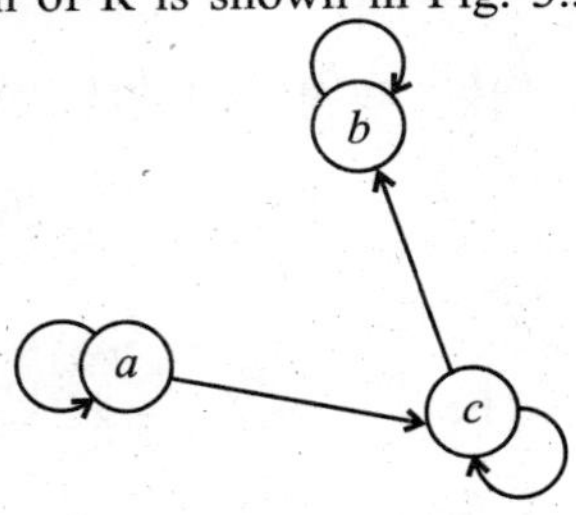

Fig. 3.5

	a	*b*	*c*
Indegree	1	2	2
Outdegree	1	2	1

3.10 PATH IN RELATIONS AND DIGRAPHS

Let R be a relation on a set A. The *path of length n* in R from *a* to *b* is a finite sequence: $a, x_1, x_2, \ldots, x_{n-1}, b$ such that

$$a \, R \, x_1, \; x_1 \, R \, x_2, \ldots, x_{n-1} \, R \, b.$$

Obviously, a path of length *n* involves *n* + 1 elements of A which may not be distinct.

A path in a digraph is the succession of edges and its length is the number of edges in the path, where the vertices need not all be distinct.

Illustration Consider the digraph in Fig. 3.6. The sequence 1, 2, 3, 4, is a path of length 3 from vertex 1 to 4, and the sequence 1, 5, 4, 1, is a path of length 3 from vertex 1 to 4 itself.

A path that begins and ends at the same vertex is called a *cycle*. In Fig. 3.6, the sequence 1, 5, 4, 1 is the cycle of length 3.

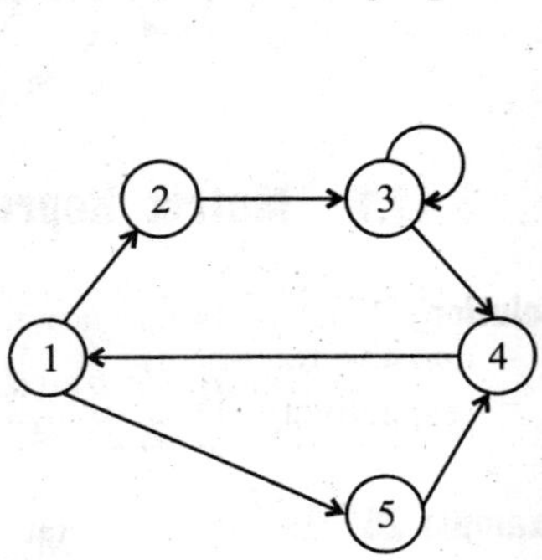

Fig. 3.6

If n is a fixed positive integer, then relation on the set A is defined as: $a \, R^n \, b$. This implies that there is a path of length n from a to b in R.

The relation R^n consists of all vertices that can be reached from a by means of a path in R of length n. The relation defined as $a \, R^\infty \, b$ is called the *connectivity relation* for R. The length of such path will depend, in general, on a and b.

3.11 COMPOSITION OF RELATIONS

Let R_1 be a relation from A to B and R_2 be the relation from B to C, then the *composition relation*, denoted as $(R_2 \circ R_1)$ is the relation from A to C defined by

$$a \, (R_1 \circ R_2) \, c \text{ if and only if } a \, R \, b \text{ and } b \, R \, c \text{ for some } b \in B.$$

In other words,

$$R_1 \, 0 \, R_2 = \{(a, c) : a \in A, c \in C \text{ and there exists } b \in B \text{ with } (a, b) \in R_1, (b, c) \in R_2\}.$$

The composition of R_1 and R_2 is sometimes also denoted by $R_1 \, R_2$.

Illustration Let A = {1, 2, 3}, B = $\{a, b, c, d\}$ and C = $\{x, y, z, w\}$ be three sets. The relation R_1 from A to B and R_2 from B to C is defined as

$$R_1 = \{(1, a), (1, b), (3, c), (3, d)\}$$
$$R_2 = \{(a, x), (b, x), (b, z), (d, w)\}$$

These relations are shown in Fig. 3.7.

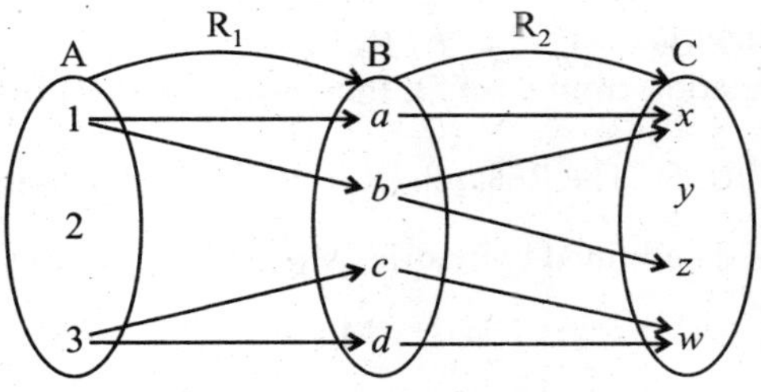

Fig. 3.7 Composition of Relations

In Fig. 3.7, it may be noted that an arrow from 1 to b is followed by an arrow from b to z, similarly an arrow from 3 to d is followed by an arrow from d to w. These arrows connect element $1 \in A$ to $x \in C$, as well as $3 \in A$ to $w \in C$. Thus, we may say that $1 \, (R_1 \, 0 \, R_2) \, x$ and $3 \, (R_1 \, 0 \, R_2) \, w$. Such connectivity of set A to C through relations R_1 and R_2 is called a *path*. The composition of R_1 and R_2 can now be written as

$$R_1 \, 0 \, R_2 = \{(1, x), (1, z), (3, w)\}.$$

Power of R Let R be a relation in a set A. The power of R is defined as:

(*a*) $R^1 = R$.

(*b*) $R^{n+1} = R \, 0 \, R^n$ for all positive integers n.

For example, if $R = \{(1, 2), (1, 3), (2, 4), (3, 2)\}$ is a relation in a set A = {1, 2, 3, 4}. then

$$R^2 = \{(1, 4), (1, 2), (3, 4)\}; \; R^3 = \{(1, 4)\} \text{ and for } n \geq 4, \; R^n = \Phi.$$

3.11.1 Matrix Representation of Composition of Relations

Let A = {1, 2, 3}, B = $\{a, b, c, d\}$ and C = $\{x, y, z, w\}$ be three finite sets. The relation, R_1 from A to B and the relation R_2 from B to C can be represented by the matrices $M_{R_1} = [m_{ij}]_{3 \times 4}$ and $M_{R_2} \, [n_{ij}]_{4 \times 4}$, respectively. These matrices are defined as:

$$M_{ij} = \begin{cases} 1 & \text{if } (i, j) \in R \\ 0 & \text{otherwise} \end{cases}$$

$$M_{R_1} = \begin{array}{c} \\ 1 \\ 2 \\ 3 \end{array}\begin{array}{c} \begin{array}{cccc} a & b & c & d \end{array} \\ \begin{bmatrix} 1 & 1 & 0 & 0 \\ 0 & 0 & 0 & 0 \\ 0 & 0 & 1 & 1 \end{bmatrix} \end{array} \quad \text{and} \quad M_{R_2} = \begin{array}{c} \\ a \\ b \\ c \\ d \end{array}\begin{array}{c} \begin{array}{cccc} x & y & z & w \end{array} \\ \begin{bmatrix} 1 & 0 & 0 & 0 \\ 1 & 0 & 1 & 0 \\ 0 & 0 & 0 & 0 \\ 0 & 0 & 0 & 1 \end{bmatrix} \end{array}$$

Further, the composite relation R_1 o R_2 from set A to C can also be obtained by multiplying two matrices M_{R_1} and M_{R_2} in the usual manner as shown below:

$$M_{R_1R_2} = M_{R_1} M_{R_2} = \begin{array}{c} \\ 1 \\ 2 \\ 3 \end{array}\begin{array}{c} \begin{array}{cccc} x & y & z & w \end{array} \\ \begin{bmatrix} 2 & 0 & 1 & 0 \\ 0 & 0 & 0 & 0 \\ 0 & 0 & 0 & 1 \end{bmatrix} \end{array}$$

The non-zero entries in matrix $M_{R_1R_2}$ indicate the relationships of particular elements in R_1 o R_2.

From this illustration we can see that it is often easier to compute $M_{R_1R_2}$ by computing $M_{R_1} M_{R_2}$ instead of searching the digraph of R_1 o R_2 for all vertices that can be joined by a path.

Standard Results

1. If R is a relation on A = $\{a_1, a_2, \ldots, a_n\}$, then $M_{R^2} = M_R \cdot M_R$
2. For $n \geq 2$, and R a relation on a finite set, we have $M_{R^n} = M_R \cdot M_R \cdot \cdots \cdot M_R$ (n factors)

Theorem 3.1 Let A, B, C, D be sets, R_1 is relation from A to B, R_2 is a relation for B to C and R_3 is a relation from C to D, then prove that $(R_1 \text{ o } R_2) \text{ o } R_3 = R_1 \text{ o } (R_2 \text{ o } R_3)$ *[UPTU., MCA., 2003]*

Proof The Relation $(R_1 \text{ o } R_2) \text{ o } R_3$ is determined by matrix $M_{(R_1 \text{ o } R_2) \text{ o } R_3}$ and then relation $R_1 \text{o } (R_2 \text{ o } R_3)$ is determined by matrix $M_{R_1 \text{ o } (R_2 \text{ o } R_3)}$. Now

$$M_{(R_1 \text{ o } R_2) \text{ o } R_3} = M_{(R_1 \text{ o } R_2)} \odot M_{R_3} = (M_{R_1} \odot M_{R_2}) \odot M_{R_3}$$

and

$$M_{R_1 \text{ o } (R_2 \text{ o } R_3)} = M_{R_1} \odot M_{R_2 \text{ o } R_3} = M_{R_1} \odot (M_{R_2} \odot M_{R_3})$$

Since matrix multiplication is associative, therefore from above results, we conclude that $(R_1 \text{ o } R_2) \text{ o } R_3 = R_1 \text{ o } (R_2 \text{ o } R_3)$

Theorem 3.2 If R^{-1} and S^{-1} are the inverse of the relations R and S respectively, then

$$(R_1 \text{ o } R_2)^{-1} = R_2^{-1} \text{ o } R_1^{-1}$$

Proof Let A, B and C be three non-empty sets such that R_1 is a relation form A to B and R_2 is a relation from B to C. Thus

$$R_1 \subseteq A \times B,\ R_2 \subseteq B \times C \text{ and } R_1 \text{ o } R_2 \subseteq A \times C. \text{ Also } (R_1 \text{ o } R_2)^{-1} \subseteq C \times A$$

Let $(z, x) \in (R_1 \text{ o } R_2)$. Then $(x, z) \in R_1 \text{ o } R_2$. This implies that there exists an element $y \in$ B such that $(x, y) \in R_1$ and $(y, z) \in R_2$. Hence,

$$(x, z) \in R_1^{-1} \text{ o } R_2^{-1} \quad \text{or} \quad (y, x) \in R_1^{-1} \text{ and } (z, y) \in R_2^{-1}$$

or

$$(z, y) \in R_1^{-1} \text{ and } (y, x) \in R_2^{-1} \quad \text{or} \quad (z, x) \in R_2^{-1} \text{ o } R_1^{-1}$$

It means that if $(x, z) \in (R_1 \text{ o } R_2)^{-1} \Rightarrow (z, x) \in R_2^{-1} \text{ o } R_1^{-1}$. Thus $(R_1 \text{ o } R_2)^{-1} \subseteq R_2^{-1} \text{ o } R_1^{-1}$

Similarly, we can prove that $R_2^{-1} \text{ o } R_1^{-1} \subseteq (R_1 \text{ o } R_2)^{-1}$

Hence, these results prove that $(R_1 \text{ o } R_2)^{-1} = R_2^{-1} \text{ o } R_1^{-1}$

3.12 HASSE DIAGRAMS

A diagram that is used to describe partial order relation associated with a set is called *Hasse diagram.* It is different than digraphs drawn earlier. However, both are used to represent different types of digraphs. The following tips are necessary to read Hasse diagram for a relation, R defined on a set A.

(*i*) Since a partial order is reflexive, therefore every vertex is related to itself. So the arrows from a vertex to itself in a digraph are removed. Thus, the digraph shown in Fig. 3.8(*a*) would be drawn as shown in Fig. 3.8(*b*).

(*ii*) Since a partial order is anti-symmetric, therefore all arrows connecting two vertex are removed. Thus, if $x \leq y$ and $y \leq x$ is not necessary unless $x = y$. In this case direction of arrows between x and y is not necessary as shown Fig. 3.8(*c*).

(*iii*) Since a partial order is transitive, all edges that are implied by this property are removed. Thus, if $x \leq y$ and $y \leq z$, then it implies that $x \leq z$, so the edge from x to z is removed as shown in Fig. 3.8(*d*).

(*iv*) Circles in the digraph representing vertices (or nodes) are replaced by dots (or points). Thus, the Hasse diagram shown in Fig. 3.8(*e*) is final form of the digraph shown in Fig. 3.8(*a*).

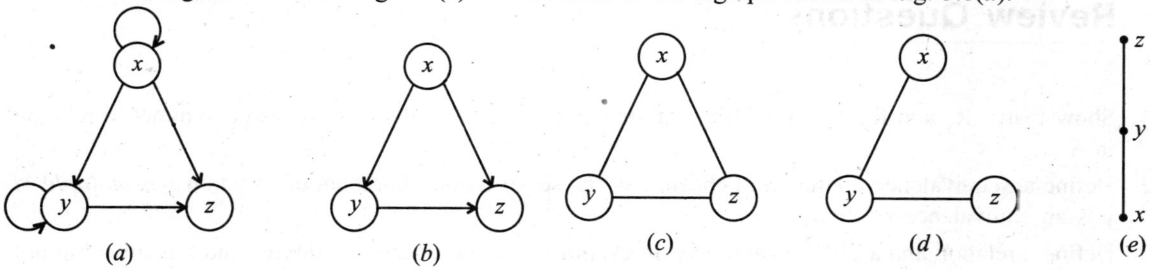

Fig. 3.8 Hasse Diagrams

Example 27 Let P(A) = {Φ, {*a*}, {*b*}, {*a*, *b*}} be the power set of the set A = {*a*, *b*}. Consider the partial order of subset (⊆) on A. Draw the Hasse diagram of the poset (A, ⊆).

Solution The digraph of the poset is shown in Fig. 3.9(*a*). The Hasse diagram can now be drawn as shown in Fig. 3.9(*b*).

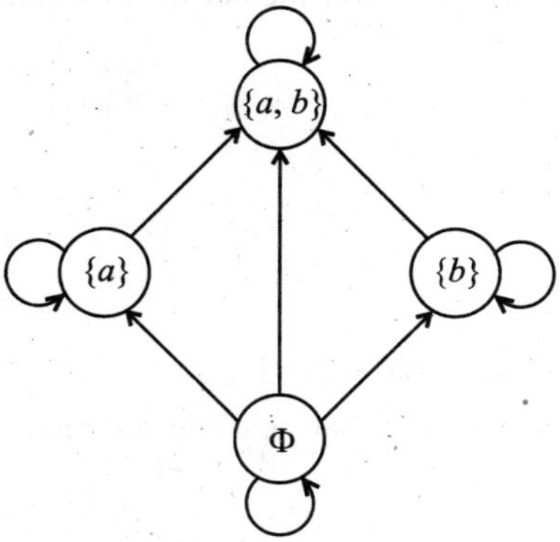

Fig. 3.9(*a*) Digraph

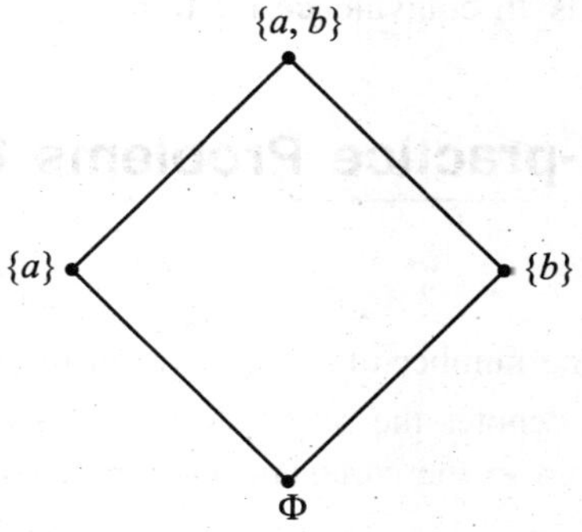

Fig. 3.9(*b*) Hasse Diagram

Example 28 Let A = {2, 3, 4, 6, 12, 36, 48} be a non-empty set and R be the partial order relation of divisibility defined on A. That is, if $a, b \in A$, then a divides b. Draw Hasse diagram of R.

Solution The digraph and Hasse diagram of poset A are shown in Figs. 3.10(a) and (b), respectively.

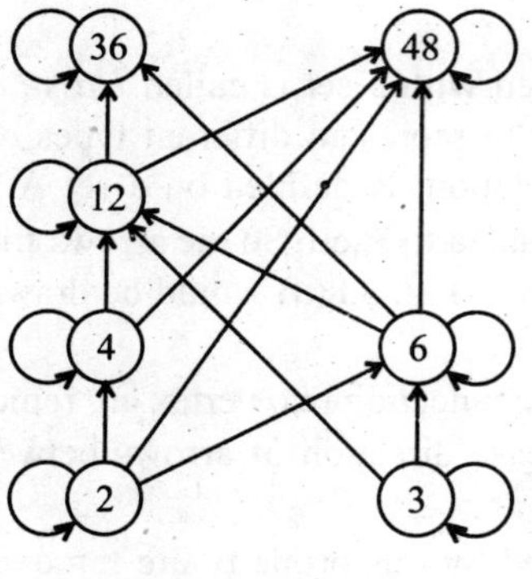

Fig. 3.10(a) Digraph

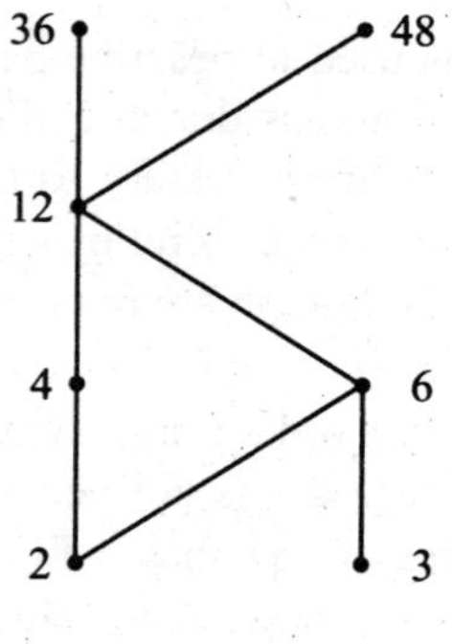

Fig. 3.10(b) Hasse Diagram

Review Questions

1. Show that if R_1 and R_2 be symmetric relations in a set A, then $R_1 \cap R_2$ is also a symmetric relation in A.
2. Define an equivalence relation and check if in the set of natural numbers the relation x *is multiple of* y is an equivalence relation.
3. Define a relation and a function and give an example to show which is reflexive and transitive but not symmetric.
4. Define a relation and give an example of a relation which is symmetric and transitive but not reflexive.
5. Show that in the set of all numbers, the relation *less than* is transitive but not reflexive.
6. Define reflexive, symmetric and transitive relations. Give one example of each.
7. Define a relation on a set. What do you mean by the domain and range of a relation? Give examples.
8. What is an equivalence relation? Show that the relation of *similarity* on the set of all triangles in a plane is an equivalence relation.

Self-practice Problems II

12. Find the number of relations from the set A = {a, b, c} to the set B = {1, 3}.
13. Let A denotes the set of shirts and B the set of pants a man owns. What possible interpretation can be given to the cartesian product A × B ?

14. Let A = {4.00, 4.30, 5.00, . . ., 7.30, 8.00} denotes the set of nine half-hour periods in the evening. Let B = {5, 20, 25, 32} denotes the set of four local television channels. Let R_1 and R_2 be two binary relations from A to B. What possible interpretations can be given to the binary relations R_1, R_2, $R_1 \cup R_2$, $R_1 \cap R_2$, $R_1 \oplus R_2$, and $R_1 - R_2$?

15. Let A = {*a*, *b*, *c*, *d*} be a set of students and B = {*x*, *y*, *z*, *w*} be a set of companies that came to the campus interviews of IT students. Let R_1 and R_2 be two binary relations from A to B. What interpretations can be given to the binary relations R_1, R_2, $R_1 \cap R_2$, $R_1 \cup R_2$, $R_1 \oplus R_2$ and $R_1 - R_2$?

16. If A = {1, 2, 3, 4, 6} = B, find the domain, range and draw digraph of R when $x \, R \, y$ if and only if x is multiple of y.

17. Given A = {1, 2, 3, 4} and B = {*x*, *y*, *z*}. Let R be the relation from A to B defined as:

$$R = \{(1, x), (2, z) (3, x), (3, y), (3, z)\}$$

(*a*) Determine the matrix of the relation.

(*b*) Find the inverse relation R^{-1} of R.

(*c*) Determine the domain and range of R.

18. Let A = {1, 2, 3}, B = {*a*, *b*, *c*} and C = {*x*, *y*, *z*} and let

$$R_1 = \{(1, a), (2, c), (3, a), (3, c)\} \text{ and } R_2 = \{(b, x), (b, z), (c, y)\}$$

(*a*) Find the composition relation $R_1 \text{ o } R_2$.

(*b*) Find the matrices M_{R_1}, M_{R_2} and $M_{R_1R_2}$ of respective relations, R_1, R_2 and $R_1 \text{ o } R_2$ and compare $M_{R_1R_2}$ with $M_{R_1} \cdot M_{R_2}$.

19. Determine whether the relation R is a partial order relation on the set of positive integers, I_+

(*a*) $a \, R \, b$ if an only if $a = 4b$

(*b*) $a \, R \, b$ if and only if $a = b^k$, $k \in I_+$

(*c*) $a \, R \, b$ if and only if $a \geq b$.

20. Let A = {1, 2, 3, 6}. Show that the relation 'divides' is a partial ordering on A. Draw the Hasse Diagram.

21. Let B = {*a*, *b*, *c*} and A = P(B) be the power set of B. Draw the Hasse diagram for ⊆ on poset A.

22. Describe the ordered pairs in the relation determined by the Hasse diagrams shown in following figures on the set A = {*a*, *b*, *c*, *d*}.

(*a*)

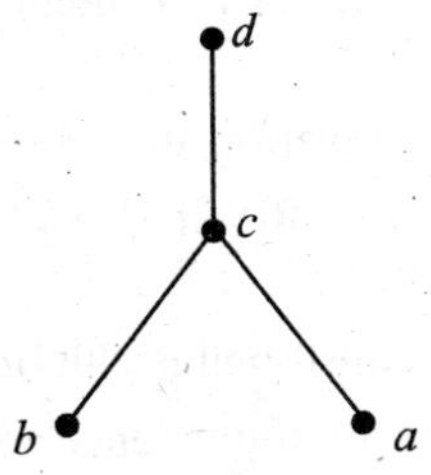

(*b*)

23. Determine the matrix of the partial order relation whose Hasse diagrams are given below:

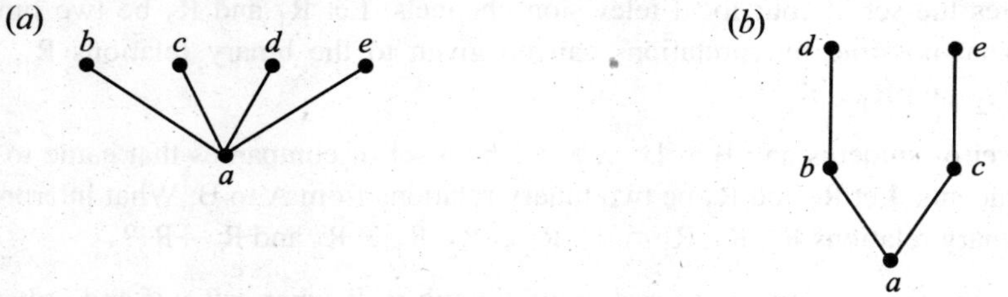

24. Given A = {x, y, z, w}. Let R be the relation on A and is defined as

$$R = \{(x, x), (y, y), (y, z), (z, y), (w, y), (w, w)\}$$

(*a*) Draw its digraph. (*b*) Is R equivalence relation?

(*c*) Is R anti-symmetric? (*d*) Find $R^2 = R \text{ o } R$.

25. Let A = {a, b, c} be a set and the relation R on A is defined as:

$$R = \{(a, a), (a, b), (b, c), (c, c)\}$$

Is R (*a*) reflexive, (*b*) symmetric, and (*c*) transitive?

26. Prove that if R is an equivalence relation on a set A, then R^{-1} is also an equivalence relation on A.

27. Let A = {1, 2, 3, . . ., 14, 15} and let R be the equivalence relation on A defined as: $(x - y)$ is divisible by 5, i.e. $x \equiv y \pmod 5$. Find the partition of A induced by R, i.e. the quotient set A/R.

28. Let A = {1, 2, 3, . . ., 11, 12} and let R be the relation on A × A defined by (a, b) R (c, d) if and only if $a + d = b + c$.

(*a*) Prove that R is an equivalence relation.

(*b*) Find the equivalence class of (2, 5).

29. Show that the union of two equivalence relations is not necessarily an equivalence relation.

30. Show that the relation of 'similarity' on the set of all triangles in a plan is an equivalence relation.

31. Which of the following relations in the set of real numbers are equivalence relations:

(*i*) $a \text{ R } b$ if $|a| = |b|$ (*ii*) $a \text{ R } b$ if $|a| \geq |b|$ (*iii*) $a \text{ R } b$ if $a - b \geq 0$.

32. Let α be a mapping of a set S onto a set T. Then if we define $a \text{ R } b$ for $a \in$ S and $b \in$ S, provided $a\alpha = b\alpha$, prove that R is an equivalence relation.

33. Show that the identity relation on a set S is an equivalence relation.

34. If R is a relation in the set of natural numbers N defined by the open sentence '$x - y$' is divisible by 5, that is, R = {$(x, y) : x \in$ N, $y \in$ N, $(x - y)$ is divisible by 5}, prove that R is an equivalence relation.

35. Show that the relation, R = {$(a, b) : a - b$ = even integer and $a, b \in$ I} in the set I of integers is an equivalence relation.

36. If R is an equivalence relation, then prove that R^{-1} is also an equivalence relation in the set A.

37. Let I be the set of all integers excluding zero and R be a relation defined by $x \text{ R } y$ if $x^y = y^x$, where $x, y \in$ I. Is the relation R an equivalence relation?

38. If R_1 and R_2 are equivalence relations in a set A, show that $R_1 \cap R_2$ is also an equivalence realtion in A.

39. Prove that in the set of natural numbers, the relation 'x is multiple of y' is an equivalence relation.

40. Define an equivalence class, and prove that two equivalence classes are either identical or disjoint.

Hints and Answers

12. The number of elements in A × B are 3 (2) = 6. Hence, there are $2^6 = 64$ subsets of A × B, and so there are 64 relations from A to B.

17. (*a*) $B = \begin{array}{c} \\ 1 \\ 2 \\ 3 \\ 4 \end{array}\!\!\begin{array}{c} \begin{array}{ccc} x & y & z \end{array} \\ \begin{bmatrix} 1 & 0 & 0 \\ 0 & 0 & 1 \\ 1 & 1 & 1 \\ 0 & 0 & 0 \end{bmatrix} \end{array}$

Here it may be observed that entry in the matrix corresponding to the elements of A and B is 1 if they are related, 0 otherwise.

(*b*)

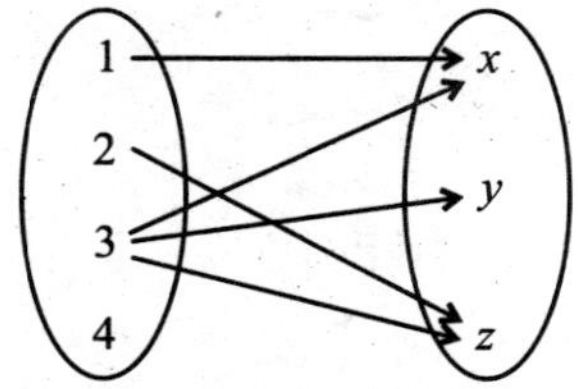

Here arrow sign indicates that $(a, b) \in R$, for all $a \in A$ and $b \in B$.

(*c*) Reverse the ordered pairs of R to obtain $R^{-1} = \{(x, 1), (z, 2), (x, 3), (y, 3), (z, 3)\}$.

(*d*) Domain (R) = {1, 2, 3} and Range (R) = $\{x, y, z\}$.

18. (*a*)

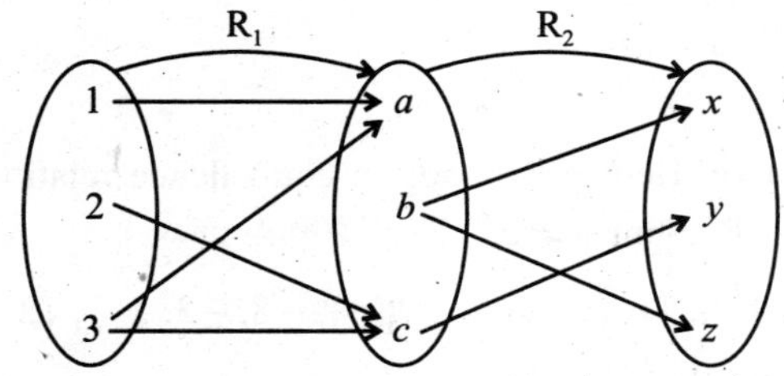

In this diagram, elements 2 and 3 of A are 'connected' to element y of C by the path $2 \to c \to y$ and $3 \to c \to y$. Thus $(2, y), (3, y) \in R_1 \circ R_2$.

(*b*) $M_{R_1} = \begin{array}{c} \\ 1 \\ 2 \\ 3 \end{array}\!\!\begin{array}{c} \begin{array}{ccc} a & b & c \end{array} \\ \begin{bmatrix} 1 & 0 & 0 \\ 0 & 0 & 1 \\ 1 & 0 & 1 \end{bmatrix} \end{array}$; $M_{R_2} = \begin{array}{c} \\ a \\ b \\ c \end{array}\!\!\begin{array}{c} \begin{array}{ccc} x & y & z \end{array} \\ \begin{bmatrix} 0 & 0 & 0 \\ 1 & 0 & 1 \\ 0 & 1 & 0 \end{bmatrix} \end{array}$; $M_{R_1R_2} = \begin{array}{c} \\ 1 \\ 2 \\ 3 \end{array}\!\!\begin{array}{c} \begin{array}{ccc} x & y & z \end{array} \\ \begin{bmatrix} 0 & 0 & 0 \\ 0 & 1 & 0 \\ 0 & 1 & 0 \end{bmatrix} \end{array}$

It may also be observed that $M_{R_1R_2} = M_{R_1} M_{R_2}$.

19. (*a*) No (*b*) Yes (*c*) Yes

20.

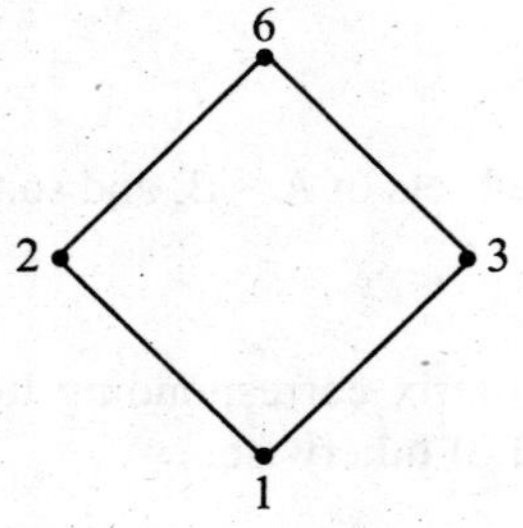

21.

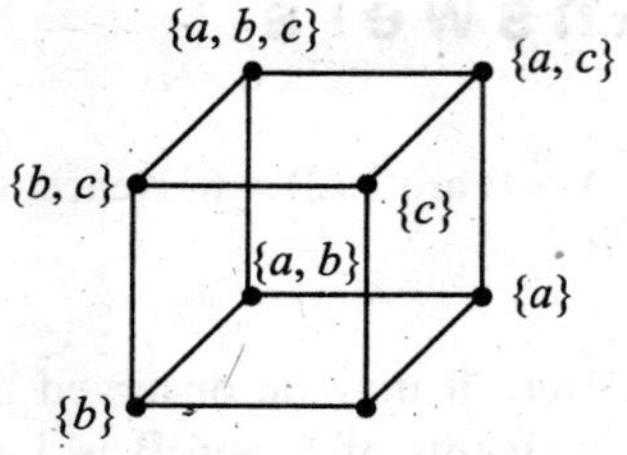

22. (*a*) $\{(a, a), (b, b), (c, c), (d, d), (a, c), (a, d), (b, c), (b, d), (c, d)\}$

(*b*) $\{(a, a), (b, b), (c, c), (d, d), (a, b), (a, c), (a, d), (b, c), (b, d), (c, d)\}$

23. (*a*)
$$\begin{array}{c} \\ a \\ b \\ c \\ d \\ e \end{array}\begin{array}{c} \begin{array}{ccccc} a & b & c & d & e \end{array} \\ \begin{bmatrix} 1 & 1 & 1 & 1 & 1 \\ 0 & 1 & 0 & 0 & 0 \\ 0 & 0 & 1 & 0 & 0 \\ 0 & 0 & 0 & 1 & 0 \\ 0 & 0 & 0 & 0 & 1 \end{bmatrix} \end{array}$$
(*b*) Same as (*a*)

24. (*a*)

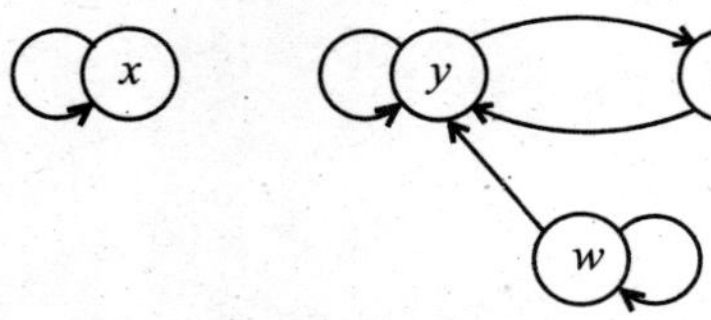

(*b*) R is reflexive and symmetric but not transitive. Hence, R is not an equivalence relation.

(*c*) R is not antisymmetric because y R z and z R y but $y \neq z$.

27. $A_0 = \{-15, -5, 0, 5, 15\}$; $A_1 = \{\ldots, -10, -5, 1, 6, 11, \ldots\}$; $A_2 = \{\ldots, -8, -3, 2, 7, 12 \ldots\}$
$A_3 = \{\ldots, -7, -2, 3, 8, 13, \ldots\}$, $A_4 = \{\ldots, -6, -1, 4, 9, 14, \ldots\}$
The equivalence classes are disjoint and $A = A_0 \cup A_1 \cup A_2 \cup A_3 \cup A_4$.

28. (*b*) $\{(1, 4), (2, 5), (3, 6), (4, 7), (5, 8)\}$

29. This relation is not symmetric.

30. The relation of divisibility in the set of natural numbers is reflexive and transitive but is not symmetric.

Chapter 4

Functions

PREVIEW

This chapter introduces various types of functions such as floor/ceiling function, Fibonacci sequence, Hamming distance function, mod function, etc. with their applications in mathematics and computer science.

LEARNING OBJECTIVES

After studying this chapter, you should be able to

- determine whether a binary relation is a function,
- create composite functions,
- differentiate between a relation and a function; between a function and its value,
- know various functions that are used in computer science.

4.1 INTRODUCTION

In this chapter, we will discuss a special kind of relation called a *function.* The knowledge of functions is very useful both in mathematics and computer science. Functions can also be used for counting and knowing the cardinality of sets. The terms such as *mapping*, *transformation,* etc., are also used for functions to depict a relation between two discrete objects.

4.2 DEFINITION AND NOTATION OF A FUNCTION

Let A and B be any two non-empty sets, not necessarily distinct. If there exists a *rule* or *correspondence*, denoted by f, which relates (or associates) each element $x \in$ A, to a unique (one and only one) element

$y \in B$, then such a rule is called a *function (or mapping)* from A to B. Symbolically, the function f is expressed as:

$$f : A \to B \text{ such that } y = f(x), \text{ for } x \in A \text{ and } y \in B.$$

However, a relation R from A to B may not relate an element of A to an element of B or more than one element of B may be related to an element of A. In other words, the relation f can be described as the set of elements written as:

$$\{(x, f(x)) : x \in A, f(x) \in B\}.$$

4.2.1 Range and Domain of a Function

The set A is called the *domain* of the function f and B is called the *co-domain* of f. The element $x \in A$ is called the argument of function f and $f(x) \in B$ is called the value of the function for x and is also called the *image* or *f-image* of x under f. Figure 4.1 represents the functional relationship between two sets A and B. Figure 4.1 suggests that x may be considered as an *input* that is transformed by f into corresponding *output* $f(x) = y$.

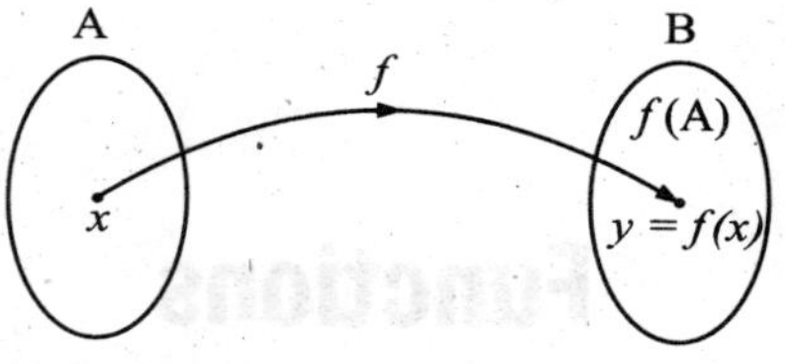

Fig. 4.1 Functional Relationship between Sets A and B

The range of f is the set of those elements in B which appear as the second components in the ordered pair of f, i.e. set of elements of B which appear as the image of at least one element in A. There can be more than one element in A which have the same image in B. The range of f is denoted by

$$f(A) = \{f(x) : x \in A\} = \{y : y \in B, y = f(x) \text{ for some } x \in A\}.$$

The sets A and B are called the *domain* and *co-domain* of f, respectively. Obviously, $f(A) \subseteq B$.

Remark Each element of the first set, A has a unique image in B and each element of B need not appear as the image of an element in A as well as two or more elements of B should not be associated to any one element of A. There can be more than one element of A which have the same image in B.

Illustrations

1. Let A be the set of books in a library and N be the set of natural numbers. Let f be the rule which assigns to each book the number of pages contained in it. Here some natural number will be assigned to each book and no book can be assigned to more than one natural number. Therefore f is a function of A to N.
2. Let A = {1, 2, 3, 4} and B = {1, 2, 3}. Then $f = \{(1, 2), (2, 3), (3, 3), (4, 2)\}$ is a function from A to B. Since to each element in A we have assigned a unique element in B, therefore $f(1) = 2$, $f(2) = 3$, $f(3) = 3$, $f(4) = 2$. Also the range of the function is $f(A) = \{2, 3\}$.
3. Let R be the set of real numbers. Then the rule

$$f : R \to R : f(x) = x^2 \text{ for all } x \in R$$

is a function from R to R, because square of a real number is unique. For example, $f(1) = 1$, $f(-1) = 1$, $f(2) = 4$, $f(-2) = 4$ and so on. Thus domain of f is R = {– 1, 1, 2, –2} and range of $f(R) = \{1, 4\}$. However, the rule defined as

$$f : R_+ \to R : f(x) = \sqrt{x} \text{ for all } x \in R_+$$

is not a function, where R_+ is the set of all real numbers. In this case for a single element, say $1 \in R_+$, there are two elements, –1 and 1 of R associated with it.

Similarly, a function $f : R \rightarrow R$ of the form

$$f(x) = a_0 x^n + a_1 x^{n-1} + \ldots + a_n$$

where a_i are real number, is a polynomial function.

4. Let A = {1, 2, 3} and B = {x, y, z}. Consider the relations

$$R_1 = \{(1, x), (2, x)\} \text{ and } R_2 = \{(1, x), (1, y), (2, z), (3, y)\}$$

The relation R_2 is not a function because $R_2(1) = \{x, y\}$. But relation R_1 is a function with Domain = {1, 2} and Range = {x}.

5. Let I be the set of integers and A = {0, 1}. The relation between I and A defined as $f : I \rightarrow A$ such that

$$f(x) = \begin{cases} 0 & \text{if } x \text{ is even} \\ 1 & \text{if } x \text{ is odd} \end{cases}$$

is a function because each set $f(x)$ consists of a single element.

4.2.2 Function as Sets of Ordered Pairs

A function f from a non-empty A to B is also defined as the set of ordered pairs such that $f \subseteq A \times B$ satisfying the following conditions:

(*i*) For each $x \in A$, there is a unique element $y \in B$ such that an ordered pair $(x, y) \in f$.

(*ii*) If $(x_1, y_1) \in f$ and $(x_1, y_2) \in f$ implies that $y_1 = y_2$.

The first condition ensures that to each $x \in A$, f associates an element $y \in B$ and the second condition guarantees that y is unique. Thus, the function f is also defined as:

$$f = \{(x, y) : x \in A, y \in B, (x, y) \in f\}$$

4.2.3 Difference between Relation and Function

Suppose A and B are two sets. Let f be a function from A to B. Then by the definition of function, f is the subset of A × B in which each $x \in A$ appears in one and only one ordered pair.

On the other hand, every subset of A × B is a relation from A to B. Thus, *every function is a relation but every relation is not a function.* If R is a relation from A to B, the domain of R may be a subset of A. But if f is a function from A to B, then the domain of f is equal to A. In a relation from A to B an element of A may be related to more than one element in B. Also there may be some elements in A which may not be related to any element in B. But, in a function from A to B each element of A must be associated to one and only one element of B.

In other words, if f is the subset of A × B, then

(a) There cannot be two or more distinct second coordinates corresponding to any first coordinate in the set of ordered pairs A × B. For example, (x, y_1) and (x, y_2) cannot be the elements of A × B. But in relation an element in A can have more than one f-image in B.

(b) Every element in A must have an image in B. This condition is not necessary for a relation. Thus every function is a relation but every relation is not a function.

(c) If R is a relation from A to B, then domain of R may be subset of A. But if f is a function from A to B, then domain of f is equal to A.

Illustration Let $A = \{a, b, c, d\}$, $B = \{1, 2, 3\}$ and $R = \{(a, 1), (b, 1), (c, 2), (d, 2)\}$. Then R is a function from A to B. Clearly, R is also a relation from A to B. But, consider the subset C of $A \times B$ given by

$$C = \{(a, 1), (b, 2), (a, 3), (c, 1), (d, 2)\}.$$

Here C is a relation from A to B. But C is not a function from A to B. The obvious reason is that the element $a \in A$ is associated to two different elements $1, 3 \in B$.

4.2.4 Difference between a Function and its Value

If f is the set of all ordered pairs of elements of A and B such that image of first element is the second element, i.e. $(x, y) \in f$, then y is called the value of f at the element x and is denoted by $f(x)$. Hence, f is an element of $A \times B$ and $f(x)$ is an element of B. For example,

$$f = \{x, y : y = 2x^2 + 4x + 5\}$$

defines the function. The numbers x and y are called variables. The domain of the function is the set of all possible values of x and the range is the set of all values of y. In this case the domain is the set of all real numbers which can be denoted with interval notation as $(-\infty, \infty)$, but the range of f is set of all positive numbers greater than or equal to 3, i.e. $(3, \infty)$ because the lowest value that y can assume is 3 for $x = -1$.

Illustration If $A = \{1, 2, 3\}$ and $B = \{2, 3, 4\}$, then the relation f defined by $f(1) = 2$, $f(2) = 3$, $f(3) = 4$ is a function from set A to B. We may also write $f = \{(1, 2), (2, 3), (3, 4)\}$.

Example 1 If $A = \{1, 2, 3\}$ and $B = \{2, 3, 4\}$ are two sets, then show that $\{(1, 2), (1, 3), (3, 4)\}$ is not a function from A to B.

Solution The relation, $f = \{(1, 2), (1, 3), (3, 4)\}$ implies that $f(1) = 2$, $f(1) = 3$ and $f(3) = 4$. Clearly f is not a function for two reasons:

(*i*) The element 2 of A has no image in B.

(*ii*) The element 1 of A has two images, viz., 2 and 3 in B.

4.3 TYPES OF FUNCTIONS

Equal functions Two functions $f : A \to B$ and $g : A \to B$ are said to be equal if and only if $f(x) = g(x)$ for every $x \in A$ and are written as $f = g$.

Constant function A function $f : A \to B$ is called a constant function, if some element $y \in B$, is assigned to every element of A, i.e. $f(x) = y$ for every $x \in A$. In other words, $f : A \to B$ is a constant function if the range of f consists of only one element.

Onto and Into functions The function $f : A \to B$ is said to be an

(*a*) *Onto (or surjective) function,* if every element of B is f-image of at least one element in A. In this case, the range of f is equal to B, that is, for all $y \in B$, $y = f(x)$ for some $x \in A$ or $f(A) = B$.

(*b*) ***Into function,*** if there is at least one element in B which is not the f-image of any element in A. But every element of A has f-image in B. In this case the range of f is proper subset of B, that is, $f(A) \subseteq B$ and $f(A) \neq B$.

Many-one and one-to-one functions The function $f: A \to B$ is said to be a

(*a*) ***Many-one function,*** if two or more different elements in A have same f-image in B, i.e. $f(x_1) = f(x_2)$ even if $x_1 \neq x_2$, for all $x_1, x_2 \in A$.

(*b*) ***One-to-one (or injective) function,*** if different elements in A have different f-image in B, i.e. $x_1 \neq x_2$ implies $f(x_1) \neq f(x_2)$ for all $x_1, x_2 \in A$ or equivalently, $f(x_1) = f(x_2) \Rightarrow x_1 = x_2$ for all $x_1, x_2 \in A$.

If f is one-to-one with A and B finite, then we must have $|A| \leq |B|$.

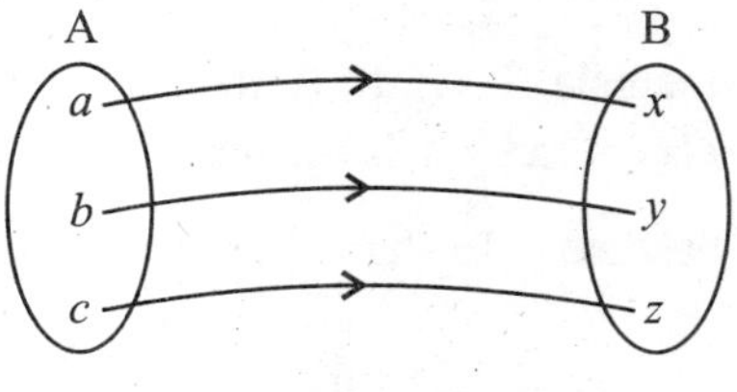

Fig. 4.2 One-One Onto Function

One-one onto (bijective) function The function $f: A \to B$ is said to be *one-one onto function* if to each element of A there corresponds one and only one element of B and every element of B have one and only image in A. One-one and onto function are also called *bijective* mapping as shown in Fig. 4.2.

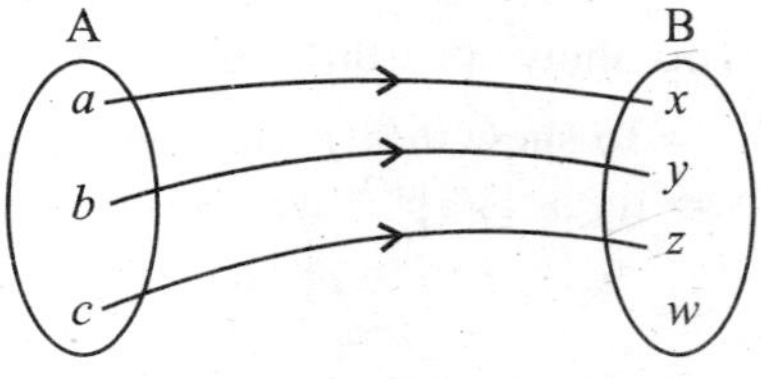

Fig. 4.3 One-One into Function

One-one into function The function $f: A \to B$ is said to be *one-one into* function if to each element of A there corresponds one element of B, but there are some element of B which do not correspond to any of the elements of A as shown in Fig. 4.3.

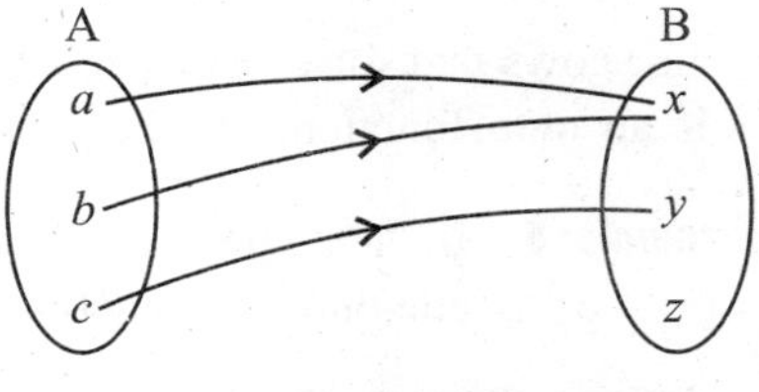

Fig. 4.4 Many-One into Function

Many-one into function The function, $f: A \to B$ is said to be *many-one into* function if two or more elements of a set A corresponds to the same element of B and there are some elements in B which do not correspond to any of the elements of A as shown in Fig. 4.4.

Many-one onto function The function, $f: A \to B$ is said to be *many-one onto* function if each element in B is joined to at least one element in A, and two or more elements in A are joined to the same element in B as shown in Fig. 4.5.

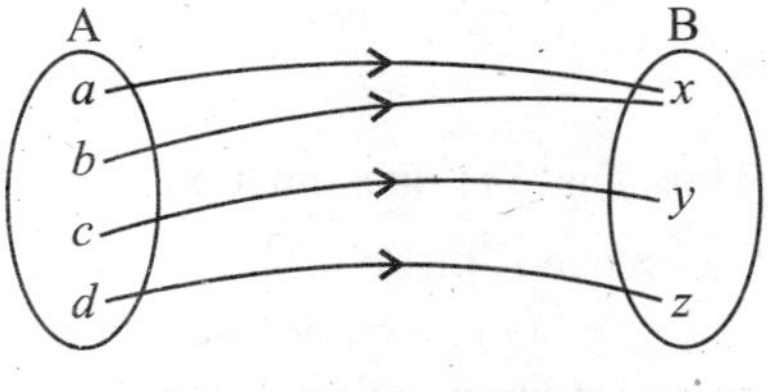

Fig. 4.5 Many-One Onto Function

Illustrations Let A be the set of students sitting on chairs in a classroom and let B be the set of chairs in the classroom. Let f be the correspondence which associates to each student the chair on which he sits.

Since every student has some chair to sit and no student can sit on two or more than two chairs, therefore for the function defined as $f: A \to B$, the following cases may arise:

1. If every student gets a separate chair and no chair is left vacant, then this is a case of a *one-one onto* function.
2. If every student gets a separate chair and still some chairs are left vacant, it is a *one-one into* function.
3. If every student does not sit on a separate chair, i.e. more than one student sit on a chair, no chair is left vacant, it is a *many-one onto* function.

Remarks To check a function to be one-one onto we must verify the following:

1. Let $f: A \to B$ such that $f(x) = y$, then for function f to be one-one

   either $f(x_1) = f(x_2) \Rightarrow x_1 = x_2$, for all $x_1, x_2 \in A$

   or $x_1 \neq x_2 \Rightarrow f(x_1) \neq f(x_2)$, for all $x_1, x_2 \in A$.

2. Function, f will be onto if for each $y \in B$ there exists an element $x \in A$ such that $y = f(x)$.
3. Let A = {1, 2, 3} and B = {1, 2, 3, 4}. The function g = {(1, 1), (2, 3), (3, 3) is not one-to-one because $g(2) = g(3)$ but $2 \neq 3$, whereas the function f = {(1, 1), (2, 3), (3, 4)} is one-to-one function from A to B.

Example 2 If R is the set of real numbers, then show that the function, $f: R \to R$ defined by $f(x) = 5x^3 - 1$ is one-one onto function.

Solution To show f to be a one-one function, let $x_1, x_2 \in R$ (domain set) be two different elements. Then

$$x_1 \neq x_2 \Rightarrow 5x_1^3 - 1 \neq 5x_2^3 - 1$$
$$\Rightarrow f(x_1) \neq f(x_2).$$

This shows that different elements in R have different f-images in R. Hence f is one-one.

To show that f is an onto function, let $y = 5x^3 - 1$ be any arbitrary element in R. Then we may write $x = \{(y + 1)/5\}^{1/3}$, which is a real number for all values of $y \in R$. Now

$$f[\{(y+1)/5\}^{1/3}] = 5[\{(y+1)/5\}^{1/3}]^3 - 1 \qquad [\text{since } f(x) = 5x^3 - 1]$$
$$= 5\{(y+1)/5\} - 1$$
$$= (y + 1) - 1 = y$$

This shows that for every element $y \in R$ (range set) there exists per-image in the domain set R of f. Thus f is an onto function, and hence f is one-one onto function.

Example 3 If N is the set of natural numbers, then show that the function, $f: N \to N$ defined by $f(x) = x^2$ is one-one into function.

Solution To show f to be a one-one function, let $x_1, x_2 \in N$ (domain set) be two different elements. Then

$$x_1 \neq x_2 \Rightarrow x_1^2 \neq x_2^2$$
$$\Rightarrow f(x_1) \neq f(x_2)$$

Thus f is one-one function.

Since f-image of every element $x \in N$ (domain set) is its square, i.e. x^2, therefore, we have $f(2) = 4$, $f(3) = 9$, and so on. However, there are natural numbers in the range set N such as 1, 3, 5, 6, 7, etc., which are not f-image of any natural number in the domain set N of f. Thus, f is an into function and hence f is one-one into function.

Example 4 If R is a set of real numbers, then show that the function, $f: R \to R$ defined by $f(x) = -\sin x$, is neither one-one nor onto.

Solution Let $\pi/6$ and $5\pi/6$ be two different elements in domain set R of function f. Now

$$f(\pi/6) = -\sin(\pi/6) = -1/2 \text{ and } f(5\pi/6) = -\sin(5\pi/6) = -1/2.$$

This implies that $f(x_1) = f(x_2)$ even if $x_1 \neq x_2$, then f is not one-one function.

Since value of $f(x) = -\sin x$ oscillates between -1 and 1, i.e. $-1 \leq -\sin x \leq 1$, therefore, range set of f is $\{-1, 1\}$, which is the subset of the set R. Hence, f is not onto.

Example 5 List all possible function X = $\{a, b, c\}$ to Y = $\{0, 1\}$ and indicate in each case whether the function is one to one, is onto and is one-one onto. *[UPTU, BTech, 2007]*

Solution Given X = $\{a, b, c\}$ and Y = $\{0, 1\}$. All possible functions from X to Y are

$$f_1 = \{(a, 0)\ (b, 0)\ (c, 0)\} \quad f_2 = \{(a, 1)\ (b, 1)\ (c, 1)\}$$
$$f_3 = \{(a, 0)\ (b, 0)\ (c, 1)\} \quad f_4 = \{(a, 1)\ (b, 1)\ (c, 0)\}$$
$$f_5 = \{(a, 0)\ (b, 1)\ (c, 1)\} \quad f_6 = \{(a, 1)\ (b, 0)\ (c, 0)\}$$
$$f_7 = \{(a, 0)\ (b, 1)\ (c, 0)\} \quad f_8 = \{(a, 1)\ (b, 0)\ (c, 1)\}$$

The total numbers of function from X $\to$ Y are $2^3 = 8$.

Functions f_3, f_4, f_5, f_7, f_8, are onto functions. None of these function is one-one and one-one-onto.

Example 6 Let N be the set of natural numbers including zero. Determine which of the following functions are one-to-one, which are onto and which are one-to-one onto.

(*a*) $f: \text{N} \to \text{N}$, where $f(j) = j^2 + 2$

(*b*) $f: \text{N} \to \text{N}$, where $f(j) = \begin{cases} 1, & \text{if } j \text{ is odd} \\ 0, & \text{if } j \text{ is even} \end{cases}$ *[UPTU, BTech, 2007]*

Solution (*a*) For each value of $j \in$ N its image $f(j) = j^2 + 2 \in$ N. If $j = a_1, a_2 \in$ N, then $f(a_1) = a_1^2 + 2$ and $f(a_2) = a_2^2 + 2$ as $a_1 \neq a_2$. Thus $a_1^2 + 2 \neq a_2^2 + 2$, i.e., $f(a_1) \neq f(a_2)$, so f is one-one mapping.

There exists values of $f(j) = 0$, 1 for which the pre-images $\sqrt{-2}$ and $\sqrt{-1}$ do not belong to N

(*b*) All odd values of $j \in N$ have the same image (= 1) $\in$ N so f is not one-one mapping.

Example 7 Let $f: \text{R} \to \text{R}$ and $g: \text{R} \to \text{R}$, where R is the set of real numbers. Find *fog* and *gof*, where $f(x) = x^2$ and $g(x) = x + 4$. State whether these functions are injective, surjective and bijective. *[UPTU, BTech, 2006]*

Solution We know that $fog(x) = f(x + 4) = (x + 4)^2$, and $gof(x) = g(x^2) = x^2 + 4$.

Function $f(x) = x^2$ is not one-one (injective) as $f(-1) = f(+1) = 1$. It is also onto (surjective) because corresponding to negative values of $f(x)$, the value of x shall be imaginary, which does not belong to R.

Function $g(x) = x + 4$ is one-one (injective). It is also onto (surjective). Hence, it is bijective (one-one onto mapping).

Function $fog(x) = (x + 4)^2$ is not one-one (injective) because for $x = \pm i \in$ R, the value of $fog(x)$ will be the same. It is also not onto (surjective) because corresponding to negative value of $fog(x)$ (which belongs to R), there will not be a value of $x \in$ R.

Function $gof(x) = x^2 + 4$ is not one-one (injective) because for $x = \pm i$, the value of $gof(x)$ will be the same. It is not one-one (surjective) because corresponding to negative values of $gof(x)$, the value of x shall be imaginary (which does not belong to R).

4.4 INVERTIBLE FUNCTIONS

The domain and range of any function may be interchanged to form a new function. It is obtained by interchanging the position of elements in the ordered pair of original function.

Let $f : A \to B$ be an one-one onto function, and let $y \in B$ be an arbitrary element. Then the inverse relation

$$f^{-1} : B \to A$$

is called the *inverse function* of f if and only if the original function f is such that to each element of its range there exists one and only one element in its domain. That is, to each element $y \in B$, a unique element $x \in A$ exists, such that $f(x) = y$.

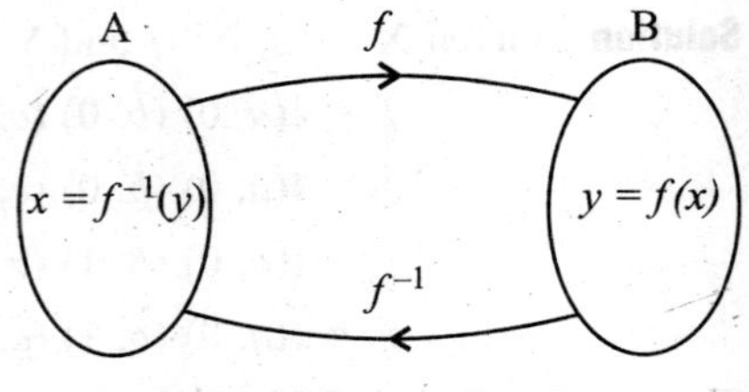

Fig. 4.6 Inverse Function

If f^{-1} is the inverse of f, then $f[f^{-1}(x)] = x$ for all x in the domain of f^{-1} and $f^{-1}[f(x)] = x$ for all x in the domain of f. In other words $f(x) = y \Leftrightarrow x = f^{-1}(y)$.

A function and its inverse can be represented diagrammatically in terms of sets as shown in Fig. 4.6.

Illustrations

1. Let $f : A \to B$ such that y or $f(x) = x + 1$, for all $x \in A$ and $y \in B$ be a function.
   Since f is everywhere defined as well as one-one onto, therefore f is invertible. Thus f^{-1} is also one-one onto function from set B to A.
2. Let $f : R \to R$ such that y or $f(x) = x^2$, for all $x \in A$ and $y \in B$ be a function.
   Since R is the set of real numbers, where $f(-2) = f(2) = 4$, therefore f is not one-one. Hence f is not invertible.
3. Let $f : A \to B$ be a function defined everywhere. Let A and B be two finite sets containing equal number of elements. In such a case
   (*a*) If f is one-one, then f is onto.
   (*b*) If f is into, then f is one-one.

Moreover, when A = B we need to prove only that a function is one-one onto to show that it is bijection.

Remarks

1. A function which possesses an inverse is said to be *invertible.*
2. Only one-one and onto functions are invertible.
3. The inverse function of the given function f is unique.
4. Inverse function f^{-1} is onto if and only if f is everywhere defined.

Example 8 Let A and B be two non-empty sets. If f is one-one onto function from A to B, then prove that $f^{-1} : B \to A$ is also one-one onto.

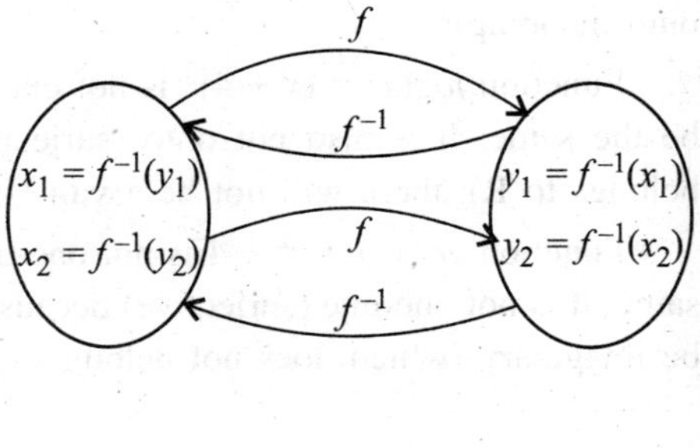

Solution To prove f^{-1} to be one-one, let y_1, y_2 be any two elements of B. Since f is already one-one function, therefore there exist unique elements x_1 and x_2 in A such that $y_1 = f(x_1)$ and $y_2 = f(x_2)$.

Suppose $x_1 = f^{-1}(y_1)$ and $x_2 = f^{-1}(y_2)$. If f^{-1} is one-one, then by definition, we have

$$f^{-1}(y_1) = f^{-1}(y_2) \Rightarrow x_1 = x_2$$

$$\Rightarrow f(x_1) = f(x_2) \qquad \text{[since } f \text{ is one-one]}$$

$$\Rightarrow y_1 = y_2$$

Therefore f^{-1} is one-one.

To prove f^{-1} is also onto, let x be any arbitrary element of A. Since f is a function from A to B, therefore there exits an unique element $y \in$ B such that $f(x) = y$ or $x = f^{-1}(y)$. This mean that f^{-1} image of the element $y \in$ B is $x \in$ A. Hence, the function f^{-1} is also onto.

Example 9 Let A and B be two non-empty sets. If f is one-one onto function from A to B, then prove that inverse function $f^{-1} : B \to A$ is unique.

Solution Let $g : B \to A$ and $h : B \to A$ be two inverse functions of f. Then we need to prove that $g = h$. Let y be any arbitrary element in B, and $g(y) = x_1$ and $h(y) = x_2$.

Since g is the inverse function of f, therefore $g(y) = x_1 \Rightarrow f(x_1) = y$. Also h is the inverse function of f, therefore $h(y) = x_2 \Rightarrow f(x_2) = y$.

Since f is one-one function, therefore

$$f(x_1) = y \text{ and } f(x_2) = y \Rightarrow x_1 = x_2$$

$$\Rightarrow g(y) = h(y)$$

Hence, $g = h$.

Example 10 If R is the set of all real numbers, then discuss the type of function defined by

$$f : R \to R \text{ such that } f(x) = x^2 \text{ for all } x \in R.$$

What is the domain and range of this function?

Solution Since square of any real number is also real and unique, therefore each element $x \in$ R has a unique f image in R.

Since there is no real number whose square is negative, therefore any negative number in R is not the f-image of any element in R. That is, for $y = -1$ in R, $\sqrt{-1}$ is not a real number. Thus $-1 \in$ R has no pre-image in R and so f is into function.

Let 2 and –2 be two different real numbers such that $(2)^2 = 4$ and $(-2)^2 = 4$ or $f(2) = 4$ and $f(-2) = 4$. Thus, two different real numbers have the same f-image in R. Hence, f is a many-one into function.

Example 11 Show that the function f defined by

$$f : Q \to Q \text{ such that } f(x) = 3x + 4, \text{ for all } x \in Q$$

is one-one onto, where Q is the set of all rational numbers. Also find a formula that defines the inverse function f^{-1}.

Solution Let $x_1, x_2 \in$ Q be two different elements. Then,

(*a*) $$x_1 \neq x_2 \Rightarrow 3x_1 + 4 \neq 3x_2 + 4$$

$$\Rightarrow f(x_1) \neq f(x_2)$$

This shows that different elements in Q have different f-images in Q. Hence, f is one-one.

(*b*) Let $y = 3x + 4$ be any arbitrary element in Q. Then we may write $x = (y - 4)/3$ which is a rational number for all values of $y \in$ Q. Thus, the element $(y - 4)/3$ in Q is the f-image of the element $(y - 4)/3$ in Q. Hence, f is onto.

(*c*) Since $f : Q \to Q$ is one-one onto function, therefore f has an inverse function $f^{-1} : Q \to Q$.

Let $y = 2x + 4$ be the image of x under function f. Then $x = (y - 4)/2$ will be the image of y under inverse function f^{-1}, i.e. $x = f^{-1}(y)$. Thus the formula defining the inverse function as:

$$f^{-1} : Q \to Q \text{ such that } f^{-1}(y) = (y - 4)/3 \text{ for all } y \in Q.$$

Example 12 Let R_0 be the set of all non-zero real numbers. Show that if $f : R_0 \to R_0$ such that $f(x) = 1/x$, for all $x \in R_0$, is one-one onto function.

Solution Since each non-zero real number has a unique reciprocal number, therefore for $x_1, x_2 \in R_0$ we have

$$x_1 \neq x_2 \Rightarrow \frac{1}{x_1} \neq \frac{1}{x_2}$$
$$\Rightarrow f(x_1) \neq f(x_2).$$

Thus different non-zero real numbers have different f-images in R_0. Hence, f is one-one.

If $y \in R_0$ is any arbitrary element, then $y = 1/x$ or $x = 1/y$, which is a non-zero real number. Thus $f(1/y) = y$, i.e. each element $y \in R_0$ has its f-image $f(1/y)$ in R_0. Hence, f is one-one onto.

Example 13 Let $A = \{x : x \in R, \text{ and } -\frac{\pi}{2} \leq x \leq \frac{\pi}{2}\}$, and

$B = \{y : y \in R, \text{ and } -1 \leq y \leq 1\}$.

Show that the function

$$f : A \to B \text{ such that } f(x) = \sin x, \text{ for all } x \in A$$

is one-one onto. Also find the inverse function, f^{-1}.

Solution Let $x_1, x_2 \in R$ be any two real numbers in the closed interval $[-\pi/2, \pi/2]$. Therefore,

$$x_1 \neq x_2 \Rightarrow \sin x_1 \neq \sin x_2$$
$$\Rightarrow f(x_1) \neq f(x_2).$$

Thus f is one-one.

If $y \in R$ is any real number in the closed interval $[-1, 1]$, then there exists a real number $x \in A$ in the closed interval $[-\pi/2, \pi/2]$ such that $\sin x = y$, i.e. $f(x) = y$. Thus each element $y \in B$ is the f-image of an element $x \in A$. Hence, f is onto.

Since f is one-one onto, therefore it is invertible. That is,

$$f(x) \text{ or } y = \sin x \Leftrightarrow x = \sin^{-1} y$$
$$\Leftrightarrow f^{-1}(y) = \sin^{-1} y \qquad [\text{since } f(x) = y \Leftrightarrow x = f^{-1}(y)]$$

Thus, inverse function is defined as:

$$f^{-1} : B \to A \text{ such that } f^{-1}(y) = \sin^{-1} y, \text{ for all } y \in B.$$

Example 14 If $f : X \to Y$ and A, B are two subsets of Y, then prove that

(*a*) $f^{-1}(A \cup B) = f^{-1}(A) \cup f^{-1}(B)$

(*b*) $f^{-1}(A \cap B) = f^{-1}(A) \cap f^{-1}(B)$

Solution (*a*) Let x be an arbitrary element of $f^{-1}(A \cup B)$. Then

$$\begin{aligned} x \in f^{-1}(A \cup B) &\Leftrightarrow f(x) \in A \cup B \\ &\Leftrightarrow f(x) \in A \text{ or } f(x) \in B \\ &\Leftrightarrow x \in f^{-1}(A) \text{ or } x \in f^{-1}(B) \\ &\Leftrightarrow x \in [f^{-1}(A) \cup f^{-1}(B)] \end{aligned}$$

Since each element of $f^{-1}(A \cup B)$ is also the element of $f^{-1}(A) \cup f^{-1}(B)$, therefore result is proved.

(*b*) Let x be an arbitrary element of $f^{-1}(A \cap B)$. Then

$$\begin{aligned} x \in f^{-1}(A \cap B) &\Leftrightarrow f(x) \in A \cap B \\ &\Leftrightarrow f(x) \in A \text{ and } f(x) \in B \\ &\Leftrightarrow x \in f^{-1}(A) \text{ and } x \in f^{-1}(B) \\ &\Leftrightarrow x \in [f^{-1}(A) \cap f^{-1}(B)]. \end{aligned}$$

Since both $f^{-1}(A \cap B)$ and $f^{-1}(A) \cap f^{-1}(B)$ have same common elements, therefore result is proved.

Example 15 Let R and C be the set of all real and complex numbers, respectively. Prove that the function defined as:

$$f : C \to R \text{ such that } f(z) = |z|, \text{ for all } z \in C$$

is neither one-one nor onto.

Solution Let $z = x + iy$ be any complex number, where $x, y \in R$ and $i = \sqrt{(-1)}$. Then modulus of z is $|z| = \sqrt{x^2 + y^2}$. Also $|z|$ is always a non-negative real number, i.e. $|z| \geq 0$.

Let m be any negative real number. Obviously we cannot find a complex number z such that $|z| = m$ (because $|z| \geq 0$). Thus no negative number like m can be f-image of any complex number. Hence, f is not onto.

Let $z_1 = (2 + 3i)$ and $z_2 = (2 - 3i)$ be two complex numbers. Then $|z_1| = |z_2| = \sqrt{13}$. Although $z_1 \neq z_2$ but $|z_1| = |z_2|$, i.e. $f(z_1) = f(z_2)$. This means two different complex numbers have the same f-image in R. Hence, f is not one-one but many-one.

Example 16 If S and T are non-empty sets, then prove that there exists a one-to-one correspondence between (S × T) and (T × S).

Solution Consider the function defined as

$$f : (S \times T) \to (T \times S) \text{ such that } f(a, b) = (b, a), \text{ for all } (a, b) \in (S \times T).$$

(*a*) To prove f is one-one, let $(a, b) \in (S \times T)$ and $(c, d) \in (S \times T)$. Then

$$\begin{aligned} f(a, b) = f(c, d) &\Rightarrow (b, a) = (d, c) \quad \text{[by definition of } f\text{]} \\ &\Rightarrow b = d \text{ and } a = c \\ &\Rightarrow (a, b) = (c, d). \end{aligned}$$

Since $f(a, b) = f(c, d) \Rightarrow (a, b) = (c, d)$, therefore f is one-one.

(*b*) To prove f is onto, let $(b, a) \in (T \times S)$. Obviously $b \in T$ and $a \in S$. Thus for $(b, a) \in (T \times S)$, there exists a number $(a, b) \in (S \times T)$ such that $f(b, a) = (a, b)$. Hence, f is onto.

4.5 COMPOSITION OF FUNCTIONS

Let $f : A \to B$ and $g : B \to C$ be two functions. Then the *composition of functions* f and g denoted by $(g\,0\,f)$ is a function from A to C and is written as

$$(g\,0\,f) : A \to C$$

such that $(g\,0\,f)(x) = g(f(x))$ for each $x \in A$.

Obviously domain of $(g\,0\,f)$ is set A and range of $(g\,0\,f)$ is set of C. The composition function $(g\,0\,f)$ of f and g is shown in the Fig. 4.7.

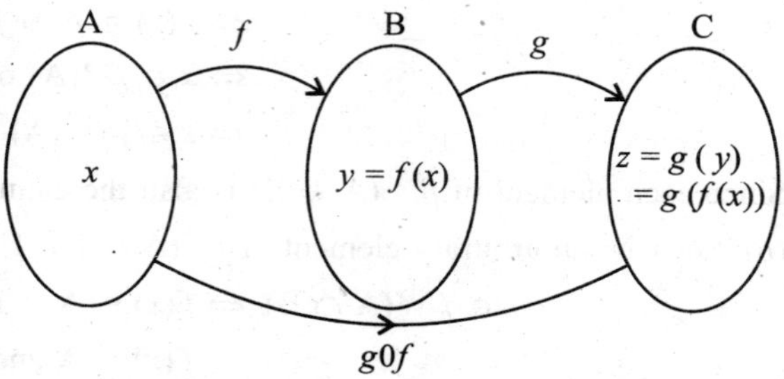

Fig. 4.7 Composition of Functions

Example 17 Show that the composition of functions obeys associative law.

Or

If A, B, C and D are four sets and f, g and h are three functions (or mappings) defined as:

$$f : A \to B \;;\; g : B \to C \text{ and } h : C \to A$$

then prove that $(h\,0\,g)\,0\,f = h\,0\,(g\,0\,f)$.

Solution Let x be an arbitrary element of set A. Then

$$[(h\,0\,g)\,0\,f](x) = (h\,0\,g)\,[f(x)]$$
$$= (h\,0\,g)(y), \text{ because } y = f(x), \text{ for } x \in A, y \in B$$
$$= h[g(y)] = h(z), \qquad [\text{since } z = g(y), \text{ for all } y \in B, z \in C]$$
$$= x \in A.$$

Again, let $[h\,0\,(g\,0\,f)](x) = h[(g\,0\,f)(x)]$

$$= h[g\{f(x)\}] = h[g(y)]$$
$$= h(z) = x \in A.$$

Hence $(h\,0\,g)\,0\,f = h\,0\,(g\,0\,f)$.

Example 18 Let A = {1, 2, 3, 4}, B = {a, b, c, d} and C = {x, y, z}. Consider the function $f : A \to B$ and $g : B \to C$ defined by

$$f = \{(1, a), (2, c), (3, b), (4, a)\}$$

and $$g = \{(a, x), (b, x), (c, y), (d, y)\}.$$

Find the composition function $(g\,0\,f)$.

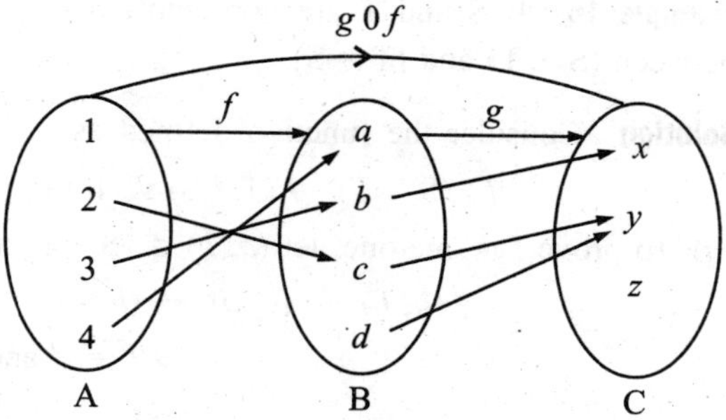

Fig. 4.8

Solution The composition function $(g \, 0 \, f)$ is computed as shown in Fig. 4.8.

$$(g\,0\,f)\,(1) = g\,(f\,(1)) = g\,(a) = x\,; \quad (g\,0\,f)\,(2) = g\,(f\,(2)) = g\,(c) = y$$
$$(g\,0\,f)\,(3) = g\,(f\,(3)) = g\,(b) = x\,; \quad (g\,0\,f)\,(4) = g\,(f\,(4)) = g\,(a) = x$$

Thus, the composition function $(g\,0\,f)$ from the set A to C is given by

$$g\,0\,f = \{(1, x), (2, y), (3, x), (4, x)\}$$

Example 19 Consider the functions defined by

$$f : A \to B \text{ such that } y \text{ or } f(x) = x + 1$$

and $$g : B \to C \text{ such that } z \text{ or } g\,(y) = 2y$$

where A, B and C are the set of even integers. Find $g\,0\,f$.

Solution We have

$$(g\,0\,f)(x) = g\,(f\,(x)) = g\,(x + 1) = 2(x + 1).$$

4.5.1 Important Results on Composition of Functions

1. Composite functions required that the co-domain of f = domain of g. If range of f is the subset of g, it yields the composite function $g\,0\,f : A \to C$.
2. If the function f from A to B is one-one and onto so that the inverse function f^{-1} from B to A exists, then the composition function
$$(f^{-1}\,0\,f) : A \to A$$
is the *identity function* on A, and the composition function
$$(f^{-1}\,0\,f) : B \to B$$
is the identity function on B.
3. Let f be the function from A to B and g be the function from B to A. Then g is the inverse function of f, i.e. $g = f^{-1}$, provided the composition function $(g\,0\,f) : A \to A$ is the identity function on A and $(f\,0\,g) : B \to B$ is the identity function on B.

4.6 IDENTITY FUNCTION

The function $f : A \to A$ defined by $f(x) = x$ for each $x \in A$ is called the identity function of A and is denoted by I_A.

If a function $f : A \to B$ is invertible and a function $g : B \to A$ satisfy conditions $g\,0\,f = I_A$ and $f\,0\,g = I_B$, then function g is unique.

If $f : A \to A$, then $f^1 = f, f^2 = f\,0\,f \ldots f^{n+1} = f\,0\,(f^n)$ for all $n \in I_+$. For example, for A = {1, 2, 3, 4} and $f : A \to A$ defined by f = {(1, 2), (2, 2), (3, 1), (4, 3)}, we have

$$f^2 = f\,0\,f = \{(1, 2), (2, 2), (3, 2), (4, 1)\},$$
$$f^3 = f\,0\,f^2 = f\,0\,f\,0\,f$$
$$= \{(1, 2), (2, 2), (3, 2), (4, 2)\} \text{ and so on.}$$

Theorem 4.1 If $f : A \to B$ and $g : B \to C$ are two one-one onto functions, then prove that

(*a*) $g\,0\,f : A \to C$ is one-one onto, and

(*b*) $g\,0\,f$ is invertible, i.e. $(g\,0\,f)^{-1} = (f^{-1}\,0\,g^{-1}) : C \to A$.

Proof (*a*) Let x_1 and x_2 be any two different elements in the set A. Then

$$x_1 \neq x_2 \Rightarrow f(x_1) \neq f(x_2), \quad \text{[since } f \text{ is one-one]}$$
$$\Rightarrow g[f(x_1)] \neq g[f(x_2)] \quad \text{[since } g \text{ is one-one]}$$
$$\Rightarrow (g\,0\,f)(x_1) \neq (g\,0\,f)(x_2)$$

This shows that the function $(g\,0\,f)$ establishes functional relationship from two different elements of the set A to two different elements of the set C. Thus, $(g\,0\,f)$ is one-one function.

To prove that $g\,0\,f$ is onto function, let z be an arbitrary element of C. Since g is an onto function from set B to set C, therefore there exists an arbitrary element $y \in$ B such that $g(y) = z$.

Also, f is an onto function from set A to B, therefore there exists an arbitrary element $x \in$ A such that $f(x) = y$.

Hence, for every arbitrary element $z \in$ C, there exists an element $x \in$ A such that $z = g(y) = g[f(x)] = (g\,0\,f)(x)$. This implies that every element of C has image under $g\,0\,f$. Therefore, $g\,0\,f$ is an one-one onto function from A to C.

(*b*) Since $g\,0\,f$ is an one-one onto function, therefore it is invertible. To prove it, let $z \in$ C be an arbitrary element such that $z = g(y)$, for all $y \in$ B. As g is a one-one onto function, therefore $g^{-1}(z) = y$.

Also, let $y \in$ B be an arbitrary element, such that $y = f(x)$, for all $x \in$ A. As f is a one-one onto function, therefore $f^{-1}(y) = x$.

Since $g\,0\,f$ is one-one onto function, therefore

$$(g\,0\,f)(x) = z \Rightarrow (g\,0\,f)^{-1}(z) = x$$

Also $$(f^{-1}\,0\,g^{-1})(z) = f^{-1}[g^{-1}(z)]$$
$$= f^{-1}(y) = x.$$

Thus, $(g\,0\,f)^{-1}(z) = (f^{-1}\,0\,g^{-1})(z)$, for all $z \in$ C.

Hence, the functions $(g\,0\,f)^{-1}$ and $f^{-1}\,0\,g^{-1}$ are equal.

Theorem 4.2 Let f : A $\rightarrow$ B be any function. Then product of any function with the identity function is the function itself. That is,

(*a*) $I_B\,0\,f = f$

(*b*) $f\,0\,I_A = f$

Proof (*a*) Let $x \in$ A be any arbitrary element such that $y = f(x)$ for an unique element $y \in$ B. Since

$$f : A \rightarrow B \text{ and } I_B : B \rightarrow B,$$

therefore, $$(I_B\,0\,f) : A \rightarrow B$$

such that $$(I_B\,0\,f)(x) = I_B[f(x)] = I_B(y) = y \quad \text{[since } I_B \text{ is identity function]}$$

Also, since $f(x) = y$, therefore for every $x \in$ A, we have $(I_B\,0\,f)(x) = f(x)$ and hence $I_B\,0\,f = f$.

(*b*) Since I_B : A $\rightarrow$ A and f : A $\rightarrow$ B, therefore

$$(f\,0\,I_A) : A \rightarrow B$$

such that $$(f\,0\,I_A)(x) = f(x), \text{ and hence } f\,0\,I_A = f.$$

Theorem 4.3 If $f : A \to B$ is a one-one and onto function, then

(*a*) $f^{-1} 0 f = I_A$, and

(*b*) $f 0 f^{-1} = I_B$

where I_A and I_B are the identity functions of the set A and B, respectively.

Proof (*a*) For every $x \in A$,

$$I_A(x) = x = f^{-1}[f(x)]$$
$$= (f^{-1} 0 f)(x)$$

Thus, $f^{-1} 0 f$ is the function from set A to itself.

(*b*) Similarly, for every $y \in B$,

$$I_B(y) = y = f[f^{-1}(y)]$$
$$= (f 0 f^{-1})(y)$$

Thus, $f 0 f^{-1}$ is the function from set B to itself.

Theorem 4.4 If $f : A \to B$ and $g : B \to C$ are invertible functions, then $g 0 f : A \to C$ is also invertible and $(g 0 f)^{-1} = f^{-1} 0 g^{-1}$.

Proof Since f and g are one-to-one functions, therefore we have

$$f(x_1) = f(x_2) \Rightarrow x_1 = x_2 \text{ for all } x_1, x_2 \in R$$
$$g(y_1) = g(y_2) \Rightarrow y_1 = y_2 \text{ for all } y_1, y_2 \in R$$
$$(g 0 f)(x_1) = (g 0 f)(x_2) \Rightarrow g(f(x_1)) = g(f(x_2))$$
$$\Rightarrow f(x_1) = f(x_2)$$
$$\Rightarrow x_1 = x_2$$

Also as f and g are both onto functions, so for each $z \in C$, there exists $y \in B$ such that $g(y) = z$ and $f(x) = y$ for each $x \in A$. Thus $z = g(y) = g(f(x) = (g 0 f)(x)$. Hence each element $z \in C$ has pre-image under $g 0 f$ and therefore $g 0 f$ is an onto function. Since it has been proved that $g 0 f$ is one-one onto function, therefore $(g 0 f)$ exists. This implies that if $g 0 f : A \to C$, then $(g 0 f)^{-1} : C \to A$ or $g^{-1} 0 f^{-1}$: $C \to A$ (See Theorem 4.1).

Now $(g 0 f)^{-1}(z) = x \Leftrightarrow (g 0 f)(x) = z$

$$\Leftrightarrow \quad g(f(x)) = z \text{ or } g(y) = z$$
$$\Leftrightarrow \quad y = g^{-1}(z)$$
$$\Leftrightarrow \quad f^{-1}(y) = f^{-1}(g^{-1}(z)) = (f^{-1} 0 g^{-1})(z)$$
$$\Leftrightarrow \quad x = (f^{-1} 0 g^{-1})(z).$$

Hence $\quad (g 0 f)^{-1}(z) = (f^{-1} 0 g^{-1})(z) \quad \text{or} \quad (g 0 f)^{-1} = (f^{-1} 0 g^{-1}).$

Example 20 Let X and Y be two nonempty sets and let $f : X \to Y$ is an into mapping and also $A \subseteq X$, $B \subseteq X$, then prove that

(*a*) $f(A \cap B) \subseteq f(A) \cap f(B)$, and

(*b*) $f^{-1}(A \cap B) = f^{-1}(A) \cap f^{-1}(B)$

Solution (*a*) Let $y \in f(A \cap B)$ be an arbitrary element. Then this implies that

$$y = f(x) \text{ for some } x \in (A \cap B)$$
$$y = f(x) \text{ for } x \in A \text{ and } x \in B$$
$$y = f(x) \text{ such that } f(x) \in f(A) \text{ and } f(x) \in f(B)$$
$$y = f(x) \text{ such that } f(x) \in f(A) \cap f(B)$$
$$y \in f(A) \cap f(B).$$

Therefore $f(A \cap B) \subseteq f(A) \cap f(B)$

(b) Let $x \in f^{-1}(A \cap B)$ be an arbitrary element. This implies that

$$f(x) \in (A \cap B) \text{ or } f(x) \in A \text{ and } f(x) \in B$$
$$x \in f^{-1}(A) \text{ and } x \in f^{-1}(B)$$
$$x \in f^{-1}(A) \cap f^{-1}(B),$$
$$x \in f^{-1}(A \cap B) \text{ or } x \in f^{-1}(A) \cap f^{-1}(B)$$
$$f^{-1}(A \cap B) \subseteq f^{-1}(A) \cap f^{-1}(B) \qquad ...(i)$$

Again, let $y \in f^{-1}(A) \cap f^{-1}(B)$. Then $y \in f^{-1}(A)$ and $y \in f^{-1}(B)$, i.e. $f(y) \in A$ and $f(y) \in B$.

$$\Rightarrow \qquad f(y) \in A \cap B \Rightarrow y \in f^{-1}(A \cap B)$$

It means that, if $y \in f^{-1}(A) \cap f^{-1}(B)$, then also $y \in f^{-1}(A \cap B)$. Thus

$$f^{-1}(A) \cap f^{-1}(B) \subseteq f^{-1}(A \cap B) \qquad ...(ii)$$

Hence, from result (*i*) and (*ii*), we have $f^{-1}(A \cap B) = f^{-1}(A) \cap f^{-1}(B)$

Example 21 Prove that $f^{-1}og^{-1} = (gof)^{-1}$, where $f : Q \to Q$ such that $f(x) = 2x$ and $g : Q \to Q$ such that $g(x) = x + 2$ are two functions. [*UPTU, MCA, 2007*]

Solution Since $f(x) = 2x$, therefore $f^{-1}(x) = \dfrac{1}{2} f(x)$

and $\qquad g(x) = x + 2$, therefore $g^{-1}(x) = g(x) - 2$

Now $\qquad gof = gof(x) = go(2x) = g(2x) = 2x + 2$. Hence, $(gof)^{-1} = \dfrac{1}{2}(gof - 2)$

Similarly, $\qquad fog = fog(x) = f(x + 2) = 2(x + 2) = 2x + 4$

$$= 2x + 2 + 2 = (2x + 2) + 2 = (gof) + 2,$$

Now, $\qquad (f^{-1} og^{-1}) = f^{-1}og^{-1}(x) = f^{-1}[g(x) - 2]$

$$= \frac{1}{2} f[g(x) - 2] = \frac{1}{2}[fog(x) - f(2)]$$
$$= \frac{1}{2}[gof(x) - 2 - f(2)]$$
$$= \frac{1}{2}[gof(x) + 2 - 2 \times 2], \text{ as } f(x) = 2x$$
$$= \frac{1}{2}(gof - 2) = (gof)^{-1}$$

Self-practice Problems I

1. Discuss the nature of the function f defined by

(a) $f: \mathrm{N} \rightarrow \mathrm{N}$ such that $f(n) = n - (-1)^n$, $n \in \mathrm{N}$

(b) $f: \mathrm{R} \rightarrow \mathrm{R}$ such that $f(x) = x^3 - x$

(c) $f: \mathrm{I} \rightarrow \mathrm{I}$ such that $f(x) = -x + 4$.

2. If the functions f and g are given by $f = \{(1, 2), (3, 5), (4, 1)\}$ and $g = \{(2, 3), (5, 1), (1, 3)\}$, then write down pairs in the functions $(f\, 0\, g)$ and $(g\, 0\, f)$.

3. Examine the validity of the following statements and justify your answers.

(a) If $x \notin \mathrm{B} \Rightarrow x \notin \mathrm{A}$, then $\mathrm{A} \subseteq \mathrm{B}$.

(b) There exists a one-one function from a set consisting of 10 elements to the set consisting of 7 elements.

(c) $-|-3| = -3$.

4. Decide whether or not the following are functions from A to B, where

$$\mathrm{A} = \{1, 2, 3, 4, 5\} \text{ and } \mathrm{B} = \{a, b, c, d, e\}.$$

If they are functions, give the range of each. If they are not, then discuss why?

(a) $f = \{(1, a), (2, b), (3, b), (5, e)\}$

(b) $g = \{(1, e), (5, d), (3, a), (2, b), (1, d), (4, a)\}$

(c) $h = \{(5, a), (1, e), (4, b), (3, c), (2, d)\}$.

5. Each of the following formula defines a function from R to R. Find the range of each function.

(a) $f(x) = x^2$ (b) $f(x) = \sin x$ (c) $f(x) = 1 - |x|$ (d) $1/(1 - x^2)$

6. Let $\mathrm{A} = \{1, 0\}$ and let N be the set of natural numbers. Show that $f: \mathrm{N} \rightarrow \mathrm{A}$ defined by $f(1) = 1$, $f(2x) = 0$, $f(2x + 1) = 1$, for all $x \in \mathrm{N}$ is a many-one onto function of N to A. What is the domain and the range of function f?

7. Let $f(x) = \begin{cases} 1 & \text{if } x \text{ is rational number} \\ 0 & \text{if } x \text{ is irrational number} \end{cases}$

be a function from R to R. Find $f(1/3)$ and $f\left(\sqrt{7}\right)$.

8. Let N be the set of all natural numbers and A be the set of even natural numbers. Consider

$$f: \mathrm{N} \rightarrow \mathrm{A} \text{ such that } f(x) = 2x, \text{ for all } x \in \mathrm{N}.$$

Show that f is one-one onto. Also define the inverse function f^{-1}.

9. Let A be a subset of the set B. Let

$$f(x) = \begin{cases} 1 & \text{if } x \in \mathrm{A} \\ 0 & \text{if } x \notin \mathrm{A} \end{cases}$$

be a function defined as, $f: \mathrm{B} \rightarrow \{0, 1\}$. Find the function f when

(a) $\mathrm{A} = \{a, b\}$ and $\mathrm{B} = \{a, b, c\}$

(b) $\mathrm{A} = \{a, c, e\}$ and $\mathrm{B} = \{a, b, c, d, e\}$.

10. Find the ranges of the following functions on the set of integers.

(*a*) $f = \{(x, 4x + 1) : x \in I\}$

(*b*) g = largest integer that is greater than or equal to $\sqrt{|x|}$.

11. Let A = {*a*, *b*, *c*, *d* } and B = {1, 2, 3}. Determine whether the relation *f* from A to B is a function. If yes, give its range.

(*a*) $f = \{a, 1\}, \{b, 2\}, (c, 1), (d, 2)\}$

(*b*) $f = \{(a, 1), (b, 2), (a, 2), (c, 1), (d, 2)\}$

12. Let $f : A \to B$ such that $f(x) = x - 1$ and $g : B \to C$ such that $g(y) = y^2$. Find

(*a*) $(f \, 0 \, g)(2)$

(*b*) $(g \, 0 \, f)(2)$

(*c*) $(f \, 0 \, g)(x)$

(*d*) $(g \, 0 \, f)(x)$

(*e*) $(f \, 0 \, f)(y)$

(*f*) $(g \, 0 \, f)(y)$

13. Let $f : A \to B$. Determine which of the following functions are one-one or onto or both or neither.

(*a*) A = {1, 2, 3, 4,} = B and f = {(`1, 1}, (2, 3), (3, 4), (4, 2)}

(*b*) A = {1, 2, 3}, B = {*a*, *b*, *c*, *d*} and f = {(1, *a*), (2, *a*), (3, *c*)}

(*c*) A = {1/2, 1/3, 1/4}, B = {*x*, *y*, *z*, *w*} and f = {(1/2, *x*), (1/4, *y*), (1/3, *w*)}

14. Let $f : N \to I_{10}$ such that $f(x)$ is the remainder after dividing 10 by x.

(*a*) What is $f(23)$?

(*b*) Is f an one-one (injective) function?

(*c*) Is f an onto (surjective) function?

15. Which of the following functions are one-one onto or one-one into on the set of real numbers?

(*a*) $f(x) = -2x$ (*b*) $g(x) = x^2 - 1$ (*c*) $g(x) = x$, when $x < 0$ and x^2 when $x \geq 0$

16. Let $f : A \to B$ such that $f(x) = 2x + 1$ and $g : B \to C$ such that $g(y) = y/3$ be two functions defined on the set of real numbers. Verify that $(g \, 0 \, f)^{-1} = f^{-1} \, 0 \, g^{-1}$.

17. (*a*) If the set A has *n* elements, how many functions are there from A to itself?

(*b*) If the set A has *n* elements, how many bijections are there from A to itself?

18. Let $f : A \to B$ and $g : B \to C$ be two functions defined on the set of real numbers.

(*a*) If f and g are one-one onto functions, then $g \, 0 \, f$ is one-one onto function.

(*b*) If $g \, 0 \, f$ is one-one function, then f is one-one function.

(*c*) If $g \, 0 \, f$ is onto function, then g is onto function.

19. Let A = {1, 2, 3}, B = {*a*, *b*, *c*, *d*} and C = {7}. The two functions f and g are defined as:

$f : A \to B$ such that $f(1) = c$, $f(2)$ and $f(3) = a$

$g : B \to C$ such that $g(y) = 7$ for all $y \in B$

(*a*) Find $g \, 0 \, f$ (*b*) Does f^{-1} exist? (*c*) Does g^{-1} exist?

20. Let A = {1, 2, 3}. Define $f : A \to A$ such that $f = \{(1, 2), (2, 1), (3, 3)\}$. Find f^{-1}, f^2 and f^3.

21. Let f be a function from R to R, where R is the set of real numbers so that $f(x) = x^2$. Determine $f^{-1}(0)$, $g^{-1}(-1)$ and $g^{-1}(4)$.

22. Let f, g and h be functions from N to N, where N is the set of natural numbers, so that $f(x) = x + 1$, $g(x) = 2x$ and $h(x) = 1$ Determine (*a*) $f \, 0 \, f$, (*b*) $f \, 0 \, g$, (*c*) $g \, 0 \, f$, (*d*) $g \, 0 \, h$, (*e*) $h \, 0 \, g$.

23. Let f, g and h be functions from I to I, where I is the set of integers, so that $f(x) = x + 5$ and $g(x) = x - 2$ and $f(x) = x^3$. Define (*a*) $f \, 0 \, g$ and (*d*) $f \, 0 \, h$.

24. Let $f: \mathrm{R} \to \mathrm{R}$ such that $f(x) = 2x - 1$. Since f is one-one onto function, find the formula for f^{-1}.

Hints and Answers

9. (*a*) $f = \{(a, 1), (b, 1), (c, 0)\}$ (*b*) $f = \{(a, 1), (b, 0), (c, 1)\ (d, 0), (e, 1)\}$

10. (*a*) $\{4x + 1: x \in \mathrm{I}\}$ (*d*) $\mathrm{N} = \{0, 1, 2, 3, \ldots\}$

11. (*a*) R is the function; Range (R) = {1, 2} (*b*) R is not the function

12. (*a*) 3 (*b*) 1 (*c*) $x^2 - 1$ (*d*) $(x - 1)^2$ (*e*) $y - 2$ (*f*) y^4

13. (*a*) Both (*b*) Neither (*c*) One-one

15. (*a*) One-one into (*b*) Neither (*c*) One-one onto

16. $(g \, 0 \, f)(x) = (2x + 1)/3$, $(g \, 0 \, f)^{-1}(z) = (3z - 1)/2$;
$f^{-1}(y) = (y - 1)/2$; $g^{-1}(z) = 3z$
$(f^{-1} \, 0 \, g^{-1})(z) = f^{-1}(3z) = (3z - 1)/2$.

20. (*a*) $f^{-1} = \{(2, 1), (1, 2), (3, 3)\}$ (*b*) $f^2 = (f \, 0 \, f)(1) = f[f(1)] = f(2) = 1$
Similarly, $(f \, 0 \, f)(2) = 2$, $(f \, 0 \, f)(3) = 3$; $f^2 = \{(1, 1), (2, 2), (3, 3)\}$
(*c*) $f^3 = f \, 0 \, f \, 0 \, f = f \, 0 \, f^2$.
$(f \, 0 \, f)^2(1) = f[f^2(1)] = f(1) = 2$; $(f \, 0 \, f^2)(2) = 1$; $(f \, 0 \, f^2)(3) = 3$
$\therefore f^3 = \{(1, 2), (2, 1), (3, 3)\}$.

21. $f^{-1}(0) = \{0\}$; $f^{-1}(-1) = \phi$; $f^{-1}(4) = \{2, -2\}$.

22. (*a*) $(f \, 0 \, f)(x) = f[f(x)] = f(x + 1) = (x + 1) + 1 = x + 2$
(*b*) $(f \, 0 \, g)(x) = f[g(x)] = f(2x) = 2(x + 1) = 2x + 2$
(*c*) $(g \, 0 \, f)(x) = g[f(x)] = g(x + 1) = 2(x + 1) = 2x + 2$
(*d*) $(g \, 0 \, h)(x) = g[h(x)] = 1$ when x is odd and 0 when x is even
(*e*) $(h \, 0 \, g)(x) = h[g(x)] = h(2x) = 1$

24. Given $y = f(x) = 2x - 1$. As x will be the image of y under inverse function f^{-1}, therefore we have $x = (y - 1)/2$. Thus $f^{-1}(y) = (y - 1)/2$. Replacing y by x to get the formula $f^{-1}(x) = (x - 1)/2$.

4.7 FUNCTIONS FOR COMPUTER SCIENCE

Few mathematical functions that are often used in the application of computer science are as follows:

4.7.1 Floor and Ceiling Functions

Let x be any real number. The *floor function*, $f(x)$ defined for x is the greatest integer less than or equal to x. This function $f: R \to I$ is given by $f(x) = \lfloor x \rfloor$ = greatest integer less than or equal to x. Thus, if x is a real number and n is an integer, then $\lfloor x \rfloor = n \Leftrightarrow n \le x < n + 1$ as shown below.

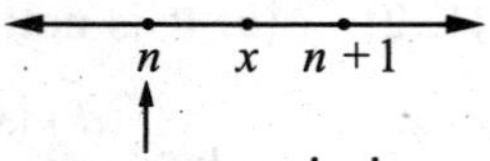

Floor of $x = \lfloor x \rfloor$

The floor function is different than the truncation function which merely truncates any fraction. If x is positive, then floor function and truncation function are same. But if x is negative, then these functions are different, for example $\lfloor -3.75 \rfloor = -4$ not -3. Examples of floor function are as follows:

(*i*) $f(4.25) = \lfloor 4.25 \rfloor = 4$, (*ii*) $f(-3.75) = \lfloor -3.75 \rfloor = -4$,

(*iii*) $f(\sqrt{7}) = \lfloor \sqrt{7} \rfloor = 2$ (*iv*) $f(-3) = \lfloor -3 \rfloor = -3$

A *ceiling function,* $c(x)$ defined for x is the smallest (least) integer greater than or equal to x. The notation $\lceil x \rceil$ is used for $c(x)$.

When $c(x) = x$ for $x \in I$ and $c(x)$ is the integer to the immediate right of x on the real number line for $x \in R - I$ as shown below:

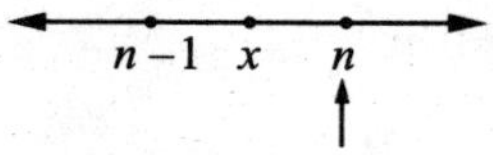

Floor of $x = \lceil x \rceil$

Symbolically, if x is real number and n is integer, then $\lceil x \rceil = n \Leftrightarrow n - 1 < x \le n$.

Definition of floor function and ceiling function implies that the floor of a real number x 'rounds x down' while the ceiling of x 'rounds x up'. For example, Trunc $(3.75) = \lfloor 3.75 \rfloor = 3$; Trunc $(-3.75) = \lceil -3.75 \rceil = -3$.

Few examples of ceiling function are as follows:

(*i*) $c(4.25) = \lceil 4.25 \rceil = 5$ (*ii*) $c(-3.75) = \lceil -3.75 \rceil = -3$

(*iii*) $c(\sqrt{7}) = \lceil \sqrt{7} \rceil = 3$ (*iv*) $c(-3) = \lceil -3 \rceil = -3$

Properties of Floor and Ceiling Functions Few useful properties of floor and ceiling functions are as follows:

1. (*a*) $\lfloor x \rfloor = n \Leftrightarrow n \le x < n + 1$, where n is an integer.

   (*b*) $\lfloor x \rfloor = n \Leftrightarrow n - 1 < n \le x$, where n is an integer.

   (*c*) $\lceil x \rceil = n \Leftrightarrow n \le x \le n + 1$, where n is an integer.

   (*d*) $\lceil x \rceil = n \Leftrightarrow n - 1 < x \le n$, where n is an integer.

2. (*a*) If $a = qb + r$ for $q, r \in$ I and $0 \le r < b$, then $q = \lfloor a/b \rfloor$ and $r = a - \lfloor a/b \rfloor b$.
   (*b*) $\lceil n/k \rceil = \lfloor (n-1)/k \rfloor + 1$, where n and k are positive integers.
   (*c*) $\lfloor \lceil n \rceil / n \rfloor = 1$ and $\lceil \lfloor n \rfloor / n \rceil = 1$, where n (≥ 1) is a positive integer.
   (*d*) $\lceil nx \rceil = n\lceil x \rceil$, where n (> 1) is a positive integer.
3. $x - 1 < \lfloor x \rfloor \le x \le \lceil x \rceil < x + 1$
4. (*a*) $\lfloor -n \rfloor = -\lfloor n \rfloor$
   (*b*) $\lfloor -n \rfloor = -\lfloor n \rfloor$
5. (*a*) $\lfloor m+n \rfloor = \lfloor m \rfloor + n$, where n is an integer
   (*b*) $\lceil m+n \rceil = \lceil m \rceil + n$, where n is an integer

Remark In general, for all real numbers x and y, $\lfloor x \rfloor + \lfloor y \rfloor \ne \lfloor x+y \rfloor$. For example,

$$\lfloor x \rfloor + \lfloor y \rfloor = \lfloor 7.7 \rfloor + \lfloor 8.4 \rfloor = 7 + 8 = 15$$
$$\lfloor x+y \rfloor = \lfloor 7.7+8.4 \rfloor = \lfloor 16.1 \rfloor = 16$$

However, $\lfloor x \rfloor + \lfloor y \rfloor = \lfloor 7.1 \rfloor + \lfloor 8.2 \rfloor = 7 + 8 = 15$

$$\lfloor x+y \rfloor = \lfloor 7.1+8.2 \rfloor = \lfloor 15.3 \rfloor = 15$$

Example 22 How many bytes are required to encode n bits of data when (*a*) $n = 175$ and (*b*) $n = 2009$ where each byte is made up of 8 bits.

Solution (*a*) The number of bytes required is the smallest integer number that is greater than or equal 175/8. Thus the required number of bytes are $\lceil 175/8 \rceil = \lceil 21.87 \rceil = 22$.

(*b*) Similarly, required number of bytes are $\lceil 2009/8 \rceil = \lceil 251.125 \rceil = 252$.

Example 23 For all real numbers x, prove that

(*a*) $\lfloor x+m \rfloor = \lfloor x \rfloor + m$, where m is any integer.

(*b*) $\lfloor 2x \rfloor = \lfloor x \rfloor + \left\lfloor x + \frac{1}{2} \right\rfloor$.

Solution (*a*) Suppose $n = \lfloor x \rfloor$. Then by definition of floor function, we have

$$n \le x < n + 1$$

or $$n + m \le x + m < n + 1 + m, \text{ by adding } m \text{ to all sides.}$$

Consequently by definition of floor function, we can write $\lfloor n+m \rfloor = n + m$ because $n + m$ is an integer. But $n = \lfloor x \rfloor$, so we have $\lfloor x+m \rfloor = x + m$.

(*b*) Let $n = \lfloor x \rfloor$. Then by definition of floor function, we may write $n \le x \le n + 1$. If $n + \frac{1}{2} < x$, then $\lfloor 2x \rfloor = 2n + 1$ and $\left\lfloor x + \frac{1}{2} \right\rfloor = n + 1$. Hence, $\lfloor 2x \rfloor = \lfloor x \rfloor + \left\lfloor x + \frac{1}{2} \right\rfloor$.

Further, if $x < n + \frac{1}{2}$, then $\lfloor 2x \rfloor = 2n$ and $\lfloor x \rfloor = \left\lfloor x + \frac{1}{2} \right\rfloor = n$. Since identity holds in both the cases, the result also holds.

Example 24 Let x (≥ 1) any positive real number. Then prove that

(*a*) $\lfloor \lceil x \rceil / x \rfloor = 1$ and (*b*) $\lceil \lfloor x \rfloor / x \rceil = 1$

Discuss the results in parts (*a*) and (*b*) when $0 < x < 1$.

Solution (*a*) If $x \in I_+$, then $\lceil x \rceil = x$ and $\lfloor \lceil x \rceil / x \rfloor = \lfloor 1 \rfloor = 1$. But if $x \notin I_+$, then we can write $x = n + m$ where $n \in I_+$ and $0 < m < 1$. Thus

$\lceil x \rceil / x = (n + 1)/(n + m) = 1 + (1 - m)/(n + m)$

where $0 < (1 - m)/(n + m) < 1$. Hence, $\lfloor \lceil x \rceil / x \rfloor = \lfloor 1 + (1 - m)/(n + m) \rfloor = 1$.

(*b*) Consider $x = 0.1$. Then

(*i*) $\lfloor \lceil x \rceil / x \rfloor = \lfloor 1/0.1 \rfloor = \lfloor 10 \rfloor = 10$ ($\neq 1$), and

(*ii*) $\lceil \lfloor x \rfloor / x \rceil = \lceil 0/0.1 \rceil = 0$ ($\neq 1$).

However, (*ii*) is false for all $0 < x < 1$, because $\lceil \lfloor x \rfloor / x \rceil = 0$ for all such values of x. In case of (*i*) when $0 < x \leq 0.5$, it follows that $\lceil x \rceil / x \geq 2$ and $\lfloor \lceil x \rceil / x \rfloor \geq 2 \neq 1$. But, for $0.5 < x < 1$, $\lceil x \rceil / x = 1/x$ where $1 < 1/x < 2$ and so $\lfloor \lceil x \rceil / x \rfloor = 1$ for $0.5 < x < 1$.

4.7.2 Sequence (or Discrete) Functions

A *sequence* is a function from the set of natural numbers, N = {1, 2, 3, . . .} into the set A. We can write a sequence as (a_n) or $(a_n : n \in N)$.

A finite sequence is a function from the set {1, 2, . . ., n} into A and is usually denoted as: $a_1, a_2, \ldots, a_n$. Such sequence is sometimes called an n-tuple or a list.

Sometimes the domain of a sequence is different from natural numbers. The most common variation is $(n : n \geq k)$ for some fixed number k. For example, {1, 2, 3, . . .} or {. . ., –3, –2, –1, 0, 1, 2, 3, . . .} could be the domain of a sequence.

If the elements listed in the sequence stops after n steps, $n \in N$, then such sequence is called *finite sequence*, otherwise *infinite sequence*. The elements in a sequence may all be different or some may be repeated.

In order to describe a sequence, it is desirable to have a compact notation. There are two ways of describing a sequence:

(*a*) *The sequence in which successive terms are produced with the help of preceding terms.* Such method is called *recursive* method as it starts from starting place to describe the sequence.

(*b*) *The sequence in which terms are described by using its position number.* Such method is called *explicit method* because it indicates what value any particular term of the sequence has.

Illustrations

1. The sequence 1, 0, 0, 1, 0, 1, 0, 0, 1, 1, 1 is a finite sequence with repeated digits.
2. The sequences such as (*a*) 2, 7, 12, 17, 22, . . . and (*b*) 1, 4, 9, 16 . . . are infinite sequences.

3. The sequence 2, 6, 10, 14, 18, . . . may be described by the recursive formula, $a_n = a_{n-1} + 4, 2 \leq n \leq \infty$.
4. The sequence 4, 8, 16, 32, 64, 138 may be described by the recursive formula, $a_1 = 4$, $a_n = 2a_{n-1}$, $2 \leq n \leq 6$.
5. The sequence –2, 4, – 8, 32, . . . may be described by the explicit formula, $a_n = (-2)^n$, $1 \leq n \leq \infty$.
6. The sequence 117, 112, 107, 102, 97, 92 may be described by the explicit formula, $a_n = 122 - 5n$, $1 \leq n \leq 6$.

4.7.3 Strings

Given a finite set A of *symbols* (also called *alphabet*), we can construct another set B consisting of all sequences of letters (or symbols) in A written without commas, are called *strings* or words from A, and it is usually written in the form $a_1\ a_2\ \ldots\ a_n$. The number n of alphabet in the string is called its *length.* If set A contains no symbol, then the set B contains *empty sequence* or *string.* Strings over an alphabet A and certain operations on these strings will be discussed in detail later.

Illustrations

1. The word 'Wednesday' (say) can be regard as a finite sequence *w, e, d, n, e, s, d, a, y* composed of English alphabet.
2. The sentence 'now is the time for the rest' can be viewed as a finite sequence of letters: now, is, the, time, for, the, rest. In this case the elements of the sequence are themselves words of varying length, so usually in such cases we use spaces among words instead of commas.

Remarks

1. The sequence defined as

$$a_n = \begin{cases} 1 & \text{if } n \text{ is even} \\ -1 & \text{if } n \text{ is odd} \end{cases}$$

has no limit. The sequence looks like, –1, 1, –1, 1, ... This sequence oscillates between 1 and –1 and hence does not converge to any fixed number.

2. In the applications of computer science sometimes a sequence is also called a *linear array* or *list.* But the slight but useful distinction between these two is as follows. Since in the given sequence location of its elements is fixed, therefore any change in their location will produce another sequence. For example, if we begin with the finite sequence, A_n : 1, 0, 0, 2, 3, 1, 1, 2 and change the element 2 to an element 5, we will be getting another sequence A'_n : 1, 0, 0, 5, 3, 1, 1, 2.

However, an array considered as a sequence of positions as shown in Fig. 4.9 is represented by boxes.

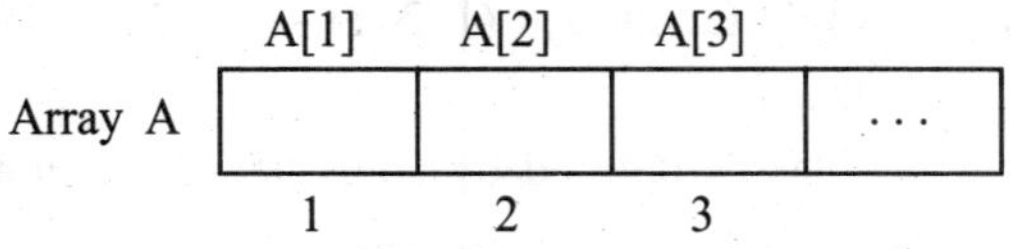

Fig. 4.9 Representation of Array

Depending on the size of an array, the positions of elements form a finite and infinite array. The element assigned to position n will be denoted by A[n] instead of simply A_n. The sequence A[1], A[2], A[3], . . . is called the *sequence of values* of the array A. In other words, an array may be defined as a well-defined object, even if the positions have not been assigned values or changed the values.

4.7.4 Fibonacci Sequence

The sequence in which each succeeding term is the sum of the two preceding terms is called *Fibonacci sequence*, and its terms are usually denoted by $F_1, F_2, F_3, \ldots$.

The formal definition of a sequence is

(*i*) If $n = 0$ or 1, then $F_n = n$

(*ii*) If $n > 1$, then $F_n = F_{n-2} + F_{n-1}$.

This definition uses recursive formula to describe the sequence, where

(*i*) The base (initial) values are 0 and 1, and

(*ii*) The value of F_n is defined in terms of smaller values of n which are close to the base values.

Illustration The infinite sequence 0, 1, 1, 2, 3, 5, 8, 13, . . . is the Fibonacci sequence, where $F_0 = 1$ and $F_1 = 1$. It may also be described as $F_n = F_{n-2} + F_{n-1}$, for $n \geq 2$.

4.7.5 Ackermann's Function

A Ackermann's function is defined as follows:

(*a*) If $m = 0$, then $A(m, n) = n + 1$

(*b*) If $m \neq 0$ but $n = 0$, then $A(m, n) = A(m - 1, 1)$

(*c*) If $m \neq 0$ and $n \neq 0$, then $A(m, n) = A[m - 1, A(m, n - 1)]$

$A(m, n)$ is explicitly given only when $m = 0$, to evaluate any such function, it is reduced to the form $A(0, n)$.

Example 25 Evaluate A(2, 0)

Solution $A(2, 0) = A(1, 1) = A[0, A(1, 0)] = A[0, A(1, 0)]$

$A(1, 0) = A(0, 1) = 1 + 1 = 2$

Again proceeding in backward direction, we have

$A(1, 0) = A(0, 1) = 2$

$A(1, 1) = A[0, A(1, 0)] = A(0, 2) = 2 + 1 = 3$

$A(2, 0) = A(1, 1) = 3$

Example 26 Let F be the collection of all function $f: (1, 2, 3) \rightarrow \{1, 2, 3\}$. If f and g define an equivalence relation by $f \sim g$, if and only if $f(3) = g(3)$

(*a*) Find the number of equivalence classes defined by equivalence relation

(*b*) Find the number of elements in each equivalence class. [*UPTU, MCA, 2005*]

Solution Since F is the collection of all function, $f: \{1, 2, 3\} \rightarrow \{1, 2, 3\}$, then total number of functions in f are : $3^3 = 27$. Given that f and g are any two functions such that $f \sim g$ if and only if $f(3) = g(3)$.

(*a*) number of equivalence classes defined by the relation are 3.

(*b*) number of elements in each class are 9.

Example 27 Suppose A = $\{a, b, c, d\}$ and $\pi_1 = \{\{a, b, c\}\ \{d\}\}$ is the partition of A.

(*a*) List the ordered pairs of the equivalence relations induced by π_1.

(*b*) Draw the graph of the such equivalence relations. [*UPTU, MCA, 2005*]

Solution Given, A = $\{a, b, c, d\}$. The partition of A is given by $\pi_1 = \{\{a, b, c\}\{d\}\}$.

(*a*) The ordered pairs of the equivalence relation in duced by π_1 are given by

$$R = \{(a, a), (a, b), (b, a), (b, b), (b, c), (c, b), (c, c), (c, a), (a, c), (d, d)\}$$

(*b*) Graph of the this relation is shwon below is the figure

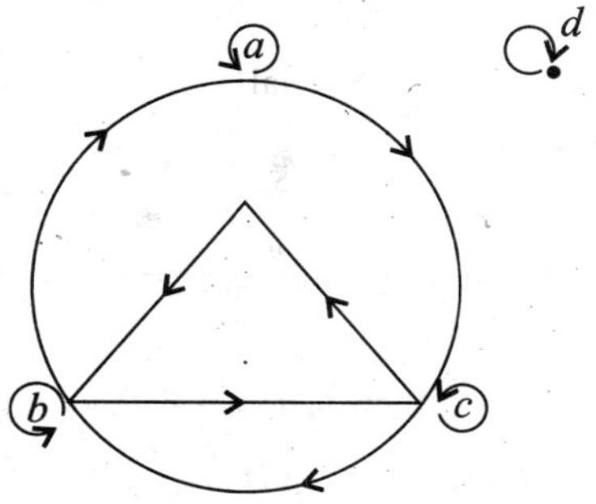

Example 28 Let A be the set {1, 2, 3}, define the following types of binary relation on A

(*a*) A relation that is both symmetric and anti-symmetric.

(*b*) A relation that is neither symmetric nor anti-symmetric. [*UPTU, MCA, 2005*]

Solution Given, A = (1, 2, 3). Then,

(*a*) For a binary relation R on A to be symmetric, it should implies that $aRb \Leftrightarrow bRa$, for all $a, b \in A$, i.e. if $(a, b) \in R$, then $(b, a) \in R$

For a binary relation R on A to be anti-symmetric, it should implies that

$$aRb \text{ and } bRa \Rightarrow a = b, \text{ i.e. } (a, b) \in R \text{ and } (b, a) \in R \text{ only when } a = b$$

which means that if $a \neq b$, then either $a \not R b$ or $b \not R a$

(*b*) A relation R on A shall neither be symmetric, i.e. $(a, b) \in R$ but $(b, a) \notin R$, nor anti-symmetric, i.e. $(a, b) \in R$ and $(b, a) \in R$ even when $a \neq b$.

4.7.6 Characteristic Function

Let A be a non-empty subset of the universal set $U = \{u_1, u_2, \ldots, u_n\}$. Then the characteristic function of A is defined as:

$$f_A : U \to [0, 1]$$

such that $$f_A(u_i) = \begin{cases} 1 & \text{when } u_i \in A \\ 0 & \text{when } u_i \notin A. \end{cases}$$

Illustration If A = {2, 5, 7} and U = {1, 2, 3, . . ., 10}, then $f_A(1) = 0$, $f_A(2) = 1$, $f_A(3) = 0$, $f_A(5) = 1$ and so on. It may also be observed that f_A is everywhere defined and onto, but is not one-one.

Theorem 4.5 If A and B are non-empty subsets, then for all x the characteristic functions of these subsets satisfy the following properties:

(*a*) $f_{A'}(x) = 1 - f_A(x)$

(*b*) $f_{A\cap B}(x) = f_A(x) \cdot f_B(x)$, for all x.

(*c*) $f_{A\cup B}(x) = f_A(x) + f_B(x) - f_A(x)\, f_B(x)$, for all x.

(*d*) $f_{A\oplus B}(x) = f_A(x) + f_B(x) - 2f_A(x)\, f_B(x)$, for all x.

Proof (*a*) $f_A(x) = 1 \Leftrightarrow x \in A' \Leftrightarrow x \notin A \Leftrightarrow f_A(x) = 0 \Leftrightarrow 1 - f_A(x) = 1$.

Thus $f_{A'}(x) = 1 - f_A(x)$.

(*b*) Since $f_A\, f_B = \begin{cases} 1 & \text{when } x \in A \cap B \\ 0 & \text{otherwise} \end{cases}$

therefore $f_A(x)\, f_B(x) = 1$ if and only if both $f_A(x)$ and $f_B(x)$ are equal to 1. It is possible only if $x \in A \cap B$. Hence, $f_{A\cap B}(x) = f_A(x)\, f_B(x)$.

(*c*) From result (*b*) we know that if $x \in A$, then $f_A\ (x) = 1$. This implies that

$$f_A(x) + f_B(x) - f_A(x)\, f_B(x) = 1 + f_B(x) - f_B(x) = 1$$

Similarly, if $x \in B$, then $f_B(x) = 1$. Thus

$$f_A(x) + f_B(x) - f_A(x)\, f_B(x) = 1.$$

However, if $x \notin A \cap B$, then both $f_A(x) = 0$ and $f_B(x) = 0$ and hence

$$f_A(x) + f_B(x) - f_A(x)\, f_B(x) = 0.$$

This shows that $f_A(x) + f_B(x) - f_A(x)\, f_B(x) = 1$ when $x \in A \cap B$ and 0 otherwise and hence equal to $f_{A\cup B}(x)$.

(*d*) Left as an exercise for the reader.

4.7.7 Exponential Functions

The limit of the sequence (a_n), where

$$a_n = \left(1 + \frac{1}{n}\right)^n, \quad n = 1, 2, \ldots$$

as $n \to \infty$ exists and is equal to a value between 2 and 3 and is approximately equal to 2.718. This number is denoted by e. Thus

$$e = \lim_{n \to \infty} \left(1 + \frac{1}{n}\right)^n$$

The exponential function $f(x)$ is defined as

$$f(x) \text{ or } e^x = \lim_{n \to \infty} \left(1 + \frac{1}{n}\right)^n, \quad -\infty < x < \infty.$$

Another exponential function $f(x) = a^x$ for all real numbers is defined as

$$a^x = \lim_{n \to \infty} a^r, \ r \text{ is a rational number.}$$

Properties of Exponentiation If a is non-negative element, and n and m are integers, then

1. $(a^n)^{-1} = (a^{-1})^n$
2. $a^n \times a^m = a^{n+m}$
3. $(a^n)^m = a^{nm}$
4. $a^{m/n} = \sqrt[n]{a^m} = \left(\sqrt[n]{a}\right)^m$
5. $a^0 = 1$
6. $a^m = a \cdot a \cdot \ldots \cdot a$ (m times)

Illustration A model proposed by Vidale and Wolfe for sales revenue rate as: $R(t) = R(0)\, e^{-\lambda t}$ where $R(0)$ is the sales rate at time $t = 0$ and α is a positive constant (also called the *sales decay constant*).

The value of $R(t)$ decreases as t increases and λ is the rate at which it decreases.

4.7.8 Logarithmic Functions

Let a be a positive number. The exponential function $y = a^x$ can be expressed as the logarithmic function as: $x = \log_a y$. The subscript a is called the *base of the logarithm*.

Similarly, the *natural logarithm* for the base e is defined in the same way as: $\log_e y$. If $x = \log_e y$, then $y = e^x$. When $x = 1$, we have $\log_e e = 1$. The domain of the logarithmic function: $x = \log_e y$ is $(0, \infty)$.

Another types of logarithms are: logarithm to base 2, written as: $\log_2 x$ and is called *binary logarithm* and logarithm to base 10, written as: $\log_{10} x$ and is called *common logarithm*.

Often the term $\log x$, usually means $\log_{10} x$, but also means $\log_2 x$ in computer science applications. For applications purpose, we require along the floor or the ceiling of a binary logarithm. For example,

$$\lceil \log_2 100 \rceil = 6 \text{ because } 2^6 = 64 \text{ and } 2^7 = 128$$

$$\lceil \log_2 1000 \rceil = 9 \text{ because } 2^8 = 512 \text{ and } 2^9 = 1024$$

Properties of Natural Logarithms Given a and b as positive integers,

1. $\log (a\, b) = \log a + \log b$
2. $\log (a^x) = x \log a$, x is any real number
3. $\log \left(\frac{a}{b}\right) = \log a - \log b$
4. $\log 1 = 0$
5. $\log e = 1$
6. $\log_e 10 = \dfrac{1}{\log_{10} e}$
7. $\log_e a = \log_e 10 \times \log_{10} a = \dfrac{\log_{10} a}{\log_{10} e}$
8. $\log_b a = x \Leftrightarrow b^x = a,\ x > 0$
9. $\log_b a = \dfrac{\log_e a}{\log_e b}$.

The property (7) is useful to compute logarithm of number to base e from logarithm tables available for base 10.

4.7.9 Mod Functions

For each positive integer number, m the mod-m function is defined as:

$$f_m(a) = b \text{ or } a \equiv b \pmod{m};\ 0 \le b < m,\ a \text{ is non-negative integer}$$

The term *mod* is used for mathematical congruence relation and is defined as:

$$a \equiv b \pmod{m} \text{ if and only } m \text{ divides } b - a.$$

Here m is called the *modulus* and $a \equiv b \pmod{m}$ is read as: *a is congruent to b modulus m*.

If $m \neq 0$ and a are non-negative integers, then we can write $a = km + r$, $0 \leq r < m$. In other words, when a is divided by m to obtain remainder r, we may also write it as: a (mod m).

Remark The arithmetic modulo 12 is sometimes called *clock* arithmetic.

Illustrations

1. (*a*) $f_6(39) = 3$ because $39 = 6 \times 6 + 3$ and $39 \equiv 3 \pmod 6$

   (*b*) $f_6(3) = 3$ because $3 = 0 \times 6 + 3$ and $3 \equiv 3 \pmod 6$
2. If the time now is 6 o'clock, then the time at 100 hours from now will be 10 o'clock, because here $a = 12$, $b = 6 + 100 = 106$ and $106 = 8 \times 12 + 10$, therefore at 100 hours it will be 10 o'clock.

4.7.10 Hamming Distance Function

The concept of hamming distance function developed by Wesley Hamming is useful for the error detecting and error correcting capabilities of a coding scheme. This function provides a measure of difference between two binary words which have the same measure of difference between them and have the same length.

Definition For each element $x = x_1x_2 \ldots x_n \in I_2^n$, where $n \in I_+$, the weight of x denoted by $r(x)$ is the number of components x_i of x, for $1 \leq i \leq n$, where $x_i = 1$. If $y \in I_2^n$, then the distance between x and y, is given by:

$d(x, y)$ = number of components x_i and y_i of x and y, respectively where $x_i \neq y_i$, for $1 \leq i \leq n$, i.e. number of positions in which x and y have different bits

$$= \sum_{i=1}^{n} d(x_i, y_i) \text{ where } x, y \in I_2, \text{ and for each } 1 \leq i \leq n,\ d(x_i, y_i) = \begin{cases} 0 \text{ if } x_i = y_i \\ 1 \text{ if } x_i \neq y_i \end{cases}.$$

Illustration For $n = 4$, let $x = 0100$ and $y = 1110$. Then $r(x) = 1$ and $r(y) = 3$ and $d(x_i, y_i) = 2$ because x and y differ in 2 positions.

Further $x + y = 1010$, so $r(x + y) = 2$. For each $1 \leq i \leq 4$, $x_i + y_i$ contributes a count of 1 to $r(x + y)$ $\Leftrightarrow x_i \neq y_i \Leftrightarrow x_i, y_i$ contribute a count of 1 to $d(x_i, y_i)$. In general, for all $n \in I_+$, $r(x + y) = d(x, y)$ for all $x, y \in I_2^n$.

Lemma For all $x, y \in I_2^n$, $r(x + y) \leq r(x) + r(y)$.

For example, in the above illustration, we found that $r(x + y) = r(1010) = 2 \leq 1 + 3$

$= r(0100) + r(1110) = r(x) + r(y)$.

Important results The Hamming distance function defined on $I_2^n \times I_2^n$ satisfies the following for all $x, y, z \in I_2^n$.

(*i*) $d(x, y) \geq 0$ (*ii*) $d(x, y) = 0 \Leftrightarrow x = y$

(*iii*) $d(x, y) = d(y, x)$ (*iv*) $d(x, z) \leq d(x, y) + d(y, z)$

4.7.11 Time-complexity Function

To measure the time or memory space required for the execution of an algorithm on a problem we need to compare relative sizes of the function $f(x)$, called the time-complexity function of the algorithm. The main aim in dealing with any algorithm is to know how the algorithm performs for large values of x.

Definition Let f and g be functions from the set of positive integers to the set of real numbers. The function f is said to be *dominated* by g if there exist positive integer constants m and k such that $|f(x)| \leq m|g(x)|$ for all $x \in I_+$ and $x \geq k$.

This definition implies that, if g dominates f, then $|f(x)/g(x)| \leq m$, i.e. the size of quotient $f(x)/g(x)$ is bounded by m for all $x \in I_+$ where $x \geq k$ and $g(x) \neq 0$.

If the function f is dominated by g, then f is of order (at most) g, denoted by $O\,g(x)$. This is read as *order* $g(x)$ or $f(x)$ *is big-oh of* $g(x)$.

If $f \in O(g)$, then $g(x)$ is upper bound for $f(x)$ but if $g \in O(f)$, then $g(x)$ is lower bound for $f(x)$. The big-o 'notation' compares the rate of growth of functions instead of their values.

Example 29 Let $f(x) = 5x^3 + 3x + 9$ and $g(x) = x^3$ for all real numbers $x > 1$. Use big-*oh* notation to express $f(x)$ and $g(x)$.

Solution $|f(x)| = |5x^3 + 3x + 9| = 5x^3 + 3x + 9 \leq 5x^3 + 3x^3 + x^3 = 9x^3 = 9|g(x)|$. Hence, for all $x > 1$ $(= k)$, $|f(x)| \leq m|g(x)|$ for any $m \geq 9$.

Order of Power Functions It is observe that, if

$$1 < x \Rightarrow x < x^2 \Rightarrow x^2 < x^3 \text{ and so on.}$$

In general, for any rational numbers r and s, if $x > 1$ and $r < s$, then $x^r < x^s$. Thus for any rational numbers r and s if $r < s$, then x^r is $O(x^s)$.

Remark Above result can also be used to derive a polynomial inequality to find the order of a polynomial.

Example 30 Use the definition of order to show that $3x^3 + 7x^2 - 4x + 2$ is $O(x^3)$.

Solution For all real numbers x, the functions f and g are defined as: $f(x) = 3x^3 + 7x^2 - 4x + 2$ and $g(x) = x^3$. Then for all real numbers $x > 1$,

$$\begin{aligned}|3x^3 + 7x^2 - 4x + 2| &\leq |3x^3| + |7x^2| + |-4x| + |2|\\ &\leq 3x^3 + 7x^3 + 4x^3 + 2x^3\\ &= 16x^3 = 16|g(x)|.\end{aligned}$$

So with $m = 16$ and $k = 1$, we find that $|f(x)| \leq m|g(x)|$ and hence $f \in O(x^3)$ or $f \in O(g)$.

Example 31 Let $f(x) = x^2 + x$ and $g(x) = x^3/2$ for all real numbers x. Show that $f \in O(g)$ but $g \notin O(f)$.

Solution To show $f \in O(g)$, let $k = 1$ and $m = 4$. Then for all $x \geq k$, $|f(x)| = x^2 + x \leq x^2 + x^2 = 2x^2 \leq 2x^3 = 4[(1/2)\,x^3] = 4|g(x)|$. Hence, f is $O(g)$.

To show that g is $O(f)$ or not, it is necessary to determine whether there are constants k and m such that $x^3/2 \leq m\,(x^2 + x)$ for $x > k$. Now if

$$|(1/2)\,x^3| \leq m\,|x^2 + x| \quad \text{or} \quad (1/2)\,x^2 \leq m\,(x + 1),$$

then

$$\frac{x^2}{2(x+1)} \leq m \Rightarrow 0 < \frac{x^2}{4x} < \frac{x^2}{2(x+1)} \leq m$$

$$\Rightarrow \frac{x}{4} \leq m \quad \text{or} \quad x \leq 4m$$

which is a contradiction because x is variable and m constant. Hence, $g \notin O(f)$.

Theorem 4.6 Let $f(x) = a_n x^n + a_{n-1} x^{n-1} + \ldots + a_2 x^2 + a_1 x + a_0$, where $a_0, a_1, a_2, \ldots, a_n$ are real numbers and $a_n \neq 0$ for $n \in N$. Then prove that $f(x)$ is $O(x^n)$.

Proof Using the triangle inequality, if $x > 1$, we have

$$\begin{aligned}|f(x)| &= |a_n x^n + a_{n-1} x^{n-1} + a_{n-2} x^{n-2} + \ldots + a_2x^2 + a_1x + a_0| \\ &\le |a_n x^n| + |a_{n-1} x^{n-1}| + |a_{n-2} x^{n-2}| + \ldots + |a_2x^2| + |a_1x| + |a_0| \\ &= |a_n|x^n + |a_{n-1}|x^{n-1} + |a_{n-2}|x^{n-2} + \ldots + |a_2|x^2 + |a_1|x + |a_0| \\ &\le |a_n|x^n + |a_{n-1}|x^n + |a_{n-2}|x^n + \ldots + |a_2|x^n + |a_1|x^n + |a_0|x^n \\ &= x^n(|a_n| + |a_{n-1}| + \ldots + |a_2| + |a_1| + |a_0|).\end{aligned}$$

This shows that $|f(x)| \le m|g(x)|$ for all $x \ge k$ where $m = |a_n| + |a_{n-1}| + \ldots + |a_2| + |a_1| + |a_0|$ and $k = 1$. Hence, f is dominated by g or f is $O(x^n)$.

Remark The result of Theorem 4.6 provides the following results on summation.

1. Let $f(n) = 1 + 2 + 3 + \ldots + n$. Then by the formula for the sum of the first n integers, we have

$$f(n) = \frac{1}{2}n(n + 1) = \left(\frac{1}{2}\right)n^2 + \left(\frac{1}{2}\right)n.$$

   Hence, by the theorem on polynomial orders, $f(n)$ is $O(n^2)$.

2. Let $f(n) = 1^2 + 2^2 + \ldots + n^2 = \frac{1}{6}n(n + 1)(2n + 1)$

$$= \left(\frac{1}{3}\right)n^3 + \left(\frac{1}{2}\right)n^2 + \left(\frac{1}{6}\right)n$$

   Hence by the theorem on polynomial orders $f(n)$ is $O(n^3)$.

Big-omega and Big-theta Notations The following is the analogous to 'big-oh' notation discussed earlier.

1. For f and g functions from the set of positive integers to the set of real numbers, we say that $f(x)$ is big-omega of $g(x)$ or $f \in \Omega(g)$ if there exists constants m and k such that $|f(x)| \ge m|g(x)|$ for all $x \in I_+$, and $x \ge k$.

   For example, if $f(x) = 7x^2 + 5x$ and $g(x) = x^2$, then for all $x \ge 1, f(x) = 7x^2 + 5x > x^2 = g(x)$. So with $m = 1$ and $k = 1$, we have $|f(x)| \ge m|g(x)|$ for all $x \ge k$ and it follows that $f \in \Omega(g)$.

2. For f and g functions from the set of positive integers to the set of real numbers, we say that $f(x)$ is 'big-theta of g' or $f \in -\Theta(g)$, if there exist constants m_1, m_2 and k such that $m_1|g(x)| \le |f(x)| \le m_2|g(x)|$ for all $x \in I_+$ and $x \ge k$.

Self-practice Problems II

25. Write the set corresponding to the sequence

(*a*) 1, 2, 1, 2, 1, 2 (*b*) *a b c c d d e e* (*c*) *a a b b c c* . . . *z z*

26. Write the first four terms (begins with $n = 1$) of the sequence whose general term is given

(*a*) $a_n = 4^n$ (*b*) $a_n = n^2 - n$ (*c*) $a_n = a^{n-1} + 2.5$; $a_1 = 1.5$

27. Write a formula for the nth term of the sequence. Identify your formula as recursive or explicit.
(*a*) 1, 4, 7, 10, 13, 16, . . . (*b*) 1, –1, 1, –1, 1, –1, . . . (*c*) 2, 5, 8, 11, 14, 17 (*d*) 2, 4, 6, 8, . . .

28. Let n be a positive integer. Suppose a function f is defined recursively as follows:
(*a*) $f(0) = 0,\ f(1) = 1,\ f(n + 2) = 2f(n) + f(n + 1)$, for $n \geq 0$,
write the first six values of f.

(*b*) If $f(n) = \begin{cases} f(\lfloor n/2 \rfloor) & ; n > 1 \\ 0 & ; \text{otherwise} \end{cases}$

then find the value of $f(23)$ and describe the nature of this function.

29. Solve the following

(*a*) $f(n) = 3 + f(\lfloor n/2 \rfloor),\ f(0) = 0$; (*b*) $f(n) = 1 + f(\lfloor n/8 \rfloor),\ f(0) = 0$

30. Prove that if $n \geq 0$, $\lfloor n/2 \rfloor + \lceil n/2 \rceil = n$.

31. Find (*a*) $\lfloor 4.5 \rfloor$, $\lfloor -4.5 \rfloor$, $\lceil -5 \rceil$ (*b*) $\lceil 4.5 \rceil$, $\lceil -4.5 \rceil$, $\lceil -5 \rceil$ (*c*) $\lfloor -0.15 \rfloor$, $\lceil -0.15 \rceil$

32. Evaluate (*a*) $\log_2 16$ (*b*) $\log_2 128$ (*c*) $\log_{10} 100$ (*d*) $\log_{10} 0.001$ (*e*) $\log_3 27$

33. Show that if seven numbers from 1 to 12 are chosen, then two of them will add up to 13.

34. How many friends you must have to gurarantee that at least five of them will have birthdays in the same month?

35. Twenty cards numbered 1 through 20 are placed face down on a table. Cards are selected one at a time and turned over until 12 cards have been chosen. If two of the cards add up to 21, the player losses. Is it possible to win this game?

36. Determine each of the following

(*a*) $\lfloor 2.3 - 1.6 \rfloor$ (*b*) $\lfloor 2.3 \rfloor - \lfloor 1.6 \rfloor$

(*c*) $\lceil 3.4 \rceil \lfloor 6.2 \rfloor$ (*d*) $\lfloor 3.4 \rfloor \lceil 6.2 \rceil$

37. Find all real numbers x such that

(*a*) $9\lfloor x \rfloor = \lfloor 9x \rfloor$ (*b*) $\lfloor 9x \rfloor = 9$

(*c*) $\lfloor x + 9 \rfloor = x + 9$ (*d*) $\lfloor x + 7 \rfloor = \lfloor x \rfloor + 7$

38. For $n, k \in I_+$, prove that $\lceil n/k \rceil = \lfloor (n-1)/k \rfloor 1$.

39. Prove that if 151 integers are selected from the set A = {1, 2, 3, . . 300}, then the selection must include two integers x and y such that x/y or y/x.

40. Show that if any 14 integers are selected from the set A = {1, 2, 3, . . ., 25}, then there are at least two integers whose sum is 26.

41. Let ABCD be a square with AB = 1. Show that if five points in the interior of this square are selected, then there are at least two points whose distance apart is less than $1/\sqrt{2}$.

42. Let A be a set of five positive integers the maximum of which is at most 9. Prove that the sum of the elements in all the non-empty subsets of A cannot all be distinct.

43. If k_{n+1} pigeons occupy n pigeonholes, then prove that at least one pigeonhole has $k + 1$ or more pigeons mosting in it, where $n, k \in I_+$.

44. Determine 'big-oh' form for each of the following:

(*a*) $f(x) = x^3 - 5x^2 + 25x - 165$ (*b*) $f(n) = n^2 + (n-1)^2$

(*c*) $f(n) = 2 + 4 + 6 + \ldots 2n$ (*d*) $f(n) = \dfrac{n(n+1)(n+2)}{n+3}$.

45. Use definition of O-notation to show that g dominates *f*.

(*a*) $f(x) = 2^x$, $g(x) = 2^{2x} - 1000$ (*b*) $f(x) = 3x^2$, $g(x) = 2^x + 2x$

Hints and Answers

25. (*a*) {1, 2} (*b*) {A, B, C, D, E} (*c*) $\{a, b, c, \ldots, z\}$

26. (*a*) {4, 16, 64, 256} (*b*) {0, 2, 6, 12} (*c*) {1.5, 4, 6.5, 9}

27. (*a*) $a_n = 1 + 3(n-1)$, explicit (*b*) $a_n = (-1)^{n+1}$, explicit (*c*) $a_n = a_{n-1} + 3$, $a_1 = 2$, recursive

(*d*) $a_n = a_{n-1} + 2$, $a_1 = 2$, recursive

28. (*a*) 0, 1, 1, 3, 5, 11

(*b*) $f(20) = f(11) + 1 = [f(5) + 1] + 1 = f(5) + 2$

$= [f(2) + 1] + 2 = f(2) + 3 = [f(1) + 1] + 3$

$= f(1) + 4 = 0 + 4 = 4.$

Each time *n* is divided by 2, the value of *f* is increased by 1. Thus *f* is the greatest integer such that $f(n) = \log_2 n$.

29. (*a*) $f(n) = 3\lfloor \log_2 n \rfloor + 1$ (*b*) $f(n) = \lfloor \log_8 n \rfloor + 1$

30. Consider the cases both for even and odd values of *n*.

31. (*a*) $\lfloor 4.5 \rfloor = 4$, $\lfloor -4.5 \rfloor = -5$; $\lfloor -5 \rfloor = -5$ (*b*) $\lceil 4.5 \rceil = 5$, $\lceil -4.5 \rceil = -4$, $\lceil -5 \rceil = -5$ (*c*) – 1, 0

32. (*a*) $\log_2 16 = 4$, since $2^4 = 16$ (*b*) $\log_2 128 = 7$ since $2^7 = 128$ (*c*) $\log_{10} 100 = 2$, since $10^2 = 100$
(*d*) $\log_{10} 0.001 = -3$, since $10^{-3} = 0.001$ (*e*) $\log_3 27 = 3$, since $3^3 = 27$

34. (*a*) At least 49 friends are needed. (*b*) No, at least one pair of the 12 cards must add up to 21.

36. (*a*) $\lfloor 2.3 - 1.6 \rfloor = \lfloor 0.7 \rfloor = 0$ (*b*) $\lfloor 2.3 \rfloor - \lfloor 1.6 \rfloor = 2 - 1 = 1$

(*c*) $\lceil 3.4 \rceil \lfloor 6.2 \rfloor = 4 \times 6 = 24$ (*d*) $\lfloor 3.4 \rfloor \lceil 6.2 \rceil = 3 \times 7 = 21$

37. (*a*) $\ldots \cup [-1, -6/7) \cup [0, 1/7) \cup [1, 8/7) \cup [2, 15/7) \cup \ldots$

(*b*) [1, 8/7]

(*c*) Set of all integer numbers.

(*d*) Set of all real numbers.

39. For each $x \in \{1, 2, 3, \ldots, 300\}$, we may write $x = 2^n m$, where $n \geq 0$ and gcd $(2, m) = 1$. Then $m \in S = \{1, 2, 5, \ldots, 299\}$, where $|S| = 150$. Since 151 integers are selected from A, then there must be two numbers of the form $x = 2^s m$, $y = 2^t m$ for some $m \in S$. If $x < y$, then $x|y$. otherwise $y < x$ and $y|x$.

40. The integers 1, 2, . . ., 25 are considered pigeons and the pigeonholes are 13 sets {1, 25}, {2, 24}, . . . {11, 15}, {12, 14}, {13}. In selecting 14 integers, we get elements in at least one two-element subset and that sum to 26.

41. Divide the interior of the square ABCD into four smaller congruent squares each having diagonal of length $1/\sqrt{2}$. If five points are chosen in the interior of ABCD, then at least two are in R_1 for some $1 \le i \le 4$ and these are within $1/\sqrt{2}$ (units) of each other.

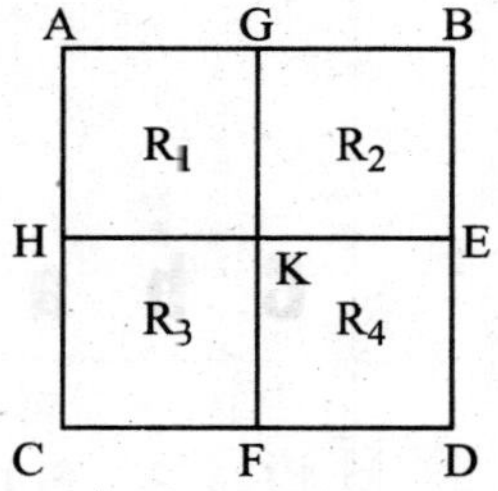

42. Consider the subset B of A where $1 \le B \le 3$. Since $|A| = 5$, there are ${}^5C_1 + {}^5C_2 + {}^5C_3 = 25$ such subsets B. The sum of elements in B will be between 1 and 24 (7 + 8 + 9). Hence, there are two subsets of A whose elements yield the same sum.

44. (*a*) $f \in O(x^3)$ (*b*) $f \in O(x^3)$ (*c*) $f \in O(x^2)$ (*d*) $f \in O(x^2)$

45. (*a*) For $x = 6$, $2^x = 64 < 3096 = 4096 - 1000 = 2^{12} - 1000 = 2^{2n} - 1000$. Assuming that $2^k < 2^{2k} - 1000$ for $n = k \ge 6$, we find that $2 < 2^2 \Rightarrow 2(2^k) < 2^2(2^{2k} - 1000) < 2^2, 2^{2k} - 1000$ or $2^{k+1} < 2^{2(k+1)} - 1000$, so $f(x) < g(x)$ for all $x \ge 6$. Thus with $k = 6$ and $m = 1$, we find that $n \ge k$ $|f(x)| \le m|g(x)$ and hence $f \in O(g)$.

Chapter 5

Mathematical Induction

PREVIEW

Mathematical induction is a deductive approach– a technique to prove a conjecture that is perhaps formulated by inductive reasoning. In other words, mathematical induction is not an exploratory proof technique– it can only confirm a correct conjecture.

The first principle of mathematical induction is an implication. That is,

- Basis step (on only basis): P(1) is true
- Inductive assumption (or hypothesis): If $P(k)$ true, then $P(k + 1)$ is also true, for all k.

Both of these implies that $P(n)$ is true for all positive integers n.

LEARNING OBJECTIVES

After studying this chapter, you should be able to

- carry out a proof by induction using either the first or second principle of induction.
- develop proofs of conjections by using either the direct proof, proof by contra-position, or proof by contradiction.

5.1 INTRODUCTION

One of the most important techniques to prove many mathematical statements or formulae which cannot be easily derived by direct methods is sometimes derived by using the principle of *mathematical induction.* In this principle, the word induction is associated with the *inductive reasoning*, by which a conclusion is drawn from a large number of special cases. Actually mathematical induction is deductive in nature, for it leads to a definite conclusion. It is usually employed in proving the validity of a statement involving all positive integer values of the number n.

5.2 PRINCIPLE OF MATHEMATICAL INDUCTION

Let P (n) be a statement or proposition defined by the set of positive integers, I_+ such that it is either true or false for all $n \in I_+$. For the given statement P (n), if we can prove that

(*a*) P (n) is true for $n = n_0$ (the smallest integer), and

(*b*) P (n) is true for $n = k + 1$, assuming that it is true for $n = k$ $(k \geq n_0)$.

then we can conclude that P (n) is true for all natural numbers $n \geq n_0$. Here the first proposition (*a*) is usually referred to as the *basis of induction.* Since assuming that P $(k + 1)$ would be true if P (k) is true is not the same as assuming that P (k) is true for some value of k, therefore the proposition (*b*) is usually referred to as the *induction step.* Also the assumption that P (n) is true for $n = k$ in (*b*) is referred to as the induction hypothesis.

The steps of the principle of mathematical induction are summarised as follows:

1. **Basis of induction** Assuming the validity of the statement for the smallest integral value of n (i.e. $n = 1, 2, 3$) for which it is true.
2. **Induction step** If the statement is true for $n = k$, where k denotes any value of n, then it is also true $n = k + 1$.
3. **Conclusion** The statement is true for all integral values of n equal to or greater than that for which it was verified in Step 1.

Illustration One of the methods of finding a square root on an ordinary computing or adding machine is based on the fact that the sum of n odd integers is equal to n^2, specifically,

$$1 = 1 = 1^2$$
$$1 + 3 = 4 = 2^2$$
$$1 + 3 + 5 = 9 = 3^2$$
$$1 + 3 + 5 + 7 = 16 = 4^2$$

Example 1 Prove by the principle of mathematical induction

$$\text{P}(n): 1 + 3 + 6 + \ldots + \frac{n(n+1)}{2} = \frac{n(n+1)(n+2)}{6}$$

Solution *Basis of induction:* For $n = 1$, the left hand side of statement, P (n) is $\frac{1(1+1)}{2} = 1$ and right hand side of P (n) is $\frac{1(1+1)(1+2)}{6} = 1$. Hence, P (n) is true for $n = 1$.

Induction step: Assuming that P (n) is true for $n = k$. Thus, we get

$$\text{P}(k): 1 + 3 + 6 + \ldots + \frac{k(k+1)}{2} = \frac{k(k+1)(k+2)}{6}$$

Adding the term $\frac{(k+1)(k+2)}{2}$ to both sides of P (k), we get

$$1 + 3 + 6 + \ldots + \frac{k(k+1)}{2} + \frac{(k+1)(k+2)}{2} = \frac{k(k+1)(k+2)}{6} + \frac{(k+1)(k+2)}{2}$$
$$= \frac{(k+1)(k+2)(k+3)}{6}$$

This shows that if P (n) is true for $n = k$, then it is also true for $n = k + 1$. Hence, by mathematical induction it is true for every value of n.

Example 2 Prove by the principle of mathematical induction

$$P(n) : 1 \cdot 2 + 2 \cdot 2^2 + \ldots + n \cdot 2^n = (n - 1)2^{n+1} + 2$$

Solution *Basis of induction:* If $n = 1$, then left hand side of the statement P (n) is $1 \cdot 2^1 = 2$ and right hand side of P (n) is $(1 - 1)2^{1+1} + 2 = 2$. Hence, P (n) is true for $n = 1$.

Induction step: Assuming that P (n) is true for $n = k$. Thus, we get

$$P(k) : 1 \cdot 2 + 2 \cdot 2^2 + \ldots + k \cdot 2^k = (k - 1)2^{k+1} + 2$$

Adding the term $(k + 1)2^{k+1}$ to both sides of P (k), we get

$$\begin{aligned} 1 \cdot 2 + 2 \cdot 2^2 + \ldots + k \cdot 2^k + (k + 1) \cdot 2^{k+1} &= (k - 1) \cdot 2^{k+1} + 2 + (k + 1) \cdot 2^{k+1} \\ &= 2^{k+1}\{k - 1 + k + 1\} + 2 \\ &= 2^{k+1} \cdot 2k + 2 = k \cdot 2^{k+1+1} + 2 \\ &= [(k + 1) - 1] \cdot 2^{k+1+1} + 2 \end{aligned}$$

This shows that if P (n) is true for $n = k$, then it is also true for $n = k + 1$. Hence, by mathematical induction P (n) is true for every integral value of n.

Example 3 Prove by the principle of mathematical induction

$$P(n) : \frac{1}{1 \cdot 2} + \frac{1}{2 \cdot 3} + \frac{1}{3 \cdot 4} + \ldots + \frac{1}{n(n+1)} = \frac{1}{n+1}$$

Solution *Basis of induction:* If $n = 1$, then left hand side of P (n) is $\frac{1}{1(1+1)} = \frac{1}{2}$ and right hand side is $\frac{1}{1+1} = \frac{1}{2}$. Hence, P ($n$) is true for $n = 1$.

Induction step: Assuming that P (n) is true for $n = k$. Thus, we get

$$P(k) : \frac{1}{1 \cdot 2} + \frac{1}{2 \cdot 3} + \frac{1}{3 \cdot 4} + \ldots + \frac{1}{k(k+1)} = \frac{1}{k+1}$$

Adding the term $\frac{1}{(k+1)(k+2)}$ to both sides of P (k), we get

$$\begin{aligned} \frac{1}{1 \cdot 2} + \frac{1}{2 \cdot 3} + \frac{1}{3 \cdot 4} + \ldots + \frac{1}{k(k+1)} + \frac{1}{(k+1)(k+2)} &= \frac{1}{(k+1)} + \frac{1}{(k+1)(k+2)} \\ &= \frac{k^2 + 2k + 1}{(k+1)(k+2)} = \frac{(k+1)(k+1)}{(k+1)(k+2)} \\ &= \frac{k+1}{k+2} = \frac{k+1}{(k+1)+1} \end{aligned}$$

This shows that if P (n) is true for $n = k$, then it is also true for $n = k + 1$. Hence, by mathematical induction P (n) is true for every integral value of n.

Example 4 Prove by the principle of mathematical induction the inequality

$$P(n) : (a + 1)^n \geq 1 + na, \text{ for } a > -1;\ n = 2, 3, 4$$

Solution *Basis of induction:* If $n = 2$, then left hand side of the given is inequality is $(c + 1)^2 = a^2 + 2a + 1$ and right hand side is $1 + 2a$.

Since left hand side is more than the right hand side, therefore inequality is ture for $n = 2$.

Induction step: Assuming that the inequality is true for $n = k$. Thus, we get

$$(a + 1)^k \geq 1 + ka, \text{ for } a > -1 \qquad \text{...(i)}$$

Since
$$\begin{aligned}(a + 1)^{k+1} &= (a + 1)^k (a + 1)\\ &\geq (1 + ka)(a + 1); \quad \text{From (i)}\\ &\geq 1 + (k + 1)a + ka^2\\ &\geq 1 + (k + 1)a, \text{ since } ka^2 > 0\end{aligned}$$

This shows that if the given inequality is true for $n = k$, then it is also true for $n = k + 1$. Hence, by mathematical induction the inequality is true for every integral value of n.

Example 5 Prove by the principle of mathematical induction that

$$P(n) : 1^2 + 2^2 + 3^2 + \ldots + n^2 = \frac{n(n+1)(2n+1)}{6}$$ [UPTU, MCA, 2006]

Solution *Basis of induction:* If $n = 1$, then left hand side of the statement P (n) is $1^2 = 1$ and right hand side is $\frac{1(1+1)(2\cdot 1+1)}{6} = 1$. Hence P ($n$) is true for $n = 1$.

Induction step: Assuming that P (k) is true for $n = k$. Then, we get

$$P(k) : 1^2 + 2^2 + 3^2 + \ldots + k^2 = \frac{k(k+1)(2k+1)}{6}$$

Adding the term $(k + 1)^2$ to both sides of P (k), we get

$$\begin{aligned}1^2 + 2^2 + 3^2 + \ldots + k^2 + (k + 1)^2 &= \frac{k(k+1)(2k+1)}{6} + (k + 1)^2\\ &= \frac{(k+1)}{6}\{k(2k + 1) + 6(k + 1)\}\\ &= \frac{(k+1)}{6}\{2k^2 + 7k + 6\}\\ &= \frac{(k+1)(k+2)(2k+3)}{6}\\ &= \frac{(k+1)\{(k+1)+1\}\{(2k+1)+1\}}{2}\end{aligned}$$

This shows that if P (n) is true for $n = k$, then it is also true for $n = k + 1$. Hence, by mathematical P (n) is true for every integral value of n.

Example 6 Using the principle of mathematical induction, prove that

$$P(n) : 1 + 3 + 5 + \ldots + (2n - 1) = n^2$$

Solution *Basis of induction:* If $n = 1$, the left hand side of P (n) is 1 and right hand side is 1. Hence, P (n) is true for $n = 1$.

Induction step: Assuming that P (n) is true for $n = k$. Then we get

$$P(k) : 1 + 3 + 5 + \ldots + (2k - 1) = k^2$$

Adding the term $(2k + 1)$ to both sides of P (k), we get

$$1 + 3 + 5 + \ldots + (2k - 1) + (2k + 1) = k^2 + (2k + 1) = (k + 1)^2$$

This shows that if P (n) is true for $n = k$, then it is also true for $n = k + 1$. Hence, by mathematical induction P (n) is true for every integral value of n.

Example 7 Using the principle of mathematical induction, prove that

$$P(n) : 1\cdot3 + 2\cdot4 + 3\cdot5 + \ldots + n(n + 2) = \frac{n}{6}(n + 1)(2n + 7)$$

Solution *Basis of induction:* If $n = 1$, then left hand side of P (n) is $1\cdot(1 + 2) = 3$ and right hand side is $\frac{1}{6}(1 + 1)(2\cdot1 + 7) = 3$. Hence, P (n) is true for $n = 1$.

Induction step: Assuming that P (n) is true for $n = k$. Then, we get

$$P(k) : 1\cdot3 + 2\cdot4 + 3\cdot5 + \ldots + k\cdot(k + 2) = \frac{k}{6}(k + 1)(2k + 7)$$

Adding the term $(k + 1)(k + 3)$ to both sides of P (k), we get

$$\begin{aligned} 1\cdot3 + 2\cdot4 + 3\cdot5 + \ldots + k(k + 2) + (k + 1)(k + 3) &= \frac{k}{6}(k + 1)(2k + 7) + (k + 1)(k + 3) \\ &= (k + 1)\left\{\frac{k}{6}(2k + 7) + (k + 3)\right\} \\ &= (k + 1)\left\{\frac{2k^2 + 13k + 18}{6}\right\} \\ &= \frac{(k + 1)(k + 2)(2k + 9)}{6} \end{aligned}$$

This shows that if P (n) is true for $n = k$, then it is also true for $n = k + 1$. Hence, by mathematical induction P (n) is true for every integral value of n.

Example 8 Prove by the principle of mathematical induction that

$$P(n) : \frac{1}{2} + \frac{1}{2^2} + \frac{1}{2^3} + \ldots + \frac{1}{2^n} = 1 - \frac{1}{2^n}$$

Solution *Basis of induction:* If $n = 1$, then left hand side of P (n) is $\frac{1}{2}$ and right hand side is $1 - \frac{1}{2} = \frac{1}{2}$. Hence P (n) is true for $n = 1$.

Induction step: Assuming that P (n) is true for $n = k$. Then, we get

$$P(k) : \frac{1}{2} + \frac{1}{2^2} + \frac{1}{2^3} + \ldots + \frac{1}{2^k} = 1 - \frac{1}{2^k}$$

Adding the term $\frac{1}{2^{k+1}}$ to both sides of P (k), we get

$$\frac{1}{2}+\frac{1}{2^2}+\frac{1}{2^3}+\ldots+\frac{1}{2^k}+\frac{1}{2^{k+1}}=1-\frac{1}{2^k}+\frac{1}{2^{k+1}}=1-\left[\frac{1}{2^k}-\frac{1}{2^{k+1}}\right]=1-\frac{1}{2^{k+1}}$$

This shows that if P (n) is true for $n = k$, then it is also true for $n = k + 1$. Hence, by mathematical induction P (n) is true for every integral value of n.

Example 9 Use the principle of mathematical induction to prove that

$$P(n): \sum_{n=1}^{n} \frac{1}{(2n-1)(2n+1)}=\frac{n}{2n+1}, \text{ for all } n \in N$$

or

$$\frac{1}{1\cdot 3}+\frac{1}{3\cdot 5}+\frac{1}{5\cdot 7}+\ldots+\frac{1}{(2n-1)(2n+1)}=\frac{n}{2n+1}$$

Solution *Basis of induction:* If $n = 1$, the LHS of P (n) is $\sum_{n=1}^{1} \frac{1}{(2n-1)(2n+1)}=\frac{1}{1\cdot 3}=\frac{1}{3}$ and RHS is $\frac{1}{2\cdot 1+1}=\frac{1}{3}$. Hence, P (n) is true for $n = 1$

Induction step: Assuming that P (n) is true for $n = k$. Then, we get

$$P(k): \sum_{r=1}^{k} \frac{1}{(2k-1)(2k+1)}=\frac{k}{2k+1}$$

Adding the term $\frac{1}{(2k+1)(2k+3)}$ to both sides of P (k), we get

$$\sum_{r=1}^{k} \frac{1}{(2k-1)(2k+1)}+\frac{1}{(2k+1)(2k+3)}=\frac{k}{2k+1}+\frac{1}{(2k+1)(2k+3)}$$

or

$$\sum_{r=1}^{k+1} \frac{1}{(2k-1)(2k+1)}=\frac{k(2k+3)+1}{(2k+1)(2k+3)}=\frac{(k+1)(2k+1)}{(2k+1)(2k+3)}=\frac{k+1}{2k+3}$$

This shows that if P (n) is true for $n = k$, then it is also true for $n = k + 1$. Hence, by mathematical induction P (n) is true for every given value of n.

Example 10 By the principle of induction, show that $3^{4n+2} + 5^{2n+1}$ is a multiple of 14, for all positive integral value of n including zero.

Solution Let the given expression P (n) : $3^{4n+2} + 5^{2n+1}$ be the multiple of 14.

Basis of induction: If $n = 1$, then we have P (1) : $3^{4\cdot 1+2} + 5^{2\cdot 1+1} = 854 = 14 \times 61$ which is a multiple of 14. Similarly, for $n = 2$,

$$P(2): 3^{4\cdot 2+2} + 5^{2\cdot 2+1} = 62174 = 14 \times 4441$$

is a multiple of 14.

Induction step: Assuming that the result is true for $n = k$, then

$$P(k) = 3^{4k+2} + 5^{2k+1} = 14 \times t\ ;\ t \in I$$

is multiple of 14.

Replacing k by $k + 1$ in P (k), we get

$$\begin{aligned}3^{4(k+1)+2} + 5^{2(k+1)+1} &= 3^{4k+6} + 5^{2k+3}\\ &= 3^{4k+2}.3^4 + 5^{2k+1}.5^2\\ &= 3^{4k+2}(11+70) + 5^{2k+1}(11+14)\\ &= 11(3^{4k+2} + 5^{2k+1}) + 70\cdot 3^{4k+2} + 14\cdot 5^{2k+1}\\ &= 11\cdot 14\, t + 14\{5\cdot 3^{4k+2} + 5^{2k+1}\}\\ &= 14\{11\, t + 5\cdot 3^{4k+2} + 5^{2k+1}\}\end{aligned}$$

which is a multiple of 14. Hence, the result is true for $n = k + 1$.

Moreover, for $n = 0$, we have

$$P(0) : 3^{4\cdot 0+2} + 5^{2\cdot 0+1} = 14\cdot 1$$

which is a multiple of 14. Hence, P (n) also holds true for $n = 0$.

Example 11 (*a*) If nth term of A.P. is $a + (n - 1)\,d$, then show by the principle of mathematical induction that the sum of n terms of A.P. is $\frac{n}{2}\{2a + (n-1)\,d\}$. That is, by the principle of mathematical induction, prove that

$$P(n) : a + (a + d) + (a + 2d) + \ldots + \{a + (n-1)\,d\} = \frac{n}{2}\{2a + (n-1)\,d\}$$

(*b*) Prove by the principle of mathematical induction the result

$$P(n) : a + ar + ar^2 + \ldots + ar^{n-1} = a\cdot\frac{r^n - 1}{r - 1}, \text{ if } r \neq 1$$

Solution (*a*) *Basis of induction:* For $n = 1$, LHS of P (n) is a and RHS is $\frac{1}{a}\{2a + (1-1)d\} = a$. Hence, P ($n$) is true for $n = 1$.

Assuming that P (n) is true for $n = k$. Then, we get

$$P(k) : a + (a + d) + (a + 2d) + \ldots + \{a + (k-1)d\} = \frac{k}{2}\{2a + (k-1)\,d\}$$

Adding the term $(a + kd)$ to both sides of P (k), we get

$$\begin{aligned}&a + (a + d) + (a + 2d) + \ldots + \{a + (k-1)d\} + (a + kd)\\ &= \frac{k}{2}\{2a + (k-1)d\} + (a + kd)\\ &= ak + \frac{k(k-1)}{2}\cdot d + a + kd = a(k+1) + \frac{k(k-1+2)}{2}\cdot d\\ &= a(k+1) + \frac{k(k+1)}{2}\cdot d = \frac{k(k+1)}{2}\{2a + (k+1-1)\cdot d\}\end{aligned}$$

This shows that if P (n) is true for $n = k$, then it is also true for $n = k + 1$. Hence, by mathematical induction P (n) is true for every positive value of n.

(*b*) *Basis of induction:* If $n = 1$, then LHS of P (n) is a and RHS is $a\left(\frac{r-1}{r-1}\right) = a$. Hence, P ($n$) is true for $n = 1$.

Induction step: Assuming that P (n) is true for $n = k$. Then we get

$$P(k) : a + ar + ar^2 + \ldots + ar^{k-1} = a \cdot \frac{r^k - 1}{r - 1}, \ r \neq 1.$$

Adding the term ar^k to both sides of P (k), we get

$$a + ar + ar^2 + \ldots + ar^{k-1} + ar^k = a \cdot \frac{r^k - 1}{r - 1} + ar^k$$

$$= a \cdot \frac{(r^k - 1 + r^{k+1} - r^k)}{r - 1} = \frac{a(r^{k+1} - 1)}{r - 1}, \ r \neq 1$$

This shows that if P (n) is true for $n = k$, then it is also true for $n = k + 1$. Hence, by mathematical induction P (n) is true for every integral value of n.

Example 12 Using principle of mathematical induction, show that

$$P(n) : 1 + p + p^2 + \ldots + p^{n-1} = \frac{p^n - 1}{p - 1} \ (p > 1)$$

for all natural numbers n.

Solution *Basis of induction:* For $n = 1$, LHS of P (n) is 1 and RHS is also 1. Hence, P (n) is true for $n = 1$.

Induction step: Assuming that P (n) is true for $n = k$. Then we have

$$P(k) : 1 + p + p^2 + \ldots + p^{k-1} = \frac{p^k - 1}{p - 1} \ (p > 1)$$

Adding the term p^k to both sides of P (k), we get

$$1 + p + p^2 + \ldots + p^{k-1} + p^k = \frac{p^k - 1}{p - 1} + p^k = \frac{p^k - 1 + p^{k+1} - p^k}{p - 1} = \frac{p^{k+1} - 1}{p - 1} \ (p > 1)$$

This shows that if P (n) is true for $n = k$, then it is also true for $n = k + 1$. Hence, by mathematical induction P (n) is true for every positive integral value of n.

Example 13 Let $\mathbf{A} = \begin{bmatrix} 1 & a \\ 0 & 1 \end{bmatrix}$ be a matrix. Prove by mathematical induction that $\mathbf{A}^n = \begin{bmatrix} 1 & an \\ 0 & 1 \end{bmatrix}$.

Solution *Basis of induction:* For $n = 2, 3$, by actual multiplication of two matrices, we have

$$\mathbf{A}^2 = \mathbf{A} \cdot \mathbf{A} = \begin{bmatrix} 1 & a \\ 0 & 1 \end{bmatrix} \cdot \begin{bmatrix} 1 & a \\ 0 & 1 \end{bmatrix} = \begin{bmatrix} 1 & 2a \\ 0 & 1 \end{bmatrix}$$

and $$\mathbf{A}^3 = \mathbf{A}^2 \cdot \mathbf{A} = \begin{bmatrix} 1 & 2a \\ 0 & 1 \end{bmatrix} \cdot \begin{bmatrix} 1 & a \\ 0 & 1 \end{bmatrix} = \begin{bmatrix} 1 & 3a \\ 0 & 1 \end{bmatrix}.$$

Hence, the result is true for $n = 2, 3$.

Induction step: Assuming that $\mathbf{A}^n$ is true for $n = k$, so that $\mathbf{A}^k = \begin{bmatrix} 1 & ka \\ 0 & 1 \end{bmatrix}$.

Multiplying both sides of $\mathbf{A}^k$ with matrix $\mathbf{A}$ once again, we get

$$\mathbf{A}^k \cdot \mathbf{A} = \mathbf{A}^{k+1} = \begin{bmatrix} 1 & ka \\ 0 & 1 \end{bmatrix} \cdot \begin{bmatrix} 1 & a \\ 0 & 1 \end{bmatrix} = \begin{bmatrix} 1 & a(k+1) \\ 0 & 1 \end{bmatrix}.$$

This shows that if $\mathbf{A}^n$ is true for $n = k$, then it is also true for $n = k + 1$. Hence, by mathematical induction $\mathbf{A}^n$ is true for every positive integral value of n.

Example 14 Prove by the principle of mathematical induction that P (n): $10^n + 3\cdot 4^{n+2} + 5$ is divisible by 9.

Solution *Basis of induction:* For $n = 1$, we have

$$10^1 + 3\cdot 4^{1+2} + 5 = 207 = 9 \times 23$$

This is divisible by 9.

Induction step: Assuming that P (n) is true for $n = k$, so that $10^k + 3\cdot 4^{k+2} + 5$ is divisible by 9.

Replacing k by $k + 1$ in P (k), we get

$$\begin{aligned} 10^{k+1} + 3\cdot 4^{k+1+2} + 5 &= 10\cdot 10^k + 3\cdot 4\cdot 4^{k+2} + 5 \\ &= (9 + 1)\cdot 10^k + 3\,(3 + 1)\cdot 4^{k+2} + 5 \\ &= 9\cdot 10^k + 10^k + 3\cdot 3\cdot 4^{k+2} + 3\cdot 4^{k+2} + 5 \\ &= 9\cdot 10^k + 9\cdot 4^{k+2} + (10^k + 3\cdot 4^{k+2} + 5) \\ &= 9\cdot 10^k + 9\cdot 4^{k+2} + 9m \qquad \text{(where } 9m = 10^k + 3.4^{k+2} + 5) \\ &= 9(10^k + 4^{k+2} + m) = 9t\,;\ \text{for } t = 10^k + 4^{k+2} + m \end{aligned}$$

which is divisible by 9.

This shows that if P (n) is true for $n = k$, then it is also true for $n = k + 1$. Hence, P (n) is true for all positive integral values of n.

Example 15 Prove by the principle of mathematical induction

$$P(n) : 1^3 + 2^3 + 3^3 + \ldots + n^3 = \frac{n^2 (n+1)^2}{4}$$

Solution *Basis of induction:* For $n = 1$, the LHS of P (n) is $1^3 = 1$ and RHS is also $\dfrac{1^2 (1+1)^2}{4} = 1$. Hence, P (n) is true for $n = 1$.

Induction step: Assuming that P (n) is true for $n = k$. Then we get

$$P(k) : 1^3 + 2^3 + 3^3 + \ldots + k^3 = \frac{k^2 (k+1)^2}{4} \qquad \ldots(ii)$$

Adding the term $(k + 1)^3$ to both sides of P (k), we get

$$\begin{aligned} 1^3 + 2^3 + 3^3 + \ldots + k^3 + (k + 1)^3 &= \frac{k^2 (k+1)^2}{4} + (k + 1)^3 \\ &= \frac{(k+1)^2}{4} \{k^2 + 4\,(k + 1)\} \\ &= \frac{(k+1)^2\ (k+2)^2}{4} = \frac{(k+1)^2\ \{(k+1)+1\}^2}{4} \end{aligned}$$

This shows that if P (n) is true for $n = k$, then it is also true for $n = k + 1$. Hence, by mathematical induction P (n) is true for every positive integral value of n.

Example 16 Show that $\sqrt{2}$ is an irrational number.

Solution Suppose $\sqrt{2}$ is not an irrational number. Then let $\sqrt{2} = p/q$, where p and q are integers having no common factor. Since $\sqrt{2} = p/q$, therefore $p^2 = 2q^2$. This implies that p^2 is even, i.e. p is even (as square of an odd number is always an odd number).

Let $p = 2n$, n being an integer. Then $2q^2 = p^2 = (2n)^2 = 4n^2$ or $q^2 = 2n^2 \Rightarrow q^2$ is even number. So q is even number. Since p and q both are even numbers, they will have common factor 2 which contradicts the assumption of p and q have no common factors. Therefore the assumption is false and $\sqrt{2}$ is an irrational number.

5.3 PRINCIPLE OF STRONG MATHEMATICAL INDUCTION

Let P(n) be a statement (or proposition) which may be either true or false for each integral value of n. The P(n) is true for all positive integers provided for n_0, $n_1 \in I_+$, where $n_1 \geq n_0$. Thus two cases may arise:

(*a*) $P(n_0), P(n_0 + 1), ..., P(n_1 - 1), P(n_1)$ are all true.

(b) For some $k \in I_+$, where $k \geq n_1$, the assumption that $P(n_0), P(n_0 + 1), ..., P(k - 1), P(k)$ are true implies that $P(k + 1)$ is also true. Then P(n) is true for all $n \geq n_0$.

Example 17 Show that for Fibonacci sequence defined as

$$F_1 = 1, F_2 = 1 \text{ and } F_n = F_{n-1} + F_{n-2}, \text{ for all } n \geq 3,$$

the nth Fibonacci number $F_n < 2^m$ of each positive integer n.

Solution Let the statement P(n) be given by: P(n) : $F_n < 2^n$.

Basis of induction: For $n = 1$, the RHS of P(n) is $F_1 = 1(< 2) = 2^1$, and hence, P(n) is true for $n = 1$.

Induction Step: Assuming that P(n) is true for $n = k$. Then P(1), P(2), P(3), ..., P(k) are true for some $k \in I_+$ where $k \geq 2$.

To show that if P(n) is true for $n = 1, 2, ..., k$, then P($k + 1$) is also true, i.e. $F_{m+1} < 2^{m+1}$. Now

$$\begin{aligned} F_{m+1} &= F_m + F_{m-1} < 2^m + 2^{m-1} \\ &= 2^{m-1}(2 + 1) < 2^{m-1}(2 + 2) \\ &= 2^{m-1} . 2^2 < 2^{m+1} \end{aligned}$$

i.e., $F_{m+1} < 2^{m+1}$. This shows that if P(n) is true for $n = 1, 2, 3, ..., k$, then it is also true for $n = k + 1$

Hence, by the principle of strong mathematical induction P(n) is true for all positive integer n.

Example 18 If F_n is the nth Fibonacci number, prove that

$$F_n = \frac{1}{\sqrt{5}}\left[\left(\frac{1+\sqrt{5}}{2}\right)^n - \left(\frac{1-\sqrt{5}}{2}\right)^n\right]$$

for all positive integral values of $n \geq 1$.

Solution *Basis of induction:* For $n = 1$, the LHS for F_n is 1, i.e. $F_1 = 1$ and RHS is

$$\frac{1}{\sqrt{5}}\left[\left(\frac{1+\sqrt{5}}{2}\right) - \left(\frac{1-\sqrt{5}}{2}\right)\right] = 1.$$

Hence, F_n is true for $n = 1$.

Induction Step: Assuming that F_n is true for $n = k$. Then, we get

$$F_k = \frac{1}{\sqrt{5}}\left[\left(\frac{1+\sqrt{5}}{2}\right)^k - \left(\frac{1-\sqrt{5}}{2}\right)^k\right]$$

Substituting k for $k + 1$ to both side of this equation, we get

$$F_{k+1} = \frac{1}{\sqrt{5}}\left[\left(\frac{1+\sqrt{5}}{2}\right)^{k+1} - \left(\frac{1-\sqrt{5}}{2}\right)^{k+1}\right]$$

Suppose $\alpha = \dfrac{1+\sqrt{5}}{2}$ and $\beta = \dfrac{1-\sqrt{5}}{2}$. Then

$$F_{k+1} = F_k + F_{k-1} \qquad \text{[by definition of Fibonacci sequence]}$$

$$= \frac{1}{\sqrt{5}}[\alpha^k - \beta^k] + \frac{1}{\sqrt{5}}[a^{k-1} - \beta^{k-1}]$$

$$= \frac{1}{\sqrt{5}}[\alpha^{k-1}(\alpha+1) - \beta^{k-1}(\beta+1)]$$

where $\alpha + 1 = \dfrac{1+\sqrt{5}}{2} + 1 = \dfrac{3+\sqrt{5}}{2} = \alpha^2$, and $\beta + 1 = \dfrac{1-\sqrt{5}}{2} + 1 = \dfrac{3-\sqrt{5}}{2} = \beta^2$

Putting these values in F_{k+1}, we get

$$F_{k+1} = \frac{1}{\sqrt{5}}[\alpha^{k-1}.\alpha^2 - \beta^{k-1}.\beta^2] = \frac{1}{\sqrt{5}}[\alpha^{k+1} - \beta^{k+1}]$$

$$F_{k+1} = \frac{1}{\sqrt{5}}\left[\left(\frac{1+\sqrt{5}}{2}\right)^{k+1} - \left(\frac{1-\sqrt{5}}{2}\right)^{k+1}\right]$$

This Shows that if F_n is ture for $n = k$, then it also true for $n = k + 1$.

Hence, by the principle of mathematical induction, F_n is true for all integer value of $n \geq 1$.

Example 19 If n is an integer greater than 1, then show that n can be written as the product of primes.

Solution Let the statement or proposition is defined as: $P(n) = n$, where n can be written as the product of primes.

Basic of induction: Since $n > 1$ already, taking $n = 2$. As 2 can be written as the product of one prime (itself). Thus P(2) is true for $n = 2$.

Induction step: Assuming that $P(n)$ is true for $n = k$, i.e. k can be written as product of primes where k is a positive integer, $k \leq n$.

Let $P(n)$ is true for all positive integers k, such that $k \leq n$. Then to prove that the result is also true for $P(n + 1)$ we need to show that $(n + 1)$ can be written as product of primes. Following two cases may arise.

(*a*) $(n + 1)$ is prime, and (*b*) $(n + 1)$ is composite.

If $(n + 1)$ is composite and can also be written as the product of two positive integers p and q such that $2 \le p \le q < n + 1$.

Since both p and q can be written as product of primes, then it implies that $(n + 1)$ is composite and can be written as the product of primes consisting of all the prime factors of p and q.

Self-practice Problems

Prove the following results by using principle of mathematical induction.

1. $4 + 8 + 12 + \ldots + 4n = 2n(n + 1)$.

2. $2^1 + 2^2 + 2^3 + \ldots + 2^n = 2^{n+1} - 2$.

3. $1^3 + 3^3 + 5^3 + \ldots + (2n - 1)^3 = n^2(2n^2 - 1)$.

4. (*a*) $1\cdot 2\cdot 3 + 2\cdot 3\cdot 4 + 3\cdot 4\cdot 5 + \ldots + n(n + 1)(n + 2) = \dfrac{n(n+1)(n+2)(n+3)}{4}$.

(*b*) $1\cdot 2 + 2\cdot 3 + 3\cdot 4 + \ldots + n(n + 1) = \dfrac{n(n+1)(n+2)}{3}$.

5. $a^{2n} - b^{2n}$ is divisible by $a - b$.

6. $7^{2n} + 16n$ is divisible by 64.

7. $n(n^2 - 1)$ is divisible by 24; n is positive odd integer.

8. $1\cdot 3 + 2\cdot 4 + 3\cdot 5 + \ldots + n(n - 2) = \dfrac{n}{6}(n + 1)(2n + 7)$.

9. $(x + a)^n = x^n + {}^nC_1 x^{n-1}\cdot a + {}^nC_2 x^{n-2}\cdot a^2 + \ldots + {}^nC_r x^{n-r}\cdot a^r + \ldots + {}^nC_n a^n$, n being a positive integer.

10. (*a*) $1^2 + 2^2 + 3^2 + \ldots + r^2 + \ldots + n^2 = \dfrac{n}{6}(n + 1)(2n + 1)$.

(*b*) $1^2 - 2^2 + 3^2 + \ldots + (-1)^{n+1} n^2 = \dfrac{(-1)^{n+1} n(n+1)}{2}$.

11. $\dfrac{1}{2\cdot 5} + \dfrac{1}{5\cdot 8} + \dfrac{1}{8\cdot 11} + \ldots$ to n terms $= \dfrac{n}{2}(3n + 2)$.

12. (*a*) $\dfrac{1}{2\cdot 5} + \dfrac{1}{5\cdot 8} + \dfrac{1}{8\cdot 11} + \ldots$ to n terms $= \dfrac{n}{n+1}$.

(*b*) $\dfrac{1}{n+1} + \dfrac{1}{n+2} + \ldots + \dfrac{1}{2n} > \dfrac{13}{24}$, for $n \ge 2$.

13. (*a*) $x^n - y^n$ is divisible by $x + y$ when n is even.

(*b*) $x^n - 1$ is divisible by $x - 1$ for all positive integral value of n.

(*c*) $x^{2n} - y^{2n}$ is divisible by $x - y$, when n is a natural number. [*UPTU, BTech, 2006*]

14. $n(n + 1)(n + 2)$ is a multiple of 6, when n is a natural number.

15. $\dfrac{n^5}{5} + \dfrac{n^3}{3} + \dfrac{7n}{15}$ is a natural number; n a natural number.

16. (*a*) $10^{2n-1} + 1$ is divisible by 11 for each natural number.

(*b*) $6^{n+2} + 7^{2n+1}$ is divisible by 43 for each natural number.

(*c*) $n^3 + 2n$ is divisible by 3 for all $n \geq 1$.

17. $1^2 + 3^2 + 5^2 + \ldots + (2n-1)^2 = \dfrac{n(2n-1)(2n+1)}{3}$.

18. $1 \cdot 3 + 3 \cdot 5 + 5 \cdot 7 + \ldots + (2n-1)(2n+1) = \dfrac{n}{3}(4n^2 + 6n - 1)$.

19. (*a*) $1 + \dfrac{1}{1+2} + \dfrac{1}{1+2+3} + \ldots + \dfrac{1}{1+2+3+\ldots+n} = \dfrac{2n}{n+1}$.

(*b*) $\dfrac{1^2}{1 \cdot 3} + \dfrac{2^2}{3 \cdot 5} + \ldots + \dfrac{n^2}{(2n-1)(2n+1)} = \dfrac{n(n+1)}{2(2n+1)}$.

20. $\log_a x^n = n \log_a x$, n being a positive integer.

21. $2 + 6 + 10 + \ldots + (4n - 2) = 2n^2$.

22. $2 \cdot 3 + 4 \cdot 5 + 6 \cdot 7 + \ldots$ up to n terms $= \dfrac{n(n+1)(4n+5)}{3}$.

23. (*a*) $n(n+1)(n+2)$ is multiple of 6.

(*b*) $n(n+1)(2n+1)$ is divisible by 6.

24. $\dfrac{1}{5}n^5 + \dfrac{1}{3}n^3 + \dfrac{7}{15}n$ is a natural number for all n.

Chapter 6

Combinatorics

PREVIEW

Combinatorics is the branch of mathematics that deals with counting. This chapter introduces some of the basic tools of combinatorics like sum and product rules, permutations, combinations, and pigeon hole principle. Next it deals with multinomial coefficient, principle of inclusion and exclusion and lastly derangements. The knowledge of combinatorics is important whenever there are finite resources and/or interested in efficiency of an algorithm.

LEARNING OBJECTIVES

After studying this chapter, you should be able to

- use the formulae of both permutations and combinations of r objects, with or without repetitions, from a set of n district (or not all district) objects,
- use the pigeon hole principle to decide when common events must occur,
- use binomial theorem to expand the expression $(a + b)^n$.

6.1 INTRODUCTION

In this chapter counting techniques which are useful in the development of several important concepts in mathematics and computer science particularly permutations and combinations will be discussed. The concept of permutations and combinations supports pigeonhole principle and is useful in the analysis of probability of an event.

6.2 SUM AND PRODUCT RULES

There are two fundamental principles of combining or selection of numbers. The first is the *principle of product* (or *multiplication*) in which outcomes of a random experiment occur independently one after the other, and second is the *principle of sum* (or *addition*) in which any of the outcomes can occur. Sometimes, these two principles have to be combined depending on the nature of the problem under analysis. When outcomes of several experiments are combined, the rules stated below are followed:

Rule of product If two tasks must be performed and if the first task can be performed independently in n_1 different ways, and second task can be performed independently in n_2 different ways, then the total number of different ways in which two tasks can be performed together are $n_1 n_2$. The extended rule of product is stated as:

> *Suppose that k tasks T_1, T_2, T_3, . . ., T_k are to be performed in this sequence. If T_1 can be performed in n_1 ways, and for each of these ways T_2 can be performed in n_2 ways, and for each of these $n_1 n_2$ ways of performing $T_1 T_2$ in this sequence, T_3 can be performed in n_3 ways, and so on, then all k tasks can be performed in the sequence $T_1 T_2 \ldots T_k$ in exactly $n_1 n_2 \ldots n_k$ different ways.*

Illustration In particular, let $k = 2$. If the first task T_1 can be performed in $n_1 = 2$ ways, then each such choice will result in different sequences of performing T_1. Now for each sequence of performing T_1, we may choose $n_2 = 3$ ways (say) of performing task T_2. Thus, there will be $n_1 n_2$ ways, in all of performing the tasks T_1 and T_2 in the sequence $T_1 T_2$, as shown in Figs. 6.1(*a*), (*b*) and (*c*).

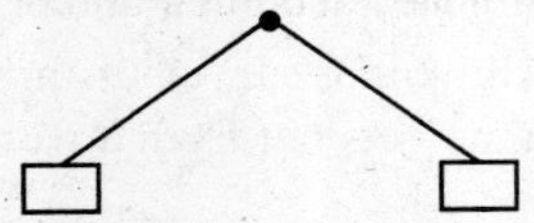

Fig. 6.1(*a*) $n_1 = 2$ Ways of Performing T_1

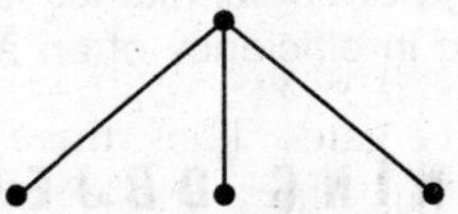

Fig. 6.1(*b*) $n_2 = 3$ Ways of Performing T_2

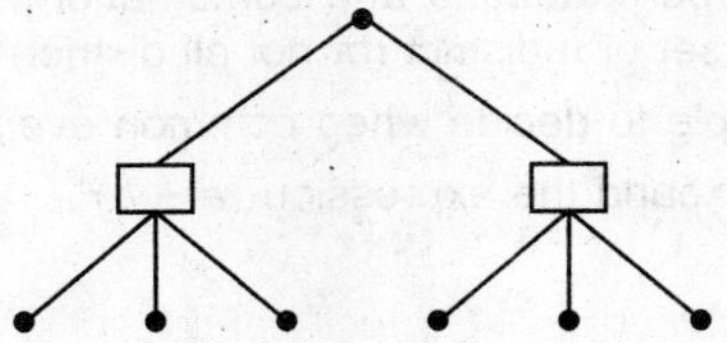

Fig. 6.1(*c*) $n_1 n_2 = 6$ Ways of Performing Sequence of Tasks $T_1 T_2$

This result can also be proved by using the principle of mathematical induction.

Rule of sum If one task can be performed independently in n_1 different ways and second task can be performed independently in n_2 different ways, then total number of ways in which exactly one of these tasks can be performed are $n_1 + n_2$.

This rule can also be extended to include more than two tasks in the same manner.

Illustration Suppose there are 5 different optional papers to select in third semester and 4 different optional papers to select in fourth semester by the MCA students. Thus, as per the *rule of product* there will be 5×4 choices for students who want to select one paper in third semester and one in the fourth semester. On the other hand, as per the *rule of sum*, students will have $5 + 4$ choices to select only one paper.

Example 1 A computer program consists of one letter followed by three digits. If repetition are allowed, then how many ways different label identifiers are possible?

Solution There are 26 English alphabet and 10 digits from 0 to 10. Thus, each sequence of three digits can be formed in 10 ways. Hence, total number of ways in which different label identifiers are possible are $26 \times 10 \times 10 \times 10 = 26{,}000$.

Example 2 Let $A = \{a_1, a_2, \ldots, a_n\}$ be a set with n elements. Find the subsets of A.

Solution From the definition of characteristic function f_A of set A in we know that

$$f_A(x) = \begin{cases} 1 & \text{if } x \in A \\ 0 & \text{if } x \notin A \end{cases}$$

The function f_A represents an array of zeros and ones having length n. Each place of the array can be filled in two ways with 0s and 1s, Thus by the *rule of product*, the number of ways of filling array of length n will be $2 \cdot 2 \cdot \ldots \cdot n$ times $= 2^n$. Hence, total number of subsets of A are 2^n.

Example 3 A football stadium has five gates on the eastern boundary and four gates on the western boundary.

(*a*) In how many ways can a person enter through an east gate and leave by a western gate?

(*b*) In how many different ways in all can a person enter and get out through different gates?

Solution (*a*) Since there are five gates on eastern side, the person can choose to enter through anyone gate out of these five gates. Thus, there are 5 ways of entrance. Again, he gets out from a west gate which are 4 in number. Thus, there are 4 ways in which he can go out.

Hence, the total number of ways in which he can enter from an eastern gate and go out from a western gate is $5 \times 4 = 20$.

(*b*) In this case the person has a choice to enter through any one of 9 gates. So there are 9 ways in which he can enter and he can get out from any one of the remaining 8 gates as he cannot go out from the gate through which he had entered.

Hence, the total number of ways in which he can enter and go out is $9 \times 8 = 72$.

Example 4 In how many different ways 3 rings of a lock can be combined when each ring has 10 digits 0 to 9? If the lock opens with only one combination of 3 digits, how many unsuccessful events are possible?

Solution The ways in which 3 rings can be combined are $10 \times 10 \times 10 = 1000$. But the lock opens with only one combination of 3 digits, therefore the unsuccessful events (attempts) will be $1000 - 1 = 999$.

Example 5 How many seven digits telephone numbers are possible, if

(*a*) only odd digits may be used?

(*b*) the number must be a multiple of 100?

(*c*) the first three digits are 481?

Solution (*a*) The odd digits are 1, 3, 5, 7 and 9. Each of the 7 places can be filled in 5 ways by odd digits to form a seven digit number. Hence, there can be $5 \times 5 \times 5 \times 5 \times 5 \times 5 \times 5 = (5)^7$ different numbers.

(*b*) To form such numbers from all the digits 0, 1, 2, . . ., 9 which must be a multiple of 100, the last digit in each number must be 0. Thus, while forming such numbers we fill the first five place by these 10 digits in different ways. The first five places can be filled in $10 \times 10 \times 10 \times 10 \times 10 = 10^5$ ways.

(*c*) Since the first three digits out of seven are fixed, thus, we have only to rearrange the remaining 4 digits. Hence, the possible ways can be: $10 \times 10 \times 10 \times 10 = 10^4$.

Example 6 Find the numbers consisting of distinct even digits between 100 and 999.

Solution There are five even digits: 0, 2, 4, 6 and 8 between 0 to 10. The number between 100 and 999 shall consist of three digits. The first position from the left can have any one of the four digits 2, 4, 6 and 8. The second position from the left can have any one of the five digits 0, 2, 4, 6 and 8 which has not been placed at the first position (to make the digits distinct). Similarly the third position from the left can have anyone of the five digits none of which have been placed either at the first position or the second position. It this way the first position can be occupied in 4 ways, the second in 5 ways and the third also in 5 ways. Hance the total number of ways can be: $4 \times 5 \times 5 = 100$.

6.2.1 Kramp's Factorial Notation

The product of first n natural numbers, viz., 1, 2, 3, . . ., n is denoted by $n!$ and it is read as *factorial n* or *n factorial*. Thus,

$$n! = 1 \cdot 2 \cdot 3 \cdot 4 \cdot \ldots \cdot (n-1) \cdot n$$

It is usually written as

$$\begin{aligned} n! &= n\,(n-1)\,(n-2) \ldots 3 \cdot 2 \cdot 1 \\ &= n\,(n-1)! \\ &= n\,(n-1)(n-2)! \\ &= n\,(n-1)\,(n-2) \ldots (n-r+1)\,(n-r)! \end{aligned}$$

From here it may be observed that the above definition of $n!$ is recursive. Thus

(*i*) If $n = 0$, then $n! = 1$

(*ii*) If $n > 0$, then $n! = n\,(n-1)!$

6.3 PERMUTATIONS

If there are n objects, then a sequence (or order) of r different objects is called *permutation* of n objects taken r at a time. The total number of permutations of n objects taken r at a time is denoted by ${}^nP_r$.

Theorem 6.1 Let A = $\{a_1, a_2, a_3, \ldots, a_n\}$ be a set of n objects. Then the number of permutations of n different objects taken r $(1 \le r \le n)$ at a time is

$$ {}^nP_r = n\,(n-1)\,(n-2) \ldots (n-r+1) = \frac{n!}{(n-r)!}.$$

Proof The total number of permutations of n different objects taken r at a time is equal to the number of ways in which r places can be filled up by n different objects.

The first place can be filled in n ways as anyone of the n given objects can be put in it, i.e. ${}^nP_1 = n$.

After filling the first place in anyone of the n ways, the second place can be filled up by any one of the remaining $(n - 1)$ objects, i.e. corresponding to each of the n ways of filling the first place, there are $(n - 1)$ ways of filling the second place. Thus, the first two places can be filled up in $n \times (n - 1)$ ways, i.e. ${}^nP_2 = n(n - 1)$.

Similarly, there remains $(n - 2)$ ways to fill up the third place. Thus, the first three places can be filled in $n(n - 1)(n - 2)$ ways, i.e. ${}^nP_3 = n(n - 1)(n - 2)$.

Proceeding in the same way, there remains $n - (r - 1) = n - r + 1$ ways to fill up the rth place. Thus, the number of ways in which r places can be filled up will be

$$\begin{aligned} {}^nP_r &= n(n-1)(n-2)\ldots \text{ to } r \text{ factors} \\ &= n(n-1)(n-2)\ldots(n-r+1) \end{aligned}$$

6.3.1 Important Deductions

1. ${}^nP_r = \dfrac{n!}{(n-r)!}$

   We know that ${}^nP_r = n(n - 1)(n - 2) \ldots (n - r + 1)$. Multiplying and dividing righthand side by $(n - r)! = (n - r)(n - r + 1) \ldots 3\cdot2\cdot1$, we get

$$ {}^nP_r = \frac{[n(n-1)(n-2)\ldots(n-r+1)][(n-r)(n-r+1)\ldots3\cdot2\cdot1]}{[(n-r)(n-r+1)\ldots3\cdot2\cdot1]}$$

$$= \frac{n(n-1)(n-2)\ldots3\cdot2\cdot1}{(n-r)(n-r+1)\ldots3\cdot2\cdot1} = \frac{n!}{(n-r)!}$$

2. ${}^nP_r = n!$

   LHS $= {}^nP_r = n(n - 1)(n - 2) \ldots (n - n + 1)$

   $= n(n - 1)(n - 2) \ldots 3\cdot2\cdot1 = n! =$ RHS

3. ${}^nP_{n-1} = {}^nP_n$

   LHS $= {}^nP_{n-1} = n(n - 1)(n - 2) \ldots [n - (n - 1) + 1]$

   $= n(n - 1)(n - 2) \ldots 2\cdot1 = n! = {}^nP_n =$ RHS

4. $0! = 1$

Example 7 If ${}^{56}P_{r+6} : {}^{54}P_{r+3} = 30{,}800 : 1$, find the value of r.

Solution Since ${}^nP_r = n!/(n - r)!$, therefore

$${}^{56}P_{r+6} = 56!/(56 - r - 6)! = 56!/(50 - r)!$$

and $${}^{54}P_{r+3} = 54!/(54 - r - 3)! = 54!/(51 - r)!$$

Now $$\frac{{}^{56}P_{r+6}}{{}^{54}P_{r+3}} = \frac{56!}{(50-r)!} \times \frac{(51-r)!}{54!} = \frac{56\cdot55\cdot54!}{(50-r)!} \times \frac{(51-r)(50-r)!}{54!} = 56\cdot55\,(51 - r)$$

But $\dfrac{^{56}P_{r+6}}{^{54}P_{r+3}} = \dfrac{30{,}800}{1}$, therefore

$$56 \cdot 55\,(51 - r) = 30{,}800$$

$$(51 - r) = 10, \text{ i.e. } r = 41.$$

6.3.2 Permutations with Repetition of Objects

Theorem 6.2 Let A = $\{a_1, a_2, \ldots, a_n\}$ is the set of n elements. Then number of permutations of elements of A taken r at a time, allowing repetition is n^r. Each such permutation is also called a *sequence of length r*.

Proof As shown below, any order arrangement of r elements of set A can be formed by filling its elements in r boxes (places) in the order

1	2	3	...	$r-1$	r

Since any one element of A can occupy the first place in n ways, while second place can also be filled up in n ways because all n elements are available to be filled at the second place as elements may be repeated. Therefore, first two places can be filled up in $n \times n = n^2$ ways. In the same way, rth place can be filled up in n ways. Hence, the total number of ways in which all r elements can be arranged is $n \times n \times n \times \ldots$ to r terms $= n^r$.

Theorem 6.3 The number of permutations of n objects of which p objects are of one type, q are of second type, r are of third type, and remaining objects are different is, $x = \dfrac{n!}{p! \times q! \times r!}$.

Proof Let x be the required number of permutations. If any one of these permutations having p objects not all different is replaced by p new objects different from each other, then $p!$ new permutations can be obtained without changing the positions of all other objects. Hence, this type of change in x number of permutations will produce $x \times p!$ permutations.

Again, if the q number of objects be replaced by q new objects different from each other and also different from the rest in each of $x \times p!$ permutations, then we will get $x \times p! \times q!$ number of permutations out of $n \times p!$ permutations.

Similarly, if r number of objects be replaced by r new objects different from each other and also different from the rest in each of $x \times p! \times q!$ permutations, then we will get $n \times p! \times q! \times r!$ number of permutations.

Now the remaining objects are different and the permutations of n different objects taken all at a time is $n!$ Therefore

$$x \times p! \times q! \times r! = n! \quad \text{or } x = \frac{n!}{p! \times q! \times r!}$$

Example 8 Find out how many 5-digit numbers greater than 30,000 can be formed from the digits 1, 2, 3, 4, 5?

Solution In order to find the 5-digit numbers greater than 30,000 from the digits 1, 2, 3, 4, 5, the first digit can be taken as 3, 4 or 5. But, the first digit can be chosen in $^3P_1 = 3$ ways and the remaining 4 digits can be any of the 4 digits taken in $^4P_4$ ways. Hence, the total number of 5-digit numbers greater than 30,000 will be $^3P_1 \times {}^4P_4 = 3 \times 4!$

Example 9 Find the number of permutations which can be made with the letters of the word ENGINEERING.

Solution The word ENGINEERING consists of 11 letters of which there are 3Es, 3Ns, 2Gs, 2Is and rest all are different. Thus the possible permutations of the word are

$$\frac{11!}{3!\times 3!\times 2!\times 2!} = \frac{11\cdot 10\cdot 9\cdot 8\cdot 7\cdot 6\cdot 5\cdot 4\cdot 3!}{3!\times 3!\times 2!\times 2!} = 2{,}77{,}200.$$

Example 10 How many permutations can be made with the letters of the word MISSISSIPPI taken all together? In how many of these will be vowels occupying the even places?

Solution (*i*) The word MISSISSIPPI consists of 11 letters of which there are 4Ss, 4Is, 2Ps and single M. Hence, the possible permutations of the word are

$$\frac{11!}{4!\times 4!\times 2!} = \frac{11\cdot 10\cdot 9\cdot 8\cdot 7\cdot 6\cdot 5\cdot 4!}{4!\times 4!\times 2!} = 34{,}650.$$

(*ii*) Out of 11 letters, there are 5 even places which can be filled up by the 4 vowels, (4Is) in 5!/4! ways only. The remaining 7 places can be filled up by 6 consonants of which there are 4Ss, 2Ps and single M. Hence, possible number of ways of filling 7 places with 6 consonants $\frac{5!}{4!}\times\frac{7!}{4!2!} = 525.$

Example 11 How many permutations can be made with letters of the word CONSTITUTION and

(*a*) in how many ways vowels occur together?

(*b*) in how many ways consonants and vowels occur alternatively?

(*c*) how many of these will have the letter N both at the beginning and at the end?

Solution The word CONSTITUTION consists of 12 letters of which there are 2Os, 2Ns, 3Ts, 2Is and remaining 3 are all different. Thus, the total possible number of permutations will be

$$\frac{12!}{2!\times 2!\times 3!\times 2!} = \frac{12\times 11\times 10\times 9\times 8\times 7!}{2\times 2\times 3\times 2\times 2} = 99{,}79{,}200$$

(*a*) There are seven consonants CNNSTTT and five vowels OOIIU. To find the number of permutations in which 5 vowels occur together, consider them as one letter. Thus, there remain 7 + 1 = 8 letters for the permutations of which there are 2Ns, 3Ts. Thus, total number of permutations will be

$$\frac{8!}{2!\times 3!} = 3{,}360.$$

But 5 vowels which were considered as single letter consist of 2Os, 2Is can be permutated themselves in 5!/(2! × 2!) = 30 ways. Hence, the total number of permutations are

$$\frac{8!}{2!\times 3!}\times\frac{5!}{2!\times 2!} = 1{,}00{,}800.$$

(*b*) The vowels and consonants can occur alternatively if vowels occur at 2nd, 4th, 6th, 8th and 10th places only. Thus, the number of permutations of these vowels will be 5!/(2! × 2!) = 30.

Further, seven consonants can be put at the remaining seven places in 7!/(2! × 3!) different ways.

Hence, the required number of ways in which vowels and consonants occur alternatively are

$$\frac{5!}{2!\times 2!}\times\frac{7!}{2!\times 3!} = 12{,}600.$$

(*c*) Fix up one N each at the beginning and at the end. Then the number of words that can be formed with remaining 10 letters, $10!/(2! \times 2! \times 3!) = 1{,}51{,}200$.

Example 12 How many numbers greater than a million can be formed with the digits 1, 7, 1, 0, 7, 3, 7?

Solution The total number of digits are 7, of which there are 2 ones, 3 sevens and rest all are different. Each number greater than a million must consist of 7 or more digits. Thus, the total numbers will be $7!/(2! \times 3!) = 420$. Out of these numbers there are some numbers which begin with zero and will naturally by less than a million, therefore, such type of numbers are undesirable. The numbers beginning with zero are $6!/(2! \times 3!) = 60$. Hence, the required numbers are $420 - 60 = 360$.

Example 13 In how many ways can the letters of the word ALLAHABAD be arranged? How many of these permutations are there in which

(*a*) two Ls come together?

(*b*) two Ls do not come together?

(*c*) two Ls and the two As come together?

Solution The word ALLAHABAD consists of 9 letters of which there are 4As, 2Ls and rest are different. Hence, the total number of ways in which the letters can be arranged among themselves are $9!/(4! \times 2!) = 7{,}560$.

(*a*) The number of arrangements in which 2Ls come together can be obtained by treating 2Ls as one letter. Thus, there remains only 8 letters and there are 4As. So the total number of permutations will be $8!/4! = 1{,}680$.

(*b*) The number of permutations in which 2Ls do not come together can be obtained by subtracting from the total number of permutations, those in which 2Ls come together. Thus, the required number is $7{,}560 - 1{,}680 = 5{,}880$.

(*c*) The number of permutations in which 2Ls and 4As come together can be obtained by treating 2Ls and 4As as a single letter. Thus, there are 7 letters which are all different and so the number of arrangements are $7! = 5{,}040$.

6.3.3 Circular Permutations

Theorem 6.4 If objects are arranged in a circular order, then the circular permutations of n different objects is $(n - 1)!$.

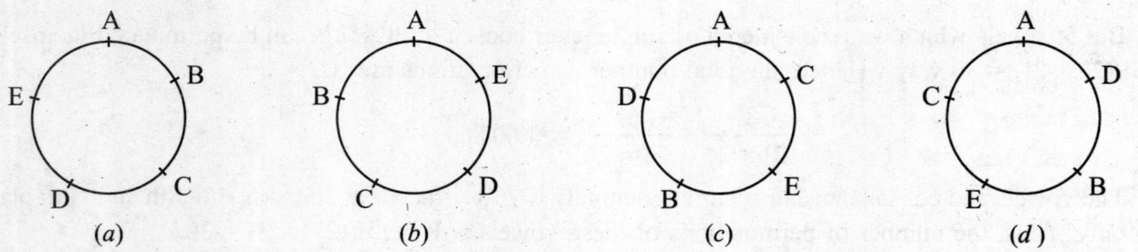

Fig. 6.2 Circular Permutations

Proof If the arrangement of objects is to be done around a circle, then arrangements remain unchanged unless the position of objects change. Let us consider arrangements of 5 objects as shown in Figs. 6.2(*a*) and (*b*). It may be noted that in these arrangements all the 5 objects have the same neighbours. But, it is not so when the position (order) change as shown in Figs. 6.2(*c*) and (*d*).

Thus, it follows that in the case of circular arrangements, the relative position of objects depends upon the position of an object placed first, and then the arrangement of the remaining objects. Hence, it can be concluded that the number of permutations of n objects taken all at a time around a circle will be $(n - 1)!$

In this illustration the circular arrangements of 5 objects will be in $4! = 24$ different ways.

Example 14 In how many ways can 5 programmers and 3 software engineers sit around a table so that no two software engineers are together?

Solution Let the programmers be seated first, then they can be asked to sit in $(5 - 1)! = 4!$ ways. Now 3 software engineers have to sit in between the programmers so that no two software engineers sit together. This means there are 3! ways for software engineers to fill up the 3 places between 5 programmers. Thus, the total number of ways in which both programmers and software engineers can sit are $4! \times 3! = 144$.

Example 15 The CEO of 18 software companies meet to discuss the problem of recession in software industry in India. In how many ways can they sit themselves around a table so that CEO of GE, HCL and Infosys, choose to sit together?

Solution First of all fix up the position of one of the CEO and then assign seats to other 17 CEOs. But, CEO of GE, HCL and Infosys are to sit together so that consider them as one. Thus, there remains 15 CEOs who can arrange themselves in 15! ways.

Further, the CEO of GE, HCL and Infosys can arrange themselves in 3! ways. Hence, the total number of ways in which CEOs can sit around a table will be $15! \times 3!$.

Remarks

1. In the circular arrangements, the clockwise and anti-clockwise arrangements do not make any difference, since turning of a given arrangement will make it otherwise.
2. If we have to arrange n different objects around a table such that no two similar objects are neighbour, then the number of permutations will be $(n - 1)!/2$.

6.3.4 Restricted Permutations

1. The number of permutations of n different objects taken r at a time in which p particular objects do not occur is ${}^{n-p}P_r$.

   Since p objects do not occur in any arrangement, therefore r places can be filled up with the remaining $n - p$ objects. Thus, the required number of such permutations will be ${}^{n-p}P_r$.
2. The number of permutations of n different objects taken r at a time in which p particular objects are present is: ${}^{n-p}P_{r-p} \times {}^{r}P_p$.

Since p objects are to be present in each arrangement, therefore the remaining $n - p$ objects are to be arranged taking $r - p$ objects at a time. Thus, the required number of permutations will be ${}^{n-p}P_{r-p}$.

Now in each of these permutations introduce remaining p particular objects one by one. Since in each permutation there are $r - p$ objects, first object out of p can be introduced in $r - p + 1$ $(= r - p - 1 + 1 + 1)$ ways. Second object in $r - p + 2$ ways, and so on, the pth object can be introduced in $r - p + p = r$ ways. Therefore, the number of ways in which p objects can be introduced is $(r - p + 1)(r - p + 2) \ldots r$ ways, which is equal to ${}^{r}P_p$.

Hence, the required number of permutations are ${}^{n-p}P_{r-p} \times {}^{r}P_p$.

Example 16 Prove that

(*a*) $^{2n}P_n = 2^n\,[1\cdot 3\cdot 5\ldots(2n-1)]$ (*b*) $^{n}P_r = n \times {}^{n-1}P_{r-1}$.

Solution (*a*) $$^{2n}P_n = \frac{(2n)!}{(2n-n)!} = \frac{(2n)!}{n!} = \frac{2n(2n-1)(2n-2)\ldots 4\cdot 3\cdot 2\cdot 1}{n!}$$

$$= \frac{[2n(2n-2)\ldots 6\cdot 4\cdot 2][(2n-1)(2n-3)\ldots 5\cdot 3\cdot 1]}{n!}$$

$$= 2^n\,n!\frac{[(2n-1)(2n-3)\ldots\cdot 5\cdot 3\cdot 1]}{n!} = 2^n\,[1\cdot 3\cdot 5\cdot\ldots\cdot(2n-1)]$$

(*b*) $$\text{RHS} = n \times {}^{n-1}P_{r-1} = n \times \frac{(n-1)!}{[(n-1)-(r-1)]!} = n \times \frac{(n-1)!}{(n-r)!} = \frac{n!}{(n-r)!} = {}^{n}P_r.$$

Example 17 If $^{2n+1}P_{n-1} : {}^{2n-1}P_n = 3 : 5$, find the value of n.

Solution $^{2n+1}P_{n-1} : {}^{2n-1}P_n = 3 : 5$ implies that $\dfrac{^{2n+1}P_{n-1}}{^{2n-1}P_n} = \dfrac{3}{5}$

or $$\frac{\dfrac{(2n+1)!}{(2n+1-n+1)!}}{\dfrac{(2n-1)!}{(2n-1-n)!}} = \frac{3}{5} \quad\text{or}\quad \frac{\dfrac{(2n+1)!}{(n+2)!}}{\dfrac{(2n-1)!}{(n-1)!}} = \frac{3}{5}$$

or $$\frac{(2n+1)!}{(n+2)!} \times \frac{(n-1)!}{(2n-1)!} = \frac{3}{5}$$

$$\frac{(2n+1)2n}{(n+2)(n+1)n} = \frac{3}{5} \quad\text{or}\quad \frac{4n+2}{n^2+3n+2} = \frac{3}{5}$$

$$20n + 10 = 3n^2 + 9n + 6 \quad\text{or}\quad 3n^2 - 11n - 4 = 0$$

$$(3n+1)(n-4) = 0, \text{ i.e. } n = 4,\ -1/3.$$

But, $n = -1/3$ is undesirable. Hence, $n = 4$.

Example 18 Six papers are set in an examination, of which two are mathematical. In how many different orders can the papers be arranged so that

(*a*) two mathematical papers are together?

(*b*) two mathematical papers are not consecutive?

Solution The total number of different orders in which 6 papers can be set is $6! = 720$.

(*a*) The number of permutations in which 2 mathematical papers are together can be obtained by treating 2 mathematical papers as one. Thus, there remains 5 papers for arrangement and which can be arranged in 5! ways. But in each of these permutations, 2 mathematical papers can be arranged themselves in 2! ways. Hence, the total number of permutations in which two mathematical papers come together are $5! \times 2! = 240$.

(*b*) The number of permutations in which 2 mathematical papers do not come together can be obtained by subtracting the total number of permutations in which 2 papers do not come together from the total number of permutations. Thus, the required numbers are $6! - 5! \times 2! = 480$.

Example 19 How many different numbers can be formed from the digits 0, 2, 3, 4, 5, 6 lying between 100 and 1000 and how many of these will be divisible by 5?

Solution The total number of different ways in which 6 digits can be arranged lying between 100 to 1000 is ${}^5P_3$. But, the numbers having 0 at the first place is to be rejected. Thus, the rest 5 numbers can be arranged taken 2 at a time in ${}^5P_2$ ways. Hence, the required number of ways is ${}^6P_3 - {}^5P_2 = 120 - 20 = 100$.

The 3-digit number between 100 and 1000 divisible by 5 must end in 0 or 5. Thus, the required numbers are $2 \times ({}^5P_2 - {}^4P_1) = 72$.

Example 20 How many ways 4-digit numbers can be formed with 0, 4, 5, 6, 7, 8, 9; no digit being repeated? How many of them are not divisible by 5?

Solution Since all 7 digits are different. Thus, these can be arranged among themselves in ${}^7P_4$ ways.

The required numbers can be obtained by subtracting the numbers divisible by 5 from total numbers. Numbers divisible by 5 must end in 0 or 5. Thus,

(*a*) when 5 is at unit place, we are left with digits 0, 4, 6, 7, 8, 9 for permutation taking 3 at a time. These can be arranged in ${}^6P_3$ ways. But the numbers having 5 at its unit place and 0 on the extreme-left are of no use. Thus, the total numbers of such type will be ${}^5P_2$. Hence, total numbers ending with 5 without having 0 on the extreme left are ${}^6P_3 - {}^5P_2$.

(*b*) when 0 is at the unit place, we are left with 6 digits to take 3 at a time. Thus, the total number of ways in which 0 occurs at the unit place is ${}^6P_3$. Hence, the required numbers not divisible by 5 are ${}^7P_4 - ({}^6P_3 - {}^5P_2) - {}^6P_3 = 30$.

Example 21 (*a*) How many 5-digit numbers can be formed from the digits 1, 2, 3, 4, 5 using the digit once? (*b*) How many of them are even? (*c*) How many are even and greater than 30,000?

Solution (*a*) With the five digits 1, 2, 3, 4, 5, the five-digit numbers can be formed in 5! different ways.

(*b*) Since the number is required to be even, therefore, it must have either 2 or 4 at its unit place. When 2 is at unit's place, we are left with four digits 1, 3, 4, 5 which have to be arranged among themselves to make a 5-digit number. This can be done in 4! ways.

Similarly, when 4 is at the unit's place, an even number can be obtained in 4! different ways. Hence, the total number of ways in which we can have even numbers are $2 \times 4!$.

(*c*) The number greater than 30,000 amongst $2 \times 4!$ must have 3, 4 or 5 at its extreme left place and either 2 or 4 on the extreme places. Thus, we are only left with digits 1, 2, 4, 5 to be arranged. But, the second place can be filled up in 3! ways with 2 as the extreme left number. Similarly, when 4 is at unit's place third place can be filled up in 3! ways. Hence, the total numbers are even and greater than 30,000 are $2 \times 3! + 2 \times 3! + 1 \times 3! = 30$.

Example 22 How many permutations can be formed with the letters of the word DELHI? How many of them begin with D and how many do not and in how many words LH will be together?

Solution The word DELHI consists of 5 letters. Thus, numbers of different words out of 5 taken all at a time will be ${}^5P_5 = 5! = 120$ words.

If we keep D at the beginning, then the number of words that can be formed with remaining 4 letters is $4! = 24$. Now the words where D is not the first letter will be $120 - 24 = 96$.

Treating LH as one letter, there are only 4 letters which can be arranged in $4! = 24$ ways. Also LH can be arranged in $2! = 2$ ways. Hence, the number of different permutations will be $4! \times 2! = 48$.

Example 23 How many different words can be framed with the letters of the word BHARAT? In how many of these B and H are never together? How many of these begin with B and end with T? (Note: The words need not convey any meaning.)

Solution (*a*) The word BHARAT consists of 6 letters in which two letters As are similar. Thus, number of different words out of 6 taken 2 at a time will be ${}^6P_2 = 6!/2! = 360$.

(*b*) When the letters B and H are taken together, there are 5 letters left in which two letters As are similar. Thus, number of different words out of 5 taken 2 at a time will be ${}^5P_2 = 5!/2! = 60$.

Since letters B and H can change their position, therefore, the total number of such arrangements along with 5 other letters will $2 \times {}^5P_2 = 2 \times 60 = 120$. Hence, the total number of arrangements in which B and H are not together equals $360 - 120 = 240$.

(*c*) When the words are to begin with B and end with T, the remaining letters to be arranged are 4 in which two letters (A) are similar. Thus, the required number of such words will be ${}^4P_2 = 4!/2! = 12$.

Example 24 How many different words containing all the letters of the word TRIANGLE can be formed, so that

(*a*) consonants are never separated,

(*b*) consonants never come together, and

(*c*) vowels occupy odd places?

Solution The word TRIANGLE consists of 8 letters. Then number of different words which can be formed is ${}^8P_8 = 8!$

(*a*) Since the consonants are never separated, treating all the consonants as single letter, so there remains only 3 letters I, A, E and these can be arranged themselves in 3! ways. But, 5 consonants T, R, N, G, L can again be arranged themselves in 5! ways. Thus, the total number of arrangement will be $3! \times 5! = 720$.

(*b*) The number of arrangements in which the consonants do not come together can be obtained by subtracting from the number of arrangements, the numbers of arrangements in which the consonants come together. Thus, the number of arrangements in which the consonants do not come together is $8! - 5! \times 3! = 39{,}600$.

(*c*) The number of vowels in the word are 3 only and the odd places are 4. Thus, vowels I, A and E occupy odd places in ${}^4P_3$ ways. After having placed vowels on odd places the remaining words T, R, N, G, L will fill up the remaining 5 places in ${}^5P_5$ ways. Hence, the total number of arrangements will be ${}^4P_3 \times {}^5P_5 = 4 \times 120 = 480$.

6.3.5 Permutation of Objects not all Different

Theorem 6.5 It there are n objects of r distinct types, with m_1 (being identical among themselves) belonging to type 1, m_2 (being identical among themselves) belonging to type 2, ... and m_r (being identical among themselves) belonging to type r such that $m_1 + m_2 + ... + m_r = n$, then the number of ways in which these n objects can be arranged is given by

$$P(n\,; m_1, m_2, ..., m_r) = \frac{n!}{m_1!\, m_2! \ldots m_r!}$$

Proof: Let the required number of permutations be x. The objects of anyone of the type say type i (containing m_i objects) can be arranged in $(m_i!)$ ways provided each of them is considered distinct. This is true for all values of $i = 1, 2, ..., r$.

Thus by multiplication rule the total number of permutation of n distinct objects taken all at a time are $n! = x \times m_1! \times xm_2! \times .... \times m_r!$ or

$$x = \frac{n!}{m_1!m_2!...m_r!}$$

Example 25 Find the number of permutations of the letters in the word ALLAHABAD.

Solution In the given word we have A four times and L two times and rest are one time. Thus the number of permutations becomes.

$$\frac{9!}{4!\times 2!\times 1!\times 1!\times 1!} = \frac{9\times 8\times 7\times 6\times 5\times 4!}{4!\times 2!} = 7560.$$

Example 26 How many numbers of 7 digits can be formed with the digits 0, 2, 2, 5, 6, 6, 6.

Solution The total number of digits are 7 where 2 appears twice, 6 appears thrice and the rest only once. Thus the total number of permutations become: $\frac{7!}{2!3!} = 420$ Since few of these numbers may have zero at the beginning, such numbers are $\frac{6!}{2!\ 3!} = 60$

Hence, total number with 7 digits will be 420 – 60 = 340.

Self-practice Problems I

1. Find the value of n, if
   (i) ${}^{n}P_5 : {}^{n-1}P_4 = 8 : 1$ (ii) ${}^{n+3}P_6 : {}^{n+2}P_n = 14 : 1$ (iii) ${}^{n}P_4 = 12 \cdot {}^{n}P_2$
   (iv) ${}^{10}P_{n-1} : {}^{11}P_{n-2} = 30 : 11$ (v) ${}^{7}P_n = 2 \cdot {}^{7}P_{n-2}$ (vi) ${}^{n}P_2 = 12$
2. Prove that ${}^{n}P_r = {}^{n-1}P_r + r \cdot {}^{n-1}P_{r-1}$.
3. A committee consisting of a president, vice-president and secretary is to be formed out of 18 members. In how many ways this can be done?
4. There are 20 buses running between cities A and B. In how many ways can a person go to B and come back to A? In how many ways he will come back by different buses?
5. There are 50 stations on a railway line. How many different kinds of single first class tickets must be printed so as to enable a passenger to go from one station to another?
6. How many different numbers can be formed from 6 digits 5, 4, 3, 2, 1, 0? How many of these have 0 at ten's place?
7. How many different numbers can be formed from the letters of the word MASTER and TRIANGLE? How many of these arrangements begin with A and end with E?

8. How many different words can be formed with the letters of the word DOGMATIC? How many of these

   (*i*) begin with D? (*ii*) begin with D or T?

   (*iii*) begin with D and end with T? (*iv*) begin and end with D or T?

   (*v*) letter G occupies 5th place? (*vi*) letter G will occupy odd places?

   (*vii*) vowels come together? (*viii*) vowels occupy even places?

   (*ix*) letters D, O, G occupy only first three places?

9. Find how many words can be formed from the letters of the word COLLEGE; so that vowels always come together.

10. Prove that the number of permutations of n objects taken r at a time which begin and end with specified objects is ${}^{n-2}P_{r-2}$.

11. Find the number of permutations of n different objects taken r at a time in which

    (*i*) a specified objects does not occur. (*ii*) a specified objects does occur.

12. In how many ways 10 examination papers can be arranged so that the best and the worst paper never come together?

13. In how many ways can the letters of the word VIOLENT be arranged so that vowels may occupy only even places?

14. How many numbers each lying between 100 and 1000 can be formed with the digits 2, 0, 3, 4, 5 (no digit being repeated)? How many of these are odd?

15. Show that the number of ways in which n books can be arranged in a shelf, so that two particular books shall not be together is $(n - 2)(n - 1)!$.

16. How many 4-digit numbers can be formed from the digits 1, 2, 3, 4? How many will be greater than 3,000?

17. A train consists of 12 carriages of which 5 are AC-I, 4 AC-II and rest 3 are AC-III. In how many different ways can these carriages be arranged?

18. How many numbers greater than a million can be formed with the digits 4, 5, 5, 0, 4, 5, 3?

19. How many words can be formed by the letters of the word SERIES? How many of these

    (*i*) begin with R and S? (*ii*) vowels and consonants occur alternatively?

20. In how many ways can 4 students win the six prizes assuming that

    (*i*) one student gets only one prize? (*ii*) one may get all the six prizes?

21. A guard of 15 men is formed from a group of soldiers in all possible ways. Find the

    (*i*) number of times three particular soldiers A, B and C are together on guard, and

    (*ii*) number of times two particular soldiers D and E are together on guard. Also find the group of soldiers if it is found that A and E are three times as often together as A, D and E are.

22. In how many ways can n balls of which w are white, r red and y yellow and rest all different be arranged in a line?

23. How many arrangements can be made with the letters of the word MATHEMATICS and in how many of them vowels occur together?

24. If all the permutations of the letters of the word CHALK be written down as in a dictionary, what is the rank of this word?

25. In how many ways can the letters of the word STRANGE be arranged so that

(*i*) vowels are never separated? (*ii*) vowels never come together?

(*iii*) vowels occupy only the odd positions?

26. There are one old man, 6 adults and 4 children in a family. They are to sit in a row for dinner so that children occupy two seats at each end and the old man refuses to have a child on either side of him. In how many ways can the seating arrangement be made for the dinner?

27. In how many ways the letters of the word PETROL can be arranged? How many of these do not begin with P? How many begin with P but do not end with L? Also find the number of words which can be formed if O and L are never together.

28. (*a*) How many 6-digit numbers can be formed from the digits 1, 2, 3, 4, 5, 6 (no digit being repeated)? How many of these are not divisible by 5?

(*b*) How many numbers greater than 7,000 can be formed with the digits 3, 5, 7, 8, 9 (no digit being repeated)?

29. In how many ways can the letters of the word ARRANGE be arranged? How many of these arrangements are there in which

(*i*) two Rs come together? (*ii*) two Rs do not come together?

(*iii*) two Rs and the two As come together?

Hints and Answers

1. (*i*) 8, (*ii*) 9, (*iii*) 6, (*iv*) 7, (*v*) $n = 7, 10$, (*vi*) $n(n-1) = 4 \cdot 3 \Rightarrow n = 4$

3. $18 \times 17 \times 16$

4. 20×20; 20×19

5. 2,450

6. 600, 120

7. (*i*) 720, 24 (*ii*) 40320, 720

8. 8!, (*i*) 7!, (*ii*) $2! \times 7!$, (*iii*) 6!, (*iv*) $2 \times 6!$, (*v*) 7!, (*vi*) $4 \times 7!$, (*vii*) $6 \times 3!$, (*viii*) $3! \times 5!$, (*ix*) $3! \times 5!$

9. 720

11. (*i*) ${}^{n-1}P_r$, (*ii*) $r \times {}^{n-1}P_{r-1}$

12. (*i*) 2,90,340, (*ii*) $6 \times 720 \times 6 \times 2 = 51{,}840$

13. 144

14. ${}^5P_3 - {}^4P_2 = 48$

16. 4!; $2 \times 3!$

17. 27,720

18. 360

19. $\dfrac{6!}{2!}$, (*i*) $\dfrac{5!}{2! \times 2!}$ (*ii*) $\dfrac{5!}{2!} \times \left(\dfrac{3!}{2!}\right)^2$

20. (*a*) (*i*) 360, (*ii*) $(6)^4$, (*b*) (*i*) 4320, (*ii*) 720

22. $\dfrac{n!}{w! \times r! \times y!}$

23. $\dfrac{8!}{2! \times 2!} \times \dfrac{4!}{2!} = 1{,}20{,}960$ ways

24. $(4! + 3! + 1! + 1)$th or 32nd

25. (*i*) 1,440, (*ii*) 4,800, (*iii*) 1,440

27. (*i*) ${}^6P_6 = 720$ ways.

(*ii*) If P is fixed, then ${}^5P_5 = 120$ ways. Hence, letters do not begin with P = 720 – 120 = 600.

(*iii*) Keeping P and L fixed, ${}^4P_4 = 24$ ways, number of words begin with P and end with L = 24. Also number of words begin with P but do not end with L = 120 – 24 = 96.

(*iv*) Letters O and L can be arranged among themselves in ${}^2P_2 = 2$ ways. In these cases O and L can be arranged among themselves in ${}^2P_2 = 2$ ways. Hence, numbers of ways where O and L are always together is 120 × 2 = 240 and where these two are together is 720 – 240 = 480.

28. (*a*) Number of arrangements, ${}^6P_6 = 720$; numbers ending with 5 are divisible by 5. Number ending with 5 = ${}^5P_5 = 120$. Hence, the numbers not divisible by 5 are 720 – 120 = 600.

(*b*) A 4-digit number greater than 7,000 must begin either with 7, 8 or 9. Thus digits can be arranged in ${}^4P_3 + {}^4P_3 + {}^4P_3 = 24 + 24 + 24 = 72$ ways.

All 5-digit numbers greater that 7000 are ${}^5P_5 = 120$. Therefore, the numbers greater than 7000 are 72 + 120 = 192.

29. (*i*) $\frac{6!}{2!} = 360$, (*ii*) $\frac{7!}{2!} \times 2! - \frac{6!}{2!} = 1260 - 360 = 900$, (*iii*) 5! = 120 ways.

6.4 COMBINATIONS

In contrast to a permutation, in which an ordered arrangement of different objects is important, *a combination is a set or collection of objects where order does not matter.* Thus, by the combinations of n different objects taken r at a time ($r \le n$) we mean all possible grouping of r different objects from the n objects without giving any importance to the order of arrangement of objects forming the group. The number of combination of n objects taken r at a time is denoted by ${}^nC_r$ or C (n, r) or $\binom{n}{r}$.

The number of permutations and combinations of n objects taken r at a time are related to each other. The following theorem will clarify this relationship.

Theorem 6.6 Let A = $\{a_1, a_2, \ldots, a_n\}$ be a set of n objects. Then the number of combinations of n different objects of A taken r ($1 \le r \le n$) at a time is

$$ {}^nC_r = \frac{n!}{r!\,(n-r)!}. $$

The relationship between ${}^nC_r$ and ${}^nP_r$ is: $r! \times {}^nC_r = {}^nP_r$.

Proof Let ${}^nC_r$ be the required number of combinations of n different objects taken r at a time. Then each of these combinations is a group of r different objects which can be rearranged among themselves in $r!$ ways.

Thus, all the ${}^nC_r$ combinations will produce $r! \times {}^nC_r$ permutations of n objects taken r at a time which are ${}^nP_r$ in number. That is,

$$ r! \times {}^nC_r = {}^nP_r \Rightarrow {}^nC_r = \frac{{}^nP_r}{r!} = \frac{n!}{r!(n-r)!}. $$

Example 27 If ${}^nC_x = 56$ and ${}^nP_x = 336$, find n and x.

Solution Dividing ${}^nP_x$ by ${}^nC_x$, we get

$$\frac{{}^nP_x}{{}^nC_x} = \frac{n!}{(n-x)!} \times \frac{x!(x-x)!}{n!} = \frac{336}{56} = 6.$$

This implies that $x! = 6 = 3!$ or $x = 3$.

Substituting value of x in ${}^nP_x$, we get

$${}^nP_x = \frac{n!}{(n-x)!} = 336$$

or $$\frac{n(n-1)(n-2)(n-3)!}{(n-3)!} = 336 \quad \text{or} \quad n(n-1)(n-2) = 338, \text{ i.e. } n = 8.$$

Example 28 A candidate is required to answer 6 out of 10 questions which are divided into two groups containing 5 questions each and he is permitted to attempt not more than four questions from any group. In how many ways can he select the questions?

Solution The questions may be answered in the following ways:

(*a*) 2 questions from first group and 4 questions from second group,

(*b*) 3 questions from first group and 3 questions from second group,

(*c*) 4 questions from first group and 2 questions from second group.

The total number of ways of selecting questions from both groups are obtained as follows:

For (*a*), number of choices : ${}^5C_2 \times {}^5C_4 = 10 \times 5 = 50$

For (*b*), number of choices : ${}^5C_3 \times {}^5C_3 = {}^5C_2 \times {}^5C_2 = 100$

For (*c*), number of choices : ${}^5C_4 \times {}^5C_2 = {}^5C_1 \times {}^5C_2 = 5 \times 10 = 50.$

Hence, the total number of ways of selecting questions are: 50 + 100 + 50 = 200.

Example 29 From 6 boys and 4 girls, 5 are to be selected for admission for a particular course. In how many ways can this be done if there must be exactly 2 girls?

Solution Since there are 6 boys and 4 girls and selection of 5 is to be made in such a way that it includes 2 girls, therefore 2 girls can be selected out of 4 in ${}^4C_2$ ways, and 3 boys can be selected out of 5 in ${}^6C_3$ ways.

Thus, the required number of ways of selection for admission will be

$${}^4C_2 \times {}^6C_3 = \frac{4!}{2! \times 2!} \times \frac{6!}{3! \times 3!} = 120.$$

Example 30 In how many ways can a cricket team of eleven be chosen out of a batch of 14 players? How many of them will

(*a*) include a particular player?

(*b*) exclude a particular player?

Solution The number of ways of selecting 11 players out of 14 are

$${}^{14}C_{11} = \frac{14!}{11!\,3!} = \frac{14 \cdot 13 \cdot 12}{3 \cdot 2 \cdot 1} = 364.$$

(*a*) Since a particular player is always to be selected, then 10 players are selected out of the remaining 13 players and this can be done in

$$^{13}C_{10} = \frac{13!}{10!\,3!} = \frac{13\cdot 12\cdot 11}{3\cdot 2\cdot 1} = 286 \text{ ways.}$$

(*b*) Since a particular player is always to be excluded, then 11 players are selected out of 13 players only and this can be done in

$$^{13}C_{11} = \frac{13!}{11!\,2!} = \frac{13\cdot 12}{2\cdot 1} = 78 \text{ ways.}$$

Example 31 In how many ways can four students be selected out of twelve students, if

(*a*) two particular students are not included at all?

(*b*) two particular students are included?

Solution The number of ways in which 4 students can be selected out of 12 student are

$$^{12}C_4 = \frac{12!}{4!\,8!} = 495.$$

(*a*) When two particular students are not included, 4 are selected out of 10 in

$$^{10}C_4 = \frac{10!}{6!\,4!} = 210 \text{ ways.}$$

(*b*) When two particular students are included, they can be selected only in one way. Thus, the number of ways of selecting 2 out of 10 students are $^{10}C_2 = \frac{10!}{2!\,8!} = 45.$

Hence, the total number of ways are $= 1 \times {}^{10}C_2 = 45.$

Example 32 Out of ten electric bulbs, three are defective but it is not known which are those. In how many ways these

(*a*) three bulb can be selected?

(*b*) selections will include at least one defective bulb?

Solution (*a*) The 3 bulbs can be selected out of 10 bulbs in $^{10}C_3 = 120$ ways.

(*b*) The number of ways the selection will include one defective bulb is equal to the number of 2 defective bulbs out of 7 good bulbs and 1 defective bulb out of 3 bulbs. Thus, we have

$$^{7}C_2 \times {}^{3}C_1 = \frac{7!}{2!\,5!} \times \frac{3!}{1!\,2!} = \frac{7\cdot 6}{2\cdot 1} \times \frac{3}{1} = 63.$$

Number of ways the selection will include 2 defective bulbs and 1 good bulb are

$$^{7}C_1 \times {}^{3}C_2 = \frac{7!}{1!\,6!} \times \frac{3!}{2!\,1!} = 21.$$

Number of ways the selection will include all the 3 defective bulbs are $^{3}C_3 = 1.$

Hence, the total number of selections which include at least one defective bulb will be $63 + 21 + 1 = 85.$

Example 33 There are 7 men and 3 women. Find the number of ways in which a committee of 6 persons can be formed if the committee is to have (*a*) exactly 4 men, and (*b*) at least 2 women.

Solution (*a*) Since a committee of 6 persons has to include exactly 4 men, then rest will be women. Thus, 4 men can be selected out of 7 in $^{7}C_4$ ways and 2 women in $^{3}C_2$ ways. Hence, a committee consisting exactly 4 men can be formed in

$$^7C_4 \times {}^3C_2 = \frac{7!}{4!\,3!} \times \frac{3!}{2!\,1!} = 35 \times 3 = 105 \text{ ways.}$$

(*b*) The committee consisting at least 2 women can be formed as follows:

(*i*) 4 men and 2 women, i.e. $^7C_4 \times {}^3C_2 = 105$ ways.

(*ii*) 3 men and 3 women, i.e. $^7C_3 \times {}^3C_3 = 35$ ways.

Hence, a committee of 6 persons consisting of at least 2 women can be formed in 105 + 35 = 140 ways.

Example 34 Out of 4 officers and 10 clerks in an organisation, a committee of five consisting of 2 officers and 3 clerks is to be formed. In how many ways can this be done if

(*a*) any officer and any clerk can be included?

(*b*) one particular clerk must be included in the committee?

(*c*) one particular officer cannot be included in the committee?

Solution (*a*) The number of ways in which a committee of any 2 officers (selected from 4) and any 3 clerks (selected from 10) can be formed including any officer and any clerk are

$$^4C_2 \times {}^{10}C_3 = \frac{4!}{2!\,2!} \times \frac{10!}{3!\,7!} = \frac{4 \cdot 3}{2 \cdot 1} \times \frac{10 \cdot 9 \cdot 8}{3 \cdot 2 \cdot 1} = 720.$$

(*b*) The number of ways in which a committee of any 2 officers (selected from 4) and any 2 clerks (selected from 9) *plus* one particular clerk are

$$^4C_2 \times {}^9C_2 = \frac{4!}{2!\,2!} \times \frac{9!}{2!\,7!} = \frac{4 \cdot 3}{2 \cdot 1} \times \frac{9 \cdot 8}{2 \cdot 1} = 216.$$

(*c*) The number of ways in which a committee of any 2 officer (selected from 3 officers excluding one particular officer) and any 3 clerks (selected from 10) are

$$^3C_2 \times {}^{10}C_3 = \frac{3!}{1!\,2!} \times \frac{10!}{3!\,7!} = \frac{3}{1} \times \frac{10 \cdot 9 \cdot 8}{3 \cdot 2 \cdot 1} = 360.$$

Example 35 For an examination, a candidate has to select 7 subjects from 3 different sections A, B and C. The three sections A, B and C contain 4, 5 and 6 subjects, respectively. In how many different ways can a candidate make his selection if he has to select at least 2 subjects from each section?

Solution The following different ways are possible in which a candidate can make a selection of 7 subjects so as to have at least 2 from each group.

(*a*) 2 from section A, 2 from section B and 3 from section C,

$$^4C_2 \times {}^5C_2 \times {}^6C_3 = 6 \times 10 \times 20 = 1{,}200 \text{ ways.}$$

(*b*) 2 from section A, 3 from section B and 2 from section C,

$$^4C_2 \times {}^5C_3 \times {}^6C_2 = 6 \times 10 \times 15 = 900 \text{ ways.}$$

(*c*) 3 from section A, 2 from section B and 2 from section C,

$$^4C_3 \times {}^5C_2 \times {}^6C_2 = 4 \times 10 \times 15 = 600 \text{ ways.}$$

Hence, the total number of ways in which a candidate can made his selection is:

$$1{,}200 + 900 + 600 = 2{,}700.$$

Example 36 An examination paper consists of 12 questions is divided into two parts: A and B. Part A contains 7 questions and Part B contains 5 questions. A candidate is required to answer 8 questions electing at least 3 questions from each part. In how many ways can the candidate select the questions?

Solution A candidate can select 8 questions from two parts in the following ways:

(*a*) 3 questions from Part A and 5 questions from Part B,

$$^7C_3 \times {^5C_5} = \frac{7!}{3!\,4!} \times 1 = 35 \text{ ways.}$$

(*b*) 4 questions from Part A and 4 questions from Part B,

$$^7C_4 \times {^5C_4} = {^7C_3} \times {^5C_1} = 35 \times 5 = 175 \text{ ways.}$$

(*c*) 5 questions from Part A and 3 questions from Part B,

$$^7C_5 \times {^5C_3} = {^7C_2} \times {^5C_2} = \frac{7 \times 6}{2 \times 1} \times \frac{5 \times 4}{2 \times 1} = 210 \text{ ways.}$$

Example 37 Out of 7 consonants and 4 vowels how many words can be made each containing 3 consonant and 2 vowels?

Solution Out of consonants, 3 can be chosen in $^7C_3$ ways and out of 4 vowels, 2 can be chosen in $^4C_2$ ways.

Since each of first group can be associated with each of the second group, therefore, the number of combined group each containing 3 consonants and 2 vowels is $^7C_3 \times {^4C_2}$.

Further, each of these groups contains 5 letters which may be arranged among themselves in 5! ways. Hence the required number of words will be

$$^7C_3 \times {^4C_2} \times 5! = \frac{7!}{3!\,4!} \times \frac{4!}{2!\,2!} \times 5! = \frac{7 \cdot 6 \cdot 5}{3 \cdot 2 \cdot 1} \times \frac{4 \cdot 3}{2 \cdot 1} \times 5! = 25{,}200.$$

Example 38 How many triangles can be made by joining 12 points in a plane, given that 7 are in one line?

Solution Since a triangle can only be formed by joining three points out of 12 points in a plane. Thus, the total number of triangles can be formed in $^{12}C_3$ ways.

But 7 points are in one line, therefore out of these 7 points, a triangle can be formed with 3 points. Thus, the total number of triangles can be formed in $^7C_3$ ways. Hence, the required number of triangles are

$$^{12}C_3 - {^7C_3} = \frac{12 \cdot 11 \cdot 10}{3 \cdot 2 \cdot 1} - \frac{7 \cdot 6 \cdot 5}{3 \cdot 2 \cdot 1} = 185.$$

Example 39 There are 10 professors and 20 students out of whom a committee of 8 professors and 14 students is to be formed. In how many ways a committee can be formed if 2 particular professors refused to join the same committee?

Solution The number of ways in which professors can be chosen for committee are $^{10}C_8$ and the students can be chosen in $^{20}C_{14}$ ways. Hence, a committee consisting of 8 professors and 14 students can be formed in $^{10}C_8 \times {^{20}C_{14}}$ ways.

But out of 10 professors 2 refused to join the committee, therefore, the selection is to be made only from 8 professors to select 6. This can be done in $^8C_6 \times {^{20}C_{14}}$ ways.

Hence, the required number of ways in which a committee can be formed is

$$^{10}C_6 \times {^{20}C_{14}} - {^8C_6} \times {^{20}C_{14}} = \frac{10!}{6!\,4!} \times \frac{20!}{14!\,6!} - \frac{8!}{6!\,2!} \times \frac{20!}{14!\,6!} = 6{,}78{,}920.$$

Example 40 Out of 5 males and 6 females, a committee of 5 is to be formed. Find the number of ways in which it can be formed so that among the persons chosen in the committee there are: (*a*) exactly 3 males and 2 females, (*b*) at least 2 males and one female.

Solution (*a*) The number of ways in which exactly 3 males and 2 females can be selected to form a committee of 5 out of 5 males and 6 females will be $^5C_3$ and $^6C_2$, respectively. Hence, the total number of ways in which the committee can be formed is:

$$^5C_3 \times {}^6C_2 = \frac{5!}{3!\,2!} \times \frac{6!}{2!\,4!} = 10 \times 15 = 150.$$

(*b*) A committee of 5 members with exactly 3 males and 2 females can be formed by selecting manner:

(*i*) 4 males and 1 females in $^5C_4 \times {}^6C_1$ ways.

(*ii*) 3 males and 2 females in $^5C_3 \times {}^6C_2$ ways.

(*iii*) 2 males and 3 females in $^5C_2 \times {}^6C_3$ ways.

Hence, the total number of different ways to form a committee of 5 is:

$$^5C_4 \times {}^6C_1 + {}^5C_3 \times {}^6C_2 + {}^5C_2 \times {}^6C_3 = \frac{5!}{4!\,1!} \times \frac{6!}{1!\,5!} + \frac{5!}{3!\,2!} \times \frac{6!}{2!\,4!} + \frac{5!}{2!\,3!} \times \frac{6!}{3!\,3!}$$

$$= \frac{5}{1} \times \frac{6}{1} + \frac{5 \cdot 4}{2 \cdot 1} \times \frac{6 \cdot 5}{2 \cdot 1} + \frac{5 \cdot 4}{2 \cdot 1} \times \frac{6 \cdot 5 \cdot 4}{3 \cdot 2 \cdot 1}$$

$$= 30 + 150 + 200 = 380.$$

Example 41 A guard of 12 policemen is formed from a group of n policemen present in the police line in all possible ways. How many times

(*a*) two particular policemen A and B are together?

(*b*) three particular policemen C, D and E are together?

(*c*) policemen A and B remain on guard together 2 times as often C, D and E are together, and hence find n.

Solution (*a*) There remains $(n - 2)$ policemen to select 10 policemen to form a guard of 12 men because 2 policemen A and B have to remain together. Hence, the possible number of ways of forming guard in which A and B are together is $^{n-2}C_{10}$.

(*b*) Similarly, the possible number of ways of forming a guard in which C, D and E are together is $^{n-2}C_9$.

(*c*) $^{n-2}C_{10} = 2 \times {}^{n-3}C_9$

or
$$\frac{(n-2)!}{10!\,(n-12)!} = 2 \times \frac{(n-3)!}{9!\,(n-12)!}$$

$$\frac{(n-2)(n-3)!}{10 \cdot 9!\,(n-12)!} = 2 \times \frac{(n-3)!}{9!\,(n-12)!}$$

$$\frac{(n-2)}{10} = \frac{2}{1}$$

$$n - 2 = 20 \text{ or } n = 22.$$

Example 42 Find the number of combinations that can be obtained by the letters of the word COMMERCE taking 4 at a time.

Solution The word COMMERCE consists of 8 letters. Then all possible number of combinations are $^8C_4$ taking 4 letters at a time. But, there are three pairs of letters of same kind, e.g. (C, C), (M, M), (E, E). Thus, one pair can be chosen in $^3C_1$ ways. Now there remains 2 letters to select from remaining 7 letters to form a word of 4 letters. This can be done in $^7C_2$ ways. Hence, the total number of combinations of this type will be $^3C_1 \times {}^7C_2$.

Moreover, two pairs of similar letters can be chosen to form a word of 4 letters, and this can be done in $^3C_2$ ways. Hence, the total number of possible combinations are

$$^8C_4 + {}^3C_1 \times {}^7C_2 + {}^3C_2 = \frac{8!}{4!\,4!} + \frac{3!}{1!\,2!} \times \frac{7!}{2!\,5!} + \frac{3!}{2!\,1!}$$

$$= \frac{8 \cdot 7 \cdot 6 \cdot 5}{4 \cdot 3 \cdot 2 \cdot 1} + \frac{3}{1} \times \frac{7 \cdot 6}{2 \cdot 1} + \frac{3}{1} = 136.$$

6.4.1 Restricted Combinations

1. The number of combination of n objects taken r at a time in which p particular objects always occur is: $^{n-p}C_{r-p}$.
   Since p objects do not occur in any group, therefore, the groups of r objects can be made from the remaining $(n-p)$ objects. The required number of such combinations will be $^{n-p}C_r$. With each of these groups we combine p particular objects to get all the combinations in which p particular objects are always included. Hence, the required number of combinations will be $^{n-p}C_{r-p}$.
2. The number of combinations of n objects taken r at a time in which p particular objects never occur is: $^{n-p}C_r$.
   If p particular objects are not included, then there remains $(n-p)$ objects of which r objects may be chosen in $^{n-p}C_r$ ways.
3. The number of combinations in which n objects can be divided into three groups containing p, q and r objects is

   $$\frac{n!}{p!\,q!\,r!}, \quad \text{when the group are distinct,}$$

   and $$\frac{n!}{p!\,q!\,r!\,3!}, \quad \text{when the groups are not distinct,}$$

   where $p + q + r = n$. Further, if $p = q = r$, then above results will become

   $$\frac{n!}{(p!)^3}, \quad \text{when groups are distinct,}$$

   and $$\frac{n!}{(p!)^3\,3!}, \quad \text{when groups are not distinct.}$$

Example 43 If $^nP_r = 240$ and $^nC_r = 120$, then find the value of n and r.

Solution Given that $^nP_r = 240 \Rightarrow \dfrac{n!}{(n-r)!} = 120$ ...(*i*)

$$^nC_r = 120 \Rightarrow \frac{n!}{r!(n-r)!} = 120 \qquad \text{...}(ii)$$

Dividing Eq. (*ii*) by Eq. (*i*), we get

$$\frac{1}{r!} = \frac{120}{240} = \frac{1}{2}, \text{ i.e. } r! = 2 \text{ or } r = 2$$

Again from Eq. (*i*), we have $\frac{n!}{(n-r)!} = 240$

or $$\frac{n(n-1)(n-2)!}{(n-2)!} = 240 \quad [\text{For } r = 2]$$

$$n(n-1) = 240 \text{ or } n^2 - n - 240 = 0$$

$$(n-16)(n+15) = 0, \text{ i.e. } n = 16, -15.$$

Since $n = -15$ is undesirable, therefore $n = 16$.

Example 44 A job placement agency in Mumbai has to send ten MCA students to five companies, two to each. Two of the companies are in Mumbai and others are outside. Two of the students prefer to work in Mumbai while three prefer to work outside. In how many ways can the assignment be made if preferences are to be satisfied?

Solution Four students are required for Mumbai. Since 2 have already preferred Mumbai, therefore, 2 more must be selected out of the 5 having no preference, and this can be done in ${}^5C_2 = 10$ ways.

Now these 4 students for Mumbai can be assigned to the two companies in ${}^4C_2 = 6$ ways. Thus, the remaining 6 (3 having no preference and 3 having preference for outside Mumbai) students are available for assignment outside Mumbai. They can be assigned to 3 companies in $6!/(2! \times 2! \times 2!) = 90$ ways.

Hence, the number of ways in which assignment can be made is: $10 \times 6 \times 90 = 5{,}400$.

6.4.2 Combinations of Objects not all Different

Theorem 6.7 The total number of combinations which can be made of *n* different objects taken some or all at a time is ${}^nC_1 + {}^nC_2 + {}^nC_3 + \ldots + {}^nC_n = 2^n - 1$.

Proof The first object can be dealt in 2 different ways, whether it either be left or taken. Similarly, second object can also be dealt in 2 ways. Thus, the number of ways of dealing with the first object can be same as the ways of dealing with the second object. Hence, the total number of ways of dealing with first two objects is $2 \times 2 = 2^2$.

Proceeding in the same manner for *n* different objects, the total number of ways of dealing with *n* objects are 2^n. But these 2^n ways include one case when none of the objects is taken.

Hence, the required number of combinations will be $2^n - 1$.

Example 45 Determine all possible ways in which a student may attempt one or more questions from 8 given questions having an alternative.

Solution The first question may be attempted in 3 ways including itself, or its alternative may be attempted or none of them may be attempted.

In the same manner, second question may also be attempted in 3 ways. Thus, the total number of ways in which first two questions may be attempted is $3 \times 3 = 3^2$.

Proceeding in the same manner, all the 8 questions may be attempted in 3^8 ways. But, this number also includes the case in which no question is attempted. Hence, the required number of ways are $3^8 - 1$.

Example 46 In order to pass MCA first semester examination minimum marks have to be secured in each of the 7 subjects. In how many cases can a student fail?

Solution The possibility of fail or pass in each subject can be considered in 2 ways. Thus, possibility of pass or fail in all the 7 subjects can be considered in 2^7 ways. But, this number includes the case in which students passes in all the 7 subjects.

Hence, the required number of combinations in which the student can fail are $2^7 - 1 = 127$.

Self-practice Problems II

30. Find value of n when (*i*) ${}^{n+2}C_8 : {}^{n-2}C_4 = 57 : 16$. (*ii*) ${}^{n}C_9 : {}^{n}C_8 = 2 : 9$

31. (*i*) If ${}^{10}C_r = {}^{10}C_{r+4}$, find ${}^{5}C_r$. (*ii*) If ${}^{n}C_3 = {}^{n}C_5$, find ${}^{2n}C_2$.

32. (*i*) If ${}^{n}P_r = 3{,}024$ and ${}^{n}C_r = 126$, find n and r. (*ii*) If ${}^{n}P_r = 6{,}84{,}800$ and ${}^{n}C_r = 120$, find n and r.

33. From 9 boys and 5 girls a committee of 7 members is to be formed. In how many ways can this be done if the committee contains (*a*) 2 girls, (*b*) at least 3 girls?

34. How many diagonals are there in an n-sided polygon?

35. Find out the number of ways in which a cricket team of 11 players can be selected from 14 players. Also find out how many of these (*a*) will include captain, (*b*) will not include the captain.

36. Four positions of programmers are vacant in an office. How many different groups of candidates can be chosen out of 35 candidates who have applied? In how many ways can selection be made, (*a*) if a particular candidate is included? (*b*) if one particular candidate is always excluded?

37. In how many ways can you choose 6 out of 9 questions? In how many of these ways (*a*) first question is excluded? (*b*) first and second questions always excluded?

38. There are 6 engineers and 8 programmers out of whom a committee of 8 is to be formed. In how many ways can it be formed so that engineers may not be out numbered?

39. A cricket team is to be formed consisting of 5 batsmen, 4 bowlers and 2 wicket keepers form a group of 11 batsmen, 8 bowers and 4 wicket keepers. In how many ways a cricket team of 11 players can be formed?

40. In an examination a candidate is required to answer 5 out of 11 questions. If question number one is compulsory, in how many ways can he answer 6 questions? Further, if question number one is compulsory and remaining 10 questions are divided into two groups each containing 5 questions, then in how many ways he can answer 6 questions if at least 2 questions are to be attempted from each group?

41. Out of 5 males and 6 females, a committee of 5 is to be formed. Find the number of ways in which it can be formed so that among the persons chosen in the committee there are
(*a*) 3 males and 2 females (*b*) at least one female
(*c*) not more than 3 males.

42. The staff of a bank consists of a manager, a deputy manager and 10 other officers. A committee of r members is to be formed. Find the number of ways in which this can be formed so as to always include
(*a*) manager (*b*) manager but not the deputy manager
(*c*) neither the manager nor the deputy manager.

43. In an election every voter may vote for any number of candidates but not greater than to be elected. There are 5 candidates and 3 are to be elected. In how many ways can a voter vote?

44. (*a*) In how many ways can 52 playing cards be distributed among 4 players.
(*b*) In how many ways 52 cards be distributed in 4 groups.

45. Out of 6 consonants and 3 vowels, how many different words can be formed each containing 4 consonants and 2 vowels?

46. There are 20 points in a plane of which 7 are collinear (lying in one straight line). Find how many
(*a*) straight lines can be formed (*b*) maximum number of triangles by joining them.

47. How many of the combinations of the letters of the word NUMBERS taken 3 at a time
(*a*) contain M? (*b*) do not contain M?
(*c*) contain M and R? (*d*) contain neither M nor R?

48. In a crossword puzzle twenty words are to be guessed of which 8 words have each an alternative solution also. Find the numbers of possible solution.

6.5 PIGEONHOLE PRINCIPLE

The pigeonhole principle also known as Dirichlet Drawer Principle or Shoe Box Principle is another useful method of counting.

Definition If n pigeonholes (categories of desired characteristics) are occupied by $k + 1$ or more pigeons, when k is a positive integer, then at least one hole is occupied by $k + 1$ or more pigeons. In other words, if k pigeonholes are occupied by $k + 1$ pigeons, then at least one hole is occupied by more than one pigeon.

The pigeonhole principle is not only easy to use but also helpful in proving interesting consequences. To use this principle we must identify the pigeons (objects) and pigeonholes (categories of the desired characteristics) and be able to count the number of pigeons as well as pigeonholes. For example, using this principle, we could prove that, given two rational numbers p and q, there is a rational number between them by showing that $(p + q)/2$ is in between p and q.

Remark The pigeonhole principle can also be stated in the set theoretic terms as follows:

Let A and B be any two finite sets. If $|A| > |B|$, then any function $f : A \rightarrow B$ cannot be one-to-one, i.e. these exists at least 2 distinct elements x_1 and x_2 in A such that $f(x_1) = f(x_2)$.

Example 47 Show that if four numbers from 1 to 7 are chosen, then two of them will add up to 8.

Solution Make three sets, each containing two numbers whose sum is 8 as: $A_1 = \{1, 7\}$, $A_2 = \{2, 6\}$, $A_3 = \{3, 5\}$. The four numbers chosen must belong to one of these sets. As there are only three sets, two of the chosen numbers belong to the same set whose sum is 8.

Example 48 Show that at least 2 people out of 13 must have their birthday in the same month when they are assembled in a room.

Solution In this case we may consider birth months of people as pigeons and the calendar months ($n = 12$) as the pigeonholes. Thus there are 13 pigeons and 12 pigeonholes. So according to the pigeonhole principle, $k + 1 = 2$ or $k = 1$. Thus the minimum number of people who may have their birthday in the same month are $kn + 1 = 1{\times}12 + 1 = 13$.

Example 49 If any 12 numbers from 1 to 23 are chosen, then show that one of them is a multiple of another.

Solution Consider each of the number from 1 to 23 as a pigeonhole ($n = 23$). Assign each of the selected 12 numbers from 23 pigeonholes with its odd part. There are only 11 such pigeonholes. Therefore out of 11 numbers, two must have the same odd part, one of which is a multiple of the other.

Example 50 If a bag contains many white, red and green balls, then find the number of balls to the chosen to be sure of getting a pair of balls with same colour.

Solution The number of balls (pigeon) to be chosen are $kn + 1$. Considering number of balls as pigeon and number of colours the pigeonholes ($n = 3$). Then, number of colours represents the number of pigeonholes. So, $k + 1 = 2$ or $k = 1$

Thus the required number of balls are: $kn + 1 = 1\times3 + 1 = 4$.

Example 51 Suppose a set of positive integers contains 37 elements. Then prove that the set contains two elements that have the same remainder upon division by 36.

Solution Consider 37 positive integers in the set A as pigeons. We know from the division algorithm that when any positive integer n is divided by 36, there exists a unique quotient q and unique remainder r, such that

$$n = 36q + r, 0 \leq r \leq 36$$

Here, the 36 possible values of r are regarded as pigeonholes.

Example 52 Prove that if 101 integers are selected from the set A = {1, 2, 3 . . ., 2000}, then there are two integers such that one divides the other.

Solution For each element $x \in$ A, we may write $x = 2^k y$, where gcd(2, y) = 1 and $k \geq 0$ (See Fundamental Theorem of Arithmetic). It follows that $y \in$ B = {1, 3, 5 . . ., 199} where |B| = 100. Since 100 integers are selected from the set A, then by the pigeonhole principle there exist two distinct integers of the form $a = 2^m b$, $b = 2^n y$ for some $y \in$ B. If $m < n$, then a divides b. But if $m > n$, then b divides a.

Example 53 Let n be a positive integer. Prove that there exists a positive integer n such that n divides $2^m - 1$.

Solution Consider the $n + 1$ positive integers $2^1 - 1, 2^2 - 1, 2^3 - 1, \ldots, 2^n - 1, 2^{n+1} - 1$. By the pigeonhole principle and the division algorithm there exists two positive integer numbers a and b with $1 \leq a < b \leq n + 1$, where $2^a - 1$ and $2^b - 1$ have the same remainder upon division by n. Hence $2^a - 1 = q_1 n + r$ and $2^b - 1 = q_2 n + r$ for $q_1, q_2 \in N_1$ and $(2^b - 1) - (2^a - 1) = (q_2 n + r) - (q_1 m + r)$, so $2^b - 2^a = (q_2 - q_1)\, n$. But $2^b - 2^a$ $2^a (2^{b-a} - 1)$ and gcd(2^a, n) = 1 because n is odd. Hence, n divides $(2^{b-a} - 1)$ and $m = b - a$.

In this definition the sets A and B can be considered as the set of pigeons and set of pigeonholes respectively. Thus, we assign a pigeon $x \in$ A to a pigeonhole $y = f(x) \in$ B.

6.5.1 Generalised Pigeonhole Principle

If there are n pigeons and m pigeonholes ($n > m$), few pigeonholes must contain at least $\left[\frac{n-1}{m}\right] + 1$ pigeons, where the symbol $[x]$ represents the largest integer less than x.

Example 54 Suppose 26 candidates appear in a competitive examination. Then show that there exists at least two among them whose roll number differ by a multiple of 25.

Solution Let A be the set of roll numbers of 2000 candidates, and B be the set of remainders when any positive integer is divided by 19, i.e. B = {1, 2, . . ., 24}. Now we define a function

$f : A \to B$ such that $f(x)$ = remainder when x is divided by 25.

Since $|A| = 26$ and $|B| = 25$, therefore by the pigeonhole principle, function f cannot be one-to-one. Thus, there exist two distinct roll numbers x_1 and x_2 such that $f(x_1) = f(x_2)$. By Euclidean algorithm, we may write

$$x_1 = 25a + f(x_1) \text{ and } x_2 = 25b + f(x_2)$$

where a and b are positive integers. This gives

$$x_1 - x_2 = 25(a - b) \text{ because } f(x_1) = f(x_2)$$

Since $a - b$ is an integer, $x_1 - x_2$ is the multiple of 25. Hence, there are at least two candidates whose roll numbers differ by a multiple of 25.

6.6 BINOMIAL THEOREM

If x and y are variables and n is a positive integer, then

$$(x + y)^n = \sum_{r=0}^{n} {}^nC_r \, x^r y^{n-r}.$$

Here, the term ${}^nC_r$ is often referred to as *binomial coefficient.* The binomial coefficient is also written as $C(n, r)$ or $\binom{n}{r}$. The symbol ${}^nC_r$ has two meanings:

(*i*) *Combinatorial meaning* where it represents the number of ways of choosing r objects from given n distinct objects,

(*ii*) *Algebraic meaning* where it is expressed as

$$ {}^nC_r = \frac{n!}{r!(n-r)!}.$$

Corollary When n is a positive integer,

(*a*) $(1 + x)^n = \sum_{r=0}^{n} C(n, r) x^r$

(*b*) $(1 + x)^{-n} = \sum_{r=0}^{n} (-1)^r C(n + r - 1, r) x^r$

(*c*) $(1 - x)^{-n} = \sum_{r=0}^{n} C(n + r - 1, r) x^r$.

Combinatorial Indentities For each $n > 0$

1. (*a*) $C(n, 0) = C(n, n) = 1$
   (*b*) $C(n, 1) = C(n, n - 1) = n$
   (*c*) $C(n, r) = \frac{n}{r} C(n - 1, r - 1)$
   (*d*) $C(n, 0) + C(n, 1) + \ldots + C(n, n) = 2^n$
   (*e*) $C(n, 1) + C(n, 3) + \ldots = C(n, 0) + C(n, 2) + \ldots = 2^{n-1}$
   (*f*) $C(n, r) = C(n, n - r)$, for $r \le n$
   (*g*) If $C(n, r) = C(n, p)$, then $r = p$ or $r + p = n$
   (*h*) $C(n, r) + C(n, r - 1) = C(n + 1, r)$

2. Newton's identity

   C (n, r).C (r, k) = C (n, k).C $(n - k, r - k)$ for integers $n \geq r \geq k \geq 0$
3. Pascal identity

   C $(n + 1, r)$ = C (n, r) + C $(n, r - 1)$
4. Vandermonde's identity

   C $(n + m, r)$ = C $(n, 0)$.C (m, r) + C $(n, 1)$.C $(m, r - 1)$ + . . . + C (n, r).C $(m, 0)$

   $$= \sum_{k=0}^{r} C(m, r-k).C(n, k) \text{ for integers } m, n \geq r \geq 0.$$

Pascal's formula provides a recurrence relation for the computation of Binomial coefficients without any multiplication, given the initial values C $(n, 0)$ = C $(n, n) = 1$, for all n. The triangular array of numbers called *Pascal's triangle*. The first five rows of Pascal's triangle are shown in Fig. 6.3.

$(n = 0)$						C (0, 0)						
$(n = 1)$					C (1, 0)		C (1, 1)					
$(n = 2)$				C (2, 0)		C (2, 1)		C (2, 2)				
$(n = 3)$			C (3, 0)		C (3, 1)		C (3, 2)		C (3, 3)			
$(n = 4)$		C (4, 0)		C (4, 1)		C (4, 2)		C (4, 3)		C (4, 4)		
$(n = 5)$	C (5, 0)		C (5, 1)		C (5, 2)		C (5, 3)		C (5, 4)		C (5, 5)	

Fig. 6.3 Pascal's Triangle

The nth row of the triangle, which contains $n + 1$ numbers, can be formed from the preceding row by using following rules:

(*a*) The first $(r = 0)$ and the last $(r = n)$ entries are both equal to 1.

(*b*) For $1 \leq r \leq n - 1$, the rth entry in the nth row is the sum of the $(r - 1)$th and kth entries in the $(n - 1)$ rows.

Illustrations

1. The coefficient of $x^5 y^2$ in the expansion of $(x + y)^7$ is ${}^7C_5 = {}^7C_2 = 21$.
2. To obtain the coefficient of a^5b^2 in the expansion of $(2a - 3b)^7$, substitute x for $2a$ and y for $- 3b$. Then the coefficient of x^5y^2 in the expansion of $(x + y)^7$ is ${}^7C_5 = 21$, and ${}^7C_5x^5y^2 = {}^7C_5(2a)^5 (- 3b)^2 = 21(2)^5 (- 3)^2 a^5b^2 = 6048\, a^5b^2$.

6.7 MULTINOMIAL COEFFICIENT

For non-negative integers $n_1, n_2, \ldots, n_k$ such that $n_1 + n_2 + \ldots + n_k = n$ for $n \in$ N, the multinomial coefficient in the expansion of $(n_1 + n_2 + \ldots + n_k)^n$ is given by

$$\binom{n}{n_1, n_2, \ldots, n_k} = \frac{n!}{n_1!\, n_2! \ldots n_k!}$$

where $0 \leq n_i \leq n$ for all $1 \leq i \leq k$.

This coefficient denotes the number of distinguishable arrangements of n numbers, in which there are n_1 objects of type 1, n_2 objects of type 2, . . ., n_k objects of type k.

The term 'multinomial coefficient' also comes from considering algebraic expressions in the same way as the term 'Binomial coefficient'. For example,

$$(x_1 + x_2 + \ldots + x_k)^n = (x_1 + x_2 + \ldots + x_k)(x_1 + x_2 + \ldots + x_k) \ldots (x_1 + x_2 + \ldots + x_k)$$

while performing the multiplication of terms on RHS and collecting like terms, we get a typical term which has the form: $x_1^{n_1} x_2^{n_2} \ldots x_k^{n_k}$. The coefficient of this term is the number of ways of selecting x_1 factors from n_1 of the n factors, x_2 factors from n_2 of the $n - n_1$ remaining factors, x_3 factors from n_3 of the remaining $n - n_1 - n_2$ factors . . . and x_k factors from n_k of the last $n - n_1 - n_2 \ldots n_{k-1}$ remaining factors. This can be carried out as follows:

$$\binom{n}{n_1}\binom{n-n_1}{n_2}\binom{n-n_1-n_2}{n_3}\cdots\binom{n-n_1-n_2-\ldots n_{k-1}}{n_k} = \frac{n!}{n_1!\, n_2! \ldots n_k!} = \binom{n}{n_1\, n_2 \ldots n_k}$$

Example 55 Determine the coefficient of

(*a*) x^9y^3 in $(x + y)^{12}$ and (*b*) xyz^2 in $(x + y + z)^4$

Solution (*a*) Consider the expression

$$(x + y)^{12} = (x + y)(x + y) \ldots (x + y).$$

Selection of x from 9 brackets and y from 3 brackets in the above expansion, can be done as follows:

$$\frac{12!}{9!\,3!} = 220.$$

(*b*) Consider the expression

$$(x + y + z)^4 = (x + y + z)(x + y + z) \ldots (x + y + z).$$

Now selection of x from 1 bracket, a y from 1 bracket and a z from 2 brackets can be done as follows:

$$\frac{4!}{1!\,1!\,2!} = 12.$$

6.8 NUMBER OF ONTO FUNCTIONS

We know that, in general, if there are two non-empty sets A and B such that $|A| = n \geq 2$ and $|B| = 2$, then there are $|B|^{|A|} - 2 = 2^n - 2$ onto functions from A to B. For example, if A = $\{a, b, c\}$ and B = $\{x, y\}$, then all functions f : A → B are onto except two constant functions $f_1 = \{(a, x), (b, x), (c, x)\}$ and $f_2 = \{(a, y), (b, y), (c, y)\}$. Thus, there are $|B|^{|A|} - 2 = 2^3 - 2 = 6$ onto functions from A to B.

Theorem 6.8 For finite sets A and B with $|A| = m$ and $|B| = n$ with $m \geq n$ there are

$${}^nC_n\, n^m - {}^nC_{n-1}(n-1)^m + {}^nC_{n-2}(n-2)^m - \ldots + (-1)^{n-2}\, {}^nC_2\, 2^m + (-1)^{n-1}\, {}^nC_1\, 1^m$$

$$= \sum_{k=0}^{n-1} (-1)^k\, {}^nC_{n-k}(n-k)^m = \sum_{k=0}^{n} (-1)^k\, {}^nC_{n-k}(n-k)^m$$

onto functions from A to B.

Example 56 Let A = {1, 2, 3, 4} and B = {x, y, z}. Find the number of onto functions from A to B.

Solution Applying the general formula with m = 4 and n = 3, we get

$$^3C_3(3)^4 - {}^3C_2(2)^4 + {}^3C_1(1)^4 = \sum_{k=0}^{2}(-1)^k\, {}^3C_{3-k}(3-k)^4$$

$$= 36 \text{ onto functions from A to B.}$$

Example 57 In how many ways six jobs can be assigned to three different employees so that each of them works on at least one job?

Solution Applying the general formula with |A| = m = 6 and |B| = n = 3, we find that there are

$$\sum_{k=0}^{2}(-1)^k\, {}^3C_{3-k}(3-k)^6 = 3^6 - {}^3C_1(2)^6 + {}^3C_2(1)^6$$

$$= 729 - 3 \times 64 + 3 = 540$$

ways to assign the jobs to the employees so that each of them works on at least on job.

6.9 PRINCIPLE OF INCLUSION AND EXCLUSION

Let $A_1, A_2, \ldots, A_s$ be the subset of a set with |A| = n and let $p_1, p_2, \ldots, p_s$ be a collection of properties or conditions satisfied by some or all, of the subsets of A. It is possible that some subsets of A may satisfy more than one of these conditions, whereas others may not satisfy any of them. The following notations are used in the statement of following theorem on the principle of inclusion and exclusion.

$N(p_i)$ = number of subsets in A that satisfy property p_i, for all $1 \le i \le s$.

$N(p_i p_j \ldots p_k)$ = number of subsets in A that satisfy the properties p_i, $p_j \ldots, p_k$ for all $i, j, k \in \{1, 2, \ldots, s\}$ where $i \ne j \ne k$. Here $N(p_i p_j \ldots p_k)$ does not count the subsets of A that satisfy only $p_i, p_j, \ldots, p_k$.

$N(\overline{p}_i)$ = $N - N(p_i)$ = number of subsets in A that do not satisfy property p_i for each $1 \le i \le s$.

$N(\overline{p}_i\, \overline{p}_j)$ = number of subsets in A that do not satisfy either of the properties p_i or p_j where $1 \le i, j \le s$ with $i \ne j$.

$= N - [N(p_i) + N(p_j)] + N(p_i p_j)$, it is shown in Venn diagram Fig. 6.4.

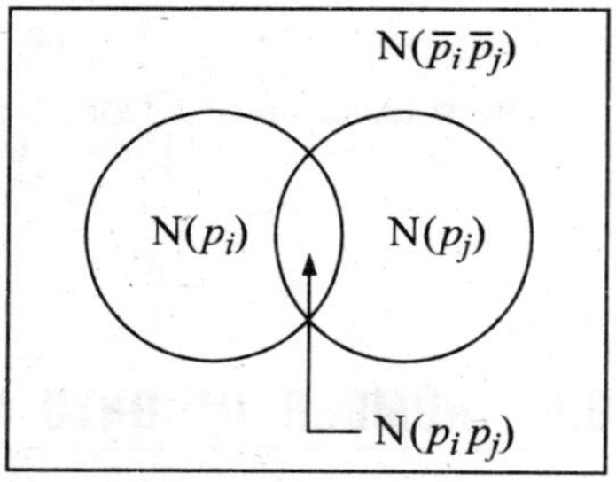

Fig. 6.4 Venn Diagram

Similarly, as shown in Fig. 6.4, we have

$$N(\overline{p}_i\, \overline{p}_j\, \overline{p}_k) = N - [N(p_i) + N(p_j) + N(p_k)] + [N(p_i p_j) + N(p_j p_k) + N(p_k p_i)] - N(p_i p_j p_k).$$

Theorem 6.9 Consider a set A containing n elements and the properties p_i for all $1 \le i \le s$ satisfied by some of the elements of A. The number of elements of A that satisfy none of the properties p_i, $1 \le i \le s$ is denoted by

$$\overline{N} = N(\overline{p}_1\, \overline{p}_2 \ldots \overline{p}_s)$$

$$= N - [N(p_1) + N(p_2) + \ldots + N(p_s)] + [N(p_1 p_2) + N(p_2 p_3)$$

$= + \ldots + (p_{s-1}\, p_s)] - [N(p_1\, p_2\, p_3) + N(p_1\, p_2\, p_4) + \ldots + N(p_1\, p_2\, p_s) + N(p_1\, p_3\, p_4)$
$+ \ldots + N(p_1\, p_3\, p_s) + \ldots + N(p_{s-2}\, p_{s-1}\, p_s)] + \ldots + (-1)^s N(p_1\, p_2\, p_3 \ldots p_s)$

$$= N - \sum_{1 \le i \le s} N(p_i) + \sum_{1 \le i \le j \le s} N(p_i p_j) - \sum_{1 \le i \le j \le k \le s} N(p_i p_j p_k) + \ldots$$

$$+ (-1)^s N(p_1\, p_2\, p_3 \ldots p_s)$$

Proof We shall prove the formula by showing that elements of the set A are counted exactly once on the right hand side of the equation.

Suppose $x \in A$ satisfy exactly m of the properties for $1 \le m \le s$. This implies that x does not contribute to $\overline{N}$. But on the right hand side of the equation, x is counted

(*i*) one time in N.

(*ii*) m times in $\sum_{1 \le i \le s} N(p_i)$, i.e. once for each of the m properties.

(*iii*) ${}^mC_2$ times in $\sum_{1 \le i \le j \le s} N(p_i p_j)$, i.e. once for each pair of conditions selected from m conditions it satisfies

⋮

${}^mC_m = 1$ time in $\Sigma N(p_{i1}\, p_{i2} \ldots p_{im})$, i.e. summation is taken over all selections of size m from s conditions.

This implies that in the right hand side of the equation in the statement of the theorem, the element x is counted

$${}^mC_0 - {}^mC_1 + {}^mC_2 - {}^mC_3 + \ldots + (-1)^m\, {}^mC_m = 1 - m + {}^mC_2 - {}^mC_3 + \ldots + (-1)^m\, {}^mC_m = [1 + (-1)]^m = 0 \text{ times.}$$

Remark If A_i is the subset having property p_i, then

(*a*) $N(p_{i1}\, p_{i2} \ldots p_{im}) = |A_{i1} \cap A_{i2} \ldots \cap A_{im}|$

(*b*) $\overline{N} = N - [N(p_1) + N(p_2) + \ldots + N(p_m) = N - |A_1 \cup A_2 \cup \ldots \cup A_m|$.

Alternative Statements of the Theorem on Principle of Inclusion and Exclusion

1. If $A_1, A_2, \ldots A_s$ are any finite sets, then

$$|A_1 \cup A_2 \cup A_3 \ldots \cup A_s| = \sum_{i=1}^{s} |A_i| + \sum_{1 \le i < j \le s} |A_i \cap A_j| + \Sigma |A_i \cap A_j \cap A_k) + \ldots + (-1)^s |A_1 \cap A_2 \cap \ldots \cap A_s|$$

where $|A_i|$ denotes the number of elements in the set A_i.

2. If $A_1, A_2, \ldots, A_n$ are subsets of any finite set A having n elements, then

$$|\overline{A} \cap \overline{A}_2 \cap \ldots \cap \overline{A}_n| = |A| - \sum_{i=1}^{n} |A_i| + \sum_{1 \le i < j \le n} |A_i \cap A_j| + \ldots + (-1)^n |A_1 \cap A_2 \cap \ldots A_n|$$

Example 58 Find the number of non-negative integer solutions to the equation: $x_1 + x_2 + x_3 + x_4 = 13$, with the extra condition that $x_i \le 5$, for all $1 \le i \le 4$.

Solution Consider the set A of all integer solutions of the given equation with each $x_i \ge 0$ for all $1 \le i \le 4$. The number of integer solutions will be

$|\overline{A}_1 \cap \overline{A}_2 \cap \overline{A}_3 \cap \overline{A}_4|$ with $x_i \le 5$, for all $1 \le i \le 4$.

So $|A| = N = {}^{4+13-1}C_{13} = {}^{16}C_{13} = 560$

and $|A_1| = N(p_i)$, number of solutions satisfying the condition $x_i > 5$ or $x_i \geq 6$

$= {}^{(13-6)+4-1}C_{13-6} = {}^{10}C_7 = 120$

$|A_1 \cap A_2| = N(p_1 p_2)$, number of solutions satisfying the conditions $x_1 \geq 6$ and $x_2 \geq 6$.

$= {}^{(13-6-6)+4-1}C_{13-6-6} = {}^{4}C_1 = 4$

$|A_1 \cap A_2 \cap A_3| = N(p_1 p_2 p_3)$, number of solutions satisfying the conditions $x_1 \geq 6$, $x_2 \geq 6$, $x_3 \geq 6$.

$= 0$ because sum of three x_i's exceeds 13.

Since $N(p_1 p_2 p_3) = 0$ for every selection of three conditions and so $N(p_1 p_2 p_3 p_4) = 0$. Hence, total number of non-negative integer solutions of the given equation satisfying the given conditions are

$$|\overline{A}_1 \cap \overline{A}_2 \cap \overline{A}_3 \cap \overline{A}_4| = N(\overline{p}_1 \overline{p}_2 \overline{p}_3 \overline{p}_4) = |A| - \sum_{i=1}^{4} |A_i| + \sum_{1 \leq i < j \leq 4} |A_i \cap A_j|$$

$$= 560 - {}^4C_1 \times 120 + {}^4C_2 \times 4$$

$$= 560 - 4 \times 120 + 24 = 104.$$

Example 59 Find the number of positive integers between 1 and 150 that are divisible by 2, 3 and 7.

Solution Here $A = \{1, 2, \ldots, 150\}$ and $|A| = N = 150$. Let

A_1 = set of all integers in the given range that are divisible by 2.

A_2 = set of all integers in the given range that are divisible by 3

A_3 = set of all integers in the given range that are divisible by 7.

Now we are to find $|A_1 \cup A_2 \cup A_3| = N(p_1 p_2 p_3)$

$|A_1| = N(p_1) = \lfloor 150/2 \rfloor = 75$ because there are 50 positive integers 2, 4, 6, 8, . . ., 148 (= 2 × 74), 150 (= 2 × 75) that are divisible by 2.

$|A_2| = N(p_2) = \lfloor 150/3 \rfloor = 50$ because there are 50 positive integers 3, 6, 9, 12, . . ., 147 (= 3 × 49), 150 (= 3 × 50) that are divisible by 3

$|A_3| = N(p_3| = \lfloor 150/7 \rfloor = 21$ because there are 21 positive integers 7, 14, 21, 28, 140 (= 7 × 20), 147(= 7 × 21) that are divisible by 7.

$|A_1 \cap A_2| = N(p_1 p_2)$ = positive integers that are divisible by both 2 and 3 and so divisible by lcm (2, 3) = 2 × 3 = 6 = $\lfloor 150/6 \rfloor = 25$

Similarly, $|A_2 \cap A_3| = \lfloor 150/3 \times 7 \rfloor = 7$ and $|A_3 \cap A_1| = \lfloor 150/7 \times 2 \rfloor = 10$

and $|A_1 \cap A_2 \cap A_3| = \lfloor 150/2 \times 3 \times 7 \rfloor = 3$. Hence,

$$|A_1 \cup A_2 \cup A_3| = N(p_1 p_2 p_3)$$

$$= N - [N(p_1) + N(p_2) + N(p_3)] + [N(p_1 p_2) + N(p_2 p_3) + N(p_3 p_1)] - N(p_1 p_2 p_3)$$

$$= |A| - [|A_1| + |A_2| + |A_3|] + [(A_1 \cap A_2) + |A_2 \cap A_3| + |A_3 \cap A_1|] - |A_1 \cap A_2 \cap A_3|$$

$$= 150 - [75 + 50 + 21] + [25 + 7 + 10] - 3 = 34.$$

6.10 DERANGEMENTS

A derangement is a permutation of objects in such a way that no object remains in its natural position. For example, consider a sequence of numbers 1, 2, 3, 4, 5. Now we would like to know in how many ways these numbers can be arranged so that 1 is not in first place (its natural position), 2 is not in second place (its natural position), . . ., 5 is not in fifth place (its natural position). Such arrangements are called the *derangements* of numbers.

Theorem 6.10 The number of derangement of n objects is equal to

$$D_n = n! \cdot \left[1 - \frac{1}{1!} + \frac{1}{2!} + \ldots + (-1)^n \frac{1}{n!}\right].$$

Proof Let P_i be the set of all permutations in which object i remains in its natural position. Then position of remaining $(n-1)$ objects can be changed among themselves keeping object i fixed at its natural position in $(n-1)!$ ways, i.e. $|P_i| = (n-1)!$. Similarly, $|P_i \cap P_j| = (n-2)!$, i.e. keeping objects i and j in their natural position, position of remaining $(n-2)$ objects can be changed among themselves in $(n-2)!$ ways. In general,

$$|P_{i_1} \cap P_{i_2} \cap \ldots \cap P_{i_n}| = (n-r)!$$

Since ${}^nC_r$ represents the ways of selecting r objects from n objects, therefore we can write

$$\Sigma\,|P_i| = {}^nC_1(n-1)!$$

$$\Sigma\,|P_i \cap P_j| = {}^nC_2(n-2)!$$

$$\vdots$$

$$\Sigma\,|P_{i_1} \cap P_{i_2} \cap \ldots \cap P_{i_r}| = {}^nC_r(n-r)!$$

Since the number of derangements is the number of permutations in which no object remains in its natural position, therefore due to principle of inclusion and exclusion, we have

$$D_n = N - \Sigma\,|P_i| + \Sigma\,|P_i \cap P_j| - \ldots + (-1)^n\,|P_1 \cap P_2 \cap \ldots \cap P_n|$$

$$= n! - {}^nC_1\,(n-1)! + {}^nC_2\,(n-2)! + \ldots + (-1)^n\,{}^nC_n\,0!$$

$$= n! - {}^nC_1\,(n-1)! + {}^nC_2\,(n-2)! + \ldots + (-1)^n$$

$$= n!\left[1 - \frac{1}{1!} + \frac{1}{2!} - \ldots + \frac{(-1)^n}{n!}\right]$$

Example 60 Suppose 4 letters are to be placed in addressed envelope. Find the number of ways such that no letter is placed in the right envelope.

Solution Here the number of objects are $n = 4$. Thus the required number of derangements in which letters can be placed are

$$D_4 = 4!\left[1 - \frac{1}{1!} + \frac{1}{2!} - \frac{1}{3!} + \frac{1}{4!}\right]$$

$$= 24\left[1 - 1 + \frac{1}{2} + \frac{1}{6} + \frac{1}{24}\right] = 9.$$

Example 61 How many derangements are there for 1, 2, 3, 4?

Solution Here $n = 4$. Thus, the required derangements are

$$D_4 = 4!\left[1 - \frac{1}{1!} + \frac{1}{2!} - \frac{1}{3!} + \frac{1}{4!}\right] = 9.$$

The possible 9 derangements are: 2143 2341 2413 3142 3412 3421 4123 4312 4321

Self-practice Problems III

49. Prove the following results
(*a*) ${}^nC_r = {}^nC_{n-r}$ (*b*) ${}^nC_r + {}^nC_{r-1} = {}^{n+1}C_r$

50. Find the value of ${}^rC_5$ when ${}^{18}C_r = {}^{18}C_{r+2}$.

51. If ${}^{2n}C_3 : {}^nC_2 = 44 : 33$, find value of n.

53. Find the coefficient of (*a*) x^5 in $(1-2x)^{-7}$, (*b*) x^{15} in $(x^2+x^3+x^4+\ldots)^4$, and (*c*) x^7 in $(1+x+x^2+x^3+\ldots)^{15}$.

54. Find prime numbers not exceeding 100.

55. Find the number of non-negative integer solutions to equation $x_1 + x_2 + x_3 = 13$ with the extra conditions that $x_1 \leq 4$, $x_2 \leq 5$ and $x_3 \leq 6$.

56. Determine the number of positive integers n, $1 \leq n \leq 200$ that are divisible by 2, 3 or 5.

57. Find the number of onto functions from the set A to B where A = {1, 2, 3, 4} and B = {a, b, c}.

Hints and Answers

50. ${}^rC_5 = {}^8C_5 = 56$

51. $n = 6$

52. (*a*) $(1-2x)^{-7} = \sum_{r=0}^{\infty} {}^{-7}C_r (2x)^r = {}^{-7}C_5(-2)^5 = (-1)^5\, {}^{7+5-1}C_5(-32) = 14,784$

(*b*) $(x^2+x^3+\ldots)^2 = [x^2(1+x+x^2+\ldots)]^4 = [x^2/(1-x)]^4 = x^8/(1-x)^4$. Coefficient of x^{15} = Coefficient x^7 in $(1-x)^{-4} = {}^{-4}C_7(-1)^7 = (-1)^7\, {}^{4+7-1}C_7 = 120$

(*c*) ${}^{21}C_7$

53. $N(\bar{p}_1\bar{p}_2\bar{p}_3\bar{p}_4) = N - [N(p_1) + N(p_2) + N(p_3) + N(p_4)] + [N(p_1p_2) + N(p_1p_3) + N(p_1p_4) N(p_2p_3) + N(p_2p_4) + (p_3p_4)] - [N(p_1p_2p_3) + N(p_1p_2p_4) + N(p_1p_3p_4) + N(p_2p_3p_4)] + N(p_1p_2p_3p_4)$

$= 99 - (50 + 33 + 20 + 14) + (16 + 10 + 7 + 6 + 4 + 2) - (3 + 2 + 1 + 0) + 0 = 21$

54. $N(\bar{p}_1\bar{p}_2\bar{p}_3) = N - [N(p_1) + N(p_2) + N(p_3)] + [N(p_1p_2) + N(p_2p_3) + N(p_3p_1)] - N(p_1p_2p_3)$

$= {}^{3+13-1}C_{13} - [{}^{3+8-1}C_8 + {}^{3+7-1}C_7 + {}^{3+6-1}C_6] + [{}^{3+2-1}C_2 + {}^{3+0-1}C_0 + {}^{3+1-1}C_1]$

$= 105 - [45 + 36 + 28] + [6 + 1 + 3] - 0 = 6.$

Chapter 7

Fundamentals of Probability

PREVIEW

Probability had its beginning in gambling, where combinatorics (counting) techniques are used. Many experiments (or actions) when performed repeatedly do not yield exactly the identical results. Such experiments are called probabilistic whose outcomes are not known. Thus to know such outcomes that are of special interest to us, probability theory is used.

LEARNING OBJECTIVES

After studying this chapter, you should be able to,

- understand fundamental concepts of probability theory,
- calculate probability of an event using three different approaches,
- use rules of probability to add and multiply two or more than two events,
- use Baye's theorem to revise probability of an event based an additional information.

7.1 INTRODUCTION

The decision-makers always face some degree of risk while selecting a particular decision (course of action or strategy) to solve a decision problem. It is because each strategy can lead to a number of different possible outcomes (or results). Thus, it is necessary for the decision-makers to enhance their capability of grasping the probabilistic situation so as to gain a deeper understanding of the decision problem and base their decisions on rational considerations. For this the knowledge of the concept of probability, probability distributions and various related statistical terms is needed. The knowledge of probability and its various types of distributions helps the development of probabilistic decision models. This chapter is devoted to

(*a*) explain the fundamentals of probability and related concepts, and

(*b*) illustrate the application of these concepts to decision problems.

7.2 CONCEPTS OF PROBABILITY

7.2.1 Random Experiment

The random experiment (also called *act, trial, operation* or *process*) is an *activity* whose *consequence* (*result* or *outcome*) is not likely to be known until after its completion. For example, if we toss a coin, the outcome will not be known with certainty until either the head or the tail is observed. The number of outcomes may be finite or infinite depending on the nature of the experiment. For example, in the experiment of tossing a coin, the outcomes are finite and are represented by the head and tail, whereas in the experiment of measuring the time between successive failures of an electronic device, the outcomes are infinite and are represented by the time of failure.

The outcome of an experiment may be expressed in numerical or non-numerical value. For example,

(*a*) counting the number of arrivals at a service window (numerical outcome), and
(*b*) payment made by cash, cheque or credit card (non-numerical outcome).

Each experiment such as tossing a coin leads to the occurrence of a *random event.* In certain decision environment such events are random in nature.

7.2.2 Sample Space

The set of all possible distinct outcomes for a random experiment is called the *sample space* (or *event space)* provided.

(*a*) no two or more of these outcomes can occur simultaneously;
(*b*) exactly one of the outcomes must occur, whenever the experiment is performed.

The sample space is denoted by S.

Consider the 'experiment' of tossing two coins. If we ask the question 'whether each coin falls heads (H) or tails (T)', then there are four possible answers (outcomes). That is

$$S = \{(H, H), (H,T), (T, H), (T,T)\}$$

Each of the pairs of letters in S is called a *sample point* or *an elementary event.* On the other hand, if we are interested in knowing whether the coins fall alike (A) or different (D), then the possible outcomes are only two.

A sample space consisting of finite (or an infinite but countable) number of sample point is called a *discrete sample space,* otherwise called a *continuous sample space.*

7.2.3 Types of Events

An event is referred to the individual outcomes (also called *points*) of a sample space associated with an 'experiment'. For example, for the above defined sample space S, the subset {(H, T), (T, H)} is the event which indicates that either head or tail occurs. A single possible outcome of an experiment is called a simple (or elementary) event. The term *event* is used synonymously with the words: *result, outcome* and *consequence.*

A subset of outcomes which corresponds to some specific event is called the *event space.*

Mutually exclusive events If any event occur and no other event of the sample space can occur at the same time, i.e. two or more events cannot together happen in a single trial, then such events are called mutually exclusive events. In other words, two events are mutually exclusive if happening of one of them prevents or precludes the happening of the other. For example, the numbers 2 and 3 cannot occur simultaneously on the roll of a die. Symbolically a set of events $\{A_1, A_2, \ldots, A_n\}$ is mutually exclusive if $A_i \cap A_j = \Phi\ (i \neq j)$. This means the intersection of two events is a null set (Φ); it is impossible to observe an event that is common in both A_i and A_j.

Collectively exhaustive events If no other events than those enumerated can occur, then such events are called collectively exhaustive events. Symbolically, a set of events $\{A_1, A_2, \ldots, A_n\}$ is called collectively exhaustive if the union of these events is identical with the sample space S. That is,

$$S = \{A_1 \cup A_2 \cup \ldots \cup A_n)$$

For example, the number 7 cannot come upon the uppermost face of a die and in the experiment of rolling a die the number of faces uppermost has the sample space

$$S = \{1, 2, 3, 4, 5, 6\}$$

Independent and dependent events When the outcome of one event does not affect, and is not affected by the other event, then such events are said to be *independent*. The outcomes of successive tosses of a coin are independent of its preceding toss. Increase in the population (in per cent) per year in India is independent of increase in wheat production (in per cent) per year in USA.

However, if the occurrence or non-occurrence of an event in any trial of a random experiment affects the probability of other events in other trials, then such events are said to be *dependent*. For example, the probability of drawing of a card (say a queen) from a pack of playing cards without replacement reduces the probability of drawing a queen in the subsequent draws.

Compound events When two or more events occur in connection with each other, then their simultaneous occurrence is called a compound event. These events may be (*i*) independent, or (*ii*) dependent.

Equally likely events Two events are said to be equally likely if both have equal chance to occur. That is, one of them cannot be expected to occur in preference to the other. For example, each number may be expected to occur the same number of times in the long run on the uppermost face of a rolling die.

Complementary events If E is any subset of the sample space, then its complement denoted by $\overline{E}$ (read as E-bar) contains all the points of the sample space that are not the part of E. If S denotes the sample space then

$$\begin{aligned}\overline{E} &= S - E\\ &= \{\text{All sample points not in E}\}.\end{aligned}$$

For example, if E represents companies with sales less than or equal to Rs 25 lakh, written as $E = \{x : x \leq 25\}$ then this set is complement of the set, $\overline{E} = \{x : x > 25\}$. Obviously such events must be mutually exclusive and collective exhaustive.

7.3 PROBABILITY DEFINED

A general definition of probability states that the probability is a numerical measure of the likelyhood or chance of occurrence of an uncertain event. However, it does not tell us how to compute the probability. In this section, we shall discuss different conceptual approaches of calculating the probability.

7.3.1 The Classical Approach

This approach of defining the probability is quite old and happens to be the earliest. Here the definition is based on the assumption that all the possible outcomes (finite in number) of an experiment are mutually exclusive and equally likely.

For example, if a fair die is rolled, then on any trial each event is equally likely to occur since there are six equally likely exhaustive events, each will occur 1/6 of the time, and therefore the probability of any one event occurring is 1/6. Similarly for the process of selecting a card at random, each event or card is mutually exclusive, exhaustive, and equiprobable. The probability of selecting anyone card on a trial is equal to 1/52, since there are 52 cards. Hence, in general, for a random experiment with n mutually exclusive, exhaustive, equiprobable events, the probability of any of the events is equal to $1/n$. In other words, if there are a possible outcomes where the favourable event A occurs and b possible outcomes where the event A does not occur, and all these possible outcomes are mutually exclusive, exhaustive and equiprobable, then the probability that event A will occur is defined as

$$P(A) = \frac{a}{a+b} = \frac{\text{Number of favourable outcomes}}{\text{Total number of possible outcomes}} = \frac{c(A)}{c(S)}$$

The classical probability is also called a *priori probability* because the definition can be used to determine probabilities in advance (or before the fact or a priori) of any trial of tossing a fair coin, rolling a fair die or drawing a card from a shuffled deck of cards. This means, we do not have to perform random experiments to find the probability of occurrence of an event. This also implies that no experimental data are required for computation of probability.

This approach of calculating probability is useful when the pattern of outcomes is countable. This definition is based on the assumption that all the possible outcomes are equally likely. This assumption may lead to wrong calculation of probability in case some outcomes are more or less frequent in occurrence. For example, if we classify two children in a family according to their sex, then possible outcomes in terms of number of boys in the family are 0, 1, 2. Thus, according to classical approach, the probability for each the outcomes should be 1/3. However, it has been calculated that the probabilities are approximately 1/4, 1/2 and 1/4 for 0, 1, 2 boys, respectively.

Since probability calculation is based entirely upon logical reasoning and hence it is known as *priori* or *mathematical probability*.

7.3.2 The Relative Frequency Approach

In situations where the outcomes of a random experiment are not all equally likely or when it is not known whether outcomes are equally likely, application of classical approach may not give correct value. For example, it is not possible to state in advance, without repetitive trials of the experiment, the probabilities in the cases like (*i*) a number greater than 3 will appear when die is rolled, (*ii*) a lot of 100 items will include 10 defective items.

Since the probability of an event is determined by repeating an experiment (observations) large number of times under the same conditions where trials are independent to each other and observing the outcome to know the ratio of the number of occurrence of an event to the number of trials, therefore, this approach is called a *relative frequency approach.* For example, if die is tossed n times and s denotes the number of times the event A (i.e. numbers 4, 5 or 6) occurs, then the ratio $P(A) = s/n$ gives the proportions of times the event A occurs in n trial, and are also called relative frequencies of the event in n trials.

Now let us repeat the experiment of rolling a die 100 times and calculate the probability of event A each time. Although our estimate about P (A) may changed after every trial but we will find that the proportions (s/n) tend to cluster around a unique central value as number of trials n becomes even larger. This unique central value (also called probability of event A) is defined as:

$$P(A) = \lim_{n \to \infty} \left(\frac{s}{n}\right).$$

Since the probability of an event is determined objectively by repetitive empirical observations, it is also known as *empirical probability.*

7.3.3 The Subjective Approach

The subjective approach of calculating probability is the best estimate in the opinion of an individual after considering what (if any) evidence is at hand. Such estimates are always based on degree of beliefs, convictions, experience and background. It is, thus, a way to quantify individual's beliefs, assessment, and judgement of the situation and then to express in the quantitative way the occurrence of an outcome. Probability assigned for the occurrence of an event may be based on just guess or having some idea about the relative frequency of past occurrence of the event. This approach must be used when either sufficient data are not available or sources of information giving different results are not known.

7.3.4 Probability Axioms

The another approach of defining probability is based on the following axioms. Let S be the sample space of an experiment that is partitioned into mutually exclusive and exhaustive events $A_1, A_2, \ldots, A_n$ which may be elementary or compound. The probability of any event A in S is then a real number associated with A such that the following axioms are satisfied.

(*i*) The probability of an event is restricted to the range *zero* to *one* inclusive, where zero represents an impossible event and one represents a certain event. Symbolically, we write $0 \leq P(A) \leq 1$.

For example, probability of number seven, $P(7) = 0$ in rolling a die because this number is an impossible event for this experiment.

(*ii*) The probability sum of all the simple events constituting the sample space is equal to one, that is $P(S) = 1$. This also implies that if a random experiment is conducted, one of its outcomes in its sample space is certain to occur. Similarly, the probability of an impossible event or an empty set is zero. That is $P(\Phi) = 0$.

(*iii*) If events A_1 and A_2 are two events in S and if occurrence of A_1 implies that A_2 occurs, i.e. if A_1 is a subset of A_2, then the probability of A_1 is less than or equal to the probability of A_2. That is, $P(A_1) \leq P(A_2)$.

Glossary of probability terms If A and B are two events, then

$A \cup B$ = an event which represents the occurrence of either A or B or both.

$A \cap B$ = an event which represents the simultaneous occurrence of A and B.

$\overline{A}$ = complement of event A and represents non-occurrence of A.

$\overline{A} \cap \overline{B}$ = both A and B do not occur.

$\overline{A} \cap B$ = event A does not occur but event B occurs.

$A \cap \overline{B}$ = event A occurs but event B does not occur.

$(A \cap \overline{B}) \cup (\overline{A} \cap B)$ = exactly one of the two events A and B occurs.

Example 1 A bag contains 6 red and 8 green balls.

(*a*) If one ball is drawn at random, then what is the probability of the ball being green?

(*b*) If two balls are drawn at random, then what is the probability that one is red and the other green?

Solution (*a*) Since the bag contains 6 red and 8 green balls, therefore it contains 6 + 8 = 14 equally likely outcomes, i.e. S = {*r*, *g*}. But one ball out of 14 balls can be drawn in $^{14}C_1$ ways, i.e.

$$^{14}C_1 = \frac{14!}{1!\,(14-1)!} = 14 \text{ ways.}$$

Let A be the event of drawing a green ball. Then out of these 8 green balls, one green ball can be drawn in $^{8}C_1$ ways:

$$^{8}C_1 = \frac{8!}{1!\,(8-1)!} = 8.$$

Hence, $$P(A) = \frac{c(A)}{c(S)} = \frac{8}{14}.$$

(*b*) All exhaustive number of cases, $c(S) = {}^{14}C_2 = \dfrac{14!}{2!\,(14-2)!} = 91.$

Also out of 6 red balls, one red ball can be drawn in $^{6}C_1$ ways and out of 8 green balls, one green ball can be drawn in $^{8}C_1$ ways. Thus, the total number of favourable cases is

$$c(B) = {}^{6}C_1 \times {}^{8}C_1 = 6 \times 8 = 48.$$

Hence, $$P(B) = \frac{c(B)}{c(S)} = \frac{48}{91}.$$

Example 2 Tickets are numbered from 1 to 100. They are well shuffled and a ticket is drawn at random. What is the probability that the drawn ticket has

(*a*) an even number?

(*b*) a number 5 or multiple of 5?

(*c*) a number which is greater than 75?

(*d*) a number which is a square?

Solution Since any of the 100 tickets can be drawn, therefore, exhaustive number of cases are $c(S) = 100$.

(*a*) Let A be the event of getting even numbered tickets. Then $c(A) = 50$, and hence

$$P(A) = 50/100 = 1/2.$$

(*b*) Let B be the event of getting a ticket bearing a number 5 or multiple of 5, i.e.

$$B = [5, 10, 15, 20, \ldots, 95, 100]$$

which are 20 in number, $c(B) = 20$. Thus $P(B) = 20/100 = 1/5$.

(*c*) Let C be the event of getting a ticket bearing a number greater than 75, i.e.

$$C = \{76, 77, \ldots, 100\}$$

which are 25 in number, $c(C) = 25$. Thus $P(C) = 25/100 = 1/4$.

(*d*) Let D be the event of getting a ticket bearing a number which is a square, i.e.

$$D = \{1, 4, 9, 16, 49, 64, 81, 100\}$$

which are 10 in number, $c(D) = 10$. Thus $P(D) = 10/100 = 1/10$.

7.4 RULES OF PROBABILITY AND ALGEBRA OF EVENTS

In probability we use set theory notations to simplify the presentation of ideas. As discussed earlier in this chapter, the probability of the occurrence of an event A is expressed as:

$$P(A) = \text{Probability of the occurrence of event A}$$

Such single probabilities are called *marginal* (or *unconditional*) *probabilities* because it is the probability of single event occurring. In a coin tossing example, the marginal probability of a tail or head in a toss can be stated as P (T) or P (H).

7.4.1 Addition Rules

Mutually exclusive events If we are interested in calculating the probability that one event or another will occur, then the following addition rule can be applied provided events are mutually exclusive (disjoint), exhaustive and equally probable.

The rule may be stated as: *If two events A and B are mutually exclusive, exhaustive and equiprobable, then the probability of the occurrence of either A* or *B is equal to the sum of their individual probabilities*. That is

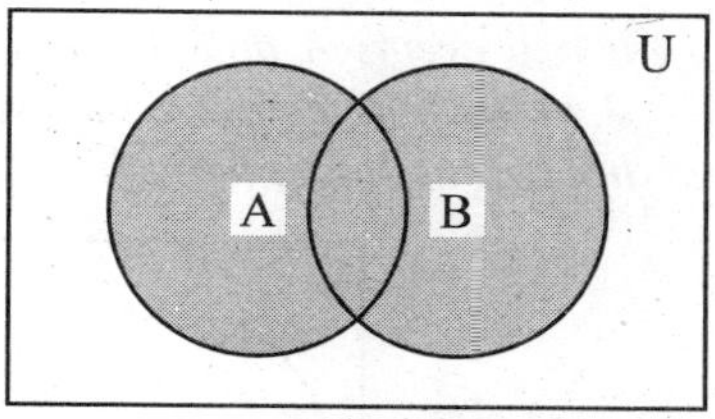

Fig. 7.1 Union of Two Events

$$P(A \text{ or } B) = P(A \cup B) = \frac{n(A \cup B)}{n(S)} = \frac{n(A) + n(B)}{n(S)}$$

$$= \frac{n(A)}{n(S)} + \frac{n(B)}{n(S)} = P(A) + P(B) \qquad ...(i)$$

where $A \cup B$ is read as A *union* B denotes the union of two events A and B and it is the set of all sample points that belongs to either event A or event B (or both). $A \cup B$ also means that either A or B occurs or both A and B occur, i.e. at least one of A and B. This rule can also be illustrated by the Venn diagram as shown in Fig. 7.1.

Illustration Consider the pattern of arrival of customers at a service counter during first hour it is open with the following probabilities

No. of persons :	0	1	2	3	4 or more
Probability :	0.1	0.2	0.3	0.3	0.1

To understand the probability that either 2 or 3 persons will be there during the first hour, we shall have

$$P(2 \text{ or } 3) = P(2) + P(3) = 0.3 + 0.3 = 0.6$$

The Eq. (*i*) can be expanded to include more than two events. In particular if there are n events in a sample space that are mutually exclusive, then the probability of the union of these events is given by

$$P(A_1 \cup A_2 \cup \ldots \cup A_n) = P(A_1) + P(A_2) + \ldots + P(A_n) \qquad ...(ii)$$

For example, if we are interested in the question, 'what is the probability that there will be two or more persons during first hour', then using Eq. (*ii*), we have

$$P(2 \text{ or more}) = P(2, 3, 4 \text{ or more}) = P(2) + P(3) + P(4) = 0.3 + 0.3 + 0.1 = 0.7.$$

An important special case of Eq. (*i*) is for complementary events. Let A be any event and $\overline{A}$ be the complement of A. Obviously A and $\overline{A}$ are exclusive and exhaustive events. Thus, either A happens or it does not, is given by

$$P(A \text{ or } \overline{A}) = P(A) + P(\overline{A}) = P(A) + \{1 - P(A)\} = 1$$

or $$P(A) = 1 - P(\overline{A}). \quad ...(iii)$$

Non-mutually exclusive (or partially overlapping) events: When events A and B are not mutually exclusive it is possible for both events to occur simultaneously. This means these events have sample points in common. Such events are also called *joint* (or *overlapping*) *events*. The sample points in common (belong to both events) represent the joint event $A \cap B$ (read as A intersection B). This addition rule in this case is stated as:

If two events A and B are not mutually exclusive, then the probability of occurrence of either A or B or both is equal to the sum of their individual probabilities minus the probability of A and B happening together.

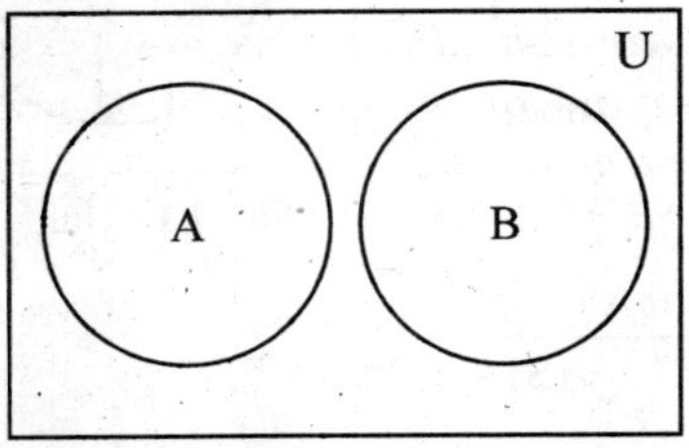

Fig. 7.2 (*a*) Mutually Exclusive Events

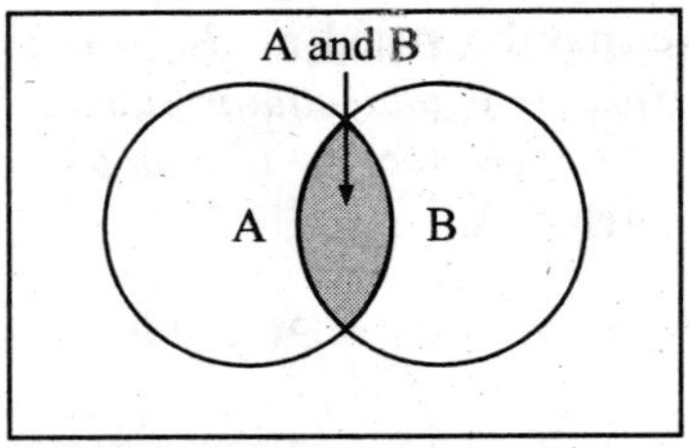

Fig. 7.2 (*b*) Joint Events

Symbolically it can be written as:

$$P(A \text{ or } B) = P(A) + P(B) - P(A \text{ and } B)$$

or $$P(A \cup B) = P(A) + P(B) - P(A \cap B). \quad ...(iv)$$

This addition rule can also be illustrated by the Venn diagram as shown in Figs 7.2 (*a*) & (*b*).

Illustration Suppose 70 per cent of all tourists who come to India will visit Agra, 60 per cent of them will visit both Agra and Goa. The probability that a tourist will visit either Goa or Agra or both is obtained by applying Eq. (*iv*) as follows:

$$P(\text{Agra or Goa}) = P(\text{Agra}) + P(\text{Goa}) - P(\text{Agra and Goa})$$
$$= 0.70 + 0.60 - 0.50 = 0.8.$$

Consequently, the probability that a tourist will visit neither Agra nor Goa is given by

$$P(\text{neither Agra nor Goa}) = 1 - P(\text{Agra or Goa}) = 1 - 0.80 = 0.20.$$

Equation (*iv*) can be expanded to include more than two events. In particular, if there are three events that are not mutually exclusive, then the probability of the union of these events is given by

$$P(A \cup B \cup C) = P(A) + P(B) + P(C) - P(A \cap B) - P(B \cap C)$$
$$- P(C \cap A) + P(A \cap B \cap C) \quad ...(v)$$

Remark The additive rules are applicable for calculating probability in case of simultaneous trails.

Example 3 From a sales force of 150 persons, one will be selected to attend a special sales meeting. If 52 of them are unmarried, 72 are college graduates and 3/4 of the 52 that are unmarried are college graduates, find the probability that a sales person selected at random will be neither single nor a college graduate.

Solution Let A and B be event that a sales person selected is married and that he is a college graduate, respectively. Then it is given that

$$P(A) = 52/150,\ P(B) = 72/150;\ P(A \text{ and } B) = (3/4)(52/150) = 39/150.$$

The probability that a sales person selected at random will be neither single nor a college graduate is:

$$P(\bar{A} \cap \bar{B}) = 1 - P(A \cup B) = 1 - \{P(A) + P(B) - P(A \cap B)\}$$

$$= 1 - \left\{\frac{52}{150} + \frac{72}{150} - \frac{39}{150}\right\} = \frac{13}{30}.$$

Example 4 From a computer tally based on employer's records, the personnel manager of a large manufacturing firm finds that 15 per cent of the firm's employees are supervisors and 25 per cent employees are college graduates. He also discovers that 5 per cent employees are both supervisors and college graduates. Suppose that an employee is selected at random from the firm's personnel record, what is the

(*a*) probability of selecting a person who is both a college graduate and a supervisor.

(*b*) probability of selecting a person who is neither a supervisor nor a college graduate?

Solution Let A and B be the event that the person selected is supervisor and that he is a college graduate, respectively. Given that

$$P(A) = 15/100;\ P(B) = 25/100;\ P(A \text{ and } B) = 5/100.$$

(*a*) Probability of selecting a person who is both a college graduate and supervisor is:

$$P(A \text{ and } B) = 5/100 = 0.05.$$

(*b*) Probability of selecting a person who is neither a supervisor nor a college graduate is:

$$P(\bar{A} \text{ and } \bar{B}) = 1 - P(A \text{ or } B) = 1 - [P(A) + P(B) - P(A \text{ and } B)]$$

$$= 1 - \left(\frac{15}{100} + \frac{25}{100} - \frac{5}{100}\right) = \frac{65}{100} = 0.65.$$

Example 5 The odds against student X solving a Business Statistics problem are 8 to 6, and odds in favour of student Y solving the problem are 14 to 16.

(*a*) What is the chance that the problem will be solved if they both try independently of each other?

(*b*) What is the probability that none of them is able to solve the problem?

Solution Let A = the event that the first student solves the problem,

B = the event that the second student solves the problem.

$$P(A) = \frac{6}{8+6} = \frac{6}{14} \text{ and } P(B) = \frac{14}{14+16} = \frac{14}{30}.$$

(*a*) Probability that the problem will be solved

$$= P(\text{at least one of them solves the problem})$$

$$= P(A \text{ or } B) = P(A) + P(B) - P(A \text{ and } B)$$

$$= P(A) + P(B) - P(A) \times P(B) \qquad \text{[since events are independent]}$$

$$= \frac{6}{14} + \frac{14}{30} - \frac{6}{14} \times \frac{14}{30} = \frac{73}{105} = 0.695.$$

(*b*) Probability that neither A nor B solves the problem

$$P(\overline{A} \text{ and } \overline{B}) = P(\overline{A}) \times P(\overline{B})$$

$$= [1 - P(A)] \times [1 - P(B)] = \frac{8}{14} \times \frac{16}{30} = \frac{32}{105} = 0.305.$$

7.4.2 Multiplication Rules

Statistical independent events When the occurrence of an event does not affect and is not affected by the probability of occurrence of any other event, the events are said to be *statistically independent.* There are three types of probabilities under statistical independence: *marginal*, *joint* and *conditional.*

(*i*) *Marginal Probability:* The term 'marginal' refers to one and only one event. The concept of marginal probability remains the same in any case.

(*ii*) *Joint Probability:* The probability that two or more independent events occur together or in succession is called the *joint probability.* The joint probability of two or more independent events is equal to the product of their respective marginal probabilities. In particular, if A and B are independent events, the probability that both A and B will occur is given by

$$P(AB) = P(A \cap B) = P(A) \times P(B). \qquad ...(vi)$$

Suppose, we toss a coin twice. The probability that both the times the coin will turn up head is given by

$$P(H_1.H_2) = P(H_1) \times P(H_2) = 1/2 \times 1/2 = 1/4.$$

Equation (*vi*) is applied here because the probability of any outcome is not affected by any preceding outcomes, in other words, outcomes are independent.

(*iii*) *Conditional Probability:* For statistically independent events, A and B, the conditional probability denoted by $P(A \mid B)$ [or $P(B \mid A)$] of event A (or B), given that event B (or A) has already occurred is simply the probability of event A (or B). Symbolically, it is written as

$P(A \mid B) = P(A)$, probability of A when it is known that event B has already occurred.

or $P(B \mid A) = P(B)$, probability of B when it is known that event A has already occurred. ...(*vii*)

Since, in the case of independent events the probability of occurrence of either of the events does not depend or affect the occurrence of other, therefore, in the coin tossing example, the probability of a head occurrence in the second toss given that head resulted in the first toss is still 0.5. That is, $P(H_2 \mid H_1) = 0.5 = P(H_2)$. It is because of the fact that probabilities of heads and tails are same for every toss and in no way influenced by whether it was a head or tail which occurred in the previous toss.

Statistical dependent events When the occurrence of an event is affected and is also affected by the probability of occurrence of any other event, the events are said to be statistically dependent. There are three types of probabilities under statistical dependence: *joint*, *conditional* and *marginal.*

(*i*) *Joint Probability:* If A and B are dependent events, then the joint probability as discussed under statistical independence case is no longer equal to the product of their respective probabilities. That is, for dependent events

$$P(A \text{ and } B) = P(A \cap B) \neq P(A) \times P(B).$$

Accordingly, $P(A) \neq P(A \mid B)$ and $P(B) \neq P(B \mid A)$.

For dependent events A and B the joint probability of A and B occurring together or in succession is given by

$$P(A \cap B) = P(A) \times P(B \mid A) \qquad \text{...}(i)$$

or

$$P(A \cap B) = P(B) \times P(A \mid B). \qquad \text{...}(ii)$$

(*ii*) *Conditional Probability:* Equations (*i*) and (*ii*) refer to as joint probability under condition of statistical dependence. Now these equations can also be written as

$$P(B \mid A) = \frac{P(A \cap B)}{P(A)}. \qquad \text{...}(iii)$$

This gives the conditional probability of B, given that event A has occurred likewise, the conditional probability of A, given that event B has occurred, is

$$P(A \mid B) = \frac{P(A \cap B)}{P(B)}. \qquad \text{...}(iv)$$

(*iii*) *Marginal Probability:* The marginal probability of an event under statistical dependence is the same as the marginal probability of an event under statistical independence.

The marginal probability of events A and B can be written as:

$$P(A) = P(A \cap B) + P(A \cap \overline{B})$$

and

$$P(B) = P(A \cap B) + P(\overline{A} \cap B)$$

Example 6 A piece of equipment will function only when the three components A, B and C are working. The probability of A failing during one year is 0.15, that of B failing is 0.05 and that of C failing is 0.10. What is the probability that the equipment will fail before the end of the year?

Solution Given that

$$P(A \text{ failing}) = 0.15; \quad P(A \text{ does not fail}) = 1 - P(A) = 0.85$$

$$P(B \text{ failing}) = 0.05; \quad P(B \text{ does not fail}) = 1 - P(B) = 0.95$$

$$P(C \text{ failing}) = 0.10; \quad P(C \text{ does not fail}) = 1 - P(C) = 0.90.$$

Since all the three events are independent, therefore, the probability that the equipment will work is given by

$$P(\overline{A}) \times P(\overline{B}) \times P(\overline{C}) = 0.85 \times 0.95 \times 0.90 = 0.726.$$

Hence, the probability that the equipment will fail before the end of the year is given by

$$P(A \cup B \cup C) = 1 - P(\overline{A} \cup \overline{B} \cup \overline{C}) = 1 - P(\overline{A} \cap \overline{B} \cap \overline{C})$$

$$= 1 - P(\overline{A}).\ P(\overline{B}).\ P(\overline{C})$$

$$= 1 - \{0.15 \times 0.05 \times 0.10\} = 1 - 0.726 = 0.274.$$

Example 7 Suppose a company selected both MBA and non-MBA for the same type of jobs. After a period of employment, some out of each group are promoted and some are not. Table on the following page shows the proportions of company's managers among the said classes:

Proportional Status	*Managers*		*Total*
	MBA (A)	*Non-MBA* ($\bar{A}$)	
Promoted (B)	0.39	0.21	0.60
Non-promoted ($\bar{B}$)	0.26	0.14	0.40
Total	0.65	0.35	1.00

Calculate P (A | B) or P (B | A) and verify whether qualification (A) and promotion (B) are independent.

Solution The table shows that the probability of selecting an MBA is 0.65 and that of selecting a non-MBA is 0.35. The probability of finding a promoted and a non-promoted manager are 0.60 and 0.40, respectively.

Probabilities appearing in the column total and row total are called *marginal probabilities*. These are the sum of probabilities in which events A and B are involved.

Since each of the joint probabilities in the given table is numerically equal to the product of the marginal (or individual) probabilities, we may say that qualification is independent of promotion.

It may be observed from the table that

P (AB) = P (A ∩ B) = Probability that a person was promoted and is an MBA (*a joint probability*)

= (Probability that a person was promoted) × (Probability that a person is an MBA)

= P (A) × P (B) = 0.65 × 0.60 = 0.39.

Also from the table, we know that, P (AB) = 0.39. Therefore, P (AB) = P (A) × P (B).

Similarly.

P (MBA and Non-promoted) = P (MBA) × P (Non-promoted) = 0.65 × 0.40 = 0.26

P (Non-MBA and Promoted) = P (Non-MBA) × P (Promoted) = 0.35 × 0.60 = 0.21

P (Non-MBA and Non-promoted) = P (Non-MBA) × P (Non-promoted)

= 0.35 × 0.40 = 0.14.

Since each of the joint probabilities is equal to the product of the probabilities of the individual event, therefore, we may conclude that the qualification and promotion are independent of each other.

Example 8 A market research firm is interested in surveying certain attitudes in a small community. There are 125 households broken down according to income, ownership of a telephone and ownership of a TV.

	Households with Annual Income of Rs 8,000 or less		*Households with Annual Income above Rs 8,000*		*Total*
	Telephone Subscriber	*No Telephone*	*Telephone Subscriber*	*No Telephone*	
Own TV set	27	20	18	10	75
No TV set	18	10	12	10	50
Total	45	30	30	20	125

(*a*) What is the probability of obtaining a TV owner in drawing at random?

(*b*) If a household has income over Rs 8,000 and is a telephone subscriber, what is the probability that he has a TV?

(*c*) What is the conditional probability of drawing a household that owns a TV, given that the household is a telephone subscriber?

(*d*) Are the events 'ownership of a TV' and 'telephone subscriber' statistically independent? Comment.

Solution (*a*) Probability of drawing a TV owner at random,

$$P\text{ (TV owner)} = 75/125 = 0.6.$$

(*b*) There are 30 (= 18 + 12) persons whose household income is above Rs 8,000 and are also telephone subscribers. Out of these, 18 own TV sets. Hence, the probability of this group of persons having a TV set is: 18/30 = 0.6.

(*c*) Out of 75 (= 27 + 18 + 18 + 12) households who are telephone subscribers, 45 (= 27 + 18) households have TV sets. Hence, the conditional probability of drawing a household that owns a TV given that the household is a telephone subscriber is: 45/75 = 0.6.

(*d*) Let A and B be the events representing TV owners and telephone subscribers, respectively. Probability of a person owning a TV, P (A) = 75/125.

Probability of a person being a telephone subscriber, P (B) = 75/125.

Probability of a person being a telephone subscriber as well as TV owner is:

$$P\text{ (A and B)} = 45/125 = 9/25.$$

But

$$P(A) \times P(B) = (75/125)\,(75/125) = 9/25.$$

Since P (AB) = P (A) × P (B), therefore, we conclude that the events 'ownership of a 'TV' and 'telephone subscriber' are statistically independent.

Example 9 A company has two plants to manufacture scooters. Plant I manufactures 80 per cent of the scooters and Plant II manufactures 20 per cent. In plant I only 85 out of 100 scooters are considered of standard quality. In plant II, only 65 out of 100 scooters are considered of standard quality. What is the probability that a scooter selected at random came from plant I, if it is known that it is of standard quality?

Solution Let A = The scooter purchased is of standard quality.

B = The scooter is of standard quality and came from plant I.

C = The scooter is of standard quality and came from plant II.

D = The scooter came from plant I.

The percentage of scooters manufactured in plant I that are of standard quality is 85 per cent of 80 per cent, i.e. 0.85 × (80/100) = 68 per cent or P(B) = 0.68.

The percentage of scooters manufactured in plant II that are of standard quality is 65 per cent of 20 per cent, i.e. 0.65 × (20/100) = 13 per cent or P(C) = 0.13.

The probability that a customer obtains a standard quality scooter from the company is: 0.68 + 0.13 = 0.81.

The probability that the scooters selected at random came from plant I, if it is known that it is of standard quality is given by:

$$P(D|A) = \frac{P(D \text{ and } A)}{P(A)} = \frac{0.68}{0.81} = 0.84.$$

Example 10 A husband and wife appear in an interview for two vacancies in the same post. The probability of husband's selection is 1/7 and that of wife's selection is 1/5. What is the probability that

(*a*) both of them will be selected?

(*b*) only one of them will be selected?

(*c*) none of them will be selected?

Solution Let A and B be the event of husband's and wife's selection, respectively. Given that P (A) = 1/7 and P (B) = 1/5.

(*a*) Probability that both of them will be selected

$$P(A \text{ and } B) = P(A)\,P(B) = (1/7) \times (1/5) = 1/35 = 0.029.$$

(*b*) Probability that only one of them will be selected

$$P[(A \text{ and } \overline{B}) \text{ or } (B \text{ and } \overline{A})] = P(A \text{ and } \overline{B}) + P(B \text{ and } \overline{A})$$

$$= P(A)\,P(\overline{B}) + P(B)\,P(\overline{A})$$

$$= P(A)\,[1 - P(B)] + P(B)\,[1 - P(A)]$$

$$= \frac{1}{7}\left(1 - \frac{1}{5}\right) + \frac{1}{5}\left(1 - \frac{1}{7}\right) = \left(\frac{1}{7} \times \frac{4}{5}\right) + \left(\frac{1}{5} \times \frac{6}{7}\right) = \frac{10}{35} = 0.286.$$

(*c*) Probability that none of them will be selected

$$= P(\overline{A}) \times P(\overline{B}) = \frac{6}{7} \times \frac{4}{5} = \frac{24}{35} = 0.686.$$

Example 11 The data for the promotion and academic qualification for a company is given below:

Promotional Status	*Academic Qualification*		
	MBA(A)	*Non-MBA* ($\overline{A}$)	*Total*
Promoted (B)	0.14	0.26	0.40
Non-promoted ($\overline{B}$)	0.21	0.39	0.60
Total	0.35	0.65	1.00

(*a*) Calculate the conditional probability of promotion after an MBA has been identified.

(*b*) Calculate the conditional probability that he is an MBA when a promoted employee has been chosen.

(*c*) Find the probability that employee promoted was an MBA.

Solution It is given that, P (A) = 0.35, P ($\overline{A}$) = 0.65, P (B) = 0.40, P ($\overline{B}$) = 0.60 and P (A ∩ B) = 0.14.

(*a*) The probability of 'promoted' after an MBA employee has been identified is given by

$$P(B \mid A) = \frac{P(A \cap B)}{P(A)} = \frac{0.12}{0.35} = 0.34$$

(*b*) If a promoted employee has been chosen, then the probability that the person is an MBA is:

$$P(A \mid B) = \frac{P(A \cap B)}{P(B)} = \frac{0.12}{0.40} = 0.30.$$

(*c*) The probability that an employee promoted was an MBA is:

$$P(A \cap B) = P(A) \times P(B \mid A) = 0.35 \times 0.34 = 0.12$$

or

$$P(A \cap B) = P(B) \times P(A \mid B) = 0.40 \times 0.30 = 0.12.$$

Example 12 The probability that a trainee will remain with a company is 0.6. The probability that an employee earns more than Rs 10,000 per month is 0.5. The probability that an employee is a trainee who remained with the company or who earns more than Rs 10,000 per month is 0.7. What is the probability that an employee earns more than Rs 10,000 per month given that he is a trainee who stayed with the company?

Solution Let A and B be the events that a trainee will remain with the company and the event that an employee earns more than Rs 10,000, respectively. Given that

$$P(A) = 0.6, P(B) = 0.5 \text{ and } P(A \text{ or } B) = P(A \cup B) = 0.7.$$

Here we need to calculate,

$$P(B|A) = \frac{P(A \text{ and } B)}{P(A)} = \frac{P(A \cap B)}{P(A)}.$$

We know that $P(A \cup B) = P(A) + P(B) - P(A \cap B)$,

or $P(A \cap B) = P(A) + P(B) - P(A \cup B) = 0.6 + 0.5 - 0.7 = 0.4.$

Hence the required probability is:

$$P(B \mid A) = \frac{(A \cap B)}{P(A)} = \frac{0.4}{0.6} = 0.667.$$

Example 13 Two computers A and B are to be marketed. A salesman who is assigned the job of finding customers for them has 60 per cent and 40 per cent chances, respectively of succeeding in case of computer A and B. The two computers can be sold independently. Given that he was able to sell at least one computer, what is the probability that computer A has been sold?

Solution Let us define the events

E_1 = Computer A is marketed and E_2 = Computer B is marketed.

It is given that $P(E_1) = 0.60$, $P(E_2) = 0.40$

$P(E_1 \text{ and } E_2)$ or $P(E_1 \cap E_2) = P(E_1) \cdot P(E_2) = 0.60 \times 0.40 = 0.24.$

Hence, the probability that computer A has been sold given that the salesman was able to sell at least one computer is given by

$$P(E_1 \mid E_1 \cup E_2) = \frac{P\{E_1 \cap (E_1 \cup E_2)\}}{P(E_1 \cup E_2)} = \frac{P(E_1)}{P(E_1 \cup E_2)}$$

$$= \frac{P(E_1)}{P(E_1) + P(E_2) - P(E_1 \cap E_2)} = \frac{0.60}{0.60 + 0.40 - 0.24}$$

$$= \frac{0.60}{0.76} = 0.789.$$

7.5 PROBABILITY TREE DIAGRAM

Decision-makers at times face difficulty in constructing the joint probability table. Thus, they prefer to use *probability tree* to calculate probability calculations. The probability tree diagram for tossing a coin at three consecutive times is shown in Fig. 7.3.

The use of probability tree diagram is convenient for calculating joint probabilities when events occur at different times or stages. In probability trees, time moves from left to right. The complete tree exhibits each outcome as a *single path* from beginning to end. Each path corresponds to distinct joint event. A joint event is represented by a path through the tree and its probability is determined by multiplying all the individual branch probabilities for its path.

In this example of three tosses of a coin, at each toss the probability of either event's (H or T) occurring remains the same, i.e. the events are independent. The joint probabilities of events occurring in succession are calculated by multiplying the probabilities of each event.

The events emanating from a single breaking point are mutually exclusive and collectively exhaustive, so that exactly one must occur. All the probabilities on branches within the same fork must there fore sum to 1.

In this case, the results in the diagram should not be confused with conditional probabilities. The probability of a head and then two tails occurring on three consecutive tosses in computed prior to any tosses taking place. If the first two tosses have already occurred, then the probability of getting a tail on the third toss is still 0.5, i.e. P (T | HT) = P (T) = 0.5.

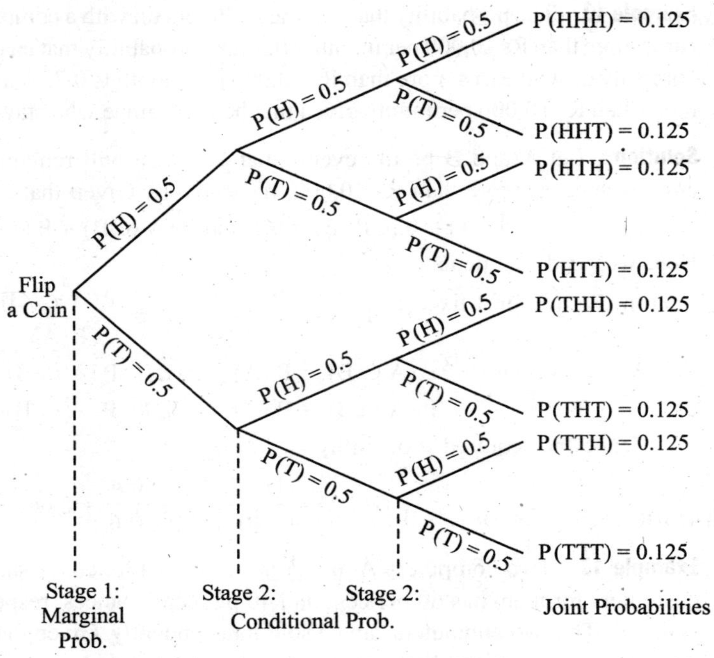

Fig. 7.3 Probability Tree Diagram

7.6 THE BAYE'S THEOREM

The Baye's formula is useful in revising the original probability estimates as we gain additional information about the events or outcomes. The prior probabilities when changed in the light of new information are called *revised* or *posterior probabilities.*

Suppose $A_1, A_2, \ldots, A_n$ are mutually exclusive and collectively exhaustive events with prior marginal probabilities $P(A_1), P(A_2), \ldots, P(A_n)$. Let B be an arbitrary event with $P(B) \neq 0$ for which conditional probabilities $P(B \mid A_1), P(B \mid A_2), \ldots, P(B \mid A_n)$ are also known. Given the information that outcome B has happened, we want to calculate the revised (or posterior) probabilities $P(A_i \mid B)$, for all i.

The conditional probability of occurrence of event A_i $(i = 1, 2, \ldots, n)$ given that event B has happened is given by

$$P(A_i \mid B) = \frac{P(A_i \cap B)}{P(B)}. \qquad \text{...}(i)$$

Since events $A_1, A_2, \ldots, A_n$ are mutually exclusive and collectively exhaustive, the event B is bound to occur with either $A_1, A_2, \ldots, A_n$. That is,

$$B = (A_1 \cap B) \cup (A_2 \cap B) \cup \ldots \cup (A_n \cap B).$$

As $(A_1 \cap B), (A_2 \cap B), \ldots, (A_n \cap B)$ are mutually exclusive. Thus, we get

$$P(B) = P(A_1 \cap B) + P(A_2 \cap B) + \ldots + P(A_n \cap B)$$
$$= \sum_{i=1}^{n} P(A_i \cap B)$$
$$= P(A_1)\, P(B \mid A_1) + P(A_2)\, P(B \mid A_2) + \ldots + P(A_n)\, P(B \mid A_n)$$
$$= \sum_{i=1}^{n} P(A_i)\, P(B|A_i).$$

From Eq. (*i*) for a fixed *i*, we have

$$P(A_i \mid B) = \frac{P(A_i \cap B)}{P(B)} = \frac{P(B|A_i)\,.\,P(A_i)}{P(B)}$$
$$= \frac{P(B|A_i)\,.\,P(A_i)}{P(A_1)\,.\,P(B|A_1) + P(A_2)\,P(B|A_2) + \ldots + P(A_n)\,P(B|A_n)}.$$

Remark Result of Baye's theorem can also be written as

$$P(A_i \mid B) = \frac{P(A_i \cap B)}{\sum_{i=1}^{n} (A_i \cap B)}.$$

Example 14 Suppose an item is manufactured by three machines X, Y and Z. All the three machines have equal capacity and are operated at the same rate. It is known that percentages of defective items produced by X, Y and Z are 2, 7 and 12 per cent, respectively. All the items produced by X, Y and Z are put into one bin. From this bin, one item is drawn at random and is found to be defective. What is the probability that this item was produced by Y?

Solution Let A be the defective item. We know the prior probability of defective items produced by X, Y and Z, i.e. P (X) = 1/3; P (Y) = 1/3 and P (Z) = 1/3. We also know that

$$P(A \mid X) = 0.02,\ P(A \mid Y) = 0.07,\ P(A \mid Z) = 0.12.$$

Now having known that the item drawn is defective, we want to know the probability that it was produced by Y. That is

$$P(Y \mid A) = \frac{P(A|Y)\,.\,P(Y)}{P(X)\,.\,P(A|X) + P(Y)\,.\,P(A|Y) + P(Z)\,.\,P(A|Z)}$$
$$= \frac{(0.07)\,.\,(1/3)}{(1/3)\,(0.02) + (1/3)\,(0.07) + (1/3)\,(0.12)} = 0.33.$$

Example 15 Assume that a factory has two machines. Past records show that machine 1 produces 30 per cent of the items of output and machine 2 produces 70 per cent of the items. Further 5 per cent of the items produced by machine 1 were defective and only 1 per cent produced by machine 2 were defective. If a defective item is drawn at random, what is the probability that the defective item was produced by machine 1 or machine 2?

Solution Let A_1 = Event of drawing an item produced by machine 1,
A_2 = Event of drawing an item produced by machine 2,
and D = Event of drawing a defective item produced either by machine 1 or machine 2.

From the data in the problem, we know that

$$P(A_1) = 0.30, \quad P(A_2) = 0.70, \quad P(D \mid A_1) = 0.05, \quad P(D \mid A_2) = 0.1.$$

The data of the problem can be summarised as under:

Event	*Prior Probability* $P(A_i)$	*Conditional Probability Event* $P(D \mid A_i)$	*Joint Probability* $P(A_i \text{ and } D)$	*Posterior (Revised) Probability* $P(A_i \mid D)\ P(A_i \text{ and } D)$
(1)	(2)	(3)	(2) × (3)	
A_1	0.30	0.05	0.015	0.015/0.022 = 0.682
A_2	0.70	0.01	0.007	0.007/0.022 = 0.318

Here $\quad P(D) = \sum_{i=1}^{2} P(D \mid A_i)\, P(A_i) = 0.05 \times 0.30 + 0.01 \times 0.70 = 0.22.$

From table, the probability that the defective item was produced by machine 1 is 0.682 or 68.2 per cent and that by machine 2 is only 0.318 or 31.8 per cent. We may now say that the defective item is more likely drawn from the output produced by machine 1.

Example 16 In a bolt factory machines A, B and C manufacture respectively 25 per cent, 35 per cent and 40 per cent. Of the total of their output 5, 4 and 2 per cent are defective bolts. A bolt is drawn at random from the product and is found to be defective. What is the probability that it was manufactured by machines A, B and C?

Solution Let A_i ($i = 1, 2, 3$) be the event of drawing a bolt produced by machines A, B and C, respectively. From problem data we know that

$$P(A_1) = 0.25, \ P(A_2) = 0.35 \text{ and } P(A_3) = 0.40.$$

From the additional information, we know that

B = the event of drawing a defective bolt.

Thus, $P(B \mid A_1) = 0.05$, $P(B \mid A_2) = 0.04$, and $P(B \mid A_3) = 0.02$.

The table of posterior probabilities can be prepared as under:

Event	*Prior Probability* $P(A_i)$	*Conditional Probability* $P(B \mid A_i)$	*Joint Probability* (2) × (3)	*Posterior Probability* $P(A_i \mid B) = \dfrac{P(A_i \cap B)}{P(B)}$
(1)	(2)	(3)	(4)	(5)
A_1	0.25	0.05	0.0125	0.0125/0.0345 = 0.362
A_2	0.35	0.04	0.0140	0.014/0.0345 = 0.406
A_3	0.40	0.02	0.0080	0.008/0.0345 = 0.232
	1.00		0.0345	1.000

The above table shows the probability that the item was defective and produced by machine A is 0.362, by machine B is 0.406 and machine C is 0.232.

Review Questions

1. Discuss different schools of thought on the interpretation of probability. How does each school define probability?
2. Explain what do you understand by the term probability. Discuss its importance in business decision-making.
3. (*a*) Give the classical and statistical definition of probability and state the relationship, if any, between the two definitions.

   (*b*) Critically examine 'a priori' definition of probability showing clearly the improvement which the empirical version of probability makes over it.
4. Define independent and mutually exclusive events. Can two events be mutually exclusive and independent simultaneously? Support your answer with an example.
5. Explain the meaning of each of the following terms:

   (*a*) Random phenomenon (*b*) Statistical experiment
   (c) Random event (*d*) Sample space
6. Distinguish between the two concepts in each of the following pairs:

   (*a*) Elementary event and compound events.
   (*b*) Mutually exclusive events and overlapping events.
   (c) Sample space and sample point.
7. (*a*) Define the terms: joint probability, marginal probability and conditional probability.

   (*b*) By comparing the three kinds of probabilities (joint, conditional and marginal), explain what information is provided by each.
8. Suppose an entire shipment of 1,000 items is inspected and 50 items are found to be defective. Assume the defective items are not removed from the shipment before sent to a retail outlet for sale. If you purchase one item from this shipment, what is the probability that it will be one of the defective items?
9. Suppose you are told that the price of a particular stock will increase with a probability of 0.7.

   (*a*) How is this probability interpreted?
   (*b*) Assuming the definition of probability in terms of long-run relative frequencies, how would you find the probability that a stock price will increase?

Self-practice Problems

1. Three unbiased coins are tossed. What is the probability of obtaining

   (*a*) all heads? (*b*) two heads? (c) one head?
   (*d*) at least one head? (*e*) at least two heads? (*f*) all tails?
2. A card is drawn from a well-shuffled deck of 52 cards. Find the probability of drawing a card which is neither a heart nor a king.

3. In a single throw of two dice, find the probability of getting (*a*) a total 11, (*b*) a total 8 or 11 (*c*) same number on both the dice.

4. Five men in a company of 20 are graduates. If 3 men are picked out of the 20 at random, what is the probability that they are all graduates? What is the probability of at least one graduate?

5. A bag contains 25 balls numbered 1 through 25. Suppose an odd number is considered a 'success'. Two balls are drawn from the bag with replacement. Find the probability of getting

   (*a*) two successes (*b*) exactly one success (*c*) at least one success (*d*) no success

6. Mr X has 2 shares in lottery in which there are 2 prizes and 5 blanks. Mr Y has 1 share in a lottery in which there is 1 prize and 2 blanks. Show that chance of Mr X's success to that of Mr Y's is 15 : 7.

7. Explain whether or not each of the following claims could be correct:

   (*a*) A businessman claims the probability that he will get contract A is 0.15 and that he will get contract B is 0.20. Further more, he claims that the probability of getting A or B is 0.50.

   (*b*) A market analyst claims that the probability of selling ten million rupees of plastic A or five million rupees of plastic B is 0.60. He also claims that the probability of selling ten million rupees of A and five million rupees of B is 0.45.

8. The probability that a contractor will get a plumbing contract is 2/3 and the probability that he will not get an electric contract is 5/9. If the probability of getting at least one contract is 4/5, what is the probability that he will get both the contracts?

9. An MBA applies for employment to companies X and Y. The probability of his being selected by X is 0.7, and being rejected by Y is 0.5. The probability of at least one of his applications being rejected is 0.60. What is the probability that he will then be employed by at least one of the companies.

10. The probability is 0.3 that a management accountant's job applicant has a postgraduate degree, 0.7 that he has had some work experience as a chief financial accountant, and 0.2 that he has both. Out of 300 applicants, approximately, what number would have either a postgraduate degree or some professional work experience?

11. A can hit a target 3 times in 5 shots; B, 2 times in 5 shots; C, 3 times in 4 shots. They fire a valley. What is the probability that 2 shots hit?

12. The odds that a person X speaks the truth is 3 : 2 and the odds that a person Y speaks truth is 5 : 3. In what percentage of cases are they likely to contradict each other on an identical point?

13. A problem in business statistics is given to five students, A, B, C, D and E. Their chances of solving it are 1/2, 1/3, 1/4. 1/5 and 1/6, respectively. What is the probability that the problem will be solved?

14. The odds against student X solving a business statistics problem are 8 : 6 and odds in favour of student Y solving the same problem are 14 : 16,

    (*a*) What is the chance that the problem will be solved if both try?

    (*b*) What is the probability that they both, working independently of each other, solve the problem?

    (*c*) What is the probability that neither solves the problem?

15. A husband and wife appear in an interview for two vacancies for the same post. The probability of husband's selection is 1/7 and that of wife's selection is 1/5. What is the probability that

    (*a*) only one of them will be selected? (*b*) both of them will be selected?

    (*c*) none of them will be selected?

16. A candidate is selected for interview for three posts. For the first, there are 3 candidates; for the second, there are 4 and for the third, there are 2. What is the probability of his getting at least one post?

17. A manufacturing firm produces steel pipes in three plants with daily production volumes of 500, 1,000 and 2,000 units, respectively. According to past experience, it is known that the fractions of defective output produced by the three plants are 0.005, 0.008 and 0.010, respectively. If a pipe is selected from a day's total production and found to be defective, find out (*i*) from which plant the pipe comes, (*ii*) what is the probability that it came from the first plant?

18. Three persons A, B and C are being considered for the appointment as Vice-Chancellor of a university whose chance of being selected for the post are in the proportion 14 : 2 : 3, respectively. The probability that A is selected, will introduce democratisation in the university structure is 0.3, and the corresponding probabilities for B and C doing the same are 0.5 and 0.8, respectively. What is the probability that democratisation would be introduced in the university?

19. There are three brands, say X, Y and Z of an item available in the market. A consumer chooses exactly one of them for his use. He never buys two or more brands simultaneously. The probabilities that he buys brands X, Y and Z are 0.20, 0.16 and 0.45.
   (*a*) What is the probability that he does not buy any of the brands?
   (*b*) Given that a customer buys some brand, what is the probability that he buys brand X?

20. There is 50-50 chance that contractor's firm A will bid for the construction of a multi-storeyed building. Another firm B submits a bid and the probability is 3/5 that it will get the job, provided that firm A does not submit bid. If firm A submits a bid, the probability that firm B will get the job is only 2/3. What is the probability that firm B will get the job?

21. Plant I of a manufacturing organisation employs 5 production and 3 maintenance foremen, another plant II of same organisation employs 4 production and 5 maintenance foremen. From any one of these plants, a single selection of two foremen is made. Find the probability that one of them would be production and the other maintenance foreman.

22. Two sets of candidates are competing for the positions on the board of directors of the company. The probability that the first and second sets will win are 0.6 and 0.4, respectively. If the first set wins, the probability of introducing a new product is 0.8 and the corresponding probability if the second set wins is 0.3.
   (*a*) What is the probability that the new product will be introduced?
   (*b*) If the new product was introduced, what is the probability that the first set won as directors?

23. In a post office, three clerks assigned to process incoming mail. The first clerk, A processes 40 per cent, the second clerk, B processes 35 per cent and the third clerk, C processes 25 per cent of the mail. The first clerk has an error rate of 0.04, the second has an error rate of 0.06 and the third has an error rate of 0.03. A mail selected at random from a day's output is found to have an error. The postmaster wishes to know the probability that it was processed by clerk A or clerk B or clerk C.

24. If a machine is correctly installed, it will produce 90 per cent acceptable items. If it is incorrectly installed, it will produce 40 per cent acceptable items. Past experience shows that 80 per cent of the installations are correctly done. If after a certain installation, the machine produces 2 acceptable items as the first 2 pieces, find the probability that the machine is correctly installed.

25. A firm plans to bid Rs 300 per tonne for a contract to supply 1,000 tonnes of a metal. It has two competitors A and B and it assumes that the probability that A will bid less than Rs 300 per tonne is 0.3 and that B will bid less than Rs 300 per tonne is 0.7. If the lowest bidder gets all the business and the firms bid independently, what is the expected value of the contract to the firm?

26. An investment consultant predicts that the odds against the price of a certain stock will go up during the next week are 2 : 1 and odds in favour of the price remaining the same are 1 : 3. What is the probability that the price of the stock will go down during the next week?

27. An article manufactured by a company consists of two parts: A and B. In the process of manufacture of part A, 9 out of 100 are likely to be defective. Similarly, 5 out of 100 are likely to be defective in the manufacture of part B. Calculate the probability that the assembled part will not be defective.

28. A bag contains 5 white and 8 red balls. Two drawings of 3 balls are made such that (*a*) the balls are replaced before the second trial, and (*b*) the balls are not replaced before the second trial. Find the probability that the first drawing will give 3 white and the second 3 red balls in each case.

29. Three groups of workers contain 3 men and one woman, 2 men and 2 women, and 1 man and 3 women, respectively. One worker is selected at random from each group. What is the probability that the group selected consists of 1 man and 2 women?

30. What is the probability that a leap year, selected at random, will contain 53 Sundays?

31. A university has to select an examiner from a list of 50 persons, 20 of them women and 30 men, 10 of them knowing Hindi and 40 not, 15 of them being teachers and the remaining 35 not. What is the probability of the university selecting a Hindi-knowing woman teacher?

32. A product is assembled from three components X, Y and Z, the probability of these components being defective is 0.01, 0.02 and 0.05, respectively. What is the probability that the assembled product will not be defective?

33. The daily production of a machine producing a very complicated item gives the following probabilities for the number of items produced: P (1) = 0.20, P (2) = 0.35 and P (3) = 0.45. Furthermore, the probability of defective items being produced is 0.02. Defective are assumed to occur independently. Determine the probability of no defectives during a day's production.

34. The personnel department of a company has records which show the following analysis of its 200 engineers:

Age (Years)	*Bachelor's Degree Only*	*Master's Degree*	*Total*
Under 30	90	10	100
30 to 40	20	30	50
Over 40	40	10	50
Total	150	50	200

If one engineer is selected at random from the company, find:

(*a*) the probability he has only a bachelor's degree;

(*b*) the probability he has a master's degree given that he is over 40;

(*c*) the probability he is under 30 given that he has only a bachelor's degree.

35. In a certain town, males and females form 50 per cent of the population. It is known that 20 per cent of the males and 5 per cent of the females are unemployed. A research student studying the employment situation selects unemployed persons at random. What is the probability that the person selected is (*a*) male, (*b*) female?

36. You note that your officer is happy in 60 per cent cases of your calls. You have also noticed that if he is happy, he accedes to your requests with a probability of 0.4, whereas if he is not happy, he accedes to your requests with a probability of 0.1. You call on him one day and he accedes to your request. What is the probability of his being happy?

37. A certain production process produces items that are 10 per cent defective. Each item is inspected before supplying to customers but the inspector incorrectly classifies an item 10 per cent of the time. Only items classified as good are supplied. If 820 items have been supplied in all, how many of them are expected to be defective?

Hints and Answers

1. (*a*) P (all heads) = 1/8 (*b*) P (two heads) = 3/8 (*c*) P (one head) = 3/8
(*d*) P (at least one head) = 7/8 (*e*) P (at least two heads) = 4/8 = 1/2
(*f*) P (all tails) = 1/8.

2. $P\text{ (neither a heart nor a king)} = \frac{^{36}C_1}{^{52}C_1} = \frac{36}{52}$

3. c (S) = 36; P (total of 11) = 2/36 P (total of 9 or 11) = 7/36

4. $P\text{ (all graduate)} = \frac{^{5}C_3 \times {}^{15}C_0}{^{20}C_3} = \frac{10 \times 1}{1{,}140} = \frac{1}{114}$

$P\text{ (no graduate)} = \frac{^{15}C_3 \times {}^{5}C_0}{^{20}C_2} = \frac{455 \times 1}{1{,}140} = \frac{91}{28}$

$P\text{ (at least one graduate)} = 1 - \frac{91}{228} = \frac{137}{228}$

5. (*a*) $P\text{ (two success)} = \frac{13}{25} \times \frac{13}{25} = \frac{169}{625}$ (*b*) $P\text{ (exactly one success)} = \frac{13}{25} \times \frac{12}{25} + \frac{13}{25} \times \frac{12}{25} = \frac{312}{625}$

(*c*) $P\text{ (at least one success)} = P\text{ (exactly one success)} + P\text{(two success)} = \frac{312}{625} + \frac{169}{625} = \frac{481}{625}$

(*d*) $P\text{ (no success)} = \frac{12}{25} \times \frac{12}{25} = \frac{144}{625}$

6. Considering Mr X's chances of success.

A = event that 1 share brings a prize and 1 share goes blank.

B = event that both the shares bring prizes.

C = event that X succeeds in getting at least one prize = $A \cup B$

Since A and B are mutually exclusive, therefore

$$P(C) = P(A \cup B) = P(A) + P(B) = \frac{^{2}C_1 \times {}^{5}C_1}{^{7}C_2} + \frac{^{2}C_2 \times {}^{5}C_0}{^{7}C_2}.$$

Similarly, if D denotes the event that Y succeeds in getting a prize, then we have

$$P(D) = \frac{{}^1C_1}{{}^3C_1} = \frac{1}{3}.$$

X's chance of success : Y's chance of success = 15/21 : 1/3 = 15 : 7.

7. (*a*) $P(A \cap B) = -0.15$ (*b*) $P(A) + P(B) = 1.05$

8. $P(A \cap B) = 14/45$; A = plumbing contract; B = electric contract

9. $P(X \cup Y) = 1 - P(X \cap Y) = 1 - P(X) - P(Y) - P(X \cup Y) = 0.8$

10. Let A = Applicant has PG degree; B = Applicant has work experience;
Given, $P(A) = 0.3$; $P(B) = 0.7$ and $P(A \cap B) = 0.2$. Therefore

$$300 \times P(A \cup B) = 300[P(A) + P(B) - P(A \cap B)] = 240.$$

11. The required event that two shots may hit the target, can happen in the following mutually exclusive cases.

(*i*) A and B hit and C fails to hit the target.

(*ii*) A and C hit and B fails to hit the target.

(*iii*) B and C hit and A fails to hit the target.

Hence, the required probability that any two shots hit is given by, P = P (*i*) + P (*ii*) + P (*iii*).

Let E_1, E_2 and E_3 be the event of hitting the target by A, B and C, respectively. Therefore

$$P(i) = P(E_1 \cap E_2 \cap \overline{E}_3) = P(E_1) \cdot P(E_2) \cdot P(\overline{E}_3) = (3/5) \cdot (2/5)(1 - 3/4) = 6/100$$

$$P(ii) = P(E_1 \cap E_3 \cap \overline{E}_2) = (3/5)(3/4)(1 - 2/5) = 27/100$$

$$P(iii) = P(\overline{E}_1 \cap E_2 \cap E_3) = (1 - 3/5)(2/5)(3/4) = 12/100.$$

Since all the three events are mutually exclusive events, hence the required probability is given by

$$P(i) + P(ii) + P(iii) = 6/100 + 27/100 + 12/100 = 9/20.$$

12. A, B = events that X and Y speak the truth, respectively.

P (X and Y contradict on an identical point)

$$= P[(A \cap \overline{B}) \cup (\overline{A} \cap B)] = P(A \cap \overline{B}) + P(\overline{A} \cap B)$$

$$= P(A).P(\overline{B}) + P(\overline{A}).P(B) = \frac{3}{5} \cdot \frac{3}{8} + \frac{2}{5} \cdot \frac{5}{8} = 0.475.$$

That is, in 47 per cent cases A and B contradict each other.

13. P (problem will be solved) = 1 – P (problem is not solved)

= 1 – P (all students fail to solve the problem)

$$= 1 - P(\overline{A} \cap \overline{B} \cap \overline{C} \cap \overline{D} \cap \overline{E})$$

$$= 1 - P(\overline{A})\,P(\overline{B})\,P(\overline{C})\,P(\overline{D})\,P(\overline{E})$$

$$= 1 - \left(1 - \frac{1}{2}\right)\left(1 - \frac{1}{3}\right)\left(1 - \frac{1}{4}\right)\left(1 - \frac{1}{5}\right)\left(1 - \frac{1}{6}\right) = 1 - \frac{1}{6} = \frac{5}{6}.$$

14. (*a*) $P(X \cap Y) = P(X) + P(Y) - P(X \text{ and } Y) = P(X) + P(Y) - P(X) \cdot P(Y)$

(*b*) $P(X \text{ and } Y) = P(X)\,P(Y) = (8/14) \times (14/30) = (4/15)$

(*c*) $P(\overline{X} \text{ and } \overline{Y}) = P(\overline{X}) \cdot P(\overline{Y}) = (8/35)$

15. P (only one of them will be selected) $= P(H \cap \overline{W}) \cup (\overline{H} \cap W)$

$$= P(H \cap \overline{W}) + (\overline{H} \cap W) = P(H)\,P(\overline{W}) + P(\overline{H})\,P(W)$$

$$= \frac{1}{7}\left(1-\frac{1}{5}\right)+\left(1-\frac{1}{7}\right)\frac{1}{5} = \frac{2}{7}.$$

(*b*) P (both of them will be selected) = P (H ∩ W) = P (H) · P (W) = 1/35.

(*c*) P (none of them will be selected) $= P(\overline{H} \cap \overline{W}) = P(\overline{H}) \cdot P(\overline{W})$ = 24/35.

17. $P(A_1) = 500/3{,}500$; $P(A_2) = 100/3{,}500$; $P(A_3) = 2{,}000/3{,}500$ where A_1, A_2 and A_3 are production volume of first, second and third plant. If E is a defective item, then $P(E \mid A_1) = 0.005$, $P(E \mid A_2) = 0.008$, and $P(E \mid A_3) = 0.010$.

18. P (D) = Probability of the event that democratisation would be introduced.

$= P[(A \cap D) \cup [(B \cap D) \cup [(C \cap D)$

$= P[(A \cap D) + P(B \cap D) + P(C \cap D)$

$= P(A) \cdot P(D \mid A) + P(B) \cdot P(D \mid B) + P(C) \cdot P(D \mid C)$

$= 0.3(4/9) + 0.5(2/9) + 0.8(3/9) = 0.51.$

19. (*a*) P (customer does not buy any brand) $= P[\overline{X} \cap \overline{Y} \cap \overline{Z}] = 1 - P[(X \cup Y \cup Z]$

$= 1 - [P(X) + P(Y) + P(Z)]$

$= 1 - [0.20 + 0.16 + 0.45] = 0.19.$

(*b*) P (customer buys brand X) = $P[X \mid (X \cup Y \cup Z)]$

$$= \frac{P[X \cap (X \cup Y \cup Z)]}{P(X \cup Y \cup Z)}$$

$$= \frac{P(X)}{P(X)+P(Y)+P(Z)-P(X\cap Y)-P(Y\cap Z)-P(X\cap Z)+P(X\cap Y\cap Z)}$$

$$= \frac{0.2}{0.2+0.16+0.45-0-0-0-0} = 0.247.$$

20. P (A) = 1/2, P (B | A) = 2/3 and $P(B \mid \overline{A}) = 3/5$.

$$P(B) = P(A \cap B) + P(\overline{A} \cap B) = P(A) \cdot P(B \mid A) + P(\overline{A}) \cdot P(B \mid \overline{A}) = \frac{1}{2}\cdot\frac{2}{3}+\frac{1}{2}\cdot\frac{3}{5} = \frac{19}{30}.$$

21. Let E_1, E_2 be the events that Plants I and II is selected, respectively. Then the probability, of the event E that a batch of 2, one is production and the other is maintenance for men is

$$P(E) = P(E_1 \cap E) + P(E_2 \cap E) = P(E_1) \cdot P(E \mid E_1) + P(E_2) \cdot P(E \mid E_2)$$

$$= \frac{1}{2}\cdot\frac{{}^5C_1 \cdot {}^3C_1}{{}^8C_2} + \frac{1}{2}\cdot\frac{{}^4C_1 \cdot {}^5C_1}{{}^9C_2} = \frac{1}{2}\cdot\frac{15}{28}+\frac{1}{2}\cdot\frac{5}{9} = \frac{275}{504}.$$

24. Let A = event that item is acceptable;

B_1 and B_2 = events that machine is correctly and incorrectly installed, respectively.

Given, $P(A \mid B_1) = 0.9$; $P(A \mid B_2) = 0.4$; $P(B_1) = 0.8$ and $P(B_2) = 0.2$.

Then $P(B_1 \mid A) = 0.9.$

25. There are two competitors A and B and the lowest bidder gets the contract.

Value of plan $= 300 \times 1{,}000 = 3{,}00{,}000$

Contractor A : P (Bid < 300) = 0.3; P (Bid ≥ 300) = 0.7

Contractor B : P (Bid < 300) = 0.7; P (Bid ≥ 300) = 0.3.

(*i*) If both bids are less than Rs 300, probability is $0.3 \times 0.7 = 0.21$. Therefore, the expected value of contract is: $3{,}00{,}000 \times 0.21 =$ Rs 63,000.

(*ii*) If A bids less than Rs 300 and B bids more than Rs 300, probability is $0.3 \times 0.3 = 0.09$. Therefore, the expected value of contract is $3{,}00{,}000 \times 0.09 =$ Rs 27,000.

(*iii*) B bids less than Rs 300 while A bids more than Rs 300, probability is $0.7 \times 0.7 = 0.49$. Therefore, the expected value of contract is $3{,}00{,}000 \times 0.49 =$ Rs 1,47,000.

Therefore, expected value of contract is = 63,000 + 27,000 + 1,47,000 = Rs 2,37,000.

26. P (price of a certain stock not going up) = 2/3

P (price of a certain stock remaining same) = 1/4

The probability that the price of the stock will go down during the next week

= P (price of the stock not going up and not remaining same)

= P (price of the stock not going up) × P (price of the stock not remaining same)

$= (2/3) \times (1 - 1/4) = (2/3) \times (3/4) = 1/2 = 0.5.$

27. The assembled part will be defective if any of the part is defective.

The probability of the assembled part being defective

= P (any of the part is defective)

$= P(A \cup B) = P(A) + P(B) - P(AB)$

$= 9/100 + 5/100 - (9/100) \times (5/100) = 0.1355.$

∴ The probability that assembled part is not defective $= 1 - 0.1355 = 0.8645$.

28. (*a*) *When balls are replaced*

Total number of balls in the bag = 5 + 8 = 13.

3 balls can be drawn from 13 in $^{13}C_3$ ways; 3 white balls can be drawn from 5 in $^5C_3$ ways.

3 red balls can be drawn from 8 in $^8C_3$ ways. Therefore, the

$$\text{Probability of 3 red balls in the second trial} = \frac{^5C_3}{^{13}C_3} = \frac{5}{143}$$

$$\text{Probability of 3 red balls in the second trial} = \frac{^5C_2}{^{12}C_2} = \frac{28}{143}$$

$$\text{Probability of the compound event} = \frac{5}{143} \times \frac{28}{143} = \frac{140}{20449} = 0.007.$$

(*b*) *When balls are not replaced*

At the first trial 3 white balls can be drawn in $^5C_3$ ways. Therefore, the

$$\text{Probability of drawing three balls at the first trials} = \frac{^5C_3}{^{13}C_3} = \frac{5}{143}.$$

When the white balls have been drawn and not replaced, the bag contains 2 white and 8 red balls.

Therefore, at the second trial 3 balls can be drawn from 10 in ${}^{10}C_3$ ways and 3 red balls can be drawn from 8 in ${}^{5}C_3$ ways.

$$\text{The probability of 3 red balls in the second trial} = \frac{{}^{8}C_3}{{}^{10}C_3} = \frac{7}{15}.$$

$$\text{The probability of the compound event} = \frac{5}{142} \times \frac{7}{15} = \frac{7}{429} = 0.016.$$

29. There are three possibilities in this case:

(*a*) Man is selected from the 1st group and women from 2nd and 3rd groups; or

(*b*) Man is selected from the 2nd group and women from the 1st and 3rd groups; or

(*c*) Man is selected from the 3rd group and women from 1st and 2nd groups.

The probability of selecting a group of 1 man and 2 women is:

$$\left(\frac{3}{4} \times \frac{2}{4} \times \frac{3}{4}\right) + \left(\frac{2}{4} \times \frac{1}{4} \times \frac{3}{4}\right) + \left(\frac{1}{4} \times \frac{1}{4} \times \frac{2}{4}\right) = \frac{9}{32} + \frac{3}{32} + \frac{1}{32} = \frac{13}{32}.$$

30. A leap year consists of 366 days, therefore it contains 52 complete weeks and 2 days extra. These 2 days may make the following 7 combinations:

1. Monday and Tuesday
2. Tuesday and Wednesday
3. Wednesday and Thursday
4. Thursday and Friday
5. Friday and Saturday
6. Saturday and Sunday
7. Sunday and Monday

Of these seven equally likely cases only the last two are favourable. Hence, the required probability is 2/7.

31. Probability of selecting a woman = 20/50; Probability of selecting a teacher = 15/50.

Probability of selecting a Hindi-knowing candidate = 10/50.

Since the events are independent, the probability of the university selecting a Hindi-knowing woman teacher will be: (20/50) × (15/50) × (10/50) = (3/125).

32. Let A, B and C denote the respective probabilities of components X, Y and Z being defective.

$$P(A) = 0.01, \; P(B) = 0.02, \; P(C) = 0.05$$

$$\begin{aligned} P(A \text{ or } B \text{ or } C) &= P(A) + P(B) + P(C) - P(AB) - P(BC) - P(AC) + P(ABC) \\ &= 0.01 + 0.02 + 0.05 - 0.0002 - 0.0010 - 0.0005 + 0.00001 = 0.0784. \end{aligned}$$

Hence, the probability that the assembled product will not be defective = 1 – 0.0784 or 0.9216.

33. Let A be the event that no defective item is produced during a day. Then

$$P(A) = P(1) \cdot P(A \mid 1) + P(2) \cdot P(A \mid 2) + P(3) \cdot P(A \mid 3).$$

The probability that a defective item is produced = 0.02, Probability that a non-defective item is produced = 1 – 0.02 = 0.98. Also defectives are assumed to occur independently, therefore

$$\begin{aligned} P(A \mid 1) &= 0.98, \; P(A \mid 2) = (0.98)(0.98) \text{ and } P(A \mid 3) = (0.98)(0.98)(0.98) \\ P(A) &= (0.20(0.98) + (0.35)(0.98)^2 + (0.45)(0.98)^3 \\ &= 0.1960 + 0.3361 + 0.4322 = 0.9643. \end{aligned}$$

Hence, the probability of no defectives during a day's production is 0.9643.

34. A : an engineer has a bachelor's degree only ; B : an engineer has a master degree;
C : an engineer is under 30 years of age ; D : an engineer is over 40 years of age.

(*a*) P (A) = 150/200 = 0.75

(*b*) $P(B \mid D) = \dfrac{P(B \cap D)}{P(D)} = \dfrac{10/200}{50/200} = 0.20$

(*c*) $P(C \mid A) = \dfrac{P(C \cap A)}{P(A)} = \dfrac{90/200}{150/200} = 0.60$

35. Given that

	Employed	*Unemployed*	*Total*
Males	0.40	0.10	0.50
Females	0.475	0.025	0.50
Total	0.875	0.125	1.00

Let M and F be the male and female chosen, respectively.
U = Male, female chosen is unemployed.

(*a*) $P(M \mid U) = \dfrac{P(M \cap U)}{P(U)} = \dfrac{0.10}{0.125} = 0.80$

(*b*) $P(F \mid U) = \dfrac{P(F \cap U)}{P(U)} = 0.20.$

36. The probability that officer is happy and accedes to requests = 0.6 × 0.4.
The probability the officer is unhappy and accedes to requests = 0.4 × 0.1 = 0.04.
Total probability of acceding to requests = 0.24 + 0.04 = 0.28.
The probability of his being happy if he accedes to requests = 0.24/0.28 = 0.875.

37. P (D) = Probability of a defective item = 0.1 ; P (classified as good | defective) = 0.1
∴ P (G) = Probability of a good item = 1 – P (D) = 1 – P (D) = 1 – 0.1 = 0.9
P (classified as good | good) = 1 – P (classified as good | defective) = 1 – 0.1 = 0.9

$$\therefore\ P(\text{defective} \mid \text{classified as good}) = \frac{P(D) \cdot P(\text{classified as good} \mid \text{defective})}{P(D) \cdot P(\text{classified as good} \mid D) + P(G)\, P(\text{classified as good} \mid G)}$$

$$= \frac{0.1 \times 0.1}{0.1 \times 0.1 + 0.9 \times 0.9} = \frac{0.01}{0.82} = \frac{1}{82} = 0.012.$$

Chapter 8

Recurrence Relations and Generating Functions

PREVIEW

If an item also becomes the part of its definition, then such a definition is called a recursive (or inductive) definition. A recursive definition has two parts:

(a) Initial condition (s) where few simple cases of an item being defined are explicitly given

(b) Inductive (or recursive) process where further cases of an item being defined are already given in previous cases.

Recurrence relations arise in many combinatorics (counting) problems and in analyzing programming problems.

LEARNING OBJECTIVES

After studying this chapter, you should be able to,

- understand recursive definitions for certain sequences,
- write recursive algorithms for sequences with constant coefficients,
- derive solution of non-homogeneous recursive relation(s),
- find solution of recursive relations using generating functions.

8.1 INTRODUCTION

We already have discussed iterative methods to find an explicit formula for a recursively defined sequence. For example, the sequence $a_n = a_{n-1} + 4$ for $n \geq 2$ and $a_1 = 3$ (initial condition) of real numbers, is an equation relating all but a finite number of terms a_r of the sequence (for any r) to its previous terms is called *recurrence* and the relation is called a *recurrence relation* (or *formula*) with given initial values about the beginning of the sequence. The initial values are also called *initial conditions*, *boundary conditions* or *basis* for the sequence.

In this chapter, we will discuss *finite order linear recurrence relations with constant coefficients*. There is no general algorithm (or method) for solving all recurrence relations, however, generating functions will be introduce to solve recurrence relations.

8.2 RECURRENCE RELATIONS

A recurrence relation is also called a *difference equation* and these two names will be used interchangeably. This means a recurrence relation may also be defined alternately as an equation relating the values of function $y = f(x)$ with the independent variable x and one or more of its differences Δy, $\Delta^2 y$, . . . of various orders, where $\Delta y = y_{k+1} - y_k$. For example, the equation

$$4 f(x+3) - f(x+2) + 11 f(x+1) - 6 f(x) = 0$$

is a difference equation. It can also be written in either of the two forms:

$$4y_{n+3} - y_{n+2} + 11y_{n+1} - 6y_n = 0$$

$$4a_{n+3} - a_{n+2} + 11a_{n+1} - 6a_n = 0.$$

Illustrations The following are the examples of recurrence relations.

1. The recurrence relation, ${}^nC_r = {}^{n-1}C_r + {}^{n-1}C_{r-1}$, for all $n\ (\geq r) > 0$ defines binomial coefficients with initial conditions, ${}^nC_0 = 1$ and ${}^nC_n = 1$.
2. The Fibonacci sequence is defined by the recurrence relation, $a_n = a_{n-1} + a_{n-2}$ with initial conditions, $a_0 = 1$ and $a_1 = 1$.
3. A arithmetic progression is an infinite sequence of numbers such as 2, 5, 8, 11, ... This sequence is defined by the recurrence relation $a_n = a_{n-1} + 3$ with initial condition, $a_1 = 2$.
4. A geometric progression is an infinite sequence of numbers such as 5, 15, 45, 135, ... This sequence is defined by the recurrence relation $a_{n+1} = 3a_n$, $n \geq 0$ with initial condition $a_0 = 5$.

Any recurrence relation (or formula) in which the value of any term of the sequence can be determined with the help of its position number is called *explicit formula*. Few explicit formule are:

1. $a_n = n^2$, $n > 1$ 2. $a_n = (-3)^n$, $1 \leq n \leq \infty$ 3. $a_n = 50 - 2n$, $1 \leq n \leq 1$

Order of a recurrence relation The order of a linear recurrence relation (also called linear difference equation) is the difference between the largest and the smallest subscript appearing in the relation.

Values such as a_0 or a_1 given in addition to the recurrence relation are called *boundary conditions*. The expression $a_0 = \text{A}$, where A is a constant, is also referred to as an *initial condition*. The recurrence relation $a_n = a_{n-1} f_1(n) + a_{n-2} f_2(n) + ... + a_{n-k} f_{n-k}(n) + g(n)$ represents the general form of a linear recurrence relation of order k.

Degree of a recurrence relation The degree of any recurrence relation is the degree of the function, f in an equation like $a_n = f(a_{n-1}, a_{n-2}, ..., a_{n-k}, n)$ where $a_{n-1}, a_{n-2}, ..., a_{n-k}$ are all variables and n is any constant. Hence, the function, f needs to be a polynomial, otherwise no degree is assigned to the given recurrence relation.

In general, the first order linear recurrence relation with constant coefficients has the form $f(n) = a_{n+1} + c\,a_n$, for $n \geq 0$ where c is a constant and $f(n)$ is a function on the set of non-negative integers. If $f(n) = 0$ for all $n \in \text{I}_+$, then the relation is called *homogeneous*, otherwise called *non-homogeneous*. For example,

(*i*) recurrence relation $a_n = -2a_{n-1}$ is of order one.
(*ii*) recurrence relation $a_n = 3a_{n-1} - 2a_{n-2}$ is of order 2.
(*iii*) $a_n = 5\,a_{n-1} + n^2$ is non-homogeneous of order 1 and degree 1.
(*iv*) $a_n = n\,a_{n-2} + 2^n$ is non-homogeneous of order 2 and degree 1.
(*v*) $a_n = + a^2_{n-2}$ is homogeneous of order 2 and no degree.
(*vi*) $a_n = a_{n-1} + a_{n-2} + \ldots + a_0$ is homogeneous of no order and degree 1.
(*vii*) $a_n = a^2_{n-1} + a_{n-2}\,a_{n-3}\,a_{n-4}$ is homogeneous of order 4 and degree 3.

8.2.1 General Solution of Recurrence Relation

The general solution of the recurrence relation $a_{n+1} = c\,a_n$, where $n \geq 0$, $a_0 = A$ and c is constant is unique and is given by $a_n = A\,d^n$, $n \geq 0$.

The solution to any recurrence relation refers to find an explicit expression for a_r (general term) as a function of n only which does not involve any other term of the series.

(*i*) Any recurrence relation of order k has *initial condition*(*s*) provided the values of one or more of the terms $a_0, a_1, \ldots, a_{k-1}$ are known.

(*ii*) A function $f(n)$ is said to be a *general solution* to the homogeneous recurrence relation if it satisfies the *recurrence equation.*

(*iii*) A function $g(n)$ is said to be a *particular solution* to the recurrence relation if it satisfies the recurrence equation together with the *initial conditions*.

Illustrations

1. The general solution of the recurrence relation $a_n = 9\,a_{n-1}$ where $n \geq 1$ and $a_2 = 50$ is of the form: $a_n = a_0\,(9^n)$. Since $a_2 = 50 = a_0\,(9^2)$, therefore $a_0 = 2$ and $a_n = 2(9^n)$, $n \geq 0$ is the unique solution.
2. The general solution of the recurrence relation, $a_n = a_{n-1}$ is $a_n = c$ (any constant) and the particular solution when $a_0 = 1$ is $a_n = 1$ $(n \geq 0)$.
3. The general solution of the recurrence relation, $a_n = a_{n-1} + 1$ is $a_n = c + n$ and the particular solution when $a_0 = 0$ is $a_n = n$ $(n \geq 0)$.
4. The general solution of the recurrence relation, $a_n = ka_{n-1}$ is $a_n = ck^n$ and the particular solution when $a_0 = 1$ is $a_n = k^n$ $(n \geq 0)$.
5. The general solution of the recurrence relation, $a_n = a_{n-1} + a_{n-2}$ is $a_n = c_1\,((1 + \sqrt{5})/2)^n + c_2\,((1 - \sqrt{5})/2)^n$ and the particular solution for $a_1 = 1$, $a_2 = 3$ is $a_n = ((1 + \sqrt{5})/2)^n + ((1 - \sqrt{5})/2)^n$, $n \geq 1$.
6. If a person deposits Rs 100 in his savings bank account at an interest rate of 6 per cent per year compounded monthly, then amount in his account at the end of n months will be

$$a_{n+1} = a_n\left(1 + \frac{6}{100 \times 12}\right)^n = a_n\,(1.005)^n$$

where a_0 = Rs 100. The solution to this relation is $a_n = a_0\,(1.005)^n = 100\,(1.005)^n$.

7. The solution of the recurrence relation $a_n = n\,.\,a_{n-1}$ where $n \geq 1$ and $a_0 = 1$ is $a_n = n!$, which counts the number of permutations of n objects, $n \geq 0$.

8.3 LINEAR RECURRENCE RELATIONS WITH CONSTANT COEFFICIENTS

A recurrence relation is called *linear* because each subscripted term appears to the first power. In a linear relation no term like $a_n\, a_{n-1}$ appears.

In general, the linear recurrence relation of order k with constant coefficients is expressed as:

$$c_0\, a_n + c_1\, a_{n-1} + c_2\, a_{n-2} + \ldots + c_k\, a_{n-k} = f(n)$$

where c_i ($i = 0, 1, 2, \ldots, k$) are constants and non-zero and $f(n)$ is the function of independent variable n only.

Illustrations

1. $2a_n + 5a_{n-1} = 3^n$ is the linear recurrence relation of order one with constant coefficients.
2. $3a_{n+3} - 9a_{n+2} + y_{n+1} - 6a_n = 2n + 1$ is the linear recurrence relation of order 3 with constant coefficients.
3. $5a_n - 3a_{n-1} + 4a_{n-2} = n^2 + 7$ is the linear recurrence relation of order 2 with constant coefficients.

Remarks 1. The recurrence relation (or difference equation) defined above is said to be *linear homogeneous relation* (or equation) of order k if $f(n) = 0$. Otherwise it is non-homogeneous relation.

2. We may use the phrase *linear difference equation with constant coefficients* instead of *linear recurrence relations with constant coefficients.*

8.3.1 Characteristic Equation of the Recurrence Relation

Consider a linear homogeneous recurrence relation of order k with constant coefficients

$$c_0\, a_n + c_1\, a_{n-1} + c_2\, a_{n-2} + \ldots + c_k\, a_{n-k} = 0.$$

Assume that the solution of this relation has the form, $a_n = \mathrm{A}\alpha^k$. Substituting this into the relation, we find

$$\mathrm{A}\,\alpha^n + c_1\,\mathrm{A}\,\alpha^{n-1} + c_2\,\mathrm{A}\,\alpha^{n-2} + \ldots + c_k\,\alpha^{n-k} = 0$$

$$\mathrm{A}\,\alpha^{n-k}\,[\alpha^k + c_1\,\alpha^{k-1} + c_2\,\alpha^{k-2} + \ldots + c_k] = 0.$$

The term in bracket is an nth order polynomial, which is also known as the *characteristic equation* or *auxiliary equation* of recurrence relation (or difference equation). Here α is called the *characteristic root* and A is a constant determined by the initial conditions.

Since we are interested only in non-trivial solution, we shall disregard solution corresponding to $\alpha = 0$. Assuming the roots are real and distinct, we obtain k roots $\alpha_1, \alpha_2, \ldots, \alpha_k$. The general solution of the recurrence relation will be some linear combination of these terms

$$a_n = \mathrm{A}_1\,\alpha_1^n + \mathrm{A}_2\,\alpha_2^n + \ldots + \mathrm{A}_k\,\alpha_k^n$$

where $\mathrm{A}_1, \mathrm{A}_2, \ldots, \mathrm{A}_k$ are constants determined by the given the initial conditions.

8.4 METHODS OF SOLVING RECURRENCE RELATIONS WITH CONSTANT COEFFICIENTS

The following methods of solving recurrence relations discussed in this chapter are:

1. Substitution method
2. Characteristic roots method
3. Generating functions method.

Substitution (or iterative) Method

In this method anyone of the following two approaches may be used to solve a recurrence relation.

(*i*) Starting with the nth term of the given recurrence relation and terminate at the first term. This approach is known as *back tracking*.

(*ii*) Starting with the first term of the given recurrence relation and terminate at the nth term.

Back tracking This approach starts by first considering nth term and terminates at the first term to find an explicit formula for the sequence defined by a recurrence relation. For example, let recurrence relation be $a_n = a_{n-1} + 5$ with initial condition, $a_1 = 1$. Starting with the term a_n and tracking back by substituting value for $a_{n-1}, a_{n-2}, \ldots$ and so on until a pattern is obtained.

$$\begin{aligned} a_n &= a_{n-1} + 5 & &= (a_{n-2} + 5) + 5 \\ &= a_{n-2} + 2 \times 5 & &= a_{n-2} + 3 \times 5 \\ &\vdots \qquad \vdots \qquad \vdots & &\vdots \\ &= a_{n-(n-2)} + (n-1) \times 5 & &= a_1 + (n-1) \times 5 \\ &= 1 + (n-1) \times 5 \text{ or } 5n - 4 \end{aligned}$$

This is an explicit formula for the sequence.

Example 1 Solve the recurrence relation, $a_n = ka_{n-1} + f(n)$ where k is a constant.

Solution Using back tracking approach as follows:

$$\begin{aligned} a_n &= ka_{n-1} + f(n) \\ &= k[k\,a_{n-2} + f(n-1)] + f(n) \\ &= k^2a_{n-2} + k\,f(n-1) + f(n) \\ &= k^2[ka_{n-3} + f(n-2)] + kf(n-1) + f(n) \\ &= k^3(a_{n-3} + k^2f(n-2) + kf(n-1) + f(n) \\ &\vdots \qquad \vdots \qquad \vdots \qquad \vdots \\ &= k^na_{(n-n)} + k^{n-1}f(1) + k^{n-2}f(2) + \ldots + kf(n-1) + f(n) \\ &= k^na_0 + k^{n-1}f(1) + k^{n-2}f(2) + \ldots + kf(n-1) + f(n) \end{aligned}$$

Alternate Approach: Substituting $n = 1, 2, 3, \ldots, n$ in the given recurrence relation, we get

$$\begin{aligned} a_n &= ka_{n-1} + f(n) \\ a_1 &= ka_0 + f(1) \\ a_2 &= ka_1 + f(2) = k[ka_0 + f(1)] + f(2) \\ &= k^2a_0 + kf(1) + f(2) \\ a_3 &= k\,a_2 + f\,(3)\;kf(2) + f(3) \\ &= k[k^{n-1}\,a_0 + k^{n-2}f(1) + \ldots + k\,f(n-1)] + f(n) \\ &= k^na_0 + k^{n-1}f(1) + \ldots + k\,f(n-1) + f(n) \end{aligned}$$

Characteristic Roots Method

There are two types of solutions of a linear recurrence relation with constant coefficients.

1. Homogeneous Solution

It is the solution which satisfies the given recurrence relation when right hand side of it is set to zero, i.e. $f(n) = 0$ for all $n \geq 0$. The solution contains k arbitrary constants equal to the order of the relation.

The general homogeneous solution of a recurrence relation with constant coefficients is of the form $a_{nh} = \sum A_i \alpha_i^n$, where α_i $(i = 1, 2, \ldots, k)$ are the *characteristic roots* and A_i are the constants determined by the initial conditions.

While finding k characteristic roots α_i $(i = 1, 2, \ldots, k)$ of the *characteristic* (*auxiliary*) *equation*

$$c_0 a^k + c_1 a^{k-1} + c_2 a^{k-2} + \ldots + c_k = 0 \qquad \ldots(i)$$

of degree k of recurrence relation

$$c_0 a_n + c_1 a_{n-1} + c_2 a_{n-2} + \ldots + c_k a_{n-k} = 0, \qquad \ldots(ii)$$

four different cases may arise:

Case 1 The roots are real and unequal.

Let all the roots $\alpha_1, \alpha_2, \ldots, \alpha_k$ of the characteristic equation (*i*) are real and unequal. Then $A_1\alpha_1^n$, $A_2\alpha_2^n$, $\ldots$, $A_k\alpha_k^n$ are all solutions of Eq. (*i*). Thus, the sum of homogeneous solutions of the recurrence relation is also a solution and is given by

$$a_{nh}^* = A_1\alpha_1^n + A_2\alpha_2^n + \ldots + A_k\alpha_k^n.$$

Case 2 The roots are real and equal (repeated).

If two roots α_1 and α_2 are equal, i.e. $\alpha_1 = \alpha_2$, then $(A_1 + A_2 n)\,\alpha_1^n$ is the homogeneous solution of the recurrence relation.

Similarly, if three roots α_1, α_2 and α_3 are equal, i.e. $\alpha_1 = \alpha_2 = \alpha_3$, then $a_n = (A_1 + A_2 n + A_3 n^2)\alpha_1^n$ is the homogeneous solution of recurrence relation.

In general, if root α_1 is repeated k times, then

$$a_{nh}^* = (A_1 + A_2 n + A_3 n^2 + \ldots + A_k n^{k-1})\alpha_1^k$$

is the homogeneous solution of the recurrence relation, where A_i $(i = 1, 2, \ldots, k)$ are constants determined by initial conditions.

Case 3 The roots are non-repeated complex roots.

Let $\alpha \pm i\beta$ be the imaginary roots of the characteristic equation, where α and β are real. Then $(\alpha + i\beta)^n$ and $(\alpha - i\beta)^n$ are solutions implying

$$a_{nh}^* = A_1(\alpha + i\beta)^n + A_2(\alpha - i\beta)^n$$

is the homogeneous solution of the recurrence relation, where A_1 and A_2 are constants determined by initial conditions.

Case 4 The roots are repeated complex roots.

Let $\alpha \pm i\beta$ be the complex roots repeated twice, where α and β are real. Then

$$a_{nh}^* = (A_1 + A_2 n)(\alpha + i\beta)^n + (A_3 + A_4 n)(\alpha - i\beta)^n$$

is the homogeneous solution of the recurrence relation, where A_1, A_2, A_3 and A_4 are constants determined by initial conditions.

Example 2 For the Fibonacci sequence (1, 1, 2, 3, 5, 8, . . .) defined by the recurrence relation, $a_n = a_{n-1} + a_{n-2}$ $n \geq 2$ with initial conditions, $a_0 = 1$ and $a_1 = 1$, show that

$$a_n = \frac{1}{\sqrt{5}}\left[\left(\frac{1+\sqrt{5}}{2}\right)^n - \left(\frac{1-\sqrt{5}}{2}\right)^n\right]$$

is the explicit formula for the sequence.

Solution The characteristic equation: $\alpha^2 - \alpha - 1 = 0$ of the recurrence relation $a_n = a_{n-1} + a_{n-2}$ has two distinct roots: $\alpha_1 = (1+\sqrt{5})/2$ and $\alpha_2 = (1-\sqrt{5})/2$, so that the general solution of the recurrence relations is:

$$a_n = A_1\alpha_1^n + A_2\alpha_2^n = A_1\left(\frac{1+\sqrt{5}}{2}\right)^n + A_2\left(\frac{1-\sqrt{5}}{2}\right)^n$$

where A_1 and A_2 are constants whose values depend on the initial conditions, $a_0 = 1$ and $a_1 = 1$ for the sequence.

For the given initial conditions, $a_0 = a_1 = 1$, we have $A_1 = 1/\sqrt{5}$ and $A_2 = -1/\sqrt{5}$. Thus, the explicit formula for Fibonacci sequence is

$$a_n = \frac{1}{\sqrt{5}}\left(\frac{1+\sqrt{5}}{2}\right)^n - \frac{1}{\sqrt{5}}\left(\frac{1-\sqrt{5}}{2}\right)^n.$$

This proves the result.

Example 3 Solve the recurrence relation: $a_{n+2} - 3a_{n+1} + 2a_n = 0$.

Solution The characteristic equation of the given recurrence relation is

$$\alpha^2 - 3\alpha + 2 = 0 \quad \text{or} \quad (\alpha - 2)(\alpha - 1) = 0, \text{ i.e. } \alpha = 1, 2.$$

Since roots are real and unequal, homogeneous solution of the recurrence relation is

$$a^*_{nh} = A_1 + A_2 (2)^n.$$

Example 4 Solve the recurrence relation: $a_{n+2} - 2a_{n+1} + 4a_n = 0$.

Solution The characteristic equation of the given recurrence relation is

$$\alpha^2 - 2\alpha + 4 = 0$$

or
$$\alpha = \frac{2 \pm \sqrt{4-16}}{2} = \frac{2 \pm i2\sqrt{3}}{2} = 1 \pm i\sqrt{3}.$$

Since roots are complex and non-repeated, homogeneous solution of the recurrence relation is

$$a^*_{nh} = A_1\left[1 + i\sqrt{3}\right]^n + A_2\left[1 - i\sqrt{3}\right]^n.$$

Example 5 Solve the recurrence relation: $4a_n - 20a_{n-1} + 17a_{n-2} - 4a_{n-3} = 0$.

Solution The characteristic equation of the given recurrence relation is

$$4\alpha^3 - 20\alpha^2 + 17\alpha - 4 = 0 \text{ or } (\alpha - 4)(4\alpha^2 - 4\alpha + 1) = 0.$$

On solving this equation, we get $a = 1/2, 1/2, 4$. Since roots are real as well as two roots are equal, homogeneous solution of the recurrence relation is

$$a^*_{nh} = (A_1 + A_2 n)\left(\frac{1}{2}\right)^n + A_3 (4)^n.$$

Example 6 Solve the recurrence relation: $a_n + 6a_{n-1} + 12a_{n-2} + 8a_{n-3} = 0$.

Solution The characteristic equation of the given recurrence relation is

$$\alpha^3 + 6\alpha^2 + 12\alpha + 8 = 0 \text{ or } (\alpha + 2)(\alpha^2 + 4\alpha + 8) = 0.$$

On solving this equation, we get $a = -2, -2, -2$. Since all three roots are real but equal, homogeneous solution of the recurrence relation is

$$a^*_{nh} = (A_1 + A_2 n + A_3 n^2)(-2)^n.$$

Example 7 Solve the recurrence relation: $a_{n+4} + 4a_{n+3} + 8a_{n+2} + 8a_{n+1} + 4a_n = 0$.

Solution The characteristic equation of the given recurrence relation is

$$\alpha^4 + 4\alpha^3 + 8\alpha^2 + 8\alpha + 4 = 0 \text{ or } (\alpha^2 + 2\alpha + 2)(\alpha^2 + 2\alpha + 2) = 0.$$

On solving this equation, we get $\alpha = -1 \pm i, -1 \pm i$. Since roots are complex and repeated, homogeneous solution of the recurrence relation is

$$a^*_{nh} = (A_1 + A_2 n)(-1 + i)^n + (A_3 + A_4 n)^n (-1 - i)^n.$$

2. Particular Solution

It is the solution obtained from homogeneous solution by giving particular values to constant coefficients, and hence satisfies the given recurrence relation with $f(n)$ on the right hand side.

Having found the homogeneous solution of the given homogeneous recurrence relation of order k, we can find its particular solution by substituting initial conditions in the homogeneous solution.

Example 8 Solve the recurrence relation: $a_n - 8a_{n-1} + 16a_{n-2} = 0$ with initial conditions, $a_2 = 16$ and $a_3 = 80$.

Solution The characteristic equation of the given recurrence relation is

$$\alpha^2 - 8\alpha + 16 = 0 \quad \text{or} \quad (\alpha - 4)^2 = 0, \text{ i.e. } \alpha = 4, 4.$$

Since roots of the equation are repeated, therefore general homogeneous solution is

$$a^*_{nh} = (A_1 + A_2 n)(4)^n \qquad ...(i)$$

where A_1, A_2 are arbitrary constants determined by initial conditions.

Putting $n = 2$ and 3 in Eq. (*i*), we get

$$\alpha_2 = (A_1 + 2A_2)(4)^2 \text{ or } 16 = (A_1 + 2A_2)\,16 \qquad ...(ii)$$

$$\alpha_3 = (A_1 + 3A_2)(4)^3 \text{ or } 80 = (A_1 + 3A_3)\,64. \qquad ...(iii)$$

Solving simultaneous Eqs. (*ii*) and (*iii*), we get $A_1 = 1/2$ and $A_2 = 1/4$. Hence, the particular solution of the recurrence relation is:

$$a^*_{np} = \left\{\frac{1}{2} + \frac{n}{4}\right\}(4)^n = (2 + n)(4)^{n-1}.$$

Example 9 Solve the recurrence relation: $a_n - 7a_{n-2} + 6a_{n-3} = 0$ with initial conditions, $a_0 = 8$, $a_1 = 6$ and $a_2 = 22$.

Solution The characteristic equation of the given recurrence relation is

$$\alpha^3 - 7\alpha + 6 = 0 \quad \text{or} \quad (\alpha - 1)(\alpha^2 + \alpha + 6) = 0, \text{ i.e. } \alpha = 1, 2, -3.$$

The general homogeneous solution to the recurrence relation is

$$a^*_{nh} = A_1(1)^n + A_2(2)^n + A_3(-3)^n \qquad ...(i)$$

where A_1, A_2 and A_3 are arbitrary constants determined by initial conditions.

Putting $n = 0$, 1 and 2 in Eq. (*i*), we get

$$a_0 = A_1 + A_2 + A_3 \quad \text{or} \quad 8 = A_1 + A_2 + A_3 \qquad ...(ii)$$
$$a_1 = A_1 + 2A_2 - 3A_3 \quad \text{or} \quad 6 = A_1 + 2A_2 - 3A_3 \qquad ...(iii)$$
$$a_2 = A_1 + 4A_2 + 9A_3 \quad \text{or} \quad 22 = A_1 + 4A_2 + 9A_3 \qquad ...(iv)$$

Solving simultaneous Eqs, (*ii*), (*iii*) and (*iv*), we get $A_1 = 5$, $A_2 = 2$ and $A_3 = 1$. Hence, the particular solution is:

$$a^*_{np} = 5 + 2(2)^n + (-3)^n = 5 + 2^{n+1} - 3^n.$$

Example 10 Find explicit formula for the sequence defined by $c_n = 3c_{n-1} - 2c_{n-2}$ with initial conditions $c_1 = 5$, $c_2 = 3$ by using following approaches.

(*a*) Characteristic equation

(*b*) Generating function

Solution (*a*) The characteristic equation of the given sequence, $c_n = 3c_{n-1} - 2c_{n-2}$ is:

$$x_n = 3x_{n-1} - 2x_{n-2}$$

or

$$x^2 - 3x + 2 = (x - 2)(x - 1) = 0$$

The roots of this equation are: $x = 1$ and 2. Hence, the general solution becomes:

$$c_n = A_1 (2)^n + A_2 = A_1 2^n + A_2$$

Since $c_1 = 5$ (initial condition), substituting $n = 1$ in the general solution, we get

$$5 = A_1 2 + A_2$$

Also, $c_2 = 3$ (initial condition), substituting $n = 2$ in the general solution, we get

$$3 = 4A_1 + A_2$$

Solving both equations, we have $A_1 = -1$ and $A_2 = 7$. Hence, the solution of the given recurrence relation becomes: $c_n = -(2)^n + 7$.

(*b*) To find the value of c_0, substituting $n = 2$ in the given recurrence relation as:

$$c_2 = 3c_1 - 2c_0 \text{ or } c_0 = (3c_1 - c_2)/2 = [3(5) - 3]/2 = 6$$

where $c_1 = 5$ and $c_2 = 5$ (Initial conditions).

To solve the recurrence relation with initial conditions, $c_0 = 6$ and $c_1 = 5$, multiplying both sides of the relation by z^n and summing from $n = 2$ to ∞, we have

$$\sum_{n=2}^{\infty} c_n z^n = 3\sum_{n=2}^{\infty} c_{n-1} z^n - 2\sum_{n=2}^{\infty} c_{n-2} z^n$$

or

$$A(z) - c_0 - c_1 z = 3z[A(z) - c_0] - 2A(z)$$
$$A(z) = (c_0 + c_1 z - 3c_0 z)/(1 - 3z + 2z^2)$$
$$= (6 - 13z)/(1 - 2z)(1 - z), \text{ where } c_0 = 6 \text{ and } c_1 = 5$$
$$= -1/(1 - 2z) + 7/(1 - z) \qquad \text{[using partial fraction]}$$

Hence,

$$c_n = \text{Coefficient of } z^n \text{ in } A(z)$$
$$= \text{Coefficient of } z^n \text{ in } (-1)(1 - 2z)^{-1} + \text{Coefficient of } z^n \text{ in } 7(1 - z)^{-1}$$
$$= (-1)(2)^n + 7(1)^n = 7 - 2^n$$

which is the $(n + 1)^{th}$ term of the sequence c_0, c_1, c_2, c_3 and nth term of the sequence c_1, c_2

Example 11 Use (*a*) generating function and (*b*) Characteristic roots approaches to solve the recurren relation

$$u_n = u_{n-1} + u_{n-2}, \text{ for } n \geq 3,\ u_2 = 1,\ u_2 = 3.$$

Solution (*a*) Given recurrence relation is: $u_n = u_{n-1} + u_{n-2}$, where $n \geq 3$, $u_1 = 1$, $u_2 = 3$.

For $n = 2$, we have, $u_2 = u_1 + u_0$ or $u_0 = u_2 - u_1 = 3 - 1 = 2$. Then recurrence relation becom $u_n = u_{n-1} + u_{n-2}$, where $n \geq 2$, $u_0 = 2$, $u_1 = 1$

Multiplying both sides of the given relation by z^n and summing from $n = 2$ to ∞, we get

$$\sum_{n=2}^{\infty} u_n z^n = \sum_{n=2}^{\infty} u_{n-1} z^n + \sum_{n=2}^{\infty} u_{n-2} z^n$$

or
$$\sum_{n=2}^{\infty} u_n z^n = z \sum_{n=2}^{\infty} u_{n-1} z^{n-1} + z^2 \sum_{n=2}^{\infty} u_{n-2} z^{n-2}$$

$$A(z) - u_0 - u_1 z = z[A(z) - u_0] + z^2 A(z)$$

or
$$A(z)[1 - z - z^2] = u_0 + u_1 z - u_0 z$$

$$A(z) = (2 - z)/(1 - z - z^2) \quad \text{(substituting values of } u_0 \text{ and } u$$

$$= (2 - z) / (1 - \alpha z)(1 - \beta z), \quad \text{where } \alpha, \beta \text{ are roots of the Eq. } 1 - z - z^2 =$$

Let $\alpha = (1 + \sqrt{5})/2$, $\beta = (1 - \sqrt{5})/2$. Then $\alpha + \beta = 1$, $\alpha\beta = -1$

$$A(z) = \frac{2\alpha - 1}{(\alpha - \beta)(1 - \alpha z)} + \frac{2\beta - 1}{(\beta - \alpha)(1 - \beta z)} \quad \text{(by partial fraction}$$

$$= \frac{1}{\alpha - \beta}[(2\alpha - 1)(1 - \alpha z)^{-1} - (2\beta - 1)(1 - \beta z)^{-1}]$$

$$= \frac{1}{\alpha - \beta}[\text{Coeff. of } z^n \text{ in } (2\alpha - 1)(1 - \alpha z)^{-1} - \text{Coeff. of } z^n \text{ in } (2\beta - 1)(1 - \beta z)^{-1}]$$

$$= \frac{1}{\alpha - \beta}[(2\alpha - 1)\alpha^n - (2\beta - 1)\beta^n]$$

$$= \frac{1}{\alpha - \beta}[\sqrt{5}\,[(1 + \sqrt{5})/2]^n + \sqrt{5}\,[(1 - \sqrt{5})/2]^n \quad \text{[substituting values of } \alpha \text{ and } \beta$$

$$= [(1 + \sqrt{5})/2]^n + [(1 - 5)/2]^n$$

This is n^{th} term of the sequence $u_1, u_2, u_3, \ldots$

(*b*) The characteristic equation is: $x^n = x^{n-1} + x^{n-2}$ or $x^2 - x - 1 = 0$

The characteristic roots of this equation are: $(1 + \sqrt{5})/2$ and $(1 - \sqrt{5})/2$. Hence, the general solution becomes:

$$u_n = A_1 [(1 + \sqrt{5})/2]^n + A_2 [(1 - \sqrt{5})/2]^n$$

$$= A_1 a^n + A_2 \beta^n, \text{ where } \alpha = (1 + \sqrt{5})/2 \text{ and } \beta = (1 - \sqrt{5})/2$$

Putting values $u_1 = 1$, i.e. $n = 1$ and $u_2 = 3$, i.e. $n = 2$ and $u_n = 3$, we have

$$1 = A_1\alpha + A_2\beta \text{ and } 3 = A_1\alpha^2 + A_2\beta^2$$

Solving these equations, we have

$A_1 = (\beta - 3) / (\alpha\beta - \alpha^2) = 1$ [substituting values of α and β]

d $A_2 = (1 - A_1\alpha) / \beta = 1$

nce, $u_n = [(1 + \sqrt{5})/2]^n + [(1 - \sqrt{5})/2]^n$

ample 12 Use generating function to solve the recurrence relation $a_n = a_{n-1} + a_{n-2}$; $n \geq 3$, $= 1$, $a_2 = 1$ directly without calculating a_0.

lution The given sequence is: $a_n = a_{n-1} + a_{n-2}$; $n \geq 3$

Multiplying both sides by z^n and summing up from $n = 3$ to $n = \infty$, we get

$$\sum_{n=3}^{\infty} a_n z^n = \sum_{n=3}^{\infty} a_{n-1} z^n + \sum_{n=3}^{\infty} a_{n-2} z^n$$

$$= z\sum_{n=3}^{\infty} a_{n-1} z^{n-1} + z^2\sum_{n=3}^{\infty} a_{n-2} z^{n-2} \qquad (i)$$

e term on the LHS of Eqn. (*i*) is:

$$\sum_{n=3}^{\infty} a_n z^n = a_3z^3 + a_4z^4 + \ldots$$

$$= (a_1z + a_2z^2 + a_3z^3 + a_4z^4 + \ldots) - a_1z - a_2z^2$$

$= A(z) - a_1z - a_2z^2$, since first term is a_1 instead of a_0.

rst term on RHS of Eqn. (*i*) is:

$$z\sum_{n=3}^{\infty} a_{n-1} z^{n-1} = z[a_2z^2 + a_3z^3 + \ldots]$$

$$= z[(a_1z + a_2z^2 + a_3z^3 + a_4z^4 + \ldots) - a_1z]$$

$$= z[A(z) - a_1z] = z\,A(z) - a_1z^2$$

Second term on RHS of Eqn. (*i*) is:

$$z^2\sum_{n=3}^{\infty} a_{n-2} z^{n-2} = z^2[a_1z + a_2z^2 + \ldots] = z^2 A(z)$$

Putting all values in Eqn. (*i*), we have

$$A(z) - a_1z - a_2z^2 = z\,A(z) - a_1z^2 + z^2 A(z)$$

$$A(z)[1 - z - z^2] = a_1z + a_2z^2 - a_1z^2$$

$$A(z) = (a_1z + a_2z^2 - a_1z^2)/(1 - z - z^2)$$

$= z/(1 - z - z^2)$, [substituting values of a_1 and a_2]

The roots of the equation: $1 - z - z^2 = 0$ are:

$$\alpha = \frac{-1+\sqrt{5}}{2} \quad \text{or} \quad \frac{1}{\alpha} = \frac{2}{\sqrt{5}-1} = \frac{2(\sqrt{5}+1)}{(\sqrt{5}-1)(\sqrt{5}+1)} = \frac{1+\sqrt{5}}{2}$$

and $$\beta = \frac{-1-\sqrt{5}}{2} \quad \text{or} \quad \frac{1}{\beta} = \frac{2}{-\sqrt{5}-1} = \frac{2(-\sqrt{5}+1)}{(-\sqrt{5}-1)(-\sqrt{5}+1)} = \frac{1-\sqrt{5}}{2}$$

Hence, $$A(z) = \frac{-z}{(z-\alpha)(z-\beta)} = \left[\frac{\alpha}{(\alpha-\beta)(z-\alpha)} + \frac{\beta}{(\beta-\alpha)(z-\beta)}\right]$$

$$= \frac{-1}{\alpha-\beta}\left[\frac{\alpha}{z-\alpha} - \frac{\beta}{z-\beta}\right] = \frac{-1}{\alpha-\beta}\left[\left(1-\frac{z}{\beta}\right)^{-1} - \left(1-\frac{z}{\alpha}\right)^{-1}\right]$$

Taking coefficient of $z^n = \frac{-1}{\alpha-\beta}\left[\frac{1}{\beta^n} - \frac{1}{\alpha^n}\right] = \frac{1}{\sqrt{5}}\left[\left(\frac{1-\sqrt{5}}{2}\right)^n - \left(\frac{1+\sqrt{5}}{2}\right)^n\right]$

which is the nth term of the sequence $a_1, a_2, a_3, \ldots$

Example 13 Solve the following recurrence relation

$$a_n = 5\, a_{n-1} + 6\, a_{n-2}\ ;\ n \geq 2,\ a_0 = a_1 = 3$$ [*UPTU, BTech, 200*

Solution Given: $a_n = 5\, a_{n-1} + 6\, a_{n-2}$, where $n \geq 2$, $a_0 = a_1 = 3$

The solution of characteristic equation: $m^2 - 5m - 6 = 0$, is: $m_1 = 6$, $m_2 = -1$,

and the general solution of this equation is: $a_n = A_1(6)^n + A_2(-1)^n$, where $a_0 = 3$ and $a_1 = 3$

Thus, $$3 = A_1(6)^0 + A_2(-1)^0 = A_1 + A_2 \text{ and } 3 = 6\,A_1 - A_2$$

On solving these equations, we get $A_1 = 6/7$ and $A_2 = 15/7$. Thus,

$$a_n = \frac{6}{7}(6)^n + \frac{15}{7}(-1)^n = \frac{1}{7}(6)^{n+1} + \frac{15}{7}(-1)^n$$

Example 14 Solve the following difference equation: [*UPTU, MCA, 200*

$$a_{r+2} - 7a_{r+1} - 8a_r = 2^r r^2$$

Solution The given equation is: $a_{r+2} - 7a_{r+1} - 8a_r = 2^r r^2$

The characteristic equation becomes:

$$m^2 - 7m - 8 = (m-8)(m+1) = 0, \quad \text{i.e.} \quad m = 8, \text{ and } m = -1$$

The homogeneous solution of the given equation becomes:

$$c_1(8)^r + c_2(-1)^r$$

Particular solution is given by

$$\text{PI} = \frac{2^r \cdot r^2}{E^2 - 7E - 8} = 2^r \frac{r^2}{(2E)^2 - 7(2E) - 8} = 2^r \frac{r^2}{4E^2 - 14E - 8}$$

$$= 2^r \frac{r^2}{4(1+\Delta)^2 - 14(1+\Delta) - 8} = 2^r \frac{r^2}{4\Delta^2 - 6\Delta - 18}$$

$$= 2^{r-1} \cdot \frac{r^2}{(3+2\Delta)(\Delta-3)} = 2^{r-1} \frac{1}{3\cdot(-3)}\left(1+\frac{2}{3}\Delta\right)^{-1}\left(1-\frac{\Delta}{3}\right)^{-1} r^2$$

$$= \frac{1}{9}2^{r-1}\left[\left(1-\frac{2}{3}\Delta+\frac{(-1)(-2)}{2!}\frac{4}{9}\Delta^2+\ldots\right)\left(1+\frac{1}{3}\Delta+\frac{(-1)(-2)}{2!}\cdot\frac{\Delta^2}{9}+\ldots\right)\right]r^2$$

$$= \frac{1}{9}2^{r-1}\left[1+\frac{1}{3}\Delta+\frac{\Delta^2}{9}-\frac{2}{3}\Delta-\frac{2}{9}\Delta^2\right]r^2 \quad \text{(neglecting higher power of } \Delta)$$

$$= \frac{1}{9}2^{r-1}\left[1+\frac{1}{3}\Delta+\frac{1}{9}\Delta^2\right]r^2 = -\frac{1}{9}2^{r-1}\left[r^2-\frac{2}{3}r-\frac{2}{9}\right]$$

$$= -\frac{1}{9}2^{r-1}\left(r^2-\frac{2}{3}r-\frac{2}{9}\right) = -\frac{1}{3^4}2^{r-1}(9r^2-6r-2)$$

Hence, the homogeneous solution is

$$a_r = [c_1 8^r + c_2(-1)^r] = -\frac{2^{r-1}}{3^4}(9r^2-6r-2)$$

Example 15 Solve the following difference equation by the method of generating function

$$a_r - 7a_{r-1} + 10a_{r-2} = 3^r,\ r \geq 2.$$

with the boundary conditions: $a_0 = 0$ and $a_1 = 1$. [*UPTU, MCA, 2005*]

Solution The given equation is: $a_r - 7a_{r-1} + 10a_{r-2} = 3^r,\ r \geq 2.$

Multiplying both sides by z^r and summing each term from $r = 2$ to ∞, we have

$$\sum_{r=2}^{\infty} a_r z^r - 7\sum_{r=2}^{\infty} a_{r-1} z^r + 10\sum_{r=2}^{\infty} a_{r-2} z^r = \sum_{r=2}^{\infty} 3^r z^r$$

$$\sum_{r=2}^{\infty} a_r z^r - 7z\sum_{r=2}^{\infty} a_{r-1} z^{r-1} + 10z^2\sum_{r=2}^{\infty} a_{r-2} z^{r-2} = 3^2 z^2 + 3^3 z^3 + \ldots$$

$$[A(z) - a_0 - a_1 z] - 7z[A(z) - a_0] + 10z^2 A(z) = \frac{3^2 z^2}{1-3z}$$

$$(1 - 7z + 10z^2)\, A(z) = a_0 + a_1 z - 7a_0 z + \frac{9z^2}{1-3z}$$

$$(1 - 5z)(1 - 2z)\, A(z) = z + \frac{9z^2}{1-3z}, \text{ where } a_0 = 0 \text{ and } a_1 = 1$$

$$(1 - 5z)(1 - 2z)\, A(z) = \frac{z+6z^2}{1-3z}$$

$$A(z) = \frac{z+6Z^2}{(1-3z)(1-5z)(1-2z)}$$

$$= \frac{-9}{2(1-3z)} + \frac{11}{6(1-5z)} + \frac{8}{3(1-2z)} \quad \text{[by partial fractions]}$$

$$= -\frac{9}{2}(1-3z)^{-1} + \frac{11}{6}(1-5z)^{-1} + \frac{8}{3}(1-2z)^{-1}$$

Hence, the general solution shall be

a_r = Coefficent of z^r in A(z)

$$= -\frac{9}{2}\text{ Coeff. of } z^r \text{ in } (1-3z)^{-1} + \frac{11}{6}\cdot\text{Coeff. of } z^r \text{ in } (1-5z)^{-1} + \frac{8}{3}\text{Coeff. of } z^r \text{ in } (1-2z)^{-1}$$

$$= -\frac{9}{2}({}^{1-1+r}C_r)(3^r) + \frac{11}{6}\cdot({}^{1-1+r}C_r)(5^r) + \frac{8}{3}({}^{1-1+r}C_r)(2^r)$$

[Coeff. of z^r in the expansion of $(1 - az)^{-n} = ({}^{n-1+r}C_r)(a^r)$]

Example 16 Express Fibonacci sequence of numbers 1, 1, 2, 3, 5, 8, 13, 21, 34, ... in terms of
(*a*) general expression for the *r*th number a_r and
(*b*) generating function. [*UPTU, MCA, 2004*]

Solution (*a*) Expressing the given sequence in terms of recurrence relation as follows:

$$a_n = a_{n-1} + a_{n-2}, \text{ where } a_0 = 1 \text{ and } a_1 = 1$$

The solution of the recurrence relation then shall be

$$a_n = \frac{1}{\sqrt{5}}[(1+\sqrt{5})/2]^n - \frac{1}{5}[(1-\sqrt{5})/2]^n$$

(*b*) Multiplying both sides of the recurrence relation by z and summing both sides from $n = 2$ to ∞, we get

$$\sum_{n=2}^{\infty} a_n z^n = \sum_{n=2}^{\infty} a_{n-1} z^n + \sum_{n=2}^{\infty} a_{n-2} z^n$$

$$= z\sum_{n=2}^{\infty} a_{n-1} z^{n-1} + z^2\sum_{n=2}^{\infty} a_{n-2} z^{n-2}$$

$$[A(z) - a_0 - a_1 z] = z[(A(z) - a_0] + z^2 A(z)$$

$$A(z)(1 - z - z^2) = a_1 z = z, \text{ since } a_1 = 1$$

$$A(z) = \frac{z}{1-z-z^2}$$

which is the required generating function corresponding to the given sequence.

Example 17 Show that

$$a^r = \frac{1}{4}r^2 + \frac{13}{24}r + \frac{71}{288}$$

is a particular solution for the recurrence relation

$$a_r + 5a_{r-1} + 6a_{r-2} = 3r^2 - 2r + 1$$ [*UPTU, MCA, 2004*]

Solution If the recurrence relation is of the type

$$a_r + c_1 a_{r-1} + c_2 a_{r-2} + \ldots + c_n a_0 = g(n)$$

where $g(n)$ is a polynomial of order k and 1 is not the characteristic root of the corresponding homogeneous recurrence relation, then the particular solution is given by

$$a_r = A_0 + A_1 r + A_2 r^2 + ... + A_k r^k + ...$$

Since 1 is not the characteristic root of the recurrence relation and $g(n)$ is given by

$$g(n) = 3r^2 - 2r + 1$$

then the particular solution is given by

$$a_r = A_0 + A_1 r + A_2 r^2$$

Substituting value of a_r in the given recurrence relation, we get

$$(A_0 + A_1 r + A_2 r^2) + 5[A_0 + A_1(r-1) + A_2(r-1)^2] + 6[A_0 + A_1(r-2) + A_2(r-2)^2] = 3r^2 - 2r + 1$$

or
$$(A_0 + A_1 r + A_2 r^2) + 5(A_0 + A_1 r - A_1 + A_2 r^2 - 2A_2 r + A_2)$$
$$+ 6(A_0 + A_1 r - 2A_1 + A_2 r^2 - 4A_2 r + 4A_2) = 3r^2 - 2r + 1$$

$$12A_2 r^2 + (12A_1 - 34A_2)r + 12A_0 - 17A_1 + 29A_2 = 3r^2 - 2r + 1$$

Equating coefficients of equal powers of r on both the sides, we have

$$\left.\begin{aligned} 12A_2 &= 3, \\ 12A_1 - 34A_2 &= -2, \\ 12A_0 - 17A_1 + 29A_2 &= 1 \end{aligned}\right\}$$

Solving these equations, we have

$$A_2 = \frac{1}{4}, A_1 = \frac{13}{24} \text{ and } A_0 = \frac{71}{288}$$

Putting values of A_0, A_1 and A_2 in the particular solution, we get

$$a_r = \frac{1}{4}r^2 + \frac{13r}{24} + \frac{71}{288}$$

8.5 NON-HOMOGENEOUS RECURRENCE RELATIONS

The general form of linear non-homogeneous recurrence relations is:

$$u_n = c_1 u_{n-1} + c_2 u_{n-2} + . . . + c_k u_{n-k} + g(n); n > k \quad ...(i)$$

where $c_1, c_2, ..., c_k$ are constants.

The corresponding homogeneous recurrence relation is

$$u_n = c_1 u_{n-1} + c_1 u_{n-2} + . . . + c_k u_{n-k} \quad ...(ii)$$

The more general type of non-homogeneous recurrence relation is of the form:

$$u_n = f_1(n)u_{n-1} + f_2(n) u_{n-2} + . . . + f_k(n) u_{n-k} + g(n) \quad ...(iii)$$

and the corresponding homogeneous recurrence relation is

$$u_n = f_1(n)u_{n-1} + f_2(n) u_{n-2} + . . . + f_k(n) u_{n-k} \quad ...(iv)$$

where $f_1(n), f_2(n), ..., f_k(n)$ and g are functions of n.

The procedure of obtaining particular solution for a non-homogeneous recurrence relation depends on the type of right hand side of the relation. The following are the two methods used to find particular solution of non-homogeneous recurrence relations:

(*a*) Undetermined Coefficients Method
(*b*) E and Δ Operator Method.

8.5.1 Undetermined Coefficients Method

This method is useful in finding the particular solution of non-homogeneous recurrence relations where the right hand side, $f(n)$ consists of terms of special forms.

In this method, we consider a trial solution according to the form of $f(n)$ containing a number of unknown constant coefficients, determined by initial conditions. The trial solutions according to the form of $f(n)$ are shown in Table 8.1 to find exact solution of the given recurrence relation. Here A, B, A_0, A_1, . . . are the unknown constant coefficients which have to be determined.

Table 8.1 Trial Solutions

Form of f (n)	*Corresponding Trial Solution*
• β^n, β is a constant.	$A\beta^n$
• $\beta^n f(n)$; $f(n)$ is polynomial of degree k in r, and β is a constant.	$\beta^k [A_0 \beta^k + A_1\beta^{k-1} + \ldots + A_k]$
• $f(n)$, a polynomial of degree k.	$A_0 \beta^k + A_1 \beta^{k-1} + \ldots + A_k$
• β, a constant	A

Example 18 Find the particular solution of the recurrence relation: $a_{n+2} - 3a_{n+1} + 2a_n = b^n$, where b is some constant.

Solution The characteristic equation of the recurrence relation is given by

$$\alpha^2 - 3\alpha + 2 = 0 \quad \text{or} \quad (\alpha - 1)(\alpha - 2) = 0, \text{ i.e. } \alpha = 1, 2.$$

Thus, homogeneous solution of the recurrence relation is: $a_{nh}^* = A_1 + A_2 2^n$.

Since, right hand side, $f(n)$ of the recurrence relation is of the form β^n, where β is a constant, to find particular solution, let the trial solution be $a_n^* = Ab^n$. Substituting value of a_n^* in the given recurrence relation, we get

$$A[b^{n+2} - 3b^{n+1} + 2b^n] = b^n$$

or

$$Ab^n [b^2 - 3b + 2] = b^n$$

$$A(b^2 - 3b + 2) = 1$$

$$A = \frac{1}{b^2 - 3b + 2} = \frac{1}{(b-1)(b-2)}, \text{ provided } b \neq 1 \text{ and } b \neq 2.$$

Putting the value of A in the trial solution, we get the particular solution as, $a_{np}^* = \dfrac{b^n}{(b-1)(b-2)}$.

Example 19 Find the particular solution of the recurrence relation

$$a_{n+2} - 4a_{n+1} + 4a_n = 3n + 2^n.$$

Solution The characteristic equation of the recurrence relation is given by

$$\alpha^2 - 4\alpha + 4 = 0 \quad \text{or} \quad (\alpha - 2)^2 = 0, \text{ i.e. } \alpha = 2, 2.$$

Thus, the homogeneous solution of the recurrence relation is: $a_{nh}^* = (A_1 + A_2 n)2^n$.

Since the right hand side of the recurrence relation is of mix type containing two terms $3n$ and 2^n, therefore to find particular solution, let the trial solution be $a_n^* = A_0 + A_1 n + Bn^2\ 2^n$. Here $Bn^2 2^n$ is representing the trial solution corresponding to the term 2^n and not B 2^n, because 2^n is the solution of the recurrence relation, so B 2^n is not taken as trial solution. Multiplying it by n we get $Bn\ 2^n$, which is also the solution of recurrence relation. Again multiplying with n, we get $Bn^2\ 2^n$ and take the trial solution as: $Bn^2\ 2^n$.

Substituting value of $a_n^* = 3n^2\ 2^n$ in the given recurrence relation, we get

$$[A_0 + A_1 (n + 2) + B (n + 2)^2\ 2^{n+2}] - 4 [A_0 + A_1 (n + 1) + B (n + 1)^2\ 2^{n+1}]$$
$$+ 4 [A_0 + A_1 n + B n^2\ 2^n] = 3n + 2^n.$$

or

$$(A_0 - 2A_1) + A_1 n + [4B (n + 2)^2 - 8B (n + 1)^2 + 4Bn^2]2^n = 3n + 2^n.$$
$$(A_0 - 2A_1) + A_1 n + 8B \cdot 2^n = 3n + 2^n. \quad ...(i)$$

Comparing coefficients on both sides of Eq. (*i*), we get

$$A_0 - 2A_1 = 0,\ A_1 = 3 \quad \text{and} \quad 8B = 1 \quad \text{or} \quad A_0 = 6,\ A_1 = 3 \text{ and } B = 1/8.$$

Putting the values of A_0, A_1 and B in the trial solution, we get, the particular solution as,

$$\alpha_{np}^* = 6 + 3n + \frac{1}{8} n^2 \cdot 2^n.$$

Example 20 Find the particular solution of the recurrence relation: $a_{n+2} - 3a_{n+1} + 2a_n = 1$.

Solution The characteristic equation of the recurrence relation is given by

$$\alpha^2 - 3\alpha + 2 = 0 \quad \text{or} \quad (\alpha - 2)(\alpha - 1) = 0, \text{ i.e. } \alpha = 2, 1.$$

Thus, the homogeneous solution of the recurrence relation is, $a_{nh}^* = A_1 + A_2\ 2^n$.

Since right hand side, $f(n)$ of the recurrence relation is a constant term, therefore to find the particular solution, let the trial solution be $a_n^* = A$. Substituting the value of a_n^* in the given recurrence relation, we get A – 3A + 2A = 1 or 0 = 1, which is not true.

Now, let the trial solution be $a_n^* = An$. Substituting this value in the given recurrence relation, we get

$$A (n + 2) - 3A (n + 1) + 2\ An = 1 \quad \text{or} \quad A = -1.$$

and hence, the particular solution is, $a_{np}^* = -n$.

Example 21 Find the particular solution of the recurrence relation

$$a_{n+2} - 6a_{n+1} + 8a_n = 3n^2 + 2 - 5 \cdot 3^n.$$

Solution The characteristic equation of the recurrence relation is given by

$$\alpha^2 - 6\alpha + 8 = 0 \quad \text{or} \quad (\alpha - 2)(\alpha - 4) = 0, \text{ i.e. } \alpha = 2, 4.$$

Thus, the homogeneous solution of the recurrence relation is, $a_{nh}^* = A_1\ 2^n + A_2\ 4^n$.

Corresponding to the term, $3n^2 + 2$ on the right hand side of the recurrence relation. let the trial solution be $a_n^* = B_1 n^2 + B_2 n + B_3$. Since none of the these terms occur in the homogeneous solution, therefore due to the presence of the term 5.3^n to the right hand side of the relation, the trial solution corresponding to this term be $B_4\ 3^n$.

Hence, the complete trial solution may be written as:

$$a_n^* = B_1 n^2 + B_2 n + B_3 + B_4\, 3^n.$$

Substituting this value in the recurrence relation, we get

$$a_{n+2} - 6a_{n+1} + 8a_n = B_1 (n+2)^2 + B_2 (n+2) + B_3 + B_4\, 3^{n+2} - 6\,[B_1 (n+1)^2 + B_2 (n+1) + B_3 + B_4\, 3^{n+1}] + 8\,[B_1 n^2 + B_2 n + B_3 + B_4\, 3^n]$$

or

$$3n^2 + 2 - 5.3^n = 3B_1 n^2 + (3B_2 - 8B_1)n + 3B_3 - 4B_2 - 2B_1 - B_4\, 3^n.$$

Equating coefficients of similar terms on both sides, we get

$$3B_1 = 3,\ 3B_2 - 8B_1 = 0,\ 3B_3 - 4B_2 - 2B_1 = 2.$$

On solving these equations, we get $B_1 = 1$, $B_2 = 8/3$, $B_3 = 44/9$ and $B_4 = 5$.

Hence, the particular solution is, $a_{np}^* = n^2 + (8/3)n + (44/9) + 5 \cdot 3^n$.

Example 22 Find the particular solution of the recurrence relation

$$a_{n+3} - 3a_{n+2} + 3a_{n+1} - a_n = 24(n+2).$$

Solution The characteristic equation of the recurrence relation is given by

$$\alpha^3 - 3\alpha^2 + 3\alpha - 1 = 0 \quad \text{or} \quad (\alpha - 1)^3 = 0, \text{ i.e. } \alpha = 1, 1, 1.$$

Thus, the homogeneous solution of the recurrence relation is, $a_{nh}^* = A_1 + A_2 n + A_3 n^2$.

In order to find the particular solution, let the trial solution be $a_n^* = B_1 n + B_2$ corresponding to the terms on the right hand side of the relation. Since terms in the trial solution also occur in the homogeneous solution, therefore, we multiply each term in the trial solution by a suitable term, i.e. n^3 so that terms should not be in homogeneous solution. Thus, the trial solution now is, $a_n^* = B_1 n^4 + B_2 n^3$.

Substituting the value of a_n^* in the recurrence relation, we get

$$a_{n+3} - 3a_{n+2} + 3a_{n+1} - a_n = B_1 (n+3)^4 + B_2 (n+3)^3 - 3\{B_1 (n+2)^4 + B_2 (n+2)^3\} + \{3B_1 (n+1)^4 + B_2 (n+1)^3\} - B_1 n^4 - B_2 n^3$$

or

$$24n + 48 = 24B_1 n + 36B_1 + B_2.$$

Comparing coefficients of similar terms on both sides, we get $B_1 = 1$ and $B_2 = 12$.

Hence, the particular solution is, $a_{np}^* = n^4 + 12n^3$.

8.5.2 E and D Operators Method

These methods are based on the *calculus of finite differences* that deals with the changes in the value of the function (dependent variable), $y = f(x)$ due to changes in the value of the independent variable, x. Also, the calculus of finite differences helps the study of relation of the value which the function assumes, whenever the independent variable changes by finite jumps whether of equal or unequal intervals. In this chapter, we shall study the case when independent variable changes with equal intervals.

The operator Δ

Let the consecutive values of independent variable x be $a, a+h, a+2h, \ldots, a+nh$, each different from other by h. Thus, the values of the function, $y = f(x)$ corresponding to each of these values of x are:

$$f(a),\ f(a+h),\ f(a+2h), \ldots,\ f(a+nh).$$

The differences, $f(a+h)-f(a), f(a+2h)-f(a+h), \ldots, f(a+nh)-f(a+(n-1)h)$, are called *first differences* of the function $y=f(x)$ and are denoted as $\Delta f(a), \Delta f(a+h) \ldots$, respectively, where Δ is called the *forward or descending difference operator*, and $\Delta f(a), \Delta f(a+h), \ldots$ are called *first forward differences.*

In general, the forward differences are defined as

$$\begin{aligned}\Delta f(x) &= f(x+h)-f(x), \text{ first forward difference}\\ \Delta^2 f(x) &= \Delta f(x+h)-\Delta f(x)\\ &= \{f(x+2h)-f(x+h)\}-\{f(x+h)-f(x)\}\\ &= f(x+2h)-2f(x+h)+f(x), \text{ second forward difference, and so on.}\end{aligned}$$

Important Results with respect to Operator Δ

1. As Δ is an operator, and Δ^2 is not the square of Δ, But it is the repetition of Δ
2. $\Delta c = 0$, c = constant.
3. Δ is commutative with respect to a constant, that is
$$(\Delta c) f(x) = \Delta c f(x) = c\{f(x+h)-f(x)\} = c\,\Delta f(x).$$
Thus, $\Delta c = c\Delta$.
4. Δ is distributive, that is
$$\Delta[f(x)+g(x)] = \Delta f(x) + \Delta g(x).$$
5. If $f(x)$ is the polynomial of degree n in x, then nth difference of $f(x)$ is constant and
$$\Delta^{n+1} f(x) = 0.$$
6. If the nth differences of a function are constant when values of the independent variable are taken at equal intervals, then the function is a polynomial of degree n.
7. $f(x+nh) = f(x) + {}^nC_1\,\Delta f(x) + {}^nC_2\,\Delta^2 f(x) + \ldots + {}^nC_n\,\Delta^n f(x)$, for all integral values of n.
8. In factorial notation, we have
$$\begin{aligned}x^{(n)} &= x(x-1)(x-2)\ldots,(x-\overline{n-1}), \text{ when interval of differencing is 1.}\\ &= x(x-h)(x-2h)\ldots,(x-\overline{n-1}\,h), \text{ when interval of differencing is } h.\end{aligned}$$
Thus (*a*) $x^{(n)} = \dfrac{x!}{(x-n)!}$, $x > n$

(*b*) $\Delta^n x^{(n)} = n!\,h^n$ and $\Delta^{n+1} x^{(n)} = 0$

(*c*) $x^{(r)}(x-r)^{(n)} = x^{(n+r)}$.
9. A polynomial can also be represented in factorial notation. For example,
$$\begin{aligned}4x^3-3x^2+6x-5 &\equiv A\,x^{(3)} + B\,x^{(2)} + C\,x^{(1)} + D\\ &\equiv A\,x(x-1)(x-2) + B\,x(x-1) + Cx + D\end{aligned}$$

The operator E

The operator E also known as *shift operator* gives an increment to the value of independent variable, x in the function and is defined as

$$E f(x) = f(x+h)$$

where h is the increment quantity.

Theorem 8.1 Prove that $E \equiv 1 + \Delta$

Proof We know that the operator Δ is defined as

$$\Delta f(x) = f(x + h) - f(x) = E f(x) - f(x)$$

or $$E f(x) = \Delta f(x) + f(x) = (\Delta + 1) f(x)$$

Since $f(x)$ is arbitrary, therefore cancelling $f(x)$ from both sides, we have

$$E \equiv 1 + \Delta.$$

Here 1 is an identity operator.

Theorem 8.2 Operators E and Δ are distributive.

Proof Consider a function $u(x)$ such that

$$u(x) = f(x) + g(x) + p(x) + \ldots$$

where $f(x)$, $g(x)$, $p(x)$, etc., are also the functions of x. Then

$$\begin{aligned} E\,u(x) &= f(x + h) + g(x + h) + p(x + h) + \ldots \\ &= E f(x) + E g(x) + E p(x) + \ldots \end{aligned}$$

That is, operator E is distributive. Also

$$\begin{aligned} \Delta\,u(x) &= \Delta \{f(x) + g(x) + p(x) + \ldots\} \\ &= \{f(x + h) + g(x + h) + p(x + h) + \ldots\} - \{f(x) + g(x) + p(x) + \ldots\} \\ &= \{f(x + h) - f(x)\} + \{g(x + h) - g(x)\} + \{p(x + h) - p(x)\} + \ldots \\ &= \Delta f(x) + \Delta g(x) + \Delta p(x) + \ldots \end{aligned}$$

That is, operator Δ is distributive.

Theorem 8.3 E and Δ are commutative with respect to constants. That is

$$E\,[k f(x)] = k\,E f(x),$$
$$\Delta\,[k f(x)] = k\,\Delta f(x),$$

where k is constant.

Proof We know that $E\,[k f(x)] = k f(x + h) = k\,E f(x)$, hence the result is proved; where h is the interval of differencing.

Similarly, we know that $\Delta\,[k f(x)] = k f(x + h) - k f(x) = k\,\Delta f(x)$, hence the result is proved.

Theorem 8.4 Prove that $E^n f(x) = f(x + nh)$.

Proof We know that $E f(x) = f(x + h)$. Since

$$\begin{aligned} E^n f(x) &= E\,E \ldots n \text{ times } f(x) \\ &= E^{n-1} [E f(x)] = E^{n-1} f(x + h) \\ &= E^{n-2} [E f(x)] = E^{n-2} f(x + 2h) \\ &\;\;\vdots \\ &= E f[x + (n - 1) h] = f(x + nh). \end{aligned}$$

Hence, $E^n f(x) = f(x + nh)$.

Similarly $\quad E^{-n} f(x) = f(x - nh)$.

8.5.3 Use of E and D Operators

Suppose $a_{n+k} + c_1 a_{n+k-1} + c_2 a_{n+k-2} + \ldots + c_k a_n = f(n)$ ...(*i*)

be a recurrence relation of order k. The Eqn. (*i*) can also be written as

$$E^k a_n + c_1 E^{k-1} a_n + c_2 E^{k-2} a_n + \ldots + c_k a_n = f(n). \quad ...(ii)$$

The Eqn. (*ii*) is also rewritten as

$$\{E^k + c_1 E^{k-1} + c_2 E^{k-2} + \ldots + c_k\} a_n = f(n) \quad ...(iii)$$

or simply $f(E)\, a_n = f(n)$ or $a_n = \dfrac{f(n)}{f(E)}$ ...(*iv*)

where $f(E) \equiv E^k + c_1 E^{k-1} + c_2 E^{k-2} + \ldots + c_k$.

The particular solution of the given recurrence relation will now depend on the different forms of $f(n)$. Few forms of $f(n)$ are discussed below:

Case 1 When $f(n) = A$, where A is constant.

We know that the operator E on any constant is the constant itself, i.e. E (A) = A. This property implies that

$$\begin{aligned} f(E)\, A &= \{E^k + c_1 E^{k-1} + \ldots + c_k\} A \\ &= E^k A + c_1 E^{k-1} A + \ldots + c_k A \\ &= A + c_1 A + \ldots + c_k A = f(1)\, A. \end{aligned}$$

This implies that E = 1.

Hence, the particular solution of given recurrence relation will be obtained by substituting E = 1. That is

$$a^*_{np} = \frac{f(n)}{f(E)} = \frac{A}{f(1)}, \text{ provided } f(1) \neq 0.$$

Here, it may be noted that $f(E)$ is an operator while $f(1)$ is a number.

Case 2 When $f(n) = A\beta^n$, where A and β are constants.

We have
$$\begin{aligned} f(E)\,(A\beta^n) &= \{E^k + c_1 E^{k-1} + \ldots + c_k\}\,(A\beta^n) \\ &= A\,\{\beta^{n+k} + c_1 \beta^{n+k-1} + \ldots + c_k \beta^n\} \\ &= A\,\{\beta^k + c_1 \beta^{k-1} + \ldots + c_k\}\, \beta^n \\ &= A f(\beta) \cdot \beta^n. \end{aligned}$$

Thus to obtain particular solution we may put $E = \beta$.

Hence, the particular solution of the given recurrence relation will be

$$a^*_{np} = \frac{f(n)}{f(E)} = \frac{A\beta^n}{f(\beta)}, \text{ provided } f(\beta) \neq 0.$$

Special Cases

(*a*) When A = 1, we get $a^*_{np} = \dfrac{\beta^n}{f(\beta)}, f(\beta) \neq 0.$

(*b*) If $f(\beta) = 0$, then

(*i*) for the relation $(E - \beta)\, a_n = A\beta^n$, the particular solution is,

$$a^*_{np} = \frac{A\beta^n}{E - \beta} = An\ \beta^{n-1}$$

(*ii*) for the relation $(E - \beta)^2\, a_n = A\beta^n$, the particular solution is

$$a^*_{np} = \frac{A\beta^n}{(E - \beta)^2} = A\,\frac{n\,(n-1)}{2!}\,\beta^{n-2}.$$

(*iii*) for the relation $(E - \beta)^3\, a_n = A\beta^n$, the particular solution is

$$a^*_{np} = \frac{A\beta^n}{(E - \beta)^3} = A \cdot \frac{n\,(n-1)\,(n-2)}{3!} \cdot \beta^{n-3}.$$

Case 3 When $f(n)$ is a polynomial of degree m in n.

Since $E \equiv 1 + \Delta$, therefore we can write

$$f(E) = f(1 + \Delta) \quad \text{or} \quad \frac{1}{f(E)} = \frac{1}{f(1+\Delta)},$$

Expanding $f(1 + \Delta)$ in ascending power of Δ up to Δ^m, we get

$$\text{or} \qquad \frac{1}{f(E)} = \frac{1}{f(1+\Delta)} = b_0 + b_1\,\Delta + \ldots + b_m\,\Delta^m + \ldots$$

$$\text{or} \qquad \frac{1}{f(E)}\,f(n) = (b_0 + b_1\,\Delta + \ldots + b_m\,\Delta^m + \ldots)\,f(n)$$

$$= b_0\,f(n) + b_1\,\Delta\,f(n) + \ldots + b_m\,\Delta^m\,f(n)$$

other terms will be zero as $f(n)$ is a polynomial of degree m.

Hence, the particular solution of the given recurrence relation becomes

$$a^*_{np} = b_0\,f(n) + b_1\,\Delta\,f(n) + \ldots + b_m\,\Delta^m\,f(n).$$

Case 4 When $f(n)$ is a polynomial of degree m in n and is of the form $\beta^n\,f(n)$.

We have
$$E^k\,[\,\beta^n f(n)\,] = \beta^{n+k}\,f(n + k)$$
$$= \beta^n\,\beta^k\,E^k f(n) = \beta^n\,(\beta E)^k f(n).$$

Similarly, we have

$$\frac{1}{f(E)}\,\beta^n f(n) = \beta^n\,\frac{1}{f(\beta E)}\,f(n) = \beta^n\,[\,f(\beta + \beta\Delta)\,]^{-1} f(n).$$

Hence, the particular solution of the given recurrence relation becomes

$$a^*_{np} = \beta^n\,[\,f(\beta + \beta\Delta)\,]^{-1} f(n).$$

Example 23 Find the particular solution of the recurrence relation

$$a_{n+2} - 5a_{n+1} + 6a_n = 36.$$

Solution The given recurrence relation can be written as

$$(E^2 - 5E + 6)\,a_n = 36.$$

Therefore,
$$f(E) = (\,E^2 - 5E + 6\,) = (E - 2)\,(E - 3).$$

hus, the particular solution is given by

$$a^*_{np} = \frac{A}{f(E)} = \frac{36}{(E-2)(E-3)}.$$

or E = 1, the particular solution becomes, $a^*_{np} = 36/2 = 18$.

xample 24 Find the particular solution of the recurrence relation

$$a_{n+2} - 5a_{n+1} + 6a_n = 5^n.$$

olution The given recurrence relation can be written as

$$(E^2 - 5E + 6)\, a_n = 5^n.$$

ince right hand side of the relation is of the form $A\beta^n$, whese A = 1, β = 5, therefore we have

$$f(E) = E^2 - 5E + 6$$

r $$f(5) = (5)^2 - 5\,(5) + 6 = 6\ (\neq 0).$$

Thus, the particular solution is given by

$$a^*_{np} = \frac{A\beta^n}{f(E)} = \frac{5^n}{f(5)} = \frac{5^n}{6}.$$

xample 25 Find the particular solution of the recurrence relation

$$a_{n+2} - 5a_{n+1} + 6a_n = 2^n.$$

olution The given recurrence relation can be written as

$$(E^2 - 5E + 6)\, a_n = 2^n.$$

Therefore, $f(E) = E^2 - 5E + 6 = (E-2)(E-3)$ and then $f(2) = 0$.

Since $f(E) = 0$ for the term $(E - 2)$, therefore the particular solution is given by

$$a^*_{np} = \frac{2^n}{f(E)} = \frac{n\,2^{n-1}}{(E-3)}$$

$$= \frac{n\,2^{n-1}}{(2-3)} = -\,n \cdot 2^{n-1}.$$

Example 26 Find the particular solution of the recurrence relation

$$a_{n+2} - 4a_{n+1} + 4a_n = 2^n.$$

Solution The given recurrence relation can be written as

$$(E^2 - 4E + 4)\, a_n = 2^n.$$

Therefore, $f(E) = E^2 - 4E + 6 = (E-2)^2$, and then $f(2) = 0$.

Since $f(E) = 0$ for the term $(E-2)^2$, therefore the particular solution is given by

$$a^*_{np} = \frac{2^n}{f(E)} = 2^n \cdot \frac{1}{f(2E)}$$

$$= 2^n \cdot \frac{n^{(0)}}{(2E)^2 - 4\,(2E) + 4} \qquad = 2^n \cdot \frac{n^{(0)}}{4E^2 - 8E + 4}, \quad n^{(0)} = 1$$

$$= 2^n \cdot \frac{n^{(0)}}{4(1+\Delta)^2 + 8(1+\Delta) + 4} = 2^n \cdot \frac{n^{(0)}}{4\Delta^2}$$

$$= 2^n \cdot \frac{\Delta^{-2} n^{(0)}}{4} = \frac{2^n}{4} \Delta^{-1} (n^{(1)}) = \frac{2^n}{4} \cdot \frac{n^{(2)}}{2}$$

$$= n(n-1)\, 2^{n-3}.$$

Example 27 Find the particular solution of the recurrence relation

$$a_{n+2} + a_{n+1} + a_n = n \cdot 2^n.$$

Solution The given recurrence relation can be written as

$$(E^2 + E + 1)\, a_n = n \cdot 2^n.$$

Therefore $f(E) = E^2 + E + 1 = (E + 1)^2$.

The particular solution is given by

$$a^*_{np} = \frac{1}{E^2 + E + 1} n \cdot 2^n = 2^n [(2E)^2 + (2E) + 1]^{-1} \cdot n$$

$$= 2^n [4E^2 + 2E + 1]^{-1} \cdot n = 2^n [4(1+\Delta)^2 + 2(1+\Delta) + 1]^{-1} \cdot n$$

$$= 2^n [7 + 10\Delta + 4\Delta^2]^{-1} \cdot n = \frac{2^n}{7} \left[1 + \frac{10\Delta + 4\Delta^2}{7}\right]^{-1} \cdot n$$

$$= \frac{2^n}{7} \left[1 - \frac{10\Delta + 4\Delta^2}{7}\right] \cdot n = \frac{2^n}{7} \left(n - \frac{10}{7}\right).$$

Example 28 Find the particular solution of the recurrence relation

$$a_{n+3} + a_{n+2} - 8a_{n-1} - 12a_n = 2n^2.$$

Solution The given recurrence relation can be written as

$$(E^3 + E^2 - 8E - 12)\, a_n = 2n^2.$$

Therefore $f(E) = E^3 + E^2 - 8E - 12 = (E - 3)(E^2 + 4E + 4) = (E - 3)(E + 2)^2$.

Thus, the particular solution is given by

$$a^*_{np} = \frac{2n^2}{(E-3)(E+2)^2} = \frac{2n^2}{(\Delta - 2)(\Delta + 3)^2};\quad E \equiv 1 + \Delta$$

$$= \frac{2n^2}{(\Delta^3 + 4\Delta^2 - 3\Delta - 18)} = 2n^2 (\Delta^3 + 4\Delta^2 - 3\Delta - 18)^{-1}$$

$$= -\frac{1}{18} \left[1 + \frac{3\Delta - 4\Delta^2 - \Delta^3}{18}\right]^{-1} \cdot 2n^2.$$

$$= -\frac{1}{18} \left[1 - \frac{3\Delta - 4\Delta^2 - \Delta^3}{18} + \frac{(3\Delta - 4\Delta^2 - \Delta^3)^2}{(18)^2} + \ldots\right] \cdot 2n^2$$

[since in factorial notation
$2n^2 \equiv An^{(2)} + Bn^{(1)} + C \equiv A\, n(n-1) + Bn + C$,
therefore A = B = 2, C = 0]

$$= -\frac{1}{18}\left[1-\frac{\Delta}{6}+\frac{\Delta^2}{4}\right](2n^{(2)}+2n^{(1)})$$

$$= -\frac{1}{18}\left[1-\frac{\Delta}{6}+\frac{\Delta^2}{4}\right][2n\,(n-1)+2n]$$

$$= -\frac{1}{18}\left[2n\,(n-1)+2n-\frac{4n+2}{6}+\frac{4}{4}\right]$$

$$= -\frac{n^2}{18}+\frac{n}{27}-\frac{2}{54}.$$

Example 29 Find the particular solution of the recurrence relation: $a_{n+1} - 2a_n = n + 1$.

Solution The given recurrence relation can be written as

$$(\mathrm{E}-2)\,a_n = n+1, \text{ i.e. } f(\mathrm{E}) = \mathrm{E}-2.$$

Thus, the particular solution is given by

$$a^*_{np} = \frac{1}{\mathrm{E}-2}(n+1) = \frac{1}{\Delta-1}(n+1);\ \mathrm{E} \equiv 1+\Delta$$

$$= \frac{-1}{1-\Delta}[n^{(1)}+n^{(0)}];\ n^{(1)} = n \text{ and } n^{(0)} = 1$$

$$= -(1-\Delta)^{-1}[n^{(1)}+n^{(0)}] = -(1+\Delta+\Delta^2+\ldots)[n^{(1)}+n^{(0)}]$$

$$= -(n^{(1)}+n^{(0)}+1), \text{ neglecting higher powers of } \Delta$$

$$= -(k+1+1) = -(k+2).$$

8.5.3 Solution of Non-homogeneous Recurrence Relations

As mentioned earlier that the total solution of a non-homogeneous linear recurrence relation with constant coefficients is the sum of its homogeneous solution (a^*_{nh}) and particular solution (a^*_{np}). If initial conditions are not given, then there is no need to determine values of constants in the homogeneous solution, otherwise obtain a linear equation corresponding to each initial condition. On solving these simultaneous linear equations, we get the values of constants. The total solution or general solution of the recurrence relation is obtained as: $a_n = a^*_{nh} + a^*_{np}$.

Example 30 Solve the recurrence relation

$$a_{n+3} + a_{n+2} - 8a_{n+1} - 12a_n = 2n^2 + 7.$$

Solution The characteristic equation of the recurrence relation is

$$\alpha^3 + \alpha^2 - 8\alpha - 12 = 0 \quad \text{or} \quad (\alpha-3)(\alpha+2)^2 = 0, \text{ i.e. } \alpha = 3, -2, -2.$$

Since two roots of the characteristic equation are repeated, therefore homogeneous solution is

$$a^*_{nh} = \mathrm{A}_1\, 3^n + (\mathrm{A}_2 + \mathrm{A}_3 n)(-2)^n$$

and the particular solution of the recurrence relation is

$$a^*_{np} = \frac{1}{(\mathrm{E}-3)(\mathrm{E}+2)^2}(2n^2+7) = [(\mathrm{E}-3)(\mathrm{E}+2)^2]^{-1}(2n^2+7)$$

$$= [(\Delta-2)(\Delta+3)^2]^{-1}(2n^2+7);\ \mathrm{E} = 1+\Delta$$

$$= [\Delta^3 + 4\Delta^2 - 3\Delta - 18]^{-1}(2n^2 + 7)$$

$$= -\frac{1}{18}\left[1 + \frac{3\Delta - 4\Delta^2 - \Delta^3}{18}\right]^{-1}(2n^2 + 7).$$

$$= -\frac{1}{18}\left[1 - \frac{3\Delta - 4\Delta^2 - \Delta^3}{18} + \frac{(3\Delta - 4\Delta^2 - \Delta^3)^2}{18} + \ldots\right](2n^2 + 7)$$

$$= -\frac{1}{18}\left[1 - \frac{\Delta}{6} + \frac{\Delta^2}{4}\right](2n^{(2)} + 2n^{(1)} + 7)$$

$$= -\frac{1}{18}\left[2n^{(2)} + 2n^{(1)} + 7 - \frac{4n+2}{6} + \frac{4}{4}\right]$$

$$= -\frac{n^2}{9} + \frac{n}{27} - \frac{23}{54}.$$

Hence, the total solution of the given recurrence relation is

$$a_n = [A_1 3^n + (A_2 + A_3 n)(-2)^n] - \frac{n^2}{9} + \frac{n}{7} - \frac{23}{27}.$$

Example 31 Solve the recurrence relation: $a_{n+2} - 2a_{n+1} + a_n = n^2 \cdot 2^n$

Solution Solving the characteristic equation of the recurrence relation, we get

$$\alpha^2 - 2\alpha + 1 = 0 \quad \text{or} \quad (\alpha - 1)^2, \text{ i.e. } \alpha = 1, 1.$$

Since roots are real and equal, homogeneous solution of the recurrence relation is: $a^*_{np} = A_1 + A_2 n$.

In order to find particular solution, let the trial solution corresponding to the term $n^2 \cdot 2^n$ on the right-hand side of the relation be, $2^n (A_1 + A_2 n + A_3 n^2)$.

Putting this solution in the left-hand side of the given relation, we get

$$a_{n+2} - 2a_{n+1} + a_n = 2^{n+2}[A_1 + A_2(n+2) + A_3(n+2)^2]$$
$$-2 \cdot 2^{n+1}[A_1 + A_2(n+1) + A_3(n+1)^2] + 2^n[A_1 + A_2 n + A_3 n^2]$$

or

$$n^2 \cdot 2^n = n^2 \cdot 2^n A_3 + n \cdot 2^n (A_2 + 8A_3) + 2^n (A_1 + 4A_2 + 12A_3).$$

Comparing coefficients of similar terms on both sides of the given relation, we get

$$A_3 = 1, \; 8A_3 + A_2 = 0, \; 12A_3 + 4A_2 + A_1 = 0$$

On solving these equations, we get $A_1 = 20$, $A_2 = -8$ and $A_3 = 1$. Thus, the particular solution is,

$$a^*_{np} = (n^2 - 8n + 20)2^n.$$

Hence, the total solution of the given recurrence relation is

$$a_n = [A_1 + A_2 n] + (n^2 - 8n + 20)\, 2^n.$$

Example 32 Solve the recurrence relation: $a_{n+2} - 5a_{n+1} + 6a_n = n^2$.

Solution Solving the characteristic equation of the recurrence relation, we get

$$\alpha^2 - 5\alpha + 6 = 0 \quad \text{or} \quad (\alpha - 2)(\alpha - 3) = 0, \text{ i.e. } \alpha = 2, 3.$$

Since roots are real and different, homogeneous solution of the recurrence relation is: $a^*_{nh} = A_1 2^n + A_2 3^n$.

In order to find particular solution, let the trial solution corresponding to the term n^2 on the right-hand side of the relation be, $A_1n^2 + A_2n + A_3$.

Putting this solution in the left-hand side of the given relation, we get

$$a_{n+2} - 5a_{n+1} + 6a_n = A_1(n+2)^2 + A_1(n+2) + A_3 - 5A_1(n+1)^2 - 5A_2(n+1) - 5A_3 + 6A_1n^2 + 6A_2n + 6A_3$$

$$n^2 = 2A_1n^2 + (-6A_1 + 2A_2)n + (-A_1 - 3A_2 + 2A_3).$$

Comparing coefficients of similar terms on both side of the given relation, we get

$$2A_1 = 1, -6A_1 + 2A_2 = 0 \text{ and } -A_1 - 3A_2 + 2A_3 = 0.$$

On solving these equations, we get $A_1 = 1/2$, $A_2 = 3/2$ and $A_3 = 5/2$. Thus, the particular solution is

$$a^*_{np} = \frac{1}{2}n^2 + \frac{3}{2}n + \frac{5}{2}.$$

Hence, the total solution of the given recurrence relation is

$$a_n = (A_12^n + A_23^n) + \left(\frac{1}{2}n^2 + \frac{3}{2}n + \frac{5}{2}\right).$$

Example 33 Solve the recurrence relation, $a_{n+2} - 2a_{n+1} + a_n = 2^n$ satisfying the initial condition $a_0 = 2$ and $a_1 = 1$.

Solution Solving the characteristic equation: $\alpha^2 - 2\alpha + 1 = 0$ or $(\alpha - 1)(\alpha - 1) = 0$ of the recurrence relation, we get $\alpha = 1, 1$. Since roots are real and repeated, therefore the homogeneous solution is, $a^*_{nh} = A_1 + nA_2$.

In order to find the particular solution, let the trial solution corresponding to the term 2^n on the right hand side of the relation be $A\,2^n$. Since the term $A\,2^n$ also appears in the homogeneous solution, therefore multiplying it by n, and hence, the trial solution now becomes: $An\,2^n$.

Putting the trial solution on the left hand side of the recurrence relation, we get

$$A(n+1)^2\, 2^{n+2} - 2A(n+1)\, 2^{n+1} + An\, 2^n = (4n^2 + 5n)\, A\, 2^n.$$

Comparing coefficients of similar terms with the right hand side of the recurrence relation, we get

$$A(4n^2 + 5n) = 1 \quad \text{or} \quad A = 1/(4n^2 + 5n).$$

Thus particular solution of the recurrence relation is

$$a^*_{np} = \frac{n}{4n^2 + 5n} \cdot 2^n$$

and, hence the total solution is: $a_n = A_1 + A_2n + \dfrac{n}{4n^2 + 5n} \cdot 2^n$.

Putting $n = 0$ and 1 in the total solution, we get

$$a_0 = 2 = A_1; \text{ for } n = 0$$

$$a_1 = 1 = A_1 + A_2 + \frac{1}{4+5} \cdot 2^1; \quad \text{for } n = 1$$

$$= A_1 + A_2 + \frac{2}{9}.$$

On solving these equations, we get $A_1 = 2$, $A_2 = -11$.

Hence, total solution satisfying initial conditions is: $a_n = (2 - 11n) + \frac{n}{4n^2 + 5n} \cdot 2^n$.

8.6 METHODS OF GENERATING FUNCTIONS

Methods of generating functions which are used to solve recurrence relations, among other problems, are based on the concept that these methods transform a recurrence relation for a sequence into an algebraic or differential equation of that sequence. If this equation is solvable, then original sequence can be rewritten by either differentiation or by other means to get a solution to the original problem.

Definition Let $a_0, a_1, a_2, \ldots, a_r, \ldots$ be a sequence of real numbers. Then the function defined as:

$$G(a, z) = a_0 + a_1 z + a_2 z^2 + \ldots + a_r z^r + \ldots = \sum_{n=0}^{\infty} a_n z^n$$

is called the generating function for the given sequence, where z is a variable.

Illustrations

1. The sequence of repeated constant numbers such as, 1, 1, 1, ... has the generating function

$$G(1, z) = 1 + z + z^2 + \ldots = \sum_{n=0}^{\infty} z^n = (1 - z)^{-1}$$

by using binomial theorem.

2. The sequence, a^n ($a \neq 0$, and constant) has the generating function

$$G(a, z) = 1 + az + a^2 z^2 + \ldots$$

$$= \sum_{n=0}^{\infty} a^k z^k = \frac{1}{1 - az}.$$

3. The sequence of binomial coefficients ${}^nC_0, {}^nC_1, {}^nC_2, \ldots, {}^nC_n$; $n \geq 1$ has the generating function

$$G(C_n, z) = {}^nC_0 + {}^nC_1 z + {}^nC_2 z^2 + \ldots + {}^nC_n z^n$$

$$= \sum_{k=0}^{\infty} {}^nC_k z^k = (1 + z)^n$$

by using binomial theorem.

4. The sequence, $a_n = 9.5^n$, $n \geq 0$, has the generating function, $G(a, z) = 9/(1 - 5z)$ which is an infinite geometric series with first term, $a = 9$ and common ratio, $r = 5z$.

5. The sequnce, $a_n = 3(-1)^n$, has the generating function, $G(a, z) = 3/(1 + z)$.

8.6.1 Solution of Recurrence Relations using Generating Functions

The summary of various types of sequences and the corresponding generating functions are given in Table 8.2.

Table 8.2 Sequences and the Corresponding Generating Functions

General Sequence	*Corresponding Generating Function*
• $a_k = 1$	$G(a, z) = (1 - z)^{-1}$
• $a_k = k + 1$	$G(a, z) = (1 - z)^{-2}$
• $a_k = k$	$G(a, z) = \dfrac{z}{(1-z)^2}$
• $a_k = ba^k$; a, b = constants	$G(a, z) = \dfrac{b}{1-az}$
• $a_k = k(k + 1)$	$G(a, z) = \dfrac{2z}{(1-z)^3}$
• $a_k = A^k$, A is a constant	$G(a, z) = (1 - Az)^{-1}$
• $a_k = {}^nC_k\, B^{n-k} A^k$; $(k = 0, 1, 2, \ldots, n)$, $k > n$	$G(a, z) = (B + Az)^n$
• $a_k = bka^k$	$G(a, z) = \dfrac{abz}{(1-az)^2}$
⋮	
• $a_k = \dfrac{1}{k!}$	$G(a, z) = e^z$ $(e \cong 2.718)$
• $a_k = a_{k+n}$	$G(a, z) = \dfrac{G(z) - a_0 - a_1 z - \ldots - a_{n-1} z^{n-1}}{z^n}$
• $a_k = \lambda_1 a_k^{(1)} + \lambda_2 a_k^{(2)}$	$G(a, z) = \lambda_1 G_1(z) + \lambda_2 G_2(z)$.

Theorem 8.5 If the generating function for the sequence whose general term, a_n is known then the generating function for the sequences, a_{n+1} and a_{n+2} can also be obtained.

Proof Let G (z) be the generating function for the sequence whose general term is a_n. Then we may write

$$G(z) = a_0 + a_1 z + a_2 z^2 + \ldots + a_r z^r + \ldots \quad \ldots(i)$$

or

$$\frac{G(z) - a_0}{z} = a_1 + a_2 z + \ldots + a_r z^{r-1} + a_{r+1} z^r + \ldots . \quad \ldots(ii)$$

This implies that $\dfrac{G(z) - a_0}{z}$ is the generating function for the sequence whose general term is a_{n+1}.

Similarly, using Eq. (*ii*) we may write

$$\frac{G(z) - a_0 - a_1 z}{z^2} = a_2 + a_3 z + a_4 z^2 + \ldots + a_r z^{r-2} + a_{r+1} z^{r-1} + \ldots a_{r+2} z^r + \ldots$$

which is the generating function for the sequence whose general term is a_{n+2}.

8.6.2 Addition and Multiplication of Generating Functions

If $G(z) = \sum_{k=0}^{\infty} a_k z^k$ and $H(z) = \sum_{k=0}^{\infty} b_k z^k$ are two generating functions and c is a real number, then following results hold.

(*i*) Sum of two functions, $(G + H)(z) = \sum_{k=0}^{\infty}(a_k + b_k) z^k$

(*ii*) Scalar product with the function, $(c\,G)(z) = c \sum_{k=0}^{\infty} a_k z^k = \sum_{k=0}^{\infty}(ca_k) z^k$

(*iii*) Product of two functions, $(GH)\,z = \sum_{k=0}^{\infty} a_k z^k \cdot \sum_{k=0}^{\infty} b_k z^k$.

$$= \sum_{k=0}^{\infty} d_k z^k, \text{ where } d_k = \sum_{i=0}^{\infty} a_i b_{k-i}.$$

8.6.3 Shifting of Generating Functions

(*a*) If $G(z) = \sum_{k=0}^{\infty} a_k z^k$ generates the sequence $(a_0, a_1, a_2, \ldots)$, then $z\,G(z)$ generates the sequence $(0, a_0, a_1, a_2, \ldots)$, $z^2\,G(z)$ generates the sequence $(0, 0, a_0, a_1, a_2 \ldots)$ and in general $z^n\,G(z)$ generates the sequence $(0, 0, \ldots 0, a_0, a_1, a_2, \ldots)$ which contains n zeros before a_0.

For example, $\frac{1}{1-x} = \sum_{k=0}^{\infty} x^k$ generates the sequence $(1, 1, 1, \ldots)$, i.e. $\{a_n\}$ where $a_n = 1$ for each $n \geq 0$

Thus,

$$\frac{x}{1-x} = \sum_{k=0}^{\infty} x^{k+1} = \sum_{k=1}^{\infty} x^k \text{ generates the sequence } (0, 1, 1, 1, \ldots)$$

$$\frac{x^2}{1-x} = \sum_{k=0}^{\infty} x^{k+2} = \sum_{k=2}^{\infty} x^k \text{ generates the sequence } (0, 0, 1, 1, \ldots)$$

(*b*) If $G(z) = \sum_{n=0}^{\infty} a_n z^n$ generates the sequence $(a_0, a_1, a_2, \ldots)$, then

$$G(z) - a_0 = \sum_{n=1}^{\infty} a_n z^n \text{ generates the sequence } (0, a_1, a_2 \ldots)$$

$$G(a) - a_0 - a_1 z = \sum_{n=2}^{\infty} a_n z^n \text{ generates the sequence } (0, 0, a_2, a_3, \ldots) \text{ and so on.}$$

(*c*) $\frac{1}{z}[G(z) - a_0] = \sum_{n=1}^{\infty} a_n z^{n-1}$ generates the sequence $(a_1, a_2, a_3, \ldots)$. That is, division by the variable z shifts the sequence to the left. Similarly,

$$\frac{1}{z^2}[G(z) - a_0 - a_1 z] = \sum_{n=2}^{\infty} a_n z^{n-2} \text{ generates the sequence } (a_2, a_3, \ldots)$$

$$\vdots$$

$$\frac{1}{z^k}[G(z) - a_0 - a_1 z - a_2 z^2 - \ldots - a_k z^k] = \sum_{n=k}^{\infty} a_n z^{n-k} \text{ generates the sequence } (a_k, a_{k+1}, \ldots).$$

Example 34 Solve the recurrence relation

$$a_{n+2} - 3a_{n+1} + 2a_n = 0$$

by the generating function method with initial conditions, $a_0 = 2$ and $a_1 = 3$.

Solution Let $G(a, z) = \sum_{n=0}^{\infty} a_n z^n$ be the generating function for the given recurrence relation.

Multiplying each term of the recurrence relation with z^n and summing from $n = 0$ to ∞, we have

$$\sum_{n=0}^{\infty} a_{n+2} z^n - 3 \sum_{n=0}^{\infty} a_{n+1} z^n + 2 \sum_{n=0}^{\infty} a_n z^n = 0$$

or $$(a_2 + a_3 z + a_4 z^2 + \ldots) - 3(a_1 + a_2 z + a_3 z^2 + \ldots) + 2(a_0 + a_1 z + a_2 z^2 + \ldots) = 0$$

or $$\frac{G(a, z) - a_0 - a_1 z}{z^2} - 3\left[\frac{G(a, z) - a_0}{z}\right] + 2\,G(a, z) = 0 \qquad \ldots(i)$$

[Since $G(a, z) = a_0 + a_1 z + a_2 z^2 + \ldots$]

Putting $a_0 = 2$ and $a_1 = 3$ in Eq. (*i*) and on solving for $G(a, z)$, we get

$$G(a, z) = \frac{2 - 3z}{1 - 3z + 2z^2} = \frac{2 - 3z}{(1 - z)(1 - 2z)}. \qquad \ldots(ii)$$

Using partial fractions method to get value of $G(a, z)$ in terms of z as follows:

$$G(a, z) = \frac{2 - 3z}{(1 - z)(1 - 2z)} = \frac{A}{1 - z} + \frac{B}{1 - 2z} = \frac{A(1 - 2z) + B(1 - z)}{(1 - z)(1 - 2z)}$$

or $$2 - 3z = A(1 - 2z) + B(1 - z). \qquad \ldots(iii)$$

For $z = 1$ in Eq. (*iii*), we get $A = 1$ and for $z = 1/2$, in Eq. (*iii*), we get $B = 1$. Thus

$$G(a, z) = \frac{1}{1 - z} + \frac{1}{1 - 2z}$$

$$= \sum_{n=0}^{\infty} z^n + \sum_{n=0}^{\infty} (2z)^n = \sum_{n=0}^{\infty} (1 + 2^n) z^n.$$

This implies that, $a_n = 1 + 2^n$ is the required solution of the recurrence relation.

Example 35 Solve the recurrence relation

$$a_{n+2} - 5a_{n+1} + 6a_n = 2$$

by the generating function method with initial conditions, $a_0 = 1$ and $a_1 = 2$.

Solution Let $G(a, z) = \sum_{n=0}^{\infty} a_n z^n$ be a generating function.

Multiplying each term of recurrence relation with z^n and summing from $n = 0$ to ∞. Then in the same manner as in previous example, we have

$$\frac{G(a, z) - a_0 - a_1 z}{z^2} - 5\left[\frac{G(a, z) - a_0}{z}\right] + 6\,G(a, z) = \frac{2}{1 - z}. \qquad \ldots(i)$$

Putting $a_0 = 1$ and $a_1 = 2$ in Eq. (*i*), we get

$$\frac{G(a,z)-1-2z}{z^2} - 5\left[\frac{G(a,z)-1}{z}\right] + 6G(a, z) = \frac{2}{1-z}$$

or $$G(a, z) = \frac{5z^2-4z+1}{(1-z)(1-2z)(1-3z)}. \quad ...(ii)$$

Using partial fractions method to get value of G (*a*, *z*) in terms of *z* as follows:

$$G(a, z) = \frac{5z^2-4z+1}{(1-z)(1-2z)(1-3z)} = \frac{A}{1-z} + \frac{B}{1-2z} + \frac{C}{1-3z}$$

$$= \frac{A(1-2z)(1-3z)+B(1-z)(1-3z)+C(1-z)(1-2z)}{(1-z)(1-2z)(1-3z)}$$

or $$5z^2 - 4z + 1 = A(1 - 2z)(1 - 3z) + B(1 - 3z)(1 - z) + C(1 - z)(1 - 2z). \quad ...(iii)$$

In Eq. (*iii*), for $z = 1$, we get A = 1; for $z = 1/2$, B = – 1 and for $z = 1/3$, C = 1. Thus

$$G(a, z) = \frac{1}{1-z} - \frac{1}{1-2z} + \frac{1}{1-3z} \quad ...(iv)$$

$$= \sum_{n=0}^{\infty} z^n - \sum_{n=0}^{\infty}(2z)^n + \sum_{n=0}^{\infty}(3z)^n = \sum_{n=0}^{\infty}(1-2^n+3^n)z^n.$$

Hence, the solution of the given recurrence relation is, $a_n = 1 - 2^n + 3^n$.

Example 36 Solve the recurrence relation

$$a_{n+2} - 2a_{n+1} + a_n = 2^n$$

by the generating function method with initial conditions, $a_0 = 2$ and $a_1 = 1$.

Solution Let $G(a, z) = \sum_{n=0}^{\infty} a_n z^n$ be a generating function.

Multiplying the recurrence relation with z^n and summing from $n = 0$ to ∞. Then in the same manner as discussed before we have

$$\frac{G(a,z)-a_0-a_1z}{z^2} - 2\left[\frac{G(a,z)-a_0}{z}\right] + G(a, z) = \frac{1}{1-2z}. \quad ...(i)$$

Putting $a_0 = 2$ and $a_1 = 1$ in Eq. (*i*), we get

$$\frac{G(a,z)-2-z}{z^2} - 2\left[\frac{G(a,z)-2}{z}\right] + G(a, z) = \frac{1}{1-2z}$$

or $$G(a, z) = \frac{2}{(1-z)^2} - \frac{3z}{(1-z)^2} + \frac{z^2}{(1-z)^2(1-2z)}. \quad ...(ii)$$

Applying partial fractions method, we have

$$\frac{z^2}{(1-2z)(1-z)^2} = \frac{A}{(1-2z)} + \frac{B}{(1-z)} + \frac{C}{(1-z)^2}$$

$$= \frac{A(1-z)^2 + B(1-2z)(1-z) + C(1-2z)}{(1-2z)(1-z)^2}$$

or $$z^2 = A(1-z)^2 + B(1-2z)(1-z) + C(1-2z). \qquad ...(iii)$$

In Eq. (*iii*), for $z = 1$, $C = -1$; for $z = 1/2$, $A = 1$; and after equating constant terms, we have $A + B + C = 0$, and therefore $B = 0$. Thus

$$\frac{z^2}{(1-2z)(1-z)^2} = \frac{1}{1-2z} - \frac{1}{(1-z)^2}.$$

Substituting this value in Eq. (*ii*), we get

$$G(a, z) = \frac{2}{(1-z)^2} - \frac{3z}{(1-z)^2} + \frac{1}{1-2z} - \frac{1}{(1-z)^2}$$

$$= \frac{1}{(1-z)^2} - \frac{3z}{(1-z)^2} + \frac{1}{1-2z}.$$

Hence, the required solution of the given recurrence relation is: $a_n = (n+1) - 3n + 2^n$.

Example 37 Solve the recurrence relation

$$a_n - 2a_{n-1} - 3a_{n-2} = 0,\ n \geq 2$$

by the generating function method with initial conditions, $a_0 = 3$ and $a_1 = 1$.

Solution Let $G(a, z) = \sum_{n=0}^{\infty} a_n z^n$ be a generating function.

Multiplying each term of the given recurrence relation with z^n and summing from $n = 2$ to ∞, we get

$$\sum_{n=2}^{\infty} a_n z^n - 2\sum_{n=2}^{\infty} a_{n-1} z^n - 3\sum_{n=2}^{\infty} a_{n-2} z^n = 0. \qquad ...(i)$$

Rewriting terms of the Eq. (*i*) as

$$\sum_{n=2}^{\infty} a_n z^n = \left[\left\{a_0 + a_1 z + \sum_{n=2}^{\infty} a_n z^n\right\} - a_0 - a_1 z\right]$$

$$= \sum_{n=0}^{\infty} a_n z^n - 3 - z;\ a_0 = 3,\ a_1 = 1$$

$$= G(a, z) - 3 - z$$

$$\sum_{n=2}^{\infty} a_{n-1} z^n = z\left[\sum_{n=2}^{\infty} a_{n-1} z^{n-1}\right] = z\left[\sum_{n=1}^{\infty} a_n z^n\right] = z\left[\left\{a_0 + \sum_{n=1}^{\infty} a_n z^n\right\} - a_0\right]$$

$$= z\left[\sum_{n=0}^{\infty} a_n z^n - 3\right] = z\,G(a, z) - 3z;\ a_0 = 3$$

$$\sum_{n=2}^{\infty} a_{n-2} z^n = z^2\left[\sum_{n=2}^{\infty} a_{n-2} z^{n-2}\right] = z^2\,G(a, z).$$

Substituting these values in Eq. (*i*), it becomes

$$[G(a, z) - 3 - z] - 2[z\,G(a, z) - 3z] - 3z^2\,G(a, z) = 0$$

or $$G(a, z)[1 - 2z - 3z^2] = 3 - 5z$$

or $$G(a, z) = \frac{3-5z}{1-2z-3z^2} = \frac{3-5z}{(1-3z)(1+z)}. \qquad ...(ii)$$

Using partial fractions method, we have

$$\frac{3-5z}{(1-3z)(1+z)} = \frac{A}{1-3z} + \frac{B}{1+z} = \frac{A(1-z)+B(1-3z)}{(1-3z)(1+z)} \qquad ...(iii)$$

or $$3 - 5z = A(1 + z) + B(1 - 3z). \qquad ...(iv)$$

In Eq. (*iv*), for $z = -1$, $B = 2$ and for $z = 1/3$, $A = 1$. Thus Eq. (*iii*) becomes

$$G(a, z) = \frac{1}{1-3z} + \frac{2}{1+z}.$$

Hence, the solution of the recurrence relation is, $a_n = 3^n + 2(-1)^n$.

Example 38 Find the generating function for each of the following sequence

(*a*) 0, 1, – 2, 4, – 8, . . . (*b*) 1, 0, 1, 0, . . .

(*c*) 2, – 2, 2, – 2, . . . (*d*) 0, 0, 2, 2, 2, . . .

Solution (*a*) The required generating function is of the form

$$G(z) = 0 + z - 2z^2 + 4z^3 - 8z^4 + \ldots$$
$$= z - 2z^2 + 4z^3 - 8z^4 = z(1 - 2z + (2z)^2 - (2z)^3 + \ldots)$$
$$= \frac{z}{1+2z} \text{ for } |2z| < 1.$$

(*b*) The required generating function is of the form

$$G(z) = 1 - 0z + 1z^2 - 0z^3 + 1z^4 - \ldots$$
$$= 1 + z^2 + z^4 + \ldots = \frac{1}{1-z^2}, \; |z^2| < 1.$$

(*c*) The required generating function is of the form

$$G(z) = 2 - 2z + 2z^2 - 2z^3 + \ldots$$
$$= 2(1 - z + z^2 - z^3 + \ldots) = \frac{2}{1+z}, \; |-z| < 1.$$

(*d*) The required generating function is of the form

$$G(z) = 0 + 0z + 2z^2 + 2z^3 + 2z^4 + \ldots$$
$$= 2z^2(1 + z + z^2 + \ldots) = \frac{2z^2}{1-z}, \; |z| < 1$$

Example 39 Evaluate the sum: $1^2 + 2^2 + 3^2 + \ldots + r^2$ using generating function.

Solution We know that

$$\frac{1}{1-z} = 1 + z + z^2 + \ldots + z^r + \ldots.$$

Differentiating both sides with respect to z, we get

$$\frac{1}{(1-z)^2} = 1 + 2z + 3z^2 + \ldots + r\,z^{r-1} + \ldots.$$

Multiplying both sides by z, we get

$$\frac{z}{(1-z)^2} = z + 2z^2 + 3z^3 + \ldots + r\,z^r + \ldots.$$

Differentiating both sides with respect to z, we get

$$\frac{1+z}{(1-z)^3} = 1^2 + 2^2 z + 3^2 z^2 + \ldots + r^2 z^{r-1} + \ldots.$$

Multiplying both sides by z, we get

$$\frac{z(1+z)}{(1-z)^3} = 0^2 + 1^2 z + 2^2 z^2 + 3^2 z^3 + \ldots + r^2 z^r + \ldots.$$

This shows that $z\,(1 + z)/(1 - z)^3$ is the generating function of the numeric function $0^2, 1^2, 2^2, \ldots, r^2$.

Now $\dfrac{1}{1-z} \times \dfrac{z(1+z)}{(1-z)^3} = (1 + z + z^2 + \ldots + z^r + \ldots) \times (0^2 + 1^2 z + 2^2 z^2 + 3^2 z^3 + \ldots + r^2 z^r + \ldots)$

$$= 0^2 + (0^2 + 1^2)\,z + (0^2 + 1^2 + 2^2)\,z^2 + \ldots + (0^2 + 1^2 + 2^2 + \ldots + r^2)\,z^r + \ldots.$$

Thus $0^2 + 1^2 + 2^2 + \ldots + r^2$ = coefficient of z^r in the expansion of $z\,(1 + z)/(1 - z)^4$.

But from Binomial theorem we know that the coefficient of z^r in the expansion of $1/(1 - z)^4$ is

$$\frac{(-4)(-4-1)\ldots(-4-r+1)}{r!}(-1)^r = \frac{4\cdot 5\cdot 6\cdot \ldots (r+3)}{r!}$$

$$= \frac{(r+1)(r+2)(r+3)}{1\cdot 2\cdot 3}.$$

So, the coefficient of z^r in the expansion of $z\,(1 + z)/(1 - z)^4$ is

$$\frac{r\,(r+1)(r+2)}{1\cdot 2\cdot 3} + \frac{r\,(r-1)(r+1)}{1\cdot 2\cdot 3} = \frac{r\,(r+1)(2r+1)}{6}.$$

Hence,

$$1^2 + 2^2 + 3^2 + \ldots + r^2 = \frac{r\,(r+1)(2r+1)}{6}.$$

Example 40 Determine the generating function of a numeric function a_r where

$$a_r = \begin{cases} 2^r \text{ if } r \text{ is even} \\ -2^r \text{ if } r \text{ in odd} \end{cases}$$

Solution The required generating function is:

$$A(z) = \sum_{r=0}^{\infty} a_r z^r = \sum_{r=0}^{\infty} (-1)^r \cdot 2^r \cdot z^r$$

$$= 2^0 z^0 - 2^1 z + 2^2 z^2 - 3^3 z^3 + \ldots$$

$$= 1 - 2z + 2^2 z^2 - 2^3 z^3 + \ldots$$

Example 42 Find the discrete numeric function corresponding to the generating function given by

(*a*) $A(z) = 1/(z^2 - 7z + 12)$

(*b*) $A(z) = 1/(1 - 2z)\ (1 + 3z)$

(*c*) $(1 + z)^n + (1 - z)^n$ [*UPTU, MCA, 2004*]

Solution (*a*) $A(z) = 1/(z^2 - 7z + 12) = 1/(z - 4)\ (z - 3)$

$= 1/(z - 4) - 1/(z - 3)$ [by partial fractions]

$= (z - 4)–1 - (z - 3)–1$

$= (-4)^{-1}(1 - z/4)^{-1} - (-3)^{-1}\ (1 - z/3)^{-1}$

$= -1/4[1 + z/4 + z^2/4^2 + z^3/4^3 + \dots] + 1/3[1 + z/3 + (z/3)^2 + (z/3)^3 + \dots]$

$= [-1/4 - 1/4\ z/4 - 1/4\ z^2/4^2 - 1/4\ z^3/4^3 + \dots]$

$+ [1/3 + 1/3\ z/3 + 1/3\ z^2/3^2 + 1/3\ (z^3/3^3) + \dots]$

$= [1/3 - 1/4] + [1/3\ 1/3 - 1/4\ 1/4]z + [1/3\ 1/3^2 - 1/4\ 1/4^2]z^2$

$+ [1/3\ 1/3^3 - 1/4\ 1/4^3]z^3 + \dots$

Therefore the required numeric function is:

$$\frac{1\,z^0}{12}, \frac{7\,z^1}{144}, \frac{37\,z^2}{1728}, \frac{175\,z^3}{20736}, \dots$$

(*b*) $A(z) = 1/(1-2z)\,(1+3z) = 1/5\,[2/(1-2z) + 3/(1+3z)]$

$= 1/5[2(1-2z)^{-1} + 3\,(1+3z)^{-1}]$

$= 1/5[2(1\ 2z + 2^2\ z^3 + 2^3\ z^3 + \dots) + 3\,(1-3z + 3^2\,z^2 - 3^3\,z^3 + \dots)]$

$= 1/5[(2 + 2^2\ z + 2^3\ z^2 + 2^4 z^3 + \dots) + (3 - 3^2\,z + 3^3\,z^2 - 3^4\,z^3 + \dots)]$

$= 1/5[(2+3) + (2^2 - 3^2)z + (2^3 + 3^3)z^2 + (2^4 - 3^4)z^3 + \dots]$

$= 1/5[5 - 5z + 35z^2 - 65z^3 + \dots]$

The required numeric function is: 5/5, – 5/5, 35/5, –65/5, ... or 1, –1, 7, –13, ... The $(n + 1)$th term of this sequence can be written as the coefficients of

$$z^n (= a_n) = 1/5[2^{n+1} + (-1)^n\, 3^n]$$

(*c*) Generating functions is given by $A(z) = a^0 z^0 + a^1 z + a^2 z^2 + \dots$

$$= \sum_{r=0}^{\infty} a_r\, z^r$$

The corresponding sequence is: $a_0, a_1, a_2, a_n, \dots$. Thus, the general term a_r = coefficient of z^r of the sequence = coefficient of z^r in the general function (GF). The given general function (GF) is: $A(z) = (1 + z)^n + (1 - z)^n$

The general term of generating function is given by a_r = Coefficient of z^r in the RHS of generating function.

$= {}^n c_r\,(1)^r + {}^n c_r\,(-1)^r$, using Binomial theorem

$$= \begin{cases} 0, & \text{if } n \text{ is odd} \\ 2\,({}^n c_r), & \text{if } n \text{ is even} \end{cases}$$

Example 43 Determine the discrete numeric function corresponding to the following generating function:

$$A(z) = \frac{1}{5 - 6z + z^2}$$ [*UPTU, MCA, 2006*]

Solution The given generating function is

$$A(z) = \frac{1}{5 - 6z + z^2} = \frac{1}{4(1-z)} - \frac{1}{4(5-z)}$$ (by partial fractions)

The general term of the function corresponding the given GF is given by

$$a_r = \text{Coefficient of } z^r \text{ in } A(z)$$

$$= \text{Coefficient of } z^r \text{ in the RHS of } A(z)$$

$$= \frac{1}{4}\text{Coefficient of } z^r \text{ in } (1-z)^{-1} - \frac{1}{4}\text{Coefficient of } z^r \text{ in } (5-z)^{-1}$$

$$= \frac{1}{4}\text{Coefficient of } z^r \text{ in } (1-z)^{-1} - \frac{1}{20}\text{Coefficient of } z^r \text{ in } \left(1 - \frac{z}{5}\right)^{-1}$$

$$= \frac{1}{4}\left({}^{1-1+r}C_r\right)(1)^5 - \frac{1}{20}\left({}^{1+1-r}C_r\right)\left(\frac{1}{5}\right)^r$$

[Coefficient. of z^r in the expansion of $(1 - az)^{-n} = {}^{n-1+r}C_r\,(a)$]

$$= \frac{1}{4} - \frac{1}{20}\left(\frac{1}{5}\right)^r = \frac{1}{4}\left[1 - \left(\frac{1}{5}\right)^{r+1}\right] \quad r \geq 0$$

Example 44 Find the generating function for ${}^nC_r$, the number of ways of selecting r objects from n objects with repetitions.

Solution Since each object can be selected as many times as we wish, therefore the generating function is

$$G(z) = (1 + z + z^2 + \ldots)^n$$

where each factor $1 + z + z^2 + \ldots$ corresponds to the number of times one of the objects is selected.

To find ${}^nC_r$, we find the coefficient of z^r in G (z):

$$G(z) = \left(\frac{1}{1-z}\right)^n = (1 - z)^{-n}.$$

Thus coefficient of $z^r = \dfrac{(-1)^r(-n)(-n-1)\ldots(-n-r+1)}{r!} = \dfrac{(n+r-1)!}{r!(n-r)!}$

$$= {}^{n+r-1}C_r.$$

Example 45 Use generating function to prove the following results

(*a*) ${}^nC_r = {}^{n-1}C_r + {}^{n-1}C_{r-1}$ (*b*) $\sum_{k=0}^{r} {}^mC_k \cdot {}^nC_{r-k} = {}^{m+n}C_r$.

Solution (*a*) Since ${}^nC_r$ is the coefficient of x^r in the expansion of $(1 + x)^n$. But this can be written as

$$(1 + x)^n = (1 + x)^{n-1} + x(1 + x)^{n-1}.$$

Now the coefficient of x^r in the expansion of $(1 + x)^{n-1}$ and $x(1 + x)^{n-1}$ is ${}^{n-1}C_r$ and ${}^{n-1}C_{r-1}$, respectively. Hence, the coefficient of x^r in the expansion of $(1 + x)^n$ can be written as the sum of coefficient of x^r of the two terms in the right side of above equation. This proves the result.

(*b*) Let $A(z) = (1 + z)^n$ and $B(z) = (1 + z)^m$ be two generating functions. Then coefficient of z^r in A (z) and B (z) is ${}^nC_r$ and ${}^mC_r$, respectively.

Since $A(z).B(z) = (1 + z)^{m+n}$, therefore coefficient of z^r in the expansion of $(1 + z)^{m+n}$ is ${}^{m+n}C_r$. But the coefficient of z^r in $(1 + z)^{m+n}$ is also given by $\sum_{k=0}^{r} {}^nC_k\, {}^mC_{r-k}$. Hence, the relation is proved.

Example 46 Find the coefficient of

(*a*) x^8 in $(1 + x + x^2 + \ldots)^3$

(*b*) x^{20} in $(x + x^2 + x^3 + x^4 + x^5)(x^2 + x^3 + \ldots)^5$

(*c*) x^9 in $(x^3 + x^4 + x^5 + \ldots)^3$.

Solution (*a*) We know that

$$G(x) = (1 + x + x^2 + \ldots)^3 = [(1 - x)^{-1}]^3 = (1 - x)^{-3} = \Sigma\, {}^{3+r-1}C_r x^r.$$

Since we need to find the coefficient of x^8, therefore $r = 8$ and hence, the required coefficient is ${}^{3+8-1}C_8 = {}^{10}C_8 = 45$.

(*b*) We know that

$$\begin{aligned} G(x) &= (x + x^2 + x^3 + x^4 + x^5)(x^2 + x^3 + \ldots)^5 \\ &= x(1 + x + x^2 + x^3 + x^4)\, x^{10}(1 + x + x^2 + \ldots)^5 \\ &= x^{11}[(1 - x^5)(1 - x)^{-1}][(1 - x)^{-5}] \\ &= (x^{11} - x^{16})(1 - x)^{-6} \\ &= (x^{11} - x^{16})\, \Sigma\, {}^{6+r-1}C_r x^2. \end{aligned}$$

Since, we need to find coefficient of x^{20}, $20 = r$. Thus, the required coefficient is ${}^{6+9-1}C_9 - {}^{6+4-1}C_4 = {}^{14}C_9 - {}^9C_4$.

(*c*) We know that

$$\begin{aligned} (x^3 + x^4 + x^5 + \ldots)^3 &= x^9(1 + x + x^2 + \ldots)^3 = x^9(1 - x)^{-3} \\ &= x^9\, \Sigma\, {}^{3+r-1}C_r x^r = \Sigma\, {}^{3+r-1}C_r x^{9+r}. \end{aligned}$$

Since, we need to find coefficient of x^9, $9 + r = 9$ or $9 = 0$. Thus, the required coefficient is

$$ {}^{3+r-1}C_r = {}^{3-1}C_0 = {}^2C_0 = 2.$$

Example 47 Use generating functions to find the number of ways to select 10 balls from a box containing red, white and blue balls in such a way that the selection has

(*a*) at least two balls of each colour.

(*b*) at most two red balls.

(*c*) even number of blue balls.

Solution Let x_1, x_2 and x_3 represents number of red, white and blue balls chosen from the box. Then the number of integer solution should satisfy the equation: $x_1 + x_2 + x_3 = 10$ for $x_i \geq 2$.

(*a*) The required solution is the coefficient of x^{10} in the generating function

$$(x^2 + x^3 + x^4 + \ldots)^3 = x^6(1 + x + x^2 + \ldots)^3$$

$$= x^6 [(1 - x)^{-1}]^3 = x^6 (1 - x)^{-3}$$
$$= x^6 \Sigma\, {}^{3+r-1}C_r x^r = \Sigma\, {}^{3+r-1}C_r x^{6+r}.$$

Since we need to find the coefficient of x^{10}, $10 = 6 + r$ or $r = 4$. Thus, the required number of ways are ${}^{3+4-1}C_4 = {}^6C_4 = 15$.

(*b*) The required solution is the coefficient of x^{10} in the generating function

$$(1 + x + x^2)(1 + x + x^2 + \ldots)^2 = (1 + x + x^2)(1 - x)^{-2}$$
$$= (1 - x)^{-2} + x(1 - x)^{-2} + x^2 (1 - x)^{-2}$$
$$= \Sigma\, {}^{2+r-1}C_r x^r + x \Sigma\, {}^{2+r-1}C_r x^r + x^2 \Sigma\, {}^{2+r-1}C_r x^r$$
$$= \Sigma\, {}^{2+r-1}C_r x^r + \Sigma\, {}^{2+r-1}C_r x^{r+1} + \Sigma\, {}^{2+r-1}C_r x^{2+r}.$$

The required coefficient of x^{10} in the given generating function is

$${}^{2+10-1}C_{10} + {}^{2+9-1}C_9 + {}^{2+8-1}C_8 = {}^{11}C_{10} + {}^{10}C_9 + {}^9C_8 = 30.$$

(*c*) The required solution is the coefficient of x^{10} in the generating function

$$G(x) = (1 + x^2 + x^4 + x^6 + \ldots)(1 + x + x^2 + \ldots)$$
$$= (1 + x^2 + x^4 + x^6 + \ldots)(1 - x)^{-2}.$$

The required coefficient of x^{10} is 36.

Self-practice Problems

Solve the following homogeneous recurrence relations with given initial conditions.

1. $a_n - 10a_{n-1} + 9a_{n-2} = 0;\ a_0 = 3$ and $a_1 = 11$
2. $a_n - 9a_{n-1} + 18a_{n-2} = 0;\ a_0 = 1$ and $a_1 = 4$
3. $a_n - 20a_{n-1} + 100a_{n-2} = 0;\ a_0 = 2$ and $a_1 = 30$
4. $a_n - 4a_{n-1} + 11a_{n-2} + 30a_{n-3} = 0;\ a_0 = 1,\ a_1 = -35$ and $a_2 - 85$
5. $a_{n+2} - 4a_{n+1} + 4a_n = 0;\ a_0 = 1$ and $a_1 = 3$
6. $9a_{n+2} - 6a_{n+1} + a_n = 0;\ a_0 = 0$ and $a_1 = 1$
7. $a_n - a_{n-1} + a_{n-2} = 0;\ a_0 = 0$ and $a_1 = 2$
8. $a_n - a_{n-1} - a_{n-2} = 0;\ a_0 = 1$ and $a_1 = 1$
9. $2a_n - 5a_{n-1} + 2a_{n-2} = 0;\ a_0 = 0$ and $a_1 = 1$
10. $a_n - 4a_{n-1} + 4a_{n-2} = 0;\ a_0 = 1$ and $a_1 = 6$
11. $a_n = -6a_{n-1} - 9a_{n-2};\ a_1 = 2.5$ and $a_2 = 4.7$
12. $a_{n-7}\, a_{n-1} + 16a_{n-2} - 12a_{n-3} = 0;\ n \geq 3\ a_0 = 1;\ a_1 = 4,\ a_2 = 8$ [*UPTU, MCA, 2001-02*]
13. $a_{k-9}\, a_{k-1} + 8_{k-2} = 9_{k+1}$ [*UPTU, BTech, 2008*]

Find the solution of the following recurrence relations.

14. $a_n + 6a_{n-1} + 12a_{n-2} + 8a_{n-3} = 0$
15. $4a_n - 20a_{n-1} + 17a_{n-2} - 4a_{n-3} = 0$
16. $a_{n+2} - 6a_{n+1} + 8a_n = 0$
17. $a_{n+4} - 4a_{n+3} + 6a_{n+2} - 4a_{n+1} + a_n = 0$
18. $3a_{n+2} - 6a_{n+1} + 4a_n = 0$
19. $a_{n+3} + a_{n+2} - 8a_{n+1} - 12a_n = 0$

20. $a_n - 2a_{n-1} + a_{n-2} = 2$

21. $a_n - 5a_{n-1} = 5^n$

22. $a_n - 4a_{n-1} + 4a_{n-2} = 3n + 2^n$

23. $2a_{n+1} - 5a_n = 3n + 1$

24. $2a_{n+1} - a_n = 2$

25. $a_{n+3} - 5a_{n+2} + 8a_{n+1} - 4a_n = n \cdot 2^n$

26. $a_{n+1} - 9a_n = (2n + 1)\, a^n$

27. $a_{n+2} - 5a_{n+1} + 6a_n = n^2$

28. $\alpha_{n+1} + \alpha_n = \sin(n\pi/2)$

Find the particular solution of the following recurrence relations using undetermined coefficients and operator methods.

29. $a_n + 5a_{n-1} + 6a_{n-2} = 3n^2 - 2n + 1$

30. $a_{n+2} + a_{n+1} + a_n = n \cdot 2^n$

31. $a_{n+2} - 2a_{n+1} + a_n = 5 + 3n$

32. $a_n - 4a_{n-1} + 4a_{n-2} = (n + 1)2^n$

33. $a_n + a_{n-1} = 3n \cdot 2^n$

34. $a_n - 2a_{n-1} = 5 \cdot 2^n$

35. $a_{n+2} - 2a_{n+1} + 5a_n = 2 \cdot 3^n - 4 \cdot 7^n$

36. $a_{n+2} - 4a_n = 9n^2$

37. $a_{n+1} - 3a_n = n;\ n_0 = 1$

38. $a_{n+2} - 4a_{n+1} + 4a_n = n^2 - 3n + 5$

39. $a_{n+2} - 4a_{n+1} + 4a_n = n^2 \cdot 2^n$

40. $a_{n+2} - a_{n+1} + a_n = 0;\ a_0 = 0$ and $a_1 = 1$

Solve the following recurrence relations by the method of generating functions.

41. $a_n + 2a_{n-1} - 4a_{n-2} = 0;\ n \geq 2;\ a_0 = 3$ and $a_1 = -2$

42. $a_{n+2} - a_{n+1} - a_n = 0;\ a_0 = 1$ and $a_1 = 1$

43. $a_n - 7a_{n-1} + 10a_{n-2} = 0;\ a_0 = 3$ and $a_1 = 3$

44. $\alpha_{n+2} - 2\alpha_{n+1} + \alpha_n = 2^n;\ \alpha_0 = 2$ and $\alpha_1 = 1$

45. $a_n = a_{n-2} + a_{n-1};\ a_0 = a_1 = 1$ *[UPTU, MCA, 2002]*

46. Solve the recurrence relation: $a_r + 3a_{r-1} + 2a_{r-2} = f(r)$, where $f(r) = \begin{cases} 1, & r = 5 \\ 0 & \text{otherwise} \end{cases}$

with the boundary condtions $a_0 = a_1 = 1$. *[UPTU, MCA, 2003]*

Hints and Answers

1. $a_n = 2 + 9^n$

2. $a_n = (2/3)3^n + (1/3)6^n$

3. $a_n = (2 + n)10^n$

4. $a_n = 4(-3)^n + 2^n - 5^{n+1}$

5. $a_n = (2 + n)2^{n-1}$

6. $a_n = 3^n\,(1/3)^n$

7. $a_n = i\,\dfrac{2}{\sqrt{3}}\left[\dfrac{-1 + i\sqrt{3}}{2}\right] - i\,\dfrac{2}{\sqrt{3}}\left[\dfrac{-1 - i\sqrt{3}}{2}\right]$

8. $a_n = \left[\dfrac{\sqrt{5}+1}{2\sqrt{5}}\right]\left[\dfrac{1+\sqrt{5}}{2}\right]^n + \left[\dfrac{\sqrt{5}-1}{2\sqrt{5}}\right]\left[\dfrac{\sqrt{5}-1}{2}\right]^n$

9. $a_n = -(2/3) + (1/2)(2)^n$

10. $a_n = (1 + 2n)2^n$

14. $a_n = (A_1 n^2 + A_2 n + A_3)(-2)^2$

15. $a_n = (A_1 n + A_2)(1/2)^n + A_3 4^n$

16. $a_n = A_1 2^n + A_2 4^n$

17. $a_n = A_1 + A_2 n + A_3 n^2 + A_4 n^3$

18. $a_n = A\left(2/\sqrt{3}\right)^n \cos\left(\dfrac{\pi n}{6} + B\right)$

19. $a_n = A_1\, 3^n + (A_2 + A_3 n)(-2)^n$

20. $a_n = n^2 - 10n + 25$

21. $a_n = (3 + n)5^n$

22. $a_n = 12 + 3n + (n^2 + 7n + 22)2^{n-1}$

23. $a_n = A\,(5/2)^n - n - 1$

24. $a_{np} = (1/2)^n + 2$

25. $a_n = A_1\,(1)^n + (A_2 + A_3 n)\,2^n + \left(\dfrac{2^r n^{(3)}}{24}\right)$

26. $a_n = (A + n^2)\,a^{n-1}$

27. $a_n = A_1\,2^n + A_2\,3^n + \dfrac{1}{2}\,n^2 + \dfrac{3}{2}\,n - \dfrac{5}{2}$

28. $\alpha_n = A\cos\left(\dfrac{n\pi}{2} + B\right) - \dfrac{1}{2} n \sin\dfrac{n\pi}{2}$

29. $a_{np} = n^2/4 + 13^n/24 + 71/288$

30. $a_{np} = \left(\dfrac{1}{7} - \dfrac{10}{49} n\right) 2^n$

31. $a_{np} = n^2 + n^3/3$

32. $a_{np} = n^2\left(1 + \dfrac{r}{6}\right) 2^n$

33. $a_{np} = \left(2n + \dfrac{2}{3}\right) 2^n.$

34. $a_{np} = 3n\,2^n.$

35. $a_{ps} = \dfrac{1}{4}\,3^n - \dfrac{1}{10}\,7^n$

36. $a_{np} = -\,3n^2 - 4n - (20/3)$

37. $a_{np} = -\,\dfrac{1}{4}\,(2n + 1)$

38. $a_{np} = -\,\dfrac{13n}{27} + \dfrac{7}{9}$

39. $a_{np} = \dfrac{2^n}{48}\,(n^4 - 4n^3 + 5n^2 - 2)$

40. $a_n = -\dfrac{1}{\sqrt{5}}\left(\dfrac{1-\sqrt{5}}{2}\right)^n + \dfrac{1}{\sqrt{5}}\left(\dfrac{1+\sqrt{5}}{2}\right)^n$

41. $G(a, z) = \dfrac{3 + 7z}{1 + 3z - 4z^2}$; $a_n = 2 + (-4)^n$

42. $G(a, z) = \dfrac{1}{1 - z - z^2}$

43. $G(a, z) = \dfrac{13}{1 - 5z} - \dfrac{10}{1 - 2z}$; $a_n = 13 \cdot 5^n - 10 \cdot 2^n$

44. $G(\alpha, z) = \dfrac{3}{1 - z} - \dfrac{2}{(1 - z^2)} + \dfrac{1}{(1 - 2z)}$

$\alpha_n = 1 - 2n + 2^n$

Chapter 9

Graph Theory

PREVIEW

The term 'graph' is often used informally for any representation of data. Various other forms of graph include the bar graph, picture graph, pie chart, graph of functions on rectangular coordinates, etc. Graphs are used to a number of real world situations such as: distribution of a product or service routes; chemical structure of a molecule; organization charts, communication lines, etc.

LEARNING OBJECTIVES

After studying this chapter, you should be able to

- understand and use many terms associated with a simple graph and directed graphs
- understand the use of graphs as representation tools in a variety of context.
- prove whether or not two graphs are isomorphic
- use Euler's formula for a simple, connected and plannar graph.
- use directed, undirected or multigraph graphs to represent in computer memory.

9.1 INTRODUCTION

Graph theory is used to analyse problems of combinatorial nature that arise in computer science, operations research, physical sciences and economics. The term graph is familiar to you because it has been used in the context of straight lines and linear inequalities. The graphs are used to represent a problem involving discrete arrangements of objects that are related to each other, but without taking into consideration their internal properties. In this chapter, first we will combine the concepts of graph theory with digraph of a relation to define a more general type of graph that has more than one edge between a pair of vertices. Second, we will identify basic components of a graph, its features and many applications of graphs.

9.2 BASIC TERMINOLOGY

In this section definition of graph and basic terminology with elementary properties of graphs are presented.

9.2.1 Graph

A *graph*, written as G = G (V, E) consists of two components:

(*i*) The finite set of *vertices* V, also called *points* (*nodes,* or *junctions*), and

(*ii*) The finite set of (*directed*) *edges* E, also called *lines* or *arcs* connecting pair of vertices.

In other words, in a graph there is a mapping from the set of edges E to the set of vertices V such that each $e \in E$ is associated with ordered or unordered pair of elements of V. In other words, each edge, say e_k is associated with an unordered pair of vertices (v_i, v_j) where v_i and v_j are called end vertices of the edge.

A graph is said to be the *directed graph* or *digraph* if it is required to associate a direction with each edge of the graph. That is, edges are ordered pairs of distinct vertices. A directed graph can be regarded as a one-way road where traffic flow is permitted in only one direction. A directed arc (*i, j*) permits flow only from vertex *i* to *j*.

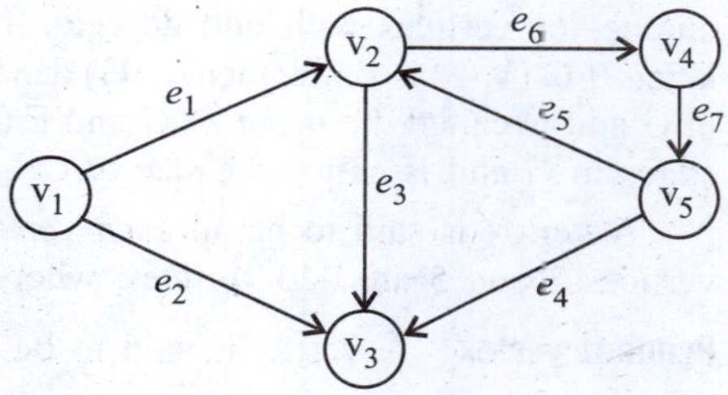

Fig. 9.1 Directed Graph

Figure 9.1 provides an example of a directed graph on $V = \{v_1, v_2, v_3, v_4, v_5\}$ with $E = \{(v_1, v_2), (v_1, v_3) (v_2, v_4), (v_2, v_3) (v_4, v_5), (v_5, v_2), (v_5, v_3)\}$. The direction of an edge is shown by placing a directed arrow on the edge.

Each pair of vertices connecting with an edge is called the *end points*. If there is only one edge between two vertices, say v_1 and v_2, then we often identify an edge, say *e*, with the set $\{v_1, v_2\}$. Thus in set theory terminology we write the set of edges, $E = \{(v_1, v_2), (v_1, v_3), (v_2, v_3), (v_5, v_3), (v_5, v_2), (v_2, v_4), (v_4, v_5)\}$ as the ordered pair of elements of the set of vertices V (also called arcs), where first element is called the *origin* or *source* of the edge (v_i, v_j) and second element is called *terminus* or *terminating vertex*. In specifying any edge of a digraph by its end points, the edge is understood to be directed from the source vertex towards the terminating vertex. But if there is more then one edge between two vertices, then different edges are identified with labels such as 0 and 1 as shown in Fig. 9.2.

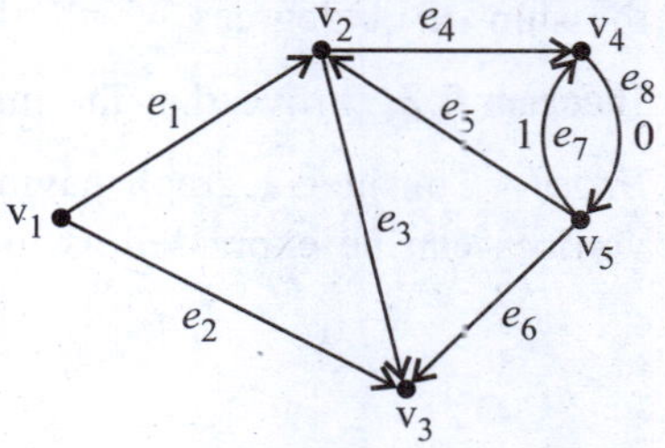

Fig. 9.2 Directed Graph

Usually in a graph G we denote the number of vertices by *n* and the number of edges by *m*. If, as in the most applications, the values of both *m* and *n* are finite, G is said to be a *finite graph*.

In place of circles, vertices in a graph are also represented by using dots. Usually names of edges in a graph are omitted because they have no intrinsic meaning. As usual, an edge between a pair of vertices is used to record information about the relationship or connections between two objects.

Indegree The indegree of a vertex v in a directed graph is the number of edges ending at it, and is denoted as, indeg (v).

Outdegree The outdegree of a vertex v in a directed graph is the number of edges beginning from it and is denoted as, outdeg (v).

Illustration In Fig. 9.2 the indegrees and outdegrees of various vertices are as index.

Vertex	:	v_1	v_2	v_3	v_4	v_5
Indegree	:	0	2	3	2	1
Outdegree	:	2	2	0	1	3

Degree of a vertex The degree of a vertex denoted by d (v) or deg (v) is the number of edges (or arcs) connected with it.

As shown in Fig. 9.3, the vertex a has degree 5, vertices c and d have degree 3 each, vertex b has degree 2.

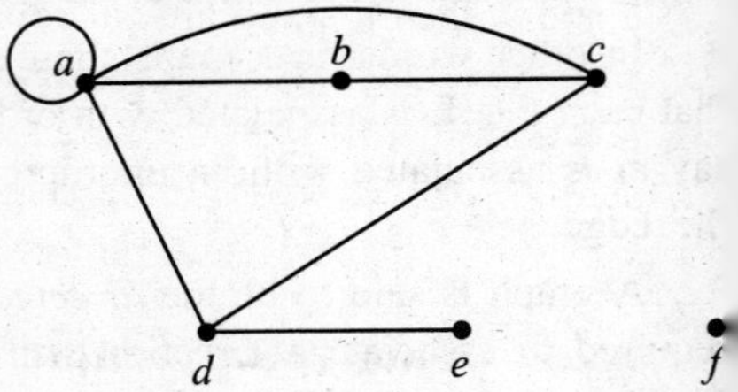

Fig. 9.3

Finite graph A graph G (V, E) is said to be *finite graph* if it contains finite number of edges and vertices (or has an even number of vertices with odd degree). This result will be proved latter. If G (V, E) is finite, then V (G) denotes the number of vertices in G and is called the *order of* G and E (G) denotes the number of edges in G and is called the *size of* G.

A vertex is said to be an *even vertex* when its degree is even, otherwise an *odd vertex*. In Fig. 9.4 vertices 3 and 5 are odd vertices, whereas vertices 1, 2 and 4 are even vertices.

Pendant vertex A vertex is said to be pendant vertex or an end vertex if its degree is one.

For example, in Fig. 9.3 we have deg (a) = 5, deg (b) = 2, deg (c) = 3, deg (d) = 3, deg (e) = 1 (pendant vertext), deg (f) = 0.

A vertext is said to be *isolated* if its degree is zero or vertex has no incident edge. For example, in Fig. 9.3, vertex f is isolated.

Theorem 9.1 If G (V, E) is a directed graph, then the sum of the outdegrees of the vertices of a diagraph equals the sum of indegrees of the vertices which equals the number of edges in a diagraph, i.e.

Σ indeg (v) = Σ outdeg (v) = e for all v $\in$ V.

Proof The proof of this theorem is obvious. As shown in Fig. 9.2, the sum of the indegrees equals the sum of outdegrees which also equals the number of edges in the diagraph.

Theorem 9.2 Prove that the number of vertices of odd degree in a graph is always even.

Proof Consider a graph having vertices of even and odd degrees. Then the sum of degrees of all vertices can be expressed as the sum of even degree vertices and odd degree vertices. That is

$$\sum_{i=1}^{n} \deg(v_i) = \sum_{\text{even}} \deg(v_j) + \sum_{\text{odd}} \deg(v_k)$$

Since the sum of degrees of all vertices is even, the left side of above equation is even. Also on the RHS, the first sum of j vertices is even. Thus for the second sum to be even, the number k must be even. Hence, the result is proved.

Illustrations In Fig. 9.3,

1. The sum of degrees of all vertices is even, i.e.

$$\deg(v) = \deg(a) + \deg(b) + \deg(c) + \deg(d) + \deg(e) + \deg(f)$$
$$= 5 + 2 + 3 + 3 + 1 + 0 = 14.$$

2. The number of vertices of odd degrees 5, 3, 3 and 1 are 4, which is even. Hence, we have even number of vertices of odd degree.

Adjacent vertices (or nodes) Two vertices are said to be adjacent if they are connected by an edge. If there is an edge, $e = \{u, v]$ connecting vertices u and v, then u and v are called adjacent to each other and also called *end points* of the edge *e*. The vertex u is said to be *adjacent to* v whereas v is *adjacent from* u. Also the edge *e* is called *incident* with the vertices u, and v it connect on its end points. In Fig. 9.4, vertices 1 and 2 are adjacent vertices. Vertices 1 and 5 are not adjacent.

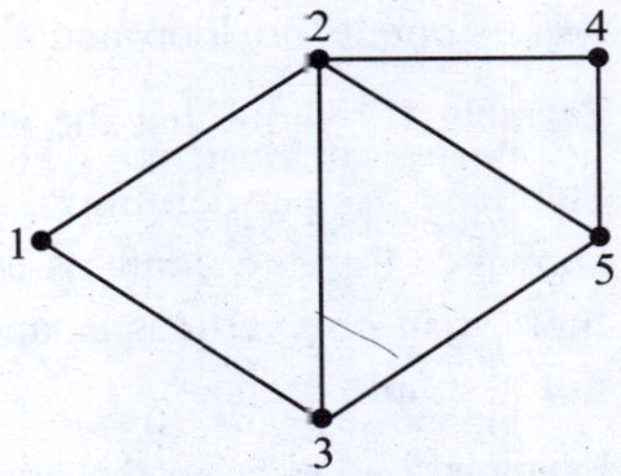

Fig. 9.4 Uncirected Graph

9.2.2 Undirected Graph

An undirected graph is also defined in the same manner as directed graph except that edges (or arcs) are unordered pairs of distinct vertices.

Figure 9.4 is an example of an undirected graph. In an undirected graph we can refer to an arc joining the vertex pair u and v as either (u, v) or (v, u). An undirected arc (u, v) can be considered as a two-way road with traffic flow permitted in both directions: either from vertex u to v or from v to u. An edge such as {u, v} stands for {(u, v), (v, u)}. Although (u, v) = (v, u) only when u = v.

Theorem (Handshaking) Theorem 9.3 If G (V, E) is an undirected graph with *e* edges, then sum of degrees of the vertices in an undirected graph is even, i.e. $\Sigma \deg(v) = 2e$ for all $v \in V$.

Proof Since degree of a vertex is defined as the number of edges (or arcs) connected with it, therefore, the sum of the degree counts the total number of times an edge is connected (incident) with a vertex. As every edge is connected with exactly two vertices, so each edge is counted twice at each of its end. This implies that the sum of the degrees equal twice the number of edges.

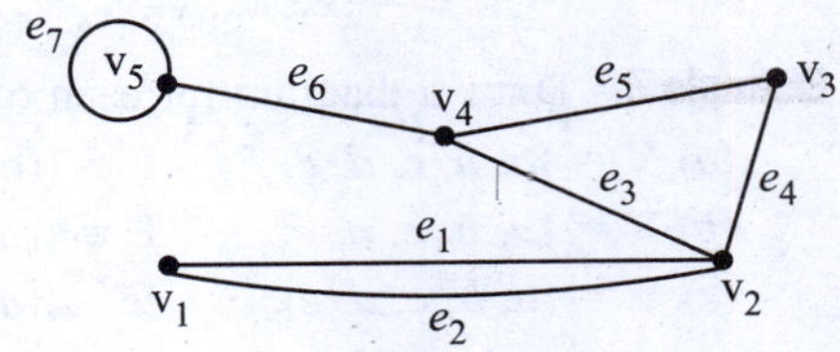

Fig. 9.5

Loop A loop is the edge in a graph whose end points are same, i.e. it is an edge from a vertex to itself. A loop contributes 2 to the degree of a vertex. In Fig. 9.5, edge e_7 from vertex 5 forms a loop, and degree of vertex 5 is 3.

9.2.3 Simple, Multigraph and Pseudo Graph

In some directed or undirected graphs, there exists certain pair of nodes that are joined by more than one edges, such edges are called *multiple* or *parallel edges.* For example, in diagraph shown in Fig. 9.5, the edges e_1 and e_2 are *multiple* or *parallel edges* because these edges connect the same end points and edge e_7 is a loop since its end points are at the same vertex. A diagram which contains multiple edges as well as loops is called a *multigraph.*

A graph in which loops and multiple edges are allowed is called a *pseudo graph.* Fig. 9.6 is an example of such graphs.

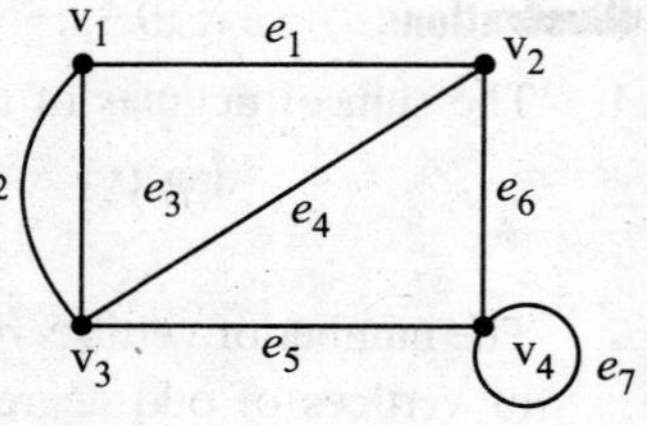

Fig. 9.6

Parallel edges Two or more edges in E of the form (v_i, v_j) and (v_r, v_s) are parallel if and only if $v_i = v_r$ and $v_j = v_s$.

A graph which has neither loops nor multiple edges is called a *simple graph.* Fig. 9.7 (*a* and *b*) are examples of simple undirected graphs because these contain no loop and the edges are all distinct.

Example 1 Show that the graph in Fig. 9.6 is not a simple graph.

Solution As per definition, a simple graph is one which does not have multiple edges or loops. The graph shown in Fig. 9.6 contains a loop formed by the edge e_7 whose end points are the same vertex v_4. Also more than one vertices e_2 and e_3 connect the same vertices v_1 and v_3. Hence, graph shown in Fig. 9.6 is not a simple graph.

Example 2 Describe the graphs shown in Fig. 9.7(a) and Fig. 9.7(*b*).

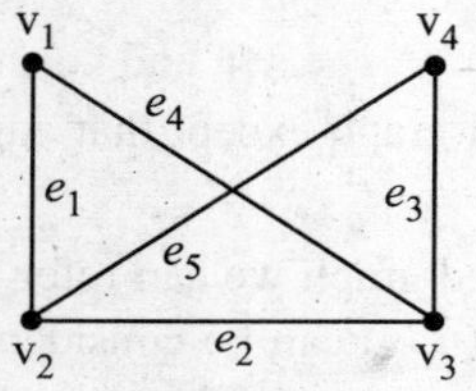

Fig. 9.7 (*a*)

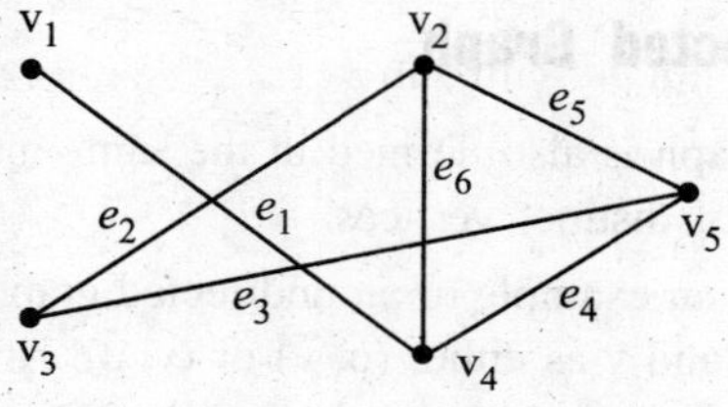

Fig. 9.7 (*b*)

Solution (*i*) The graph G (V, E) shown in Fig. 9.7(*a*) consists of 4 vertices and 5 edges as given below:

$$V = \{v_1, v_2, v_3, v_4\} \quad \text{and} \quad E = \{(v_1, v_2), (v_1, v_3), (v_2, v_3), (v_2, v_4), (v_3, v_4)\}$$

(*ii*) The graph G (V, E) shown in Fig. 9.7(*b*) consists of 5 vertices and 6 edges as given below:

$$V = \{v_1, v_2, v_3, v_4\ v_5\} \quad \text{and} \quad E = \{(v_1, v_4), (v_2, v_3), (v_2, v_5), (v_2, v_4), (v_3, v_5), (v_4, v_5)\}$$

Example 3 Draw a diagram for each of the following graph G (V, E).

(*a*) V = {*a*, *b*, *c*, *d*, *e*, *f* }, E = {(*a*, *d*), (*a*, *f*), (*b*, *c*), (*b*, *f*), (*c*, *e*)}

(*b*) V = {*a*, *b*, *c*, *d*, *e*}, E = {(*a*, *a*), (*b c*), (*b*, *d*), (*c*, *b*), (*e*, *a*), (*e*, *d*)}

(c) V = {*a*, *b*, *c*, *d*, *e*}, E = {(*a*, *d*), (*b*, *c*), (*b*, *d*), (*b*, *e*), (*d*, *e*), (*c*, *e*)}

Solution First draw vertices of the graph and then indicate edges by connecting the appropriate vertices of the graph as shown in Fig. 9.8(*a*), (*b*) and (*c*).

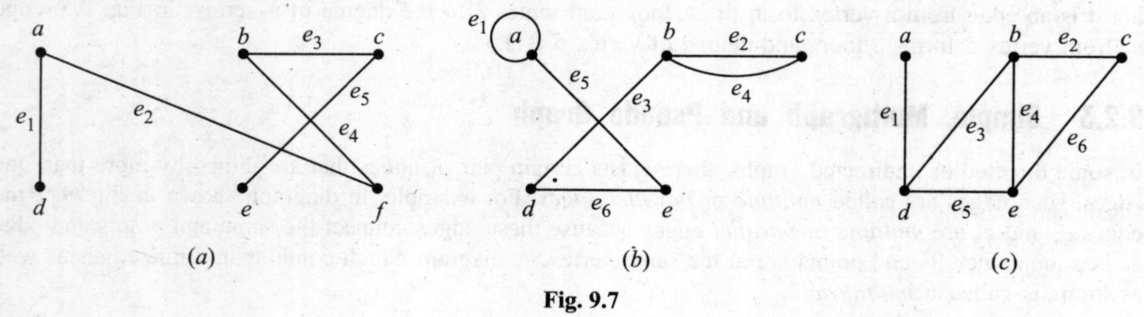

Fig. 9.7

Example 4 (*a*) Find the degree of each vertex in the multigraph shown in Fig. 9.9.

(*b*) What is the degree of an isolated vertex v_2.

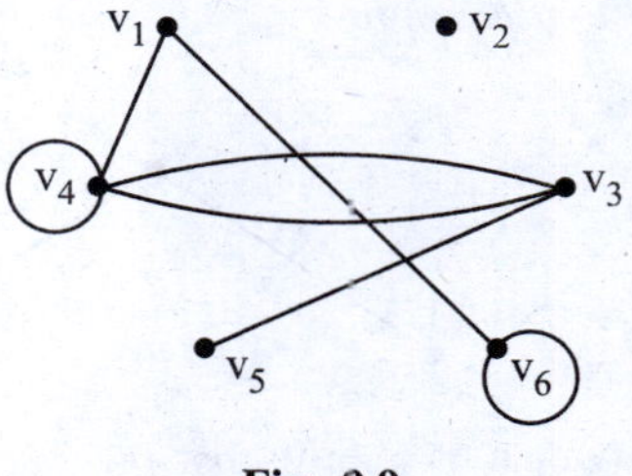

Fig. 9.9

Solution (*a*) The number of edges associated with vertex in Fig. 9.9 determine its degree. So we obtain

$\deg(v_1) = 2$, $\deg(v_3) = 3$, $\deg(v_4) = 5$, $\deg(v_5) = 1$, $\deg(v_6) = 3$.

Note that the loops at v_4 and v_6 are counted twice to determine degree of the vertex.

(*b*) Since a vertex is said to be isolated if it does not associated (or connected) with any edge, therefore $\deg(v_2) = 0$.

Example 5 Consider a graph G (V, E), where $V = \{v_1, v_2, v_3\}$ and $\deg(v_2) = 4$.

(*a*) Does such a graph exist? If not why?

(*b*) Does such a multigraph G exist? If yes, give an example.

Solution (*a*) Since there can be only one edge from vertex v_2 to each of the other two vertices and multiple edges and loops are not permitted in a graph, therefore, such a graph does not exist, and hence $\deg(v_2) \le 2$.

(*b*) A multigraph with $E = \{(v_1, v_2), (v_1, v_2)\ (v_2, v_3), (v_2, v_3)\}$ exists.

9.2.4 Walk

Let u, v be (not necessarily distinct) any two vertices in an undirected graph G (V, E). A walk u – v in G is the loop-free finite alternating sequence $u = v_0, e_1, v_1, e_2, v_2, e_3, \ldots, e_{n-1}, v_{n-1}, e_n, v_n = v$ of vertices and edges starting at vertex v_i and ending at vertex v_j involving the *n* edges $e_i = \{v_{i-1}, v_i\}$ for $1 \le i \le n$. Alternately a walk is a sequence of vertices and edges that begins at v_i and travel along edges to v_j so that no edge appears more than once. However, a vertex may appear more than once. Fig. 9.10 (*a*) and (*b*) illustrate two walks in this graph:

$$1-2-3-4 \quad \text{and} \quad 1-2-4-5-2-3.$$

In a walk, first and last vertices in the sequence are called *terminal vertices*.

Closed and open walk A walk is said to be the *closed walk* if it is possible that a walk begins and end at the same vertices, i.e. $v_i = v_j$ (and $n > 1$). Otherwise the walk is called *open*, i.e. terminal vertices are different.

9.3 PATHS, CYCLES AND CONNECTIVITY

1. A ***path*** is a walk through a sequence, $v_0, v_1, v_2, \ldots, v_n$ of vertices, each adjacent to the next, without any repetition of vertices.

   If there exists a path from vertex v_0 to v_n in an undirected graph, then there always exists a path from v_n to v_0 too. But in directed graph, it is not necessary that there exists a path from v_n tc v_0 also. The walk shown in Fig. 9.10(*a*) is also a path, but the walk shown in Fig. 9.10(*b*) is not a path because vertex 2 is repeated twice in it.

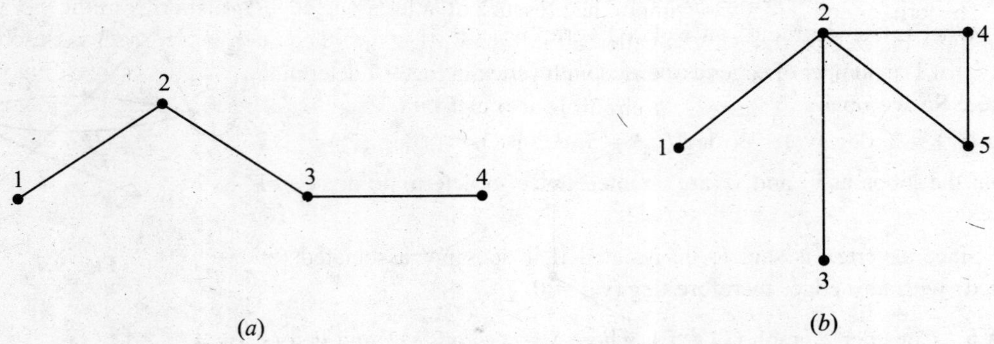

Fig. 9.10 Examples of Walk

Edges of a path can be divided into two groups: *forward edges* and *backward edges*. An edge (i, j) in a path is said to be the forward edge if the path passes through vertex i prior to passing vertex j, and is said backward edge otherwise. For example, in the path shown in Fig. 9.10(*b*), the edges (1, 2) and (2, 3) are forward edges and the edge (5, 2) is a backward edge.

2. A trail from a vertex u to v is a path that does not involve a repeated edge. Thus, a trail from u to v is a path of the form:

$$u = v_0, e_1, v_1, e_2, \ldots, e_{n-1}, v_{n-1}, e_n, v_n = v$$

where all e_i are different.

3. A ***simple path*** from u to v is the one in which no edge is repeated in the path. That is, a simple path is also a trail where all the vertices and all edges are different except the first vertex is equal to last vertex. A path is also called an *elementary path* if no vertex is repeated more than once in the path.

Length of path The number of edges in a path is called the length of the path. The sequence of $n + 1$ vertices of a graph in which each pair of vertices is an edge of the graph, is of length n.

An edge which is not a loop is a path of length one. A loop can be included in a walk but not in a path as shown in Fig. 9.11. In a path the terminal vertices are of degree 1 when other vertices (also called intermediate vertices) are of degree 2.

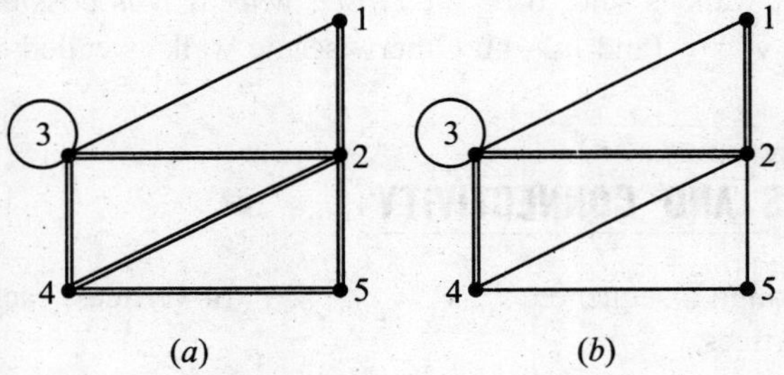

Fig. 9.11 Examples of Path

4. A ***circuit*** is a closed walk (or path) of non-zero length that contains no repeated edge where start and end vertex is same, i.e. $u = v_0, e_1, v_1, e_2, \ldots, e_n, v_n = v$ where $v_0 = v_n$ and all e_i are distinct.

The term circuit will always imply the presence of at least one edge. If no vertex of the u–v walk occurs more than once, then the walk is called an u–v path. If u = v, then such closed path is described by the term *cycle*. The term cycle will always imply the presence of at least three distinct edges. Fig. 9.12(*a*), (*b*) and (*c*) are examples of circuits. A circuit is also called *circular path, elementary cycle, polygon* or *cycle.*

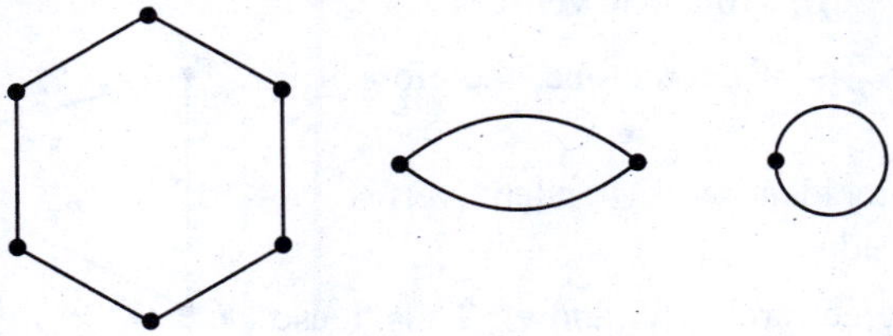

(*a*) (*b*) (*c*)

Fig. 9.12 Examples of Circuits

Table 9.1 summarises above definitions. From the comparison, it appears that a path is a trail, which in turn is an open walk. Furthermore, every cycle is a circuit and every circuit with at least two edges is a closed walk.

Table 9.1

Term	*Repeated Vertex*	*Repeated Edge*	*Starts and Ends at the Same Vertex*
Trail	Yes	No	Yes
Circuit	Yes	No	Yes
Path			
• Simple	No	No	No
• Closed	Yes	Yes	Yes
Cycle	No	No	Yes

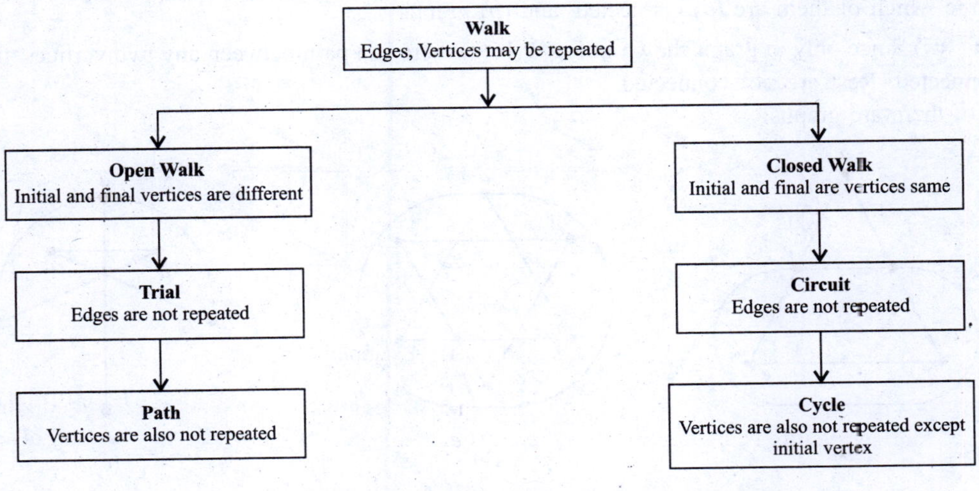

Fig. 9.13 Walk, Trial, Circuit, Path and Cycle Relationship

Example 6 Consider the graph as shown in Fig. 9.13. Determine:

(*a*) Pendent vertices (*b*) Pendent edges

(*c*) Odd vertices (*d*) Even vertices

(*e*) Incident edges (*f*) Adjacent vertices

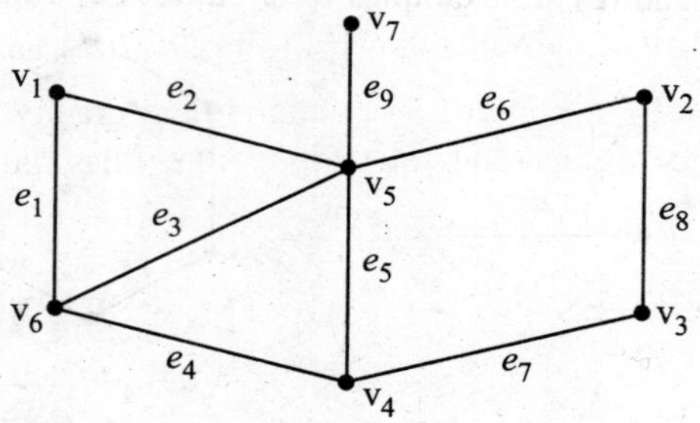

Fig. 9.14

Solution (*a*) Since vertex v_7 is of degree one, therefore it is pendent vertex.

(*b*) The edge (v_5, v_7) is incident with pendent vertex v_7, therefore it is pendent edge.

(*c*) Vertices having odd degrees are v_4, v_5 and v_6. Thus, these vertices are odd vertices.

(*d*) Vertices having even degrees are v_1, v_2, and v_3. Thus, these vertices are even vertices.

(*e*) All 9 edges in Fig. 9.14 are incident edges because each of them is connected with a pair of vertices.

(*f*) Vertex v_1 is adjacent to v_5 and v_6; Vertex v_2 is adjacent to v_3 and v_5.

Vertex v_3 is adjacent to v_2 and v_4; Vertex v_4 is adjacent to v_3, v_6.

Vertex v_5 is adjacent to v_4 and v_7; Vertex v_6 is adjacent to v_1 and v_4.

Vertex v_7 is adjacent to v_5.

Example 7 Let G be a graph shown in Fig. 9.15. Find all simple paths from vertex v_1 to v_3 and also find the length of the shortest path.

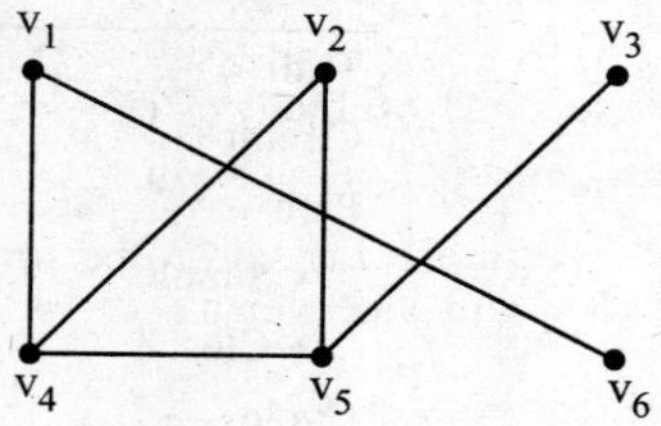

Fig. 9.15

Solution A path from vertex v_1 to v_3 will be called simple if no vertex is repeated. There are only two simple paths from vertex v_1 to v_3: $\{v_1, v_4, v_5, v_3\}$ and $\{v_1, v_4, v_2, v_5, v_3\}$.

Also, $(v_1, v_3) = 3$, since 3 is the length of the shortest path from v_1 to v_3.

Example 8 Consider the multigraphs shown in Fig. 9.16(*a*), (*b*) and (*c*). Determine which of them are (*a*) connected, and (*b*) graphs.

Solution (*a*) Since only in graph shown in Fig. 9.16(*c*) there is a path between any two vertices, therefore, it is connected. Rest are not connected.

(*b*) All of them are graphs.

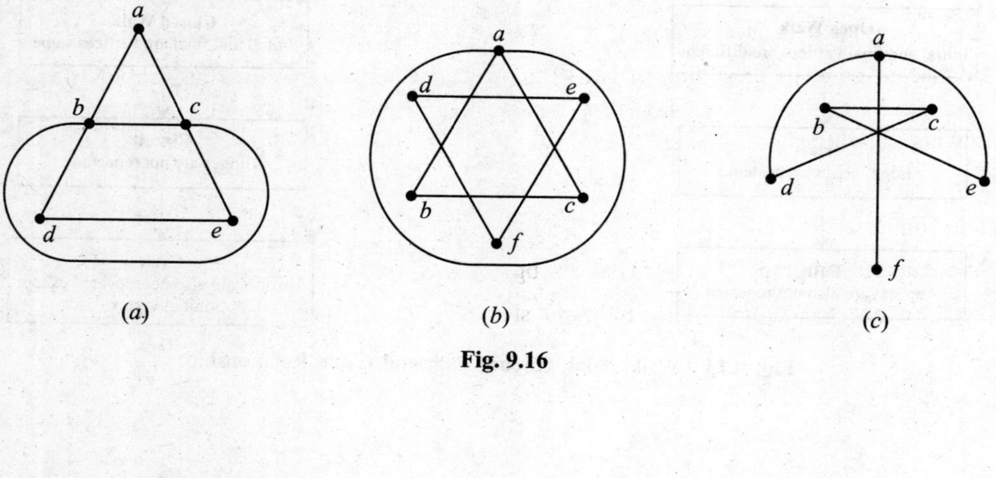

Fig. 9.16

Example 9 Consider the multigraphs shown in Fig. 9.17(*a*), (*b*) and (*c*). Determine which of them are (*a*) connected, (*b*) loop free, and (*c*) graphs.

Solution (*a*) Since only in multigraphs shown in Fig. 9.17 (*a*) and (*b*) there is a path between any two vertices, therefore these are connected.

(*b*) Multigraph shown in Fig. 9.17(*d*) has a loop, i.e. an edge with same end points.

(*c*) Only Fig. 9.17(*a*) and (*b*) are graphs. The multigraph shown in Fig. 9.17(*c*) has multiple edges, and multigraph shown in Fig. 9.17(*d*) has multiple edges and a loop.

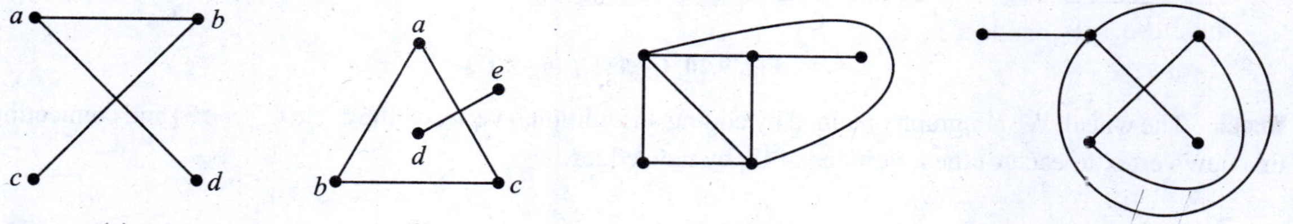

(*a*) (*b*) (*c*) (*d*)

Fig. 9.17

9.3.1 Connected Undirected Graphs

A graph is said to be *connected* if there exists at least one path between every pair of its vertices, i.e. reaching out to any vertex from any other vertex by travelling along the edges, otherwise it is *disconnected.* That is, a graph G is connected if given any vertices u and v, it is possible to travel from u to v along a sequence of adjacent edges of the graph. An example of connected graph is shown in Fig. 9.18 and a disconnected graph with three components is shown in Fig. 9.19.

A graph G is said to be *strongly connected,* if for any pair of vertices *u* and v, there exists a path from *u* to v and vice versa.

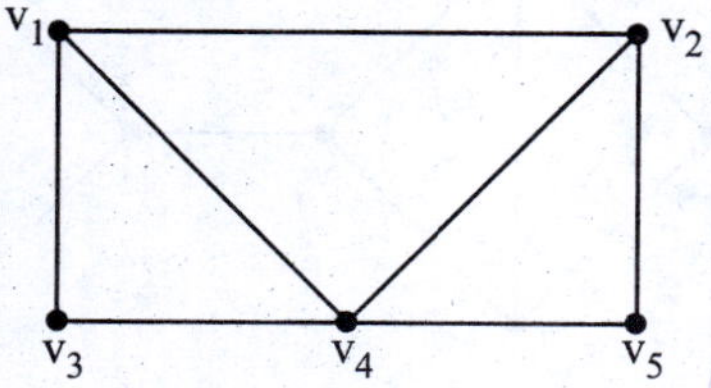

Fig. 9.18 Connected Graph

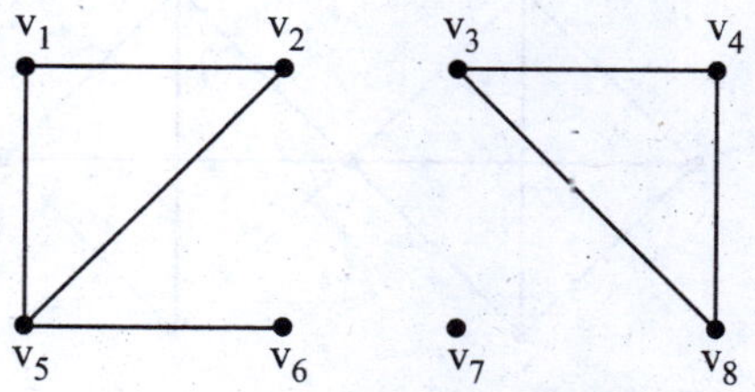

Fig. 9.19 Disconnected Graph

A disconnected graph is the union of two or more connected subgraphs each pair of which has no vertex in common. A disjoint subgraph H that is also called *connected component* of the graph G should satisfy following conditions:

(*i*) H is a subgraph of G.

(*ii*) H is connected.

(*iii*) No connected subgraph of G has H as a subgraph and contains vertices or edges that are not in H.

For example, the connected components of the graph shown in Fig. 9.18 are $\{v_1, v_2, v_5, v_6\}$, $\{v_7\}$, $\{v_3, v_4, v_8\}$.

Cycle The cycle, C_n ($n \geq 3$) is a graph of n vertices $v_1, v_2, ..., v_n$ and edges $\{v_1, v_2\}, \{v_2, v_3\}, ..., \{v_{n-1}, v_n\}, \{v_n, v_1\}$

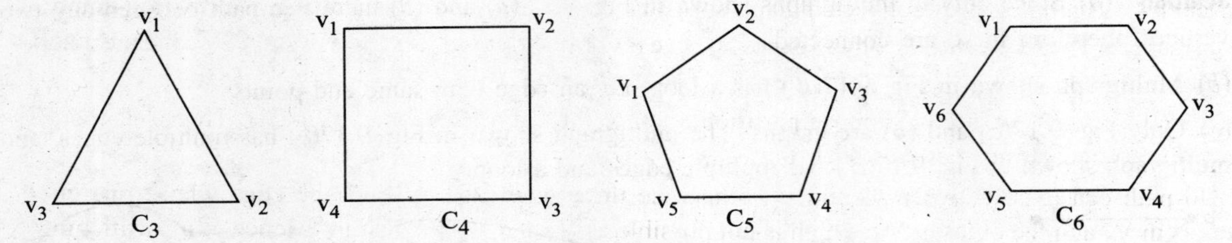

Fig. 9.20 Cycle C_n ($n \geq 3$)

Wheel The wheel, W_n is a graph obtained by adding an additional vertex to the cycle C_n ($n \geq 3$) and connecting this new vertex to each of the n vertices of C_n by new edges.

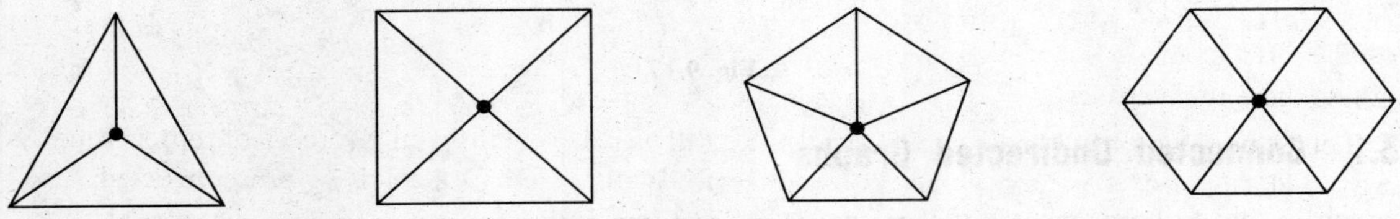

Fig. 9.21 Wheels

Distance and Diameter In a connected graph G the distance between its vertices u and v is the length of the shortest path and is denoted by d(u, v). In Fig. 9.22(a), d(1, 6) = 2 and in Fig. 9.22(b), d(1, 5) = 3.

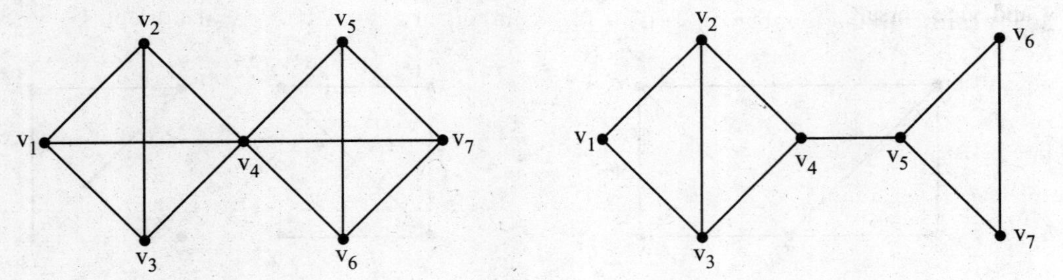

Fig. 9.22 Examples of Distance and Diameter

In a connected graph G the maximum distance between any two vertices is called *diameter* and is denoted by diam (G). In Fig. 9.22(a) and Fig. 9.22(b) the diameter is, diam (G) = 3 and diam (G) = 4, respectively.

Cut Points and Bridges Vertex v in a connected graph G is called a *cut point* if G–v is disconnected, where G–v is the graph obtained from G by deleting the vertex v and all edges connecting v. In Fig. 9.22(a), vertex 4 is the cut point.

And edge e of a connected graph G is called a *bridge* or *cut edge* if G–e is disconnected. In Fig. 9.22(b), the edge (4, 5) is a bridge.

Theorem 9.4 A graph G is disconnected if and only if its vertex set V can be partitioned into two non-empty, disjoint subsets V_1 and V_2 such that there exists no edge is G whose one end vertex is in subset V_1 and the other in subset V_2.

Proof *Necessary Condition:* Suppose partitioning of V into subsets V_1 and V_2 exists. Then, let v_1 and v_2 of G be two arbitrary vertices such that $v_i \in V_1$ and $v_2 \in V_2$.

No path can exist between v_1 and v_2, otherwise there would be at least one edge whose one end vertex is in V_1 and the other in V_2, which is not possible as V_1 and V_2 are disjoint. Hence, if a partitioning exists G is an unconnected graph.

Sufficient Condition: Let G be a disconnected graph and a vertex $v_1 \in G$. Let V_1 be the set of all vertices that are joined by path to v_1. Since G is disconnected, V_1 does not include all vertices of G. The remaining vertices will be from a non-empty set V_2. Since no vertex in V_1 is joined to any vertex in V_2 by an edge, it is possible to partition vertax set V.

Theorem 9.5 If a graph (connected or disconnected) has exactly two vertices of odd degree, there must be a path joining these two vertices.

Proof Let G be a graph with all even vertices, i.e. all vertices with even degrees, except vertices v_1 and v_2 odd degrees. If v_1 and v_2 are both connected with an edge, then these are connected.

If v_1 and v_2 belong to different components then each one of these two components shall have one vertex of odd degree, which is not possible. Hence, in graph G, v_1 and v_2 must belong to the same component and have a path between them.

Theorem 9.6 A simple graph G with n vertices and k components cannot have more than $\frac{1}{2}(n-k)(n-k+1)$ edges.

Proof If n_i represents the number of vertices in a component i, for $1 \le i \le k$ of a graph G, then the sum of all vertices in each of k components must equal the number of vertices in G, i.e. $\sum_{i=1}^{k} n_i = n$.

Any component with n_i vertices will have maximum possible number of edges $= \frac{1}{2}\, n_i(n_i - 1)$ when it is complete. Hence, the maximum number of edges in G are

$$\frac{1}{2}\sum_{i=1}^{k} n_i\left(n_i-1\right) = \frac{1}{2}\sum_{i=1}^{k} n_i^2 - \frac{1}{2}\sum_{k=1}^{k} n_i \le \frac{1}{2}\left[n^2-(k-1)(2n-k)\right]-\frac{1}{2}n$$

$$= \frac{1}{2}\left[n^2-2nk+k^2+n-k\right] = \frac{1}{2}(n-k)(n-k+1).$$

9.3.2 Connected Directed Graphs

Let G(V, E) be a directed graph. The connectedness in a directed graph is defined as follows:

(a) A directed graph G is called *connected* or *weakly connected* if it is connected as an undirected graph where each directed edge is considered as an undirected edge. Fig. 9.23(a) is an example of weakly connected graph.

(*b*) A directed graph G is called *unilaterally connected graph* if for any pair of vertices of the graph at least one of these is reachable from other vertex. Fig. 9.23 (*b*) is an example of unilaterally connected graph.

(*c*) A directed graph G is called *strongly connected graph* if for any pair of vertices of the graph both of these are reachable from other vertex. Fig. 9.23 (*c*) is an example of strongly connected graph.

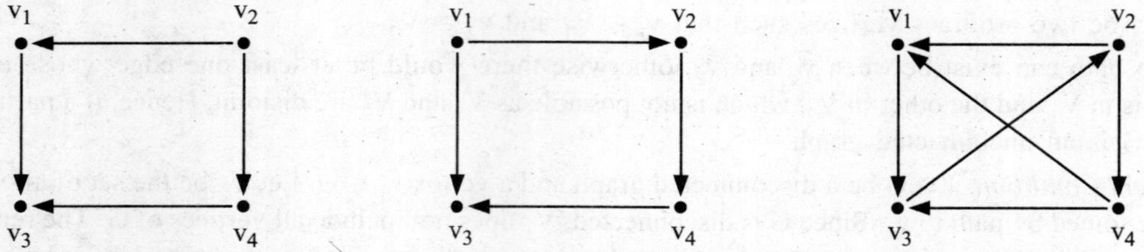

(*a*) Weakly Connected Graph (*b*) Unilaterally Connected Graph (*c*) Strongly Connected Graph

Fig. 9.23 Connected Directed Graphs

Theorem 9.7 If a simple graph G with *n* vertices has more than $\frac{1}{2}(n-1)(n-2)$ edges, then G is connected.

Proof Let G be a simple graph with *n* vertices and more than $\frac{1}{2}(n-1)(n-2)$ edges. If G has *k* components, then by Theorem 9.4, G can have atmost $\frac{1}{2}(n-k)(n-k+2)$ edges. But G has more than $\frac{1}{2}(n-1)(n-2)$ edges, therefore we have

$$\frac{1}{2}(n-k)\,(n-k+2) > \frac{1}{2}(n-1)\,(n-2)$$

which is possible only when $k = 1$. Hence, G is connected.

Example 10 Find the connected components of the graph G shown in Fig. 9.24.

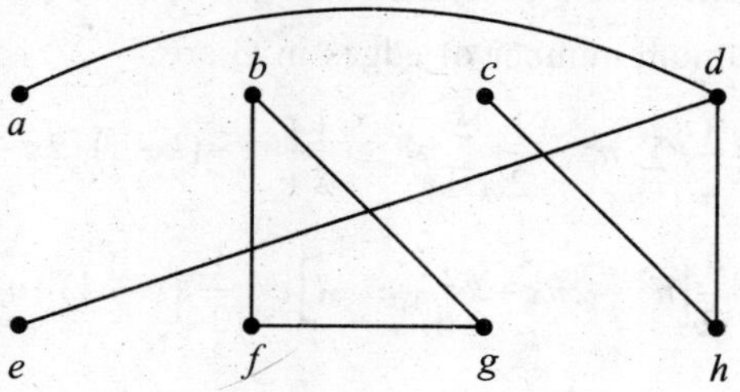

Fig. 9.24

Solution (*a*) Take any vertex, say *a*, and find all vertices connected to it. So the set of connected components is {*a*, *d*, *e*, *h*, *c*}.

Next, select a vertex not included in this component and repeat the process to obtain another component. Continue in this way until all components have been identified. In Fig. 9.24, another set of components is {*b*, *f*, *g*}. Thus, the component of the graph G are {*a*, *d*, *e*, *h*, *c*} and {*b*, *f*, *g*}.

xample 11 Consider the graph shown in Fig. 9.24. Does G has any bridge?

olution Since G – {c, h} and G – {d, h} are disconnected graphs, therefore edges {c, h} and {d, h} are ne bridges in G.

xample 12 Find the connected components of a graph G (V, E), where / = {a, b, c, d, e} and (i) E = {(a, c), (b, e), (d, c), (e, a)}, (ii) E = Φ.

olution (a) The graph G (V, E) with given set of vertices and edges is hown in Fig. 9.25. This graph is connected. Here the vertex b is connected o all vertices. Thus, G has only one set of components {b, e, a, c, d}.

b) Since E = Φ is an empty set, therefore, all vertices are isolated and hence a}, {b}, {c}, {d} and {e} are connected components of G.

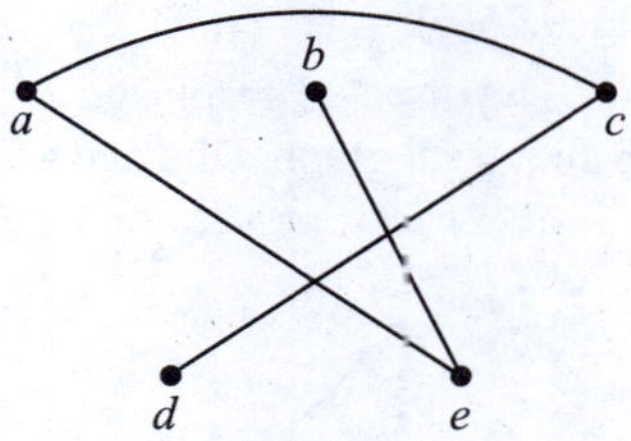

Fig. 9.25

.4 SUBGRAPHS

f G (V, E) is a directed or undirected, then the graph H (V′, E′) obtained by deleting few vertices and edges rom G is called a *subgraph* of G provided the set of vertices V′ in H contains (at least) all the end points of edges in E′, i.e. $\Phi \neq V' \subseteq V$ and $E' \subseteq E$, where each edge in E′ is incident with vertices in V′.

It is understood that the removal of an edge from G leaves its end points in place, whereas the removal of a vertex necessitates the removal of edges that vertex associated as an end point.

The graphs shown in Figs 9.26(b), (c) and (d) are few examples of a subgraph of the graph shown n Fig. 9.26(a).

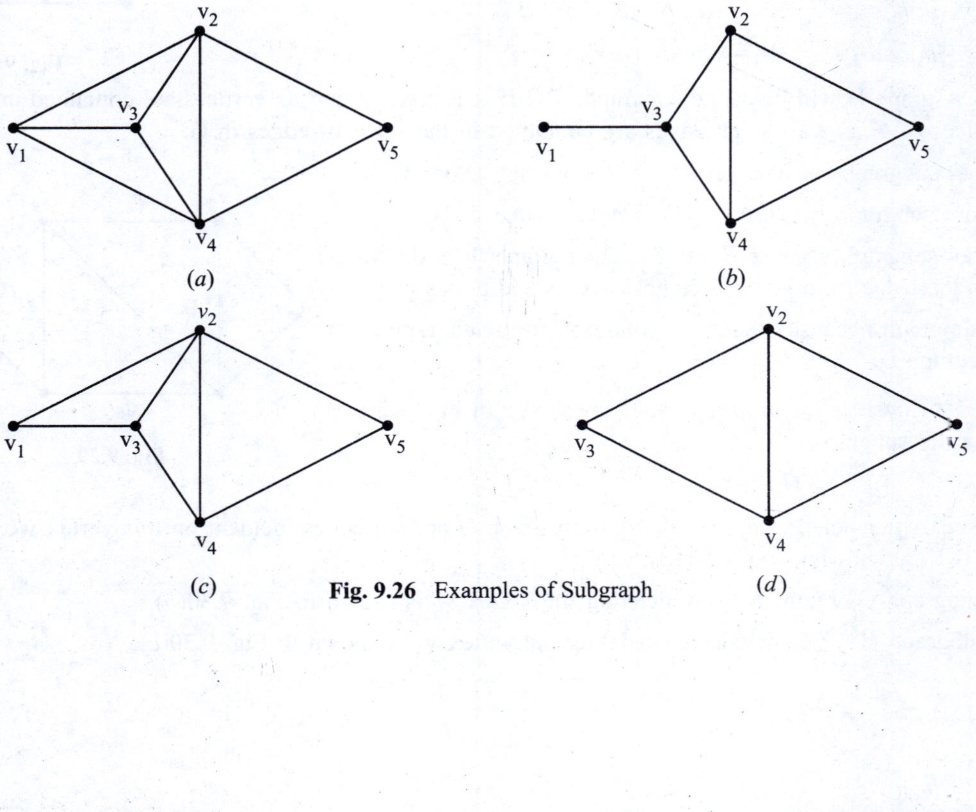

Fig. 9.26 Examples of Subgraph

9.4.1 Spanning Subgraphs

A subgraph H (V′, E′) of a given directed or undirected graph G (V, E) is called a spanning subgraph of graph G if H contains all vertices of G, i.e. V′ = V. Fig. 9.26(*b*) is the spanning subgraph of the graph shown in Fig. 9.26(*a*).

The directed graph shown in Fig. 9.27(*b*) is a subgraph of the graph shown in Fig. 9.27(*a*). But it is not a spanning subgraph. Directed graph in Fig. 9.27(*c*) and 9.27(*d*) are two of the $2^4 = 16$ possible spanning subgraphs.

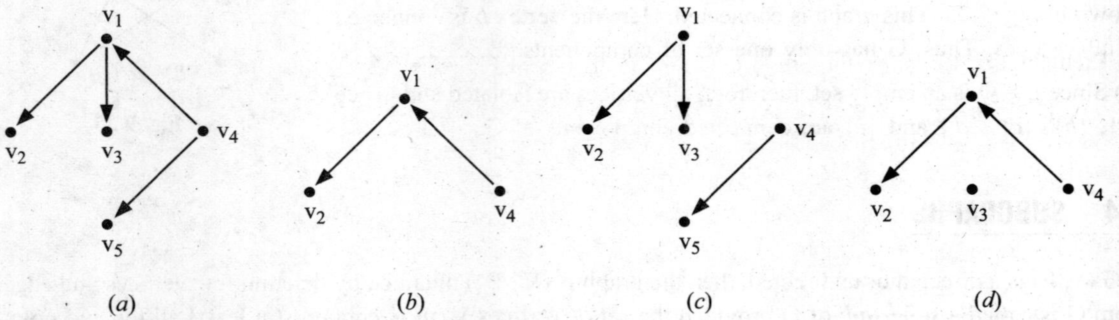

Fig. 9.27 Directed Spanning Subgraph

Example 13 Consider the graph G (V, E) shown in Fig. 9.28. Determine whether or not H (V′, E′) is a subgraph of G, where

(*a*) V′ = {*a*, *b*, *f*} , E′ = {(*a*, *b*), (*a*, *f*)}

(*b*) V′ = {*a*, *b*, *d*} , E′ = {(*a*, *b*), (*a*, *d*)}

(*c*) V′ = {*a*, *b*, *c*} , E′ = {(*a*, *b*), (*b*, *c*) (*b*, *d*)}

(*d*) V′ = {*b*, *c*, *d*} , E′ = {(*b*, *c*), (*b*, *d*)}

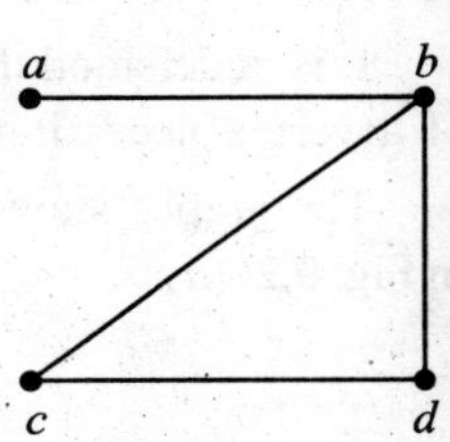

Fig. 9.28

Solution A graph H will be called subgraph of G if H is a graph and its vertices are contained in the set V of vertices of G as well as its edges are contained in the set E of edges in G.

(*a*) H is not subgraph because vertex $f \in$ V′ does not belong to V.

(*b*) H is not subgraph because (*a*, *d*) is not an edge in G.

(*c*) H is not subgraph because H itself is not a graph. It is due to the reason that edge (*b*, *d*) $\in$ E′ does not have its end points in V′.

(*d*) H is subgraph because it satisfies condition for a subgraph of the given graph G.

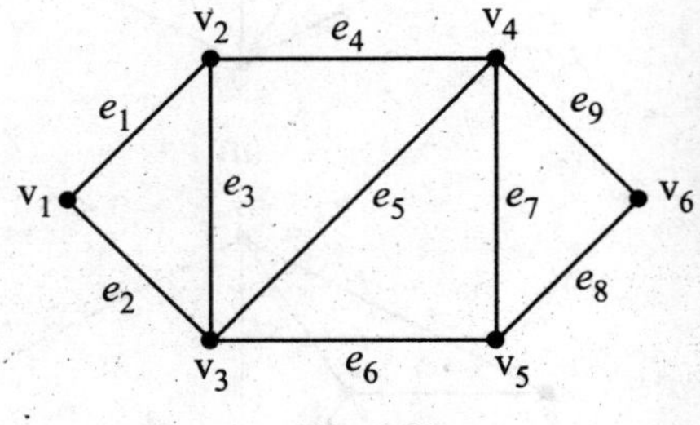

Fig. 9.29

Example 14 Consider the graph G (V, E) as shown in Fig. 9.29. Determine the subgraphs

(*a*) G–v_1 (*b*) G–v_5 (*c*) G–v_3

Solution (*a*) After deleting the vertex v_1 from graph G and all edges incident on this vertex we get the subgraph G–v_1 as shown in Fig. 9.30 (*a*).

(*b*) The subgraph G–v_5 obtained by deleting the vertex v_5 is shown in Fig. 9.30(*b*).

(*c*) The subgraph G–v_3 obtained after deleting the vertex v_3 is shown in Fig. 9.30(*c*).

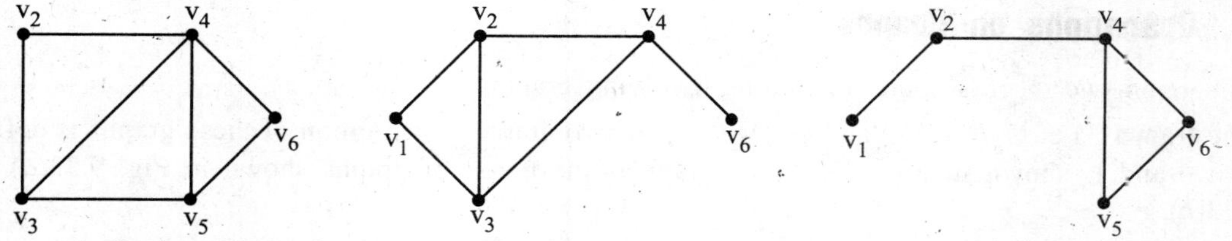

(*a*) Subgroup G–v_1 (*b*) Subgroup G–v_5 (*c*) Subgrcup G–v_3

Fig. 9.30

Example 15 Consider the graph G (V, E) as shown in Fig. 9.29. Determine the subgraphs

(*a*) G–e_1 (*b*) G–e_5 (*c*) G–e_3

Solution (*a*) After deleting the edge $e_1 = (v_1, v_2)$ from graph G, we get a subgraph G–e_1 as shown in Fig. 9.31(*a*).

(*b*) The subgraph G–e_5 obtained after deleting edge $e_5 = (v_3, v_5)$ is shown in Fig. 9.31(*b*).

(*c*) The subgraph G–e_3 obtained after deleting the edge $e_3 = (v_2, v_3)$ is shown in Fig. 9.31(*c*).

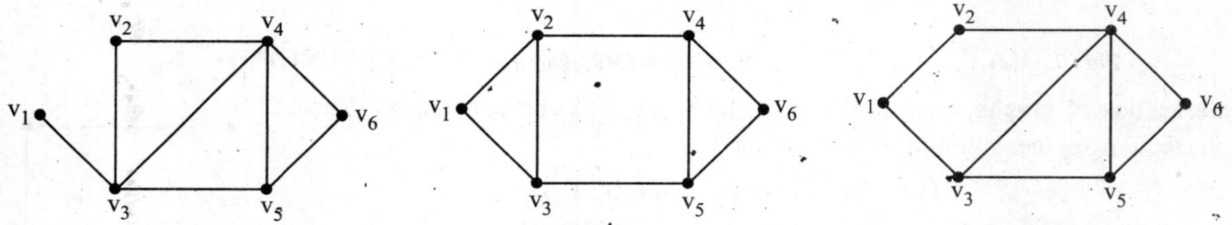

(*a*) Subgraph G–e_1 (*b*) Subgraph G–e_5 (*c*) Subgraph G–e_3

Fig. 9.31

9.4.2 Complement of a Subgraph

The complete of a subgraph H(V′, E′) of a graph G(V, E) is the subgraph $\overline{H}$ (V, E – E′). That is, the complement of a subgraph H is obtained by removing edges of H from G.

Figure 9.32(*c*) is the complement of the subgraph in Fig. 9.32(*b*) with respect to the graph in Fig. 9.32(*a*).

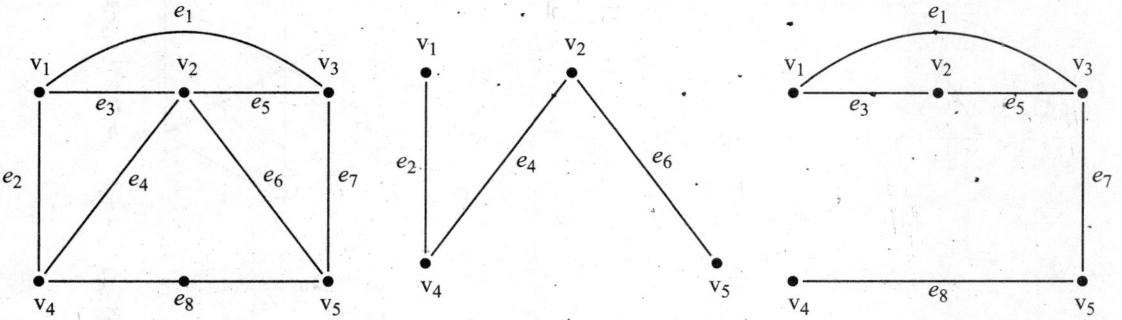

(*a*) Graph G (*b*) Subgraph H (*c*) Complement of H

Fig. 9.32 Complement of Subgraph

9.4.3 Operations on Graphs

In this section, we discuss some operations involving graphs.

Union of graphs Let $G_1 (V_1, E_1)$ and $G_2 (V_2, E_2)$ be two graphs. Then union of these graphs is defined as follows and is shown in Fig. 9.33. This graph is the union of graphs shown in Fig. 9.33(*a*) and Fig. 9.33(*b*).

$$G_1 \cup G_2 = G_3 (V_1 \cup V_2, E_1 \cup E_2).$$

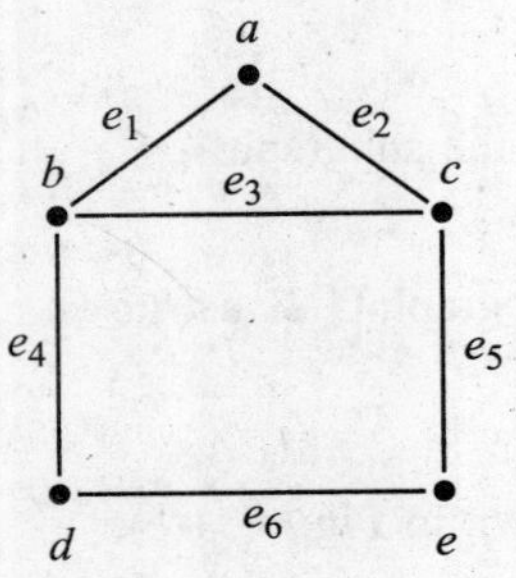

Fig. 9.33(*a*) G_1

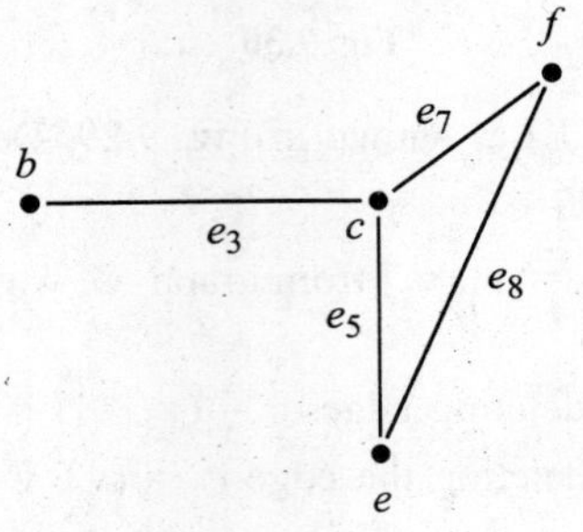

Fig. 9.33(*b*) G_2

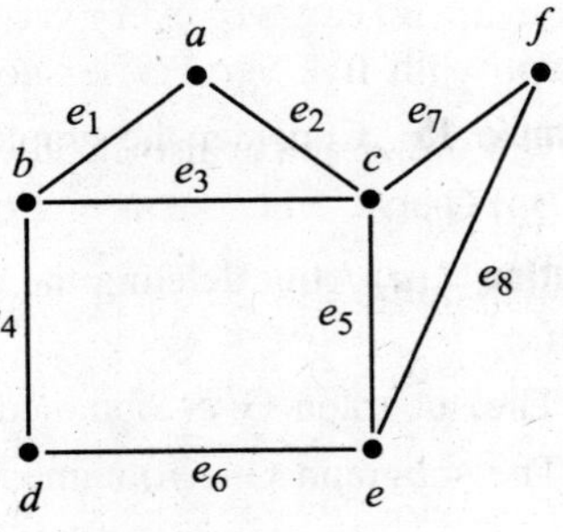

Fig. 9.33(*c*) $G_1 \cup G_2$

Intersection of graphs Let $G_1 (V_1, E_1)$ and $G_2 (V_2, E_2)$ be two graphs. Then intersection of these graphs is defined as:

$$G_1 \cap G_2 = G_3 (V_1 \cap V_2, E_1 \cap E_2)$$

The intersection of graphs G_1 and G_2 shown in Fig. 9.33(*a*) and Fig. 9.33(*b*) is given Fig. 9.34.

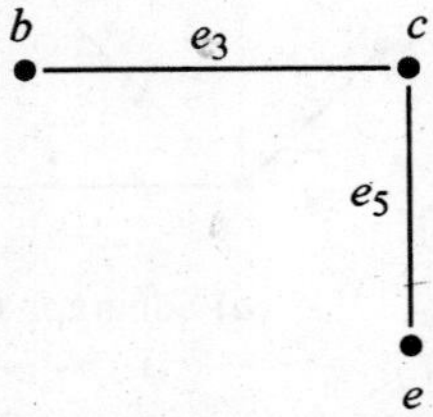

Fig. 9.34 $G_1 \cap G_2$

Complement The complement $\overline{G}$ of graph G is the graph with same vertex set as G with two adjacent vertices u and v that are not adjacent in G. Example of complement of G is shown in Fig. 9.35(*a*) and Fig. 9.35(*b*).

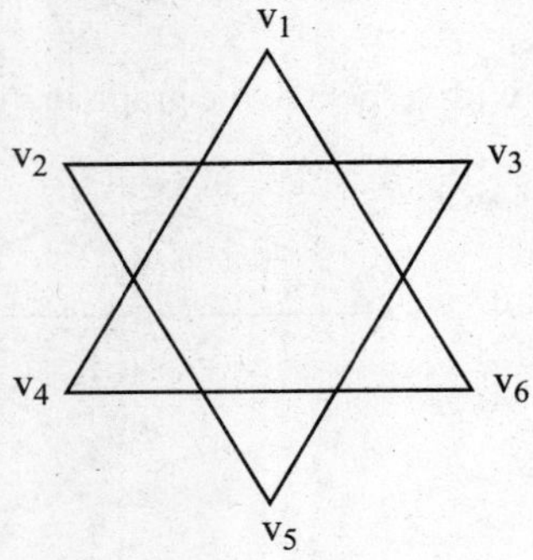

(*a*) Graph G

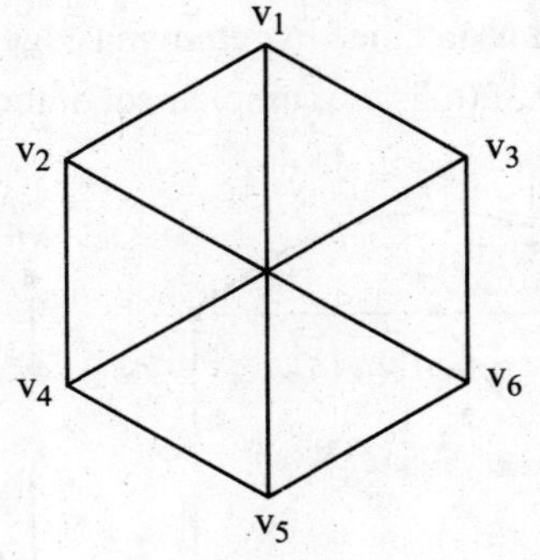

(*b*) Complement $\overline{G}$

Fig. 9.35 Example of Complement Graph

9.5 TYPES OF GRAPHS

Few important types of graphs are discussed here which are frequently used in graph theory.

Null Graph

A graph is called a *null graph* (also called totally disconnected graph) if there are no edges or every vertex is an isolated vertex. In Fig. 9.36, a null graph with five vertices is shown.

A null graph is also called *vertex graph*. Mathematically, a null graph is one whose edge set is empty, i.e. $E = \Phi$ and $V \neq \Phi$.

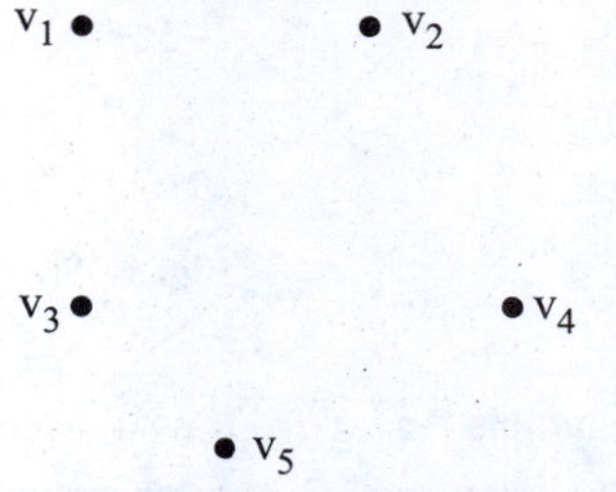

Fig. 9.36 Examp.e of Null Graph

Complete Graph

A complete graph is a loop-fee undirected graph where for all vertices $v_i, v_j \in V$, $v_i \neq v_j$, there exists an edge $\{v_i, v_j\}$. In other words, in a complete graph every vertex is connected to every other vertex, i.e. every pair of different vertices are adjacent.

A complete graph with n vertices is denoted by K_n and has $(1/2)\, n\,(n - 1)$ edges. Fig. 9.37(*a*) to Fig. 9.37(*e*) show undirected complete graph K_n, for $n = 1, 2, \ldots, 5$.

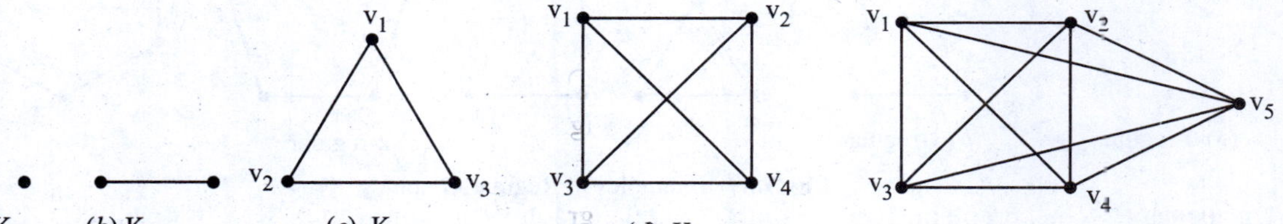

(*a*) K_1 (*b*) K_3 (*c*) K_3 (*d*) K_4 (*e*) K_5

Fig. 9.37 Examples of Undirected Complete Graph

Regular Graph

A graph in which every vertex has the same degree is called a *regular graph*. If every vertex has degree r, then graph is called a *regular graph* of degree r.

Every null graph is regular of degree zero and a complete graph K_n is regular of degree $n - 1$. Note that if a graph G has n vertices and is regular of degree r, then it has $(1/2)\, rn$ edges.

If in a regular undirected graph, every vertex has the same degree, say k, then graph is called k-regular.

Complementary Graph

Let G be a loop-free undirected graph with n vertices. The complement of G denoted as $\overline{G}$ is the subgraph of K_n consisting of the n vertices in G and all edges that are not in G. That is, if $G = K_n$, then $\overline{G}$ is a graph consisting of n vertices and no edges. In other words, $\overline{G}$ is the graph which has the same number of vertices as in G with two of them are connected if and only if they are not connected in G. That is, vertices which are adjacent in G are not adjacent in $\overline{G}$ and vice versa.

Figure 9.38 shows an undirected with 5 vertices. Its complement is shown in Fig. 9.38(*b*) where vertex v_2 is isolated.

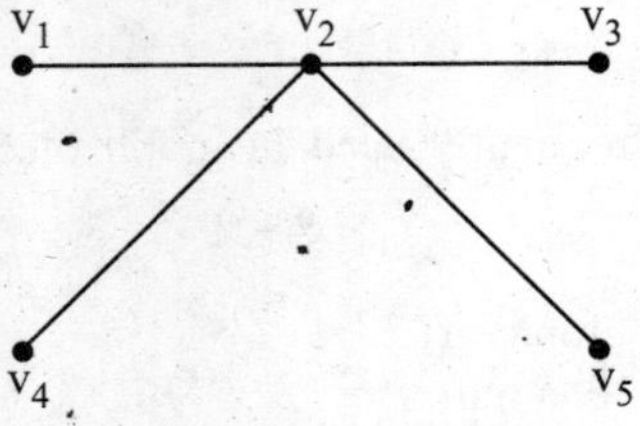

Fig. 9.38(*a*) Graph G

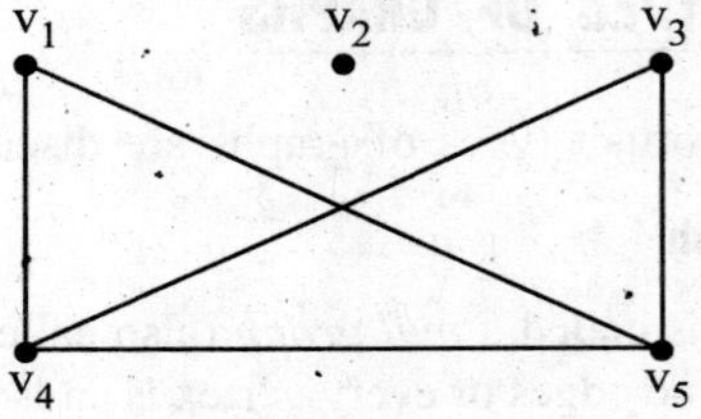

Fig. 9.38(*b*) Complement of G.

Example 16 (*a*) Draw the connected regular graphs of degree 0, 1 and 2.
(*b*) Draw two 3-regular graphs with six vertices.

Solution (*a*) The required connected regular graphs of degree 0, 1 and 2 are shown in Fig. 9.39(*a*), (*b*) and (*c*), respectively.

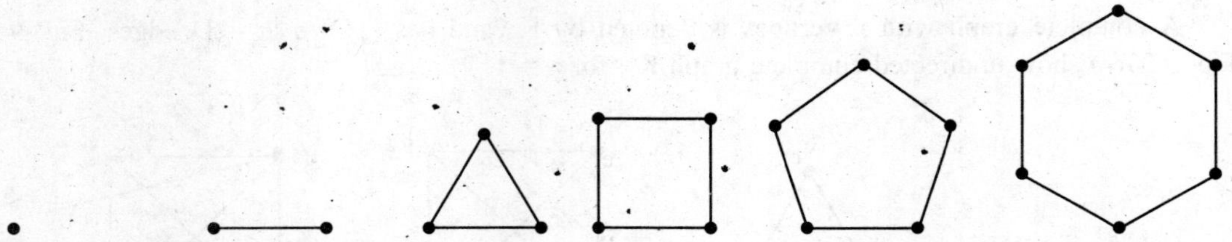

(*a*) 0-regular (*b*) 1-regular (*c*) 2-regular

Fig. 9.39 Examples of Regular Graph

(*b*) The required 3-regular graphs with six vertices are shown in Fig. 9.40(*a*) and Fig. 9.40(*b*).

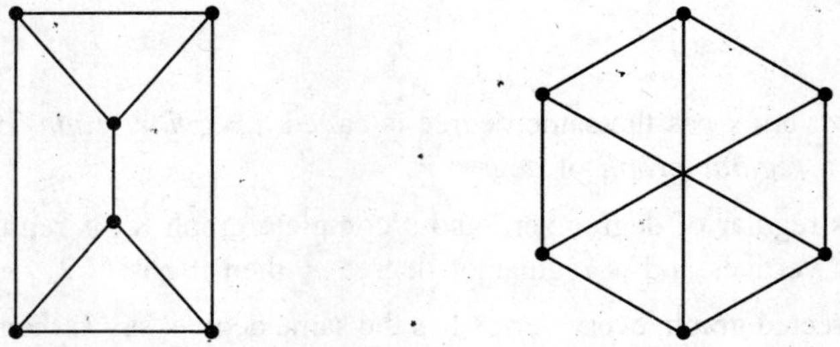

(*a*) (*b*)

Fig. 9.40 Examples of Regular Graph

Weighted Graph

A loop-free graph G is said to be a weighted graph if a positive real number called the *weight* is assigned to each of its edge $e = (v_i, v_j)$. Such a weight is denoted by wt (e) or wt (v_i, v_j). If $v_i, v_j \in V$ but $v_i, v_j \notin E$, then wt $(v_i, v_j) = \infty$.

For each $e = (v_i, v_j) \in E$, the weight may represent (*i*) the length of a road from v_i to v_j, (*ii*) the time taken to travel on this road from v_i to v_j, or (*iii*) the cost of travelling from v_i to v_j on this road.

Bipartite Graph

A loop-free graph G is called *bipartite* graph if it is possible to partition the vertex set V into two subsets V_1 and V_2 of vertices such that every edge (v_i, v_j) connects a vertex in V_1 and a vertex in V_2, i.e. there is no edge joining two vertices in set V_1 or V_2. In other words, a graph G is called *bipartite* graph when $V = V_1 \cup V_2$ and $V_1 \cap V_2 = \Phi$ and every edge of G is of the form $\{v_i, v_j\}$ with $v_i \in V_1$ and $V_j \in V_2$. The bipartite graphs $K_{2,4}$ and $K_{3,4}$ are shown in Fig. 9.41(*a*) and Fig. 9.41(*b*), respectively.

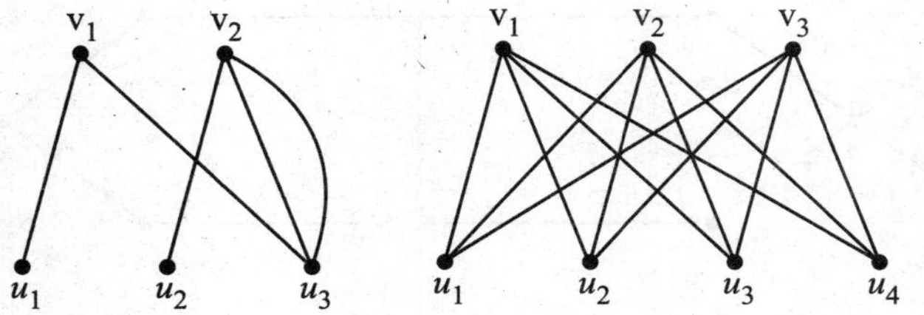

(*a*) $K_{2,4}$ (*b*) $K_{3,4}$

Fig. 9.41 Examples of Bipartite Graphs

If each vertex in V_1 is joined with every vertex in V_2, i.e. there is an edge between each pair of vertices in V_1 and V_2, then such a bipartite graph is called *complete bipartite graph.* Now if number of vertices in V_1 are *m* and number vertices in V_2 are *n*, then such a graph is denoted by $K_{m,n}$. A complete bipartite graph $K_{m,n}$ has $m + n$ vertices and *mn* edges. Fig. 9.41 is an example of complete bipartite graph.

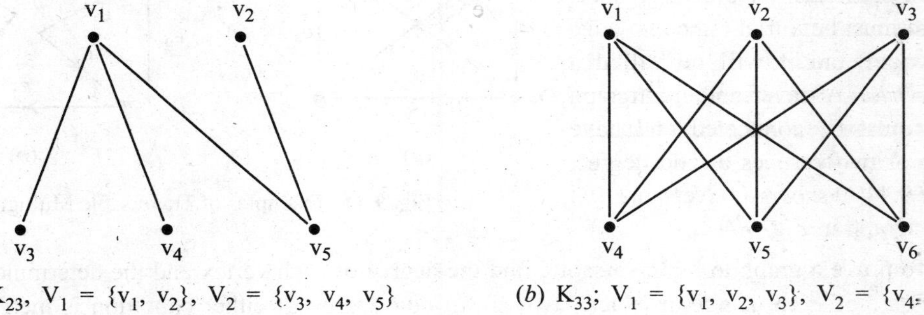

(*a*) K_{23}, $V_1 = \{v_1, v_2\}$, $V_2 = \{v_3, v_4, v_5\}$ (*b*) K_{33}; $V_1 = \{v_1, v_2, v_3\}$, $V_2 = \{v_4, v_5, v_6\}$

Fig. 9.42 Examples of Complete Bipartite Graph

Example 17 Show that the maximum number of edges in a complete bipartite graph with *n* vertices are $n^2/4$.

Solution Let G (V, E) be a bipartite graph and let V_1 and V_2 be partition of the vertex set V of G. Since G is complete bipartite graph, each vertex in V_1 is joined to each vertex in V_2 by exactly one edge. Thus, there are n_1n_2 edges in G, where $n_1 + n_2 = n$. But the maximum value of n_1n_2 subject to $n_1 + n_2 = n$ is $n^2/4$. Hence, the maximum number of edges in G are $n^2/4$.

Example 18 If the intersection of two paths in a graph G is disconnected, then their union has at least one circuit.

Solution Let P_1 and P_2 be two paths in G such that $P_1 \cap P_2$ is disconnected. Since $P_1 \cap P_2$ is disconnected, there is at least one pair of vertices u, v in $P_1 \cap P_2$ where both do not exists between them. However, u, v $\in P_1 \cap P_2$ implies that u, v belong to both paths P_1 and P_2 and therefore, there exists a path in P_1 and P_2 which connect vertices u and v.

As there is no path in $P_1 \cap P_2$ connecting vertices u and v, it may be assumed that paths in P_1 and P_2 connecting u and v are edge-disjoint. Thus, union of these two paths connecting u and v forms a circuit in $P_1 \cup P_2$ as shown in Fig. 9.43.

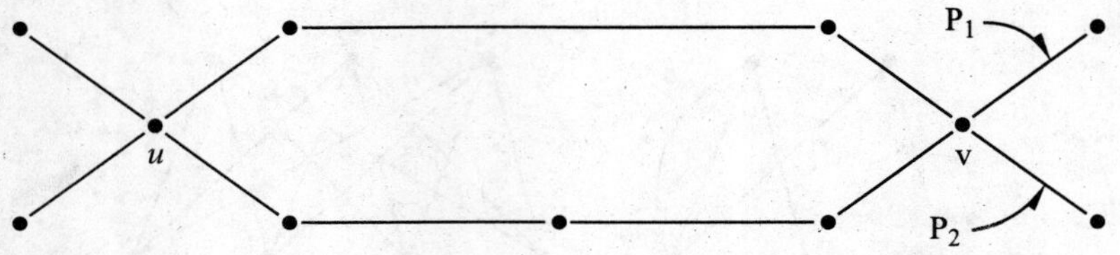

Fig. 9.43

Traversable Multigraph

A multigraph G is said to be traversable if it can be drawn without any break in the curve and without repeating any edge. That is, if there is a path which includes all vertices and uses each edge exactly once. Such a path must be a trial (since no edge is used twice) and it will be called a *traversable trial*. A traversable multigraph obviously must be connected and have either zero or two vertices of odd degree.

Figure 9.44(*b*) shows a traversable trial of the multigraph in Fig. 9.4(*a*).

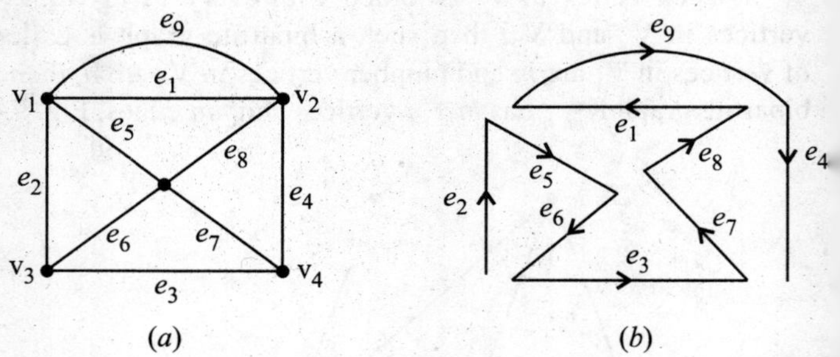

Fig. 9.44 Examples of Traversable Multigraph

Remark To prove a graph to be traversable, find the degree of each vertex and the determine whether all the vertices are of even degree or exactly two are of odd degree. If either condition is met, the graph is traversable.

Example 19 (*a*) Draw the complete graphs K_5 and K_6.

(*b*) Find the number of edges in the graphs K_{12} and K_{15}.

Solution (*a*) First draw given number of vertices and then draw edges from each vertex to every other vertex. The graphs for K_5 and K_6 are shown in Fig. 9.45(*a*) and Fig. 9.45(*b*).

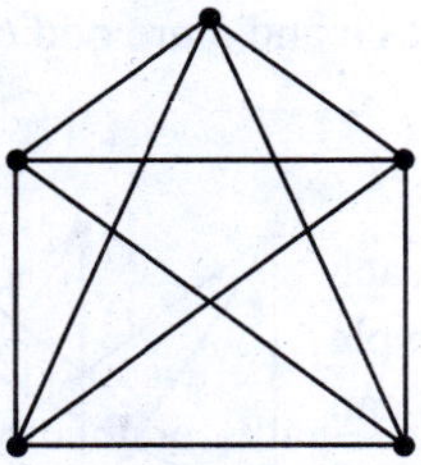

(a) K_5

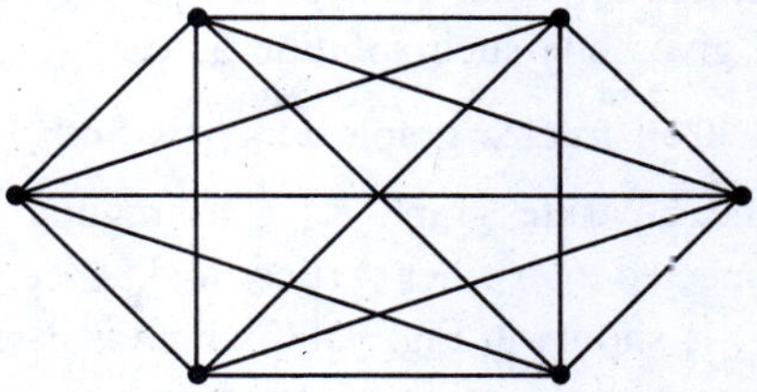

(b) K_6

Fig. 9.45

(b) Since each pair of vertices determines an edge. Thus, general formula for determining number of edges in a complete graph K_n is ${}^nC_2 = n(n-1)/2$, because there are ${}^nC_2$ ways of selecting two vertices out of n vertices. Hence, in K_{12}, $m = 12 \times 11/2 = 66$ edges and in K_{15}, $m = 15 \times 14/2 = 105$ edges.

Example 20 Find which of the graphs shown in Fig. 9.46 are traversable.

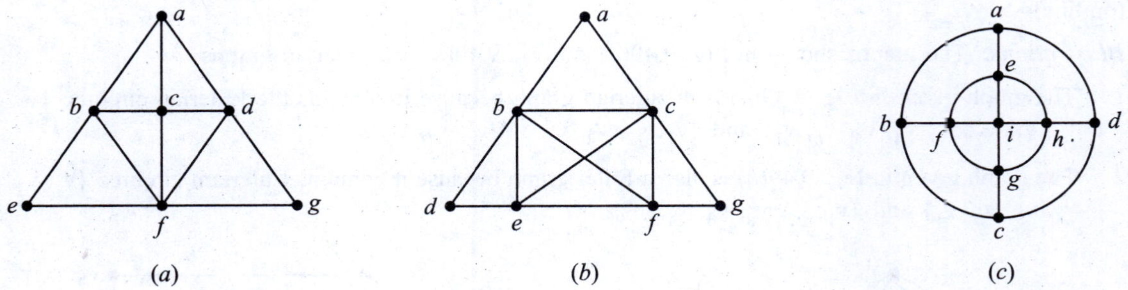

Fig. 9.46

Solution (a) Graph shown in Fig. 9.46(a) is traversable, as two of vertices a and f are of odd degree vertices.

(b) Graph shown in Fig. 9.45(b) is traversable, as two of vertices b and c are of odd degree vertices.

(c) Graph shown in Fig. 9.45(c) is not traversable, as four vertices a, b, c, and d are of odd degree.

Example 21 Find the traversable trail for the graph G (V, E), where G = {a, b, c, d} and V = {(a, c), (a, d), (b, c), (b, d), (c, d)}.

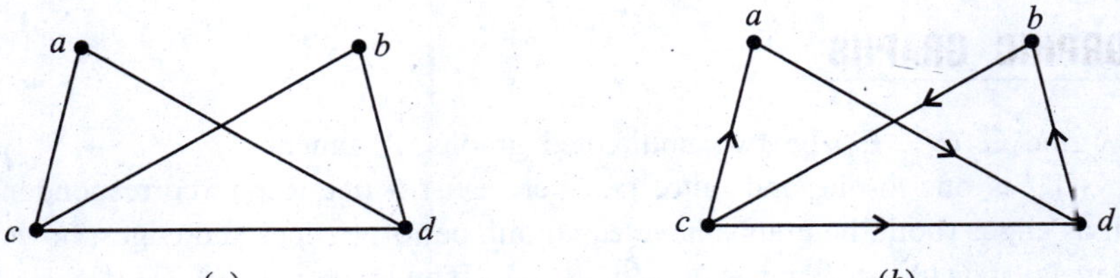

Fig. 9.47

Solution The graph G (V, E) is shown in Fig. 9.47. The traversable trial should begin from vertices of odd degree and end at the other vertices of odd degree. In Fig. 9.47(*a*), vertices *c* and *d* are odd degree vertices. Fig. 9.47(*b*) gives one such solution as $\alpha = \{c, a, d, b, c, d\}$.

Example 22 Construct a graph which is both regular and bipartite.

Solution The bipartite graph $K_{m,\,n}$ is regular of degree *m* because each vertex is connected to *m* other vertices and hence, its degree is *m*. An example of graph $K_{4,4}$ is shown in Fig. 9.48, which is 3-regular. In this graph we can delete *m* disjoint edges and each time obtain a regular graph of one degree less. These graphs may be disconnected, but in any case their connected components have the desired properties.

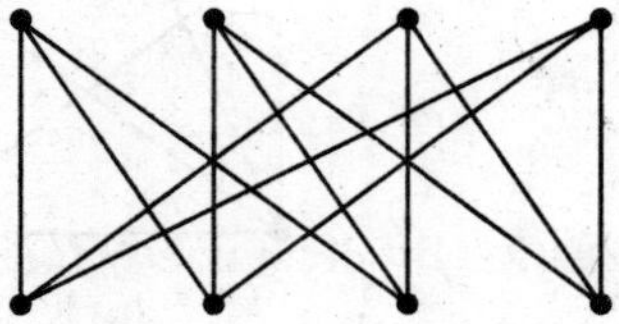

Fig. 9.48

Euler Graph

A graph is called *Eulerian or Euler graph* if it contains an Eulerian circuit (or line). A Eulerian circuit is always connected.

Eulerian trial A trial containing all the edges of the graph is called an Eulerian trial.

Euler circuit A closed trial (i.e. a circuit) containing all the edges of the graph is called an Eulerian circuit (on line).

Illustrations: The graphs shown in Fig. 9.49(*a*) and Fig.9.49(*b*) are Eulerian graphs

1. The graph given in Fig. 9.49(*a*) is a Eulerian graph because it contains the Eulerian circuits: $\{v_1, v_5, v_3, v_1, v_2, v_3, v_4, v_5, v_6, v_1\}$ and $\{v_1, v_2, v_3, v_4, v_5, v_6, v_1, v_3, v_5, v_1\}$.
2. The graph given in Fig. 9.49(*b*) is also a Euler graph because it contains Euleriam circuits: $\{v_1, v_2, v_3, v_4, v_5, v_6, v_1\}$ and $\{v_1, v_7, v_8, v_4, v_{10}, v_9, v_1\}$

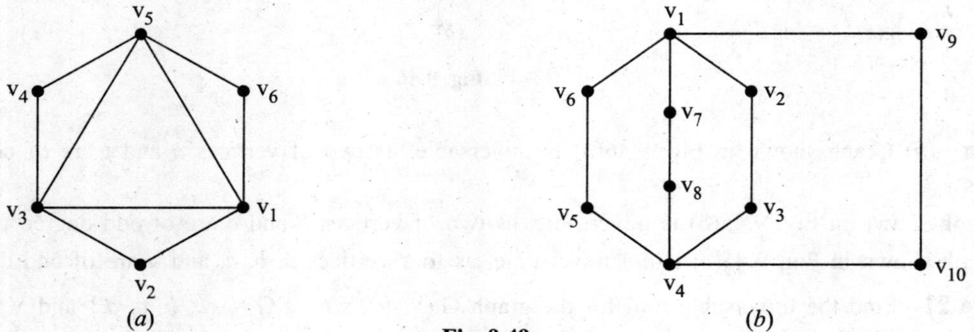

Fig. 9.49

9.6 ISOMORPHIC GRAPHS

Let $G_1 (V_1, E_1)$ and $G_2 (V_2, E_2)$ be two undirected graphs. A function $f: V_1 \to V_2$ is called a *graph isomorphism* if (*i*) *f* is one-to-one and onto, i.e. there exist a one-to-one correspondence between their vertices as well as edges (both the graphs have equal number of vertices and edges, however, vertices may have different levels) and (*ii*) for all u, v $\in V_1$, $\{u, v\} \in E_1$ if and only if $\{f(u), f(v)\} \in E_2$. If such a function exists, then graphs G_1 and G_2 are called *isomorphic graphs.* Graphs shown in Fig. 9.50(*a*) and Fig. 9.50(*b*) are examples of isomorphic graphs.

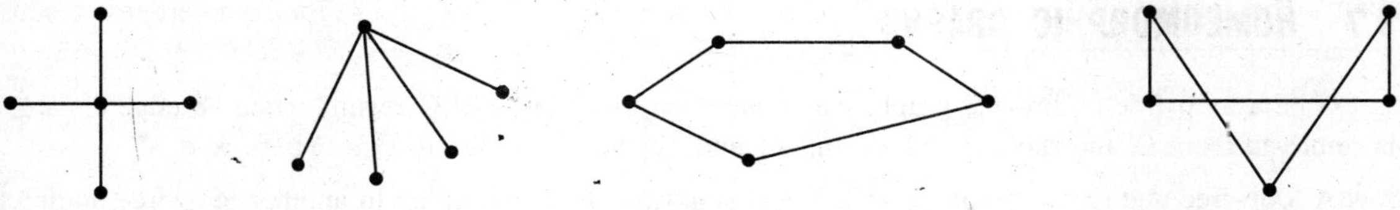

(a) (b)

Fig. 9.50 Examples of Isomorphic Graphs

Remark By definition two isomorphic graphs must have

(i) equal number of vertices and edges.
(ii) equal number of vertices with same degree.
(iii) It should be possible to start at any vertex in two graphs and find a circuit that includes every edge of the graph.

Example 23 Show that the graphs shown in Fig. 9.51(a) and Fig. 9.51(b) are isomorphic.

Solution To show that two graphs are isomorphic, we can arrange vertices from both graphs having same degrees in decreasing order of degrees. If both graphs contain vertices having same degree, then they are isomorphic, otherwise not. Note that no vertex is left out in any of the graphs which is not matching in terms of degree with the vertex in other graph. In Fig. 9.51(a) and Fig. 9.51(b), pairs of vertices in decreasing order of their degrees are follows:

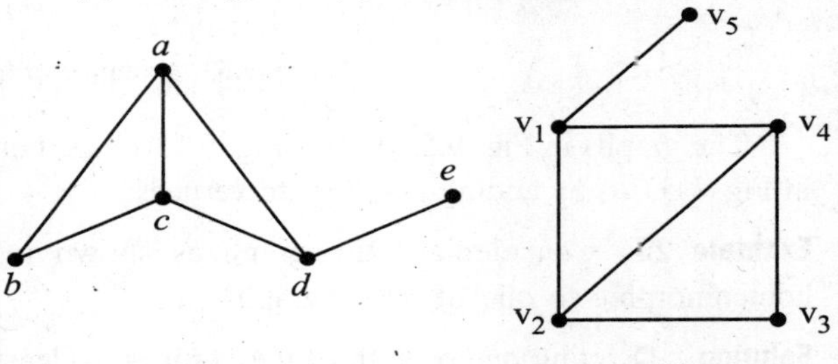

(a) (b)

Fig. 9.51

Degree 3: deg (c) = deg (v_2); deg (a) = deg (v_1); deg (d) = deg (v_4)
Degree 2: deg (b) = deg (v_3)
Degree 1: deg (e) = deg (v_1)

Since both the graphs contain vertices having same degrees, therefore, they are isomorphic.

Example 24 Show that the graphs shown in Fig. 9.52(a) and Fig. 9.52(b) are not isomorphic.

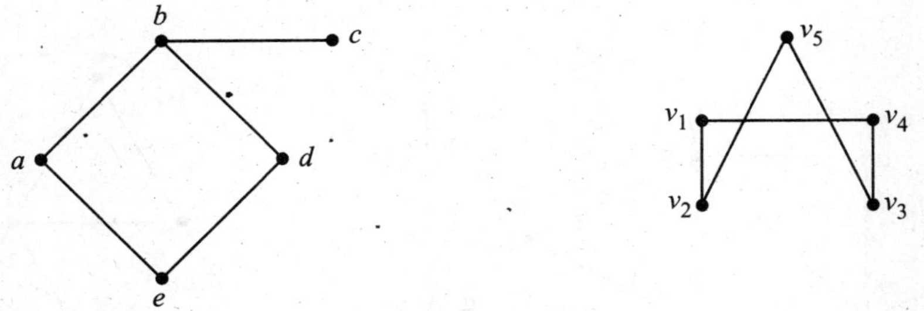

(a) (b)

Fig. 9.52

Solution The graph in Fig. 9.52(a) vertex b has degree 4, whereas graph in Fig. 9.52(b) there is no such vertex. Hence, graphs are not isomorphic.

9.7 HOMEOMORPHIC GRAPHS

Let G be a loop-free undirected graph. An *elementary subdivision* of G results when an edge $e = \{u, v\}$ is removed from G and then the edges {u, w} and {w, v} are added to G-*e*, where $w \notin V$.

A loop-free undirected graph $G_1 (V_1, E_1)$ is said to be homeomorphic to another loop-free undirected graph $G_2 (V_2, E_2)$ if they are isomorphic or if they can be obtained from the same loop-free undirected graph by a sequence of elementary subdivisions.

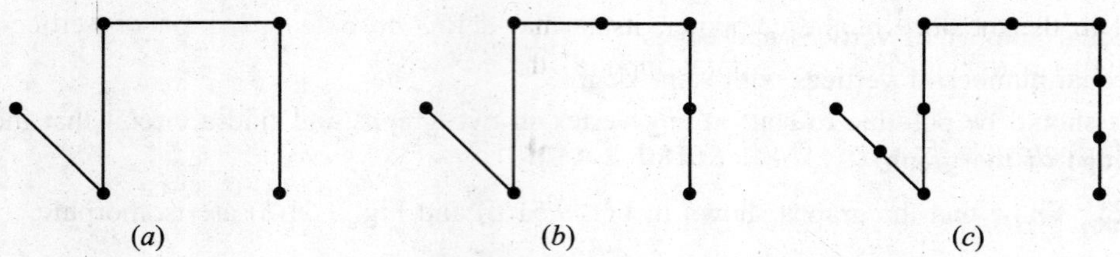

Fig. 9.53 Example of Homeomorphic Graphs

The graphs in Fig. 9.53(*b*) and Fig. 9.53(*c*) are homeomorphic, since each can be obtained from graph in Fig. 9.53(*a*) by adding appropriate vertices.

Example 25 Consider a finite graph as shown in Fig. 9.54. Show that it can be homeomorphic to one of its subgraphs.

Solution Deleting one or both of the vertices of degree 1 from G as shown in Fig. 9.54 yields a subgraph which is homeomorphic to G itself. Hence, a finite graph can be homeomorphic to one of its subgraphs other than itself.

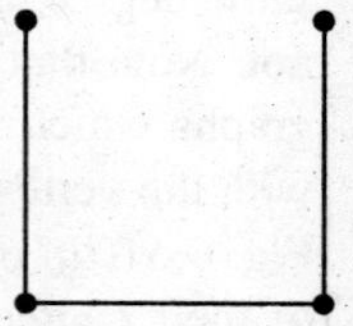

Fig. 9.54

Example 26 Verify whether two graphs in Fig. 9.55 are isomorphic.

Solution Each of the two graphs shown in Fig. 9.55 has six vertices and nine edges.

In Fig. 9.55(*b*), if we start at vertex u, then we find a circuit $u \to w \to v \to y \to w \to z \to y \to x \to v \to u$. But it is not possible to find a circuit in Fig. 9.55(*a*) where the only trails that includes each edge start at either *b* or *e* and then terminate at *e* or *b*, respectively. Hence, two graphs are not isomorphic.

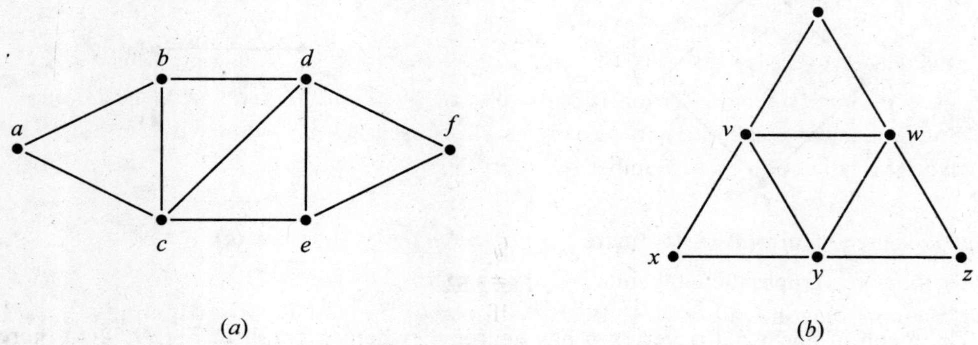

Fig. 9.55

9.8 REPRESENTATION OF GRAPHS IN COMPUTER MEMORY

Although graphs are geometric figures, their representation in computer memory requires lot of storage space. The following four representations are most common.

1. Adjacency matrix (or vertex-vertex adjacency matrix)
2. Incidence matrix (or vertex-edge incidence matrix)
3. Adjacency lists
4. Forward and reverse star representation.

In this chapter, we shall discuss only first two methods. However, irrespective of the ways a graph is maintained in the memory of the computer, its formal definition as a collection of vertices and edges (a collection of pair of vertices) is used as input to the computer.

9.8.1 Representation of Undirected Graph

(a) Adjacency matrix The adjacency matrix also called *sequential representation* method is suited to represent a graph G. A graph G with n vertices and m edges is said to be *dense* when representation of it requires kn^2, for some constant k, storage space in computer memory.

Adjacency matrix representation, stores the graph G consists of n vertices and no parallel edges, in the order $v_1, v_2, \ldots, v_n$ as an $n \times n$ matrix $A = [a_{ij}]$ defined as:

$$a_{ij} = \begin{cases} 1 & \text{if vertex } v_i \text{ is adjacent to vertex } v_j \text{ i.e., there is an edge between } v_i \text{ and } v_j \\ 0 & \text{otherwise.} \end{cases}$$

Figure 9.56 specifies 5×5 matrix representation for the graph consists of 5 vertices, shown in Fig. 9.57. Note that vertices are ordered v_1, v_2, v_3, v_4, v_5 and each edge (v_i, v_j) of G is represented twice by $a_{ij} = 1$ and $a_{ji} = 1$.

$$A = \begin{array}{c} \\ v_1 \\ v_2 \\ v_3 \\ v_4 \\ v_5 \end{array} \begin{array}{c} \begin{array}{ccccc} v_1 & v_2 & v_3 & v_4 & v_5 \end{array} \\ \begin{bmatrix} 0 & 1 & 1 & 0 & 0 \\ 1 & 0 & 1 & 1 & 0 \\ 1 & 1 & 0 & 1 & 1 \\ 0 & 1 & 1 & 0 & 1 \\ 0 & 0 & 1 & 1 & 0 \end{bmatrix} \end{array}$$

Fig. 9.56 Adjacency Matrix

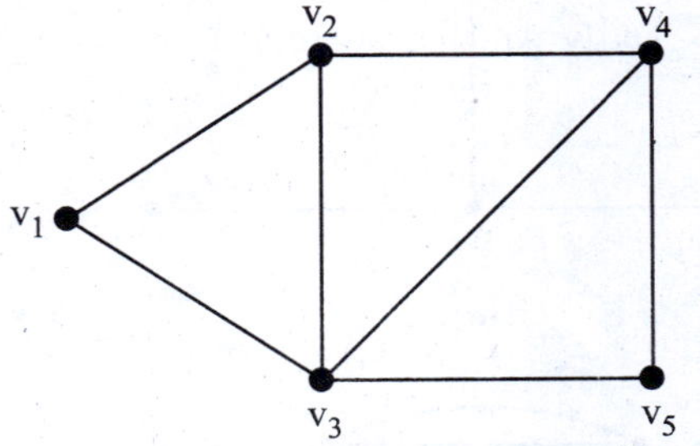

Fig. 9.57 Graph Example

Since adjacency matrix representation is based on ordering of vertices in a graph, therefore, for a graph with n vertices, we may construct as many as $n!$ different adjacency matrices because there are $n!$ different ordering of n vertices. A number of other important results about the adjacency matrix are as follows:

(*i*) The adjacency matrix A is symmetric, i.e. $a_{ij} = a_{ji}$ for all i and j.

(*ii*) For a loop-free graph, the element $a_{ij} = 1$, $i \neq j$ represents degree of a vertex. The entries along the principal diagonal are all 0's. But, a self-loop at the vertex corresponding to $a_{ji} = 1$.

(*iii*) A graph G is connected graph if and only if the matrix

$$B = A + A^2 + \ldots + A^{n-1}$$

has no zero entries off the main diagonal, where A is the adjcency matrix of G with n vertices $v_1, v_2, \ldots, v_n$.

(*iv*) The adjacent matrix has n^2 elements. Adjacency matrices are useful for determination of paths in graphs. For example, it is easy to show that there is a path of length r (i.e. having r edges) from vertex v_i to v_j if and only if $A^r = 1$, where A^r is the rth matrix product.

(*v*) The number of non-zero elements in the matrix is equal to the sum of degrees of all vertices of the graph.

Example 27 Use adjacency matrix to represent the graph shown in Fig. 9.58(*a*), (*b*), (*c*) and (*d*).

Solution (*a*) We order the vertices in Fig. 9.58(*a*) as v_1, v_2, v_3, v_4 and v_5. The adjacency matrix representing this graph will be the square matrix of order 5 as given below:

$$A = \begin{array}{c} \\ v_1 \\ v_2 \\ v_3 \\ v_4 \\ v_5 \end{array}\begin{array}{c} \begin{array}{ccccc} v_1 & v_2 & v_3 & v_4 & v_5 \end{array} \\ \begin{bmatrix} 0 & 1 & 1 & 0 & 0 \\ 1 & 0 & 0 & 1 & 0 \\ 1 & 0 & 0 & 1 & 0 \\ 0 & 1 & 1 & 0 & 1 \\ 0 & 0 & 0 & 1 & 0 \end{bmatrix} \end{array}$$

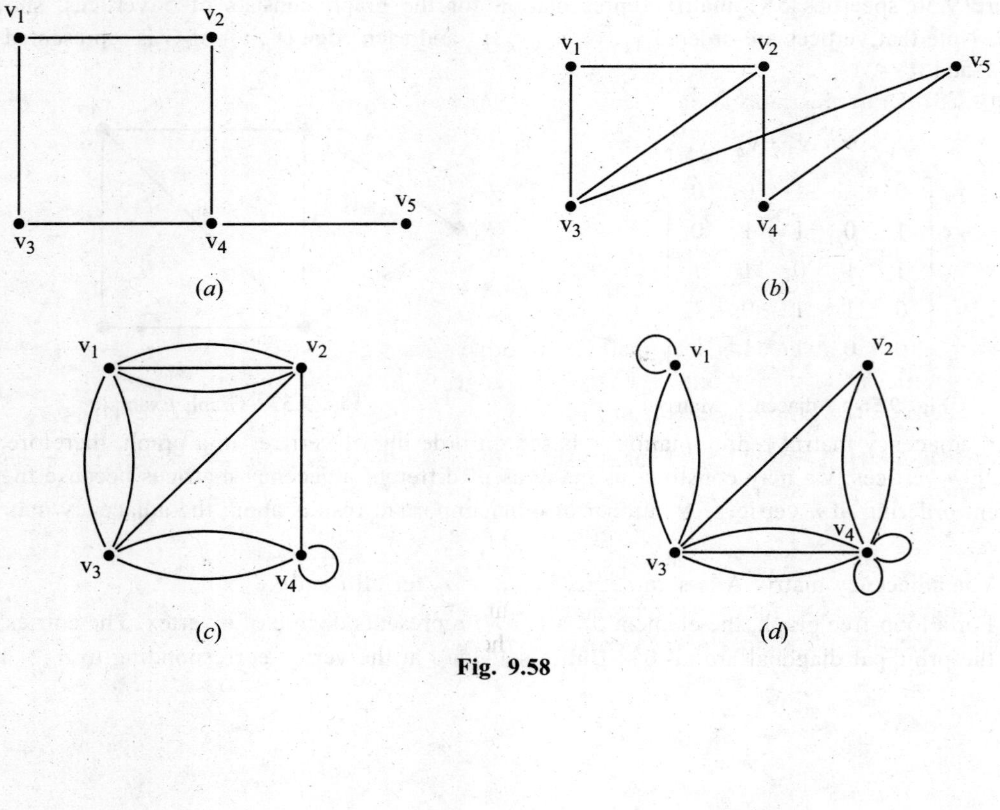

Fig. 9.58

(*b*) We order the vertices in Fig. 9.58(*b*) as v_1, v_2, v_3, v_4 and v_5. The adjacency matrix representing the graph is given below:

$$A = \begin{array}{c} \\ v_1 \\ v_2 \\ v_3 \\ v_4 \\ v_5 \end{array}\begin{array}{c} \begin{array}{ccccc} v_1 & v_2 & v_3 & v_4 & v_5 \end{array} \\ \begin{bmatrix} 0 & 1 & 1 & 0 & 0 \\ 1 & 0 & 1 & 1 & 0 \\ 1 & 1 & 0 & 0 & 1 \\ 0 & 1 & 0 & 0 & 1 \\ 0 & 0 & 1 & 1 & 0 \end{bmatrix} \end{array}$$

(*c*) We order the vertices in Fig. 9.58(*c*) as v_1, v_2, v_3 and v_4. The adjacency matrix representing this graph is given below:

$$A = \begin{array}{c} \\ v_1 \\ v_2 \\ v_3 \\ v_4 \end{array}\begin{array}{c} \begin{array}{cccc} v_1 & v_2 & v_3 & v_4 \end{array} \\ \begin{bmatrix} 0 & 3 & 2 & 0 \\ 3 & 0 & 1 & 1 \\ 2 & 1 & 0 & 2 \\ 0 & 1 & 2 & 1 \end{bmatrix} \end{array}$$

(*d*) We order the vertices in Fig. 9.58(*d*) as v_1, v_2, v_3 and v_4. The adjacency matrix representing this graph is as follows:

$$A = \begin{array}{c} \\ v_1 \\ v_2 \\ v_3 \\ v_4 \end{array}\begin{array}{c} \begin{array}{cccc} v_1 & v_2 & v_3 & v_4 \end{array} \\ \begin{bmatrix} 1 & 2 & 0 & 0 \\ 0 & 0 & 1 & 2 \\ 2 & 1 & 0 & 1 \\ 0 & 2 & 1 & 2 \end{bmatrix} \end{array}$$

Example 28 Draw diagraph G corresponding to adjacent matrix

$$A = \begin{array}{c} \\ v_1 \\ v_2 \\ v_3 \\ v_4 \end{array}\begin{array}{c} \begin{array}{cccc} v_1 & v_2 & v_3 & v_4 \end{array} \\ \begin{bmatrix} 1 & 1 & 0 & 0 \\ 0 & 0 & 1 & 0 \\ 0 & 1 & 0 & 1 \\ 1 & 0 & 0 & 1 \end{bmatrix} \end{array}.$$

Solution Since the given adjacency matrix is of order four, the graph G has four vertices v_1, v_2, v_3 and v_4. Draw an edge v_i to v_j, where $a_{ij} = 1$. The required graph is shown in Fig. 9.59.

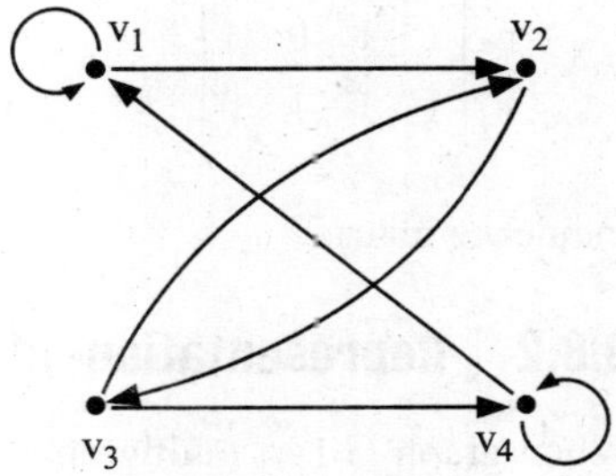

Fig. 9.59

(b) Incidence matrix An incidence matrix stores the graph G consists of *n* vertices and *m* edges as an $n \times m$ matrix $M = [m_{ij}]$ defined as:

$$m_{ij} = \begin{cases} 1 & \text{if vertex } v_i \text{ is incident on edge } e_j \\ 0 & \text{otherwise} \end{cases}$$

The matrix has a row for every vertex and a column for every edge. The non-zero elements in the matrix is equal to the sum of degrees of all vertices of the graph.

Figure 9.60 specifies 5 × 8 matrix representation for the graph consists of 5 vertices and 8 edges, shown in Fig. 9.61.

$$M = \begin{array}{c} \\ v_1 \\ v_2 \\ v_3 \\ v_4 \\ v_5 \end{array} \begin{array}{c} \begin{array}{cccccccc} e_1 & e_2 & e_3 & e_4 & e_5 & e_6 & e_7 & e_8 \end{array} \\ \begin{bmatrix} 1 & 1 & 1 & 0 & 1 & 0 & 0 & 0 \\ 1 & 0 & 0 & 1 & 0 & 0 & 0 & 0 \\ 0 & 0 & 1 & 1 & 0 & 0 & 1 & 1 \\ 0 & 0 & 0 & 0 & 1 & 1 & 0 & 1 \\ 0 & 1 & 0 & 0 & 0 & 1 & 1 & 0 \end{bmatrix} \end{array}$$

Fig. 9.60 Incidence Matrix

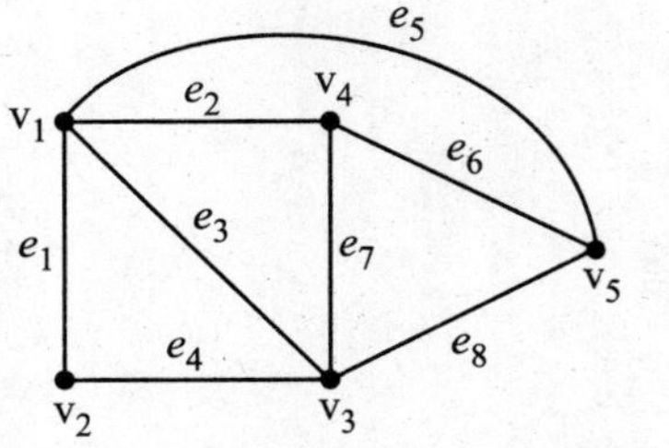

Fig. 9.61 Graph Example

We make the following observations about incident matrix M of graph G:

(*i*) Each column contains exactly two unit elements.

(*ii*) A row with all zero elements corresponds to an isolated vertex.

(*iii*) A row with a single unit element corresponds to a pendent vertex.

(*iv*) The number of unit elements in row i represents the degree of the vertex v_i.

(*v*) If graph G is connected with m vertices, then rank of incidence matrix M is $n - 1$.

Example 29 Use adjacency matrix and incidence matrix to represent the graph shown in Fig. 9.62.

Solution We order the vertices in Fig. 9.62 as v_1, v_2, v_3 and v_4, the adjacency matrix representing the graph is given below:

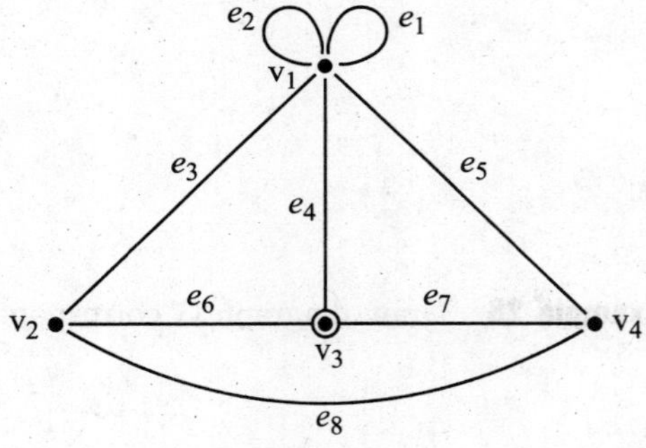

Fig. 9.62

$$A = \begin{array}{c} \\ v_1 \\ v_2 \\ v_3 \\ v_4 \end{array} \begin{array}{c} \begin{array}{cccc} v_1 & v_2 & v_3 & v_4 \end{array} \\ \begin{bmatrix} 2 & 1 & 1 & 1 \\ 1 & 0 & 1 & 0 \\ 1 & 1 & 0 & 1 \\ 1 & 1 & 1 & 0 \end{bmatrix} \end{array}, \quad M = \begin{array}{c} \\ v_1 \\ v_2 \\ v_3 \\ v_4 \end{array} \begin{array}{c} \begin{array}{cccccccc} e_1 & e_2 & e_3 & e_4 & e_5 & e_6 & e_7 & e_8 \end{array} \\ \begin{bmatrix} 1 & 1 & 1 & 1 & 1 & 0 & 0 & 0 \\ 0 & 0 & 1 & 0 & 0 & 1 & 0 & 1 \\ 0 & 0 & 0 & 1 & 0 & 1 & 1 & 0 \\ 0 & 0 & 0 & 0 & 1 & 0 & 1 & 1 \end{bmatrix} \end{array}$$

Adjacency matrix

Incidence matrix

9.8.2 Representation of Multigraph

When graph G is a multigraph, then a_{ij} denote the number of edges (v_i, v_j). Thus, only adjacency matrix is used to represent multigraphs.

Adjacency matrix For a multigraph G consists of n vertices, an $n \times n$ adjacency matrix $A = [a_{ij}]$ is defined as

$$a_{ij} = \begin{cases} m & \text{if there are one or more edges between vertex } v_i \text{ and } v_j \text{ ; } m = \text{number of edges} \\ 0 & \text{otherwise} \end{cases}$$

The following adjacency 4 × 4 matrix representation for a multigraph consists of 4 vertices is shown in Fig. 9.63.

$$A = \begin{array}{c} \\ v_1 \\ v_2 \\ v_3 \\ v_4 \end{array} \begin{array}{c} \begin{array}{cccc} v_1 & v_2 & v_3 & v_4 \end{array} \\ \begin{bmatrix} 1 & 3 & 0 & 0 \\ 3 & 0 & 1 & 1 \\ 0 & 1 & 2 & 2 \\ 0 & 1 & 2 & 0 \end{bmatrix} \end{array}$$

Adjacency Matrix

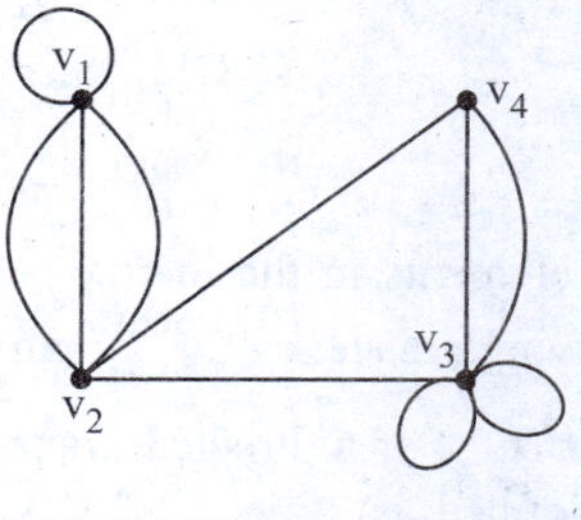

Fig. 9.63 Multigraph Example

Example 30 Determine the number of loops and multiple edges in a multigraph G from its adjacency matrix

$$A = \begin{bmatrix} 1 & 1 & 2 & 0 \\ 1 & 2 & 1 & 3 \\ 2 & 1 & 0 & 1 \\ 0 & 3 & 1 & 0 \end{bmatrix}$$

Draw the graph G and check your answer.

Solution Since adjacency matrix A is a square matrix of order 4, graph G has four vertices v_1, v_2, v_3, v_4. The diagonal of A is indicating the vertices having loops because these entries indicate the number of edges originating and terminating at the same vertex. Thus, there are three loops: one at vertex v_1 and two at vertex v_2 as shown in Fig. 9.64.

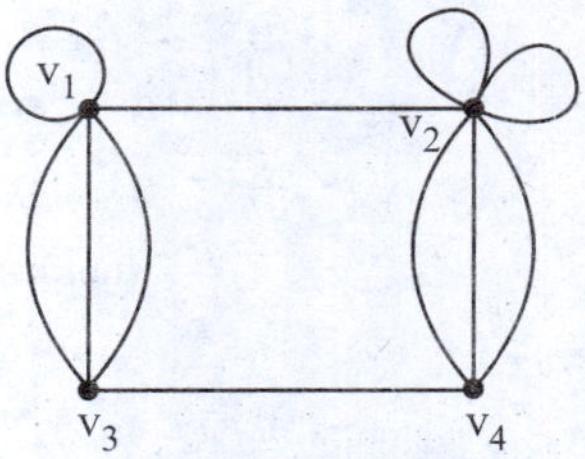

Fig. 9.64 Multigraph

To find the number of multiple edges, add the number of entries greater than 1 below the main diagonal and add the number of entries greater than 1 along the main diagonal. Main diagonal is also included since multiple loops are also multiple edges. Entries above the main diagonal are not considered because matrix is symmetric. Thus, in Fig. 9.64, there are seven multiple edges, 2 edges from v_3 to v_1; 3 edges from v_4 to v_2; and 2 loops at v_2.

9.8.3 Representation of Directed Graph

A directed graph, also called a digraph is denoted by drawing a diagram rather than explicitly listing its vertices and edges.

$$A = \begin{array}{c} v_1 \\ v_2 \\ v_3 \\ v_4 \\ v_5 \end{array} \begin{bmatrix} 0 & 0 & 1 & 1 & 1 \\ 1 & 0 & 0 & 1 & 0 \\ 0 & 1 & 0 & 0 & 0 \\ 0 & 0 & 0 & 1 & 1 \\ 0 & 0 & 1 & 0 & 0 \end{bmatrix}$$

Adjacency Matrix

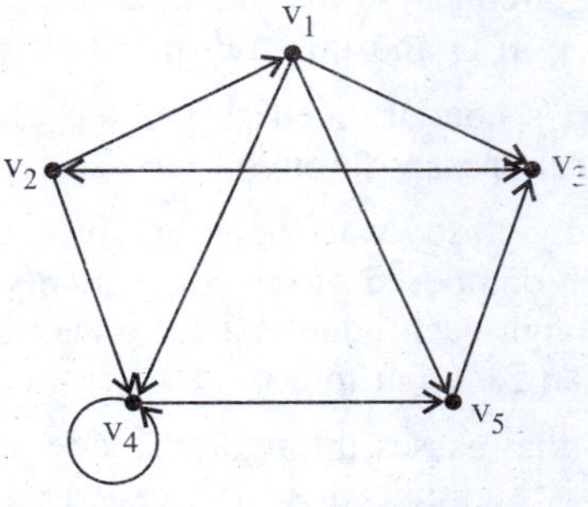

Fig. 9.65 Digraph Example

Adjacency matrix For a directed graph G consists of n vertices, an $n \times n$ adjacency matrix $A = [a_{ij}]$ is defined as:

$$a_{ij} = \begin{cases} 1 & \text{if edges beginning at vertex } v_i \text{ and ending at } v_j \\ 0 & \text{otherwise} \end{cases}$$

The non-zero elements in the matrix is equal to the number of edges in the digraph.

The following adjacency 5×5 matrix representation for the digraph is given in Fig. 9.65.

Incidence matrix For a directed graph G consists of n vertices and m edges, an $n \times m$ incidence matrix $M = [m_{ij}]$ is defined as

$$M_{ij} = \begin{cases} 1 & \text{if } v_i \text{ is initial vertex of edge } e_j \\ -1 & \text{if } v_i \text{ is final vertex of edge } e_j \\ 0 & \text{if } v_i \text{ not incident on edge } e_j \end{cases}$$

The non-zero elements in the matrix is equal to the number of edges in the digraph.

The following 5×5 incidence matrix representation for the digraph is given in Fig. 9.66.

$$M = \begin{array}{c} \\ v_1 \\ v_2 \\ v_3 \\ v_4 \\ v_5 \end{array} \begin{array}{c} \begin{array}{ccccccc} e_1 & e_2 & e_3 & e_4 & e_5 & e_6 & e_7 \end{array} \\ \begin{bmatrix} -1 & 1 & 0 & -1 & 0 & 0 & 0 \\ 1 & 0 & 1 & 0 & 1 & 0 & 0 \\ 0 & -1 & -1 & 0 & 0 & 0 & 1 \\ 0 & 0 & 0 & 1 & 0 & -1 & 0 \\ 0 & 0 & 0 & 0 & -1 & 1 & -1 \end{bmatrix} \end{array}$$

Incidence Matrix

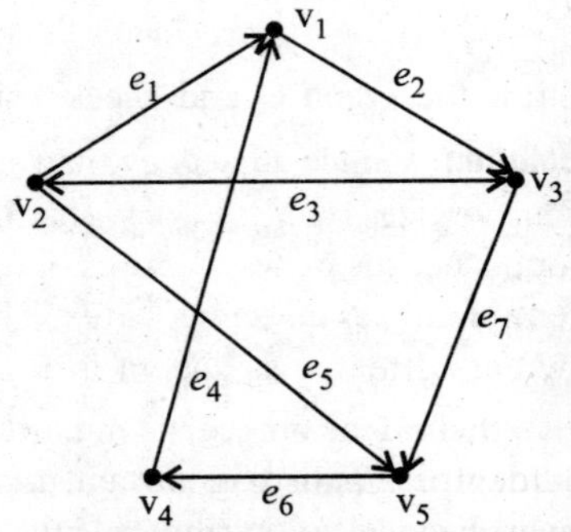

Fig. 9.66 Digraph Example

9.9 EULERIAN AND HAMILTONIAN GRAPHS

A directed graph contains a directed walk through each edge exactly once if and only if the indegree of each vertex equals its outdegree. For an undirected graph a similar result holds, which is known as *Euler's theorem*.

Definition An undirected graph or multigraph G with no isolated vertices is said to have an *Euler circuit* if there is a circuit in G that traverses every edge of the graph exactly once. If there is an open trail from vertex u to v in G and this trail traverses each edge in G exactly once, the trail is called *Euler trail*.

In Euler's honour, even graphs are often referred as *Eulerian graphs* and to closed walk that starts at some vertex, passes through each edge exactly once and returns to the starting vertex, as *Eulerian tours*.

Since any closed walk in an undirected graph enters and leaves any vertex the same number of times, the subgraph composed of the edges in any closed walk is even. Thus, if the graph contains a closed walk passing through each edge exactly once, the graph must be even. Conversely, if the graph is even, then it contains an Eulerian tour.

A path that passes through each edge exactly once but vertices may be repeated is called Euler path. If the path is a circuit, then it is called an *Euler circuit*.

Definition A graph that contains an Euler tour (path or circuit) is called *Eulerian graph.*

One of the oldest problem involving graphs is the problem of seven bridges of Konigsberg. A map of the city of Konigsberg (in East Prussia, now Kaliningrad in Russia) is shown in Fig. 9.67(*a*), which is divided into four sections by the Pregel River. Seven bridges connected these regions. The problem was to take a walk about the city so as to cross each bridge exactly once and returning to the start point. This problem of crossing each of bridges only once is equivalent to find a closed walk through a multi-graph shown in Fig. 9.67(*b*). The Swiss mathematician Leonhard Euler solved this problem in 1736. He discovered that there is no Eulerian circuit for Konigsberg bridge multi-graph in which all vertices have odd degree.

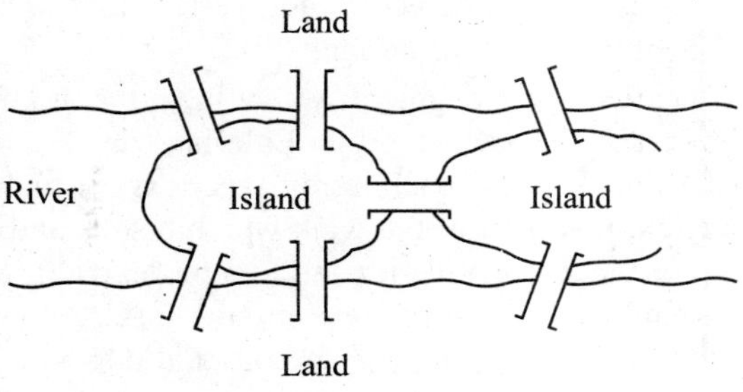

(*a*)

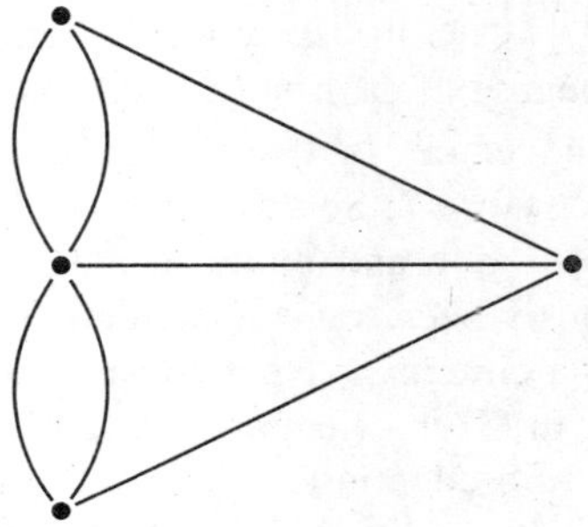

(*b*)

Fig. 9.67 Konigsberg Graph

Euler Theorem 9.6 An indirected graph G is Eulerian if and only if it is connected and each vertex has even degree.

Proof Let graph G has an Euler circuit or Eulerian. Then an Eulerian trial starts and terminates at the same vertex in G. Suppose u is the starting vertex of the Euler circuit. If we move along a trial, then each time the circuit come to a vertex v of G twice one for incoming (indegree) and one for outgoing (outdegree). Thus, the circuit traversed two edges that are incident with u or loop at v. In either case, occurrence of v contributes 2 to its degree, so deg (v) is even. This is also the case with starting vertex u, so deg (u) is even.

Conversely, let G be connected and every vertex is of even degree. If G has *n* edges, then select an edge *e* to construct Euler circuit. Suppose we begin a trial T_1 from an edge and then keep on adding edges one after the other. If T_1 is not closed, i.e. T_1 begins at vertex u and ends at vertex $v \neq u$, then there are only odd number of the edges incident on vertex v in T_1 and it can further be extended by another edge incident on v. Since the graph G is finite, we continue this process of extending the trial T_1 until we reach to the initial vertex u. If this process covers all edges of G, then there exists an Euler circuit and the trial T_1 is an Eulerian trial.

Suppose trial T_1 starting from vertex u does not ends at u, and leaves few edges uncovered. Then consider a subgraph H of G by deleting all edges of T_1. Each vertex of subgraph H has even degree because T_1 contains an even number of edges incident to any vertex.

Since graph G is connected, consider an edge *e'* of H which has terminating vertex u' in T_1. Construct another trial T_2 in H starting at vertex u' and extend it until returns to u'. It is possible because all vertices in H has even degree. Combining T_1 and T_2, we form a closed trial covering all the edges in G to obtain an Eulerian trial and so G is Eulerian.

Corollary An undirected or multigraph G with no isolated vertices has an Eulerian trial if and only if G is connected and has exactly two vertices of odd degree.

Proof Suppose G is connected and u and v are vertices of G that have odd degree. Add an addition edge $e' = \{u, v\}$ to G, so that the new graph G_1 so obtained is connected and all vertices in G_1 have even degree. Hence, G is Euler and so has Eulerian trial.

Conversely, suppose G has an Eulerial trial which is not closed. Since each vertex is associated with two edges and also there is only one edge associated with each end-vertex of the trial, all end vertices are of odd degree whereas others are of even degree.

Theorem 9.7 A non-empty connected graph G is Eulerian if and only if G is the union of some edges-disjoint circuits.

Proof Let G be the union of some edges-disjoint circuits. Since degree of each vertex in a circuit is two, degree of each vertex in G is even. Hence, by Theorem 6, G is Eulerian.

Conversely, let G be Eulerian. Then all vertices of G are of even degree. If $v \in V$ in G, that it implies that $\deg(v) \geq 2$. So at least two edges are incident at vertex v. To construct an Eulerian trial, we begin a trial T at v and move to adjacent vertices (all of even degree) reaching at a vertex u which has previously been traversed. But, if u is the first such vertex, then part of the walk which begins and ends at u is a circuit C_1. Now consider the graph $G_1 = G - C_1$. Since every vertex in C_1 is of degree 2, every vertex in G_1 is of even degree. We can continue to extend trial T and remove circuits C_2, C_3 . . . from G until no edge is left. Thus G is the union of circuits C_1, C_2, . . ., having no common edges.

Example 31 Verify that following graphs to have an Eulerian circuit.

Solution (*a*) Since all vertices in G are having even degree, therefore it has an Eulerian circuit $v_1 \rightarrow v_2 \rightarrow v_3 \rightarrow v_5 \rightarrow v_4 \rightarrow v_3 \rightarrow v_1$.

(*b*) Since all vertices in G are not of even degree, i.e. $\deg(v_2) = 3 = \deg(v_3)$, therefore, G does not contain Eulerian circuit but has a Eulerian path $e_1 \rightarrow e_2 \rightarrow e_6 \rightarrow e_7 \rightarrow e_5 \rightarrow e_3 \rightarrow e_4$.

(*c*) The graph G is connected and all the vertices are having even degree. Thus, G has a Eulerian circuit.

Important Results

1. If a graph G has a vertex of odd degree, then there can be no Euler circuit in G.
2. If a graph G is connected and each vertex has even degree, then there is an Euler circuit in G.
3. If a graph G has more than two vertices of odd degree, then there can be no Euler path in G.

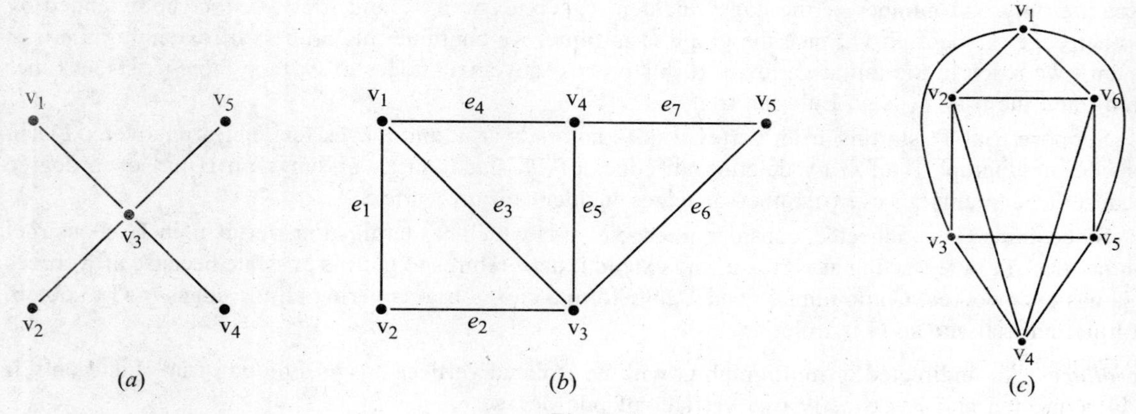

Fig. 9.68

4. If a graph G is connected and has exactly two vertices of odd degree, then there is an Euler path in G. Any Euler path in G must begin at one vertex of odd degree and end at the other.

Example 32 Determine which of the graph shown in Fig. 9.69(*a*), (*b*) and (*c*) has an Euler circuit, Euler path.

Solution (*a*) The graph in Fig. 9.69(*a*), has exactly two vertices of odd degree. Thus, graph is an Euler path but no Euler circuit.

(*b*) In Fig. 9.69(*b*) each vertices has odd degree. Thus, there is neither Euler circuit nor a path.

(*c*) In Fig. 9.69(*c*), each vertex has even degree. Thus, graph has an Euler circuit.

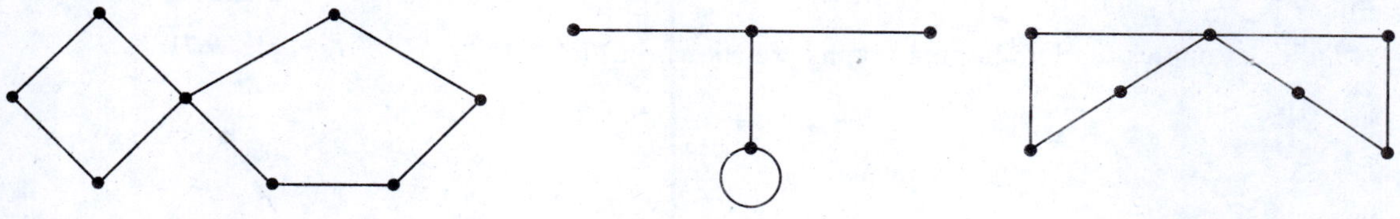

(*a*) (*b*) (*c*)

Fig. 9.69

9.9.1 Hamiltonian Graphs

Hamiltonian graphs are named after Sir William Rowan Hamilton (1805–1865), an Irish mathematician who developed the technique to find a cycle along the edges of a graph so that each vertex appears in it exactly once.

Let G be a connected graph with $|V| \geq 3$. If there is a path in G that uses each vertex of the graph exactly once, then such a path is called *Hamiltonian path.* If the path is a circuit that contains each vertex in G exactly once, except initial vertex that appears twice as the terminal vertex, then such path is called a *Hamiltonian circuit* or *cycle.*

A graph with a closed path that includes every vertex exactly once is called a *Hamiltonian graph.*

Remarks 1. Eulerian cycle uses every edge exactly once but may repeat vertices, while Hamiltonian cycle uses each vertex exactly once (except for the first and the last) but may skip edges.

2. The complete bipartite graph $K_{m,n}$ is hamiltonian if and only if $m = n$ and $n > 1$.

Illustrations Following graphs shown in Fig. 9.70(*a*), (*b*) and (*c*) illustrate comparison between Eulerian and Hamiltonian graphs.

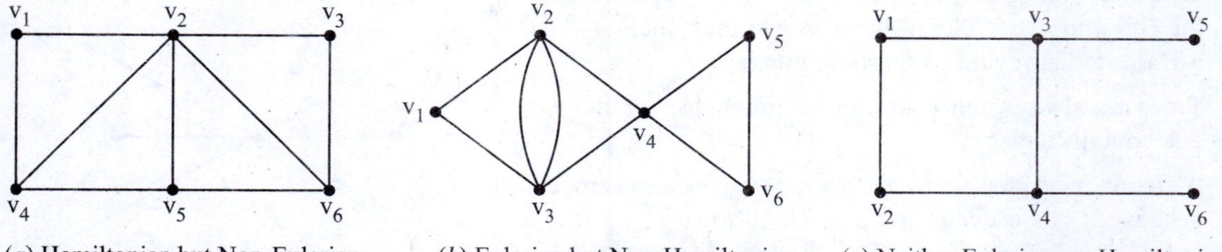

(*a*) Hamiltonian but Non-Eulerian [deg (v) ≠ even]

(*b*) Eulerian but Non-Hamiltonian [deg (v) = even]

(*c*) Neither Eulerian nor Hamiltonian

Fig. 9.70

Illustration The graph shown in Fig. 9.71 is a Hamiltonian graph, containing a Hamiltonian cycle shown by arrows, starting from vertex v_5.

Hamiltonian cycle is (v_1, v_2, v_3, v_4, v_5, v_6, v_7, v_8, v_1).

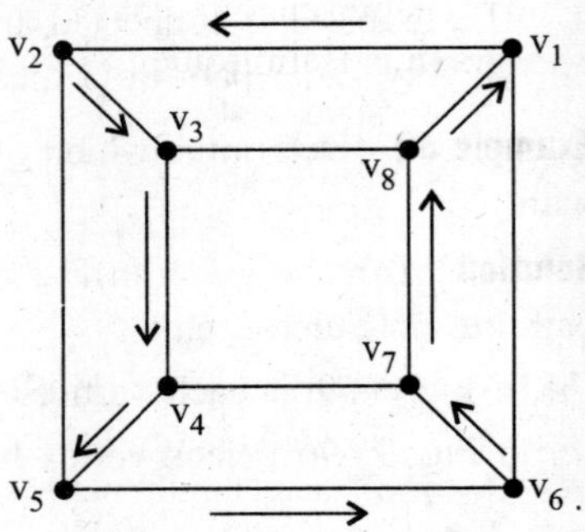

Fig. 9.71

Hamiltonian Path

The Hamiltonian path is obtained by removing any one edge from a Hamiltonian cycle which transverses all the vertices in G. The length of Hamiltonian path, (if exists) in a connected graph of n vertices is $(n - 1)$.

Other examples of Hamiltonian graphs are shown in Fig. 9.72.

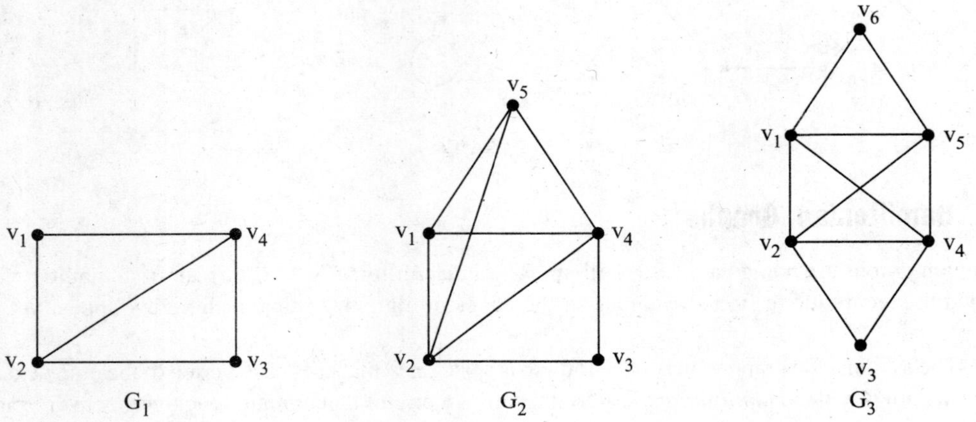

Fig. 9.72

Hamiltonian cycles in graphs G_1, G_2, and G_3 are (v_1, v_2, v_3, v_4), (v_1, v_2, v_3, v_4, v_5) and (v_1, v_2, v_3, v_4, v_5, v_6) respectively.

Remarks

1. Every cycle C is a Hamiltonian graph which includes all the vertices without repetition.
2. Any graph obtained by addition edges to a Hamiltonian graph is also Hamiltonian. Addition of edges does not change the vertices of the new graph. Hence, the Hamiltonian cycle of the original graph shall also be a Hamiltonian cycle of the new graph.
3. Trees are always non-Hamiltonian graphs, since they do not contain any cycle.
4. A Hamiltonian graph can not have cut (or pendant) vertices, because it has a cycle graph to Hamiltonian.

   If any graph has a cut (or pendant) vertax, then it can not be Hamiltonian graph as shown in Fig. 9.73 with v as cut vertex.

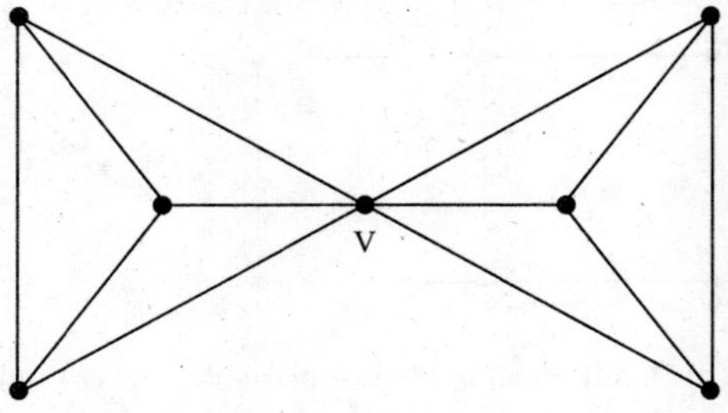

Fig. 9.73

The absence of cut (or pendant) vertices does not mean that the graph is Hamiltonian. A complete graph $K_{m,n}$ $(m, n > 2)$ has no cut (or pendant) vertices, but it is non-Hamiltonian when $m + n$ is odd.

5. A Hamiltonian cycle (or path) does not include a self loop or parallel edges because each of these transverses every vertex of G exactly once. Hence, a graph G needs to be converted into simple graph to find Hamiltonians circuit.
6. Any graph that has a Hamiltonian cycle may not have Hamiltonian path. However, conversely any graph that has Hamiltonian path does not have Hamiltonian cycle as shown in Fig. 9.74 the hamiltonian path is $\{v_1, v_2, v_6, v_5, v_7, v_3, v_4\}$ but has no Hamiltonian cycle.

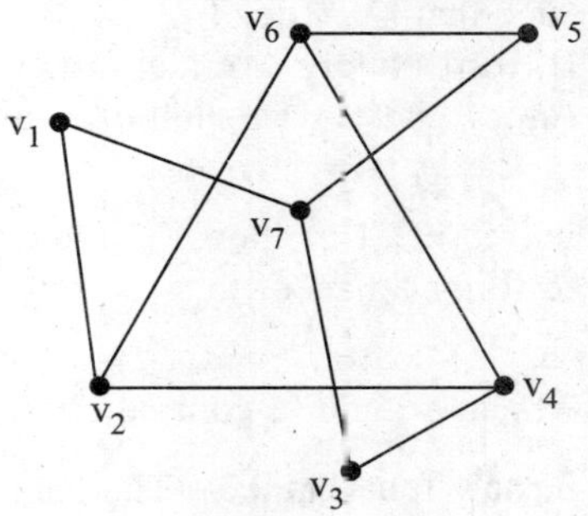

Fig. 9.74

Illustration The graph shown in Fig. 9.75 is Hamiltonian but not Eulerian, because it contains a Hamiltonian cycle $\{v_1, v_2, v_3, v_4, v_5, v_6, v_1\}$. On the other hand, the graph shown in Fig. 9.76 is Eulerian but not Hamiltonian because degree of each of its vertices in even. Also, it does no: contain any Hamiltonian cycle because it contains a cut vertex v.

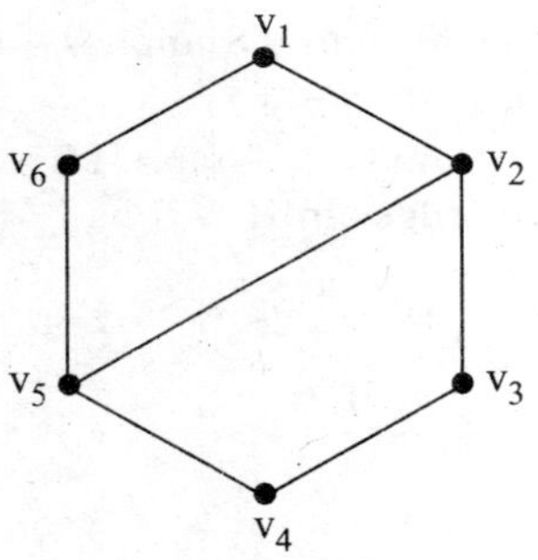

Fig. 9.75

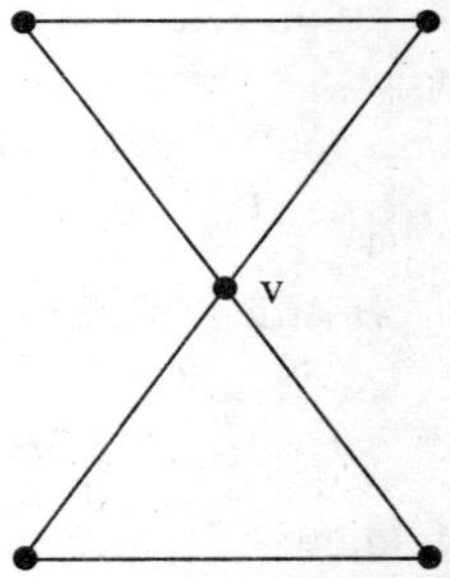

Fig. 9.76

Theorem 9.8 The connected graph G has a Hamiltonian circuit if for any two vertices u and v which are not adjacent, deg (u) + deg (v) $\geq n$, where n is the number of vertices.

Proof This theorem will be proved by contradiction. Suppose, a non-Hamiltonian graph with n vertices satisfies the condition deg (u) + deg (v) $\geq n$ for every non-adjacent vertices u and v. Choosing a graph G with maximum number of edges from all non-Hamiltonian graphs satisfying the given condition. Thus in G, there exist two non-adjacent vertices u and v such that an edge joining u and v will result in a Hamiltonian graph. It follows that there is a Hamiltonian path in G: $u = x_1, x_2, \ldots x_n = v$, where u and v are end vertices as shown in Fig. 9.77.

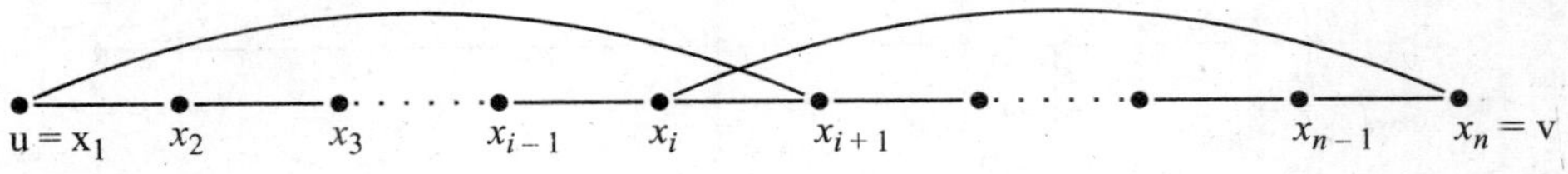

Fig. 9.77

Define two sets

$$A = \{i : x_i \text{ is adjacent to u in G}\}$$
$$B = \{i : x_i \text{ is adjacent to v in G}\}$$

such that |A| = deg (u) and |B| = deg (v) in G, where |A| and |B| denote the number of elements in A and B, respectively. We claim that $A \cap B = \Phi$ and $|A \cup B| \le n - 1$. Thus, if $x_i \in A \cap B$, then the edges (u, x_i) and (x_i, v) should be in G and the path

$$u = x_1, x_2, \ldots x_{i-1}, x_i, x_{i+1}, \ldots x_n, x_{n-1}, \ldots x_1$$

will form a Hamiltonian circuit in G, which is a contradiction. Further, since vertex $u = x_1$ is neither adjacent to u nor adjacent to v, therefore $x_1 \notin A \cup B$. Thus $|A \cup B| \le n - 1$. So

$$\deg(u) + \deg(v) = |A| + |B| = |A \cup B| \le n - 1$$

which is again a contradiction to our hypothesis. Hence, G is Hamiltonian.

Dirac's Theorem 9.9 The connected graph G with $|V| \ge n$ vertices has a Hamiltonian circuit if deg (v_i) $\ge n/2$ for each vertex n_i.

Proof The proof of this theorem follows from Theorem 9.8, because deg (u) + deg (v) $\ge n$ for every pair of non-adjacent vertices u and v.

Corollary The connected graph G with n vertices has a Hamiltonian circuit provided the number of edges in G, $e \ge \frac{1}{2}(n^2 - 3n + 6)$, for $n \ge 3$.

Proof Suppose graph G is non-Hamiltonian. Then the result of Theorem 9.8 implies that there exists a pair of non-adjacent vertices u and v such that deg (u) + deg (v) $\le n - 1$.

Let H be subgraph of G obtained by deleting vertices u and v from G. So graph H has $n - 2$ vertices and e – [deg (u) + deg (v)] edges. Thus, the maximum number of edges in H will be ${}^{n-2}C_2$ and hence

$$e - [\deg(u) + \deg(v)] \le {}^{n-2}C_2 = \frac{1}{2}(n-2)(n-3) = \frac{1}{2}(n^2 - 5n + 6)$$

or

$$e \le \frac{1}{2}(n^2 - 5n + 6) + (n - 1) = \frac{1}{2}(n^2 - 3n + 4) \le \frac{1}{2}(n^2 - 3n + 6)$$

which is a contradiction. Hence, the result of theorem is true.

Theorem 9.10 In a complete graph G with n (an odd number) ≥ 3 vertices, there are $(n - 1)/2$ edge-disjoint Hamiltonian circuits.

Proof Since a complete graph G with n vertices has $n(n - 1)/2$ edges and a Hamiltonian circuit in G contains n edges, the number of edge-disjoint Hamiltonian circuits in G cannot exceed $(n - 1)/2$. We shall now show that there are $(n - 1)/2$ edge-disjoint Hamiltonian circuits when n is old.

A subgraph H of the complete graph G with n vertices as shown in Fig. 9.78 is a Hamiltonian circuit.

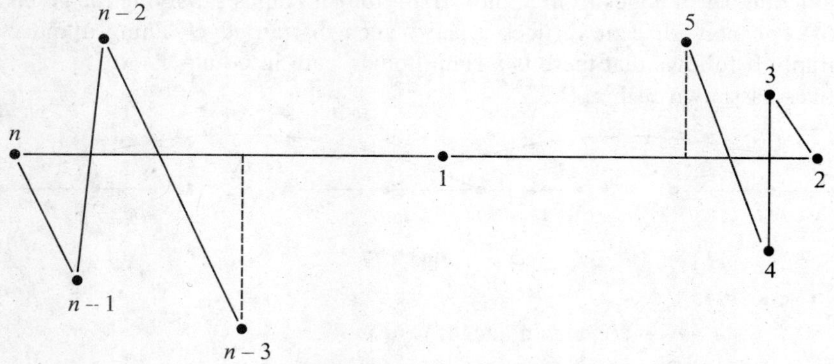

Fig. 9.78 Hamiltonian Circuit

Keeping the vertices fixed on a circle, if we rotate the polynomial pattern anti-clockwise by $\frac{2\pi}{n-1}$, $2\frac{2\pi}{n-1}$, $3\frac{2\pi}{n-1}$, . . ., $\frac{n-3}{2}.\frac{2\pi}{n-1}$ degrees, then we find that each rotation produces a Hamiltonian circuit which has no edge in common with any of the previous ones. Consequently, we obtained $(n - 3)/2$ new edge-disjoint Hamiltonian circuits. Hence, there are $1 + (n - 3)/2 = (n - 1)/2$ edge-disjoint Hamiltonian circuits.

9.10 CARTESIAN PRODUCT OF TWO GRAPHS G_1 AND G_2

If $G_1 = (V_1, E_1)$ and $G_2 = (V_2, E_2)$ are two general graphs, then cartesian product of G_1 and G_2, written as: $G(V, E) = G_1 \times G_2$ is defined as:

(*i*) $V = V_1 \times V_2 = \{(u, v) : u \in V_1 \text{ and } v \in V_1\}$

(*ii*) $E = \{[(u_1, v_1), (u_2, v_2)] : \text{either } u_1 = u_2 \text{ and } v_1v_2 \in E_2 \text{ or } v_1 = v_2 \text{ and } u_1u_2 \in E_1\}$

In other words, if $u = (u_1, u_2)$ and $v = (v_1, v_2)$ be any two elements in the vertex set V of G, then u and v are adjacent (end vertices of edge) in V provided either $u_1 = v_1$ and u_2 is adjacent to v_2, or $u_2 = v_2$ and u_1 is adjacent to v_1 in G_1 and G_2.

Illustration The product of two graphs G_1 and G_2 shown below can be obtained as follows:

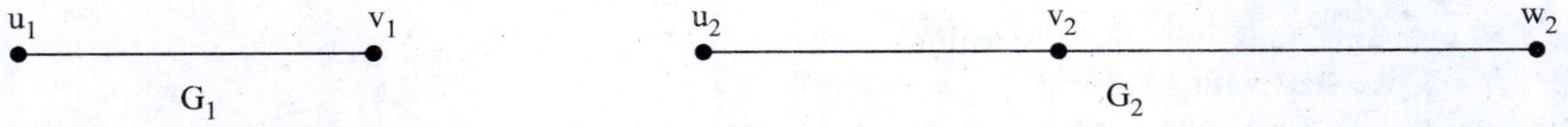

Fig. 9.79

Vertex set, $V = \{(u_1, u_2), (u_1, v_2), (u_1, w_2), (v_1, u_2), (v_1, v_2), (v_1, w_2)\}$.

The cartesian product of graphs G_1 and G_2 is shown in Fig. 9.80.

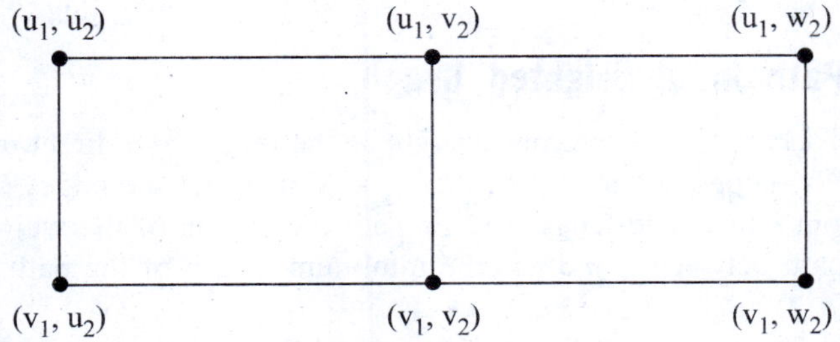

Fig. 9.80

9.11 SHORTEST PATH PROBLEMS

In this section, we shall discuss two algorithms to determine the shortest distance between a designated vertex and each of the other vertices in a loop-free connected undirected graph.

9.11.1 Shortest Path in a Unweighted Graph

The shortest distance path in a loop-free unweighted graph G (V, E) is the path between two vertices v_i and v_j that uses the least number of edges. In this section, we shall discuss *Breath First Search (BFS)* and *Back Tracking Search (BTS)* algorithms to find shortest path in an unweighted graph.

Steps of Breath First Search Algorithm

1. Set the count $i = 0$ and assign the first vertex $v_0 = s$ (say).
2. Search all other unassigned adjacent vertices of vertex $v_0 = s$ in G. If there is no such vertex or vertices, then vertex v_0 is isolated, otherwise set the counter at $i = i + 1$ and assign $v_{i+1} = t$ and find the length of a shortest path from s to t.

Steps of Back-tracking Search Algorithm

1. Set the counter at $i = \lambda(t)$ and assign the last vertex $v_i = t$.
2. Search an unassigned adjacent vertex u to v_i and assign $u = v_{i-1}$, where $\lambda(u) = i - 1$.
3. If $i = 1$, stop, otherwise decrease i to $i - 2$ and go to Step 2.

Example 33 Find the shortest path from vertex v_1 to v_6 and its length in the graph shown in Fig. 9.81.

Solution Using BFS, label v_1 as v_0 for $i = 0$, then adjacent vertices v_2 and v_3 are labelled $i = i + 1 = 0 + 1 = 1$. Vertices adjacent to v_2 and v_3, i.e. vertices v_4 and v_5 are labelled $i = i + 1 = 2$. Then vertex v_6 is labelled as $i = i + 1 = 2 + 1 = 3$. Since last vertex v_6 is labelled 3, therefore, the length of a shortest path from v_1 to v_6 is 3.

Now using back-tracking algorithm, since $\lambda(v_6) = 3$, we start with $i = 3$ and $v_i = v_6$. Vertex adjacent to v_6 are v_4 and v_5. So we choose v_4 or v_5 with $\lambda(v_5) = 2$ and assign $v_5 = 2$. Next we choose v_3 adjacent to v_5 with $\lambda(v_3) = 1$ and assign $v_3 = 1$. Finally, we take v_1 adjacent to v_3 with $\lambda(v_1) = 0$ and assign $v_1 = 0$. Hence, the shortest path from v_1 to v_6 is $v_1v_3v_5v_6$.

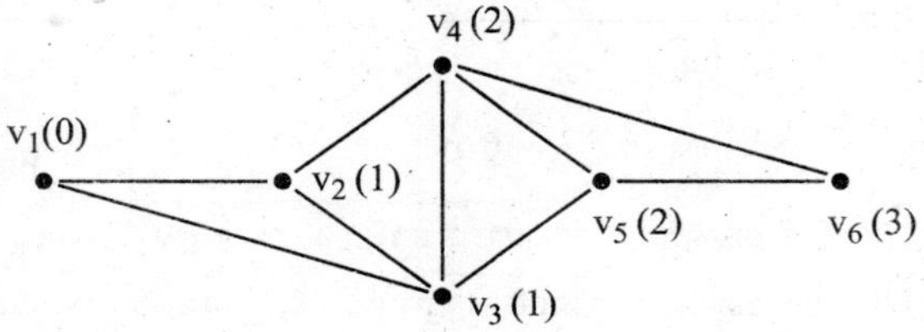

Fig. 9.81

9.11.2 Shortest Path in a Weighted Graph

Given a weighted graph G (V, E), we interpret weight as the length of a direct route from vertex v_i to v_j. For any vertices $s, t \in V$, suppose that $v_1, v_2, \ldots v_n \in V$ and that the edges $(s, v_1), (v_1, v_2), \ldots (v_n, t)$ provide a path in G from s to t. The length of this path is the sum of the weights of the edges on this path and the shortest path between s and t is the minimum length of the path. We denote the shortest path from s to t as $d(s, t)$.

The procedure to find shortest path between the first vertex and every other vertex in the weighted graph was developed by Edsger Wybe Dijkstra, a Dutch mathematician.

The computations start from a vertex v_i to an immediately succeeding vertex using a labelling procedure. Let d_i be the shortest distance (weight) from initial vertex v_i to vertex v_j and d_{ij} (≥ 0) as the length of edge (i, j). Then the label for vertex v_j is defined as: $(d_j, i) = (d_i + d_{ij}, i)$.

Labels in Dijkstra's algorithm are of two types: temporary and permanent. A temporary label can be replaced with another label if the shortest distance to the same vertex can be found. At a point when it becomes evident that no better route can be found, the status of the temporary label is changed to permanent.

Steps of the algorithm are summarised below:

1. Label the initial vertex v_i ($i = 1$) with the permanent label (0, –) and all other vertices with temporary label (∞, –). That is, initialise the optimal lengths

$$v_i = \begin{cases} 0 \text{ if } i = s(\text{initial vertex}) \\ \infty \text{ otherwise.} \end{cases}$$

2. Assign each vertex v_j that can be reached from initial vertex v_i ($i = 1$) a permanent label with

$$L(v_i) = \text{Min}\{L(v_j), L(v_i) + d_{ij}\}$$

provided vertex v_j is not permanently labelled. If vertex v_j is already labelled with $(L(v_j), v_k]$ with another vertex v_k and if $L(v_i) + d_{ij} < L(v_j)$ replace $[L(v_j) + v_k]$ with $[L(v_j), L(v_i) + d_{ij}]$.

3. If all the vertices have permanent labels, stop. Otherwise select the label with least current value v_r from all the temporary labels (break ties arbitrarily), i.e.

$$L(v_r) = \text{Min}\{L(v_i): i \text{ temporary}\}$$

set $i = r$ and return to Step 2.

Example 34 Apply Dijkastra algorithm to find the shortest path from vertex v_1 to v_5 in the graph shown in Fig. 9.82.

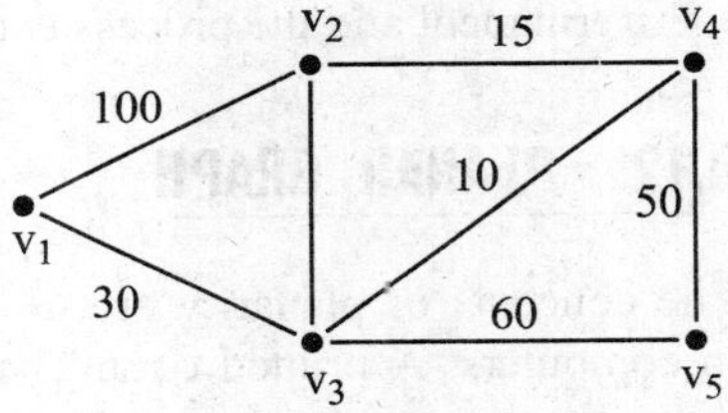

Fig. 9.82

Solution 1. Assign the permanent label (0, –) to initial (source) vertex v_1.

2. Vertices v_2 and v_3 can be reached from the last permanently labelled vertex v_1. Thus, the list of labelled vertices (temporary T and permanent P) becomes

Vertex:	v_1	v_2	v_3	v_4	v_5
Level L (v):	0	100	30	∞	∞
Status:	P	T	P	T	T

$$L(v_2) = \min[L(v_2), L(v_1) + d_{12}] = \min(\infty, 0 + 100) = 100$$

$$L(v_3) = \min[L(v_3), L(v_1) + d_{13}] = \min(\infty, 0 + 30) = 30$$

For the two temporary levels $L(v_2)$ and $L(v_3)$, vertex v_3 yields the smaller distance 30. Thus, the status of vertex v_3 is changed to permanent.

3. Vertices v_4 and v_5 can be reached from vertex v_3 and the list of labelled vertices becomes

Vertex:	v_1	v_2	v_3	v_4	v_5
L (v):	0	100	30	40	90
Status:	P	T	P	P	T

$$L(v_4) = \min[L(v_4), L(v_3) + d_{34}] = \min(\infty, 30 + 10) = 40$$

$$L(v_5) = \min[L(v_5), L(v_3) + d_{35}] = \min(\infty, 30 + 60) = 90$$

Since vertex v_4 yields the smaller distance 40, the status of vertex v_4 is changed to permanent.

4. Vertices v_2 and v_5 can be reached from vertex v_4, their labels are calculated as follows:

$$L(v_2) = \text{Min}\,[L(v_2), L(v_4) + d_{42}] = \min(100, 40 + 15) = 55$$
$$L(v_5) = \min\,[L(v_5), L(v_4) + d_{45}] = \min(90, 40 + 50) = 90$$

Since vertex v_2 yields the smaller distance 55, the status of vertex v_2 is changed to permanent. Thus, the list of labeled vertices becomes.

Vertex:	v_1	v_2	v_3	v_4	v_5
L (v):	0	55	30	40	90
Status:	P	P	P	P	T

Vertex v_2's temporary label $L(v_2) = 100$ in earlier iteration is changed to 55 in this iteration to indicate that shorter distance has been found through vertex v_4. Also, in iteration 2, vertex v_5 has two alternative labels with the same distance 90.

5. Only vertex v_3 can be reached from vertex v_2. But it is already permanently labeled. Thus, the new list of labels remains same as in iteration 2 except that the label at v_2 is permanent. This leaves vertex v_5 as the only vertex with temporary level. Since vertex 5 does not lead to other vertices, its status is converted to permanent and the process ends. Hence, the desired shortest distance is $v_1 \rightarrow v_3 \rightarrow v_4 \rightarrow v_2 = 55$.

9.12 PLANAR GRAPH

The concepts of planarity are useful application of graph theory in the design of a printed-circuit board in computers. A printed circuit board is a planar network.

Definition A graph or a multigraph G is said to be *planar* if it can be drawn in a plane with its edges do not cross (or intersect) each other except at vertices, i.e. edges are allowed to touch each other only at the vertices.

If a graph is planar, then any drawing of it with no edges crossing is called a *plane graph.* A drawing of geometric representation of G on any surface without a crossover (intersection) between its edges is called an *embedding* of G.

A graph G with crossing edges does not mean that it is non-plannar. There may be another way to draw the graph without crossing edges. Thus, before declaring a graph non-plannar, we have to redraw it in several possible manner so that none can be embedded in plane. Figures 9.83 (*a*) and (*b*); (*c*) and (*d*) are few examples of planar graphs. The graph shown in Fig. 9.83 (*a*) does not appear to be planar because edges (v_1, v_5) and (v_2, v_4) cross one another at point, which is not a vertex. Therefore, it can be redrawn in Fig. 9.83 (*b*) so that none of its edges cross.

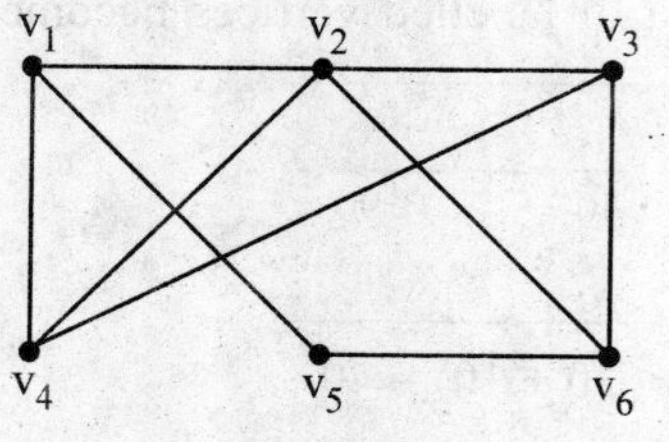

(*a*) Graph G

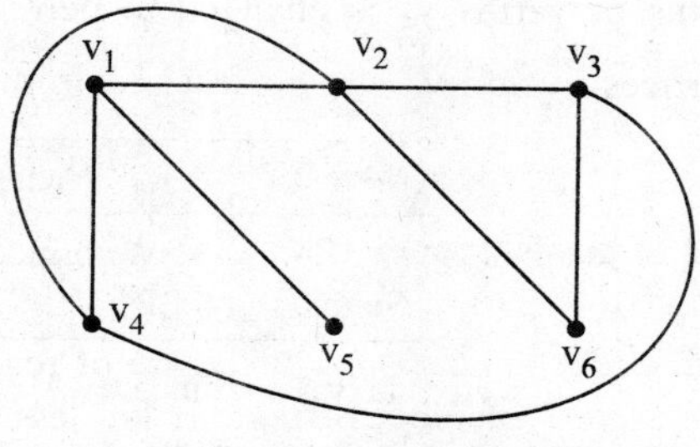

(*b*) Planar Representation of G

Fig. 9.83 Examples of Planar Graph

Example 35 Draw the planar graphs shown in Figs. 9.84(*a*) and (*c*) so that none of the edges cross.

Solution The planar representation of graphs shown in Figs. 9.84(*a*) and (*c*) without crossing edges is shown in Figs. 9.84(*b*) and (*d*) respectively.

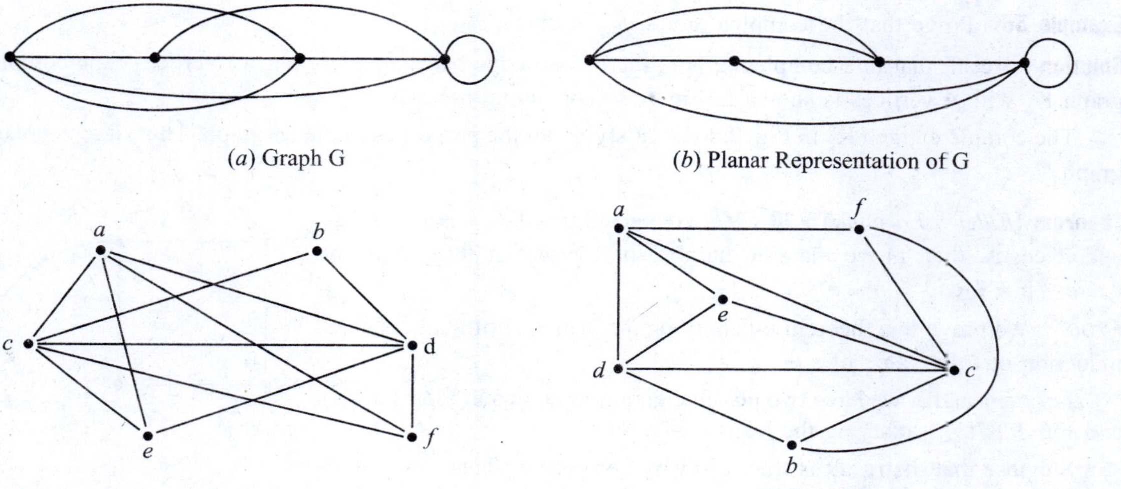

(*a*) Graph G (*b*) Planar Representation of G

(*c*) Graph G (*d*) Planar Representation of G

Fig. 9.84

9.12.1 Maps and Region of Graph

A particular planar representation of a finite planar multigraph is called a *map*. This map is said to be *connected* if the multigraph is connected.

Region The map so drawn divides the plane into various disjoint areas bounded by edges and cannot be further subdivided is called the region.

The area of the region may be finite or infinite. However a planar graph has only one infinite region. The planar graph shown in Fig. 9.85 has 3 finite regions bounded by edges and one infinite region r_1.

For each region r in the embeding of a (planar) graph or multigraph the degree of r, denoted by deg (r), is the number of edges traversed in a closed walk about (the edges in) the boundary of r. Fig. 9.85 has four regions, where

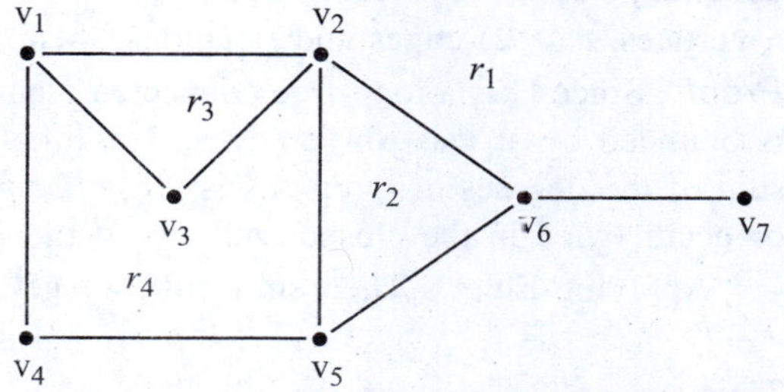

Fig. 9.85 Example of Regions of Planar Graph

$$\deg(r_1) = 7, \deg(r_2) = 3 \deg(r_3) = 3 \deg(r_4) = 5$$

Properties of planar graphs Let G be a connected planar graph having n vertices, m edges and r regions. Then

(*i*) $\Sigma \deg(r) = 2m$, sum of degree of regions is equal to twice the number of edges

(*ii*) $r \leq 2m/3$, number of regions are less than or equal to 2/3 of the number of edges.

(*iii*) $m \leq 3n$, number of edges are less than or equal to 3 times the number of vertices.

(*iv*) $3n - m \geq 6$.

(*v*) $r = m - n + 2$.

(*vi*) A connected planar graph has a vertex with degree 5 or less.

(*vii*) A complete graph $K_{m,n}$ is planar if and only if $m < 3$ or $n < 3$.

(*viii*) A complete graph K_n is planar if and only if the number of vertices, $n < 5$.

Example 36 Prove that the complete graph K_4 is planar.

Solution Recall that in a complete graph, there is an edge between every pair of vertices. The complete graph K_4 with 4 vertices is shown in Fig. 9.86 containing 6 edges.

The complete graph K_4 in Fig. 9.86 is satisfying all the properties of planar graph. Thus, it is the planar graph.

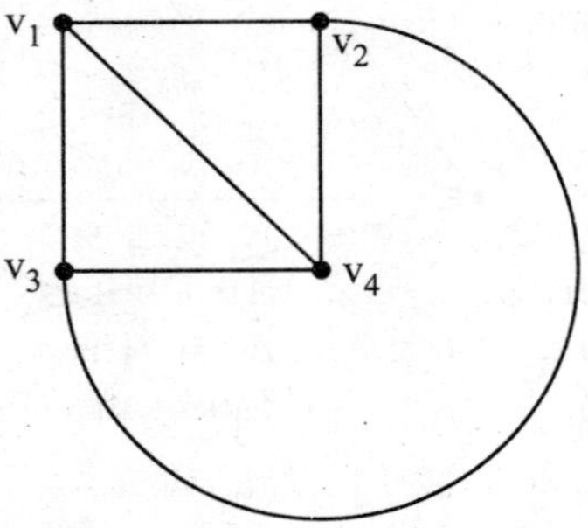

Fig. 9.86 Complete Graph K_4

Theorem (*Euler's Formula*) **9.10** If a connected graph G has n vertices and e edges, then prove that the number of regions in the graph are $r = e - n + 2$ or $r + n - e = 2$.

Proof We prove this theorem by applying the principle of mathematical induction on the value of r.

For $r = 1$ and 2, we have two possible graphs as shown in Figs. 9.87(*a*) and Fig. 9.87(*b*), satisfying the result $r = e - n + 2$.

Suppose that the result is true for every connected planar graph with $k + 1$ regions and n vertices.

Select any edge (v_i, v_j) that belongs to two regions, say r_1 and r_2. If this edge is deleted from G, then two regions r_1 and r_2 merge into a single region. Thus, resulting graph G′ has e edges, k regions and n vertices, and by the principle if induction, $k = e - n + 2$. Therefore, if we reintroduce the edge (v_i, v_j) into G′, we see that $k + 1 = (e + 1) - n + 2$. Hence, Euler's formula remains valid.

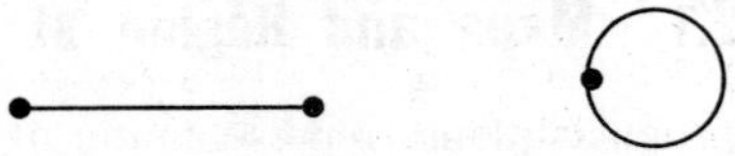

(*a*) $n = 2, m = 1$ (*b*) $n = 1, m = 1$

Fig. 9.87

Corollary If G is a loop-free connected planar graph with n vertices, e (> 2) edges and r regions, then (*a*) $3r \le 2e$ and (*b*) $e \le 3n - 6$.

Proof Since G is a loop-free connected planar graph, therefore each region (including infinite region) is bounded by at least three edges. This implies that each region is of degree ≥ 3. Consequently, the sum of the degrees of r regions is $2|E| = 2e$ and each edge is the part of the boundary of two regions or occur twice in the closed walk about the edges in the boundary for one region, so $2e \ge 3r$.

Applying Euler's Theorem result, we get

$$2 = n - e + r \le n - e + (2/3)\, e = n - (1/3)\, e$$

Thus, so $6 \le 2n - e$ or $e \le 3n - 6$.

Corollary If G is a loop-free connected planar graph with n (≥ 3) vertices, e edges and number of circuits of length 3, then $e \le 2n - 4$.

Proof Since degree of each region in a planar graph is at least 4, the total number of edges as the part of boundary of all regions is at least $4r$. In general, each edge of the planar graph is part of the boundary of two regions, and therefore, total number of edges around all the regions is $2e$. Consequently, $2e \ge 4r$ or $e \ge 2r$. Substituting this value of r in Euler's formula; we get

$$2 = n - e + r \quad \text{or} \quad r = 2 - n + e$$

$$2r = 4 - 2n + 2e \le e \quad \text{or} \quad e \le 2n - 4 \text{ (because } 2r \le e)$$

Example 37 If G is a loop-free connected graph with 4 vertices each of degree 3. In how many regions this graph can be divided.

Solution Given $n = 4$ and degree of each vertex is 3. So $\Sigma \deg(v) = 3 \times 4 = 12$. But $\Sigma \deg(v) = 2e$ implies that $2e = 12$ or $e = 6$.

By Euler's formula, we have $n - e + r = 2$ or $4 - 6 + r = 2$ or $r = 4$ regions.

9.12.2 Non-planar Graphs

For some graphs, no matter how these are drawn, some edges will always cross. Such graphs are said to be non-planar.

Graphs shown in Figs. 9.89 (*a*) and (*b*) are non-planar graphs.

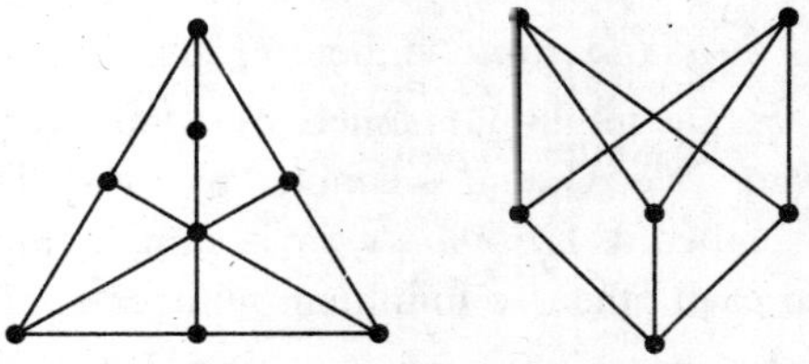

(*a*) (*b*)

Fig. 9.88 Non-planar Graphs

Remark (*Kuratowski Theorem*) A graph is non-planar if and only if it contains a subgraph homeomorphic to $K_{3,3}$ or K_5.

Example 38 Show that the graph K_5 is non-planar.

Solution The graph K_5 shown in Fig. 9.89 is the loop-free connected graph with 5 vertices and 10 edges. For a planar graph, we know that $3n - e \geq 6$ or $e \leq 3n - 6$. Here $n = 5$ and $e = 10$, therefore, $3n - e = 15 - 10 = 5$, which does not satisfy the inequality $3n - e \geq 6$ or $e \leq 3n - 6$. Thus, K_5 is non-planar graph.

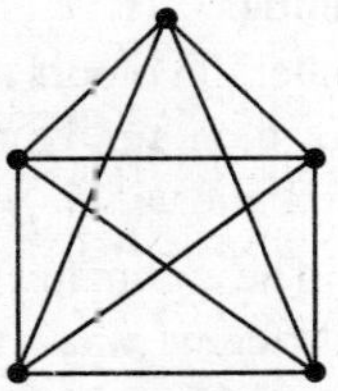

Fig. 9.89 K_5 Graph

Example 39 Identify which of the graph shown in Figs. 9.90(*a*), (*b*) and (*c*) is non-planar, i.e. which graphs have subgraph homeomorphic to $K_{3,3}$ or K_5.

Solution All the graphs shown in Figs. 9.90(*a*), (*b*) and (*c*) are non-planar.

Graph shown in Fig. 9.90(*a*) is a subgraph isomorphic to K_5. Recall that graph K_5 contains $n = 5$ vertices and $e = 10$ edges, so that $3n - e = 3 \times 5 - 10 = 5$. This does not satisfy the inequality $3n - e \geq 6$, and hence K_5 is non-planar.

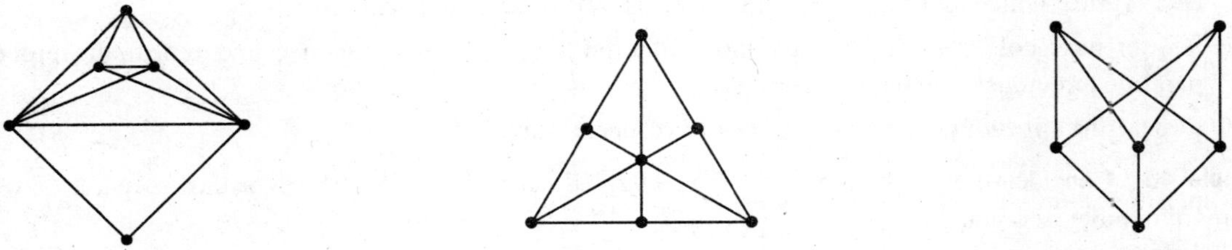

(*a*) (*b*) (*c*) $K_{3,3}$

Fig. 9.90 Examples of Nonplanar Graphs

Graphs shown in Figs. 9.90(*b*) and (*c*) are homeomorphic to $K_{3,3}$. The graph $K_{3,3}$ is loop-free and connected (it is bipartite) with $n = 6$ vertices and $e = 9$ edges, so that $3n - 6 = 18 - 6 = 12 \geq 9$ $(= e)$. Since it does notsatisfy the inequality $e \leq 3n - 6$, graph is non-planar.

9.12 GRAPH COLOURING AND CHROMATIC POLYNOMIALS

Let G (V, E) be an undirected graph with no multiple edges, and let C = $\{c_1, c_2, \ldots, c_n\}$ be the set of n colours. The *colouring of G* or *simply vertex colouring* is a function f: V → C such that adjacent vertices have different colours. In other words, any two adjacent vertices $(v_i, v_j) \in$ E have different colour, i.e. $f(v_i) \neq f(v_j)$. Thus, a colouring is *proper* when any two adjacent vertices have different colours. A graph G is *n-colourable* if there exists a colouring of G which uses *n colours*.

The minimum number of colours needed to properly colour vertices of a graph G is called the *chromatic number* of G, and is denoted by c (G). The minimum number of colours (i.e. elements) belong to the set C is called its *cardinality*. For example, in the graph shown in Fig. 9.91, vertices v_1, v_3 and v_4 are connected to each other, so minimum number of colours are needed to colour G, will be c (G) = 3.

1. Chromatic number for complete graph with n (≥ 1) vertices is $c(K_n) = n$.
2. Chromatic number for a bipartite graph is $c(K_{m,n}) = 2$.

Definition If G is the undirected graph and λ denotes the number of colours available for properly colouring the vertices of G, then a polynomial function P (G, λ) with variable λ is called *chromatic polynomial* of G.

The chromatic polynomial of G helps to determine number of different ways to colour vertices of G, using at most λ colours.

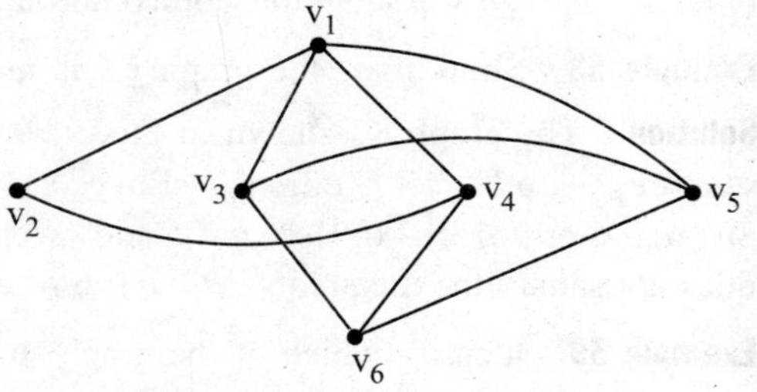

Fig. 9.91

Welch and Powell algorithm It is an algorithm developed by Welch and Powell for finding colours needed to paint a graph. However, algorithm may not always yield a minimum number of colours needed to paint a graph. Steps of the algorithm are summarised below.

(*i*) Arrange vertices in a sequence according to the descending order of their degrees.

(*ii*) Assign colour c_1 to the first vertex of the sequence and then assign c_1 in the sequential order to each vertex which is not adjacent to previous vertex assigned with colour c_1.

(*iii*) Assign next colour c_2 to the next non-coloured vertex in the sequence and repeat the process painting previously unpainted vertices.

(*iv*) Repeat the procedure until all vertices have been painted.

Example 40 Consider the graph shown in Fig. 9.92. Use the Welch-Powell algorithm to paint G with minimum number of colours.

Solution The vertices of graph G shown in Fig. 9.92 are arranged in the sequential order according to their degree as follows:

Vertex :	v_1	v_2	v_3	v_4	v_5	v_6	v_7
Degree :	5	4	4	4	3	3	3
Colour :	c_1	c_2	c_3	c_3	c_2	c_4	c_1

Assign colour c_1 to the vertex v_1 of highest degree. Other vertices in the list not adjacent to v_1 is only v_7. So assign colour c_1 to v_7 also.

Now choose next colour c_2 and assign it to v_2. Since all unpainted vertices are adjacent to v_2 or v_4. Thus, moving to colour c_3. Repeat the procedure with additional colours until all vertices are painted as shown above in the table.

Since vertices v_1, v_3, v_4 and v_6 are connected to each other and therefore, should be painted with different colours. Hence, chromatic number of G is 4.

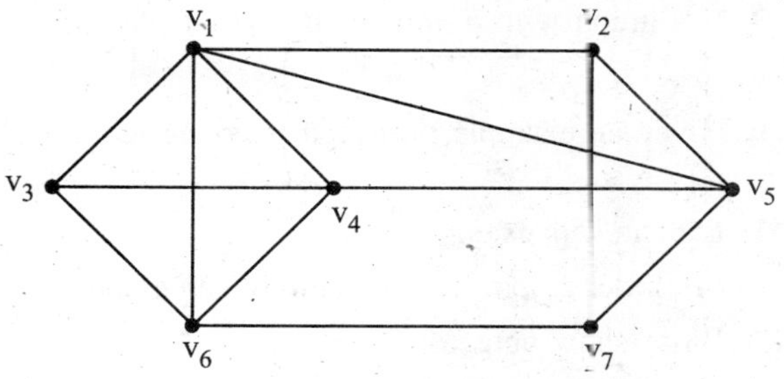

Fig. 9.92

Remarks 1. There is no limit on the number of colours that may be required for colouring an arbitrary graph. However, if the graph is a plannar graph, then regardless of the number of vertices, five colours are suffice. It is also called *five-colour theorem.*

2. The following rules may be helpful in finding chromatic number c (G) of a graph:

   (*a*) $c(G) \leq |V|$, where $|V|$ is the number of vertices in a graph.

   (*b*) If the degree of a vertex in G is r, then at most r colours are required to colour all vertices adjacent to it.

   (*c*) If a subgraph of G requires m colours, then $c(G) \geq m$.

3. For a graph G, following results are equivalent:

   (*a*) G is 2-colourable.

   (*b*) G is bipartite graph.

   (*c*) Every cycle of G is of even length.

Conceptual Review Questions

1. Differentiate with examples, between a null graph and a simple graph.
2. How graph theory can be applied to computer science?
3. Show that the number of vertices of odd degree in graph is always even.
4. What do you understand by a directed graph? Discuss the degree in directed graph and explain the types of directed graphs.
5. Define the following concepts in graph theory with suitable example

   (*i*) Complete Graph (*ii*) Cycle (*iii*) Bridge edge [*IGNOU, CS-07, Dec 02*]
6. Define the following terms in brief. Given examples :

   (*a*) Multi Graph (*b*) Undirected Graph (*c*) Complete undirected Graph
7. Prove that in a simple digraph, the length of any elementary path is less than or equal to n – 1, where n is the number of nodes in the graph. [*CCS Univ., 3CA, 2000*]
8. Show that the sum of degree of all vertices in G is twice the number of edges in G. [*UPTU., MCA. 2001*]

9. If a graph with n vertices is regular of degree r, what is sum of the degree of the vertices.

[*UPTU, MCA., 2001*]

10. Given an example of a regular connected graph with six vertices that is not complete.

[*CCS Univ. MCA., 2001*]

11. Explain with examples

(*a*) Euler graph (*b*) Hamiltonian graph

12. Write short notes on:

(*i*) Complete bipartite graph (*ii*) Hamiltonian path and circuit

(*iii*) Chromatic number of a graph (*iv*) Euler graph [*UPTU, B Tech, 2007*]

Self-practice Problems

1. Draw a diagram for each of the following graphs, G (V, E)

(*a*) V = $\{v_1, v_2, v_3, v_4\}$, E = $\{(v_1, v_2), (v_4, v_1), (v_3, v_1), (v_3, v_4)\}$

(*b*) V = $\{v_1, v_2, v_3, v_4, v_5\}$, E = $\{(v_1, v_5), (v_3, v_4), (v_2, v_3), (v_2, v_5), (v_1, v_5)\}$

(*c*) V = $\{v_1, v_2, v_3, v_4, v_5, v_6\}$, E = $\{(v_1, v_4), (v_1, v_6), (v_3, v_4), (v_4, v_4), (v_3, v_5), (v_6, v_6)\}$

2. Describe formally the graph shown in Fig. 9.93(*a*) and Fig. 9.93(*b*).

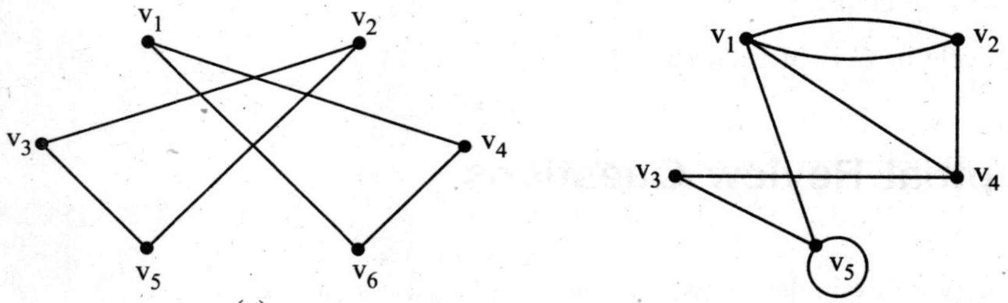

(*a*) (*b*)

Fig. 9.93

3. Find the degree of each vertex in the multigraph shown in Fig. 9.94(*b*).

4. Suppose v is an isolated vertex in a graph or multigraph G, what is its degree?

5. Suppose a graph G has five vertices. Find the maximum number of edges in G if (*a*) G is a graph, and (*b*) G is a multigraph.

6. Let G be the graph shown in Fig. 9.94. Consider the following path in G.

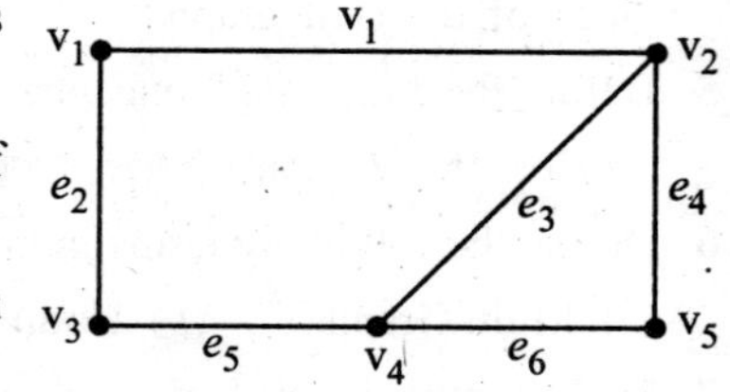

Fig. 9.94

(*a*) $\alpha = (e_1, e_4, e_6, e_5)$, and (*b*) $\beta = (e_2, e_5, e_3, e_4, e_6, e_3, e_1)$.

Convert each sequence of edges into the corresponding sequence of vertices.

7. Can a simple undirected graph of eight vertices have 40 edges excluding self-loops?

8. Let G be a graph shown in Fig. 9.95. Determine whether or not each of the following sequences of edges form a path:

(a) $\{(v_1, v_3), (v_3, v_2), (v_6, v_4), (v_4, v_3)\}$

(b) $\{(v_1, v_3), (v_3, v_4), (v_4, v_5), (v_5, v_1)\}$

(c) $\{(v_3, v_2), (v_2, v_4), (v_4, v_6)\}$

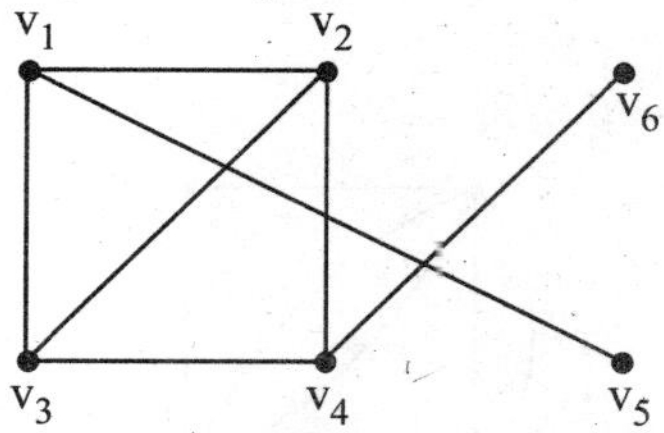

Fig. 9.95

9. Find all cycles in the graph shown in Fig. 9.95.

10. Consider the graph G (V, E) shown in Fig. 9.96. Determine whether or not H (V′, E′) is a subgraph of G, where

(a) $V' = \{v_1, v_2, v_6\}$, $E' = \{(v_1, v_2), (v_1, v_6)\}$

(b) $V' = \{v_1, v_2, v_4\}$, $E' = \{(v_1, v_2), (v_1, v_4)\}$

(e) $V' = \{v_1, v_2, v_3\}$, $E' = \{(v_1, v_2), (v_2, v_3), (v_2, v_4)\}$

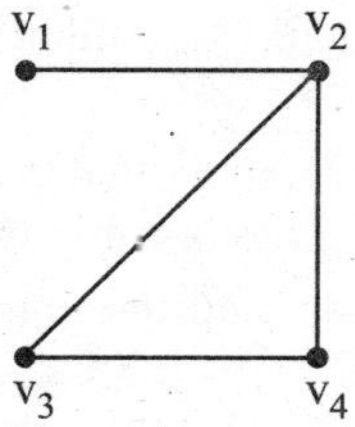

Fig. 9.96

11. Find the connected components of the graph shown in Fig. 9.97.

12. Let G be the graph in Fig. 9.96. Find:

(a) $G-(v_1, v_2)$ (b) $G-(v_2, v_3)$ (c) $G-(v_2, v_3)$ (d) $G-(v_3, v_4)$

13. Find the connected components the graph G (V, E), where $V = \{v_1, v_2, v_3, v_4, v_5, v_6\}$ and

(a) $E = \{(v_1, v_4), (v_3, v_4)\}$, (b) $E = \{v_1, v_5), (v_2, v_3), (v_6, v_5), (v_4, v_6)\}$.

14. Let G be the graph shown in Fig. 9.98. Find:

(a) $G-v_6$ (b) $G-v_4$ (c) $G-v_5$.

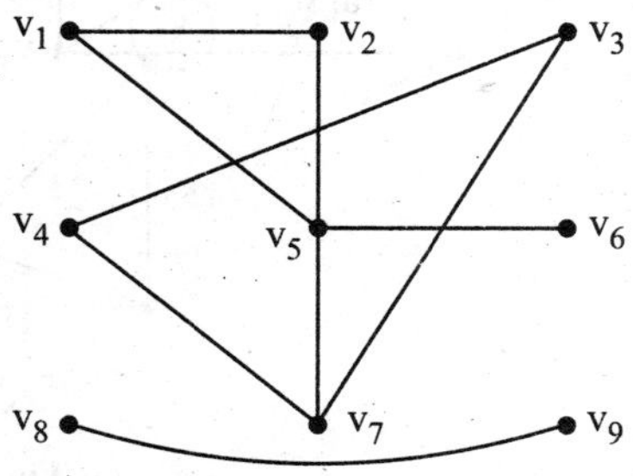

Fig. 9.97

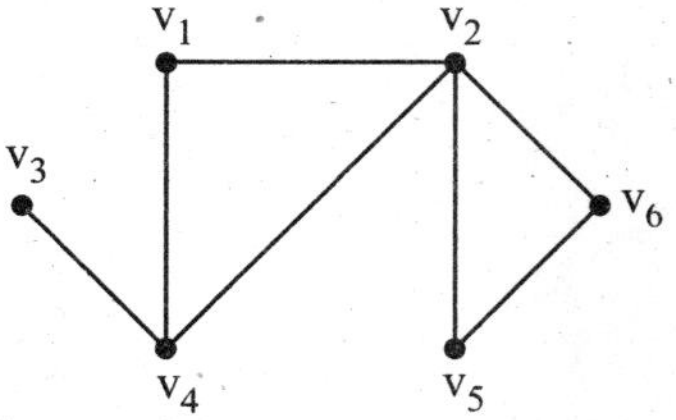

Fig. 9.98

15. In a simple undirected graph with no self-loops, what is the maximum number of edges that you can have keeping the graph unconnected. What is the maximum number of edges that will assure that the graph is connected?

16. Determine which of the graphs shown in Figs. 9.99 (*a*) to (*c*) is traversable?

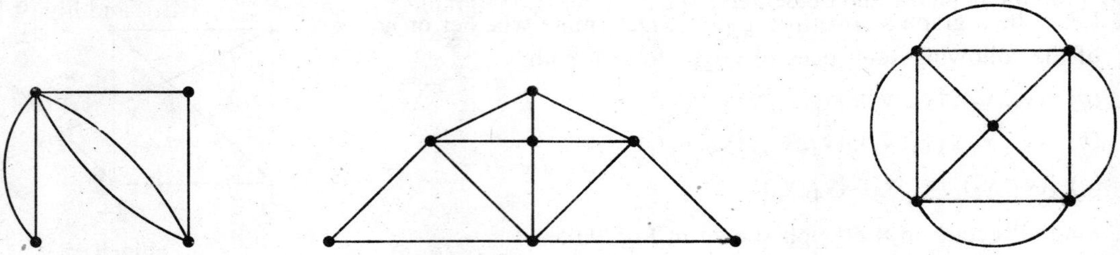

(*a*) (*b*) (*c*)

Fig. 9.99

17. Show that one can add or delete loops from a multigraph and the graph G remains traversable or non-traversable, respectively.

18. Find the number *m* of edges in the complete graph, K_n.

19. Let G be a connected graph with three vertices. Show that G is traversable.

20. (*a*) Find those values of *n*, the number of vertices for which K_n is traversable.

(*b*) Find those values of *n*, the number of vertices for which the complete graph K_n is regular.

21. Draw two regular graphs with seven vertices.

22. Draw a graph with six vertices which is Hamiltonian but not Eulerian, and (*b*) Eulerian but not Hamiltonian.

23. (*a*) Draw two 3-regular graphs with six vertices.

(*b*) Draw two 3-regular graphs with eight vertices.

24. (*a*) Draw the graph G whose adjacency matrix A = $[a_{ij}]$ is given on next page.

(*b*) Find the adjacency matrix and incidence matrix of the multigraph shown in Fig. 9.100.

$$A = \begin{bmatrix} 0 & 1 & 0 & 1 & 0 \\ 1 & 0 & 0 & 1 & 0 \\ 0 & 0 & 0 & 1 & 1 \\ 1 & 1 & 1 & 0 & 1 \\ 0 & 1 & 1 & 1 & 0 \end{bmatrix}$$

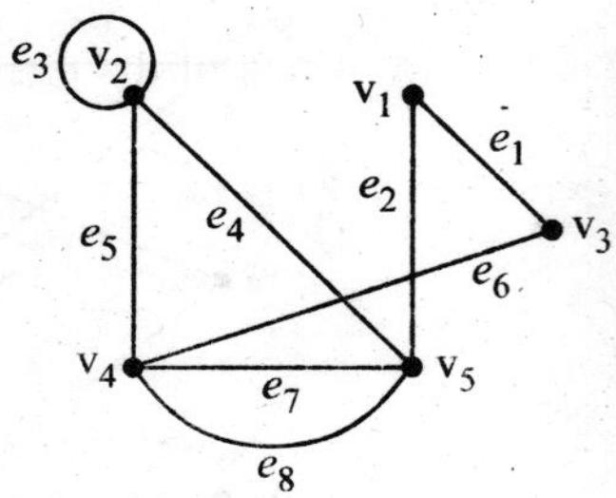

Fig. 9.100

25. Show how every connected graph G may be viewed as a (weighted) labelled graph. What is the minimum path in such a graph?

26. Suppose G and H are isomorphic graphs and G is traversable. Is H traversable? If yes, what would be traversable path for H?

27. Suppose G and H are isomorphic graphs. Find the number of connected components of H if G has eight connected components.

28. Can a finite graph G be homeomorphic to one of its subgraphs other then itself?

29. Find the indegree and outdegree of each vertex in the graphs shown in Fig. 9.101(*a*) and Fig. 9.101(*b*).

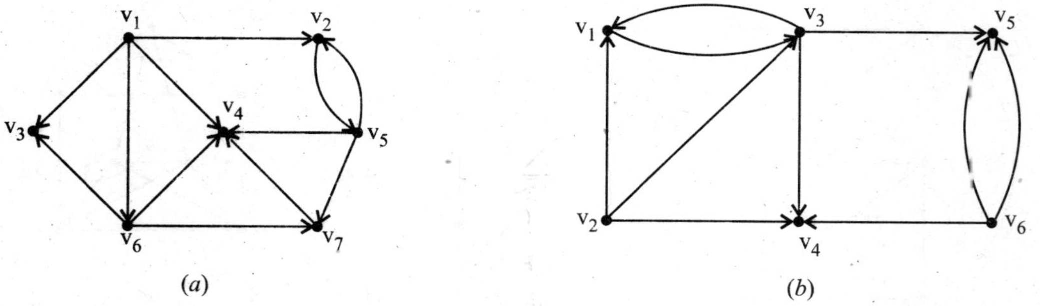

Fig. 9.101

30. Let G be the labelled graph as shown in Fig. 9.102. Find all simple paths between v_1 and v_6.

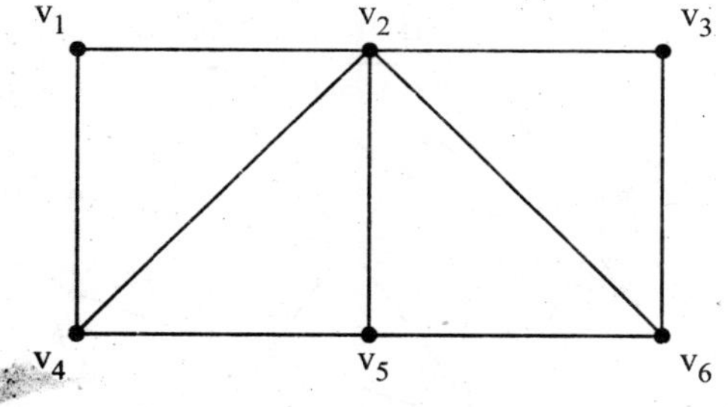

Fig. 9.102

31. Identify which of the graphs shown in Figs. 9.103 (*a*) to 9.103 (*d*) is coplanar.

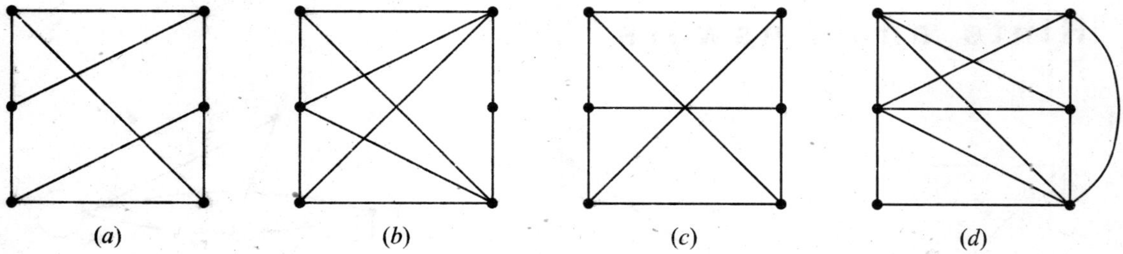

Fig. 9.103

32. The graphs shown in Fig. 9.104 has a subgraph isomorphic to $K_{3,3}$. Identify the subgraph.

33. The graph shown in Fig. 9.105 has a subgraph isomorphic to K_5. Identify the subgraph.

34. The graph shown in Fig. 9.106 has a subgraph homeomorphic to K_5. Identify the subgraph.

35. The graph shown in Fig. 9.107 has a subgraph homeomorphic to $K_{3,3}$. Identify the subgraph.

36. Determine the chromatic number of the complete graphs K_6, K_{10} and in general K_n.

37. Use the Welch-Powell algorithm to paint graphs shown in Fig. 9.108(a) and Fig. 9.108(b) and find th chromatic number n of the graph.

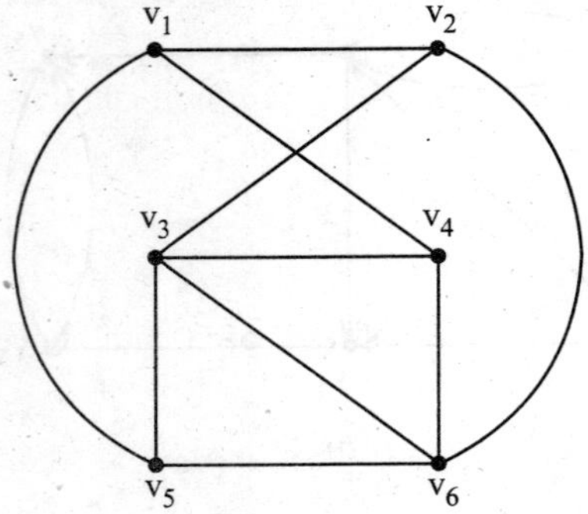

Fig. 9.104

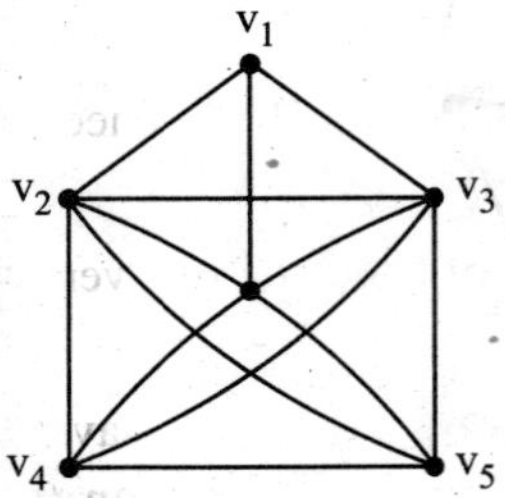

Fig. 9.105

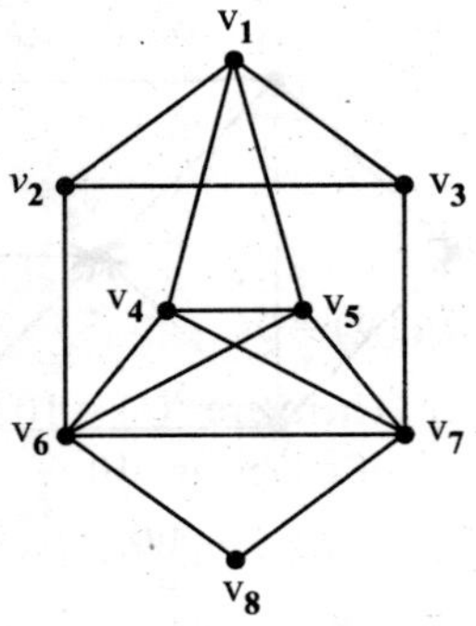

Fig. 9.106

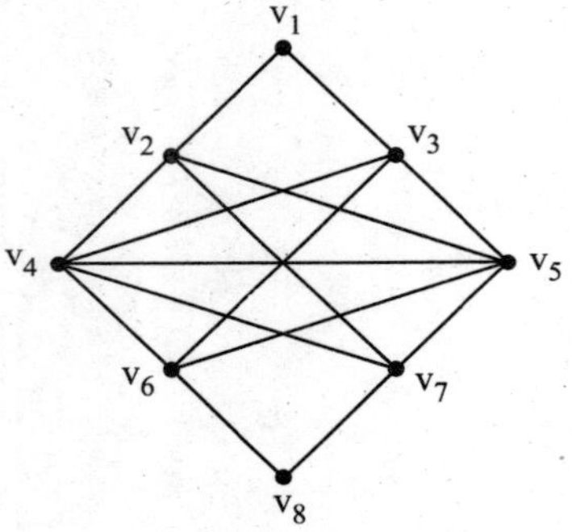

Fig. 9.107

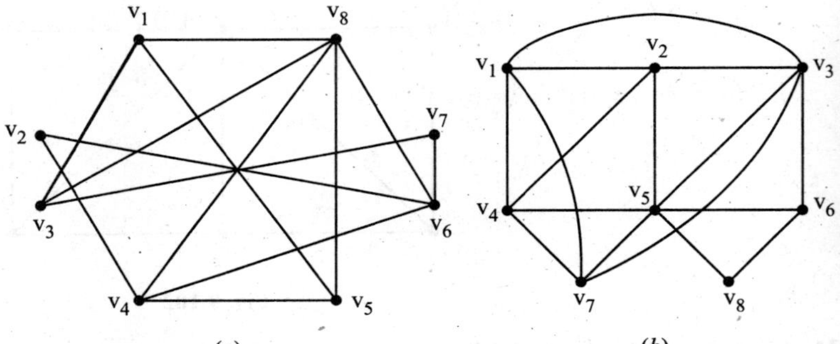

(a) (b)

Fig. 9.108

Hints and Answers

1.

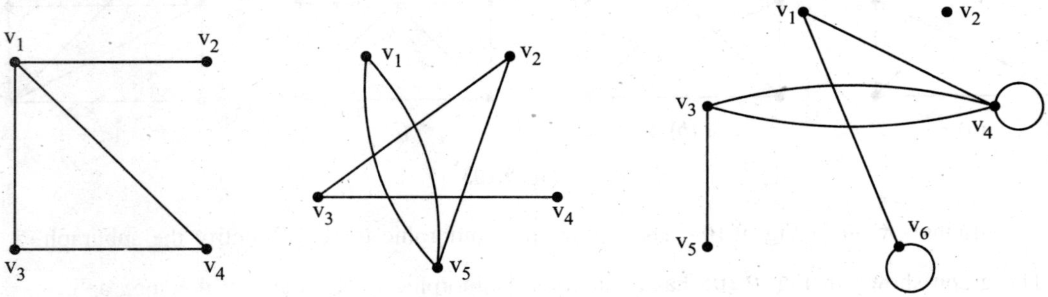

(a) (b) (c)

Fig. 9.109

2. (a) There are six vertices, $V = \{v_1, v_2, v_3, v_4, v_5, v_6\}$ and six edges and therefore six pair of vertices,

$$E = \{(v_1, v_4), (v_1, v_6), (v_4, v_6), (v_3, v_2), (v_3, v_5), (v_2, v_5)\}.$$

(*b*) There are five vertices $V = \{v_1, v_2, v_3, v_4, v_5\}$ and eight edges of which two are multiple edges and one is loop, and thus eight pairs of vertices,

$$E = \{(v_1, v_2), (v_1, v_2), (v_1, v_4), (v_1, v_5), (v_2, v_4), (v_3, v_4), (v_3, v_4), (v_3, v_5), (v_5, v_5)\}.$$

3. Deg $(v_1) = 4$; deg $(v_2) = 3$; deg $(v_3) = 2$; deg $(v_4) = 3$; deg $(v_5) = 4$ (Here the loop at v_5 is counted twice towards the degree of v_5).

4. The vertex v is isolated, if it does not connect to any edge. Thus, v is isolated if any only if deg (V) = 0.

5. (*a*) There are ${}^5C_2 = 10$ ways of choosing two vertices from set V of vertices of G, and hence number of edges are 10 in G.

(*b*) Since multiple edges are permitted, G can have any number of edges (and loops), finite or infinite, hence no such maximum number of edges in G exists.

6. List the initial vertex of the first edge followed by the terminal vertex of each edge in the sequence to obtain

(*a*) $\alpha = (v_1, v_2, v_5, v_4, v_3)$ (*b*) $\beta = (v_1, v_3, v_4, v_2, v_5, v_4, v_2, v_1)$

7. No, because maximum number of edges would be $\dfrac{n(n-1)}{2!} = \dfrac{8 \times 7}{2} = 28.$

8. (*a*) No, the edge (v_3, v_2) is not followed by the edge (v_6, v_4).

(*b*) No, the pair (v_4, v_5) is not an edge (*c*) Yes

9. There is only one cycle in G, the 3-cycle, $\alpha = (v_2, v_3, v_4, v_2)$

10. (*a*) No, the vertex v_6 is not a vertex in G.

(*b*) No, since (v_1, v_4) is not an edge in G.

(*c*) No, although $V' \subseteq V$ and $E' \subseteq E$, H is not a subgraph of G because H is not a graph. Specifically, edge (v_2, v_4) in E′ does not have its end points in V′.

11. The connected components are (v_1, v_2, v_5, v_6), (v_3, v_4, v_7), (v_8, v_9).

12.

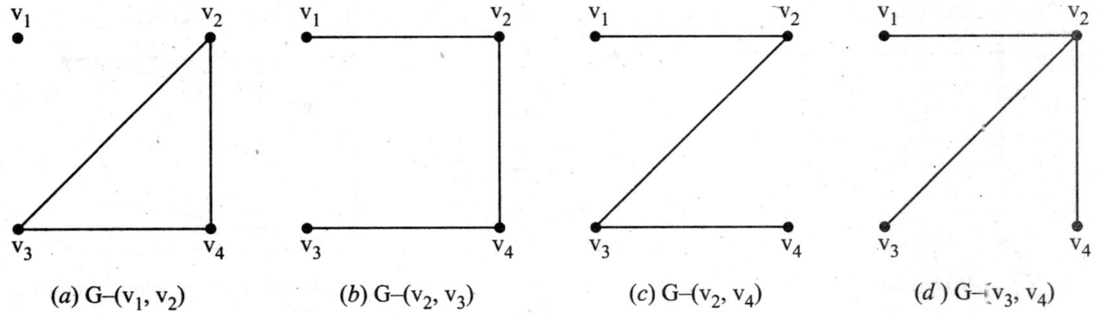

Fig. 9.110

13. (*a*) Vertex v_1 is connected to vertex v_3, and v_4; and vertices v_2, v_5 and v_6 are isolated, and hence, v_1, v_3, v_4, v_2, v_5, v_6 are the connected components of G.

(*b*) Vertices v_1, v_5, v_6 and v_4 are connected, and vertices v_2 and v_3 are connected. Thus, (v_1, v_4, v_5, v_6) and (v_2, v_3) are the connected components of G.

14.

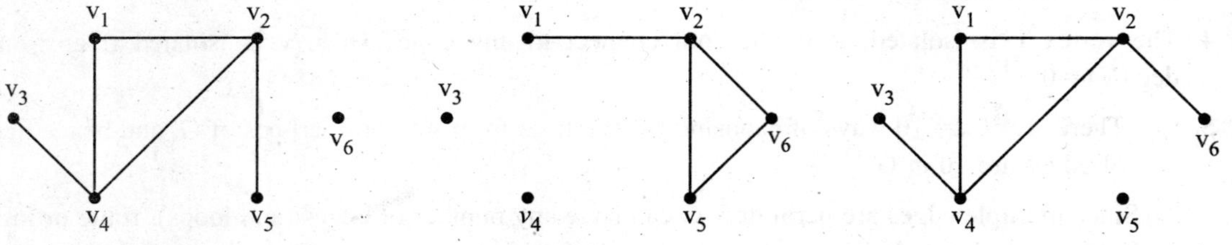

(*a*) G–v_6 (*b*) G–v_4 (*c*) G–v_5

Fig. 9.111

15. If the number of vertices is *n*, then there can be $(n-1)(n-2)/2$ vertices with one vertex not connected to any others. One more edge and connectivity is assured.

16. (*a*) Yes, since exactly two of vertices are of odd degree.

(*b*) Traversable since five vertices are of even degree and two are of odd.

(*c*) Traversable since all the vertices are even.

18. Each pair of vertices determines an edge. Thus, $m = {}^nC_2 = n(n-1)/2$, since there are ${}^nC_2$ ways of selecting two vertices out of *n* vertices.

19. The graphs G must have only even vertices or two odd vertices and one even vertex. In either case, G is traversable.

20. (*a*) The complete graph K_n is traversable for *n* odd number of vertices. For $n > 2$ and *n* even, K_n will have *n* (more than two) odd vertices and hence will not be traversable.

(*b*) Every vertex in K_n has degree $n - 1$. Thus, for every *n*, the graph K_n is regular of degree $n - 1$.

21. No such graphs exist since a 3-regular graph must have an even number of vertices.

22. (*a*) (*b*)

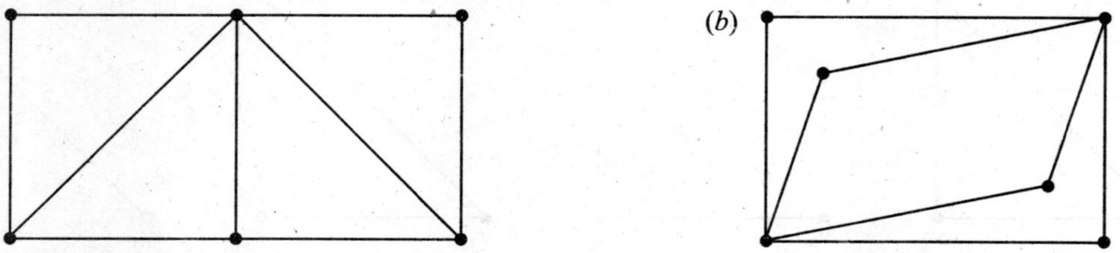

(*a*) Hamiltonian and Non-Eulerian (*b*) Euleriam and Non-Hamiltonian

Fig. 9.112

23. (*a*)

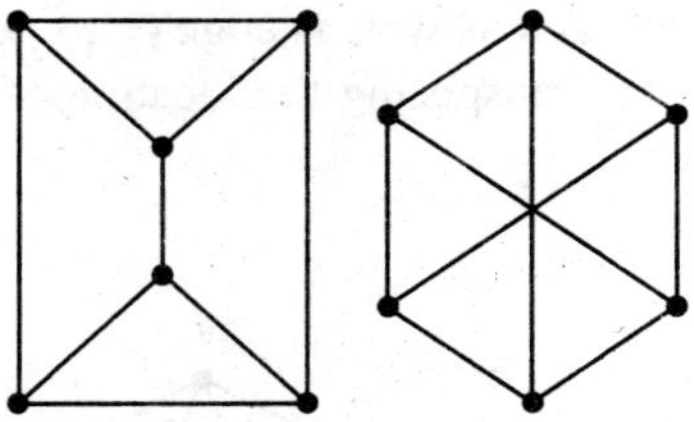

Fig. 9.113

(*b*)

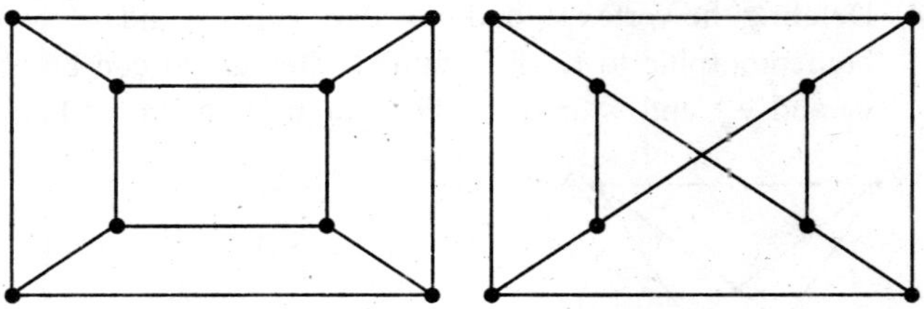

Fig. 9.114

24. (*a*)

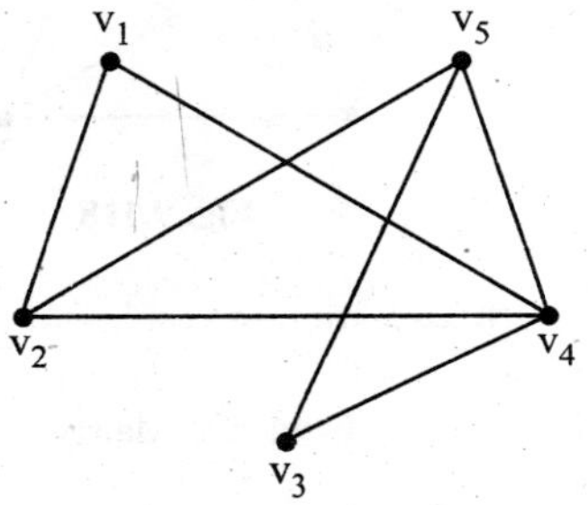

Fig. 9.115

(*b*)

$$\begin{array}{c} \\ v_1 \\ v_2 \\ v_3 \\ v_4 \\ v_5 \end{array}\begin{array}{c} \begin{array}{ccccc} v_1 & v_2 & v_3 & v_4 & v_5 \end{array} \\ \begin{bmatrix} 0 & 0 & 1 & 0 & 1 \\ 0 & 1 & 0 & 1 & 1 \\ 1 & 0 & 0 & 1 & 0 \\ 0 & 1 & 1 & 0 & 1 \\ 1 & 1 & 0 & 1 & 0 \end{bmatrix} \end{array} \quad \begin{array}{c} \\ v_1 \\ v_2 \\ v_3 \\ v_4 \\ v_5 \end{array}\begin{array}{c} \begin{array}{cccccccc} e_1 & e_2 & e_3 & e_4 & e_5 & e_6 & e_7 & e_8 \end{array} \\ \begin{bmatrix} 1 & 1 & 0 & 0 & 0 & 0 & 0 & 0 \\ 0 & 0 & 1 & 1 & 1 & 0 & 0 & 0 \\ 1 & 0 & 0 & 0 & 0 & 1 & 0 & 0 \\ 0 & 0 & 0 & 0 & 1 & 1 & 1 & 1 \\ 0 & 1 & 0 & 1 & 0 & 0 & 1 & 1 \end{bmatrix} \end{array}$$

26. Yes, H is traversable. Also, if α is traversable path for G, then the corresponding path, say α' would be traversable path for H.

27. The graph H must also have eight connected components.

28. No, since isomorphic graphs must have the same number of elements.

29. (*a*)

Vertex	v_1	v_2	v_3	v_4	v_5	v_6	v_7
Indegree	0	2	2	4	1	1	2
Outdegree	4	1	0	0	3	3	1

(*b*)

Vertex	v_1	v_2	v_3	v_4	v_5	v_6
Indegree	2	0	2	4	3	0
Outdegree	1	3	3	1	0	3

30. There are 10 simple paths between v_1 and v_6: (v_1, v_2, v_6), (v_1, v_4, v_2, v_6); (v_1, v_2, v_3, v_6); (v_1, v_4, v_5, v_6); $(v_1, v_4, v_2, v_3, v_6)$; $(v_1, v_4, v_5, v_2, v_6)$; (v_1, v_2, v_5, v_6); $(v_1, v_4, v_2, v_5, v_6)$; $(v_1, v_4, v_5, v_2, v_3, v_6)$; $(v_1, v_2, v_4, v_5, v_6)$.

31. Graphs in Fig. 9.103(*a*) and (*b*) are planar.

32. By removing edge (v_3, v_6), a graph isomorphic to $K_{3,3}$ is obtained. In other words, transpose the location of vertices v_5 and v_6 and redraw the edges as shown in the Fig. 9.116.

33. The graph is identified as shown in Fig. 9.117.

34. Deleting the vertex v_8 and the edges (v_6, v_8) and (v_7, v_8) incident on it as well as edge (v_2, v_3), a graph homeomorphic to K_5 is obtained. This graph can be obtained by transposing the location of vertices v_2 and v_4; and v_3 and v_5. The graph is shown in Fig. 9.118.

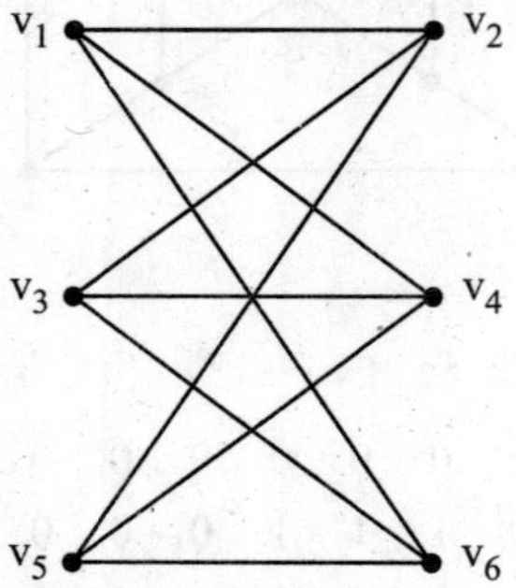

Fig. 9.116

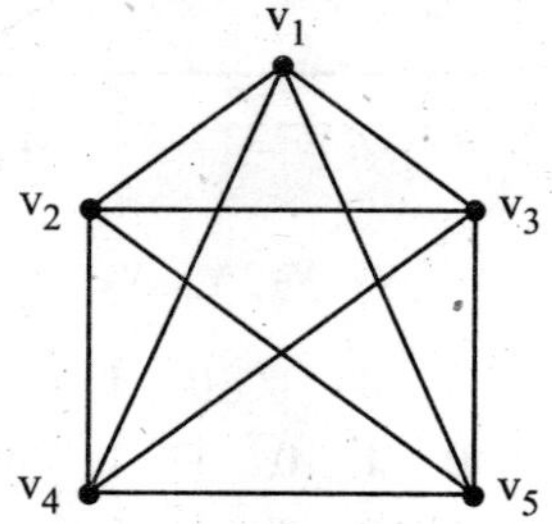

Fig. 9.117

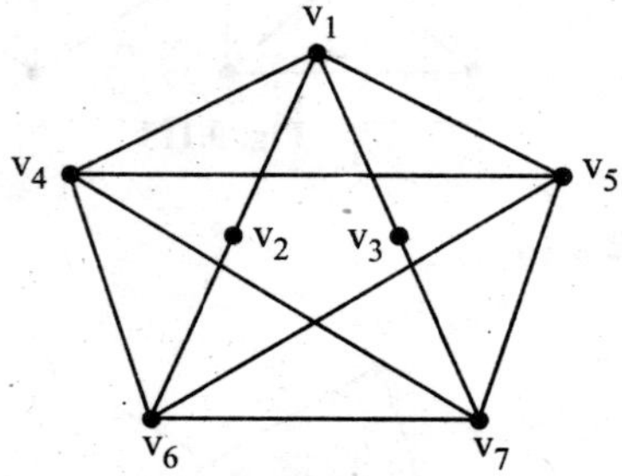

Fig. 9.118

35. The graph is identified as shown in Fig. 9.119.

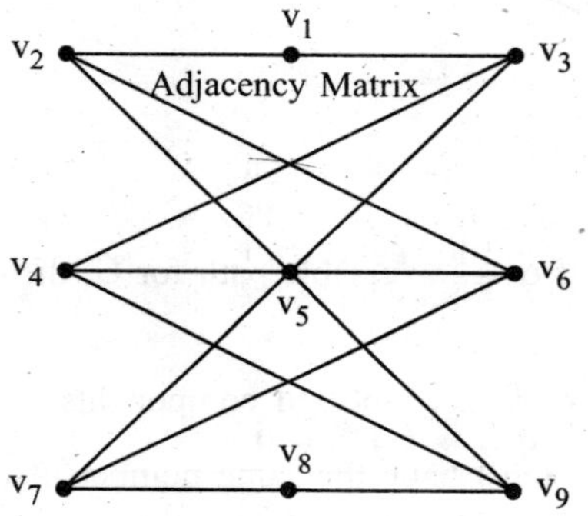

Incidence Matrix

Fig. 9.119

36. Six colours are needed to colour a K_6 graph since every vertex is adjacent to very other vertex and, therefore, for every vertex a different colour is needed. Also, ten colours are needed to colour the graph K_{10} and n colours to colour the graph K_n. Accordingly $c(K_6) = 6$, $c(K_{10}) = 10$ and, in general $c(K_n) = n$.

37. (*a*) The sequence of vertices according to decreasing order of degrees is:

Vertex :	v_8	v_1	v_4	v_6	v_2	v_3	v_5	v_7
Degree :	5	4	4	4	3	3	3	2
Colour :	c_1	c_2	c_2	c_3	c_1	c_3	c_3	c_1

Three colours are essential since vertices v_8, v_4 and v_5 must be painted with different colours because these are connected to each other. Thus, the chromatic number cannot be greater than three. Hence, $c(G) = 3$.

(*b*) The sequence of vertices according to decreasing order of degrees is

Vertex :	v_5	v_3	v_1	v_4	v_7	v_2	v_6	v_8
Degree :	6	5	4	4	4	4	3	2
Colour :	c_1	c_2	c_1	c_2	c_3	c_3	c_3	c_2

Three colours are essential since v_1, v_2 and v_3 must be painted with different colours. Hence, $c(G) = 3$.

Chapter 10

Trees

PREVIEW

A tree is a visual representation of data items in a computer and the connections between some of these items. It is a special case of a more general structure called a graph. Trees can be used to represent logic networks, finite-State machines and formal language deviations. Binary trees are used for data compression of large files.

LEARNING OBJECTIVES

After studying this chapter, you should be able to

- define and understand properties of several types of trees.
- perform preorder, inorder and postorder tree traversal
- use array and pointer representation of binary trees.
- develop infix, prefix, and postfix form of an algebraic expression.
- build a binary search tree and conduct a binary tree search.

10.1 INTRODUCTION

A tree is a connected graph without cycles (also called paths). There are several different form of a tree sand can be applied for solving a wide variety of problems in computer science applications, such as organising and sorting of data in a database, coding theory, language compliers, etc. A graph without self loop parallel edges is also a tree. In this chapter, we will discuss various types of undirected trees, that too, without certain paths called *cycles* and their properties. Obviously, then all multigraphs cannot be considered as trees.

10.2 DEFINITIONS AND PROPERTIES OF TREES

Let G (V, E) be a loop-free undirected graph. The graph G is called tree if it is connected and without cycles (loops). In Fig. 10.1, the graph G_1 is a tree, but graph G_2 is not a tree because it contains the cycle $\{v_0, v_1\}, \{v_1, v_4\}, \{v_4, v_2\}, \{v_2, v_0\}$. Trees are also defined as a non-empty finite set of elements called *nodes (or vertices)* v_i, partitioned into three disjoint subsets. The first subset contains a single element called the *root* of the tree. The other two subsets are called *left* and *right subtrees* (or *branches*) of the original tree. Any of these subtrees can be empty. The relationship among elements of the original tree satisfy the following properties:

(*i*) There exist at least two vertices with degree 1.

(*ii*) Every pair of vertices is connected by a unique path.

(*iii*) There exists one vertex v_0 (often called the *root* of the tree) with the property that there is a unique path from v_0 to every other vertex in the tree but no path from v_0 to itself.

(*iv*) Each vertex other than the root has exactly one predecessor.

(*v*) A tree with n vertices has $n - 1$ edges.

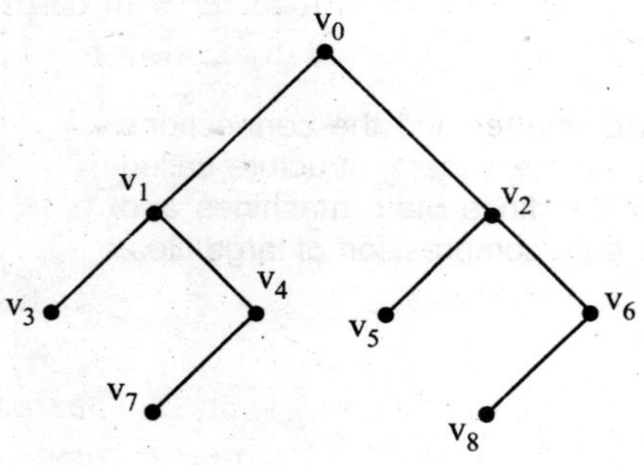

Fig. 10.1(*a*) Graph G_1 – Tree

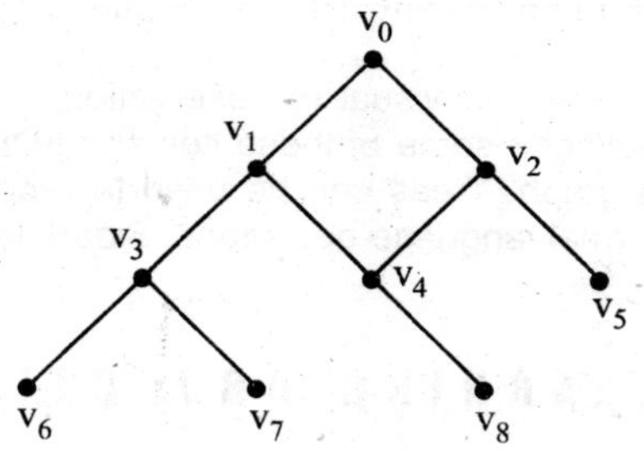

Fig. 10.1(*b*) Graph G_2 – Not a Tree

The tree in Fig. 10.1(*a*) consists of eight nodes with v_0 as its root. Its left subtree is rooted at v_1 and its right subtree is rooted at v_2. This is indicated by the two branches emanating from v_0 to v_1 on the left and to v_2 on the right. The absence of a branch indicates an empty subtree. Fig. 10.1(*b*) illustrates a structure that is not a tree.

10.2.1 Rooted Trees

A rooted tree is a tree with a designated vertex v_0 called its *root* and satisfies following properties:

(*i*) There is no cycle in the tree.

(*ii*) The root of the tree is unique.

(*iii*) Each vertex in the tree other than its root vertex, has indeg $(v_i) = 1$ and root vertex has indeg $(v_0) = 0$.

Figure 10.2 is an example of a rooted tree where vertex v_0 is the root of the tree that appears on the top of the tree (i.e. the tree grows downwards).

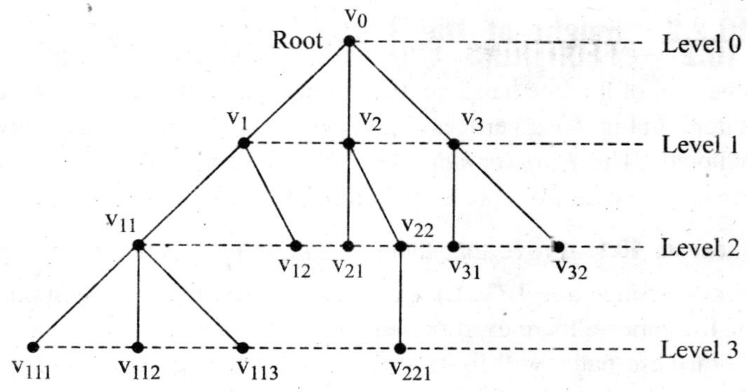

Fig. 10.2

The edges in a rooted tree define predecessor-successor (or parent-child) relationship. In Fig. 10.2, vertex v_0 is the predecessor of vertices v_1, v_2 and v_3. These vertices are called *level* 1 vertices, while v_0 is said to be at *level* 0 and v_0 is sometimes called the *parent* of these level 1 vertices. Vartices v_{11} and v_{12} are called *offspring (children)* of vertex v_1, while vertices v_{21} and v_{22} are called children of vertex v_2. The vertices having the same parent are also called *siblings* of each other. Each vertex at level 1 has no other edges entering it, but vertices v_1, v_2 and v_3 have edge or edges leaving it. The edges leaving a vertex at level 1 are drawn downward and terminate at various vertices such as v_{11}, v_{12}, v_{21}, v_{22}, v_{31} and v_{32}, which are said to be at *level* 2, vertex v_2 would be called the parent of the two offspring v_{21} and v_{22}. This process continues for as many levels as are required to complete the tree.

The *descendent* of a vertex v_i consists of the vertex itself, all its children together with descendents of those children. In Fig. 10.2, for example, vertices v_{221}, v_{21}, v_{22} are considered descendent of v_2 and v_0. In a tree all vertices are descendents of the root vertex.

The vertices of the tree that have no offspring are called the *leaves* of the tree, i.e. leaves of the tree are the vertices of degree one. A directed path from any vertex to a leaf is called a *branch.* Vertices v_{111}, v_{112}, v_{113} and v_{221} etc. are leaves in Fig. 10.2. The paths (v_0, v_1, v_{11}, v_{111}), (v_0, v_1, v_{11}, v_{112}), (v_0, v_2, v_{22}, v_{221}), etc. are branches. The *path length of a vertex* in a rooted tree is the number of edges in the path from root to the vertex.

The vertices which has no children, i.e. outdeg (v_i) = 0, are called *external* (*leaf* or *terminal*) *vertices,* otherwise *branch (non-terminal) nodes.* A vertex n_1 is called an *ancestor* of vertex n_2 (and n_2 is a descendant of n_1) if n_1 is either the parent of n_2 or the parent of some ancestor of n_2. For example, in Fig. 10.2, v_0 is an ancestor of v_{221} and v_{221} is the descendant of v_2. but v_{11} is neither an ancestor nor a descendant of v_2.

Ordered Tree In the tree T shown in Fig. 10.2, the edges (children or offspring) leaving each internal vertex are ordered from left to right. Hence, such a roted tree T is called *ordered rooted tree.* The method of labeling vertices in T is summarised below:

(*a*) Assign a label (or address) 0 to the root vertex.

(*b*) Assign positive integers 1, 2, 3, . . . to vertices at level 1 leaving root vertex from left to right.

(*c*) If a level x is assigned to internal vertex v and v_1, v_2, . . ., v_k are children of v going from left to right, then assign the labels x_1, x_2, . . ., x_k to children v_1, v_2, . . ., v_k, respectively.

For the tree shown in Fig. 10.2, this ordering yields.

0	→ 1.1.1	→ 1.2	→ 2.2.1
1	1.1.2	2.1	3.1
1.1	1.1.3	2.2	3.2

Such ordering is called *lexicographic* or *dictionary order.*

10.2.2 Height of the Tree

The root of the tree has level 0, and the level of any other vertex in the tree is one more than the level of its parent. In Fig. 10.2, vertices $v_{11}, v_{12}, v_{21}, v_{22}, v_{31}$ and v_{32} are at level 2 and vertices $v_{111}, v_{112}, v_{113}$ and v_{221} are at level 3. The *depth* (or *height*) of a tree is the maximum level of any leaf in the tree. This equals the length of the longest path from the root of any leaf. The tree in Fig. 10.2 is of height 3.

Theorem 10.1 Prove that there is unique path between each pair of vertices in a tree T (V, E).

Proof Since tree T (V, E) is connected, there exists at least one path between each pair of distinct vertices in T. Suppose, there exist two distinct paths between a pair of vertices u and v in T, then some of the edges from these paths will form a cycle. Since a tree has no cycle, therefore, T cannot be a tree.

Theorem 10.2 If in a graph G, there is one and only one path between every pair of vertices, G is a tree.

Proof In a connected graph G, there is only one path between every pair of vertices of G. Since G has only one path between every pair of it cannot have any cycle (as existence of cycle requires two or more path between a pair of vertices). Hence, G is a connected acyclic graph and is, hence it is, a tree.

Theorem 10.3 A tree T (V, E) with n vertices has $n - 1$ edges.

Proof We shall prove this theorem by mathematical induction on the number of vertices, n in T.

Suppose the theorem is true for $n = k\ (\geq 2)$, where k is some positive integer. To prove that the result is also true for $n = k + 1$, suppose T has $k + 1$ vertices. If we remove an edge with end points u and v from T, then we are left with two sub-trees T_1 (V_1, E_1) and T_2 (V_2, E_2) such that $|V| = |V_1| + |V_2|$ and $|E| = |E_1| + |E_2| + 1$. Obviously T_1 and T_2 are connected with no cycles, and having vertices less than n, i.e. $|V_1| \leq n$ and $|V_2| \leq n$. This means that we can apply induction process on T_1 and T_2 to give

$$|E_1| = |V_1| - 1;\ |E_2| = |V_2| - 1$$

$$|V| = |V_1| + |V_2| = (|E_1| + 1) + (|E_2| + 1)$$

$$= (|E_1| + |E_2| + 1) + 1 = |E| + 1$$

or $$|E| = (|V_1| - 1) + (|V_2| - 1) + 1 = (|V_1| + |V_2|) - 1 = k + 1 - 1 = k$$

This implies that T has k edges, which is the required number. Hence, theorem is proved.

Alternative statement In every tree T, number of vertices are one more than the number of edges, i.e. $|V| = |E| + 1$.

Theorem 10.4 If G is a connected graph with n vertices and $n - 1$ edges, for any positive integer n, then G is a tree.

Proof We know that a connected graph G (V, E) with no cycles is a tree T which contains n vertices and $n - 1$ edges.

Now the converse of this statement that any graph G (V, E) that contains $n - 1$ edges and no cycles, is connected need to be proved. For this, let us decompose G (V, E) into k components $G_1, G_2, \ldots, G_k$, each of which is connected with no cycles. It follows that each G_k, is a tree. Thus each of these should satisfy the condition that $\sum_{i=1}^{k} e_i = \sum_{i=1}^{k} (n_i - 1) = n - k$ or $e = n - k$, since $|E_k| = |V_k| - 1$. This implies that for $k = 1$. This gives $e = n - 1$, and hence G is a tree having only one component.

Theorem 10.5 Any connected graph with n vertices and $(n - 1)$ edges is tree.

Proof Since minimum number of edges required for a graph with n vertices to be a connected (or acyclic) graph is $n - 1$, therefore a connected graph with n vertices and $(n - 1)$ edges cannot have a cycle and hence the graph G is a tree.

Remark A minimal connected graph (a connected graph with n vertices and $n - 1$ edges). Thus the theorem can also be stated as A minimally connected graph is a tree. Thus if a graph is not minimally connected, then deletion of an edge e from G shall also keep $G - e$ as connected, but G will not be a tree.

Theorem 10.6 A connected graph G is a tree if any only if adding an edge between any two vertices of G creates an exactly one circuit.

Proof Since any pair of vertices of a tree are connected by a path, therefore addition of an edge between any two vertices of a tree shall create an additional path. Hence, a closed circuit (or cycle) is formed. So addition of an edge between any two vertices of a tree creates a cycle.

Conversely, if G is a connected graph, there exists a path between every pair of its vertices. Addition of edge between any two vertices of G creates exactly one closed circuit (or cycle) in G. Thus implies that G is a connected acyclic graph. Thus G becomes a tree.

Theorem 10.7 A acyclic graph G having n vertices, $n - 1$ edges is connected.

Proof Let G be a disconnected acyclic graph will n vertices and $n - 1$ edges. Then G consists of two or more acyclic components. Suppose G consists of two acyclic components G_1 and G_2.

If an edge e is added between a vertex v_1 in G_1 and a vertex v_2 in G_2, then G becomes connected. However, addition of an edge e between vertex $v_1 \in G_1$ and $v_2 \in G_2$ does not make G cyclic. This implies that G is a tree with n vertices and n edges. This contradicts our assumption. Thus the graph G which is acyclic with n vertices and $n - 1$ edge cannot be disconnected and is a connected graph.

Theorem 10.8 An arborescence is a tree in which every vertex other than the root has an in-degree of exactly one.

Proof Since an arborescence with n vertices can have at most $n - 1$ edges and contributes atmost $n - 1$ in-degrees to all the vertices of tree. Sum of in-degrees of all vertices in the arborescence is given by

$$d^-(v_1) + d^-(v_2) + \dots + d^-(v_n) \le n - 1$$

Only one term out of n terms on the RHS of this equation is zero, because in-degree of root is zero. The value of each of the remaining $(n - 1)$ terms must be a positive integer.

The sum of $(n - 1)$ positive integers must be equal to $(n - 1)$, which is possible only when value of each of these is 1. Other than root whose in-degree is zero, the in-degree of each of the remaining $n - 1$ vertices in one. Thus number of edge shall be $n - 1$.

Hence, an acyclic graph with n vertices and $n - 1$ edges is connected and is also a tree.

Theorem 10.9 In an arborescence, there is a directed path from the root v to every other vertex.

Proof If we consider a direct path starting from the root v to any finite distance, then it will have to terminate at a vertex with degree one (pendant vertex), otherwise a vertex with in-degree two or more exist. But it is not possible have a vertex whose indegree is two or more in an arborescence.

Also arborescence is connected, every vertex shall lie on a direct path from root v to each of the pendant vertices.

Theorem 10.10 An acyclic digraph G with n vertices is an arborescence if there is a vertex v in G such that every other vertex is reachable from v, and v is not reachable from any other vertex.

Proof Since every vertex of G is reachable from vertex v, it is connected. Also G is acyclic therefore, graph G is a tree. As v is not reachable from any other vertex, so its in-degree $d^-(v) = 0$. Also in-degree of every vertex other then v is at least one. Moreover, number of edges in G is $n - 1$ and sum of in-degree (which is an integer) of each of the $(n - 1)$ vertices is $n - 1$, then condition is satisfied only with the sign of equality, i.e. then in-degree of each of the vertices other than v is exactly one. Hence, G in an arborescence.

Remark The following are 5 different but equivalent definitions of a tree. A loop-tree undirected graph with n vertices is called a tree if

(*a*) G is connected, and no cycles.
(*b*) G is connected with $|V| = |E| + 1$.
(*c*) G contains no cycles with $|V| = |E| + 1$.
(*d*) G is connected and can be disconnected into two sub-graphs that are trees by deleting any edge from G.
(*e*) There is exactly one path between each pair of vertices in G.

Example 1 Consider the rooted tree shown in Fig. 10.3. Identify the path length from its root to each of the vertices v_1, v_6, v_8 and v_{13}.

Solution The path length in the rooted tree as shown in Fig. 10.3 from its root to each of the vertices v_1, v_6, v_8 and v_{13} is obtained by listing the vertices between these two vertices including the vertex itself as follows:

From v_0 to v_1 : (v_0, v_1) path length is one.
From v_0 to v_6 : (v_0, v_2, v_6), path length is two.
From v_0 to v_8 : (v_0, v_1, v_3, v_8), path length is three.
From v_0 to v_{13} : $(v_0, v_2, v_7, v_{12}, v_{13})$, path length is four.

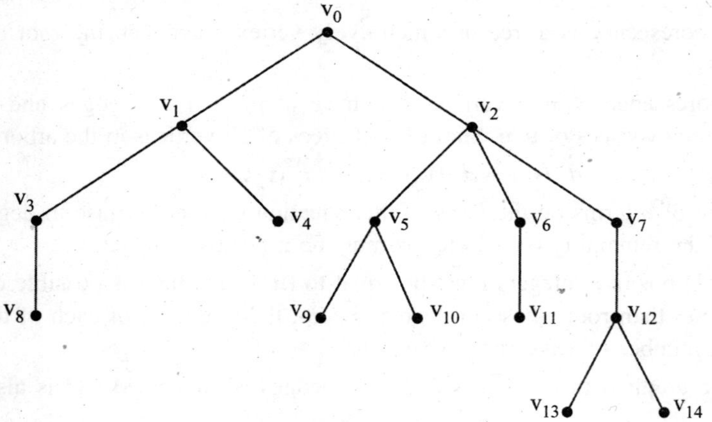

Fig. 10.3

Example 2 Consider the rooted tree shown in Fig. 10.3. Identify (*a*) all vertices which precede vertex v_{11}, (*b*) all vertices which follow vertex v_1, (*c*) the children of vertex v_2, and (*d*) the siblings of vertex v_3.

Solution (*a*) The path from the root v_0 to v_{11} includes the vertices v_0, v_2, v_6 and v_{11}. Hence, v_0, v_2 and v_6 precede v_{11}.

(*b*) All paths from vertex v_1 include the vertices v_3, v_4 and v_8, i.e. v_3, v_4 and v_8 follow v_1.

(*c*) The vertices v_5, v_6 and v_7 are the children of vertex v_2, since these vertices immediately follow v_2.
(*d*) The parent of v_3 is v_1 and the children of v_1 are v_3 and v_4, and hence v_4 is the only sibling of v_3.

10.3 BINARY TREES

A rooted tree T is called an *m-ary rooted tree* when each internal vertex has at most m children i.e., order (v) $\leq m$ for all $v \leq V$. In particular, if each internal vertex of a rooted tree has at most two children, then such tree is called 2-ary *binary tree,* i.e. order (v) = 0, 1 or 2. Figure 10.4 is an example of a binary tree.

Every binary tree is also a rooted tree having following properties:

(*i*) The number of vertices in a binary tree is always odd.

(*ii*) If the number of vertices, say p in a binary tree is of degree 1, then $n - p - 1$ is the number of vertices of degree 3.

A tree consisting of no node (null tree or empty tree) is also a binary tree.

One of the applications of binary tree as a useful data structure in computer science arises when we need to take a two-way decision at each point in a process. For example, to find all duplicates in a list of numbers, we compare each number with all those that precede it. The large number of comparisons needed in this process can be reduced by using a binary tree.

If a tree T contains unique vertex called its root, then the remaining vertices of T form an ordered pair of disjoint binary trees T_1 and T_2. However, if T does not contain root, then the two trees T_1 and T_2 are called, respectively, the *left sub-tree* and *right sub-tree* of the root. In other words, the left and right child of some vertex (node) is called left and right sub-tree of that vertex, respectively. But if T_1 is non-empty, then its roots is called the *left successor* of the root vertex. Similarly, if T_2 is non-empty, then its root is called the *right successor* of the root vertex.

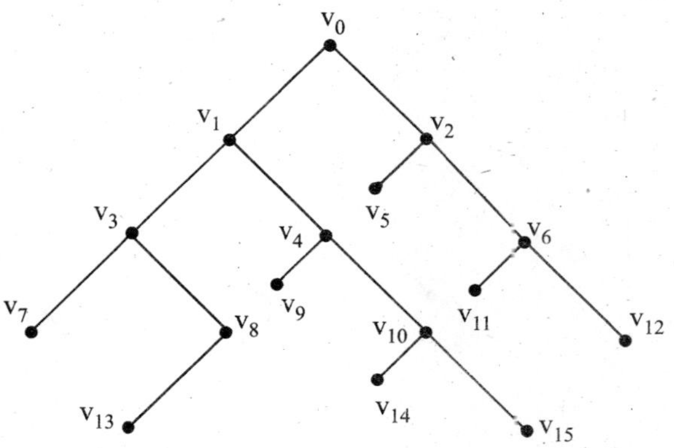

Fig. 10.4. Example of a Binary Tree

Remarks 1. A vertex of degree one is also called *pendent vertex.*

2. If every non-leaf vertex in a binary tree has non-empty left and right sub-trees, the tree is termed as a *strictly binary tree.*

Theorem 10.11 The number of vertices in a binary tree is always odd.

Proof In any tree there is only one vertex of even (= 2) degree and the remaining $(n - 1)$ vertices are of odd degree. Since number of vertices with odd degree in a tree has to be even, therefore $n - 1$ is even provided n is odd.

Theorem 10.12 The number of pendant vertices in a tree is equal to $p = (n + 1)/2$, where n is the number of vertices in a binary tree.

Proof Since total number of vertices in the given binary tree are n, therefore number of pendant vertices will be p. Also number of vertices with degree 2 is one, so number of vertices with degree 3 = $n - (p + 1)$ = $n - p - 1$. Thus number of degrees of all vertices are

= number of degrees of p vertices each of degree 1

+ number of degree of 1 vertex of degree 2

+ number of degrees of $n - p - 1$ vertices each of degree 3

$= (p \times 1) + (1 \times 2) + [(n - p - 1) \times 3] = 3n - 2p - 1$

Alternatively, the number of edges in the tree with n vertices are $n - 1$. The total number of degrees associated with these $n - 1$ edges (and hence n vertices) are $2(n - 1)$. Hence,

$$3n - 2p - 1 = 2n - 2, \text{ i.e. } p = (n + 1)/2 \qquad (1)$$

Corollary: From Eqn. (2) we have $n - p = p - 1$, i.e. in a binary tree the number of internal vertices are one less than the number of pendant vertices.

10.3.1 Forest

A graph with no cycle, where each node (vertex) has one or no predecessors is called a *forest*. A tree may be defined as a forest in which only a single node (called the *root*) has no predecessors. Any forest consists of a collection of disjoint trees obtained by deleting the root and the corresponding edges connecting the vertices from a given tree. Fig. 10.5(*a*) gives an example of a forest.

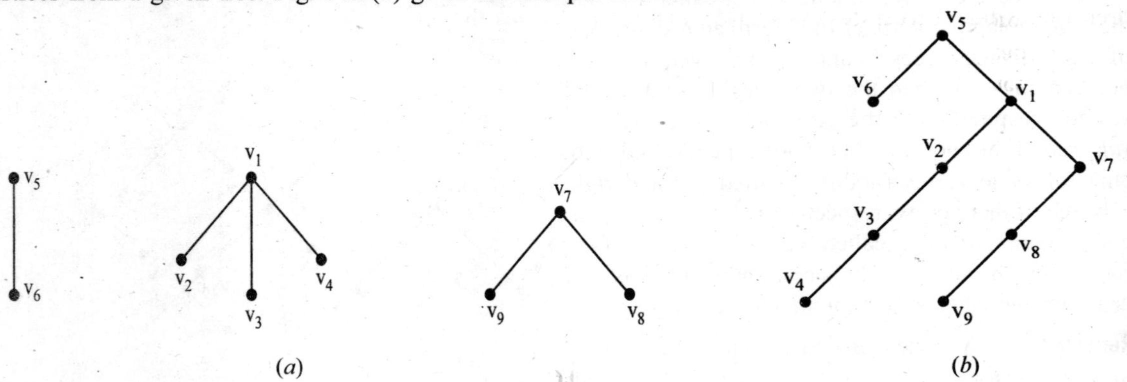

Fig. 10.5 Example of a Forest

To convert the given forest as shown in Fig. 10.5(*a*) into a binary tree, the root node of it is the root node of the first tree from left and the root node of the second tree becomes the right child of the root node in binary tree and the root node of the third tree becomes the right child of the right child in binary tree.

The process is repeated for each level till complete binary tree is obtained as shown in Fig. 10.5(*b*).

Example 3 Consider the binary tree shown in Fig. 10.4.

(*a*) Find the level of each vertex.

(*b*) Find the height or depth of each vertex.

(*c*) List the children of each vertex.

Solution (*a*) The level of each vertex is listed below:

Level	0	1	2	3	4
Vertex	v_0	v_1, v_2	v_3, v_4, v_5, v_6	$v_7, v_8, v_9, v_{10}, v_{11}, v_{12}$	v_{13}, v_{14}, v_{15}

(*b*) The height of each vertex is given below:

Height	1	2	3	4	5
Vertex	v_0	v_1, v_2	v_3, v_4, v_5, v_6	$v_7, v_8, v_9, v_{10}, v_{11}, v_{12}$	v_{13}, v_{14}, v_{15}

(*c*) Children of each vertex are given below:

Vertex	v_0	v_1	v_2	v_3	v_4	v_6	v_8	v_{10}
Children	v_1, v_2	$v_3, v_4,$	v_5, v_6	$v_7, v_8,$	$v_9, v_{10},$	v_{11}, v_{12}	$v_{13},$	v_{14}, v_{15}

Example 4 Consider three trees T_1, T_2 and T_3 shown in Fig. 10.6. Identify trees which represent (*a*) rooted tree, (*b*) ordered rooted tree, and (*c*) binary tree.

Solution (*a*) All the trees T_1, T_2 and T_3 shown in Fig. 10.6 are rooted trees, i.e. vertex v_0 is the root with children (immediate successors) v_1 and v_2, and vertex v_2 has the single child v_3.

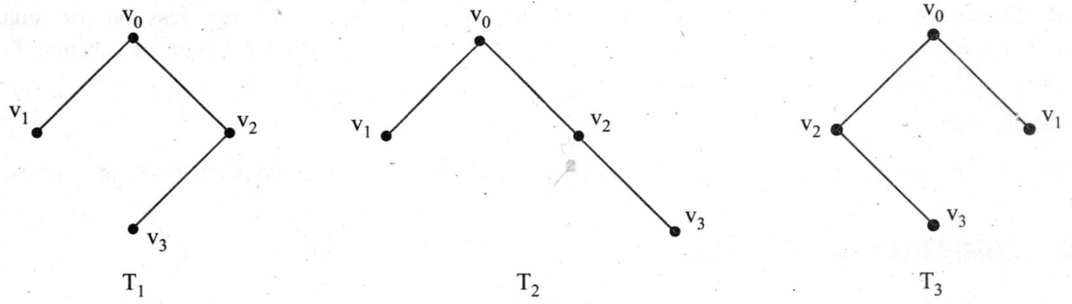

Fig. 10.6

(*b*) Trees T_1 and T_2 are ordered rooted trees but T_3 is different. For example, vertex v_1 is the first child in T_1 and T_2 but it is the second child in T_3.

(*c*) All T_1, T_2 and T_3 represent different binary trees. Because T_1 and T_2 are different since successors between left and right are different even when there is only one successor (which is not true for ordered rooted trees). Vertex v_3 is the left successor of v_2 in T_1 but right successor of v_2 in T_2.

Example 5 Distinguish between a general tree and a binary tree.

Solution The general tree is different than a binary tree due to following reasons:

(*i*) A binary tree may be an empty tree (i.e. it may not contain any node), but a general tree is non-empty.

(*ii*) In binary tree, if a vertex (nodes) has a child, then it is distinguished as a left child or right, but no such distinction exists in a general tree, for example, trees T_1 and T_2 in Fig. 10.6 are different when these are considered as binary trees, otherwise these two are same general trees.

Theorem 10.11 Let T be a complete m-ary tree with n vertices, i.e. $|V| = n$. If T has i internal vertices an l leaves, then prove that $n = mi + 1$.

Proof Since T is a complete m-ary, therefore, its each vertex has m children. Thus, the total number o vertices will be mi, expect the root vertex, where i is the internal vertex. This implies that T has $n = mi +$ vertices.

Remark The following results can easily be derived from the statement of Theorem 10.11.

(a) n vertices has $i = (n - 1)/m$ internal vertices and l $[(m - 1)n + 1]/m$ leaves.

(b) i internal vertices has $n = mi + 1$ vertices and $l = (m - 1)\, i + 1$ leaves.

(c) l leaves has $(ml - 1)/(m - 1)$ vertices and $i = (l - 1)/(m - 1)$ internal vertices.

Theorem 10.12 Let T be a complete m-ary tree of height h and l leaves. Then prove that $l \leq m^h$ an $h \geq \log_{ml}$.

Proof We shall prove this theorem by mathematical induction on h.

For $h = 1$, the tree T consists of one root and m children, each of which is a leaf. Thus, $l = m = m$ in a m-ary tree of height one. Suppose the result is true for all m-ary trees of height less than h.

Consider a m-ary tree T of height h having l leaves (at least m of the leaves at level h). The l leave of T are also the leaves of m sub-trees T_i, $1 \leq i \leq m$ obtained by deleting the edges from the root of each of the vertices at level 1. Since $m \geq 1$, therefore each of these sub-trees is of height less than or equal to $h - 1$. Thus, by the induction hypothesis, $l_i \leq m^{h(T_i)} \leq m^{h-1}$ where $h\,(T_i)$ is the height of sub-tree T_i for $1 \leq i \leq m$. Since there are almost m such trees, each with a maximum of m^{h-1} leaves, $l = \sum_{i=1}^{m} l_i \leq m\left(m^{h-1}\right)$ $= m^h$. This proves the theorem.

Since $l \leq m^h$, taking log on both sides, we have $\log_m l = \log_m (m^h) = h$. This implies that $h \geq \log_m l$ for $h \in I_+$.

10.4 COMPLETE BINARY TREE

If every vertex (node) of a rooted tree has m offsprings (children), then such a tree is said to be a complete m-ary tree.

In particular, if $m = 2$, then such a tree is called a *complete binary tree*. The height (or depth) of the complete binary tree having m vertices is, $h_m = \log_2 m + 1$.

All vertices in a binary tree having two children are called *internal vertices* (also called *non-pendent vertices*) and the vertices having no children are called *external vertices* or *leaves*. If a binary tree contains m vertices (nodes) at level r, it contains atmost $2m$ vertices at level $r + 1$. Since a binary tree can contain atmost one vertex at level 0 (the root), it can contain atmost 2^r nodes at level r.

A rooted binary tree in which all leaves are at level h is also termed as *complete binary tree* of height h, and that contains exactly 2^r vertices (nodes) at each level r. The *total number of vertices in a complete binary tree of height (h) equals the sum of the number of vertices at each level.* The significance of complete binary tree is that it is the binary tree with the maximum number of vertices at a given height. Although a complete binary tree contains many vertices, the distance from the root to any leaf (the tree's height) is relatively small.

In other words, a rooted binary tree of height, h is called *a complete binary tree* provided

(*i*) Any vertex at a level less than $h - 1$ has two children.

(*ii*) For any vertex v_h in the tree with a right descendant at level h, v_h must have a left child and every left descendant of v_h is either a leaf at level h or has two children.

A fifteen vertices, three-level binary tree is shown in Fig. 10.7. The number of vertices at levels 1, 2 and 3 are 2, 4 and 8, respectively.

Level	*Vertices*
1	2
2	4
3	8

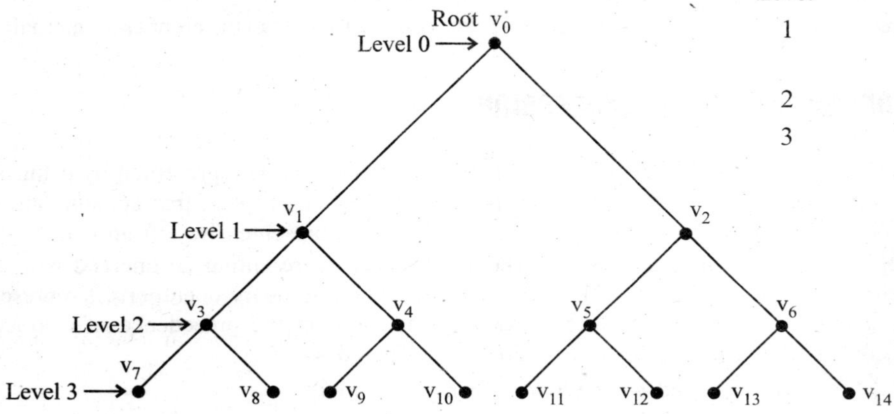

Fig. 10.7 Complete Binary Tree of Height 3

Balance Tree A rooted m-ary tree of height h is said to be balanced if all its leaves are at level h or $h - 1$.

Theorem 10.13 Prove that the maximum number of vertices on level k of a binary tree is $2^k \geq n$, $n \geq 0$.

Proof For $n = 0$, we have the only one vertex, i.e. root of the binary tree at level 0. Thus, the result is true.

Let the result is true for $n = k$. Thus, the maximum number of vertices possible in a k-level binary tree is $2^k \geq k$.

For $n = k + 1$, we have 2^{k+1}, the maximum number of vertices possible for a $(k + 1)$-level binary tree. Since each vertex in a binary tree has atmost two children, therefore maximum number of vertices at a particular level is twice the maximum number of vertices at the preceding level. Hence, $2^{k+1} = 2.2^k$ proves the result for any integer value of n.

Theorem 10.14 Every tree T with at least two vertices has at least two pendent vertices.

Proof We know that in every tree $|V| = |E| + 1$ or $|E| = n - 1$, because $|V| = n \geq 2$, so $2|E| = 2(n - 1)$ = sum of all degrees of vertices in T. Since T is connected, deg (v) ≥ 1 for all $v \in V$, But if there are pendent vertices less than 2, then either deg (v) ≥ 2 or deg (v) $= 1$ for a particular single vertex in T. But in both these cases we arrive at a contradiction because $2(n - 1) = \Sigma \deg(v) \geq 2|V| = 2n$ and $2(n - 1) = \Sigma \deg(v) \geq 1 + 2(n - 1)$.

Theorem 10.15 Prove that the maximum number of vertices possible in a binary tree of height, n is $2^n - 1$; $n \geq 1$.

Proof For $n = 1$, we have $2^1 - 1 = 1$, which is obviously the root vertex of the binary tree.

Let the result is true for $n = k$. Thus, the maximum possible number of vertices at height k of binary tree is $2^k - 1$.

For $n = k + 1$, we have $2^{k+1} - 1 = (2.2^k) - 1$. Since each vertex in a binary tree has atmost two childre therefore the maximum number of vertices at the particular height is twice the number of vertices at th preceding height. Hence, $2^{k+1} = (2.2^k) - 1$ proves the result for any integer value of n.

Remarks

1. In a binary tree, there is at least two vertices of degree one.
2. The maximum possible height of an n-vertex binary tree is, Min $\lceil \log_2 (n + 1) - 1 \rceil$, where $\lceil n \rceil$ represen the smallest integer greater than or equal to n.
3. In a binary tree, the number of external vertices (or leaves) is one more than the number of internal vertice

10.5 TREE OF AN ALGEBRAIC EXPRESSION

An algebraic expression E containing operands and binary operators can be represented by using a binar tree called *expression binary tree* (also called *strictly binary tree*). The root of tree contains an operato that is to be applied to the results of evaluating the expressions represented by left and right sub-trees A vertex representing an operator is a non-leaf, whereas a vertex representing an operand is a leaf. Fo example, an algebraic expression $E = (a + b) * ((c \div d) - e)$ involving binary operations is represented b a binary tree as shown in Fig. 10.8. In other words, depending upon the precedence of evaluation, a algebraic expression involving binary operators can be considered as:

$$\ldots \text{ left operand} \leftrightarrow \text{(operator)} \leftrightarrow \text{right operand} \ldots$$

to help in the construction of an expression binary tree.

Example 8 Consider the algebraic expression $E = (2x - y)^3 * \{a + (b \div c)\}$.

(*a*) Draw the tree which corresponds to the expression E.

(*b*) Find the sub-tree at the exponential operator.

Solution (*a*) Using symbol ↑ (arrow) and * (asterisk) for exponential and multiplication operators, respectively we obtain a tree as shown in Fig. 10.9.

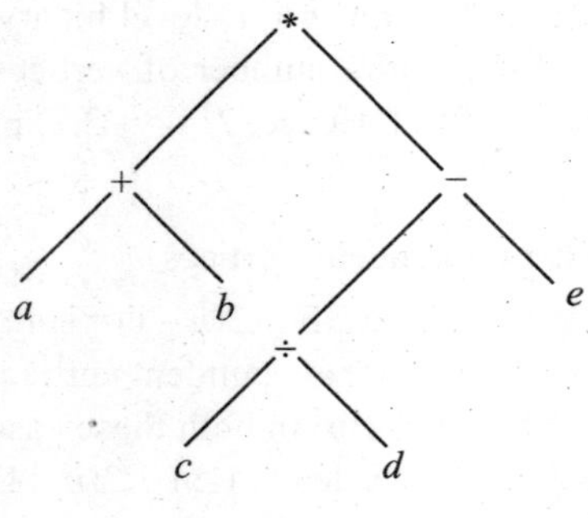

Fig. 10.8 $E = (a + b) * ((c \div d) - e)$

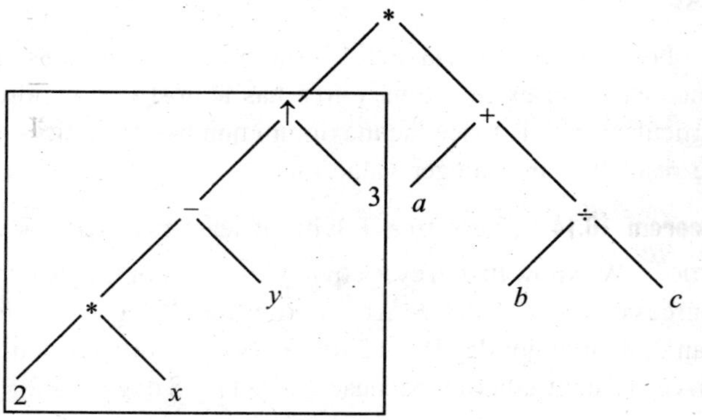

Fig. 10.9 $E = (2x - y)^3 * \{a + (b \div c)\}$

(*b*) The sub-tree at the exponential operator corresponds to expression $(2x - y)^3$ as shown in the box in Fig. 10.9.

10.5.1 Infix, Prefix and Postfix Form of an Algebraic Expression

Instead of performing scanning operations back and forth repeatedly, the computer converts algebraic expressions in the following three different forms – *Infix*, *Prefix* and *Postfix*, that are independent of parentheses. These are also known as *Polish notation* in the honour of Jan Lukasiewicz, a polish (Ukrainian) logician.

The use of such forms is important for the compilation of computer programmes and can be obtained by representing a given expression by a rooted tree. While traversing a binary tree, starting from its roots, from top to bottom and left to right, we find polish notation by writing down the labels of the vertices in the order in which they are visited.

Infix form This form of an algebraic expression is inorder traversal of binary tree representing the expression. The original algebraic expression appears with its elements and operations in the order in which the originally occurred. In this form, it is important to include parantheses in the inorder traversal as and when an operation appears. For example, the infix notation $x*y$ for a binary operation*becomes*ab.

Prefix or polish form It is an algebraic expression form without parentheses. The procedure to express an algebraic expression in polish form is as follows: Move from left to right until a string of the form Bxy is found, where B represents a binary operator such as: +, –, ÷, ×, etc., and x and y are numbers. Evaluate xBy and substitute the answer for the string Bxy. This form of an algebraic expression is named as prefix because in this form binary operator symbol precedes the arguments. For example, expression E = $(2x - y)^3$ $\{(a + (b \div c)\}$ can be expressed in prefix form as: $* - * 2xy \uparrow 3 + a \div bc$.

Postfix or reverse polish form This form of an algebraic expression is same as prefix form except that in this form binary operator comes after its arguments. For example, the expression E = $(2x - y)^3$ $(a + (b \div c))$ can be expressed in postfix form as: $2xy3 \uparrow abc \div + \times$.

Few examples on equivalent forms of these polish notations are given below:

Infix	*Prefix*	*Postfix*
$(a + b - c)/(x * y)^3$	$/ + a - bc * x \uparrow yz$	$- abc * \uparrow xyz + /$
$(a + b) * (c - d)$	$* + ab - cd$	$ab - cd + *$
$[(a - b) \uparrow 2]/(a + b)$	$/ - ab \uparrow 2 + ab$	$ab \uparrow 2 + ab$

Example 9 Determine the value of the expression represented by binary tree shown in Fig. 10.10.

Solution The expression represented by the given binary tree is:

$$(a + 5) * [\{(7 * b) + c\}/(9 + d)\}]$$

The value of this expression is

$$(a + 5)\ [(7b + c)/(9 + d)] = (7b + c)\ (a + 5)/(9 + d)$$

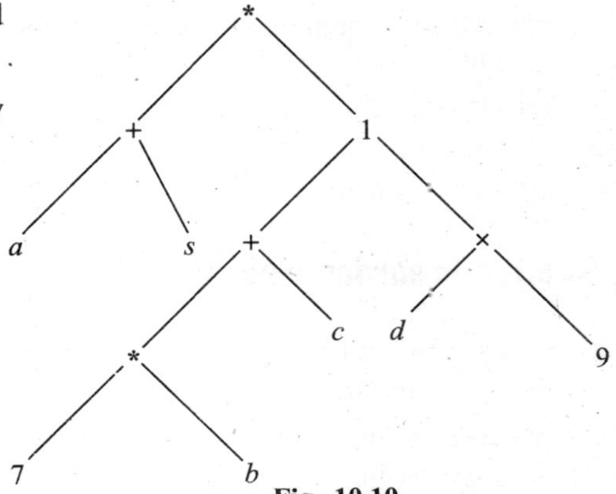

Fig. 10.10

Example 10 Write the value of the expression

(*a*) + 4/* 23 + 1 – 9 ↑ 23

(*b*) – + 5 * 62 + /921

(*c*) 56 + 25 + *

Solution (*a*) $+ 4/* 23 + 1 - 9 \uparrow 23 = + 4 / * 23 + 1 - 98$; $\uparrow 23 = 2 - 3 = 8$

$= 4 / * 23 + 11$; $- 98 = 9 - 8 = 1$

$= + 4 / * 232$; $+ 11 = 1 + 1 = 2$

$= + 4 / 62$; $* 23 = 2 * 3 = 6$

$= + 43$; $/ 62 = 6/2 = 3$

$= 7$; $+ 43 = 4 + 3 = 7$

(*b*) $- + 5 * 62 + / 921 = - + 5 * 623 = (5 + 6 * 2) - 3 = 19$

(*c*) $56 + 25 + * = * + 56 + 25 = (5 + 6) * (2 + 5)$

$= 11 * 7 = 77$

10.6 TRAVERSING BINARY TREES

Sometimes, it is desirable to pass through the tree, enumerating each of its nodes (vertices) once in some specified order so as to take some action or perform some computation appropriate to the application being represented by the tree. For example, at each vertex of a tree that represents an algebraic expression, we may desire to perform a computation indicated by the *operator* at that vertex. The process of visiting and then performing some computation at each vertex of a tree is called *searching* or performing a *tree search.* This process is also called *walking* or *traversing* the tree.

Suppose each vertex of a binary tree has exactly *n* children (offspring). Each of these children are labelled with different numbers from 1 to *n* or with English alphabet A to Z. Such a labelled tree is sometimes called *positional binary tree.* The root of the tree is not labelled, since it is not an offspring of any vertex. The positions of the offspring is also indicated by the direction of the edges as shown in Fig. 10.11. Since each vertex of a positional binary tree has two potential offspring, therefore, these offspring will be denoted as v_L (left offspring) and v_R (right offspring).

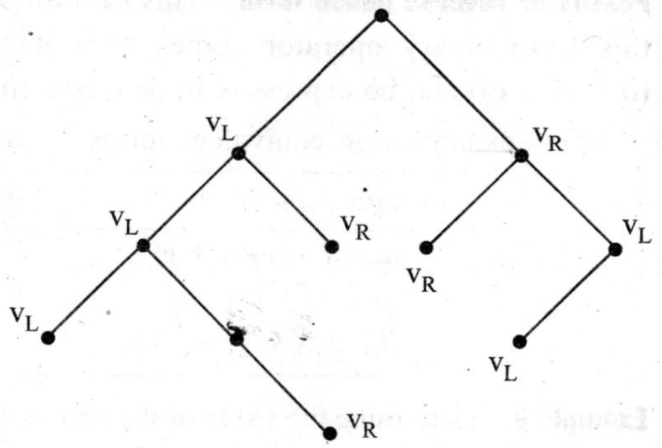

Fig. 10.11 Positional Tree

The following are three standard recursive methods (or algorithms) for searching (traversing) a positional binary tree T with root v.

1. Preorder search
2. Inorder search
3. Postorder search

10.6.1 Preorder Search

The steps of preorder search (also known as *depth-first order search*) algorithm are as follows:

(*a*) First visit the root v of the tree.

(*b*) If v_L exists, then search the left sub-tree T_L of T in preorder.

(*c*) If v_R exists, then search the right sub-tree T_R of T in preorder.

For example, let a positional binary tree T be shown in Fig. 10.12 with root vertex A. The left subtree T_L consists of vertices B, D, E, H, I and J, and its right sub-tree T_R consists of vertices C, F, G, K and L.

First, we apply preorder search on left subtree T_L numbered as 1. For this, we start preorder search of T_{L1} by first visiting its roots vertex B and then vertices D, H, E, I and J in this order.

Similarly, applying preorder search of right sub-tree T_R numbered as 7. The preorder search of T_{R7} starts by first visiting its root vertex C and then vertices F, K, G and L in this order.

The result of complete preorder search of the given tree is the string A B D H E I J C F K G L.

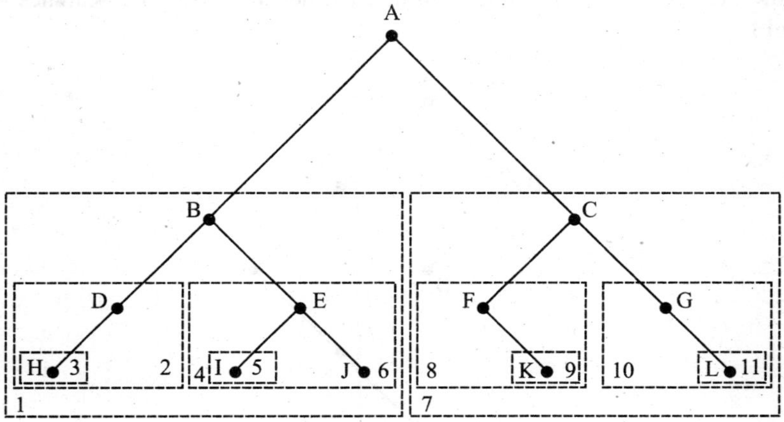

Fig. 10.12 Example of Preorder Search

10.6.2 Inorder Search

The steps of inorder search (also known as *systematic order search*) algorithm are as follows:

(*a*) Search the left sub-tree T_L in inorder.

(*b*) Visit the root of the tree.

(*c*) Search the right sub-tree T_R, in inorder.

The inorder search of T as shown in Fig. 10.12 starts with the search of left sub-tree T_L. For this, we first search root vertex A and then right sub-tree T_R. The inorder search of left sub-tree T_L starts from its root vertex B numbered as T_{L1} and then searches vertices D and H. Since search of tree with one vertex only prints the level of the vertex, therefore, the order of search is HDB. Now, we proceed to search sub-tree T_{L4} and search vertices I, E and J in this order, which gives the string IEJ. After completing search of left sub-tree T_L, we visit the root vertex A. This yields the string HDBIEJA. Proceed to search right sub-tree T_R starting from sub-tree 8, 9 and 7 and then search sub-tree 10 and 11. This gives the string FKCGL.

The result of complete inorder search of the given tree is the string: H D B I E J A F K C G L.

10.6.3 Postorder Search

The steps of postorder search algorithm are as follows:

(*a*) Search left sub-tree T_L in postorder.

(*b*) Search right sub-tree T_R, in postorder.

(*c*) Visit the root of the tree.

The postorder search of T as shown in Fig. 10.12 searches sub-tree T_L, T_R and then search root vertex A. That is, in postorder search sub-tree T_{L1} and T_{R7} will be searched before root of the tree. Similarly, sub-trees T_{L2} and T_{L4} will be searched before root of subtree T_{L1}, and so on. The postorder search of T_{L1} gives the string H D I J E B. In the same way the postorder search of T_{R7} gives the string KFLGC.

The result of complete postorder search of the given tree is the string: H D I J E B K F L G C A.

Example 11 Show the result of performing preorder, inorder and postorder searches of the binary tree shown in Fig. 10.13.

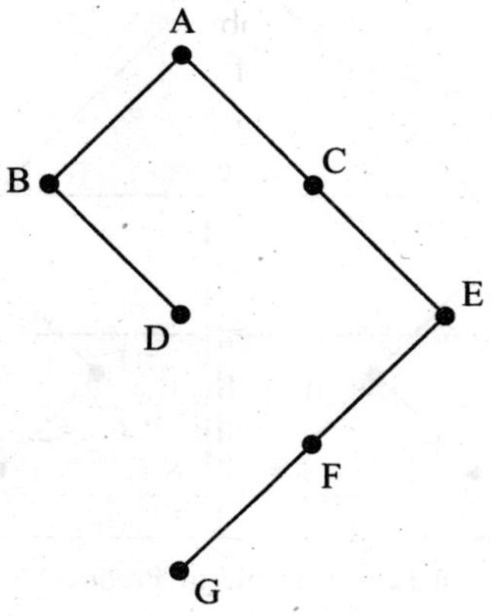

Fig. 10.13

Solution The result of preorder, inorder and postorder tree search algorithms is as follows:

Preorder : A B D C E F G

Inorder : B D A C G F E

Postorder : D B G F E C A

Example 12 Show the result of performing preorder, inorder and postorder searches of the binary tree shown in Fig. 10.14.

Solution The result of performing three tree searching algorithms is as follows:

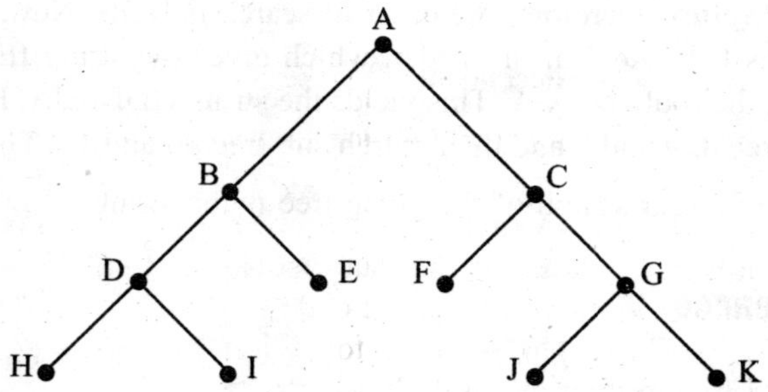

Fig. 10.14

Preorder	: A	B	D	H	I	E	C	F	G	J	K
Inorder	: H	D	I	B	E	A	F	C	J	G	K
Postorder	: H	I	D	E	B	F	J	K	G	C	A

10.6.4 Binary Tree Construction Based on String of Tree Search

Case 1 When both preorder search and inorder search of binary tree T are given.

(*i*) Identify the root vertex of T by choosing the first vertex (node) in its preorder search.

(*ii*) Find left child of the root vertex. For this first use inorder search of T to find the vertices in the left vertices sub-tree T_L of T. Obviously, all these vertices are at the left side of the root vertex and then left child of the root vertex is obtained by choosing the first vertex in the preorder search of T_L, which appears in the preorder of T.

(*iii*) Find the right subtree T_R of T by using inorder search of T. Then right child of the root vertex is obtained by choosing the first root vertex in the preorder search of T_R.

(*iv*) Repeat Steps (*ii*) and (*iii*) with each new vertex until all vertices are visited in preorder search to draw complete binary tree.

Case 2 When both postorder search and inorder search of binary tree T are given

(*i*) Identify the root vertex of T by choosing the last vertex in its postorder search.

(*ii*) Find the left child of the root vertex obtained in Step (*i*). For this first use inorder search of T to find vertices in the left subtree T_L of T, and then left child of the root vertex is obtained by selecting the last vertex in the postorder search of T_L.

(*iii*) Similarly, find the right child of the root vertex obtained in Step (*i*). For this first use inorder search of T to find the vertices in the right subtree T_R of T, and then the right child is obtained by choosing the last vertex in the postorder search of T_R.

(*iv*) Repeat Steps (*ii*) and (*iii*) with each new vertex until all vertices are visited in the postorder to draw complete binary tree.

Example 13 Suppose the preorder and inorder search of a binary tree T yield the following sequence of vertices (nodes):

Inorder : D B P H Q S E A C R K F L

↑ root node (A)

Preorder : A B D E H P Q S C F K R L

↑ root node (A)

(*a*) Draw the binary tree T.

(*b*) Find depth (height) of T and the decendants of vertex B.

Solution Since the root of the binary tree T is the first vertex in preorder search, therefore, we choose vertex A as the root vertex of T.

To find left child of root vertex A first mark in inorder search all nodes that are to the left of A. These vertices constitute a sub-tree T_L of A. The left sub-tree T_L consists of the vertices D, B, P, H, Q, S and E which are to the left of root vertex A. Then left child of A is then obtained by choosing the next vertex (root) in the preorder, i.e. vertex B. This vertex is to the left of A in inorder search. Thus B is the left child of A.

Similarly, the right sub-tree T_R of root vertex A consists of the vertices C, R, K, F and L and C is the root of T_R, i.e. C is the right child of T.

To draw the tree as shown in Fig. 10.15, repeat the above process for each new vertex until all the nodes of preorder search are considered.

(*b*) The depth (height) of T is 5 and descendents of B are D, E, H, P, Q, S.

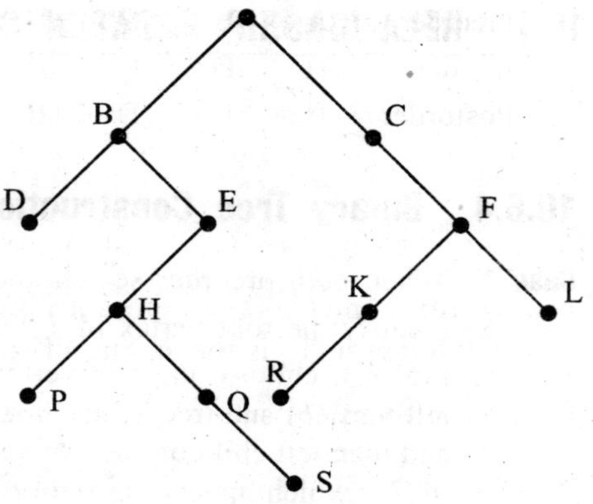

Fig. 10.15

Example 14 Suppose the preorder and inorder search of a binary tree T yield the following sequence of nodes:

Inorder : D B G E H A C F
↑ root node (A)

Preorder : A B D E G H C F
↑ root node (A)

Draw the binary tree T and find its height.

Solution Since first vertex in the preorder search of a binary tree is the root vertex, therefore vertex A is the root vertex of the binary tree to be drawn.

The binary tree drawn by using same method as discussed in Example 9 is shown in Fig. 10.16.

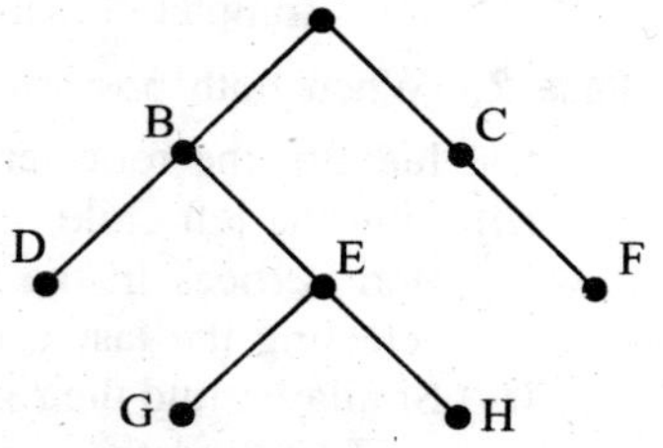

Fig. 10.16

Example 15 Suppose the postorder and inorder search of a binary tree T yield the following sequence of nodes (vertices):

Inorder : D B G E H A C F
↑ root node (A)

Postorder : D G H E B F C A
↑ root node (A)

Solution As we know that the last vertex in the postorder search is the root vertex the binary tree. Thus, vertex A is the root vertex.

In the inorder search all vertices D, B, G, E and H that are left to A constitute left subtree of A and vertices, C and F that are right to A constitute right subtree of A.

The next vertex on the left of root vertex in postorder search is vertex C. This vertex is on the right of root vertex A in inorder search. Thus, vertex C is the right child of A. Similarly, vertex F in postorder search is also on the right of root vertex A in inorder search. Thus, it is the child of vertex C which is the right child of A.

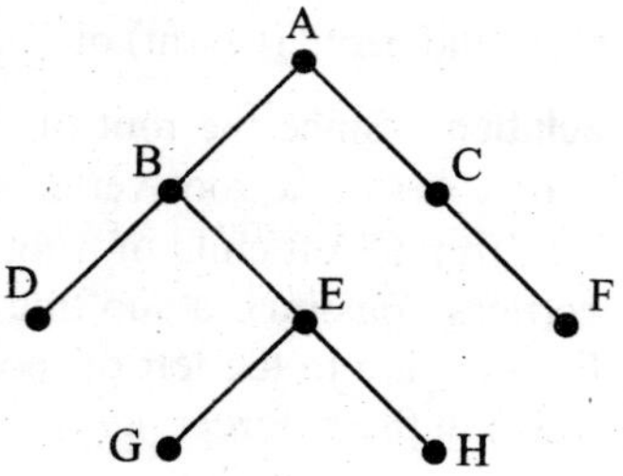

Fig. 10.17

To draw the tree as shown in Fig. 10.17 repeat the above process for each vertex until all vertices of postorder search are considered.

10.7 RELATIONSHIP BETWEEN GENERAL TREES AND BINARY TREES

A general tree T_g can be converted into a unique binary T_b tree as follows:

(*a*) The *number of vertices* (*nodes*) in a binary tree T_b is same as in the general tree T_g.

(*b*) The *root vertex* of T_b is also the root vertex of T_g.

(*c*) The first child (from left) of root vertex v_0 in T_g is the left child of the root vertex in T_b, and right child root vertex in T_b is the sibling of root vertex in T_g.

(*d*) Repeat Steps (*b*) and (*c*) for each new vertex of general tree till corresponding binary tree is drawn.

Remark It is important to understand the relationship between a general tree and a binary tree because certain algorithms that applied to binary trees such as the searching (traversing) algorithms may also be applied to general trees.

Example 16 Consider the general tree T_g as shown in Fig. 10.18(*a*). Draw the corresponding binary tree T_b.

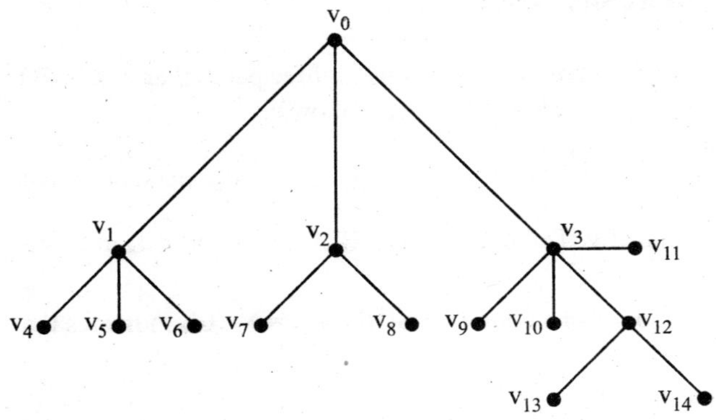

(*a*) General Tree

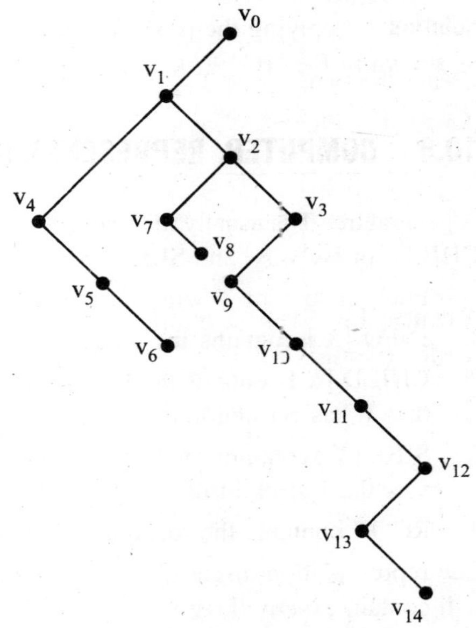

(*b*) Binary Tree

Fig. 10.18 Correspondence between General Tree and Binary Tree

Solution Applying the procedure for converting a general tree into a binary tree, we obtain a binary tree as shown in Fig. 10.18(*b*).

Example 17 Consider the general tree shown in Fig. 10.19(*a*). Draw the corresponding binary tree.

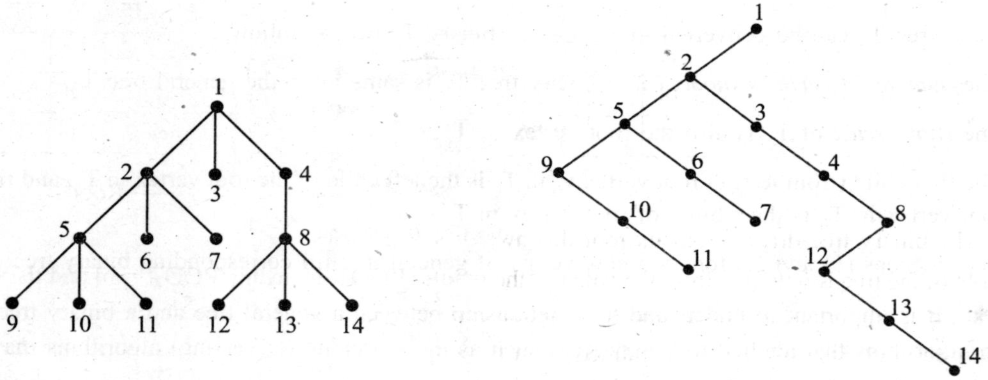

(*a*) General Tree (*b*) Binary Tree

Fig. 10.19 Correspondence between General Tree and Binary Tree

Solution Applying the procedure for converting a general tree into a binary tree, we obtain a binary tree as shown in Fig. 10.19(*b*).

10.8 COMPUTER REPRESENTATION OF GENERAL TREES

A general tree T is usually maintained in the memory of the computer by means of three parallel arrays INFO, CHILD (or DOWN) and SIBL (or HORZ), and a pointer variable ROOT as follows:

Each node v of T will correspond to a location *K* such that

1. INFO [*K*] contains the data at node v.
2. CHILD [*K*] contains the location of the first child of v. The condition CHILD [*K*] = NULL indicates that v has no children.
3. SIBL [*K*] contains the location of the next sibling of v. The condition SIBL [*K*] = NULL indicates that v is the last child of its parent.
4. ROOT contains the location of the root of the tree T.

The representation discussed above pointed out that each node of T, regardless of the number of children will contain exactly three fields.

The procedure of representing a general tree in the memory of the computer may also be extended to represent a forest F consisting of trees $T_1, T_2, \ldots, T_n$ by assuming that the roots of these trees are siblings. This means, ROOT will contain the location of the root of the first tree T_1 or when F is empty, ROOT = NULL.

The computer representation of a general tree and of the corresponding binary tree is exactly the same except that the name of the arrays CHILD and SIBL for the general tree will correspond to the name of the arrays LEFT and RIGHT for the binary tree. Such a correspondence between general trees and binary trees is important because certain algorithms that applied to binary tree such as search (traversal) algorithm may also apply to general trees as shown in Figs 10.18(*b*) and 10.19(*b*).

Example 18 Draw the diagram of the binary tree maintained in the memory of the computer as shown below:

	1	2	3	4	5	6	7	8	9	10	11	12	13	14	15	16	17	18
INFO	H	R		P	B		E		C	F	Q	S		A	K	L		D
LEFT	4	0		0	18		1		0	15	0	0		5	2	0		0
RIGHT	11	0		0	7		0		10	16	12	0		9	0	0		0
														↑ Root				

Solution The binary tree drawn from its root downward is as follows:

(*a*) The root of the tree is obtained from the value of the pointer ROOT. Note that ROOT = 14 and therefore, INFO [14] = A is the root of the tree.

(*b*) The left child of the root node is obtained from the left pointer field of the root, which is stored in LEFT [14] = 5. Thus, set INFO [5] = B as the left child of A. Since root node A has no sibling, set SIBL [14] = NULL.

(*c*) The right child of the root node is obtained from the right pointer field of the root, which is stored in RIGHT [14] = 9. Thus, set INFO [9] = C as the left child of A.

Continuing this process for each new node yields the diagram of binary tree as shown in Fig. 10.20.

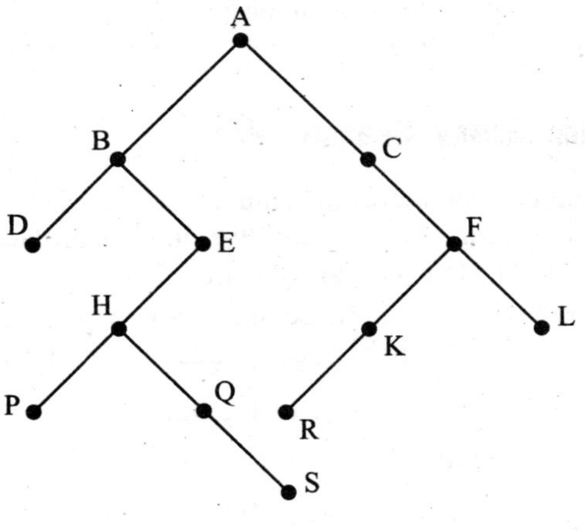

Fig. 10.20 Binary Tree

10.9 BINARY SEARCH TREE

In many applications of binary trees, it is first desirable to build a binary tree, and then traverses the tree. For example, if the input list of numbers in a input file appears as:

14 15 4 9 7 18 3 5 16 20 17

then to print them in ascending order we first read the numbers and they can be inserted into a binary tree as shown in Fig. 10.21.

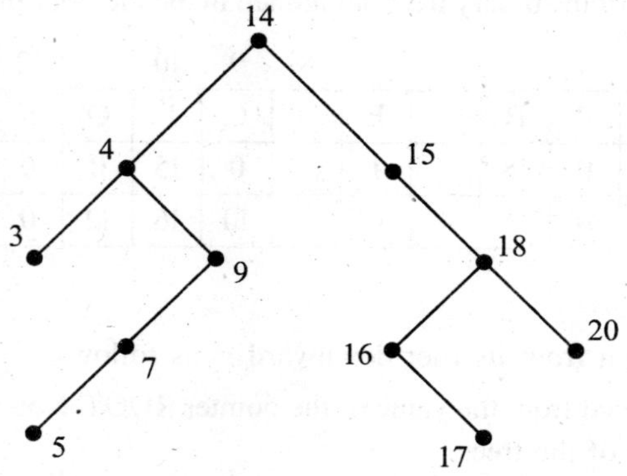

Fig. 10.21 Binary Search Tree

A binary tree T is called a *binary search tree* provided vertices in the left sub-tree of a vertex v_i are less than the contents of v_i, and all vertices in the right sub-tree of v_i are greater than or equal to the contents of v_i. In other words, in binary search tree, every vertex v satisfies following properties:

(*i*) $v > n$ for every vertex n in its left subtree.

(*ii*) $v \leq n$ for every vertex n in its right subtree.

It is, however, not necessary that a node in a binary search tree, point to the nodes whose values immediately precede and follow it. Unlike traverse algorithms used to find duplicates, duplicate values are placed in the tree. For example, tree shown in Fig. 10.21 is a binary search tree.

10.9.1 Inserting ITEM in Binary Search Tree

A binary search tree, an important data structures in computer science, helps search for and find an element with an average running time $f(n) = \log_2 n$, where n is the number of data items as well as to insert and delete an information (also called ITEM) from the data structure.

Let an ITEM of information be given to be inserted as a new node in a binary search tree T. Steps of the procedure to insert an ITEM in the binary search tree are as follows:

Step 1 Compare the ITEM with the root vertex v of the tree T.

(*a*) If ITEM < v, proceed to the left child of v.

(*b*) If ITEM > v, proceed to the right child of v and it becomes the root vertex for the right subtree.

Step 2 Repeat Step 1 until one of the case arises:

(*a*) There is no left and right sub-tree, i.e. sub-tree is empty. Then search is said to be unsuccessful and therefore insert ITEM in place of the empty sub-tree.

(*b*) A vertex v′ is found such that ITEM = v′.

10.9.2 Deleting ITEM in Binary Search Tree

Let an ITEM of information be given in a binary search tree T. Steps of the procedure to delete an ITEM from a binary search tree are as follows:

Step 1 Find the location of the vertex v_i in T which contains ITEM by the same procedure as discussed above.

Step 2 Determine the number of children of vertex v_i. The following three cases may arise depending upon the number of children of vertex v_i.

(*a*) Vertex v_i has no children. Then delete v_i from T by replacing it in the parent vertex with null vertex.

(*b*) Vertex v_i has only one child. Then delete v_i from T by replacing it in the parent vertex with its only child.

(*c*) Vertex v_i has two children. Then delete v_i from T by replacing it in the parent vertex with the vertex that is closest in the value of v_i. This vertex is called the immediate predecessor of v_i. After replacing the value of v_i with the immediate predecessor, delete the replaced vertex by using Steps 2(*a*) and 2(*b*).

Example 19 Consider the binary tree T in Fig. 9.22(*a*).

(*a*) Why is T a binary search tree?

(*b*) Suppose ITEM = 73 is added to the tree. Find the new tree.

Solution (*a*) The binary T shown in Fig. 9.22(*a*) is a binary search tree because every vertex is greater than the values in its left sub-tree and less than the values in its right sub-tree.

(*b*) Compare ITEM = 73 with the root 100. Since ITEM < 100, we proceed to the left child, 70. Since ITEM > 70, we proceed to the right child, 95. Also ITEM < 95, we proceed to left child, 75 which is also more than ITEM. But vertex 75 has no left child. Thus, insert ITEM = 73 as a left child of vertex 75 to give a new binary tree as shown in Fig. 9.22(*b*). The dotted lines indicate the path down through the tree during the insertion.

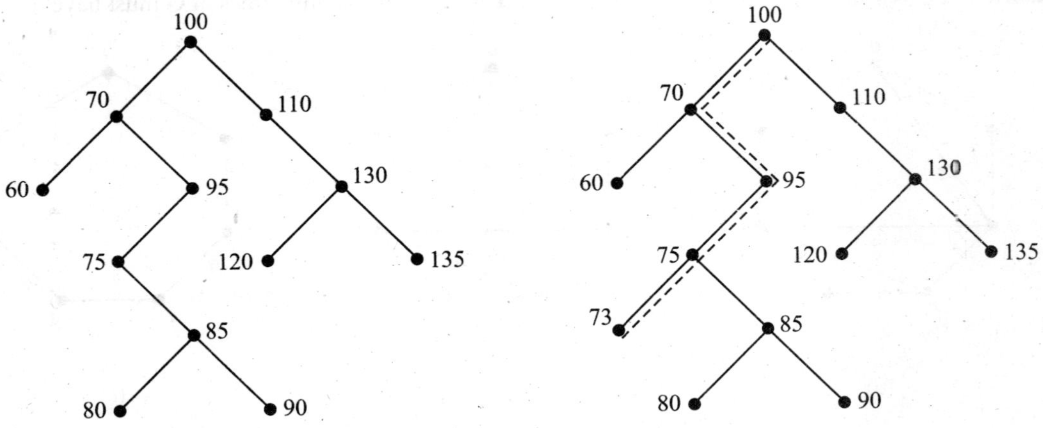

Fig. 9.22(*a*) Fig. 9.22(*b*)

Example 20 Consider the binary search tree T in Fig. 10.23(*a*). Describe the tree T when vertices B and C are deleted.

Solution The vertex C has only one child E. Thus on deleting C, the next vertex F becomes the right child of root vertex A as shown in Fig. 10.23(*b*).

The vertex B has two children. To delete the vertex B, we should find its inorder successor, which is vertex H. Thus, first delete H from T and then replace B by H as shown in Fig. 10.23(*b*).

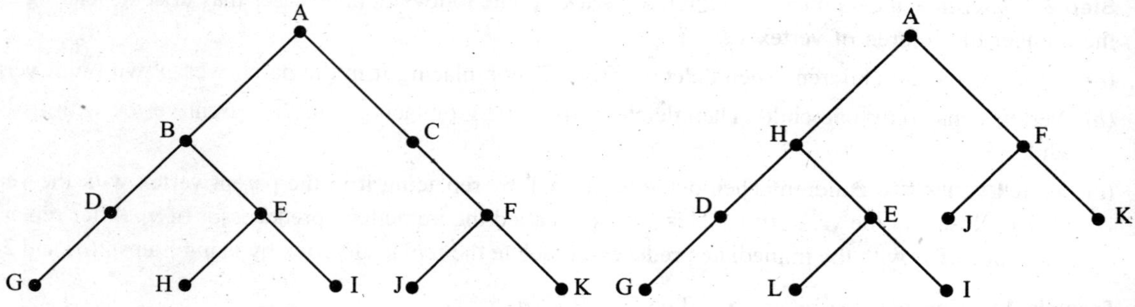

Fig. 10.23(*a*) **Fig. 10.23(*b*)**

10.11 SPANNING TREES

A subgraph T of a graph G (V, E) is called a *spanning tree* of G if T is a tree with exactly the same number of vertices as G and can also be obtained from G by deleting some of its edges. Two spanning trees for a graph in Fig. 10.24(*a*) are shown in Fig. 10.24(*b*) and Fig. 10.24(*c*).

In the section, we will discuss only undirected spanning trees for a connected graph G.

Remark If G is a finite connected graph with *n* vertices, then all spanning trees of G must have $n - 1$ edges.

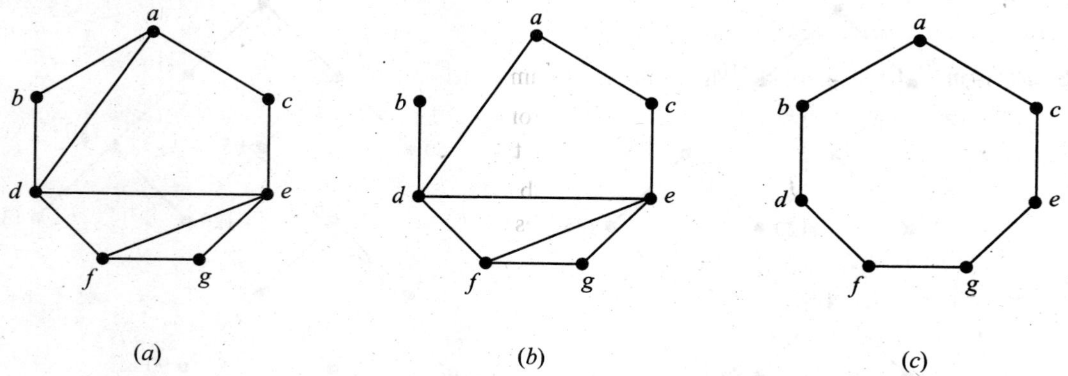

Fig. 10.24 Example of Spanning Trees

Theorem 10.16 A simple graph G is connected if and only if it is spanning tree.

Proof Let the graph G has a spanning tree (a sub-group) T. Then every vertex u, v $\in$ G that is also vertex of tree T, provides unique path between u and v. Hence, G is connected.

Conversely, if G is connected, and not a tree, then G must contain a circuit. Remove an edge e_1 from such a circuit and let $G_1 = G - e_1$. If resulting sub-group G_1 is also a connected graph because it contains all the vertices of G though one edge less than in G. If the resulting sub-group is a tree, it must contain a circuit. Again remove an edge e_2 from G_1 and let $G_2 = G_1 - e_2$. If G_2 contains no circuit, then G_2 is spanning tree of G because it contains all the vertices of G and so G_2 is loop-tree connected graph. If G_2 contains a circuit, then remove another edge e_3. Repeat this procedure until we get a loop-free connected graph that contains no circuit and contains all the vertices of G. Consequently, a spanning tree of G is obtained.

Illustration The graph G has one cycle *abc*. If we remove any edge of the cycle, the resulting graphs are all trees. These trees contain all the vertices of graph G, and are spanning trees.

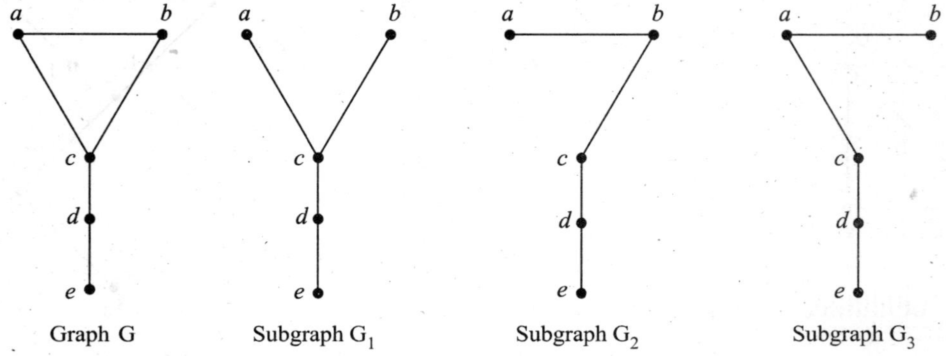

Fig. 10.25

10.10.1 Algorithms for Constructing Spanning Trees

The following two are algorithms for searching the vertices of a loop-free connected undirected graph.

1. Depth-First-Search (DFS) or Backtracking algorithm
2. Breadth-First-Search (BFS) algorithm.

DFS Algorithm The steps of the algorithm are summarised as follows:

1. Choose the smallest subscript vertex $v_1 \in V$ from the vertices of the loop-free connected undirected graph and designate this vertex as the root of the spanning tree.
2. Select an edge $\{v_1, v_i\} \in E$ with the smallest subscript i such that $2 \le i \le n$ and vertex v_i has not been visited before. If no such vertex with the smallest subscript is found, then go to Step 3. Otherwise do the following:

   (*a*) Add the edge $\{v_1, v_i\}$ to the tree T,

   (*b*) Assign v_i to the set of vertices, and go to Step 2.
3. If $v_i = v_1$, the tree T is spanning tree of specified order.
4. For $v_i \ne v_1$, backtrack from v_1. If u is the parent of vertex included in vertex set V of T, then assign u to V and return to Step 2.

Example 21 Apply DFS algorithm to find a spanning tree of the graph shown in Fig. 10.26(*a*).

Solution First we choose smallest subscript vertex v_1 as the root vertex.

Next we choose vertex v_2 such that $\{v_1, v_2\} \in E$ and v_2 has not be visited earlier. So we attach edge $\{v_1, v_2\}$ to T and repeat the process to form a path by successively adding edges incident with vertices not already visited in the path.

The resulting tree is shown in Fig. 10.26(*b*) and the path it produces is $v_1 - v_2 - v_6 - v_5 - v_3$ or $v_1 - v_2 - v_6 - v_5 - v_4$.

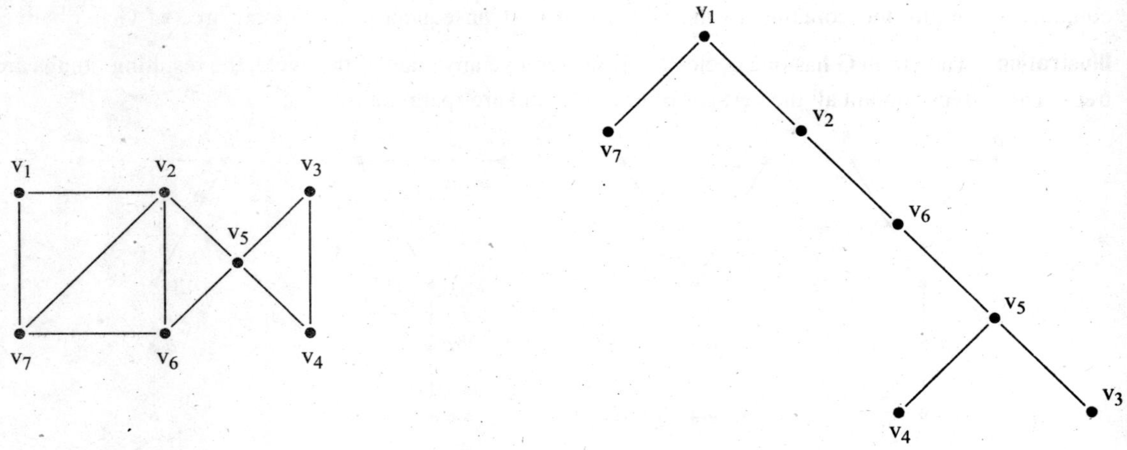

(*a*) Graph (*b*) Spanning Tree

Fig. 10.26 Spanning T

BFS Algorithm Steps of the algorithm are summarised as follows:

1. Choose a vertex v_1 as the root vertex.
2. Add edges incident with all vertices adjacent to root vertex v_1. If the edges $\{v_1, v_i\}$, $2 \leq i \leq n$ have not visited before, then the tree generated by these vertices is the rooted ordered spanning tree.
3. Add edges from the vertices v_i at level 1 (Step 2) to adjacent vertices not already included. If the edges $\{v_i, v_j\}$ at level 2 have not been added, then tree so formed is the required tree. Continue the process until all vertices are visited once.

Example 22 Apply BFS algorithm to find a spanning tree of the graph shown in Fig. 10.27.

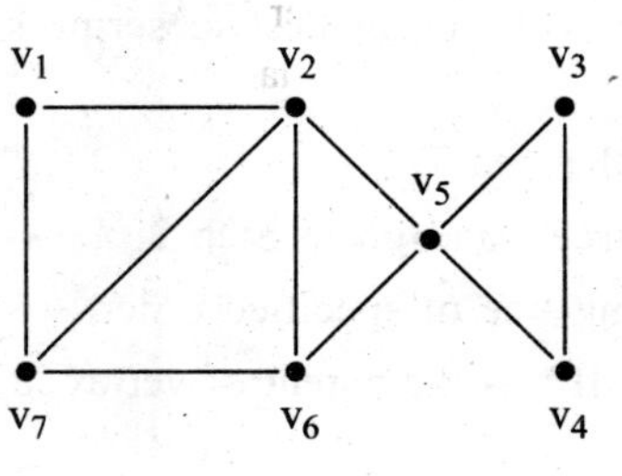

Fig. 10.27

Solution The steps of BFS algorithm are shown in Figs. 10.28(*a*) to (*e*). Let v_1 be the root of the tree. Graph shown in Fig. 10.28(*d*) and Fig. 10.29(*e*) are the required spanning trees.

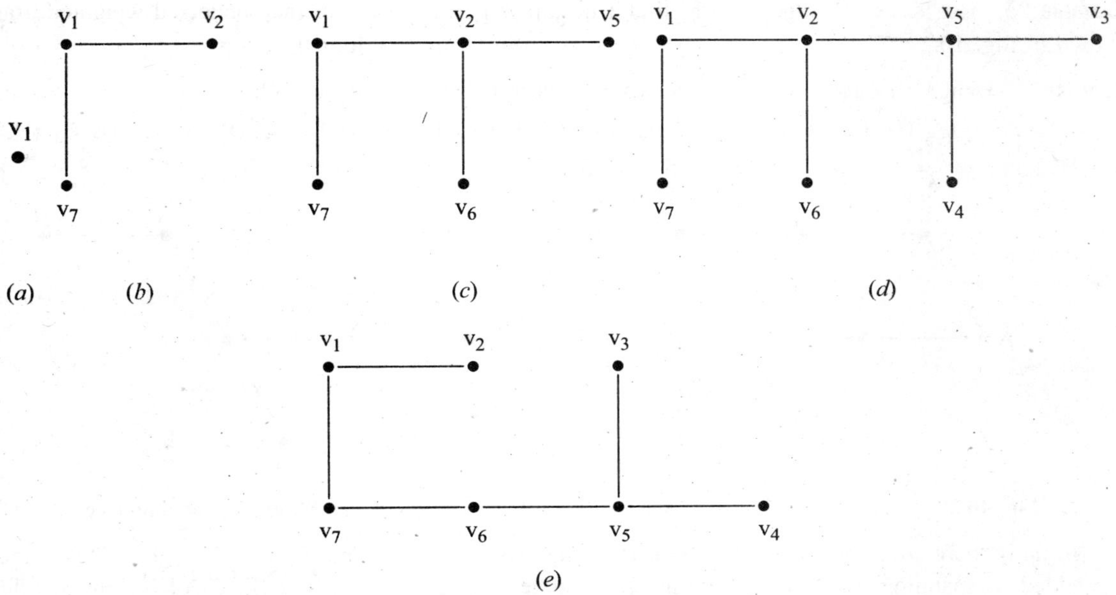

Fig. 10.28 Spanning Tree

10.10.2 Minimal Spanning Trees

A connected graph G in which each of its edge is assigned a non-negative numerical value $w(e)$ (also called weight or length) is said to be a *weighted graph* and the numerical value associated with a edge is called the *weight* of that edge. In many applications of weighted graph it is necessary to find a spanning tree for which the total weight of edges in the tree is the smallest.

A spanning tree of a weighted graph having minimum weight is called a *minimal spanning tree*. There may be more than one spanning trees whose minimum weight is same because more than one edge may have the same weight.

Two algorithms due to Joseph Kruskal and Robert Prim are used to construct an optimal spanning tree. These algorithms are *greedy*, because these help to make an optimal (minimal) choice at each step ignoring previous choices.

Kruskal's algorithm The steps of the algorithm to construct a minimal spanning tree T for a connected weighted graph G with n vertices and m edges are summarised below:

Step 1 Arrange all edges of G in an increasing order of their weights. Choose among them an edge e_i with least weight. Add this edge in T.

Step 2 Choose another edge e_j with next higher weight in such a way that these edges do not form a cycle. If edge e_j does not form a cycle, then add e_j also in T.

Step 3 Repeat Step 2 until all $n - 1$ edges are added in T and do not form a cycle.

Example 23 Use Kruskal's algorithm to find a minimal spanning tree for the connected weighted graph shown in Fig. 10.29. The weight of each edge is given in terms of kilometres.

Solution Arrange all edges in G in a increasing order of their weights as follows:

Edge :	(A, C)	(E, G)	(G, H)	(A, B)	(B, C)	(E, F)	(D, E)	(D, F)	(C, D)	(C, E)	(B, G)	(H, F)
Weight :	4	4	4	5	5	5	6	6	7	7	8	8

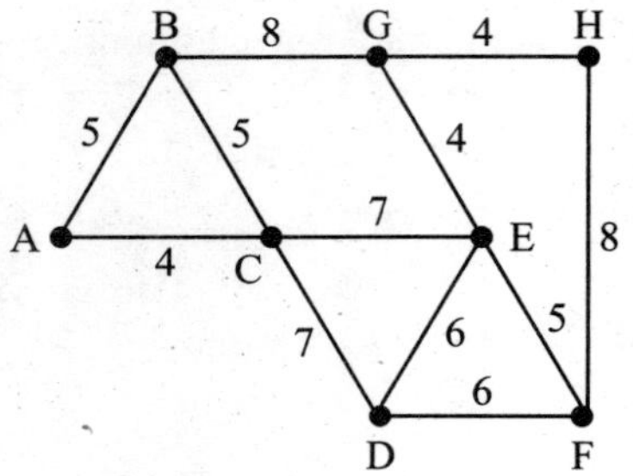

Fig. 10.29 Connected Weighted Graph

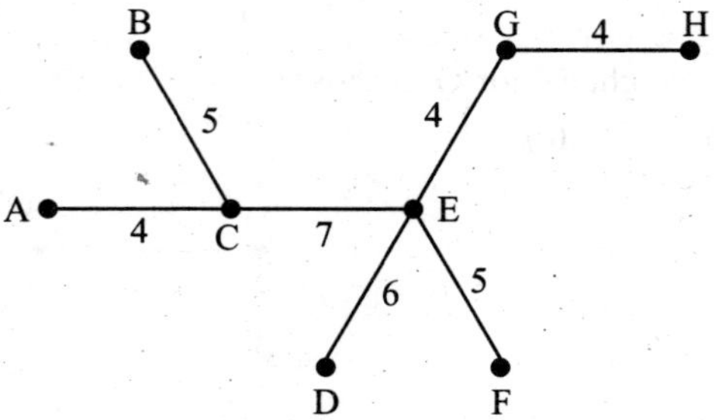

Fig. 10.30 Minimal Spanning Tree

Initially there are three edges of equal least weight (A, C), (E, G) and (G, H) of weight 4. All of them are added in spanning tree. Next, there are three edges: (A, B), (B, C) and (E, F) of weight 5. Edges (B, C) and (E, F) are added in spanning tree without forming any cycles. However, edge (A, B) is not added because edges (B, C) and (A, B) form a cycle with edge (A, C). Thus, so far 5 edges are added to T. To complete the spanning tree T, two more edges are to be added to make a total of 7 ($= n - 1$) edges.

Edges with next higher weight are (D, E) and (D, F) with weight 6. But only edge (D, E) is added in T because edge (D, F) form a cycle with edge (F, E).

Adding the edge (C, E) in T with weight 7, gives a 7-edge spanning tree for a 8-vertex graph. Hence, a minimal spanning has been found. One sequence of edge selection in Fig. 10.30 is (A, C), (E, G), (G, H), (B, C), (E, F), (D, E), (C, E) with a total weight of 35 km.

Prim's algorithm The steps of the algorithm to construct a minimal spanning tree T for a connected weighted graph G with n vertices and m edges are summarised below:

Step 1 Select an arbitrary vertex v_i and choose the edge with smallest weight in G.

Step 2 Select another edge e_j with smallest weight joining the vertex v_i.

Step 3 Repeat Step 2 until all $n - 1$ edges are added in the tree T.

Example 24 Use Prim's algorithm to find a minimal spanning tree for the connected weighted graph shown in Fig. 10.31.

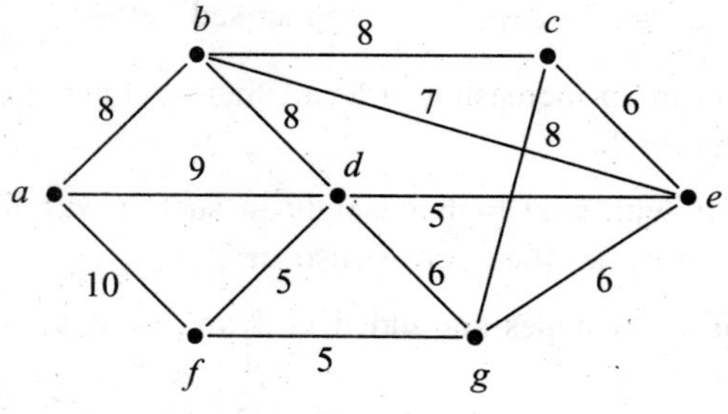

Fig. 10.31 Connected Weighted Graph

Solution Select the vertex a. Among all edges incident on a with smallest weight is $(a, b) = 8$.

Among all edges incident on vertex b with smallest weight is $(b, e) = 7$

Among all edges incident on vertex e with the smallest weight is $(e, d) = 5$.

Among all edges incident on d and have not been visited before with the smallest weight is $(d, f) = 5$.

From vertex f we cannot choose edge (f, a), so choosing the edge (f, g) with weight 5.

In the last only one edge $(g, c) = 8$ is left that has not been visited before. Thus, a minimal spanning tree of weight 38 for G is shown in Fig. 10.32.

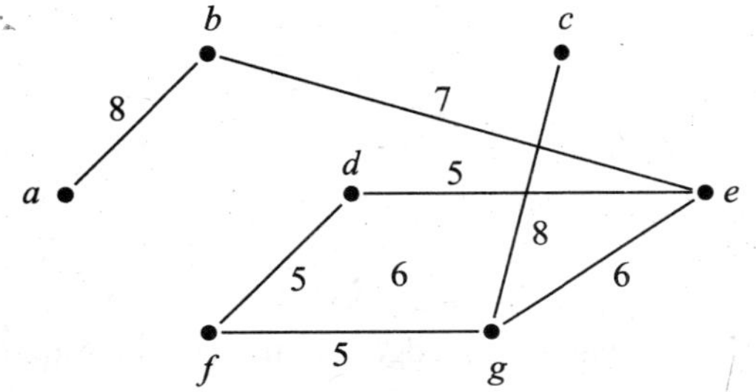

Fig. 10.32 Minimal Spanning Tree

Conceptual Review Questions

1. What is a binary rooted tree? Indicate its importance and show that the maximum height of an n vertex binary tree is $[\log_2(n + 1) - 1]$. *[Rohtak Univ., MCA, 1999]*
2. Prove that any connected graph G with n vertices and $n - 1$ edges is a tree.
3. Show that the number of simple, labelled graphs of n vertices is $2^{n(n-1)/2}$. *[Rohtak Univ., MCA., 1996]*
4. What do you mean by polish notation? Explain the same through an example. *[Rohtak Univ., MCA., 1996]*
5. Show that the maximum number of edges in a simple graph with n vertices is $n(n - 1)/2$. *[Rohtak Univ., MCA., 1999]*
6. What is arborescence of a digraph? State its significance. *[Rohtak Univ., MCA., 1999]*
7. Show that a graph is tree if and only if it is minimally connected.
8. Show that a connected graph G is a tree if adding an edge between any two vertices in G creates exactly one circuit.

9. Find the eccentricity of all vertices, radius and centre of the given graph shown in Fig. 10.33.

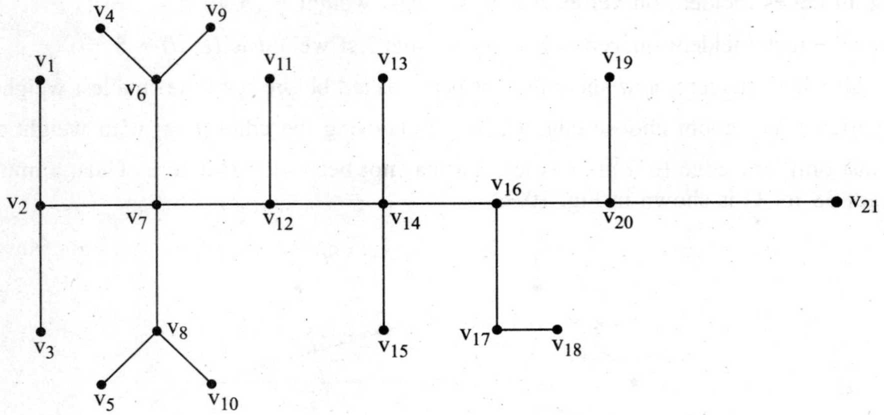

Fig. 10.33

10. Show that a tree with n vertices has $n - 1$ edges. Is the converse tree? Justify your answer. *[CCS Univ., MSc (Comp. Sci.), 1999]*

1 . (*a*) Is every tree also a graph? Give examples.
(*b*) Is every pat also a tree? Give examples. *[CCS Univ., BCA, 2000]*

12. Prove that in any tree (with two or more vertices), there at least two pendant vertices. Given examples also. *[CCS Univ., BCA., Dec. 2001]*

13. How many internal vertices does a full binary tree with h levels has? *[UPTU., MCA, 2001]*

14. State the theorems when a graph G with n vertices is said as tree. *[UPTU., MCA., 2002]*

15. Given $n = 4$ distinct vertices, what is the number of different trees that one can construct? *[UPTU., MCA, 2002]*

16. Show that an arborescence is a tree in which every vertex other than the root has an in-degree of exactly one. Also, discuss an important application of arborescence to the theory of computer algorithms. *[UPTU., MCA, 2002]*

17. Define rooted tree, binary tree and binary search tree.

Self-practice Problems

1. Is a rooted tree T with root vertex v_0 an n-tree. If yes, then for what integral value of n?
2. If T is a rooted tree with root vertex v_0, then prove that v_0 has indegree zero.
3. If T is a complete n-tree with three levels, then prove that the number of vertices of T must be $1 + kn$; $2 \leq k \leq n + 1$.
4. If G (V, E) is a tree, then prove that G (v, E – $\{e\}$) is a forest of two trees.

5. Draw the expression trees for the following expressions.

(*a*) $a + b * c$

(*b*) $(a + b) * c$

(*c*) $((2 + a) - (2a)) - (a - 2)$

(*d*) $(a - b)\ (c + (d \div e))$

(*e*) $(a \div b) \div ((3a) - (c \div 4))$

(*f*) $(2a + b)\ (5a\ - b)^2$

(*g*) $a + (b - c) * d \uparrow (e * f)$

(*h*) $((a + b) * c) \uparrow ((a + b) * c)$

6. Prove that the maximum number of vertices at level k of a binary tree must have at least $2^{k+1} - 1$ vertices.

7. Prove that in a binary tree, the number of leaves is one more than the number of internal vertices (non-leaves).

8. Let T be a binary tree stored in the memory of the computer with each vertex labelled as given below.

(*a*)

	1	2	3	4	5	6	7	8	9
INFO		D	E	C	F	B	G	A	H
LEFT	8	5	9	2				6	
RIGHT		7		3		4			

(*b*)

	1	2	3	4	5	6	7	8	9	10	11	12
INFO		A	B	C	D	E	F	G	H	I	J	K
LEFT	2	3	4	6	8							
RIGHT			5	7		10		11		9	12	

9. Consider the binary tree T in Fig. 10.34. Give arrays INFO, LEFT and RIGHT to describe the tree.

10. Consider the binary tree T in Fig. 10.35. Search T using preorder, inorder and postorder algorithms.

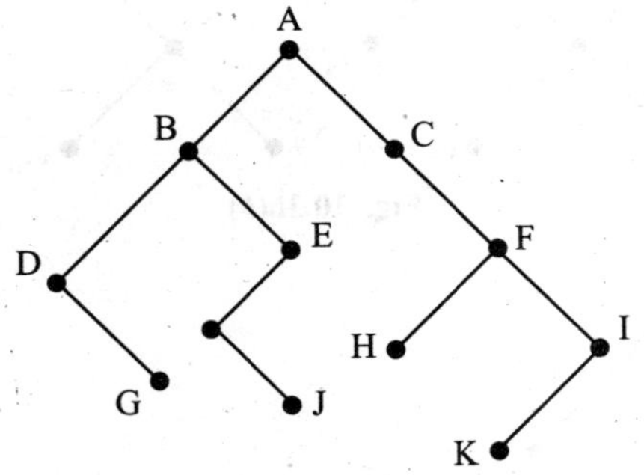

Fig. 10.34

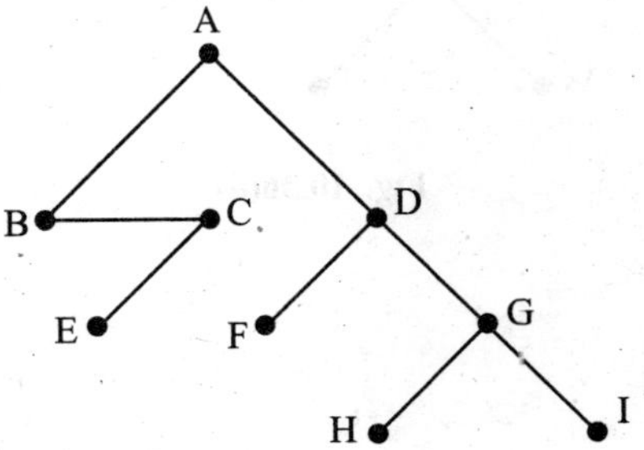

Fig. 10.35

11. Suppose the preorder and inorder searches of a binary tree T yield the following sequence of nodes. Draw the diagram of T.

(*a*) Preorder : A B D E G H C F

Inorder : D B G E H A C F

(*b*) Preorder : F G A H B C D
Inorder : G A F H D C B

(*c*) Preorder : A B D G C E H I F
Inorder : D G B A H E I C F

(*d*) Preorder : A B C E I F J D G H K L
Inorder : E I C F J B G D K H L A

12. Suppose the postorder and inorder searches of a binary tree T yield the following sequence of nodes. Draw the diagram of T.

(*a*) Inorder : G A F H D C B
Postorder : A G D C B H F

(*b*) Inorder : C D B A F E H G
Postorder : D C B F H G E A

(*c*) Inorder : D G B A H E I C F
Postorder : G D B H I E F C A

(*d*) Inorder : E I C F J B G D K H L A
Postorder : I E J F C G K L H D B A

13. Identify which of the tree is

(a) Binary tree (b) Strictly binary tree (c) Complete tree

as shown in Fig. 10.36(*a*), (*b*) and (*c*).

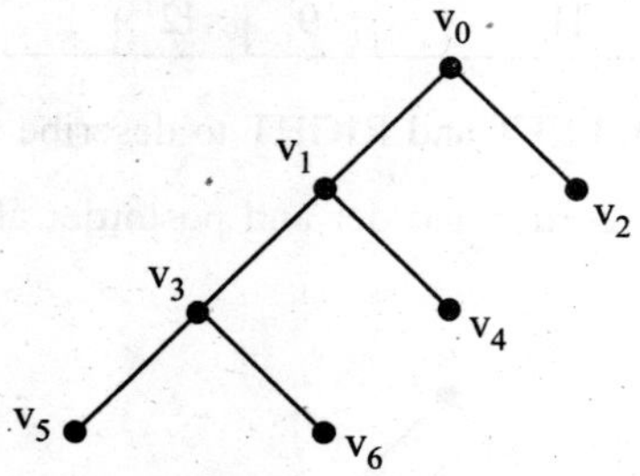

Fig. 10.36(*a*)

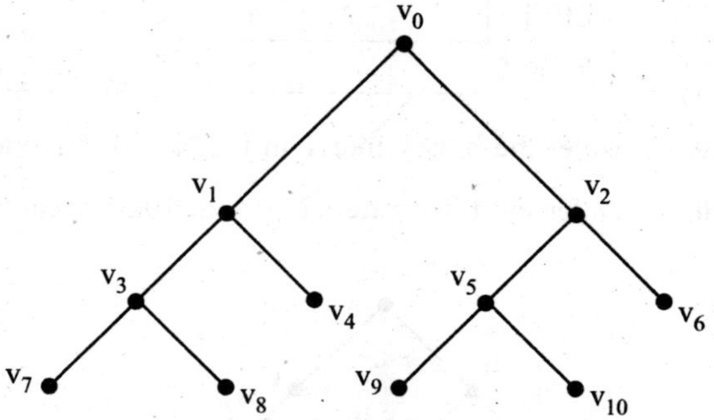

Fig. 10.36(*b*)

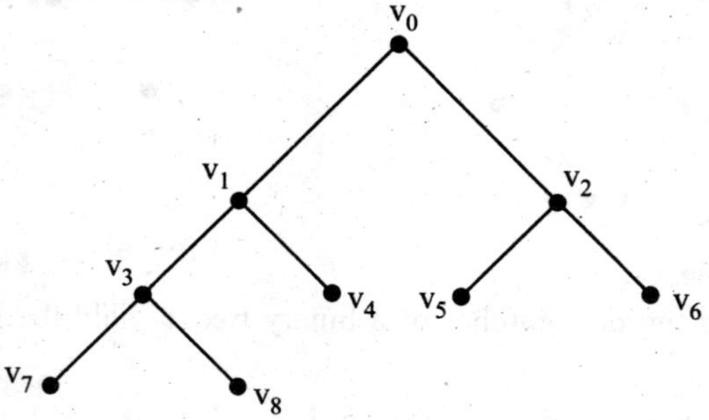

Fig. 10.36(*c*)

14. Let G be an undirected graph with n vertices. Prove that G is a tree if and only if P (G, λ) = λ (λ – 1)$^{n-1}$.

15. Let G be a loop-free undirected graph. If deg (v) ≥ 2 for all v∈ V, prove that G contains a cycle.

16. If a complete 3-ary tree T has 34 internal vertices, then how many edges and leaves does T have.

17. Traverse trees shown in Fig. 10.37(*a*) and Fig. 10.37(*b*) in preorder, inorder and postorder.

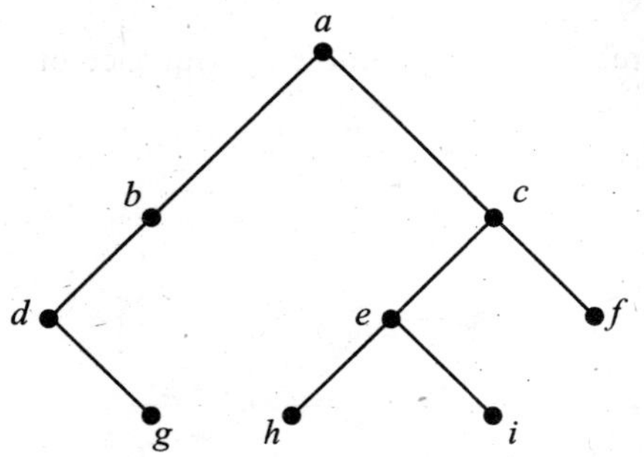

Fig. 10.37(*a*)

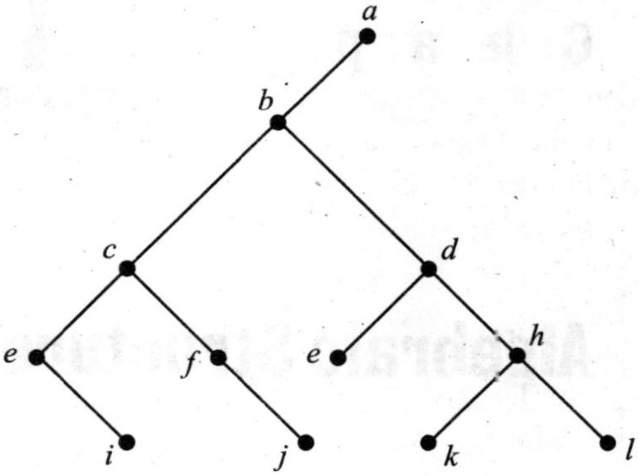

Fig. 10.37(*b*)

18. Write down a binary expression based on binary trees shown in Fig. 10.38(*a*) and Fig. 10.38(*b*).

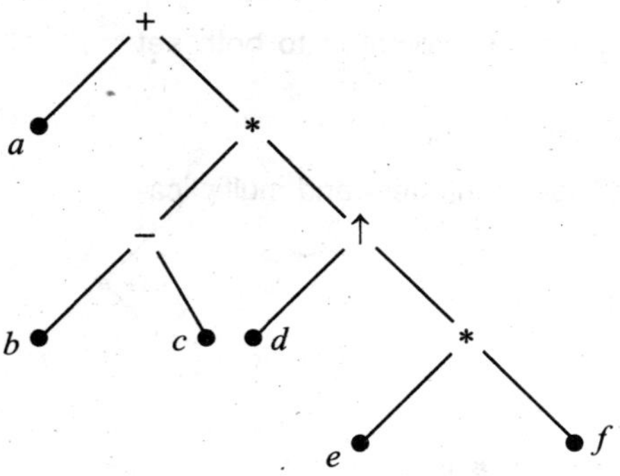

Fig. 10.38(*a*)

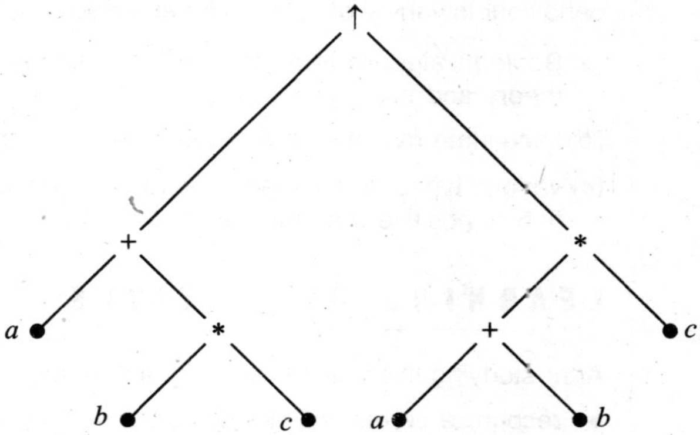

Fig. 10.38(*b*)

Chapter 11

Algebraic Structures

PREVIEW

An algebraic structure is a formal model used to represent common properties or behaviour in variety of context, for example

(*a*) Boolean algebra is a model of properties and behaviour common to both set theory and propositional logic,

(*b*) finite-state machine is a device useful for lexical analysis, and

(*c*) various types of arithmetic operations such as addition of integers and multiplication of positive real numbers.

LEARNING OBJECTIVES

After studying this chapter, you should be able to

- recognize certain well known groups, rings and field structures,
- understand congruence relations and quotient structures,
- understand what it means groups to be semi-group and monoid,
- appreciate the meaning of homomorphism and isomorphism,
- be aware of properties of rings, field and integral domain.

11.1 INTRODUCTION

A set of objects with mathematical operation or operations defined on it and the properties associated with it form an *algebraic system* or *structures.* Roughly, an algebraic system consists of following parts:

(*i*) *A set* or *universe of objects.*

(*ii*) *Definitions:* These explain the meaning and concepts that relate the given set of objects with the universe.

(*iii*) *Axioms:* These are the assertions about the properties of the universe and rules for creating and justifying more assertions. These rules include the system of logic (see Chapter 9).

(*iv*) *Theorems:* These are considered as the extension of the axioms to put additional assertions.

Few examples of algebraic structures are given below:

(*i*) In set theory, the collection of sets with the operations of *union* (∪), *intersection* (∩) and *complement* (′ or *c*) with their properties constitute a discrete algebraic structure and is denoted as [sets, ∪, ∩, *c*].

(*ii*) In matrices, the operations of *addition* (+), *multiplication* (×), and *transpose* (*t*) with their properties constitute an algebraic structure and is denoted as [matrices, +, ×, *t*].

(*iii*) In Boolean algebra, the operations of *and* (∧) *or* (∨) and *negation* (~) with their properties constitute an algebraic structure and is denoted as [propositions, ∧, ∨, ~].

An important property of any algebraic structure is its *closure property* with respect to a mathematical operation defined on it. A system is said to satisfy closure property with respect to an operation if that operation always produces another element of the collection of objects. For example, algebraic structure

(*i*) [sets, ∪, ∩, *c*] is closed with respect to *union*.

(*ii*) [3 × 5 matrices, +, ×, *t*] is not closed with respect to *transpose*.

(*iii*) [odd integers, +, ×] is not closed with respect to *addition*.

In this chapter, we will introduce few other algebraic structures such as *semi-groups*, *groups*, *rings* and *fields*. These structures shall be helpful in the study of finite-state machines and coding theory.

The relationships among the different algebraic structures can be summarized diagrammatically as shown in figure 11.1.

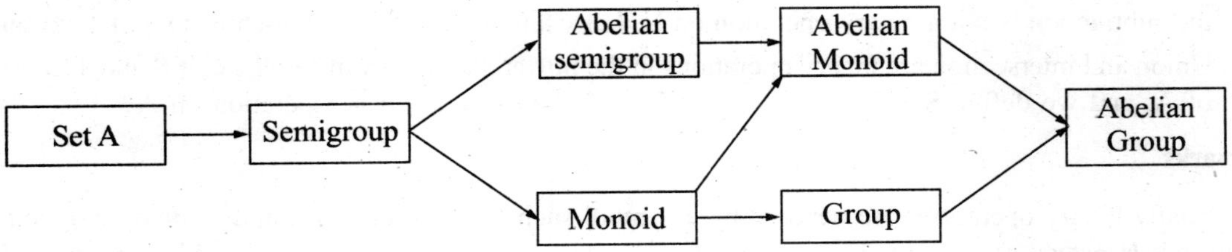

Fig. 11.1 Relationship among Different Algebraic Structures

11.2 MATHEMATICAL OPERATIONS

In the previous chapter, we already have introduced mathematical operations such as *addition* and *multiplication* of numbers; *union* and *intersection* of sets, as well as *composition of functions*, i.e.

$$a + b = c,\ a \cdot b = c,\ A \cup B = C,\ A \cap B = C, \text{ etc.}$$

In all these cases, a function is defined which assigns an element to each ordered pair of elements in a given set.

Definition An *operation*, *f* on a set A is a function of the cartesian product A × A into A, that is $f: A \times A \to A$, such that each element $f(a, b) \in A$ is associated with each ordered pair $(a, b) \in A \times A$.

11.3 BINARY OPERATIONS

An operation that specifies the way in which two objects are combined to yield a third object belonging to the same collection of objects is called a *binary operation.* A binary operation can be described using functional notation also.

Definition Let A be a non-empty set. Then $A \times A = \{(a, b) : a \in A; b \in A\}$. If $f : A \times A \rightarrow A$, then f is said to be a *binary operation* on the set A. In other words, a binary operation f on a non-empty set A is a function (mapping) which associates each ordered pair (a, b) of elements of A to a uniquely defined element, $a \in A$.

A binary operation obviously must satisfy following conditions:

(*i*) It must be defined for each ordered pair of elements of A.

(*ii*) Only one element of A is assigned to each ordered pair of its elements.

Often we use symbols such as +, ×, ·, □, ⊕, ∇, ∗, etc., to denote binary operation on a set. Thus, if we define a binary operation $a * b$ as $a + b$, then ∗ will be a binary operation on A if and only if

$$(a + b) \in A \text{ for all } a, b \in A \text{ and } (a + b) \text{ is unique.}$$

The binary operation is sometimes known as *binary composition* on the set A.

Illustrations

1. Let N be the set of natural numbers. Define $a * b$ as $a + b$. Then '∗' is a binary operation in N, since corresponding to each ordered pair (a, b) of natural numbers, $(a + b)$ is also a natural number. Therefore, N is closed with respect to addition, i.e. $(a + b) \in N$ for all $a, b \in N$. However, it may be noted that the subtraction is not a binary operation on N. For example, $2 - 5 = -3$ which is not a natural number.
2. Union and intersection are binary operations in the power set of a given set A, i.e. if S and T are subsets of A, and we define $S * T$ as $S \cap T$ and $S \cup T$, then ∗ is a binary operation on A.

Remarks

1. Usually binary operations are denoted by a symbol such as ∗, instead of f, and element assigned to (a, b) by $a * b$ instead of $*(a, b)$. This means, if elements $a, b \in A$, then $a * b \in A$. This property indicates that the set A is closed under the operation ∗.

2. (*i*) A function $f : A \rightarrow A$ is called an *unary* operation.

(*ii*) A function $g : A \times A \times A \rightarrow A$ is called a *ternary* operation.

(*iii*) A function $h : A \times A \times \ldots \times A$ (n factors) is called an *n-ary* operation.

11.3.1 Properties of Binary Operations

Commutative A binary operation '∗' on a set A is said to be *commutative* if for every $a, b \in A$.

$$a * b = b * a.$$

For example,

(*i*) Addition and multiplication of real numbers are commutative operations, i.e.

$$a + b = b + a \text{ and } a \cdot b = b \cdot a$$

(*ii*) The union and intersection in the power set of a set are both commutative, i.e.

$$A \cup B = B \cup A \text{ and } A \cap B = B \cap A$$

(*iii*) Subtraction in the set of integers is not commutative operation, i.e

$$a - b \neq b - a \text{ for all } a, b \in I.$$

Associative A binary operation '$*$' on a set A is said to be associative if for every $a, b, c \in A$,

$$a * (b * c) = (a * b) * c$$

For example,

(*i*) Addition and multiplication of real numbers are associative operations, i.e.

$$(a + b) + c = a + (b + c)$$

and $$a \cdot (b \cdot c) = (a \cdot b) \cdot c$$

(*ii*) Union and intersection of sets are associative operations, i.e.

$$(A \cup B) \cup C = A \cup (B \cup C)$$

and $$(A \cap B) \cap C = A \cap (B \cap C)$$

(*iii*) Subtraction in the set of integers, I is not associative, i.e.

$$(a - b) - c \neq a - (b - c) \text{ for all } a, b, c \in I.$$

(*iv*) In the set of real numbers R define $a * b$ as $a + 2b$. Then for all $a, b \in R$, the operation $*$ is not associative, because if

$$(a * b) * c = (a + 2b) * c = a + 2b + 2c$$

and $$a * (b * c) = a * (b + 2c) = a + 2b + 4c$$

then $$(a * b) * c \neq a * (n * c)$$

Identity element The set A is said to have an *identity element* for the operation '$*$' on itself if there exists an element $e \in A$, such that

$$e * a = a = a * e, \text{ for all } a \in A.$$

For example,

(*i*) '1' is the multiplicative identity in the set of natural numbers, rational numbers and complex numbers.

(*ii*) Consider the operation of multiplication on the set R of real numbers. Then, '1' is the identity element for this operation, because

$$a \cdot 1 = a = 1 \cdot a, \text{ for all } a \in R.$$

(*iii*) Consider the operation of addition on the set R of real number. Then, 0 (zero) is the identity element for this operation, because

$$a + 0 = a = 0 + a, \text{ for all } a \in R.$$

Inverse element Let there exists an identity element e in the set A for the operation $*$'. If

$$a * b = e = b * a, \text{ for all } a, b \in A,$$

then b is called the *inverse* of the element a and vice versa.

For example,

(*i*) Consider the operation of addition on the set of real numbers, R. Then for every real number a its negative $-a$ is its inverse, because

$$a + (-a) = 0 = (-a) + a$$

where 0 (zero) is the additive identity element.

(*ii*) Consider the operation of multiplication on the set of rational numbers and complex numbers. Then every element except zero has its multiplicative inverse.

Distributive law If '+' and '∗' are two operations defined on the set A such that

(*i*) $a + (b * c) = (a + b) * (a + c)$

(*ii*) $(b * c) + a = (b + a) * (c + a)$

In this case laws (*i*) and (*ii*) are called *left* and *right* distributive laws, respectively.

The operation of addition and multiplication on the set of integers obey the distribution laws.

11.4 GROUPS

A group is the mathematical structure consisting of a *set* together with a *binary operation* defined on it such that the elements of the given set have some relationships among themselves. The applications of group theory in studying the structure of human genes, theory of relativity, and in quantum mechanics have been appreciated.

Definition Let G be a non-empty set together with a binary operation ∗ defined on it. Then the algebraic structure (G, ∗) is said to be a group, if the binary operation satisfies the following properties (also called *group axioms*)

G_1. *Closure property* For every pair of elements

$$a, b \in G, (a * b) \in G.$$

G_2. *Associative property* For any three elements $a, b, c \in G$,

$$a * (b * c) = (a * b) * c$$

G_3. *Existence of identity* There exists an element $e \in G$ such that

$$a * e = a = e * a, \text{ for all } a \in G.$$

G_4. *Existence of inverse* For each element $a \in G$, there exists an element $b \in G$ such that

$$b * a = e = a * b$$

The element b is then called the *inverse* of a and is denoted by $b = a^{-1}$. Thus, a^{-1} is an element of G such that

$$a^{-1} * a = e = a * a^{-1}$$

Remarks

1. A group with addition binary operation is known as *additive group,* and that with multiplication binary operation is known as *multiplicative group.*
2. The number of distinct elements in a group is called *order of the group,* denoted by O (G).

11.4.1 Abelian (or Commutative) Group

A group G is said to be an *abelian* (or *commutative*) group if in addition to the above four properties the following property is also satisfied.

G_5. *Commutative property* $a * b = b * a$, for all $a, b \in G$.

Remark Commutative property is not the necessary property for G being a group.

Illustrations

1. The set of integers, I is an abelian group under addition because, $a + b = b + a$, for all $a, b \in$ I.
2. The set Q of all rational numbers forms an abelian group under addition, 0 is the additive identity and for every rational number a, the negative $(-a)$ of a is the inverse.
3. The set R – (0) of real numbers except zero is a group under multiplication, 1 is the multiplicative identity element and the inverse of a is $1/a$.

11.4.2 Finite and Infinite Groups

A group (G, $*$) is called a *finite group* if the set G contains a finite number of distinct elements, otherwise an *infinite group*.

The number of elements in a finite group is called the *order of the group* and is denoted by $|G|$.

11.4.3 Order of a Group

The number of elements in a finite group is called the order of the group. The order of an infinite group is infinite. Order of a group is denoted by o(G).

The group of the smallest order for a given operation or simply the smallest group is the group in which the set G consists of the identity element e alone, i.e. $G = \{e\}$.

Example 1 Show that the set of integers is a group under the operation of addition.

Solution (*i*) *Closure* Since sum of two integers is an integer, therefore I is closed with respect to addition.

(*ii*) *Associativity* If a, b, c are any arbitrary elements in I, then

$$(a + b) + c = a + (b + c)$$

is true, because associative law of addition holds in I.

(*iii*) *Existence of identity* The number $0 \in$ I, such that

$$a + 0 = a = 0 + a \text{ for all } a \in \text{I}.$$

Thus the integer 0 is the additive identity.

(*iv*) *Existence of inverse* If $a \in$ I, then $-a \in$ I such that

$$a + (-a) = 0 = (-a) + a.$$

Thus every element in I has its additive inverse.

Since the set I is satisfying all the properties of a group, therefore, it is a group with respect to addition. Again, it is noted that addition of integers is commutative, i.e. $a + b = b + a$ for all $a, b \in$ I.

Hence (I, +) is an abelian group of infinite order.

Example 2 Show that the set, N of natural numbers is not a group under the operation of addition.

Solution In the set of natural numbers there exists no natural number e called the *additive identity* such that $a + e = a = e + a$ for all $a \in$ N. Hence, (N, +) is not a group.

Example 3 Prove that ({1, –1}, ×), where '×' is the multiplication operation, is an abelian group of order 2.

Solution We define operation, × on the given set as shown in the table below:

×	1	–1
1	1	–1
–1	–1	1

In the above composition table, we see that multiplication is closed and associative; 1 is the multiplicative identity element and the inverse of 1 is – 1 and inverse of – 1 is 1. Multiplication is commutative also.

Hence, ({1, – 1), ×) is an abelian group of order 2.

Example 4 Show that the set, G = {1, ω, ω^2}, where 1, ω, ω^2 are cube roots of unity, form an abelian group under the operation of ordinary multiplication.

Solution We prepare a composition table as follows:

×	1	ω	ω^2
1	1	ω	ω^2
ω	ω	ω^2	1
ω^2	ω^2	1	ω

The following properties can be verified from the composition table:

(*i*) *Closure* Since all the entries in the composition table are also elements of the set G, therefore it is closed for multiplication.

(*ii*) *Associativity* $1 \cdot (\omega \cdot \omega^2) = (1 \cdot \omega) \cdot \omega^2$. Ordinary multiplication is always associative.

(*iii*) *Identity element* From the above table, we see that

$$1 \cdot 1 = 1; \; 1 \cdot \omega = \omega = \omega \cdot 1$$
$$1 \cdot \omega^2 = \omega^2 = \omega^2 \cdot 1$$

(*iv*) *Inverse element* From the above table, we see that

$$1 \cdot 1 = 1; \text{ and } \omega^2 \cdot \omega = \omega \cdot \omega^2 = \omega^3 = 1$$

Therefore, inverse of 1, ω, ω^2 are 1, ω^2, ω, respectively.

(*v*) *Commutativity* We have

$$1 \cdot \omega = \omega = \omega \cdot 1; \; \omega \cdot \omega^2 = \omega^3 = \omega^2 \cdot \omega \text{ and } \omega^2 \cdot 1 = \omega^2 = 1 \cdot \omega^2$$

Example 5 Prove that the four roots of unity 1, –1, *i*, – *i*, where $i = \sqrt{-1}$ form an abelian multiplicative group.

Solution Preparing the composition table for the set, G = {1, – 1, *i*, – *i*} with multiplication composition as given below:

×	1	–1	*i*	–*i*
1	1	–1	*i*	–*i*
–1	–1	1	–*i*	*i*
i	*i*	–*i*	–1	1
–*i*	–*i*	*i*	1	–1

he following properties are verified from the composition table:

(*i*) *Closure* Since all the elements in the composition table are also elements of G, therefore, G is closed with respect to multiplication. Hence, multiplication is a binary operation on G.

(*ii*) *Associativity* Ordinary multiplication is always associative.

(*iii*) *Identity element* 1 is the identity element, because

$$1 \cdot 1 = 1;\ 1 \cdot (-1) = -1;\ 1 \cdot (i) = i;\ 1 \cdot (-i) = i$$

e. $1 \cdot a = a$, for all $a \notin$ G.

(*iv*) *Inverse* It can be verified from the composition table that

$$1 \cdot 1 = 1;\ -1 \cdot (-1) = +1;\ (-1) \cdot (-i) = i;\ i \cdot (-1) = -i$$

e. the inverse of 1, – 1, i and $-i$ are 1, – 1, $-i$ and i, respectively.

(*v*) We see in the composition table that multiplication is commutative because all the entries in first, econd, third and fourth rows in the table are identical to those of first, second, third and fourth columns espectively, i.e. the law $a \times b = b \times a$, for all $a, b, \in$ G holds.

Hence, G is an abelian group under multiplication.

xample 6 Determine whether the set $G = \{(a, b): a, b \in R, a \neq 0\}$ under the operation defined as

$$(a, b) \cdot (c, d) = (ac, bc + d), \text{ for all } (a, b), (c, d) \in G$$

s an abelian group.

olution Verifying the following properties with respect to the given operation defined on the set A.

(*i*) *Associativity* Let (a, b), (c, d) and (e, f) be any three elements of A. Then

$$[(a, b) \cdot (c, d)] \cdot (e, f) = (ac + bc + d) \cdot (e, f)$$
$$= [(ac) \cdot e,\ (bc + d) \cdot e + f)]$$
$$= [ace,\ bce + de + f] \quad ...(i)$$

Also, $$(a, b) \cdot [(c, d) \cdot (e, f)] = (a, b) \cdot [ce, de + f]$$
$$= [a \cdot (ce),\ b \cdot (ce) + (de + f)]$$
$$= [ace,\ bce + de + f] \quad ...(ii)$$

From Eqs. (*i*) and (*ii*) we conclude that multiplication is associative.

(*ii*) *Identity element* Now if (x, y) is an element of A such that

$$(a, b).\ (x, y) = (a, b) \in A.$$

Then, $(xa, ya + b) = (a, b)$. Therefore, $xa = a$ and $ya + b = b$, which gives $x = 1$ and $y = 0$. Hence, (1, 0) is the identity element.

(*iii*) *Inverse* Let for each $(a, b) \in A$ there exists an element $(x, y) \in A$ such that

$$(x, y) \cdot (a, b) = (1, 0)$$

or $$(xa, ya + b) = (1, 0)$$

Then, $xa = 1$ and $ya + b = 0$, which gives $x = 1/a$ and $y = -b/a$.

Since $a \neq 0$, x, y are real numbers, therefore, it follows that $(1/a, -b/a)$ is the inverse of (a, b). Hence, A is a group. Now

$$(a, b) \cdot (c, d) = (ac, bc + d)$$

and $$(c, d) \cdot (a, b) = (ca, da + b)$$

In general, $$(a, b) \cdot (c, d) \neq (c, d) \cdot (a, b)$$

Hence, A is non-abelian group.

Example 7 Prove that the set, I of integers forms an abelian group with respect to the operation, $*$ defined on it as: $a * b = a + b + 1$.

Solution Verifying the following properties with respect to the given operation defined on the set I.

(*i*) *Closure* Since a, b are the elements of I, therefore $a * b = a + b + 1$.

(*ii*) *Associativity* If $a, b, c \in I$, then

$$(a * b) * c = (a + b + 1) * c = (a + b + 1) + c + 1 = a + b + c + 2.$$

Also $$a * (b * c) = a * (b + c + 1) = a + (b + c + 1) + 1 = a + b + c + 2.$$

Hence, $(a * b) * c = a * (b * c)$

(*iii*) *Identity element* Let e be an element of I such that $e * a = a = a * e$.

From Property (*ii*), we have $e * a = e + a + 1$. Thus $(e + a + 1) = a \Rightarrow e = -1$.

Now, $-1 \in I$ and for any $a \in I$, we have

$$(-1) * a = -1 + a + 1 = a.$$

It follows that -1 is the identity element.

(*iv*) *Inverse* For each $a \in I$ there exists an element $b \in I$ such that $b * a = -1$, i.e.

$$(b + a + 1) = -1 \text{ or } b = -2 - a.$$

Therefore if, $a \in I$, then $-2 - a \in I$. Also, $(-2 - a) * a = (-2 - a) + a + 1 = -1$.

Hence $-2 - a$ is the left inverse of a.

(*v*) *Commutative* $a * b = a + b + 1 = b + a + 1 = b * a$.

Hence, the set I of integers is an abelian group for the given operation.

Example 8 Prove that the set Q of rational numbers other than 1 forms an abelian group with respect to the operation $*$ defined as: $a * b = a + b - ab$.

Solution Verifying the following properties with respect to the given operation defined on the set Q.

(*i*) *Closure* $a \in Q$ and $b \in Q$ implies that a and b are rational numbers other than 1.

Now $a * b = a + b - ab$ which is also a rational number and it cannot be equal to 1.

Since $a + b - ab = 1$, therefore

$$a + b - ab - 1 = 0 \quad \text{or} \quad (a - 1)(1 - b) = 0, \text{ i.e. } a = 1, b = 1,$$

which is not true. Therefore, $a * b \in Q$ for all $a, b \in Q$. Hence, Q is closed with respect to the given operation.

(*ii*) *Associativity* If $a, b, c \in Q$, then

$$\begin{aligned}(a * b) * c &= (a + b - ab) * c \\ &= (a + b - ab) + c - (a + b - ab)c \\ &= a + b - ab + c - ac - bc + abc \\ &= a + b + c - ab - bc - ca + abc.\end{aligned}$$

Also $$\begin{aligned}a * (b * c) &= a * (b + c - bc) \\ &= a + (b + c - bc) - a(b + c - bc) \\ &= a + b + c - bc - ab - ca + abc\end{aligned}$$

Hence $(a * b) * c = a * (b * c)$, for all $a, b, c \in Q$.

(*iii*) *Identity* Let $e \in Q$ be the identity element such that

$$e * a = a = a * e; \; a \in Q$$

Now $$e * a = e + a - ea = a$$

$$e(1 - a) = 0, \text{ i.e. } e = 0, \text{ since } a \neq 1.$$

Thus $0 \in Q$, is the identity element.

(*iv*) *Inverse* For each $a \in Q$, there exists an element, $b \in Q$ such that

$$a * b = 0 = b * a.$$

Now $$a * b = 0 \text{ or } a + b - ab = 0$$

or $$b(1 - a) + a = 0, \text{ i.e. } b = a/(a - 1).$$

Thus $a/(a - 1)$ is the inverse of $a(a \neq 1)$, because

$$a * \left(\frac{a}{a-1}\right) = \left(\frac{a}{a-1}\right) + a - \left(\frac{a}{a-1}\right)a$$

$$= \frac{a + a^2 - a - a^2}{a-1} = \frac{0}{a-1} = 0, \qquad \text{[since } a \neq 1\text{]}$$

(v) *Commutative* If $a, b \in Q$, then $a * b = b * a$. Now

$$a * b = a + b - ab = b + a - ba = b * a.$$

Hence, mathematical structure (Q, *) is an abelian group.

Example 9 Prove that the set Q of rational numbers other than – 1 forms an abelian group for the operatio * defined as: $a * b = a + b + ab$.

Solution Proceed as in Example 12. In this case 0 is the identity element and the inverse of a i $a/(a + 1)$, $a \neq 1$. The mathematical structure (Q, *) is also an abelian group.

Example 10 Prove that the set Q_+ of positive rational numbers forms an abelian group with respect t the operation * defined as: $a * b = ab/2$.

Solution Verifying the following properties with respect to the given operation defined on the set Q_+

(*a*) *Closure* If $a, b \in Q_+$, then $ab/2$ is also in Q_+.

(*b*) *Associativity* If $a, b, c \in Q_+$, then

$$(a * b) * c = \frac{ab}{2} * c = \frac{1}{2}\left[\frac{ab}{2}c\right]$$

$$= \frac{1}{4}[(a \cdot b) \cdot c] = \frac{1}{4}[a \cdot (b \cdot c)]$$

$$= \frac{1}{2}\left[a \cdot \frac{bc}{2}\right] = \frac{1}{2}[a \cdot (b * c)]$$

$$= a * (b * c)$$

Hence, $(a * b) * c = a * (b * c)$.

(*c*) *Identity* The number e serves as an identity element provided $e \in Q_+$ and

$$e * a = a = a * e, \text{ for all } a \in Q_+.$$

Now $e * a = a$ or $\dfrac{ea}{2} = a$, i.e. $e = 2$.

Thus, $2 \in Q_+$ serves as an identity element, because

$$2 * a = \frac{2a}{2} = a = a * 2, \text{ for all } a \in Q_+$$

(*d*) *Inverse* Any element, $b \in Q_+$ is said to be the inverse of $a \in Q_+$ if $b * a = 2 = a * b$ [since $e = 2$]

Now $b * a = 2$ implies that $ab/2 = 2$ or $b = 4/a$. Thus, $a \in Q_+$ implies that $4/a \in Q_+$ and

$$\frac{4}{a} * a = \frac{1}{2}\left(\frac{4}{a}\right) \cdot a = \frac{4a}{2a} = 2.$$

Hence, $4/a$ is the inverse of a.

(*e*) *Commutative* If $a, b \in Q_+$, then $a * b = \dfrac{ab}{2} = \dfrac{b.a}{2}$.

Hence $(Q_+, *)$ is an abelian group.

Theorem 11.1 A group in which each element is the inverse of itself is an abelian group. In other words, if $a^2 = e$ with $a \neq e$, then G is abelian, where $a \in$ G.

Proof: Let (G, *) be a group and $a, b \in$ G. Then $a * b \in$ G (since G is a group and * is binary operation), also $a^{-1} = a$ and $b^{-1} = b$.

Now $(a * b)^{-1} = (a * b)$ because $a * b \in$ G. We have

$$(a * b)^{-1} = b^{-1} * a^{-1} = b * a$$

Since $a * b = b * a$, the operation * is commutative and the group (G, *) is an ablian group.

Theorem 11.2 Let (G, *) be a group. If an element $a \in$ G commutes with an element x of G, then a^{-1} also commutes with x.

Proof Given that $a * x = x * a$. Then we need to prove that $a^{-1} * x = x * a^{-1}$.

As $\quad a * x = x * a$

$$a^{-1} * (a * x) = a^{-1} * (x * a)$$

or $\quad (a^{-1} * a) * x = (a^{-1} * x) * a$

or $\quad e * x = (a^{-1} * x * a)$

or $\quad x = (a^{-1} * x) * a$

Hence, $x * a^{-1} = (a^{-1} * x) * (a * a^{-1}) = (a^{-1} * x) * e = a^{-1} * x$

11.4.3 Properties of a Group

1. *Cancellation Law* Let G be a group and $a, b, c \in$ G. Then $a \cdot b = a \cdot c$ implies that $b = c$.

   Since, $a^{-1} \in$ G, then

   $$a^{-1}(a \cdot b) = a^{-1} (a \cdot c) \Rightarrow (a^{-1} \cdot a) \cdot b = (a^{-1} a) \cdot c$$

   $$\Rightarrow e \cdot b = e \cdot c, \text{ i.e. } b = c$$

   This property is known as *left cancellation law.* Similarly, we can prove the *right cancellation law,* i.e. $b \cdot a = c \cdot a \Rightarrow b = c$.

2. *The Left Identity is also the Right Identity* $e \cdot a = a = a \cdot e$, for $a \in$ G.

   If a^{-1} is the inverse element of a, then

   $$a^{-1} \cdot (a \cdot e) = (a^{-1} \cdot a) \cdot e = e \cdot e = e = a^{-1} \cdot a.$$

   Therefore, $\quad a^{-1} \cdot (a \cdot e) = a^{-1} \cdot a \Rightarrow a \cdot e = a$ [Prop. 1]

   Hence, e is also the *right identity* element in group G.

3. *The Left Inverse of an Element is also the Right Inverse* $a^{-1} \cdot a = e = a \cdot a^{-1}$, for $a \in$ G.

   Let a^{-1} is the inverse of a, then

   $$a^{-1} \cdot (a \cdot a^{-1}) = (a^{-1} \cdot a) \cdot a^{-1} = e \cdot a^{-1} \text{ or } a^{-1} \cdot e.$$ [Prop. 2]

   Thus $\quad a^{-1} \cdot (a \cdot a^{-1}) = a^{-1} \cdot e \Rightarrow a \cdot a^{-1} = e$ [Prop. 1]

4. *Uniqueness of Identity* Identity element of a group is unique.

Suppose, if possible, e and e' be two identity elements of the group G, then

(*i*) $a \cdot e = a = e \cdot a$

and (*ii*) $a \cdot e' = a = e' \cdot a$, for $a \in G$

Now for $a = e'$ in (*i*), we have $e \cdot e' = e'$, and for $a = e$ in (*ii*), we have $e \cdot e' = e$.

But, $e \cdot e' = e'$ and $e \cdot e' = e \Rightarrow e = e'$. Hence, the identity element is unique.

5. *Uniqueness of Inverse* The inverse of each element of a group is unique.

Suppose, if possible b and c are two inverse of an element, $a \in G$, then

$$a \cdot b = e = b \cdot a \text{ and } a \cdot c = e = c \cdot a$$

Now $a \cdot b = e \Rightarrow a \cdot b = a \cdot c$ or $b = c$. [by cancellation law]

Hence, the inverse of a is unique.

6. *If a, b are any two Elements of a Group G, then the Equations: ax = b and ay = b, have Unique Solution in G.*

For an element, $a \in G \Rightarrow a^{-1} \in G$ such that $a^{-1} \cdot a = e = a \cdot a^{-1}$, and $a^{-1} \in G$, $b \in G$ implies due to closure property that $a^{-1}b \in G$.

Now pre-multiplying the equation $ax = b$ by a^{-1}, we have

$$a^{-1}(ax) = a^{-1}b \Rightarrow (a^{-1}a)x = a^{-1}b$$

$$\Rightarrow ex = a^{-1}b \text{ or } x = a^{-1}b.$$

Thus, $x = a^{-1}b$ is the solution of the equation $ax = b$ in G.

To prove uniqueness, let us suppose that x_1 and x_2 are two solutions of the equation $ax = b$, such that

$$ax_1 = b \text{ and } ax_2 = b \Rightarrow x_1 = x_2$$ [Left cancellation law]

Therefore, the solution is unique.

In a similar manner, it can be proved that the equation $ay = b$ has a unique solution $y = a^{-1} b$.

7. *Inverse of an Inverse of the Element of a Group is the Element Itself,* i.e. $(a^{-1})^{-1} = a$.

By the definition of inverse, we know that $a^{-1}a = e$. Pre-multiplying both sides by $(a^{-1})^{-1}$, we get

$$(a^{-1})^{-1}[a^{-1}a] = (a^{-1})^{-1}e$$

$$[(a^{-1})^{-1} a^{-1}]a = (a^{-1})^{-1} \text{ or } ea = (a^{-1})^{-1}, \text{ i.e. } a = (a^{-1})^{-1}$$ [$(a^{-1})^{-1}$ is the inverse of a^{-1}]

8. *Reversal Law* Inverse of the product of two elements of a group is the product of their inverses taken in the reverse order, i.e. $(ab)^{-1} = b^{-1}a^{-1}$, for $a, b \in G$.

Let $a, b \in G$ and a^{-1}, b^{-1} are inverses of a and b, respectively such that

$$aa^{-1} = e = a^{-1}a, \text{ and } bb^{-1} = e = b^{-1} b,$$

where e is the identity element of the group G.

Now $\quad (a, b)(b^{-1}a^{-1}) = a[b(b^{-1}a^{-1})]$

$= a[(b b^{-1}) a^{-1}]$

$= - a[ea^{-1}]$ [since $b b^{-1} = e$]

$= a a^{-1} = e$

Also, $\quad (b^{-1} a^{-1})(a b) = b^{-1}[a^{-1}(a b)]$

$= b^{-1}[(a^{-1} a) b] = b^{-1}[e b] = b^{-1} b = e.$

Thus, $\quad (a b)(b^{-1} a^{-1}) = e = (b^{-1} a^{-1})(a b).$

Hence, by the definition of inverse, we have $(ab)^{-1} = b^{-1}a^{-1}$.

Theorem 11.3 If a^{-1} is the inverse of a, then $(a^{-1})^{-1} = a$, i.e. the inverse of (a^{-1}) is a.

[UPTU, MCA, 2002]

Proof Let G be a non-empty set and $a \in$ A. Also, let e be the identity for G. Let the binary operation be $*$. Then $a^{-1} * a = e$. Operating on both sides by $(a^{-1})^{-1} \in$ G, we have

$$a^{-1} * a = e$$

$$(a^{-1})^{-1} * [a^{-1} * a] = (a^{-1})^{-1} * e$$

Applying associative property on LHS and identity on RHS, we get

$$[(a^{-1})^{-1} * (a^{-1})] * a = (a^{-1})^{-1}$$

or $\quad e * a = (a^{-1})^{-1} = a \quad [(a^{-1})^{-1}$ is inverse of $a^{-1}]$

Remark Sometimes the notation for operation is not used e.g. instead of $a * b$, we write ab in case no confusion is created by it.

Theorem 11.4 The inverse of the product of two elements of a group G is equal to the product of the inverses taken in the reverse order, i.e.

$$(ab)^{-1} = b^{-1}a^{-1}, \text{ for all } a, b \in G$$

[UPTU, MCA, 2002]

Proof Let $a, b \in$ G and a^{-1} and b^{-1} are inverse of a and b respectively. Let e be the identity element. Then $a^{-1} * a = e = a * a^{-1}$ and $b^{-1} * b = e = b * b^{-1}$

Now $\quad (a * b) * (b^{-1} * a^{-1}) = (a * b * b^{-1}) * a^{-1}$

$= [a * (b * b^{-1})] * a–1$

$= [a * e] * a^{-1}$

$= (a) * a^{-1} = e$

Again $\quad (b^{-1} * a^{-1}) * (a * b) = b^{-1} * [a^{-1} * (a * b)]$

$= b^{-1} * (a^{-1} * a) * b]$

$= b^{-1} * (e) * b$

$= b^{-1} * (e * b)$

$= b^{-1} * b = e$

From these results, we have

$$(b^{-1} * a^{-1}) * (a * b) = e = (a * b) * (b^{-1} * a^{-1})$$

Thus, by definition of inverse, $b^{-1} * a^{-1}$ is the inverse of $a * b$ or $(a * b)^{-1} = b^{-1} * a^{-1}$.

11.5 MODULO(M)

Earlier we have discussed operations such as union, intersection and complement of sets. In this chapter a new type of operation on sets known as *modulo(m)* will be discussed.

Addition Modulo(m) Let a and b be any two distinct integers and m be a fixed positive integer. Then the addition modulo(m) denoted by $a +_m b$ is defined as:

$$a +_m b = r, \quad 0 \leq r < m,$$

where r is the least non-negative remainder when the sum of a and b is divided by m. In other words, in order to determine addition modulo(m) we *add a and b in the ordinary way and then form the sum, and remove the integral multiple of m in such a way that the remainder r is either zero or a positive integer less than m.*

For example, $14 +_6 8 = 4$, since $14 + 8 = 22 = 6\,(3) + 4$, i.e. 4 is the least non-negative remainder when ordinary sum $14 + 8 = 22$ is divided by 6.

Similarly, $(-19) +_4 5 = 2$. Since $-19 + 5 = -14 = 4\,(4) + 2$.

Remark When a and b are two integers such that $a - b$ is divisible by a fixed positive integer m, then we write $a \equiv b \pmod{m}$, which is read as: *a congruent b modulo m.*

Obviously, $a +_m b \equiv a + b \pmod{m}$. For example, $15 +_5 7 = 2$ because $15 + 7 = 22$ and we write $22 \equiv 2 \pmod{5}$.

Multiplication Modulo(p) The multiplication modulo (p) of any two integers a and b is denoted by $a \times_p b$ and is defined as:

$$a \times_p b = r; \quad 0 \leq r < p,$$

where r is the least non-negative remainder, when the ordinary product of a and b is divided by p. In other words, in order to determine multiplication modulo (p) *we multiply a and b to form the product, and integral multiple of p in such a way that the remainder r is either zero or a positive integer less than p.*

For example, $8 \times_5 3 = 4$, since $8 \times 3 = 24 = 4\,(5) + 4$, i.e. 4 is the least non-negative remainder when the product $8 \times 3 = 24$ is divided by 5. Similarly, $-17 \times_5 4 = 4$, since $-17 \times 4 = -68 = 6\,(-12) + 4$.

Also, it is obvious that, $a \times_p b \equiv ab \pmod{p}$. For example, $6 \times_4 5 = 2$ because $6 \times 5 = 30$, and we write $30 \equiv 2 \pmod{4} \Leftrightarrow 30 - 2$ is divisible by 4.

Theorem 11.5 (*Additive group of integers modulo m*) Prove that the set, G = $\{0, 1, 2, ..., m - 1\}$ of m non-negative integers is a finite abelian group under the operation of ordinary addition modulo m.

Proof To prove G to be a group, we proceed to verify following properties:

(*i*) *Closure* For all $a, b \in$ G,

$$a +_m b = r \in G, \quad 0 \leq r \leq (m - 1)$$

(*ii*) *Associativity* Let $a, b, c \in G$. Then

$(a +_m b) +_m c = (a + b) +_m c$ [since $a +_m b \equiv a + b \pmod{m}$]

= least positive remainder or zero when the sum $(a + b) + c$ is divided by m.

= least positive remainder or zero when the sum $a + (b + c)$ is divided by m.

$= a +_m (b + c) = a +_m (b +_m c)$

Thus, $(a +_m b) +_m c = a +_m (b +_m c)$.

(*iii*) *Identity* For an element $0 \in G$, we have $0 +_m a = a = a +_m 0$, for all $a \in G$.

Thus, 0 is the identity element.

(*iv*) *Inverse* If $a \in G$ and $r \neq 0$, then $m - a$ is also an element of G. Also,

$$a +_m (m - a) = 0 = (m - a) +_m a.$$

Thus, $(m - a)$ is the inverse of a.

(*v*) *Commutativity* Let $a, b \in G$. Then

$a +_m b = r$ (say), when $a + b$ is divided by m

$= r$, when $b + a$ is divided by m

$= b +_m a$

Thus, $a +_m b = b +_m a$, and Hence, G is an abelian group.

Example 11 Prove that the set, G = {0, 1, 2, 3, 4, 5} is a finite abelian group under ordinary addition modulo 6 as the operation.

Solution Let us form the following composition table and verify properties for an abelian group as shown below:

$+_6$	0	1	2	3	4	5
0	0	1	2	3	4	5
1	1	2	3	4	5	0
2	2	3	4	5	0	1
3	3	4	5	0	1	2
4	4	5	0	1	2	3
5	5	0	1	2	3	4

(*i*) *Closure* Since all the entries in the composition table are also elements of the set G. Therefore, G is closed under $+_6$.

(*ii*) *Associativity* If $a, b, c \in G$, then $(a +_6 b) +_6 c = a +_6 (b +_6 c)$

Since we shall get the same remainder, when the ordinary sum is divided by 6. For example,

$(5 +_6 3) +_6 4 = 8 +_6 4 = 0$

Also $5 +_6 (3 +_6 4) = 5 +_6 7 = 0$.

Therefore, $(5 +_6 3) +_6 4 = 5 +_6 (3 +_6 4)$

(*iii*) *Identity* It is evident that the integer 0 is the identity element for the given operation.

(*iv*) *Inverse* From the table, we see that the inverses of 0, 1, 2, 3, 4 and 5 are 0, 5, 4, 3, 2 and 1, respectively. For example, $5 +_6 1 = 0 = 1 +_6 5$. This implies that 1 is the inverse of 2.

(*v*) *Commutativity* If $a, b \in G$, then $a +_6 b = b +_6 a$.

For example, $5 +_6 4 = 9 = 4 +_6 5$. Hence, G is an abelian group.

Theorem 11.6 (*Multiplicative group of integers modulo p*) Prove that the set $G = \{1, 2, 3, \ldots, p - 1\}$ of $(p - 1)$ integers, p being prime, is a finite abelian group of order $p - 1$, under the ordinary multiplication modulo p.

Proof To prove G to be a group, we proceed to verify following properties:

(*i*) *Closure* By the definition of multiplication modulo p, we have

$$a \times_p b = r, \text{ for all } a, b \in G,$$

where $1 \le a \le p - 1,\ 1 \le b \le p - 1,\ 1 \le r \le p - 1$

Since p is prime, therefore, ab is not exactly divisible by p because

$$a \times_p b = 0 \text{ implies that } p \text{ divides } ab$$

or $\quad p$ divides a or b, since p is prime

or $\quad$ either $a = 0$ or $b = 0$, since $a < p$, $b < p$.

But $a \ne 0$, $b \ne 0$. Hence, $a \times_p b \ne 0$, so closure axiom is satisfied.

(*ii*) *Associativity* If $a, b, c \in G$, then

$$a \times_p (b \times_p c) = a \times_p (bc) \qquad [\text{since } b \times_p c = bc \ (\text{mod } p)]$$

= least non-negative remainder when $a\,(bc)$ is divided by p

= least non-negative remainder when $(ab)\,c$ is divided by p

$= (ab) \times_p c$

$= (a \times_p b) \times_p c \qquad [\text{since } ab = a \times_p b \ (\text{mod } p)]$

Thus, $a \times_p (b \times_p c) \times_p c = (a \times_p b) \times_p c$.

(*iii*) *Identity* For an element, $1 \in G$, we have $1 \times_p a = a = a \times_p 1$, for all $a \in b$.

Thus 1 is the identity element.

(*iv*) *Inverse* To show that every element has a unique inverse, let s be any member of the set. Consider the following product

$$1 \times_p s,\ 2 \times_p s,\ 3 \times_p s,\ \ldots,\ (p - 1) \times_p s \qquad \ldots(i)$$

All these are elements of the set G and no two of these can be equal as shown below:

Let i, j be two unequal integers less than p with $i > j$. Then,

$$i \times_p s = j \times_p s \equiv is - js \ (\text{mod } p)$$

or $\quad is - js$ is divisible by p

$(i - j)\, s$ is divisible by p

$\Rightarrow$ either $(i - j)$ or s is divisible by p since p is prime or $i - j$ or s is zero since both $i - j$ and s are less than p.

$\Rightarrow$ $i - j = 0$, since $s \neq 0$, i.e. $i = j$

which is a contradiction because $i \neq j$. Therefore, $i \times_p s \neq j \times_p s$.

Thus, in the product (*i*) all elements are distinct and one of these elements must be equal to 1. Let $s' \times_p s = 1$. Then s' is the inverse of s.

(*v*) *Commutativity* Let $a, b \in$ G. Then

$$a \times_p b = \text{least non-negative remainder when } ab \text{ is divided by } p$$
$$= \text{least non-negative remainder when } ba \text{ is divided by } p$$
$$= b \times_p a.$$

Thus, the operation is commutative.

Hence, $[G, \times_p]$ is an abelian group.

Example 12 Prove that the set, G = {1, 2, 3, 4} is a finite abelian group under the operation multiplication modulo 5.

Solution Let us form the following composition table and verify properties for an abelian group as shown below:

$\times_5$	1	2	3	4
1	1	2	3	4
2	2	4	1	3
3	3	1	4	2
4	4	3	2	1

(*i*) *Closure* For any $a, b \in$ G, we have $a \times_5 b \in$ G. For example, $4 \times_5 3 = 5\,(2) + 2$, i.e. 2 is an element of G.

(*ii*) *Associativity* For any $a, b, c \in$ G, we have

$$a \times_5 (b \times_5 c) = (a \times_5 b) \times_5$$

$$\text{LHS} = a \times_5 (b \times_5 c) = a \times_5 (bc)$$
$$= \text{least non-negative remainder when } a\,(bc) \text{ is divided by } 5$$
$$= \text{least non-negative number when } (ab)\,c \text{ is divided by } 5$$
$$= (ab) \times_5 c$$
$$= (a \times_5 b) \times_5 c = \text{RHS} \qquad [\text{since } ab = a \times_5 b \text{ (mod 5)}]$$

For example, $(2 \times_5 3) \times_5 4 = 4 = 2 \times_5 (3 \times_5 4)$.

(*iii*) *Identity* Obviously, $1 \in$ G is the identity element.

(*iv*) *Inverse* For every element $a \in$ G, there exists an element $b \in$ G such that $a \times_5 b = 1 = b \times_5 a$. For example, $2 \times_5 3 = 1 = 3 \times_5 2$. From above table it can be seen that the inverses of 1, 2, 3 and 4, are 1, 3, 2 and 4, respectively.

(*v*) *Commutativity* The operation is commutative as the corresponding rows and columns in the composition table are identical. For example, $3 \times_5 4 = 2 = 4 \times_5 3$.

Hence, $[G, \times_5]$ is a finite abelian group.

Example 13 Let A = {0, 1, 2, 3, 4, 5, 6, 7} and * denotes "multiplication module 8, i.e. $x * y = (x\,y) \bmod 8$". Write three distinct groups (G, *), where G ⊂ A and G has two elements. *[UPTU, BTech, 2007]*

Solution Let us form the composition table as follows:

*	0	1	2	3	4	5	6	7
0	0	0	0	0	0	0	0	0
1	0	1	2	3	4	5	6	7
2	0	2	4	6	0	2	4	6
3	0	3	6	1	4	7	2	5
4	0	4	0	4	0	4	0	4
5	0	5	2	7	4	1	6	3
6	0	6	4	2	0	6	4	2
7	0	7	6	5	4	3	2	1

The required three distinct group (G, *), where G ⊂ A and G has following two elements.

(*i*)

*	1	3
1	1	3
3	3	1

(*ii*)

*	1	5
1	1	5
5	5	1

(*iii*)

*	1	7
1	1	7
7	7	1

11.5.1 Order of an Element

The order of an element a in a group is the smallest positive integer n such that $a^n = e$, and is denoted by $o(a) = e$. If no such integer exists, then order of an element a is *infinite*.

To find the order of an element in a group, we compute the sequence of products $a, a^2, a^3, \ldots$ until an identity is reached. The exponent of such a product is the order of the element a. If identity in the sequence of products does not reach, then order of a is infinite.

Example 14 Let G = $\{1, i, -1, -i\}$ be a multiplicative group. Then find the order of each element of G.

Solution The element 1 is the multiplicative identity in G. Now we develop the sequence of products as follows :

$$1^1 = 1, \text{ so } o(1) = 1$$

$$(-1)^2 = 1 \text{ but } (-1)^n \neq 1 \text{ for any positive integer } n < 2, \text{ so } o(-1) = 2$$

$$(i)^4 = 1 \text{ but } (i)^4 \neq 1 \text{ for any positive integer } n < 4, \text{ so } o(i) = 4$$

$$(-i)^4 = 1 \text{ but } (-i)^4 \neq 1, \text{ for any positive integer } n < 4, \text{ so } o(-i) = 4.$$

Hence, the order of elements $1, i, -1, -i$ is 1, 4, 2, 4, respectively.

Example 15 Let G = $\{a, a^2, a^3, a^4, a^5, a^6\}$ be a multiplicative group, where $a^6 = e$ is the multiplicative identity. Find the order of each element of G.

Solution We develop the sequence of products as follows:

$$(a^2)^3 = a^6 = e, \text{ so } o(a^2) = 3$$

$(a^3)^2 = a^6 = e$, so o $(a^3) = 2$

$(a^4)^3 = a^{12} = (a^6)^2 = e^2 = e$, so o $(a^4) = 3$

$(a^5)^6 = (a^6)^5 = e^5 = e$, so o $(a^5) = 6$, but $(a^5)^n \neq e$ for any positive integer $n < 6$

$(a^6)^1 = a^6 = e$, so o $(a^6) = 1$

Hence, the order of elements $a, a^2, a^3, \ldots, a^6$ is 6, 3, 2, 3, 6 and 1, respectively.

Theorem 11.7 The order of an element a of a group is the same as that of its inverse a^{-1}.

Proof Let n and m be the orders of a and a^{-1}, respectively. Then

$$\text{o}(a) = n \Rightarrow a^n = e \qquad \text{(identity element)}$$
$$\Rightarrow (a^n)^{-1} = e^{-1} \quad \text{or} \quad (a^{-1})^n = e$$
$$\Rightarrow \text{o}(a^{-1}) \le n \Rightarrow m \le n.$$

Also o $(a^{-1}) = m$

$$\Rightarrow (a^{-1})^m = e$$
$$\Rightarrow (a^m)^{-1} = e \quad \text{or} \quad a^m = e \qquad [\text{since } b^{-1} = e \Rightarrow b = e]$$
$$\Rightarrow \text{o}(a) \le m \Rightarrow n \le m.$$

But $m \le n$ and $n \le m$ implies that $m = n$.

If the order of a is infinite, then the order of a^{-1} cannot be finite, because o $(a^{-1}) = m$, so o $(a) \le m$ and hence o (a) is infinite. Thus if the order of a is infinite, then the order of a^{-1} must also be infinite.

Theorem 11.8 The order of an integral power of an element cannot exceed its order.

Proof Let a^k be any integral power of an element a. Let o $(a) = n$. Now

$$\text{o}(a) = n \Rightarrow a^n = e \qquad \text{(identity element)}$$
$$\Rightarrow (a^n)^k = e^k \quad \text{or} \quad a^{nk} = e$$
$$\Rightarrow (a^k)^n = e$$

Hence, o $(a^k) \le n$.

Theorem 11.9 The orders of the elements a and $x^{-1}ax$ are the same, where a, x are any two elements of a group.

Proof Let n and m be the orders of elements a and $x^{-1}ax$, respectively. Then

$$(x^{-1}ax)^2 = (x^{-1}ax)(x^{-1}ax) = x^{-1}a(xx^{-1})ax$$
$$= x^{-1}(ae)ax = x^{-1}aax = x^{-1}a^2x.$$

In general, we get $(x^{-1}ax)^n = x^{-1}a^nx = x^{-1}ex = x^{-1}x = e$ [since o $(a) = n \Rightarrow a^n = e$]

Thus o $(x^{-1}ax) \le n$ implies that $m \le n$. Again

$$\text{o}(x^{-1}ax) = m \Rightarrow (x^{-1}ax)^m = e$$
$$\Rightarrow x^{-1}a^m x = e \Rightarrow x^{-1}a^m x = x^{-1}x$$
$$\Rightarrow a^m x = x \quad \text{or} \quad a^m x = ex \qquad \text{(Left cancellation law)}$$
$$\Rightarrow a^m = e \qquad \text{(Right cancellation law)}$$

Thus o $(a) \le m$ implies that $n \le m$. Hence, $m \le n$, $n \le m \Rightarrow m = n$.

Example 16 Find the order of each element of the group $[\{0, 1, 2, 3, 4\}, +_5]$.

Solution Since identity element in the group $[\{0, 1, 2, 3, 4\} +_5]$ is 0, therefore, o (0) = 1. Now

$$1(1) = 1,\ 2(1) = 1 +_5 1 = 2,\ \ 3(1) = 1 +_5 1 +_5 1 = 3,$$

$$4(1) = 1 +_5 1 +_5 1 +_5 1 = 4,\ \ 5(1) = 1 +_5 1 +_5 1 +_5 1 +_5 1 = 0 \qquad \text{(Identity element)}$$

Thus, 5 is the least positive integer such that 5 (1) = identity of the group. So o (1) = 5

Again $1(2) = 2,\ \ 2(2) = 2 +_5 2 = 4,\ \ 3(2) = 2 +_5 2 +_5 2 = 4 +_5 2 = 1,$

$$4(2) = 2 +_5 2 +_5 2 +_5 2 = 1 +_5 2 = 3,$$

$$5(2) = 2 +_5 2 +_5 2 +_5 2 +_5 2 = 3 +_5 2 = 0 \qquad \text{(Identity element)}$$

Thus o (2) = 5.

Since $1 +_5 4 = 0 = 4 +_5 1$ and $2 +_5 3 = 0 = 3 +_5 2$, therefore, in the given group the inverse of 1 is 4 and that of 2 is 3. Thus, o (4) = o (1) = 5 and o (3) = o (2) = 5.

Hence, in the given group o (0) = 1 and each other element is of order 5.

Example 17 Find the order of each element in the multiplicative group $G = \{1, \omega, \omega^2\}$, where $\omega^3 = 1$.

Solution Since identity of the given multiplicative group $G = \{1, \omega, \omega^2\}$ is 1 therefore o (1) = 1.

Again $\omega^1 = \omega$, $\omega^2 = \omega^2$, $\omega^3 = 1$. Thus, 3 is the least positive integer such that $\omega^2 = 1$ and hence, o (ω) = 3. Finally, $(\omega^2)^1 = \omega^2$, $(\omega^2)^2 = \omega^4 = \omega^2$, $\omega = 1.\omega = \omega$, $(\omega^2)^3 = \omega^6 = (\omega^3)^2 = 1^2 = 1$ identity of (G). Thus, o (ω^2) = 3.

11.6 PERMUTATION GROUPS

A group (G, *) is called a *permutation group* of any finite set P if elements of G are permutations of G and the operation * is the composition of two functions.

The number of elements in the finite set P is known as the *degree of permutation.*

Let $P = \{a_1, a_2, a_3, \ldots, a_n\}$ be a finite set having n distinct elements and $f: P \rightarrow P$ be an one-to-one onto function such that $f(a_i) = b_i$ for all i is some arrangement of the elements of P. Then f is a *permutation of degree n*. The arrangement of elements in columns is immaterial, for example $b_1, b_2, \ldots, b_n$ represents some arrangement of n elements of P. The following notation is used to denote the permutation f on n symbols:

$$f = \begin{pmatrix} a_1 & a_2 & a_3 & \ldots & a_n \\ b_1 & b_2 & b_3 & \ldots & b_n \end{pmatrix}$$

That is, each element in the second row is the f-image of the element of the first row lying directly above it.

For example, if P = {1, 2, 3, 4} is a finite set having four elements, then

$$f_1 = \begin{pmatrix} 1 & 2 & 3 & 4 \\ 2 & 4 & 1 & 3 \end{pmatrix}, f_2 = \begin{pmatrix} 1 & 2 & 3 & 4 \\ 1 & 3 & 2 & 4 \end{pmatrix}$$

are all permutations of degree 4. Here in the permutation f, the elements 1, 2, 3, 4 have been replaced respectively by the elements 2, 4, 1, 3. Thus $f_1(1) = 2, f_1(2) = 4, f_1(3) = 1, f_1(4) = 3$.

Equality of two permutations Two permutations f and g of degree n are said to be equal if and only if $f(a) = g(a)$ for all $a \in$ P.

For example, if $f = \begin{pmatrix} 1 & 2 & 3 & 4 \\ 2 & 3 & 4 & 1 \end{pmatrix}$ and $g = \begin{pmatrix} 2 & 4 & 3 & 1 \\ 3 & 1 & 4 & 2 \end{pmatrix}$

are two permutations of degree 4. Evidently $f(1) = 2 = g(1)$, $f(2) = 3 = g(2)$, $f(3) = 4 = g(3)$ and $f(4) = 1 = g(4)$.

Total number of permutations If P is a finite set containing n distinct elements, then it will have n ! distinct arrangements of the elements of P, n ! distinct permutations of degree n and is denoted by P_n, i.e. the set P_n is called the symmetric set of permutations of degree n.

If the set P contains 3 elements, then the set P_3 of all permutations of degree 3 will have 3! = 6 elements. For example,

$$P_3 = \left\{ \begin{pmatrix} 1 & 2 & 3 \\ 1 & 2 & 3 \end{pmatrix}, \begin{pmatrix} 1 & 2 & 3 \\ 2 & 1 & 3 \end{pmatrix}, \begin{pmatrix} 1 & 2 & 3 \\ 1 & 3 & 2 \end{pmatrix}, \begin{pmatrix} 1 & 2 & 3 \\ 3 & 2 & 1 \end{pmatrix}, \begin{pmatrix} 1 & 2 & 3 \\ 2 & 3 & 1 \end{pmatrix}, \begin{pmatrix} 1 & 2 & 3 \\ 3 & 1 & 2 \end{pmatrix} \right\}$$

Identity permutation If each element of a permutation be replaced by itself, then such a permutation is called the identity permutation of degree n. For example,

$$I = \begin{pmatrix} a_1 \, a_2 \, a_2 \ldots a_n \\ a_1 \, a_2 \, a_2 \ldots a_n \end{pmatrix}$$

is the identity permutation of degree n.

Product of permutations The product of two permutations f and g of degree n, denoted by fg or gof is obtained by first carrying out the operation defined by f and then by g.

Suppose P_n is the set of all permutations of degree n, and let

$$f = \begin{pmatrix} a_1 \; a_2 \; a_3 \ldots a_n \\ b_1 \; b_2 \; b_3 \ldots b_n \end{pmatrix} \text{ and } g = \begin{pmatrix} b_1 \; b_2 \; b_3 \ldots b_n \\ c_1 \; c_2 \; c_3 \ldots c_n \end{pmatrix}$$

be any two elements of P_n.

The permutation of g is written in such a way that the first row of g coincides with the second row of f. The product of the permutation f and g is given by

$$fog = \begin{pmatrix} a_1 \; a_2 \; a_3 \ldots a_n \\ c_1 \; c_2 \; c_3 \ldots c_n \end{pmatrix}$$

To get fog, first f replaces a_1 by b_1 and then g replaces b_1 by c_1 so that fog replaces a_1 by c_1; a_2 by c_2, a_3 by c_3, . . ., a_n by c_n.

Since fog is a permutation of degree n, we conclude that the product of two permutations of degree n is also a permutation of degree n.

Example 18 Let $f = \begin{pmatrix} 1 & 2 & 3 \\ 1 & 3 & 2 \end{pmatrix}$ and $g = \begin{pmatrix} 1 & 2 & 3 \\ 2 & 3 & 1 \end{pmatrix}$ be two permutations of degree 3. Then show that fog is not commutative.

Solution $fog = \begin{pmatrix} 1 & 2 & 3 \\ 1 & 3 & 2 \end{pmatrix} \begin{pmatrix} 1 & 2 & 3 \\ 2 & 3 & 1 \end{pmatrix} = \begin{pmatrix} 1 & 2 & 3 \\ 1 & 3 & 2 \end{pmatrix} \begin{pmatrix} 1 & 3 & 2 \\ 2 & 1 & 3 \end{pmatrix} = \begin{pmatrix} 1 & 2 & 3 \\ 2 & 1 & 3 \end{pmatrix}$

and $gof = \begin{pmatrix} 1 & 2 & 3 \\ 2 & 3 & 1 \end{pmatrix}\begin{pmatrix} 1 & 2 & 3 \\ 1 & 3 & 2 \end{pmatrix} = \begin{pmatrix} 1 & 2 & 3 \\ 2 & 3 & 1 \end{pmatrix}\begin{pmatrix} 2 & 3 & 1 \\ 3 & 2 & 1 \end{pmatrix} = \begin{pmatrix} 1 & 2 & 3 \\ 3 & 2 & 1 \end{pmatrix}.$

We found that $fog \neq gof$. This shows that the product of two permutations is not in general commutative.

Cyclic permutations Suppose P_n is a permutation of degree n on a set having n distinct elements. If m elements of the set in a row are arranged in such a way that the f-image of each element in the row is the element which follows it, the f-image of the last element is the first element and the remaining $n–m$ elements of the set P are left unchanged by f. Then P_n is called a *cyclic permutation* or a *cycle of length m* or an *m-cycle.*

The length of a cycle refers to the number of objects permuted by the cycle. For example, consider the permutation

$$P = \begin{pmatrix} 1 & 2 & 5 & 3 & 6 & 4 \\ 2 & 4 & 5 & 1 & 6 & 3 \end{pmatrix}$$

This assignment of values could be presented by the cycle (1 2 4 3). This means that each element in the bracket is replaced by the element following it, the last element is replaced by the first element and the missing elements 5 and 6 are left unchanged. Thus, 1 is replaced by 2, 2 by 4, 4 by 1, 3, 3 by 1, 5 by 5 and 6 by 6. The length of this cycle is 4. Note that (1, 2, 4, 3) = (2, 4, 3, 1) = (4, 3, 1, 2) = (3, 1, 2, 4).

However, the permutation $P = \begin{pmatrix} 1 & 2 & 4 & 3 & 5 & 6 \\ 2 & 1 & 3 & 4 & 5 & 6 \end{pmatrix}$ is not cyclic.

A cycle does not change by changing the place of its elements provided their order is not changed. For example (1 2 3 4) = (2 3 4 1) = (3 4 1 2) = (4 1 2 3).

Two cycles are said to be *disjoint* if they have no symbols in common. For example, (1 3 5) and (2 6 8 9) are disjoint cycles, while (1 3 4) and (2 3 5 6) are not disjoint.

Transposition The cycle of length two which interchanges the symbols leaving all other unchanged is called transposition. For example, the cyclic permutation (1, 2) in which the image of 1 is 2, the image of 2 is 1 and the remaining missing elements are unchanged.

If the transposition (2 3) is a permutation of degree 3 on three symbols 1, 2, 3, then the corresponding permutation will be

$$\begin{pmatrix} 1 & 2 & 3 \\ 1 & 3 & 2 \end{pmatrix}$$

Multiplication of cycles Two or more cycles are multiplied by multiplying the permutations represented by them. For example, if the disjoint cycles (1 2 3) and (5 6 4 1) represent permutations of degree 6 on six symbols 1, 2, 3, 4, 5, 6, then

$$(1\ 2\ 3)\ (5\ 6\ 4\ 1) = \begin{pmatrix} 1 & 2 & 3 & 4 & 5 & 6 \\ 2 & 3 & 1 & 4 & 5 & 6 \end{pmatrix}\begin{pmatrix} 5 & 6 & 4 & 1 & 2 & 3 \\ 6 & 4 & 1 & 5 & 2 & 3 \end{pmatrix}$$

$$= \begin{pmatrix} 1 & 2 & 3 & 4 & 5 & 6 \\ 2 & 3 & 5 & 1 & 6 & 4 \end{pmatrix} = (1\ 2\ 3\ 5\ 6\ 4).$$

Since a cycle of length one represents the identity permutation, therefore, (1) (2 3 4) (6) = (2 3 4).

Inverse of a cyclic permutation The inverse of the cyclic permutation P = (1, 2, . . ., n) is obtained by writing its elements in the reverse order, i.e. $P^{-1} = (1, 2, \ldots, n-1, n)^{-1} = (n, n-1, \ldots, 2, 1)$.

In particular, every transposition is its own inverse. If (1 2) is a transposition, then $(1\ 2)^{-1} = (2\ 1) = (1\ 2)$.

Even and odd permutations A permutation is said to be an even permutation if it can be expressed as a product of an even number of transpositions, otherwise it is said to be an odd permutation.

Remarks Important results:

1. A permutation cannot be both even and odd, i.e. if a permutation P is expressed as a product of transpositions, then the number of transpositions is either even or odd.
2. A cycle of length n can be expressed as the product of $n - 1$ transpositions. Therefore, a cycle of length n will be an even permutation when n is odd and it will be an odd permutation when n is even.
3. Identity permutation is always an even permutation.
4. The product of two even permutations is an even permutation.
5. The product of two odd permutations is an even permutation.
6. The product of an even permutation and an odd permutation is an odd permutation. Similarly, the product of an odd permutation and an even permutation is an odd permutation.
7. The inverse of an even permutation is an even permutation and the inverse of an odd permutation is an odd permutation.
8. Out of $n!$ permutations on n symbols, (1/2) $n!$ are even permutations and (1/2) $n!$ are odd permutations.
9. The set P_n of all even permutations of degree n forms a finite group of order (1/2) $n!$ with respect to permutation multiplication.

Example 19 Find the inverse of the permutation:

$$f = \begin{pmatrix} 1 & 2 & 3 & 4 \\ 1 & 3 & 4 & 2 \end{pmatrix}$$

Solution To find the inverse of the permutation f, we form another permutation g by interchanging the two rows of f, such that

$$g = \begin{pmatrix} 1 & 3 & 4 & 2 \\ 1 & 2 & 3 & 4 \end{pmatrix} = \begin{pmatrix} 1 & 2 & 3 & 4 \\ 1 & 4 & 2 & 3 \end{pmatrix}$$

We have $$fog = \begin{pmatrix} 1 & 2 & 3 & 4 \\ 1 & 3 & 4 & 2 \end{pmatrix}\begin{pmatrix} 1 & 2 & 3 & 4 \\ 1 & 4 & 2 & 3 \end{pmatrix} = \begin{pmatrix} 1 & 2 & 3 & 4 \\ 1 & 2 & 3 & 4 \end{pmatrix} = I \text{ (identity)}$$

and $$gof = \begin{pmatrix} 1 & 2 & 3 & 4 \\ 1 & 4 & 2 & 3 \end{pmatrix}\begin{pmatrix} 1 & 2 & 3 & 4 \\ 1 & 3 & 4 & 2 \end{pmatrix} = \begin{pmatrix} 1 & 2 & 3 & 4 \\ 1 & 2 & 3 & 4 \end{pmatrix} = I$$

Example 20 Express following permutations as the product of disjoint cycles.

$$f = \begin{pmatrix} 1 & 2 & 3 & 4 & 5 & 6 & 7 & 8 & 9 \\ 2 & 3 & 4 & 5 & 1 & 6 & 7 & 9 & 8 \end{pmatrix}$$

Solution We express the given permutation as a product of disjoint cycles as follows:

$$\begin{pmatrix} 1 & 2 & 3 & 4 & 5 & 6 & 7 & 8 & 9 \\ 2 & 3 & 4 & 5 & 1 & 6 & 7 & 9 & 8 \end{pmatrix} = (6)\ (7)\ (1\ 2\ 3\ 4\ 5)\ (8\ 9)$$

or

$$= (1\ 2\ 3\ 4\ 5)\ (8\ 9),$$

omitting cycles of length 1 as they represent identity permutation.

Example 21 Decompose the permutation

$$f = \begin{pmatrix} 1 & 2 & 3 & 4 & 5 & 6 & 7 \\ 6 & 5 & 2 & 4 & 3 & 1 & 7 \end{pmatrix}$$

into transpositions.

Solution First we express the given permutation f as a product of disjoint cycles as:

$$(1\ 6)\ (2\ 5\ 3)\ (4)\ (7) = (1\ 6)\ (2\ 5\ 3)$$

omitting cycles of length 1 as they represent identity permutation.

Express the cycle (2 5 3) as a product of transpositions as follows:

$$(2\ 5\ 3) = (2\ 5)\ (2\ 3)$$

This gives $f = (1\ 6)\ (2\ 5)\ (2\ 3)$.

Example 22 Examine whether the following permutation is even or odd.

$$f = \begin{pmatrix} 1 & 2 & 3 & 4 & 5 & 6 & 7 & 8 & 9 \\ 2 & 5 & 4 & 3 & 6 & 1 & 7 & 9 & 8 \end{pmatrix}.$$

Solution Expressing f as a product of disjoint cycles as follows:

$$f = (1\ 2\ 5\ 6)\ (3\ 4)\ (7)\ (8\ 9) = (1\ 2\ 5\ 6)\ (3\ 4)\ (8\ 9),$$

omitting cycle of length 1 because it represents identity permutation.

Again expressing the cycle (1 2 5 6) as a product of transpositions as follows:

$$(1\ 2\ 5\ 6) = (1\ 2)\ (1\ 5)\ (1\ 6)$$

Consequently, we get

$$f = (1\ 2)\ (1\ 5)\ (1\ 6)\ (3\ 4)\ (8\ 9).$$

Since f is the product of odd transpositions, the permutation is an odd permutation.

11.7 CYCLIC GROUPS

Let G be a group. If there exists an element $a \in$ G such that every element is expressible as $b = a^n$ for some integer n, then G is called a *cyclic group,* and the element a is then called a *generator of the cyclic group* G. Symbolically, it is expressed as:

$$G = \{a^n : n \in I\}, \quad \text{for multiplication operation}$$

$$G = \{na : n \in I\}, \quad \text{for additive operation}$$

There can be more than one generator of a cyclic group. If G is a cyclic group generated by a, then we shall write G = $\{a\}$. *Every cyclic group is abelian.*

Illustration Consider the set, A = {1, 2, 3, 4} and the operation $\times_5$. Obviously, (A, $\times_5$) is a group. It may be noted that

$$2^1 = 2 \pmod 5 ; \qquad 2^2 = 4 \pmod 5$$
$$2^3 = 8 = 3 \pmod 5 ; \qquad 2^4 = 16 = 1 \pmod 5$$

that is, $2^1 = 2$, $2^2 = 4$, $2^3 = 3$, $2^4 = 1$.

Since all the elements 1, 2, 3, 4 of the group can be expressed as integral multiple of 2, therefore, 2 generates the elements or the group of 2 is the generator of the group.

Example 23 Prove that the group ({1, –1, *i*, – *i*}, ×) is cyclic and find its generator.

Solution We know that $i^1 = i$, $i^2 = -1$, $i^3 = -i$ and $(-i)^4 = 1$. It follows that elements of the set can be expressed as integral power of *i* and, therefore, it is a cyclic group generated by *i*. Thus, $-i$ is also the generator of the group.

Remark If some element *a* is the generator of cyclic group G, then its inverse will also be another generator of G.

11.8 SUBGROUP

A non-empty subset (H, *) of a group (G, *) is said to be a *subgroup of G* if (H, *) is also a group by itself.

Theorem 11.10 The necessary and sufficient condition for a non-empty subset H of group (G, *) to be a subgroup is

$$a \in H, b \in H \Rightarrow a * b^{-1} \in H$$

Proof (*Necessary condition*) Suppose H is a subgroup of G and *a* and *b* are any two elements in H.

Since each element in a subgroup has its multiplicative inverse, $b \in H \Rightarrow b^{-1} \in H$. Now

$$a \in H, b^{-1} \in H \Rightarrow a * b^{-1} \in H. \text{ (Closure property)}$$

Sufficient condition Let H be a non-empty subset of G with the condition, $a \in H, b \in H \Rightarrow a * b^{-1} \in H$. Now to show that H is a subgroup, let

$$a \in H, a^{-1} \in H \Rightarrow a * a^{-1} \in H \quad \text{or} \quad e \in H$$

Also $$e \in H, a \in H \Rightarrow e * a^{-1} = a^{-1} \in H$$

Since $a \in H$, $b^{-1} \in H$, combining these two, we get

$$a \in H, b^{-1} \in H \Rightarrow a (b^{-1})^{-1} \in H \text{ (given condition)}$$
$$\Rightarrow ab \in H \text{ (closure property)}$$

As H ⊆ G and the associative law holds for a group G also holds for H, so all conditions for a group are also satisfied for H. Hence, H is a subgroup of G.

Theorem 11.11 The intersection of any two subgroups of a group is also a subgroup of G.

[*UPTU, BTech, 2008*]

Proof Let H_1 and H_2 be any two subgroups of G. Then $H_1 \cap H_2 \neq \phi$, since at least the identity element *e* is common to both H_1 and H_2.

To prove that $H_1 \cap H_2$ is a subgroup, it is enough to prove that

$$a \in H_1 \cap H_2, b \in H_1 \cap H_2 \Rightarrow ab^{-1} \in H_1 \cap H_2.$$

Now $\quad a \in H_1 \cap H_2 \Rightarrow a \in H_1$ and $a \in H_2$,

$\quad b \in H_1 \cap H_2 \Rightarrow b \in H_1$ and $b \in H_2$.

But H_1, H_2 are subgroups, so

$$a \in H_1, b \in H_1 \Rightarrow ab^{-1} \in H_1 \text{ and } a \in H_2, b \in H_2 \Rightarrow ab^{-1} \in H_2.$$

Finally, $\quad ab^{-1} \in H_1$ and $ab^{-1} \in H_2 \Rightarrow ab^{-1} \in H_1 \cap H_2$.

Thus, we have shown that

$$a \in H_1 \cap H_2, b \in H_1 \cap H_2 \Rightarrow ab^{-1} \in H_1 \cap H_2$$

Hence, $H_1 \cap H_2$ is a subgroup of G.

Theorem 11.12 If H and K are two subgroups of G, then $H \cap K$ is also a subgroup of G.
[*UPTU, BTech, 2008*]

Proof: Let e be an identity of G. Then e will also be the identity element in H as well as in K, i.e. $e \in H \cap K$. Thus

$$e \in H \cap K \text{ and } H \cap K \neq \varnothing$$

If $a, b \in H$, then $a * b^{-1} \in H$ (H is subgroups of G)

Similarly, if $a, b \in K$, and $a * b^{-1} \in K$, then $a * b^{-1} \in H \cap K$

Hence, $H \cap K$ is subgroup of G.

11.8.1 Cosets

Suppose G is a group and H is any subgroup of G. Let a be any element of G. Then the set $Ha = \{ha : h \in H\}$ is called a *right coset* of H in G generated by a. Similarly, the set $aH = \{ah : a \in H\}$ is called *left coset of H* in G generated by a. Obviously Ha and aH are both subsets of G.

If e is the identity element of G, then $He = H = eH$. Therefore, H itself is a right as well as a left coset.

If Ha is a right coset of H in G, then $ea \in Ha$ and at least one element $a \in Ha$. This implies that right coset cannot be empty. Similarly, left coset cannot be empty.

If the group G is abelian, then $ah = ha$ for all $h \in H$. Thus, the right coset Ha will be equal to the left coset aH. However, if the group G is not abelian, then we may have $aH = Ha$ or $aH \neq Ha$.

If G is group under addition, then the cosets of H in G generated by a are defined as

Right coset, $H + a = \{H + a : h \in H\}$

Left coset, $\quad a + H = \{a + h : h \in H\}$.

Illustration Let G be the additive group of integers, i.e.

$$G = \{\ldots, -3, -2, -1, 0, 1, 2, 3, \ldots\}.$$

Let H be subgroup of G obtained on multiplying each element of G by 3. Then

$$H = \{\ldots, -9, -6, -3, 0, 3, 6, 9, \ldots\}$$

Since the group G is abelian, any right coset will be equal to the left coset. Let us form the right cosets of H in G as follows:

$$H + 0 = \{\ldots, -9 -6, -3, 0, 3, 6, 9, \ldots) \quad \text{for } 0 \in G$$

$$H + 1 = \{\ldots, -8, -5, -2, 1, 4, 7, 10, \ldots\} \quad \text{for } 1 \in G$$

$$H + 2 = \{\ldots, -7, -4, -1, 2, 5, 8, 11, \ldots\} \quad \text{for } 2 \in G$$

This shows that the right cosets H, H + 1 and H + 2 are disjoint and distinct. But

$$H + 3 = \{\ldots, -6, -3, 0, 3, 6, 9, 12, \ldots)$$

for $3 \in G$ is equal to H, also $3 \in H$.

Again $H + 4 = \{\ldots, -5, -2, 1, 4, 7, 10, 13, \ldots\}$ for $4 \in G$ is equal to H + 1, also $4 \in H + 1$.

Similarly, the right coset H + 5 coincides with H + 2, H + 6 with H, H + (– 1) with H = 2, H + (– 2) with H + 1 and so on. Thus, there exist only three distinct right cosets, H, H + 1 H + 2. Obviously, $G = H \cup (H + 1) \cup (H + 2)$.

Properties of cosets Let H be a subgroup of G and elements $a, b \in G$. Then

(*i*) $a \in a\text{H}$ (*ii*) $a\text{H} = \text{H}$ if and only if $a \in \text{H}$

(*iii*) $a\text{H} = b\text{H}$ or $a\text{H} \cap b\text{H} = \phi$ (*iv*) $a\text{H} = b\text{H}$ if and only if $a^{-1} \in \text{H}$.

Theorem (*Lagrange's Theorem*) 11.13 The order of each subgroup of a finite group G is a divisor of the order of the group G.

Proof Let G be a group of a finite order n, and let H be a subgroup of order m, i.e. $H = \{h_1, h_2, \ldots, h_m\}$.

Let $a \in G$. Then Ha is a right coset of H in G and, we have

$$\text{H}a = \{h_1 a, h_2 a, \ldots, h_m a\}.$$

This implies that Ha has m distinct members because $h_i a = h_j a \Rightarrow h_i = h_j$).

Thus each right coset of H in G has m distinct members. Any two distinct right cosets of H in G are disjoint, i.e. they have no element in common. Since G is a finite group, the number of distinct right cosets of H in G will also be finite, say k. The union of these k distinct right cosets of H in G is equal to G. Thus, if $\text{H}a_1, \text{H}a_2, \ldots, \text{H}a_k$ are the k distinct right cosets of H in G, then

$$G = \text{H}a_1 \cup \text{H}a_2 \cup \ldots \cup \text{H}a_k$$

In other words, Number of elements in G = Number of elements in Ha_1 + Number of elements in Ha_2 + . . . + Number of elements in Ha_k

Hence, order of group G, O (G) = $km \Rightarrow n = km$, i.e. $k = n \mid m$, i.e. m is a divisor of n.

So O (H) is divisor of O (G).

This proves the theorem.

Corollary If G is a finite group of order n and $a \in$ G, then $a^n = e$.

Example 24 Let G be the set of all ordered pairs (a, b) of real numbers for which $a \neq 0$. Let a binary operation in G be defined by the formula $(a, b) * (c, d) = (ac, bc + d)$. Show that (G, $*$) is a non-abelian group.

Does the subset H of all these elements of G which are of the form $(1, b)$ form a subgroup of G?

Solution It has been already proved in Example 10 that (G, ·) is a non-abelian group. Now to prove whether H is a subgroup of G, let $(1, a)$ and $(1, b)$ be any two elements of H. Then

$$(1, a) * (1, b)^{-1} = (1, a) \cdot \left(\frac{1}{1}, \frac{-b}{1}\right) = (1, a) * (1 - b) = (1, a - b) \in \text{H}$$

Hence, H is a subgroup of G.

11.9 SEMIGROUP AND MONOID

Let S be a non-empty set and $*$ be a binary operation defined on S. Then an algebraic structure (S, $*$) is is called *gropoid.* The algebraic structure (S, $*$) is said to be a *semigroup* if it satisfies the following properties:

(*i*) The binary operation $*$ is a closed operation (closure property), i.e

$$a * b \in \text{S, for all } a, b \in \text{S}$$

(*ii*) The binary operation $*$ is an associative operation (associative property), i.e.

$$a * (b * c) = (a * b) * c \text{ for all } a, b, c \in \text{S.}$$

Remark The mathematical structure (G, $*$) is said to be a *quasigroup* if binary operation $*$ satisfies closure property only.

Illustrations

1. Let A be a non-empty set, and P (A) be the power set of A. Then the mathematical structure (P (A), $\cup$) is a semigroup.
2. The mathematical structure ({0, 1}, ×} is a semigroup because it satisfies both closure and associative properties as shown in the table below.

×	0	1
0	0	0
1	0	1

Example 25 Let (S, $*$) be a semigroup. If $a * c = c * a$ and $b * c = c * b$, then show that $(a * b) * c = c * (a * b)$, for all $a, b, c \in$ S.

Solution

$$\begin{aligned} \text{LHS} &= (a * b) * c \\ &= a * (b * c) && [\text{since } * \text{ is associative}] \\ &= a * (c * b) && [\text{since } b * c = c * b] \\ &= (a * c) * b && [\text{since } * \text{ is associative}] \\ &= (c * a) * b && [\text{since } a * c = c * a] \\ &= c * (a * b) = \text{RHS} && [\text{since } * \text{ is associative}] \end{aligned}$$

Example 26 Let I be the set of positive integers and $*$ be an operation on S = I $\times$ I, defined by

(*i*) $(a, b) * (c, d) = (ac, bd)$

(*ii*) $(a, b) * (c, d) = (a + c, b + d)$

Show that S is a semigroup.

Solution Suppose $x = (a, b)$, $y = (c, d)$ and $z = (e, f)$.

(*i*) We have $(x\,y)\,z = (ac, bd) * (e, f) = [(ac)\,e, (bd)\,f]$

$x\,(y\,z) = (a, b) * (ce, df) = [a\,(ce), b\,(df)]$

Since a, b, c, d, e, f are positive integers, therefore $(ac)\,e = a\,(ce)$ and $(bd)\,f = b\,(df)$. Thus, $(xy)\,z = x\,(yz)$ and since $*$ is associative, i.e. S is a semigroup.

(*ii*) We have $(x\,y)\,z = (a+c)\,(b+d) * (e, f) = [(a+c)+e, (b+d)+f]$

$x\,(y\,z) = (a, b) * (c + e, d + f) = [a + (c + e), b + (d + f)]$

Since a, b, c, d, e, f are positive integers, therefore, $(a + c) + e = a + (c + e)$ and $(b + d) + f = b + (d + f)$. Thus $(xy)\,z = x\,(yz)$, and hence $*$ is associative, i.e. S is a semigroup.

Example 27 Let the operation $*$ is defined on the set of real number R as follows:

$$a * b = |\,a - b\,| \text{ for all } a, b \in \text{R}$$

Prove that * is a commutative binary operation on R but (R, $*$) is not a semigroup.

Solution Since $|\,a - b\,|$ is always a positive real number, $a * b \in$ R for all $a, b \in$ R. Thus, $*$ is binary operation on R. Also

$$|\,a - b\,| = |\,b - a\,| \text{ for all } a, b \in \text{R}$$

so $a * b = b * a$ for all $a, b \in$ R. Thus, $*$ is a commutative operation.

As $1 * (2 * 3) = 1 * |\,2 - 3\,| = 1 * 1 = |\,1 - 1\,| = 0$

and $(1 * 2) * 3 = |\,1 - 2\,| * 3 = 1 * 3 = |\,1 - 3\,| = 2$

Thus, $1 * (2 * 3) \neq 1 * 2) * 3$ and therefore $*$ is not associative.

Hence (R, $*$) is not a semigroup.

Sub-semigroup

Let (S, $*$) be a semigroup and let the set A be the subset of S. Then the algebraic structure (A, $*$) is called the sub-semigroup of (S, $*$), provided A is closed under the operation $*$, i.e. $a * b \in$ A all $a, b \in$ A.

For example, if N be the set of natural numbers, and A and B are set of positive even and odd integers, respectively, then the algebraic structure (A, +) is the sub-semigroup of (N, +) because the set of positive even integers, A is closed under addition. But (B, +) is not the sub-semigroup of (N, +) because the set of positive odd integers, B is not closed under addition.

11.9.1 Monoid

The semigroup (S, $*$) which has also an identity element with respect to the binary operation $*$ is said to be a *monoid* and is denoted as (M, $*$).

In other words, an algebraic structure (M, $*$) is called a monoid if the following three conditions are satisfied.

(*a*) M is closed under binary operation $*$, i.e. if $x, y \in$ M, then $x * y \in$ M.

(*b*) The binary operation $*$ is an associative operation, i.e. if any $x, y, z \in$ M, then $x * (y * z) = (x * y) + z$.

(*c*) There exists an identity element, i.e. for some $e \in$ M,

$$e * x = x * e = x, \text{ for any } x \in M$$

Illustration Let N be the set of natural numbers. Then the algebraic structures (N, +) and (N, ×) are semigroup. Moreover, (N, ×) is also a monoid because multiplication in N has the identity element 1, whereas (N, +) is not a monoid because addition in N has no additive identity element, 0.

Cyclic monoid

A monoid (M, *) is said to be cyclic monoid if there exists an element $a \in$ M such that every element of M can be written as some non-negative powers of a, i.e. a^n for $n \in$ N.

The cyclic monoid is said to be generated by the element $a \in$ M and the element a is called *generator* of the cyclic monoid.

Illustrations

1. The set M = {0, 1, 2, 3, 4, 5} with the operation $+_6$ is a cyclic monoid generated by 1 as shown below:

$$1^1 = 1,\ \ 1^2 = 1 +_6 1 = 2,\ \ 1^3 = 1^2 +_6 1 = 3,\ \ 1^4 = 1^3 +_6 = 4,$$
$$1^5 = 1^4 +_6 1 = 5 \text{ and } 1^6 = 0$$

   Thus, M = $\{1, 1^2, 1^3, 1^4, 1^5, 1^6 = 0\}$.

   It may be noted that the element $5 \in$ M is also the generator of M.

2. Let Σ^* be the set of all strings over an alphabet Σ including empty string λ. Then Σ^* is a monoid with respect to the operation of concatenation. The empty string λ is considered as an identity element. If Σ contains more than one element, then Σ^* with concatenation operation is not commutative monoid.

Sub-monoid

Let (M, $*$) be a monoid with identity element e and let the set B be nonempty subset of M. If B is closed under the binary operation, $*$ and $e \in$ B, then mathematical structure (B, $*$) is called a sub-monoid of (M, $*$).

For example, if M be the set of positive integers and (M, $*$) be a monoid under the binary operation $*$, then mathematical structure (B, $*$) is sub-monoid of (M, $*$) provided set B is defined as

$$B = \{a^i : i \in I, 0 \le i \le n \text{ and } a \in M\}.$$

11.10 CONGRUENCE RELATIONS AND QUOTIENT STRUCTURES

An equivalence relation R (also denoted as ~) defined on the semigroup (S, $*$), is called a *congruence* relation if a R b and c R d, then $(a * c)$ R $(b * d)$ or simply ac R bd.

Since equivalence relation R on semigroup S divides it into equivalence classes, therefore, we can also define an operation $*$ on equivalence classes as:

$$[a] * [b] = [a * b] \text{ or simply } [a]\,[b] = [ab]$$

where $[a]$ denotes the equivalence class containing the element a in the semigroup S. The collection of equivalence classes is denoted by S/R or S/~.

Example 28 Show that the equivalence R on the semigroup (S, +) of positive integers, defined by

$$a \, R \, b \text{ if and only if } a \equiv b \pmod 2$$

is the congruence relation.

Solution Suppose $a \equiv b \pmod 2$, and $c \equiv d \pmod 2$. Now we shall prove that $(a + c) \equiv (b + d) \pmod 2$.

Since $a \equiv b \pmod 2$ implies that 2 divides $(a - b)$ and $c \equiv d$ implies that 2 divides $(c - d)$, where m and n are remainders when $a - b$ and $c - d$ are divided by 2, respectively for all $m, n \in$ S. Adding $(a - b)$ and $(c - d)$, we get

$$(a - b) + (c - d) = 2m + 2n$$

$$(a + c) - (b + d) = 2(m + n) \text{ or } (a + c) \, R \, (b + d)$$

This proves that $(a + c) \equiv (b + d) \pmod 2$. Hence, this relation is a congruence relation on S.

Theorem 11.13 Let R be a congruence relation on the semi group (S, $*$). If ∇ is the function defined as

$$\nabla : S/R \times S/R \to S/R \text{ such that } ([a], [b]) \, R \, [a * b]$$

for all $a, b \in$ S, then prove that (S/R, ∇) is a semigroup.

Proof To prove (S/R, ∇) to be a semigroup, we shall prove that it satisfies closure as well as associative property under the operation ∇.

Since ∇ is a function defined from S/R × S/R to S/R, therefore, instead of writing an element of S/R as $\nabla([a], [b])$, we shall write $[a] \nabla [b] = [a \nabla b]$.

Closure property The relation R is a congruence relation in the semigroup (S, $*$), therefore, if a R b and c R d, then $(a * c) \, R \, (b * d)$ or $[a * c] = [b * d]$ for all $a, b, c, d \in$ S. This means ∇ is a function and also a binary operation on S/R.

Associative property Consider

$$\begin{aligned} [a] \nabla ([b] \nabla [c]) &= [a] \nabla (b * c) = [a * (b * c)] \\ &= [(a * b) * c), \qquad \text{(since $*$ is an associative operation in S)} \\ &= [a * b] \nabla [c] = ([a] \nabla [b]) \nabla [c]. \end{aligned}$$

Since both closure and associative properties are satisfied, therefore, S/R is a semigroup.

Theorem 11.14 Let R be a congruence relation on the monoid (M, $*$). If ∇ is the function defined as

$$\nabla : S/R \times S/R \to S/R \text{ such that } ([a], [b]) \, R \, [a * b]$$

for all $a, b \in$ M, then prove that (S/R, ∇) is a monoid.

Proof Left as an exercise.

Example 29 Let A = {0, 1} and A* be the set of all strings over alphabet A. Consider the free-semigroup (A*, ·), where '·' is concatenation operation. Show that the relation R defined on A* by

$$\alpha \, R \, \beta \text{ if } \alpha \text{ and } \beta \text{ have the same number of 1's,}$$

is a congruence relation.

Solution To show that R is an equivalence relation, we must show that

R is reflexive That $\alpha \, R \, \alpha$, for any $\alpha \in A*$, this result is obvious.

R is symmetric Suppose $\alpha \, R \, \beta$, for any $\alpha, \beta \in A*$. Then α and β have the same number of 1's, we may also say that β and α also have the same number of 1's. Consequently $\beta \, R \, \alpha$.

R is transitive Suppose that α R β and β R γ. Then

$$\alpha \text{ R } \beta \Rightarrow \alpha \text{ and } \beta \text{ have the same number of 1's.}$$

$$\beta \text{ R } \gamma \Rightarrow \beta \text{ and } \gamma \text{ have the same number of 1's.}$$

Thus, α R β and β R $\gamma \Rightarrow \alpha$ and γ have the same number of 1's.

$$\Rightarrow \alpha \text{ R } \gamma$$

Since all properties of equivalence relation are satisfied, R is an equivalence relation.

To show that R is congruence relation, we assume that α R α' and β R β'. Thus

$$\alpha \text{ R } \alpha' \Rightarrow \alpha \text{ and } \alpha' \text{ have the same number of 1's.}$$

$$\beta \text{ R } \beta' \Rightarrow \beta \text{ and } \beta' \text{ have the same number of 1's.}$$

We know by definition of concatenation that the, Number of 1's in $\alpha.\beta$ = Number of 1's in a + Number of 1's in β. Since α R α' and β R β', therefore

$$\text{Number of 1's in } \alpha.\beta = \text{Number of 1's in } \alpha'.\beta'$$

$$\Rightarrow (\alpha \,.\, \beta) \text{ R } (\alpha' \,.\, \beta')$$

$$\Rightarrow \text{R is a congruence relation.}$$

If R is an equivalence relation on a set S, then the quotient set S/R = $\{[a] : a \in S\}$, where $[a] = \{x \in S : a \text{ R } x\}$ is a equivalence class containing a in S.

Remark The semigroup (S/R, $\otimes$) is called *quotient semigroup* or *factor semigroup* of semigroup (S, $*$) by the congruence relation R. The operation $\otimes$ on S/R is also called *quotient binary relation.*

Corollary Let R be a congruence relation on the monoid (M, $*$). Then (M/R, $\otimes$) is a monoid, where operation $\otimes$ on M/R is defined as:

$$[a] \otimes [b] = [a * b]$$

Proof Let (M, $*$) be a monoid with identity element e. Then by Theorem 10, (M/R, $\otimes$) is a semigroup. Now we need to show that $[e]$ is the identity element (M/R, $\otimes$). Let $[a]$ be any element in M/R, where $a \in$ M. Then

$$[a] \otimes [e] = [a * e] = [a] = [e * a]$$
$$= [e] \otimes [a]$$

Thus, $[e]$ is the identity in M/R. Hence, (M/R, $\otimes$) is a monoid.

11.11 HOMEOMORPHISM AND ISOMORPHISM

Homeomorphism: Let (S_1, $*$) and (S_2, $\otimes$) be two semigroups. Then the function, f from S_1 to S_2 is homeomorphism provided.

(*a*) f is one-one onto function, and

(*b*) $f(a * b) = f(a) \otimes f(b)$.

If such a function exists, then S_1 is homeomorphic to S_2.

Illustration Let A = {0, 1} be a set and (S_1, ×) and (S_2, +) be two semigroups, where operation + is defined as

+	0	1
0	0	1
1	1	0

A function from S_1 to S_2 is defined as

$$f : S_1 \to S_2 \text{ such that for all } a \in S_1$$

$$f(a) = \begin{cases} 1 & \text{if A has an odd number of 1s} \\ 0 & \text{if A has an even number of 1s} \end{cases}$$

This definition of $f(a)$ implies that for any elements $x, y \in S_1$, we have $f(x\,y) = f(x) + f(y)$. Obviously f is onto because $f(0) = 0$ and $f(1) = 1$. Thus, f is a homeomorphism.

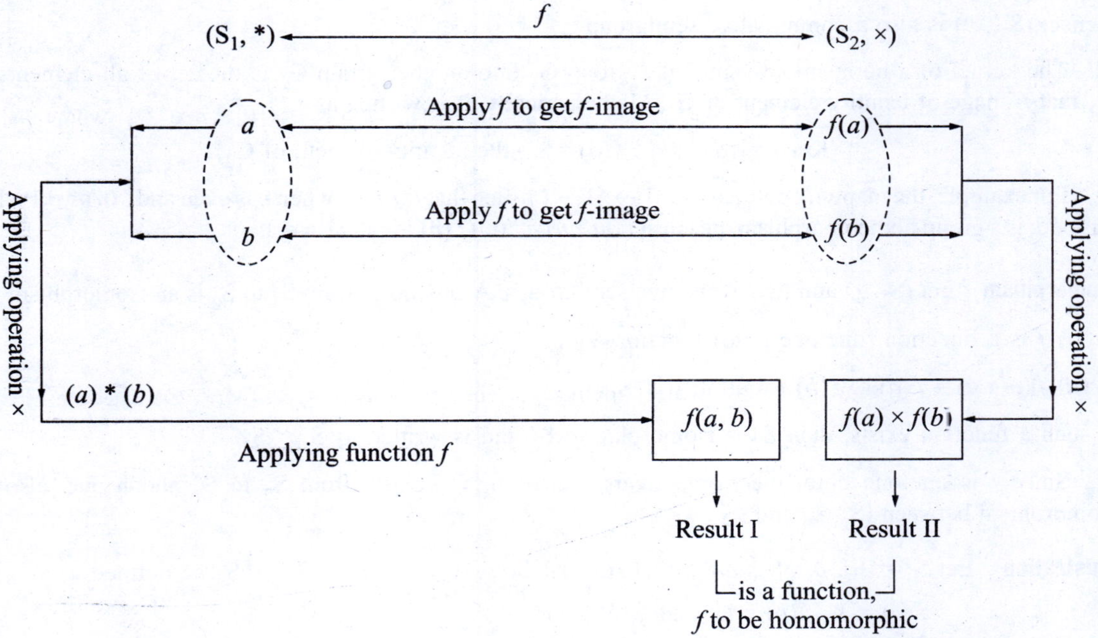

Fig. 11.2 Concept of Homomorphism

Theorem 11.15 Let (S_1, ∗) and (S_2, ⊗) be two semigroups, and f be a homeomorphism from S_1 to S_2. If T is a sub-semigroup of (S_1, ∗), then the image

$$f(T) = \{s : s \in S_2 \text{ and } s = f(t) \text{ for some } t \in T\}$$

under f, is a sub-semigroup of (S_2, ⊗).

Proof Let $s_1, s_2 \in f(T)$ be two arbitrary elements. Then for some t_1 and t_2 in T, we have $s_1 = f(t_1)$ and $s_2 = f(t_2)$.

To prove that $f(\mathrm{T})$ is a sub-semigroup, we should prove that it is closed under operation ⊗ as well as associative property holds in it. Since

$$s_1 \otimes s_2 = f(t_1) \otimes f(t_2) = f(t_1 * t_2) = f(t_3)$$

where $t_3 = t_1 * t_2 \in \mathrm{T}$. Hence, $s_1 \otimes s_2 \in f(\mathrm{T})$. This means $f(\mathrm{T})$ is closed under the operation ⊗.

Since (S_2, ⊗) is the semigroup, therefore, associative property holds in it. It follows that the associative property is also holds in $f(\mathrm{T})$, and hence, it is the sub-semigroup of (S_2, ⊗).

Theorem 11.16 If f is a homeomorphism from a commutative semigroup (S_1, ∗) onto a semigroup (S_2, ⊗), then (S_2, ⊗) is also commutative.

If s_1 and s_2 are any elements of S_2, then there exist s_1' and s_2' in S_1 such that $s_1 = f(s_1')$ and $s_2 = f(s_2')$. Thus

$$s_1 \otimes s_2 = f(s_1') \otimes f(s_2') = f(s_1' * s_2') = f(s_2' * s_1')$$
$$= f(s_2') \otimes f(s_1') = s_2' \otimes s_1'$$

Hence, (S_2, ⊗) is also a commutative semigroup.

The kernel of a homeomorphism f of a group G_1 into another group G_2 is the set of all elements of G_1 that f-image of identity element of G_2. Mathematically, it is written as

$$\mathrm{Ker}\, f = \{a \in \mathrm{G}_1 : f(a) = e_2, \text{ the identity element of } \mathrm{G}_2\}.$$

For example, the mapping defined as $f: \mathrm{I} \to \{1, -1\}$ such that $f(n) = 1$ when n is even and $f(n) = -1$ when n is odd, is a group homeomorphism, because $f(m+n) = f(m) f(n)$ for all $m, n \in \mathrm{I}$.

Isomorphism Let (S_1, ∗) and (S_2, ⊗) be two semigroups. A function f from S_1 to S_2 is an isomorphism if

(*i*) f is a bijection (one-one onto) function, and

(*ii*) $f(a * b) = f(a) \otimes f(b)$ for all $a, b \in \mathrm{S}_1$.

If such a function exists, then S_1 is isomorphic to S_2 and is written as $\mathrm{S}_1 \sim \mathrm{S}_2$.

Since f is one-one onto, therefore, inverse function, f^{-1} exists from S_2 to S_1 and hence, also an isomorphism between (S_2 ⊗) and (S_1, ∗).

Illustration Let $\mathrm{S}_1 = \{a, b, c\}$ and $\mathrm{S}_2 = \{1, \omega, \omega^2\}$ be two sets. Let $f : \mathrm{S}_1 \to \mathrm{S}_2$ be defined as

$$f(a) = 1,\ f(b) = \omega \text{ and } f(c) = \omega^2.$$

Then the following tables show that f is one-one onto. Thus f is homeomorphism as well as isomorphism between S_1 and S_2 with binary operation ∗ and ⊗, and hence (S_1, ∗) and (S_2, ⊗) are isomorphic semigroups.

∗	a	b	c
a	a	b	c
b	b	c	a
c	c	a	b

⊗	1	ω	ω^2
1	1	ω	ω^2
ω	ω	ω^2	1
ω^2	ω^2	1	ω

It may be noted that the table for S_2 is obtained by replacing elements in S_1 by their images.

Example 30 Let A be the set of 2 × 2 matrices. Show that semigroups, $A = \left[\begin{bmatrix} 1 & a \\ 0 & 1 \end{bmatrix}, \times\right]$ and (R, +) are isomorphism.

Solution We define the function $f: R \to A$ by $f(a) = \begin{bmatrix} 1 & a \\ 0 & 1 \end{bmatrix}$, for all $a \in R$. To prove f to be one-one, suppose $f(a_1) = f(a_2)$. Then

$$\begin{bmatrix} 1 & a_1 \\ 0 & 1 \end{bmatrix} = \begin{bmatrix} 1 & a_2 \\ 0 & 1 \end{bmatrix} \quad \text{or } a_1 = a_2.$$

Hence f is a one-one function.

Also, suppose that $b \in R$ is any real number. Then

$$f(a) = \begin{bmatrix} 1 & a \\ 0 & 1 \end{bmatrix} = \begin{bmatrix} 1 & b \\ 0 & 1 \end{bmatrix}$$

so f is a onto function.

For $a, b \in R$, we have

$$f(a)\,f(b) = \begin{bmatrix} 1 & a \\ 0 & 1 \end{bmatrix}\begin{bmatrix} 1 & b \\ 0 & 1 \end{bmatrix} = \begin{bmatrix} 1 & a+b \\ 0 & 1 \end{bmatrix} = f(a+b)$$

Since both properties of isomorphism are satisfied, therefore, A and R are isomorphic semigroups.

Example 31 If R is the additive group of real numbers and R_+ the multiplicative group of positive real numbers, prove that the mapping $f: R \to R_+$ defined by $f(x) = e^x$ for all $x \in R$ is an isomorphism of R onto R_+.

Solution If x is any real number, then e^x is always a positive real number, and e^x is unique. Therefore, if $f(x) = e^x$, then $f: R \to R_+$.

If $f(x_1) = f(x)_2$ for $x_1, x_2 \in R$, then $e^{x1} = e^{x2}$ or $x_1 = x_2$. Thus, f is one-to-one mapping.

Suppose $y \in R_+$. Then $\log y$ is a real number, i.e. $\log y \in R$. Now $f(\log y) = e^{\log y} = y$. Thus, for each $y \in R_+$ there exists $\log y \in R$ such that $f(\log y) = y$, so each element of R_+ is the f-image of some element of R. Thus, f is onto mapping.

Suppose x_1 and x_2 are any two elements of R. Then

$$f(x_1 + x_2) = e^{x1 + x2} = e^{x1}\,e^{x2} = f(x_1)\,f(x_2) \qquad \text{[by definition]}$$

Thus, f preserves compositions in R and R_+ and hence, f is an isomorphism of R onto R_+.

11.11.1 Properties on Homeomorphisms

Theorem 11.17 Let f be a homeomorphic (or an isomorphic) mapping of a group G_1 into a group G_2. Then

(*a*) The f-image of the identity element $e \in G_1$ is the identity of G_2, i.e. $f(e_1) = e_2$, where e_1 is the identity element in G_1 and e_2 is the identity element in G_2.

(*b*) The f-image of the inverse of an element $a \in G_1$ is the inverse of the f-image of a, i.e. $f(a^{-1}) = [f(a)]^{-1}$ for all $a \in G_1$.

(c) If H is a subgroup of G_1, then $f(H) = \{f(h) : h \in H\}$ is a subgroup of G_2. Ker f is a normal subgroup of G_1.
[UPTU, MCA, 2003]

Proof (a) Let e_1 be the identity element in G_1 and e_2 be the identity element in G_2. Let a be any element of G_1. Then $f(a) \in G_2$. Now

$$e_2 f(a) = f(a) = f(e_1 a)$$
$$= f(e_1) f(a) \quad [f \text{ is an isomorphic mapping}]$$
$$\text{or} \quad e_2 = f(e_1) \quad [\text{Right cancellation law in } G_2]$$

Thus $f(e_1)$ is the identity of G_2.

(b) Suppose e_1 is the identity element in G_1 and e_2 is the identity element in G_2. Then $f(e_1) = e_2$. Now for any $a \in G_1$, $a^{-1} \in G_1$, and $aa^{-1} = e_1$ we have

$$e_2 = f(e_1) = f(aa^{-1}) = f(a) f(a^{-1}) \quad [f \text{ is composition preserving}]$$

This implies that $f(a^{-1})$ is the inverse of $f(a)$ in G_1. Hence, $f(a^{-1}) = [f(a)]^{-1}$.

(c) If $e_1 \in H$, $f(e_1) = e_2 \in H$, so $f(H) \neq \phi$.

For $a, b \in H$, implies that $ab^{-1} \in H$ because H is a subgroup. Let $f(a), f(b) \in f(H)$. Then $f(a)(f(b))^{-1} \in f(H)$. Hence, $f(H)$ is subgroup of G_2.

Example 32 State and prove the function theorem of semigroups homomorphism.
[CCS Univ., MSc(Maths), 2005]

Solution Let $f : S_1 \to S_2$ be a homomorphism of the semigroup $(S_1, *)$ onto the semigroup $(S_2, \otimes)$. Let R be the relation on S_1 defined by $a\text{R}b$ if and only if $f(a) = f(b)$, for all $a, b \in S_1$. Then

(a) R is a congruence relation

(b) $(S_2, \otimes)$ and the quotient semigroup $(S_1/R, \otimes)$ are isomorphic.

(a) To prove that R is an equivalence relation:

Since $f(a) = f(a)$, or $a\text{R}a$, for all $a \in S_1$, i.e. R is reflexive.

$f(a) = f(b)$ or $a\text{R}b$ and $f(b) = f(a)$ or $b\text{R}a$, i.e. R is symmetric.

Also as $a\text{R}b \Rightarrow f(a) = f(b)$ and $b\text{R}c \Rightarrow f(b) = f(c)$. This means $f(a) = f(c)$, i.e. $a\text{R}c$. Thus, $a\text{R}b$ and $a\text{R}c \Rightarrow a\text{R}c$, i.e. R is transitive.

Hence, R is an equivalence relation.

Suppose $a\text{R}a_1$ and $b\text{R}b_1$. This implies that $f(a) = f(a_1)$ and $f(b) = f(b_1)$

Using operation of S_2, $f(a) \otimes f(b) = f(a_1) \otimes f(b_1)$

$$\text{or} \quad f(a * b) = f(a_1 * b_1), \text{ i.e. } (a * b)\,\text{R}\,(a_1 * b_1) \quad [\text{since } f \text{ is homomorphism}]$$

Hence, R is a congruence relation.

(b) Consider a relation g from $\dfrac{S_1}{R}$ to S_2 defined as:

$$g = \left\{[a], \frac{f(a)}{[a]} \in \frac{S_1}{R}\right\}$$

To show that g is a function, let $[a] = [a']$, i.e. aRa'. This means $f(a) = f(a')$, so g is a function.

So we can write that

$$g : \frac{S_1}{R} \rightarrow S_2 \text{ where } g[a] = f(a), \text{ for all } [a] \in \frac{S_1}{R}$$

Now to show that g is one-to-one, let

$$g([a]) = g([a']) \quad \text{or} \quad f(a) = f(a')$$

This means $aRa' \Rightarrow [a] = [a']$, so g is one-to-one.

To prove that g is also onto function, let $b \in S_2$ as f is onto, $f(a) = b$, for some $a \in S_1$. Then

$$g([a]) = f(a) = b, \text{ so } g \text{ is onto.}$$

Function g is homomorphism

$$g = ([a] \otimes [b]) = g([a * b]) = f(a * b)$$
$$= f(a) \otimes f(b) = g([a] \otimes f[b])$$

Since g is an one-one onto and is homomorphism, therefore, it is also isomorphism.

Example 33 Show that a semigroup with more than one idempotent element cannot be a group.

[UPTU, BTech, 2006]

Solution Let A = {e, 0, 1} and a binary operation denoted by * be defined as shown below:

*	e	0	1
e	e	0	1
0	0	0	0
1	1	0	1

(A, *) is a semigroup with three idempotent elements e, 0, 1 because

$$e * e = 0;\ 0 * 0 = 1, \text{ and } 1 * 1 = 1$$

In case (A, *) is a group, Then $e * 1 = 1$, and $1 * 1 = 1$. This implies that $e * 1 = 1 * 1$, i.e., $e = 1$ (right cancellation rule) which is not true. Hence, (A, *) is not group.

Example 34 If (S, $*$) and (T, $*'$) be monoids with identity elements e and e' respectively. Let $f : S \rightarrow T$ be an isomorphism, then $f(e) = e'$.

[CCS Univ., MSc (Maths), 2005]

Solution Let $b \in T$ be an arbitrary element. Since f is onto, there must be an element $a \in S$ (i.e. premage of b) such that $f(a) = b$. Thus, $a = a * e$, and $b = f(a) = f(a * e) = f(a) * f(e) = b * f(e)$

Similarly, $a = e * a$ implies that

$$b = f(a) = f(e * a) = f(e) * f(a) = f(e) * b$$

These two results show that $b * f(e) = b = f(e) * b$ which means that $f(e)$ is the identity element for the monoid (T, *). But this is equal to e'. Hence, $f(e) = e'$.

Self-practice Problems I

1. Let R be the set of real numbers. For each of the following determine whether binary operation, $*$ satisfies both commutative and associative operations.

(*a*) $a * b = a \times |b|$ (*b*) $a * b = ab/3$

(*c*) $a * b = ab + 2b$ (*d*) $a * b = \min(a, b)$

2. Let N be the set of natural numbers. For each of the following determine whether binary operation, $*$ satisfies both commutative and associative operations.

(*a*) $a * b = a + 4b$ (*b*) $a * b = a + b + 5$

(*c*) $a * b = \min(a, b + 4)$ (*d*) $a * b = \max(a, b)$

3. Let A and B be two non-empty sets. If $*$ is an operation defined on the set S such that A, B are subsets of S. If A and B are closed under binary operation, $*$, then prove that $A \cap B$ is also closed under $*$, but $A \cup B$ need not be.

4. Let a binary operation $*$ be defined on the set $A = \{a, b, c, d\}$ as follows:

$*$	a	b	c	d
a	a	b	c	d
b	b	a	b	c
c	c	d	a	a
d	d	b	a	c

Compute, (*a*) $c * d$ and $d * c$, (*b*) $b * d$ and $d * b$, (*c*) $a * (b * c)$ and $(a * b) * c$

5. Determine whether the operation, $*$ defined on the following sets is the binary operation

(*a*) A = I, where $a * b = a + b$ (*b*) A = I, where $a * b = \min(a, b)$

(*c*) A = I, where $a * b = \max(a, b)$ (*d*) A = R, where $a * b = a/b$

(*e*) $A = I_+$, where $a * b = a - b$ (*f*) A = R, where $a * b = a\sqrt{b}$.

6. Let (S, $*$) be a semigroup. Show that a binary operation $\otimes$ defined on S as $x \otimes y = x * a * y$ for all $x, y, a \in S$, is an associative operation.

7. Let (S, $*$) be a semigroup. Also for every $a, b \in S$, if $a \neq b$, then $a * b \neq b * a$.

(*a*) Show that for every $a \in A$, $a * a = a$

(*b*) Show that for every $a, b \in A$, $a * b * a = a$

(*c*) Show that for every $a, b, c \in A$, $a * b * c = a * c$

8. Let ($\{a, b\}$, $*$) be a semigroup, where $a * a = b$. Show that

(*a*) $a * b = b * a$ (*b*) $b * b = b$

9. Let (S, $*$) be a commutative semigroup. Show that if $a * a = a$ and $b * b = b$, then $(a * b) * (a * b) = a * b$.

0. Let (I, $*$) be an mathematical structure, where operation $*$ is defined as: $a * b = \max(a, b)$. Show whether (I, $*$) is a monoid?

11. Let A = $\{a, b\}$. Which of the following tables defines a semigroup as well as monoid on A?

(*a*)

$*$	a	b
a	a	b
b	a	a

(*b*)

$*$	a	b
a	a	b
b	b	a

(*c*)

$*$	a	b
a	a	a
b	b	b

(*d*)

$*$	a	b
a	b	b
b	a	a

12. Let A = $\{a, b\}$. Are there two semigroups (S_1, $*$) and (S_2, $\otimes$) that are not isomorphic?

13. If (S_1, $*$) and (S_2, $\otimes$) are commutative semigroups, then show that $S_1 \times S_2$ is also a commutative semigroup.

14. If (S_1, $*$) and (S_2, $\otimes$) are monoids, then show that $S_1 \times S_2$ is also a monoid.

15. Determine whether relation R on a semigroup is a congruence relation under the relation defined as follows:

(*a*) (I, +); a R b if and only if $a \equiv b \pmod 3$)

(*b*) (I_+, $\times$); a R b if and only if $|a - b| \leq 2$.

11.12 RINGS

While studying groups we consider sets which were closed with respect to a single operation either addition or multiplication. Here we shall study an important type of mathematical structure with above two binary operations '+' and '·' known as *ring*. A ring has its origin in the set of integers which is closed with respect to both operations of ordinary addition (+) and ordinary multiplication (·).

Definition Suppose R is a non-empty set with two binary operations, called addition (+) and multiplication (·). Then the mathematical structure (R, +, ·) is called a ring if the following conditions are satisfied.

1. (R, +) is an abelian group

(*i*) *Closure law* $a + b \in$ R, for all $a, b \in$ R.

(*ii*) *Associative law* $a + (b + c) = (a + b) + c$, for all $a, b, c \in$ R.

(*iii*) *Additive identity* There exists an element $0 \in$ R called the identity element such that

$$0 + a = a + 0, \text{ for all } a \in \text{R}$$

The element $0 \in$ R is called the *zero element* of the ring.

(*iv*) *Additive inverse* To each element $a \in$ R, there exists an element $b \in$ R such that

$$a + b = 0 = b + a$$

Then b is called the inverse of a and is denoted by $(-a)$.

(*v*) *Addition is commutative* $a + b = b + a$, for all $a, b \in$ R.

2. Closure and associative law for multiplication

(*i*) *Closure law* $a \cdot b \in$ R, for all $a, b \in$ R.

(*ii*) *Associative law* $a \cdot (b \cdot c) = (a \cdot b) \cdot c$, for all $a, b, c \in$ R.

3. Distributive law

Multiplication is distributive with respect to addition, i.e. for all $a, b, c \in R$.

$$a \cdot (b + c) = b \cdot c + c \cdot a \quad \text{[Left distributive law]}$$

and $$(b + c) \cdot a = b \cdot a + c \cdot a \quad \text{[Right distributive law]}$$

In short, a group is said to be a ring provided following conditions are satisfied.

(*i*) (R, +) is an abelian group

(*ii*) (R, ·) is a semigroup, i.e. multiplication is associative on R, and

(*iii*) Multiplication operation is distributive over addition operation, i.e.

$$a \cdot (b + c) = a \cdot b + a \cdot c$$

and $$(b + c) \cdot a = b \cdot a + c \cdot a, \text{ for all } a, b, c \in R$$

Illustration The set of integer numbers I, the set of rational numbers Q, the set R of real numbers and the set of complex numbers, C are examples of ring under ordinary addition and multiplication.

11.12.1 Properties of a Ring

If the algebraic structure (R, +, ·) is a ring, then for all $a, b, c \in R$, the following properties are satisfied

1. $a \cdot 0 = 0 \cdot a = 0$
2. $a \cdot (-b) = -(a \cdot b) = (-a) \cdot b$
3. $(-a)(-b) = a \cdot b$
4. $a \cdot (b - c) = a \cdot b - a \cdot c.$
5. $(b - c) \cdot a = b \cdot a - c \cdot a$

Proof 1. We have $a \cdot 0 = a \cdot (0 + 0)$ [Since $0 + 0 = 0$]

$= a \cdot 0 + a \cdot 0$ [Distributive law]

Adding to both sides $-(a \cdot 0)$, we have

$$-(a \cdot 0) + a \cdot 0 = -(a \cdot 0) + [a \cdot 0 + a \cdot 0]$$

$$0 = [-(a \cdot 0) + a \cdot 0] + a \cdot 0$$

$$0 = 0 + a \cdot 0 \text{ or } 0 = a \cdot 0$$

Similarly, $0 \cdot a = 0.$

2. We have, $a[(-b) + b] = a \cdot 0$

$a \cdot (-b) + a \cdot b = 0$ [Distributive law]

$-(a \cdot b) = a \cdot (-b)$ [since $a + b = 0 \Rightarrow a = -b$]

Therefore $a \cdot (-b) = -(a \cdot b)$

Similarly, we have

$$(-a) \cdot b = -(a \cdot b)$$

Hence, $a \cdot (-b) = -(a \cdot b) = (-a) \cdot b$

3. $(-a) \cdot (-b) = -[(-a) \cdot b)]$, because $a \cdot (-b) = -(a \cdot b) = -[(-a \cdot b)] = a \cdot b$

Since R is a group with respect to addition, because $-(-a) = a$ and hence

$$(-a) \cdot (-b) = a \cdot b$$

4. $a \cdot (b - c) = a \cdot [b + (-c)] = a \cdot b + a \cdot (-c)$

$= a \cdot b + [-(a \cdot c)] = a \cdot b - a \cdot c$

5. $(b - c) \cdot a = [b + (-c)] \cdot a = b \cdot a + (-c) \cdot a$

$= b \cdot a + [(c \cdot a)] = b \cdot a - c \cdot a$

Example 35 Prove that the set, S = {0, 1, 2, 3, 4} is a ring with respect to the operation of addition and multiplication modulo 5.

Solution Forming the composition tables for the two operations $+_5$ and $\times_5$ as follows:

$+_5$	0	1	2	3	4
0	0	1	2	3	4
1	1	2	3	4	0
2	2	3	4	0	1
3	3	4	0	1	2
4	4	0	1	2	3

$\times_5$	0	1	2	3	4
0	0	0	0	0	0
1	0	1	2	3	4
2	0	2	4	1	3
3	0	3	1	4	2
4	0	4	3	2	1

It is easy to prove from the first table that the structure (S, $+_5$) is an abelian group.

From the second composition table we see that S is closed with respect to the operation '$\times_5$'. Since S being a set of numbers, we have

(*i*) $(a \times_5 b) \times_5 c = a \times_5 \{(b \times_5 c)\}$ for all $a, b, c \in$ S

(*ii*) $a \times_5 \{(b +_5 c)\} = (a \times_5 b) +_5 (a \times_5 c)$

and $(b +_5 c) \times_5 a = (b \times_5 a) +_5 (c \times_5 a)$, for all $a, b, c \in$ S.

Hence, S is a ring with respect to the given operations.

11.12.2 Type of Rings

Subring A subset S of a ring (R, +, ·) is a subring of R, if and only if S is closed under addition (+) and multiplication (·) and forms a ring under these operations.

For example, the set of even integers forms a subring of the ring of the set of integers because for any even integers a and b, $a + b$ and $a \cdot b$ are both even.

Criteria for subring

A non-empty subset S of a ring (R, +, ·) will form a subring of R, if and only if the following conditions are satisfied:

(*i*) If $a, b \in$ S, then $a + b \in$ S

(*ii*) If $a, b \in$ S, then $a \cdot b \in$ S

(*iii*) $0 \in$ S

(*iv*) $a \in$ S, then $-a \in$ S

Commutative ring A ring (R, +, $\cdot$) is said to be commutative ring if it is commutative under the multiplication operation, i.e.

$$a \cdot b = b \cdot a, \text{ for all } a, b \in R.$$

Ring with unity element A ring (R, +, $\cdot$) is said to be a ring with unity element if it has a multiplicativ identity, that is, if there exists an element in R denoted by 1 such that

$$1 \cdot a = a = a \cdot 1, \text{ for all } a \in R$$

For example, the ring of all integers is a ring with unity, 1 being the unity of the ring.

Ring with or without zero divisors

(*i*) If in a ring,. R, there exist elements a and b such that $a \cdot b = 0$ when neither $a = 0$ nor $b = 0$, then the elements a and b are called *zero divisors.*

(*ii*) A ring, R is said to be a ring *with zero divisors* if the product of no two non-zero elements a and b such that $a \cdot b = 0$.

(*iii*) A ring, R is said to be a ring *without zero divisors* if the product of no two non-zero elements of R is zero, i.e. if

$$a \cdot b = 0 \Rightarrow a = 0 \text{ or } b = 0 \text{ or both } a = 0, b = 0$$

Illustrations

1. The ring of integers is a ring without zero divisors because the product of two non-zero integers cannot be equal to the zero integer.
2. The ring of integers with addition and multiplication modulo 6 is a ring with zero divisors. For example, we have $2 \times_6 3 = 0$, so that the product of two non-zero elements 2 and 3 is zero.

Boolean ring A ring R is said to be a *Boolean ring* if all its elements are indempotent, i.e.

$$a^2 = a, \text{ for all } a \in R.$$

11.12.3 Cancellation Law in a Ring

If R is a ring, then it is an abelian group with respect to addition. For operation of addition, the cancellation law holds in all integers. Now we say that cancellation law holds in a ring R if

$$ab = ac \Rightarrow b = c \ (a \neq 0)$$

and $$ba = ca \Rightarrow b = c \ (a \neq 0),$$

for all $a, b, c \in R$.

Thus, in a ring with zero divisors, it is impossible to define a cancellation law, i.e. from the equation $ab = 0$ we cannot conclude that either a or b (or both) is zero.

Example 36 A ring, R is without zero divisors if the cancellation law holds in R.

Solution Suppose that R is without zero divisors. Then we have to show that the cancellation law holds in R. For this let, $a \cdot b = a \cdot c$ $(a \neq 0)$, for all $a, b, c \in R$. Then

$$a \cdot b = a \cdot c \Rightarrow a \cdot b - a \cdot c = 0 \text{ or } a \cdot (b - c) = 0$$

Since R is without zero divisors and $a \neq 0$, therefore, $b - c = 0$, i.e. $b = c$. Thus, the left cancellation law holds in R. Similarly, we can show that the right cancellation law also holds in R.

Conversely, let the cancellation law holds in R. Then to show that R is without zero divisors, let $a \cdot b = 0$ $(a \neq 0, b \neq 0)$ or $a \cdot b = a \cdot 0$ (because $a \cdot 0 = 0$). So by cancellation law, we have $b = 0$, which is a contradiction. Hence, R is without zero divisors.

Example 37 If (R, +, •) is a ring with unity, show that, for all $a \in$ R.

(*a*) $(-1) \times a = -a$ (*b*) $(-1)(-1) = 1$ [*UPTU, BTech, 2007*]

Solution Given that (R, +, •) is a ring with unity.

(*a*) $(-1)a = -(1.a) = -a$ because R is ring with unity, i.e. $1 \in$ R such that $1 \times x = x = x \times 1$, for all $x \in$ R.

(*b*) $(-1) \cdot (-1) = -[(-1) \cdot 1] = -[-(1.1)] = 1.1 = 1$ [as $(-a) \cdot b = -(a \cdot b)$ and $-(-a) = a$]

11.13 INTEGRAL DOMAIN

A commutative ring with unity containing no zero divisors is called an *integral domain*. It is usually denoted by (D, +, ∗). In other words, a ring is also integral domain, if it

(*i*) is commutative,

(*ii*) has unit element,

(*iii*) is without zero divisors.

Illustrations

1. The set of integers, I is an integral domain. It has been earlier proved that I is a commutative ring with unity. Also, I does not have zero divisors. We know that if $a, b \in$ I such that $a \cdot b = 0$, then either a or b (or both) is zero.

2. The ring, R of numbers : $a + \sqrt{2}\, b$; $a, b \in$ I is an integral domain. R is a commutative ring with unity because $1 + \sqrt{2} \,.\, 0 = 1$. It is also without zero divisors. For example, we have

$$(a + \sqrt{2}\, b)\,(c + \sqrt{2}\, d) = (ac + 2bd) + \sqrt{2}\,(bc + ad) = 0$$

only if $ac + 2bd = 0$ and $bc + ad = 0$ and this will happen only when either $a = 0$ and $b = 0$ or $c = 0$ and $d = 0$. Hence, $(a + \sqrt{2}\, b)\,(c + \sqrt{2}\, d) = 0$ implies that either $a + \sqrt{2}\, b = 0$ or $c + \sqrt{2}\, b = 0$.

11.14 FIELDS

A ring, R with at least two elements is called the *field* if the non-zero elements of R form an abelian group under multiplication, i.e. a ring R is a field if it has at least two elements and

(*i*) is commutative,

(*ii*) has a unit element,

(*iii*) is such that each non-zero element has a multiplicative inverse.

Illustrations

1. The ring (Q, +, ·) of rational numbers is a field since it is a commutative ring with unity and each non-zero element has a multiplicative inverse. The ring (R, +, ·) of real numbers and (C, +, ·) of complex numbers are also examples of fields.

2. Consider the set {0, 1} under the addition and multiplication modulo 2. It can be easily verified that $[\{0, 1\}, +_2, \times_2]$ is a commutative ring with unit element. The only non-zero element is 1 which is its inverse. Hence, it is a field.

Remarks

1. Every field is an integral domain but the converse is not true, i.e. every integral domain is not a field. For example, the ring of integers is an integral domain which is not a field since no element except 1 and –1 has a multiplicative inverse.

2. For a field 1 and 0 are distinct elements, i.e. $1 \neq 0$. If a is any non-zero element, then a^{-1} exists and is non-zero. Now a field has no zero divisors. Therefore, $1 \neq a^{-1} \cdot a \neq 0$.

Example 38 Prove the set of all real numbers of the form $a + b\sqrt{2}$, where a and b are real numbers forms a field under the operation of addition and multiplication.

Solution *(i) Closure property* Let $R = \{a + b\sqrt{2} : a, b \in Q\}$. If $a_1 + b_1\sqrt{2} \in R$ and $a_2 + b_2\sqrt{2} \in R$, then

(*a*) $(a_1 + b_1\sqrt{2}) + (a_2 + b_2\sqrt{2}) = [(a_1 + a_2) + (b_1 + b_2)\sqrt{2}] \in R$

(*b*) $(a_1 + b_1\sqrt{2})(a_2 + b_2\sqrt{2}) = [(a_1a_2 + 2b_1b_2) + (a_1b_2 + a_2b_1)\sqrt{2}] \in R$

Thus, R is closed under addition and multiplication.

(ii) Associativity Since all the elements of R are real numbers, it follows that associative laws hold for *addition* and *multiplication.*

(iii) Identity Since $0 \in Q$, it is easy to see that $0 + 0\sqrt{2}$ is the additive identity.

(iv) Inverse The additive inverse of $a + b\sqrt{2}$ is $(-a) + (-b)\sqrt{2}$.

(v) Multiplicative identity Since $(1 + 0\sqrt{2})(a + b\sqrt{2}) = a + b\sqrt{2}$, Therefore, $1 + 0\sqrt{2}$ is the multiplicative identity.

(vi) Multiplicative inverse Let $a + b\sqrt{2}$ be an element of the ring so that both a and b are not zero. Now

$$\frac{1}{a+b\sqrt{2}} = \frac{1}{a+b\sqrt{2}} \times \frac{a-b\sqrt{2}}{a-b\sqrt{2}} = \frac{a-b\sqrt{2}}{a^2-2b^2} = \frac{a}{a^2-2b^2} - \left(\frac{b}{a^2-2b^2}\right)\sqrt{2}$$

which is of the form $c + d\sqrt{2}$ where c, d are rational numbers. Hence, the multiplicative inverse of $a + b\sqrt{2}$ is

$$\frac{a}{a^2-2b^2} + \sqrt{2}\left[-\frac{b}{a^2-2b^2}\right]$$

(vii) Commutativity Since all the elements of R are real numbers, it follows that commutative law holds for *addition* and *multiplication.*

Hence, the given system is a field.

Example 39 Give an example of ring which is not a field.

Solution The set I = {. . ., – 3, 2, – 1, 0, 1, 2, 3, . . .} of integers is an example of ring which is not a field because multiplicative inverse is not there.

Self-practice Problems II

16. Define a group. When is it abelian? Show that set of all integers, positive or negative, including zero with additive binary operation is an infinite abelian group.

17. Prove that the set, Q of non-zero rational numbers forms a group under the operation of multiplication of real numbers.

18. Show that the set of natural numbers is not a group with respect to addition.

19. Find which of the following sets are groups under multiplication (mod 11).

(i) {1, 2, 3, 4, 5, 9} *(ii)* {1, 3, 5, 7, 8}.

20. Show that the real number of the form $a + b\sqrt{2}$, where a and b are rational numbers and $a_1 + b_2 \neq 0$, constitutes a group under multiplication.

21. Define an abelian group. Show that the set, G = {–1, 1} is a finite abelian group of order 2 under multiplication as operation.

22. Show that the set, G = $\{\ldots, 2^{-4}, 2^{-3}, 2^{-2}, 2^{-1}, 1, 2, 2^2, 2^3, 2^4, \ldots\}$ forms an infinite abelian group under multiplication.

23. Show that the following sets form groups under ordinary addition as operation.

(i) $\{a + b\sqrt{2}\ ;\ a, b \in Q\}$ *(ii)* $\{a + ib\ ;\ a, b \in Q\}$.

24. Is the set of integers, I = {. . ., –3, – 2, – 1, 0, 1, 2, 3, . . .} a group for the binary operation defined as $a * b = a - b$ for all $a, b \in$ I?

25. If a, b are any two elements of a group G, then show that $(ab)^2 = a^2b^2$, if and only if G is abelian.

26. Define a cyclic group. Show that the set G = [{1, 2, 3, 4}, ×] is a cyclic group.

27. Define a subgroup. If H is any subgroup of a group G, then show that H^{-1} = H. Is the converse true?

28. Let G be the abelian group of integers {. . ., – 3, – 2, – 1, 0, 1, 2, 3, . . .}. Then, verify that H is the subgroup of even integers {– 6, – 4, – 2, 0, 4, 6, . . .}.

29. Define a ring and give two examples.

30. Prove that the following operation tables define a ring. Is it an integral domain?

$+_{10}$	0	2	4	6	8
0	0	2	4	6	8
2	2	4	6	8	0
4	4	6	8	0	2
6	6	8	0	2	4
8	8	0	2	4	6

$\times_{10}$	0	2	4	6	8
0	0	0	0	0	0
2	0	4	8	2	6
4	0	8	6	4	8
6	0	2	4	6	8
8	0	6	2	8	4

31. Define a field. Prove in general that the ring I_p of integers modulo p, where p is prime, is a field.

32. Let R be the set of ordered pairs (a, b) of real numbers. Let addition and multiplication compositions be defined as

$$(a, b) + (c, d) = (a + c, b + d)$$
$$(a, b)\,(c, d) = (ac - bd, bc + ad)$$

Show that R is a field.

33. Prove that the following operation tables define a finite field.

+	*a*	*b*	*c*	*d*
a	*a*	*b*	*c*	*d*
b	*b*	*a*	*d*	*c*
c	*c*	*d*	*a*	*b*
d	*d*	*c*	*b*	*a*

×	*a*	*b*	*c*	*d*
a	*a*	*a*	*a*	*a*
b	*a*	*b*	*c*	*d*
c	*a*	*c*	*d*	*b*
d	*a*	*d*	*b*	*c*

Hints and Answers

16. See the text **19.** (*i*) Yes **21.** Yes

24. I is not group under the given operation.

26. See text for definition; G is cyclic and 2 is the generator of G.

27. Converse is not true.

29. (*i*) The set of all integers under addition and multiplication operations constitute a ring
(*ii*) The set R of real numbers under addition and multiplication operation constitute a ring.

30. It is an integral domain.

Chapter 12

Mathematical Logic

PREVIEW

The mathematic logic is widely used to represent the statements to communicate facts or information. A statement of proposition is a sentence that is either true or false. A formal logic helps in organizing our thinking which characterizes any reasoned activity such as scientific experiment, socio-political study. Thus allows us to concentrate on underlying reasoning being applied. The circuit logic which controls computer circuitry has close analog to the statement logic.

LEARNING OBJECTIVES

After studying this chapter, you should be able to

- develop knowledge of logical connectives and compound propositions.
- use the formal symbols of propositional logic
- understand basic logical operations
- find true value of an expression in propositional logic
- find converse, inverse and contra-positive statements
- use the formal symbols to predicate logic

12.1 INTRODUCTION

In this chapter, we will introduce certain rules and techniques of mathematics that deal with the method of logical reasoning to recognise valid logical arguments. A Greek philosopher and scientist Aristotle (381–322 DC) is said to be the first person to have studied logical reasoning. Logical reasoning, for example is applied by a software engineer to verify correctness of programs; by a lawer in the courtroom to defend his client; by a scientist to draw conclusions from experiments and so on. In this chapter, few of the basic ideas about the *logic* – the algebra of propositions (also called *proposition calculus* that deals statements with true and false and their analysis) and conclusions (also called *predicate calculus* that deals with the propositions containing variables) that are useful in designing circuits of digital computers, automata theory and computability, artificial intelligence, etc. will be discussed.

We use four different types of sentences: *declarative, interrogative, exclamatory* and *imperative* to express our ideas. But in mathematics, sentences that are *declarative* and it is possible to judge, whether such sentences are true or false but not both, are used to draw conclusions and/or to prove theorems. Such sentences are called *statements.* The value associated with the truth (or falsity) of a statement is called its *truth* (or falsity) *value.*

12.2 PROPOSITIONAL CALCULUS

Propositional calculus also called *statement calculus* is the branch of mathematics that is used to describe a logical system. A logical system or structure consists of: (*i*) a universe of statements, (*ii*) truth tables as axioms for the logical operators, and (*iii*) definitions that explain equivalence and implication of statements.

A proposition (or statement) is used to describe any mathematical structure. It is a declarative sentence that is either *true* or *false*, but not both. However, a statement becomes *predicate* when variables are assigned an appropriate type of values in a formula to drawn conclusions. Questions, exclamations and commands are not propositions.

The following statements are of declarative type:

(*i*) $\{2, 3\} \subseteq \{1, 2, 3, 4\}$

(*ii*) $a^2 + b^2 = (a + b)^2 - 2ab$, for all values of a and b

(*ii*) 9 is greater that 7.

(*iv*) The earth is round.

(*v*) The price of wheat in Delhi rose on 15 August 1998.

(*vi*) Jaipur is the capital of Rajasthan.

However, following are not the statements:

(*i*) What is the exchange rate of Indian rupees to US dollars?

(*ii*) What are you doing?

(*iii*) May God bless you?

(*iv*) Who are you?

(*v*) Lahore is in Bangladesh.

12.2.1 Propositional Variables and Constants

For convenience English alphabet: $p, q, r, s, \ldots$, etc., are used to represent simple statements and are called propositional variables. For example, we can write,

$$p = \text{It is raining}, \quad q = \text{It is cold.}$$

Similarly, two letters T and F that are used to represent true and false statement are called *propositional constants.*

12.2.2 Logical Connectives and Compound Proposition

Fundamental connectives Two or more simple propositions (or statements) can be combined by means of logical operators or connectives to form a single proposition called *compound proposition* (*composite* or *molecular*). However, a proposition consisting of only one propositional variable (or constant) is called *atomic* (*primary* or *primitive*) proposition.

The words and phrases (or symbols) that are used to form a compound proposition are called *connectives*. There are five logical connectives which are frequently used for this purpose. These are shown in Table 12.1.

Table 12.1 Logical Connectives

Symbol	*Connective Word*	*Name*
$\sim$	not	negation
$\wedge$	and	conjunction
$\vee$	or	disjunction
$\Rightarrow$	implies or If . . . then	implication or conditional
$\Leftrightarrow$	If and only if	equivalence or biconditional

If p and q are any two propositions, then $\sim q$, $p \wedge q$, $p \vee q$, $p \Rightarrow$, $p \Leftrightarrow q$ $\{\equiv (p \Rightarrow q) \wedge (q \Rightarrow p)\}$ are all propositions.

A compound proposition (or statement) which does not contain any connective is called a *prime proposition.* Few propositions are composed of several simple propositions and their true value is determined by the truth value of each simple propositions and the type of connectives used. The meaning of the connectives can be summarised in the tabular form known as *truth tables*.

The compound proposition can be constructed as: $p \wedge q \wedge r$, $p \wedge (q \Rightarrow r)$, etc. The precedence conventions are in the order given above so that $\sim p_1 \wedge p_2 \vee p_3 \Rightarrow p_4 \Leftrightarrow p_5$ is taken as: $((((\sim p_1) \wedge p_2) \vee p_3) \Rightarrow p_4) \Leftrightarrow p_5$. However, it is good to insert brackets even where it is unneccessary, if this improves readability.

Truth tables A truth table is the convenient way of summarising the truth values of logical statements. It indicates the truth values of compound propositions composed of several simple propositions. A truth table consists of columns and rows. The number of columns depend upon the numbers of sample propositions and connectives used to form a compound proposition. The number of rows in a truth table are found on the basis of simple propositions. For example, if there are two simple propositions, there will be $2^2 = 4$ rows. In general, for n simple propositions, the total number of rows will be 2^n.

Since each proposition (or statement) has two possible truth values, we assign letter T to a proposition when it is true and the letter F when it is false. The truth table is useful in

(*i*) finding out the validity of an equivalence relation between functions.

(*ii*) designing and testing the electronic circuits to perform a given operation based on certain relationship.

12.3 BASIC LOGICAL OPERATIONS

As mentioned earlier that there are three basic logical operations, namely, *conjunction*, *disjuction* and *negation*, which correspond to the English words *and*, *or* and *not*, respectively. These logical operations are discussed in detail below.

12.3.1 Conjunction

When two or more statements are joined by the connective, denoted by the symbol, $\wedge$, the compound statement so formed is called a *conjunction.* Let p and q be two statements. Then $p \wedge q$ form a statement which is true if and only if both p and q are true, and is false if p is false or q is false or both are false. The statement $p \wedge q$ is read as: *p and q*, and is called the *conjunction of p and q*.

For example, let p be the statement 'Delhi is the capital of India' and q be the statement '8 + 5 = 13'. Then $p \wedge q$ which is read as 'Delhi is the capital of India and 8 + 5 = 13' is true. If, however, q is the statement '8 + 5 = 12', then $p \wedge q$ is false even though p is true.

Truth values of the statement $p \wedge q$ in terms of truth values of p and q are given in the truth table shown in Table 12.2.

Table 12.2 Truth Table for $p \wedge q$

p	q	$p \wedge q$
T	T	T
T	F	F
F	T	F
F	F	F

From Table 12.2, the following observations may be noted:

(*i*) With two statements, there are $2^2 = 4$ possible combinations.

(*ii*) The basic truth values in the first two columns are same as binary digits if you place 1 for T and 0 for F.

Remark In logic we may join two totally unrelated statements by the connective '*and*'. For example, the statement $p \wedge q$: Raju is handsome and Ramesh is selfish, is obtained by joining two statements p: Raju is handsome and q: Ramesh is selfish, which are not related.

12.3.2 Disjunction

When two or more statements are joined by the connective *or* denoted by the symbol, $\vee$, the compound statement, so formed is called a *disjunction* of two statements p and q and is written as p or q or $p \vee q$.

The connective has two different meanings: *Inclusive Disjunction* and *Exclusive Disjunction.* In the inclusive sense the compound statement $p \vee q$ is true only when at least one of the statements is true. But in the exclusive sense the compound statement $p \vee q$ is true only when either p or q but not both, is true.

The exclusive disjunction is commonly called *inequivalence* (written as $p \not\equiv q$), since it defines a statement that is true precisely when p and q have opposite or equivalent truth values. Two statements p and q are logically equivalent ($\equiv$) if they have same truth values.

Illustrations

1. Let p : 5 is a positive integer, and

   q : $\sqrt{5}$ is a rational number.

   Then $p \vee q$: '5 is a positive integer' or '$\sqrt{5}$ is a rational number' is the true disjunction because p is true even though q is false.
2. Let p : Ganga is the mountain, and

   q : Delhi is the capital of England.

   Then $p \wedge q$: Ganga is the mountain or Delhi is the capital of England is false disjunction because both p and q are false.
3. Let p : XYZ company earned 20 per cent profit per share in 1998.

   q : XYZ company paid 12 per cent dividend per share in 1998.

   The inclusive disjunction of p and q is

   $p \vee q$: XYZ company earned 20 per cent profit per share in 1998 or XYZ company paid 12 per cent dividend per share in 1998 or both.

   The exclusive disjunction of p and q is

   $p \vee q$: XYZ company earned 20 per cent profit per share in 1998 or XYZ company paid 12 per cent dividend per share in 1998 but not both.
4. In the compound statement, p : This weekend I will go to Shimla or Nainital, the use of 'or' is not clear. Because, if 'or' means inclusive disjunction, then at least one or possibly both cities will be visited. If 'or' means exclusive disjunction, the only one city will be visited.

If p and q are two simple statements, then truth values of and $p \wedge q$ and $p \vee q$ are given in the truth table shown in Tables 12.3(*a*) and (*b*).

Table 12.3(*a*) Truth Table for $p \wedge q$

p	q	$p \wedge q$
T	T	T
F	T	F
T	F	F
F	F	T

Table 12.3(*b*) Truth Table for $p \vee q$

p	q	$p \vee q$
T	T	T
F	T	T
T	F	T
F	F	F

In table 12.3(*b*), it may be noted that the statement $p \vee q$ is true if either p or q or both are true. If both p and q are false, the statement $p \vee q$ is false. The statement $p \wedge q$ is true only when both p and q are either true or false.

12.3.3 Negation

If p is a statement, then *negation* (*or not*) of p, denoted by $\sim p$ is defined to form a statement that is true when p is false, and false when p is true. The negation of statement p is read as 'not p'.

In fact, ~ (not) is not a connective because it does connect two statements. However, ~ p is a statement if p is the statement.

Illustrations

1. If $p : x = 3$, then $\sim p : x \neq 3$.
2. If p : Sun is shinning, then ~ p : sun is not shinning

The truth values of ~ p relative to p are given in Table 12.4.

Table 12.4 Truth Table for Negation (~)

p	$\sim p$
T	F
F	T

Double negative is positive and can be justified from the truth Table 12.4.

Table 12.5 True Table for Double Negation [~ (~)]

p	$\sim p$	$\sim (\sim p)$
T	F	T
F	T	F

In Table 12.5 it may be noted that double negation of any statement gives the original statement.

Negation of compound statements When a compound statement is regreted, its logical connective changes from *and* to *or* and from *or* to *and*. For example,

$$\sim (p \wedge q) = \sim p \vee \sim q$$
$$\sim (p \vee q) = \sim p \wedge \sim q$$

The following relations are based on above laws:

(*a*) $\sim (p \wedge \sim q) = \sim p \vee \sim (\sim q) = \sim p \vee q$

(*b*) $\sim (\sim p \wedge q) = \sim (\sim q) \vee \sim q = p \vee \sim q$

(*c*) $\sim (\sim p \vee \sim q) = \sim (\sim p) \wedge \sim (\sim q) = p \wedge q$

Example 1 Let p be the statement 'South-West monsoon is very good this year' and q be the statement 'Rivers are rising'. Give the verbal translations of statement (*a*) and verify the statement (*b*).

(*a*) (*i*) $p \vee \sim q$ and (*ii*) $\sim (\sim p \vee \sim q)$.

(*b*) statement $x > 1$ is false if and only if $x^2 > 1$ is false, when x is a real number.

Solution (*a*) (*i*) The South-West monsoon is very good this year or the rivers are not rising.

(*ii*) It is not true that the South-West monsoon is not very good this year and the rivers are not rising. In other words, we can also say that the South-West monsoon is very good this year but the rivers are rising.

(*b*) The statement $x > 1 \Leftrightarrow x^2 > 1$ is false because if $x^2 > 1$ means $x > 1$ or $x < 1$.

Example 2 With the help of truth tables, prove that

(*a*) $\sim p \vee \sim q = \sim (p \wedge q)$ (*b*) $\sim (p \vee q) = \sim (p \wedge \sim q)$

(*c*) $p \vee q = \sim (\sim p \wedge \sim q)$ (*d*) $p \vee \sim q = (p \vee q) \wedge \sim (p \wedge q)$

(*e*) $\sim (p \wedge q) = \sim p \vee \sim q$

Solution (*a*)

Table 12.6 Truth Table

p (1)	q (2)	$\sim p$ (3)	$\sim q$ (4)	$\sim p \vee \sim q$ (5)	$p \wedge q$ (6)	$\sim (p \wedge q)$ (7)
T	T	F	F	F	F	F
T	F	F	T	T	F	T
F	T	T	F	T	F	T
F	F	T	T	T	F	T

Since values in Columns (5) and (7) are identical, therefore, $\sim p \vee \sim q = \sim (p \wedge q)$.

(*b*)

Table 12.7 Truth Table

p (1)	q (2)	$p \vee q$ (3)	$\sim (p \vee q)$ (4)	$\sim p$ (5)	$\sim q$ (6)	$\sim p \wedge \sim q$ (7)
T	T	T	F	F	F	F
T	F	T	F	F	T	F
F	T	T	F	T	F	F
F	F	F	T	T	T	T

Since values in Columns (4) and (7) are identical, therefore, $\sim (p \vee q) = \sim p \wedge \sim q$.

(*c*)

Table 12.8 Truth Table

p (1)	q (2)	$p \vee q$ (3)	$\sim p$ (4)	$\sim q$ (5)	$\sim p \wedge \sim q$ (6)	$\sim(\sim p \wedge \sim q)$ (7)
T	T	T	F	F	F	T
T	F	T	F	T	F	T
F	T	T	T	F	F	T
F	F	F	T	T	T	F

Since values in Columns (3) and (7) identical, therefore, $p \vee q = \sim (\sim p \wedge \sim q)$.

(*d*)

Table 12.9 Truth Table

p (1)	q (2)	$p \vee \sim q$ (3)	$p \vee q$ (4)	$p \wedge q$ (5)	$\sim (p \vee q)$ (6)	$(p \vee q) \wedge \sim (p \wedge q)$ (7)
T	T	F	T	T	F	F
T	F	T	T	F	T	T
F	T	T	T	F	T	T
F	F	F	F	F	T	F

Since values in Columns (3) and (7) are identical, therefore, $p \vee \sim q = (p \vee q) \wedge \sim (p \wedge q)$.

(*e*)

Table 12.10 Truth Table

p (1)	q (2)	$p \wedge q$ (3)	$\sim(p \wedge q)$ (4)	$\sim p$ (5)	$\sim q$ (6)	$\sim p \vee \sim q$ (7)
T	T	T	F	F	F	F
T	F	F	T	F	T	T
F	T	F	T	T	F	T
F	F	F	T	T	T	T

Since values in Columns (4) and (7) are same, therefore, result is proved.

12.3.4 DERIVED CONNECTIVES

NAND It is the *Negation after ANDing* of two statements. For example, if p and q are two statements, then NANDing of p and q denoted by $p \uparrow q$ is a false statement when both p and q are true, otherwise true. The truth values of $p \uparrow q$ are given in truth table as shown in Table 12.11.

Table 12.11 Truth Values for $p \uparrow q$

p	q	$p \uparrow q$
T	T	F
T	F	T
F	T	T
F	F	T

NOR or joint denial It is the *Negation after ORing* of two statements. For example, if p and q are two statements, then NORing of p and q, denoted by $p \downarrow q$ is a true statement when both p and q are false, otherwise false. The truth values of $p \downarrow q$ are given in truth table as shown in Table 12.12.

Table 12.12 Truth Values for $p \downarrow q$

p	q	$p \downarrow q$
T	T	F
T	F	F
F	T	F
F	F	T

XOR If p and q are two statements, then *XORing* of p and q denoted by $p \oplus q$ is a true statement when either p or q is true but not both and vice versa. The truth values of $p \oplus q$ are given in truth table as shown in Table 12.13.

Table 12.13 Truth Values of $p \oplus q$

p	q	$p \oplus q$
T	T	F
T	F	T
F	T	T
F	F	F

mple 3 Make truth tables for

(*a*) $(p \downarrow q) \wedge (p \downarrow r)$ (*b*) $p \uparrow q \uparrow r$ (*c*) $p \oplus q \oplus r$

Solution (*a*) Truth table is shown in Table 12.14

Table 12.14 Truth Values of $(p \downarrow q) \wedge (p \downarrow r)$

p	q	r	$p \downarrow q$	$p \downarrow r$	$(p \downarrow q) \wedge (p \downarrow r)$
T	T	T	F	F	F
T	T	F	F	F	F
T	F	T	F	F	F
T	F	F	F	F	F
F	T	T	F	F	F
F	T	F	F	T	F
F	F	T	T	F	F
F	F	T	T	T	T

(*b*) The truth table is shown in Table 12.15.

Table 12.15 Truth Values of $p \uparrow q \uparrow r$

p	q	r	$p \uparrow q$	$p \uparrow q \uparrow r$
T	T	T	F	T
T	T	F	F	T
T	F	T	T	F
T	F	F	T	T
F	T	T	T	F
F	T	F	T	T
F	F	T	T	F
F	F	F	T	T

(*c*) The truth table is shown in Table 12.16.

Table 12.16 Truth Values of $p \oplus q \oplus r$

p	q	r	$p \oplus q$	$p \oplus q \oplus r$
T	T	T	F	T
T	T	F	F	F
T	F	T	T	F
T	F	F	T	T
F	T	T	T	F
F	T	F	T	T
F	F	T	F	T
F	F	F	F	F

Example 4 If p and q are two statements, then show that the statement $(p \uparrow q) \oplus (p \uparrow q)$ is equivalent to $(p \vee q) \wedge (p \downarrow q)$.

Solution The equivalence of two compound statements is shown in the truth Table 12.17.

Table 12.17 Truth Table for Equivalence

p (1)	q (2)	$p \uparrow q$ (3)	$(p \uparrow q) \oplus (p \uparrow q)$ (4)	$p \vee q$ (5)	$p \downarrow q$ (6)	$(p \vee q) \wedge (p \downarrow q)$ (7)
T	T	F	F	T	F	F
T	F	T	F	T	F	F
F	T	T	F	T	F	F
F	F	T	F	F	T	F

Since values in Columns (4) and (7) are same, therefore two statements are equivalent.

Example 5 If p and q are two statements, then show that $p \oplus q$ is equivalent to $(p \wedge \sim q) \vee (\sim p \wedge q)$.

Solution The equivalence of two compound statements is shown in truth Table 12.18.

Table 12.18 Truth Table for Equivalence

p (1)	q (2)	$p \oplus q$ (3)	$\sim p$ (4)	$\sim q$ (5)	$p \wedge \sim q$ (6)	$\sim p \wedge q$ (7)	$(p \wedge \sim q) \vee (\sim p \wedge q)$ (8)
T	T	F	F	F	F	F	F
T	F	T	T	T	T	F	T
F	T	T	F	F	F	T	T
F	F	F	T	T	F	F	F

Since values in Columns (3) and (8) are same, therefore, two compound statements are equivalent.

Example 6 If p and q are two statements, then show that the statement $(p \oplus q) \vee (p \downarrow q)$ is equivalent to $p \uparrow q$.

Solution The equivalence of two compound statements is shown in truth Table 12.19.

Table 12.19 Truth Table for Equivalence

p (1)	q (2)	$p \oplus q$ (3)	$(p \downarrow q)$ (4)	$(p \oplus q) \vee (p \downarrow q)$ (5)	$p \uparrow q$ (6)
T	T	F	F	F	F
T	F	T	F	T	T
F	T	T	F	T	T
F	F	F	T	T	T

Since values in Columns (5) and (6) are same, therefore, two compound statements are equivalent.

12.4 STATEMENTS GENERATED BY A SET

Let $X = \{ p, q, r\}$ be a finite set of statements p, q and r. The following are two definitions for a compound statement generated by X.

(*i*) A compound statement generated by X is any valid combination of statements p, q and r in X with logical operators: conjunction, disjunction and negation.

(*ii*) (*a*) $p \in X$ be any arbitrary element. Then p is said to be the statement generated by X.

(*b*) If A and B are two statements generated by X, then $\sim$ A, A $\wedge$ B, A $\vee$ B, etc., are also statements.

As mentioned above, the precedence conventions for interpreting statements are in the order: *negation*, *conjunction* and *disjunction*. For example, instead of writting $(p \wedge q) \wedge r$, we may write $p \wedge q \vee r$. Some of the expressions and their parenthesized versions are:

(*i*) $((p \wedge (q)) \wedge r$ for $p \wedge q \wedge r$

(*ii*) $(\sim p) \vee (\sim r)$ for $\sim p \vee \sim r$

(*iii*) $\sim (\sim p)$ for $\sim \sim p$

The truth values of any statement, $c : (p \wedge q) \vee (\sim q \wedge r)$ generated by the combination of statements p, q are r with logical connectives can be obtained as shown in Table 12.20.

Table 12.20 Truth Table for $(p \wedge q) \vee (\sim q \wedge r)$

p (1)	q (2)	r (3)	$(p \wedge q)$ (4)	$\sim q$ (5)	$(\sim q \wedge r)$ (6)	$(p \wedge q) \vee (\sim q \wedge r)$ (7)
F	F	F	F	T	F	F
F	F	T	F	T	T	T
F	T	F	F	F	F	F
F	T	T	F	F	F	F
T	F	F	F	T	F	F
T	F	T	F	T	T	T
T	T	F	T	F	F	F
T	T	T	T	F	F	F

Remarks 1. Conditional ($\Rightarrow$) and biconditional ($\Leftrightarrow$) connectives have not been included in the definition of a statement generated by a set because these can be obtained from three main connectives: conjunction, disjunction and negation.

2. A compound statement in its expression generated by a set need not include its each element. For example, in Table 12.20 the truth value of the statement $\sim q \wedge r$ is same as the truth values of the statement $(p \wedge q) \vee (\sim q \wedge r)$ generated by p, q and r.

12.5 CONDITIONAL STATEMENTS

Statements such the following occur quite frequently in mathematics and computer science, and therefore, an understanding of their nature is very important.

(*i*) If you read, then you will pass the examination.

(*ii*) If you do your homework, then you will be allowed to play.

(*iii*) If 9 + 3 = 12, then 12 – 9 = 3.

(*iv*) If it rains, then there will be no picnic.

In all these statements, two simple statements are connected by *If . . ., then* to form a single statement In other words, all such statements can be restated to fit into the form *if* (*condition . . .*), *then* (*conclusion*). Therefore, a compound statement using the connective 'If . . ., then' is called a *conditional statement* or *implication.*

A conditional statement is used as a guarantee for the conclusion to be true provided condition is true. If p and q are two statements, then compound statement *if p, then q* denoted by $p \Rightarrow q$ is an implication, and the connective *if . . ., then . . .,* is the *conditional connective.*

The statement, p is called the *hypothesis* or *antecedent.* The hypothesis p is often a relational expression such as $x > 5$ and becomes a logical statement that has the truth value 0 or 1, depending on the value of the variable x. The statement q is called the *consequent* or *conclusion*, i.e. it is an 'executable statement'. While dealing with 'if p and then q', the computer executes q and on the condition that p is true. Otherwise computer goes to the next instruction in the program sequence. For example, for the decision structure 'if p then q else r'; q is executed when p is true and r is executed when p is false. The connective *if . . ., then* is denoted by $\rightarrow$ or $\Rightarrow$ and can be read as follows:

(*i*) p implies q

(*ii*) p is sufficient for q

(*iii*) p only if q

(*iv*) q is necessary for p

(*v*) q is consequence of p

Remark Reader should not confuse with $p \Rightarrow q$ and $q \Rightarrow p$. These two conditions are completely different. The statement $p \Rightarrow q$ is ture when either both p and q are true or false expect when p is true and q is false.

Truth Table for $p \Rightarrow q$

The truth values of $p \Rightarrow q$ in terms of the truth values of p and q are shown in Table 12.21.

Table 12.21 Truth Table for $p \Rightarrow q$

p	q	$p \Rightarrow q$
T	T	T
T	F	F
F	T	T
F	F	T

Example 7 Write each of the following as conditional sentences in the *If . . ., then* form:

(*i*) Freezing water expands.

(*ii*) A racer wins the race only if he runs fast.

(*iii*) Roses are vegetables if carrots are flowers.

Solution (*i*) If water freezes, then it expands.

(*ii*) If a racer wins the race, then he runs fast.

(*iii*) If carrots are flowers, then roses are vegetables.

12.6 CONVERSE, INVERSE AND CONTRAPOSITIVE STATEMENTS

12.6.1 Converse Statements

The converse of a given conditional statement, $p \Rightarrow q$ is a new statement formed by interchanging the *phypothesis, p* and the *conclusion, q* of the given conditional statement. For example, if $p \Rightarrow q$, then converse of it is $q \Rightarrow p$.

Illustration Consider the statements:

p : You are guilty, q : You are punished

The implication $p \Rightarrow q$ states that *If you are guilty, then you are punished.* The converse of this implication, $q \Rightarrow p$ states that *If you are punished, then you are guilty.*

Obviously, the statement: *all guilty persons are punished,* is not the same as the statement: *every one who is punished is also guilty.* The truth values for $p \Rightarrow q$ and its converse are gives in Table 12.22.

Table 12.22 Truth Table of Converse

p	q	$p \Rightarrow q$	$q \Rightarrow p$
T	T	T	T
T	F	F	T
F	T	T	F
F	F	T	T

12.6.2 Inverse Statements

The inverse of a given conditional statement, $p \Rightarrow q$ is a new statement whose hypothesis is the negation of the original hypothesis and whose conclusion is the negation of the original conclusion. For example, if $p \Rightarrow q$, then inverse of it is *if not p, then not q* and is written as $\sim p \Rightarrow \sim q$.

Illustration As mentioned above, the statement $p \Rightarrow q$ means that 'If you are guilty, then you are punished.' The inverse of this statement is $\sim p \Rightarrow \sim q$ which means that 'If you are not a guilty, then you are not punished.' The truth values for $p \Rightarrow q$ and its inverse are given in Table 12.23.

Table 12.23 Truth Table of Inverse

p	q	$\sim p$	$\sim q$	$p \Rightarrow q$	$\sim p \Rightarrow \sim q$
T	T	F	F	T	T
T	F	F	T	F	T
F	T	T	F	T	F
F	F	T	T	T	T

12.6.3 Contrapositive Statements

The contrapositive of a given conditional statement, $p \Rightarrow q$ is a new statement whose conclusion is the negation of the original hypothesis. For example, if $p \Rightarrow q$, then contrapositive of it is *if not q, then not p,* and is written as $\sim q \Rightarrow \sim p$.

Illustration As mentioned above, the statement, $p \Rightarrow q$ means that 'If you are guilty, then you are punished'. The contrapositive of this statement is $\sim q \Rightarrow \sim p$, which means that 'If you are not punished then you are not guilty.' The truth values for $p \Rightarrow q$ and its contrapositive are given is Table 12.24.

Table 12.24 Truth Table of Contrapositive

p	q	$\sim p$	$\sim q$	$p \Rightarrow q$	$\sim q \Rightarrow \sim p$
T	T	F	F	T	T
T	F	F	T	F	F
F	T	T	F	T	T
F	F	T	T	T	T

In Table 12.24 it may noted that the entries under $p \Rightarrow q$ and contrapositive are same. Thus, the statement $p \Rightarrow q$ is logically equivalent to its contrapositive '$\sim q \Rightarrow \sim p$'. Also from Tables 12.22 and 12.23, it may be noted that the converse $q \Rightarrow p$ of conditional statement $p \Rightarrow q$ is logically equivalent to its inverse $\sim p \Rightarrow \sim q$. Hence, we conclude that

(*i*) $p \Rightarrow q \equiv \sim q \Rightarrow \sim p$

(*ii*) $q \Rightarrow p \equiv \sim p \Rightarrow \sim q$

In other words, given any simple conditional statement, we may obtain the following:

Statement : If $p \Rightarrow q$ is an implication

Converse : then converse of $p \Rightarrow q$ is the implication $q \Rightarrow p$, but $q \Rightarrow p \neq p \Rightarrow q$

Inverse : then inverse of $p \Rightarrow q$ is the implication $\sim p \Rightarrow \sim q$, but $\sim p \Rightarrow \sim q \neq p \Rightarrow q$

Contrapositive : then contrapositive of $p \Rightarrow q$ is the implication, $\sim q \Rightarrow \sim p$

The summary of the above definitions is given in truth Table 12.25.

Table 12.25 Summary of Converse, Inverse and Contrapositive

p	q	*Conditional* $p \Rightarrow q$	*Converse* $q \Rightarrow p$	*Inverse* $\sim p \Rightarrow \sim q$	*Contrapositive* $\sim q \Rightarrow \sim p$
T	T	T	T	T	T
T	F	F	T	T	F
F	T	T	F	F	T
F	F	T	T	T	T

Example 8 Prove that $p \Rightarrow q \equiv \sim q \Rightarrow \sim p$.

Solution Truth values for the given result is shown in Table 12.26.

Table 12.26 $p \Rightarrow q \equiv \sim q \Rightarrow \sim p$

p (1)	q (2)	$p \Rightarrow q$ (3)	$\sim q$ (4)	$\sim p$ (5)	$\sim q \Rightarrow \sim p$ (6)
T	T	T	F	F	T
T	F	F	T	F	F
F	T	T	F	T	T
F	F	T	T	T	T

Since entries in Columns (3) and (6) are identical, therefore, $p \Rightarrow q = \sim q \Rightarrow \sim p$.

Example 9 Prove by means of a truth table that $\sim (p \Rightarrow q) = p \wedge \sim q$.

Solution Truth values for the given result is shown in Table 12.27.

Table 12.27 $\sim (p \Rightarrow q) \equiv p \wedge \sim q$

p (1)	q (2)	$p \to q$ (3)	$\sim (p \to q)$ (4)	$\sim q$ (5)	$p \wedge \sim q$ (6)
T	T	T	F	F	F
T	F	F	T	T	T
F	T	T	F	F	F
F	F	T	F	T	F

Since entries in Columns (4) and (6) are identical, therefore, $\sim (p \Rightarrow q) = p \wedge \sim q$.

Example 10 Prove that conditional operation distributes over conjunction, i.e.

$$p \Rightarrow (q \wedge r) \equiv (p \Rightarrow q) \wedge (p \Rightarrow r)$$

Solution Truth values for the given result is shown in Table 12.28.

Table 12.28 $(q \wedge r) \equiv (p \Rightarrow q) \wedge (p \Rightarrow r)$

p (1)	q (2)	r (3)	$q \wedge r$ (4)	$p \Rightarrow (q \wedge r)$ (5)	$p \Rightarrow q$ (6)	$p \Rightarrow r$ (7)	$(p \Rightarrow q) \wedge (p \Rightarrow r)$ (8)
T	T	T	T	T	T	T	T
T	F	T	F	F	F	T	F
F	T	T	T	T	T	T	T
F	F	T	F	T	T	T	T
T	T	F	F	F	T	F	F
T	F	F	F	F	F	F	F
F	T	F	F	T	T	T	T
F	F	F	F	T	T	T	T

Since entries in Columns (5) and (8) are identical, therefore, the result is true.

Example 11 If p, q and r are three statements, then show that $[(p \Rightarrow q) \wedge (q \Rightarrow r)] \Rightarrow (p \Rightarrow r)$. Also write its truth set. *[UPTU, MCA, 2008]*

Solution Truth values for the given result is shown in Table 12.29.

Table 12.29 $[(p \Rightarrow q) \wedge (q \Rightarrow r)] \Rightarrow (p \Rightarrow r)$

p (1)	q (2)	r (3)	$p \Rightarrow q$ (4)	$q \Rightarrow r$ (5)	$(p \Rightarrow q) \wedge (q \Rightarrow r)$ (6)	$p \Rightarrow r$ (7)	$(6) \Rightarrow (7)$ (8)
T	T	T	T	T	T	T	T
T	T	F	T	F	F	F	T
T	F	T	F	T	T	F	T
T	F	F	F	T	F	F	T
F	T	T	T	T	T	T	T
F	T	F	T	F	T	F	T
F	F	T	T	T	T	T	T
F	F	F	T	T	T	T	T

All T entries in Column (8) indicate that the given statement is true, i.e.

$$(p \Rightarrow q) \wedge (q \Rightarrow r) \Rightarrow (p \Rightarrow r)$$

and its truth set is: {TTT, TTF, TFT, FTT, TFF, FTF, FFT, FFF}.

Example 12 A firm of chartered accountants makes the following declaration: *An article clerk passing the final C.A. examination in the first attempt will be awarded a prize of Rs 100.* Five clerks P, Q, R, S and T appeared for the first time from the firm and only P and Q could pass. The firm awards prizes not only to them but to R and S also. Is this action logically justified? T claims the prize comparing himself with R and S but the firm refuses. Is the refusal logically justified? How should the statement be worded so that only P and Q will be entitled for the prize?

Solution Let p be the statement that 'a clerk passes C.A. examination in the first attempt, and q be the statement 'awarded a prize of Rs 100.' Obviously, $p \Rightarrow q$. But it does not exclude other because the statement be true even when p is false and q is true as shown in truth Table 12.30.

Table 12.30 Truth Table for $p \Rightarrow q = \sim p \vee q$

p (1)	q (2)	$p \Rightarrow q$ (3)	$\sim p$ (4)	$\sim p \vee q$ (5)
T	T	T	F	T
T	F	F	F	F
F	T	T	T	T
F	F	T	T	T

In Table 12.30, Columns (3) and (5) are identical. Columns (1) and (2) indicate that only in one case when the statement p is true, the statement q is false, i.e. the prize is not given to those p and q who have passed the examination in first attempt but having given to others who have not passed. Therefore, the statement should be worded as: *Only those who pass the examination in the first attempt shall be awarded the prize, i.e.* $p \Leftrightarrow q$.

Example 13 Test the validity of the following arguments. 'If my brother passes the examination of the Institute of Chartered Accountants in the first attempt, I will give him a valuable prize. Either he passes the examination or I was out of station. I did not give my brother a prize this time although he passed the examination. Therefore, I was out of station.'

Solution If we use p for the statement *brother passes the examination* and q for the statement *I will give him a prize* and r for the statement *I was out of station*, then symbolically the given argument is written as: $(p \Rightarrow q) \wedge (p \vee r) \wedge (\sim q) \Rightarrow r$. The truth values for argument are shown in Table 12.31.

Table 12.31 $(p \Rightarrow q) \wedge (p \vee r) \wedge (-q) \Rightarrow r$

p (1)	q (2)	r (3)	$p \Rightarrow q$ (4)	$\sim q$ (5)	$p \vee r$ (6)	(4) ∧ (6) (7)	(7) ∧ (5) (8)	(8) ⇒ r (9)
T	T	T	T	F	T	T	F	T
T	T	F	T	F	T	T	F	T
T	F	T	F	T	T	F	F	T
T	F	F	F	T	T	F	F	T
F	T	T	T	F	T	T	F	T
F	T	F	T	F	F	F	F	T
F	F	T	T	T	T	T	T	T
F	F	F	T	T	F	F	F	T

Since entries in Column (9) are Ts only, therefore, the given statement is true or valid.

Example 14 Giving reasons in support of your answers decide if the two compound statements given below are equivalent statements:

(*a*) If Ram is 21 years old, then he has right to vote.

(*b*) Ram is not 21 years old, then he has a right to vote.

Solution Let p be the statement *Ram is 21 years old* and q is the statement, *he has a right to vote.* Then statement (*a*) is of the form $p \Rightarrow q$ and statement (*b*) is of the form $\sim p \vee q$. Now in order to show that the two compound statements are equivalent, we have to show that biconditional formed by them is true for all assignments of p and q. Truth values for the compound statement are shown in Table 12.32.

Table 12.32 Truth Table for $(p \Rightarrow q) \Leftrightarrow (\sim p \vee q)$

p (1)	$\sim p$ (2)	q (3)	$p \Rightarrow q$ (4)	$\sim p \vee q$ (5)	$p \Rightarrow q \Leftrightarrow \sim p \vee q$ (6)
T	F	T	T	T	T
T	F	F	F	F	T
F	T	T	T	T	T
F	T	F	T	T	T

Since entries in Column (6) are only Ts, therefore, the biconditional formed by p and q is true for all possible assignments. Hence, the two given compound statements are equivalent.

Example 15 Prove that $\sim (p \Rightarrow q) \equiv p \cap \sim q$.

Solution The truth values for the given result is shown in Table 12.33.

Table 12.33 $\sim (p \Rightarrow q) \equiv p \cap - q$

p (1)	q (2)	$p \to q$ (3)	$\sim (p \to q)$ (4)	$\sim q$ (5)	$p \cap \sim q$ (6)
T	T	T	F	F	F
T	F	F	T	T	T
F	T	T	F	F	F
F	F	T	F	T	F

Since entries in Columns (4) and (6) are identical, therefore, the given statement is true.

Example 16 Construct truth tables for the following statements:

(*a*) $(p \Rightarrow p) \Rightarrow (p \Rightarrow \sim p)$

(*b*) $(p \Rightarrow p) \vee (p \Rightarrow \sim p)$

(*c*) $(\sim q \Rightarrow \sim p) \Rightarrow (p \Rightarrow q)$

(*d*) $(p \Rightarrow (q \Rightarrow r)) \Rightarrow ((p \Rightarrow q) \Rightarrow (p \Rightarrow r))$

Solution (*a*) Truth values for the statement are shown in Table 12.34.

Table 12.34 $(p \Rightarrow p) \Rightarrow (p \Rightarrow \sim p)$

p	$\sim p$	$p \Rightarrow p$	$p \Rightarrow \sim p$	$(p \Rightarrow p) \Rightarrow (p \Rightarrow \sim p)$
T	F	T	F	F
F	T	T	T	T

(*b*) Truth values for the statement are shown in Table 12.35.

Table 12.35 $(p \Rightarrow p) \vee (p \Rightarrow \sim p)$

p	$\sim p$	$p \Rightarrow p$	$p \Rightarrow \sim p$	$p \Rightarrow p \vee (p \Rightarrow \sim p)$
T	F	T	F	T
F	T	T	T	T

(*c*) Truth values for the statement are shown in Table 12.36.

Table 12.36 $(\sim q \Rightarrow - p) \Rightarrow (p \Rightarrow q)$

p	q	$\sim p$	$\sim q$	$\sim q \Rightarrow \sim p$	$p \Rightarrow q$	$(\sim q \Rightarrow \sim p) \Rightarrow (p \Rightarrow q)$
T	T	F	F	T	T	T
T	F	F	T	F	F	T
F	T	T	F	T	T	T
F	F	T	T	T	T	T

(*d*) Truth values for the statement are shown in Table 12.37.

Table 12.37 $((p \Rightarrow (q \Rightarrow r)) \Rightarrow ((p \Rightarrow q) \Rightarrow (p \Rightarrow r))$

p	q	r	$q \Rightarrow r$	$p \Rightarrow q$	$p \Rightarrow r$	$p \Rightarrow (q \Rightarrow r)$	$((p \Rightarrow q) \Rightarrow (p \Rightarrow r))$
T	T	T	T	T	T	T	T
T	T	F	F	T	F	F	F
T	F	F	T	F	F	T	T
T	F	T	T	F	T	T	T
F	T	T	T	T	T	T	T
F	T	F	F	T	T	T	T
F	F	T	T	T	T	T	T
F	F	F	T	T	T	T	T

12.7 BICONDITIONAL STATEMENTS

The connective 'if and only if' is called *biconditional connective* and is denoted by $\Leftrightarrow$. If p and q are two statements for which $p \Rightarrow q$ and $q \Rightarrow p$, then we say *p if and only if q* and we write $p \leftrightarrow q$ or $p \Leftrightarrow q$ and the two statements are said to be logically equivalent and is read as *p implies and implied by q* or *p is necessary and sufficient condition for q* or *p iff q*. In other words, if p and q are statements, then the compound statement means $(p \Rightarrow q) \wedge (q \Rightarrow p)$. This statement is *true if p and q are true or both are false and is false when one of the statements is true and other is false.* The essential features of such relationship are:

(*i*) p is equivalent to q

(*ii*) p implies q and conversely

(*iii*) if p, then q and conversely

(*iv*) p if and only if q

(*v*) q if and only if p

(*vi*) p is an essential condition for q

(*vii*) q is an essential condition for p.

The connective 'if and only if' is often used in situations where an implication is used when the intension actually calls for a biconditional. For example, consider two sentences

$p \Rightarrow q$: If you do your homework then, you will be allowed to watch TV.

$q \Rightarrow q$: You will be allowed to watch TV only if you do your homework.

The truth values of $p \Leftrightarrow q$ are given in Tables 12.38 and 12.39.

Table 12.38 Truth Table for $(p \Rightarrow q) \wedge (q \Rightarrow p)$

p	q	$p \Rightarrow q$	$q \Rightarrow p$	$(p \Leftrightarrow q)$	$(p \Rightarrow q) \wedge (q \Rightarrow p)$
T	T	T	T	T	T
T	F	F	T	F	F
F	T	T	F	F	F
F	F	T	T	T	T

Thus, the compound statement $(p \Rightarrow q) \wedge (q \Rightarrow p)$ is true only when both p and q are either true or false. Moreover, $p \Leftrightarrow q \equiv (p \vee q) \Rightarrow (p \wedge q)$ as shown in Table 12.39.

Table 12.39 Truth Table $p \Leftrightarrow q \equiv (p \vee q) \Rightarrow (p \wedge q)$

p	q	$p \Leftrightarrow q$	$p \vee q$	$p \wedge q$	$(p \vee q) \Rightarrow (p \wedge q)$
T	T	T	T	T	T
T	F	F	T	F	F
F	T	F	T	F	F
F	F	T	F	F	T

12.7.1 NEGATION OF BICONDITIONAL STATEMENT

The negation of the biconditional statement is defined as follows:

$$\sim (p \Leftrightarrow q) = \sim p \Leftrightarrow q = p \Leftrightarrow \sim q$$

The validity of these statements can be proved by the Truth Table 12.40.

Table 12.40 $\sim (p \Leftrightarrow q) = \sim p \Leftrightarrow q = p \Leftrightarrow \sim q$

p	q	$p \Leftrightarrow q$	$\sim (p \Leftrightarrow q)$	$\sim p$	$\sim p \Leftrightarrow q$	$\sim q$	$p \Leftrightarrow \sim q$
T	T	T	F	F	F	F	F
T	F	F	T	F	T	T	T
F	T	F	T	T	T	F	T
F	F	T	F	T	F	T	F

Example 17 If p and q are two statements, then prove that

(a) $\sim (p \wedge q) \Leftrightarrow \sim p \vee \sim q$

(b) $p \Leftrightarrow q \equiv (p \Rightarrow q) \wedge (q \Rightarrow p)$

(c) $(\sim p \vee q) \wedge (\sim q \vee p) \equiv p \Leftrightarrow q$

(d) $\sim p \vee (p \Leftrightarrow q) \equiv p \Rightarrow q$

(e) $p \Leftrightarrow (\sim p \vee \sim q)$

(f) $\sim (p \Leftrightarrow q) \equiv p \Leftrightarrow \sim q$ or $\sim p \equiv q$.

Solution (a)

Table 12.41 Truth Table for $\sim (p \wedge q) \Leftrightarrow \sim p \vee \sim q$

p (1)	q (2)	$p \wedge q$ (3)	$\sim (p \wedge q)$ (4)	$\sim p$ (5)	$\sim q$ (6)	$\sim p \vee \sim q$ (7)
T	T	T	F	F	F	F
T	F	F	T	F	T	T
F	T	F	T	T	F	T
F	F	F	T	T	T	T

Since entries in Columns (4) and (7) are identical, therefore, the result is true.

(*b*) For proof see Tables 12.38 and 12.39.

(*c*) **Table 12.42** Truth Table for $(\sim p \vee q) \wedge (\sim q \vee p) \equiv p \Leftrightarrow q$

p (1)	q (2)	$\sim p$ (3)	$\sim p \vee q$ (4)	$\sim q$ (5)	$\sim q \vee p$ (6)	$(\sim p \vee q) \wedge (\sim q \vee p)$ (7)	$p \Leftrightarrow q$ (8)
T	T	F	T	F	T	T	T
T	F	F	F	T	T	F	F
F	T	T	T	F	F	F	F
F	F	T	T	T	T	T	T

Since entries in Columns (7) and (8) are identical, therefore, the result is true.

(*d*) **Table 12.43** Truth Table for $\sim p \vee (p \Leftrightarrow q)$

p (1)	q (2)	$\sim p$ (3)	$p \Rightarrow q$ (4)	$p \Leftrightarrow q$ (5)	$\sim p \vee (p \Leftrightarrow q)$ (6)
T	T	F	T	T	T
T	F	F	F	F	F
F	T	T	T	F	T
F	F	T	T	T	T

Since entries in Columns (4) and (6) are identical, therefore, the result is true.

(*e*) **Table 12.44** Truth Table for $p \Leftrightarrow (\sim p \vee \sim q)$

p (1)	q (2)	$\sim p$ (3)	$\sim q$ (4)	$\sim p \vee \sim q$ (5)	$p \Leftrightarrow (\sim p \vee \sim q)$ (6)
T	T	F	F	F	T
T	F	F	T	T	T
F	T	T	F	T	F
F	F	T	T	T	F

Since entries in Columns (1) and (6) are identical, therefore, the results is true.

(*f*) **Table 12.45** Truth Table for $\sim (p \Leftrightarrow q) \equiv p \Leftrightarrow \sim q$ or $\sim p \Leftrightarrow q$

p (1)	q (2)	$\sim p$ (3)	$\sim q$ (4)	$p \Leftrightarrow q$ (5)	$\sim (p \Leftrightarrow q)$ (6)	$p \Leftrightarrow \sim q$ (7)	$\sim p \Leftrightarrow q$ (8)
T	T	F	F	T	F	F	F
T	F	F	T	F	T	T	T
F	T	T	F	F	T	T	T
F	F	T	T	T	F	F	F

Since entries in Columns (6), (7) and (8) are identical, therefore, results is true.

Example 18 Make a truth table to show that $p \Rightarrow q$ is equivalent to $\sim p \vee q$, where p and q are statements.

Solution Truth values for $p \Rightarrow q \equiv \sim p \vee q$ is shown in Table 12.46.

Table 12.46 $p \Rightarrow q \equiv \sim p \vee q$

p (1)	q (2)	$p \Rightarrow q$ (3)	$\sim p$ (4)	$\sim p \vee q$ (5)
T	T	T	F	T
T	F	F	F	F
F	T	T	T	T
F	F	T	T	T

Since entries in Columns (3) and (5) are identical, therefore, the result is true.

Example 19 If $l = \sim (p \vee q)$ and $m = \sim p \wedge \sim q$, write the truth table for $l \Leftrightarrow m$.

Solution Truth values for $l \Leftrightarrow m$ are shown in Table 12.47.

Table 12.47 $\sim (p \vee q) \equiv \sim p \wedge - q$

p (1)	q (2)	$p \vee q$ (3)	$l = \sim (p \vee q)$ (4)	$\sim p$ (5)	$\sim q$ (6)	$m = \sim p \wedge \sim q$ (7)	$l \Leftrightarrow m$ (8)
T	T	T	F	F	F	F	T
T	F	T	F	F	T	F	T
F	T	T	F	T	F	F	T
F	F	F	T	T	T	T	T

Example 20 Construct the truth table for the following and write the truth set

$$p \Rightarrow [(p \vee r) \wedge \sim (p \Leftrightarrow \sim r)].$$

Solution Truth values for the given statement is shown in Table 12.48.

Table 12.48 $p \Rightarrow [(p \vee r) \wedge \sim (p \Leftrightarrow \sim r)]$

p (1)	q (2)	r (3)	$\sim r$ (4)	$q \vee r$ (5)	$p \Leftrightarrow \sim r$ (6)	$\sim (p \Leftrightarrow \sim r)$ (7)	$(q \vee r) \wedge \sim (p \Leftrightarrow r)$ (8)	$p \Rightarrow [(q \vee r) \wedge \sim (p \Leftrightarrow \sim r)$ (9)
T	T	T	F	T	F	T	T	T
T	T	F	T	T	T	F	F	F
T	F	T	F	T	F	T	T	T
T	F	F	T	F	T	F	F	F
F	T	T	F	T	T	F	F	T
F	T	F	T	T	F	T	T	T
F	F	T	F	T	T	F	F	T
F	F	F	T	F	F	T	F	T

The required truth set is: {TTT, TFT, FTT, FTF, FFT, FFF}. The elements of the set are the values of p, q and r, respectively for T values in Column (8).

12.8 TAUTOLOGIES

Earlier in this chapter we have seen how compound statements are obtained from given simple statements by using connectives. Ordinarily when a compound statement is written, the truth of such statement is not known unless we know the truth or falsity of the component statements. There are certain compound statements that always are true regardless of the truth or falsity of component statements. Compound statements which have this characteristic are called *tautologies*. In other words, *tautology is a compound statement* if it is true for all truth value assignments for its component statements. Obviously, the truth table of a tautology will contain only T entries in the last column.

Importance of tautology Tautology helps conclude some statement from some given statements. For example, if the statement $p \Rightarrow q$ is a tautology, then it is easy to conclude the truth of q from the truth of p. Thus with the help of tautology we move from some given statement to some concluding statement in a step-by-step manner which is justified within the framework of mathematical logic.

Fxample 21 Show that the statement $(p \wedge q) \Rightarrow q$ is a tautology.

Solution The truth table for the given statement is shown in Table 12.49.

Table 12.49 $(p \wedge q) \Rightarrow q$

p	q	$p \wedge q$	$p \wedge q \Rightarrow q$
T	T	T	T
T	F	F	T
F	T	F	T
F	F	F	T

Since the statement $(p \wedge q) \Rightarrow q$ has its truth value T for all its entries in the truth table. Hence, it is a tautology.

Example 22 Show that the statement $p \vee \sim p$ is a tautology.

Solution The truth table for the given statement is shown in Table 12.50.

Table 12.50 $p \vee \sim p$

p	$\sim p$	$p \vee \sim p$
T	F	T
F	T	T

Since the statement $p \vee \sim p$ has its truth value T for all its entries in the table, therefore, it is tautology. This is called the *law of exclusive middle,* i.e. either p is true or it is false. There is no middle possibility.

12.9 CONTRADICTION

If a compound statement is false for all true value assignments for its component statements, then it is called a *contradiction,* i.e. a compound statement is said to be contradicting its truth value if it is false (F) for all its entries in the truth table. If a statement is a contradiction, then its negation will be a tautology.

Example 23 Show that the statement $p \wedge \sim p$ is a tautology or contradiction.

Solution The truth table for $p \wedge \sim p$ is shown in Table 12.51.

Table 12.51 $p \wedge \sim p$

p	$\sim p$	$p \wedge \sim p$
T	F	F
F	T	F

Since the statement $p \wedge \sim p$ has its false value F for all its entries in the truth table. Hence, it is a tautology.

Example 24 Establish $\sim (p \wedge q) \Leftrightarrow (\sim p \vee \sim q)$ as a tautology.

Solution The truth table for the given statement is shown in Table 12.52.

Table 12.52 $\sim (p \wedge q) \Leftrightarrow (\sim p \vee \sim q)$

p (1)	q (2)	$p \wedge q$ (3)	$\sim (p \wedge q)$ (4)	$\sim p$ (5)	$\sim q$ (6)	$\sim p \vee \sim q$ (7)	$\sim (p \wedge q) \Leftrightarrow (\sim p \vee \sim q)$ (8)
T	T	T	F	F	F	F	T
T	F	F	T	F	T	T	T
F	T	F	T	T	F	T	T
F	F	F	T	T	T	T	T

Since Column (8) contains only Ts, therefore, the given proposition is a tautology.

Example 25 Show that the statement $(p \wedge q) \Rightarrow (p \vee q)$ is a tautology, but $(p \vee q) \Rightarrow (p \wedge q)$ is not.

Solution Truth table for $(p \vee q) \Rightarrow (p \wedge q)$ is shown in Table 12.53.

Table 12.53 $(p \vee q) \Rightarrow (p \wedge q)$

p	q	$p \vee q$	$p \wedge q$	$(p \vee q) \Rightarrow (p \wedge q)$
T	T	T	T	T
T	F	T	F	F
F	T	T	F	F
F	F	F	F	T

Since, the given statement does not have all entries as T values in the last column of the truth table, therefore it is not a tautology.

Example 26 By means of a truth table, show that the statement $p \vee \sim (p \wedge q)$ is a tautology.

Solution The truth table for the given statement is shown in Table 12.54.

Table 12.54 $p \vee \sim (p \wedge q)$

p	q	$p \wedge q$	$\sim (p \wedge q)$	$p \vee \sim (p \wedge q)$
T	T			
T	F	T		
T	F	F	T	T
F	T	F	T	T
F	F	F	T	T

Since the given statement has its truth values T for all its entries in the last column of the truth table, therefore, it is a tautology.

Example 27 Show that $(p \wedge q) \wedge \sim (p \vee q)$ is a fallacy or contradiction.

Solution Truth table for the given statement is shown in Table 12.55.

Table 12.55 $(p \wedge q) \wedge \sim (p \vee q)$

p (1)	q (2)	$p \wedge q$ (3)	$p \vee q$ (4)	$\sim (p \vee q)$ (5)	$(p \wedge q) \wedge \sim (p \vee q)$ (6)
T	T	T	T	F	F
T	F	F	T	F	F
F	T	F	T	F	F
F	F	F	F	T	F

Since the statement has its false values (F) for all its entries in the last column of the truth table, therefore, it is a fallacy.

12.10 CONTINGENCY

Contingency is a compound statement whose truth or falsity depends upon the truth values of its component statements.

Example 28 Show that the following statements are contingency.

(*a*) $p \Rightarrow (p \Rightarrow q)$ (*b*) $(p \wedge \sim q) \vee (\sim p \wedge q)$

(*c*) $\sim (p \vee q) \wedge (\sim p \vee \sim q)$ (*d*) $(p \Rightarrow (q \wedge r)) \Rightarrow \sim (p \Rightarrow q)$

Solution (*a*) Truth table for the given statement is shown in Table 12.56.

Table 12.56 $p \Rightarrow (p \Rightarrow q)$

p	q	$p \Rightarrow q$	$p \Rightarrow (p \Rightarrow q)$
T	T	T	T
T	F	F	F
F	T	T	T
F	F	T	T

Since entries in the last column of the truth table depend on statements p, q and $p \Rightarrow q$, therefore, given statement is a contingency.

(*b*) Truth table for the given statement is shown in Table 12.57.

Table 12.57 $(p \wedge \sim q) \vee (\sim p \wedge q)$

p	q	$\sim p$	$\sim q$	$p \wedge \sim q$	$\sim p \wedge q$	$(p \wedge \sim q) \vee (\sim p \wedge q)$
T	T	F	F	F	F	F
T	F	F	T	T	F	T
F	T	T	F	F	T	T
F	F	T	T	F	F	F

Since entries in the last column of the truth table depend on component statements associated with this compound statement, therefore, it is a contingency.

(*c*) Truth table for the given statement is shown in Table 12.58.

Table 12.58 $\sim (p \wedge q) \vee (\sim p \wedge \sim q)$

p	q	$\sim p$	$\sim q$	$p \wedge q$	$\sim (p \wedge q)$	$\sim p \vee \sim q$	$\sim (p \wedge q) \vee (\sim p \wedge \sim q)$
T	T	F	F	T	F	F	F
T	F	F	T	F	T	T	T
F	T	T	F	F	T	T	T
F	F	T	T	F	T	T	T

Since entries in the last column of the truth table depend on component statements associated with this compound statement, therefore, it is a contingency.

(*d*) Truth table for the given statement is shown in Table 12.59.

Table 12.59 $((p \Rightarrow (q \wedge r)) \Rightarrow \sim (p \Rightarrow q)$

p	q	r	$q \wedge r$	$p \Rightarrow (q \wedge r)$	$p \Rightarrow q$	$\sim (p \Rightarrow q)$	$((p \Rightarrow (q \wedge r)) \Rightarrow \sim (p \Rightarrow q)$
T	T	T	T	T	T	F	F
T	T	F	F	F	T	F	T
T	F	F	F	F	F	T	T
T	F	T	F	F	F	T	T
F	T	T	T	T	T	F	F
F	F	T	F	T	T	F	F
F	T	F	F	T	T	F	F
F	F	F	F	T	T	F	F

Since entries in the least column of the truth table depend on component statements associated with this compound statement, therefore, it is a contingency.

12.11 ARGUMENT

The knowledge of tautology and implication can now be used to describe a valid argument. Terms such as *theorem, premises, conclusion* used in the definition of an 'argument' are defined as follows:

- *Theorem:* It is a true statement (or proposition) derived from the axioms of a mathematical structure or system.
- *Conclusion:* It is the statement that is asserted on the basis of other proposition or propositions.
- *Premise (or hypotheses):* It is a statement (or proposition) which is assumed to be true for accepting a conclusion.

An argument (also called *proof*) is an assertion that a finite number of statements (or propositions) $p_1, p_2, \ldots, p_n$ called the *premises (or hypotheses)* yields (has a consequence) another statement q called *conclusion*. In other words, if each of $p_1, p_2, \ldots, p_n$ is a true statement, then an implication of the form:

$$p_1 \wedge p_2 \wedge p_2 \ldots \wedge p_n \Rightarrow q$$

is also true, where q may be a compound statement and hence, we say that q logically follows from $p_1, p_2, \ldots, p_n$ and q is called the conclusion. Obviously, the mathematical proof of such implication does not show that q is true, but it shows that q is true only if the p_is are all true.

Validity of an argument An argument is valid if and only if the conjunction of the premises implies the conclusion, i.e. if premises are all true, then conclusion must also be true. However, a valid argument does not necessarily lead to a true conclusion. Moreover, if any of $p_1, p_2, \ldots, p_n$ is false, then no matter what truth value q has, the implication $p_1 \wedge p_2 \wedge p_3 \wedge \ldots \wedge p_n \Rightarrow q$ is true. Consequently, if each p_i carries a truth value 1 and find that q also has the value 1, then implication $p_1 \wedge p_2 \wedge p_3 \wedge \ldots \wedge p_n \Rightarrow q$ is tautology and the argument is valid.

Method to test the validity of an argument

1. Construct a truth table showing the truth values of all premises (p_is) and conclusion (q).
2. Find rows (called critical rows) in which all the premises are true. In case no such row is found, the argument is invalid.
3. In each critical row, determine whether the conclusion is also true. If yes, then the argument is valid. Otherwise invalid. However, if at least one critical row contains false conclusion, then the argument is invalid.

An argument is called *invalid* (also called *fallacy*) if it is not a valid argument. For example, a student who attempts a problem and makes several errors but by chance arrives at the right answer, is an example of an invalid argument which implies a true conclusion.

Fundamental rules Few elementary rules of conclusion are as under:

(*i*) Modus ponens or Rule of detachment

$$\begin{array}{ll} & p \Rightarrow q \\ & p \quad \text{(Premises)} \\ \hline \therefore & q \quad \text{Conclusion} \end{array}$$

(*ii*) Modus tollens

$$\begin{array}{ll} & \sim q \\ & p \Rightarrow q \\ \hline \therefore & \sim p \end{array}$$

(*iii*) Hypothetical syllogism

$$\begin{array}{ll} & p \Rightarrow q \text{ and } q \Rightarrow r \\ \hline \therefore & p \Rightarrow r \end{array}$$

(*iv*) Constructive dilemma

$$\begin{array}{ll} & (p \Rightarrow q) \wedge (r \Rightarrow s) \\ & p \wedge r \\ \hline \therefore & q \vee s \end{array}$$

(*v*) Disjunctive syllogisms

$$\begin{array}{ll} & p \wedge q \\ & \sim p \\ \hline \therefore & q \end{array}$$

(*vi*) Absorption

$$\begin{array}{ll} & p \Rightarrow q \\ \hline \therefore & p \Rightarrow (p \wedge q). \end{array}$$

Table 12.60 contains brief summary of 'rules of inference' discussed above based on Demorgan's laws and law of contraposition.

Table 12.60 Rules of Inference

Name	*Rule of Inference*	*Tautological Form*
Addition	p / $\therefore p \vee q$	$p \Rightarrow p \vee q$
Simplification	$p \wedge q$ / $\therefore p$	$(p \wedge q) \Rightarrow p$
Conjunction	p, q / $\therefore p \wedge q$	$[(p) \wedge (q)] \Rightarrow (p \wedge q)$
Modus ponens	$p \Rightarrow q$, p / $\therefore q$	$[(p \Rightarrow q) \wedge p] \Rightarrow q$
Modus tollens	$p \Rightarrow q$, $\sim q$ / $\therefore \sim p$	$[(p \Rightarrow q) \wedge \sim q] \Rightarrow \sim p$
Hypothatical syllogism	$p \Rightarrow q$ and $q \Rightarrow r$ / $\therefore P \Rightarrow r$	$[(p \Rightarrow q) \wedge (q \Rightarrow r)] \Rightarrow (p \Rightarrow r)$
Constructive dilemma	$(p \Rightarrow q) \wedge (r \Rightarrow s)$, $p \wedge r$ / $\therefore q \vee s$	$[(p \Rightarrow q) \wedge (r \Rightarrow s) \wedge (p \wedge r)] \Rightarrow (q \vee s)$
Disjunction dilemma	$p \wedge q$, $\sim p$ / $\therefore q$	$[(p \wedge q) \wedge (\sim p)] \Rightarrow q$
Destructive dilemma	$p \Rightarrow q \wedge (r \Rightarrow s)$, $\sim q \vee \sim s$ / $\therefore \sim p \vee \sim r$	$[(p \Rightarrow q) \wedge (r \Rightarrow s) \wedge (\sim q \vee \sim s)] \Rightarrow (\sim p \vee \sim r)$
Absorption	$p \Rightarrow q$ / $\therefore p \Rightarrow (p \wedge q)$	$[(p \Rightarrow q) \Rightarrow (p \Rightarrow (p \wedge q)]$

Example 29 Examine the validity of following arguments

(*a*) $\dfrac{p \wedge q}{\therefore q}$ (*b*) $p \vee q$, $\dfrac{\sim q}{\therefore p}$ (*c*) $p \Rightarrow q$, $\dfrac{q \Rightarrow p}{\therefore p \vee q}$

(*d*) $p \Rightarrow q$, $\dfrac{\sim q}{\therefore \sim p}$ (*e*) $p \vee q$, $p \Rightarrow \sim q$, $\dfrac{p \Rightarrow r}{\therefore r}$

Solution To examine the validity of the given arguments, we construct the truth table for the premises and conclusion as given below.

(a) Conclusion Presises

p	q	$p \wedge q$
T	T	T ← Critical row
T	F	F
F	T	F
F	F	F

The given argument is valid since critical row contains true conclusion.

(*b*) Premises Conclusion

p	q	$\sim q$	$p \vee q$	p
T	T	F	T	T
T	F	T	T	T
F	T	F	T	F
F	F	T	F	F

Since critical row 2 contain true conclusion, the argument is valid.

(*c*) Premises Conclusion

p	q	$p \Rightarrow q$	$q \Rightarrow p$	$p \vee q$
T	T	T	T	T
T	F	F	T	T
F	T	T	F	T
F	F	T	T	F

Since critical row 4 does not contain true conclusion, the argument is invalid.

(*d*) Premises Conclusion

p	q	$\sim p$	$\sim q$	$p \Rightarrow q$	$\sim p$
T	T	F	F	T	F
T	F	F	T	F	F
F	T	T	F	T	T
F	F	T	T	T	T

Since the critical row 3 contains true conclusion, the argument is valid.

(*e*)

				Premises			Conclusion
p	q	r	$\sim p$	$p \vee q$	$p \Rightarrow \sim q$	$p \Rightarrow r$	r
T	T	T	F	T	F	T	T
T	T	F	F	T	F	F	F
T	F	T	F	T	T	T	T
T	F	F	F	T	T	F	F
F	T	T	T	T	T	T	T
F	T	F	T	T	T	T	F
F	F	T	T	F	T	T	T
F	F	F	T	F	T	T	F

Since one of the critical rows (i.e. row 6) do not contain true conclusion, the argument is invalid.

Example 30 Prove the validity of the following argument. If a man is bachelor, he is unhappy. If a man is unhappy, he dies young. Therefore, bachelors die young.

Solution Let

p : Man is bachelor

q : Man is unhappy

r : Man dies young

Consider the premises that the arguments: p, $p \Rightarrow q$, $q \Rightarrow r$ are true.

1. p Premises (given)
2. $p \Rightarrow q$ Premise (given)
3. $q \Rightarrow r$ Premise (given)
4. $p \Rightarrow r$ Hypothetical syllogism using 2 and 3 (conclusion)

Hence, the argument is valid.

12.12 METHODS OF PROOF

So far we were concerned only with the various truth values that could be assigned to the implication 'if p, then q' for all possible truth values of p and q. In this section, we shall show that the implication $p \Rightarrow q$ is true when p is true. Of course, if q is false whenever p is true, then the implication $p \Rightarrow q$ is false. Two proof techniques called *direct and indirect methods* will be discussed in this section.

12.12.1 Direct Proof

If a chain of premises $p_1, p_2, \ldots, p_n$ is given, then we need to arrive at a conclusion q, i.e. $p_1 \wedge p_2 \wedge p_3 \wedge \ldots \wedge p_n \Rightarrow q$, using 'direct proof' method. It is described below without proof.

Example 31 Let the following statements be true.

- It is snowing.
- If it is warm, then it is not snowing.
- If it is not warm, then I cannot go for swimming.

Show that the statement 'I cannot go for swimming' is a true statement

Solution Let p, q and r represent the statements.

p : it is snowing
q : it is warm
r : I can go for swimming

Consider premises that the statements

$$p, \quad q \Rightarrow \sim p \quad \text{and} \quad \sim q \Rightarrow \sim r$$

are true. Then we shall prove that the statement $\sim r$ is true.

Since $q = \sim p$ is true, its contrapositive $p = \sim r$ is also true. Since $p \Rightarrow \sim q$ and $\sim q \Rightarrow \sim r$, then as per the law of syllogisms $p \Rightarrow \sim r$. Since statement p is true, therefore, the statement $\sim r$: I cannot go for swimming, must be true.

Example 32 Let the following statements be true.

- If I enjoy studying, then I will study.
- I will do my homework or I will not study.
- I will not do my homework.

Show that the statement 'I do not enjoy studying' is a true statement.

Solution Let p, q and r represent the statements.

p : I enjoy study
q : I will study
r : I will do my homework

Consider premises that the statements

$$p \Rightarrow q, \quad \sim q \vee r \quad \text{and} \sim r$$

are true. Then we shall prove that the statement $\sim p$ is true.

Since $\sim r$ is true, then r is false. Also either r or $\sim q$ is true. Thus, $\sim q$ is true. Since $p \Rightarrow q$, therefore, $\sim q \Rightarrow \sim p$ is true and hence, the statement $\sim p$: I do not enjoy study, must be true.

12.12.2 Indirect Proof

In this method, the conclusion of a theorem along with the statements of other premises is proved with contradiction. For example, to prove the theorem, $p_1 \wedge p_2 \wedge \ldots \wedge p_n \Rightarrow q$, we start with the assumption that $\sim q$ is true and then proving that this assumption adding with the statements of other premises of the theorem, leads to a logical contradiction. In other words, the indirect method is based on the equivalence $(p_1 \wedge p_2 \wedge \ldots \wedge p_n) \Rightarrow q \Leftrightarrow (p_1 \wedge p_2 \wedge \ldots \wedge p_n) \wedge (\sim q)$. To prove this result, we start with $\sim q$ (negation of conclusion) of the theorem and then combine this assumption with the statement of other premises. If the contradiction can be implied from the premise $(p_1 \wedge p_2 \wedge \ldots \wedge p_n) \wedge (\sim q)$, then at least one of the statements $p_1, p_2, \ldots, p_n, \sim q$ must be false. That is, if all p_is are true, then $\sim q$ must be false, so q must be true. Thus q follows from $p_1, p_2, \ldots, p_n$.

Example 33 Let the following statements be true.

- If I am lazy, then I do not study.
- I study or I enjoy myself.
- I do not enjoy myself.

Show that the statement 'I am not lazy' is a true statement.

Solution Let p, q and r represent the statements

p : I am lazy

q : I study

r : I enjoy myself

Consider premises that the statements

$$p \Rightarrow \sim q, \quad q \vee r \text{ and } \sim r$$

are true. Then we shall prove that the statement $\sim p$ is true.

Assuming that p is true. Since p is true and $p \Rightarrow \sim q$, then $\sim q$ is true. Thus, q is false. Also, r is false. But either q or r is true. This is not possible This contradiction implies that the assumption that p is true is false and hence $\sim p$ is true.

Example 34 Show that $5 + \sqrt{2}$ is an irrational number.

Solution Let $5 + \sqrt{2} = r$ be a rational number. Then $\sqrt{2} = r - 5$.

Since r and 5 are rational numbers, therefore, $r - 5$ is also a rational number. This implies that $\sqrt{2}$ is a rational number, which is a contradiction because $\sqrt{2}$ is an irrational number. Thus, the assumption that $5 + \sqrt{2}$ must be false and hence, the given number is irrational.

12.13 EQUIVALENCE AND IMPLICATION

If two statements p and q have the same truth values in every possible case, then such statements are said to be *logically equivalent* or simply *equivalent.* Logical equivalence of statements p and q is denoted by $p \equiv q$.

12.13.1 Algebra of Statements (Propositions)

The mathematical structure [statements, $\wedge$, $\vee$, $\sim$] has many properties that are used for regulating manipulations on statements which are used in finding out new relations and equivalence among some of them. The operations for statements (or propositions) have the following properties:

Idempotent law According to this law the given statement does not change by conjunction with another statement in certain situations such as:

(a) $p \vee p \equiv p$ (b) $p \wedge p \equiv p$,

For example, in the ordinary algebra $a + b = a \equiv b = 0$, i.e. b is the additive identity. Similarly, $ab = a \equiv b = 1$, i.e. b is multiplicative identity.

Associative law According to this law a compound statement consists of three statements remains unchanged with the change in the position of three statements with the same connective. That is,

(a) $p \wedge (q \wedge r) \equiv (p \wedge q) \wedge r$ (b) $p \vee (q \vee r) \equiv (p \vee q) \vee r$

Commutative law According to this law order is irrelevant. That is,

(a) $p \vee q \equiv q \vee p$ (b) $p \wedge q \equiv q \wedge p$

Laws of disjunctive

(a) $p \Rightarrow (p \vee q)$ (b) $q \Rightarrow (p \vee q)$

(c) $(p \vee q) \wedge \sim p \Rightarrow q$ and $(p \vee q) \wedge \sim q \Rightarrow p$

Laws of conjunctive

(a) $(p \wedge q) \Rightarrow p$ (b) $(p \wedge q) \Rightarrow q$

Laws of negation

(*a*) $\sim(\sim p) \equiv p$ (*b*) $\sim(p \vee q) \equiv (\sim p) \wedge (\sim q)$

(*c*) $\sim(p \wedge q) \equiv (\sim p) \vee (\sim q)$

Law of contrapositive According to this law if *p* is a statement, then *q* if and only if negation of *q*, then negation of *p*. That is,

$$(p \Rightarrow q) \wedge \sim q \Rightarrow \sim p \text{ or } (p \Rightarrow q) \equiv (\sim q \Rightarrow \sim p)$$

DeMorgan's Laws According to this law a statement remains unchanged if we change *and* ($\wedge$) by *or* ($\vee$) and *or* ($\vee$) by *and* ($\wedge$) provided that the inverse of the constituent statement exists, i.e.

(*a*) $\sim(p \wedge q) \equiv (\sim p) \vee (\sim q)$ (*b*) $\sim(p \vee q) \equiv (\sim p) \wedge (\sim q)$

Distributive laws According to this law if connective *or* distributes over *and* from the left, then this is left distributive. Similarly, *or* distributes over *and* from the right.

(*a*) $p \wedge (q \vee r) \equiv (p \wedge q) \vee (p \wedge r)$
$p \vee (q \wedge r) \equiv (p \vee q) \wedge (p \vee r)$

(*b*) $(p \wedge q) \vee r \equiv (p \vee r) \wedge (q \vee r)$
$(p \vee q) \wedge r \equiv (p \wedge r) \vee (q \wedge r)$

Identity law In ordinary algebra 1 and 0 are multiplicative and additive identities, respectively. Similarly, in Boolean algebra identity elements are denoted by T (tautology) and F (fallacy) such that

(*a*) $p \vee T \equiv T$ (*b*) $p \wedge T \equiv p$

(*c*) $p \vee F \equiv p$ (*d*) $p \wedge F \equiv F$.

Absorption laws

(*a*) $p \vee (p \wedge q) \equiv p$ (*b*) $p \wedge (p \vee q) \equiv p$

Law of complement

(*a*) $p \vee \sim p \equiv T$ (*b*) $p \wedge \sim p \equiv F$

(*c*) $\sim(\sim p) \equiv p$ (*d*) $\sim T \equiv F$ and $\sim F \equiv T$.

The implication operation for statements have the following properties:

1. $(p \Rightarrow q) \equiv ((\sim p) \vee q)$
2. $(p \Rightarrow q) \equiv ((\sim q) \Rightarrow \sim p)$
3. $(p \Leftrightarrow q) \equiv ((p \Rightarrow q) \wedge (q \Rightarrow p))$
4. $\sim(p \Rightarrow q) \equiv (p \wedge \sim q)$
5. $\sim(p \Leftrightarrow q) \equiv ((p \wedge \sim q) \vee (q \wedge \sim p))$.

Example 35 Show that $p \vee (q \wedge r) \equiv (p \vee q) \wedge (p \vee r)$.

Solution The truth table for the given statement is shown in Table 12.61.

Table 12.61 $p \vee (q \wedge r) \equiv (p \vee q) \wedge (p \vee r)$

p (1)	q (2)	r (3)	$q \wedge r$ (4)	$p \vee (q \wedge r)$ (5)	$(p \vee q)$ (6)	$p \vee r$ (7)	$(p \vee q) \wedge (p \vee r)$ (8)
T	T	T	T	T	T	T	T
T	T	F	F	T	T	T	T
T	F	T	F	T	T	T	T
T	F	F	F	T	T	T	T
F	T	T	T	T	T	T	T
F	T	F	F	F	T	F	F
F	F	T	F	F	F	T	F
F	F	F	F	F	F	F	F

Since entries in Columns (5) and (8) are identical, therefore, two statements are logically equivalent.

Example 36 Using the truth table prove that

(*a*) $p \vee (q \vee r) \equiv (p \vee q) \vee r$ (*b*) $p \wedge (q \wedge r) \equiv (p \wedge q) \wedge r$

Solution (*a*) Truth table for the given statement is shown in Table 12.62.

Table 12.62 $p \vee (q \vee r) \equiv (p \vee q) \vee r$

p (1)	q (2)	r (3)	$q \vee r$ (4)	$p \vee (q \vee r)$ (5)	$p \vee q$ (6)	$(p \vee q) \vee r$ (7)
T	T	T	T	T	T	T
T	F	T	T	T	T	T
F	T	T	T	T	T	T
F	F	T	T	T	F	T
T	T	F	T	T	T	T
T	F	F	F	T	T	T
F	T	F	T	T	T	T
F	F	F	F	F	F	F

Since Columns (5) and (7) are identical, therefore, two statements are logically equivalent.

(*b*) Truth table for the given statement is shown in Table 12.63.

Table 12.63 $p \wedge (q \wedge r) \equiv (p \wedge q) \wedge r$

p (1)	q (2)	r (3)	$q \wedge r$ (4)	$p \wedge (q \wedge r)$ (5)	$p \wedge q$ (6)	$(p \wedge q) \wedge r$ (7)
T	T	T	T	T	T	T
T	F	T	F	F	F	F
F	T	T	T	F	F	F
F	F	T	F	F	F	F
T	T	F	F	F	T	F
T	F	F	F	F	F	F
F	T	F	F	F	F	F
F	F	F	F	F	F	F

Since entries in Columns (5) and (7) are identical, therefore, two statements are logically equivalent.

Example 37 If p, q and r are any three statements, then using the truth table prove that,

(*a*) $p \vee (q \wedge r) \equiv (p \vee q) \wedge (p \vee r)$

(*b*) $p \wedge (q \vee r) \equiv (p \wedge q) \vee (p \wedge r)$

(*c*) $(p \wedge q) \vee r \equiv (p \vee r) \wedge (q \vee r)$

Solution (*a*) Truth values for the result is shown Table 12.64.

Table 12.64 $p \vee (q \wedge r) \equiv (p \vee q) \wedge (p \vee r)$

p (1)	q (2)	r (3)	$q \wedge r$ (4)	$p \vee (q \wedge r)$ (5)	$(p \vee q)$ (6)	$p \vee r$ (7)	$(p \vee q) \wedge (p \vee r)$ (8)
T	T	T	T	T	T	T	T
T	F	T	F	T	T	T	T
F	T	T	T	T	T	T	T
F	F	T	F	F	F	T	F
T	T	F	F	T	T	T	T
T	F	F	F	T	T	T	T
F	T	F	F	F	T	F	F
F	F	F	F	F	F	F	F

Since Columns (5) and (8) are identical, therefore, two statements are logically equivalent.

(*b*) Truth table for the result is shown in Table 12.65.

Table 12.65 $p \wedge (q \vee r) \equiv (p \wedge q) \vee (p \wedge r)$

p (1)	q (2)	r (3)	$q \vee r$ (4)	$p \wedge (p \vee r)$ (5)	$(p \wedge q)$ (6)	$p \wedge r$ (7)	$(p \wedge q) \vee (p \wedge r)$ (8)
T	T	T	T	T	T	T	T
T	F	T	T	T	F	T	T
F	T	T	T	F	F	F	F
F	F	T	T	F	F	F	F
T	T	F	T	T	T	F	T
T	F	F	F	F	F	F	F
F	T	F	T	F	F	F	F
F	F	F	F	F	F	F	F

Since Columns (5) and (8) are identical, therefore, two statements are logically equivalent.

(*c*) Truth table for the result is shown in Table 12.66.

Table 12.66 $(p \wedge q) \vee r \equiv (p \vee r) \wedge (q \vee r)$

p (1)	q (2)	r (3)	$p \wedge q$ (4)	$(p \wedge q) \vee r$ (5)	$p \vee r$ (6)	$q \vee r$ (7)	$(p \vee r) \wedge (q \vee r)$ (8)
T	T	T	T	T	T	T	T
T	T	F	T	T	T	T	T
T	F	T	F	T	T	T	T
T	F	F	F	F	T	F	F
F	T	T	F	T	T	T	T
F	T	F	F	F	F	T	F
F	F	T	F	T	T	T	T
F	F	F	F	F	F	F	F

Since Columns (5) and (8) are identical, therefore, two statements are logically equivalent.

12.14 PREDICATE CALCULUS

The predicate calculus is limited to the structure of sentences in terms of component sentences and does not breakdown a sentence into parts for most purposes. For the *conclusion* that requires an analysis of sentence structure, we introduce three logical notions: *terms*, *predicates* and *quantifiers*. These logical notions are helpful in symbolising everyday and/or mathematical language in such a way that make possible an analysis of an argument.

Illustrations

1. The statement such as

   (*a*) the first president of India

   But if we let

   x = Dr Rajendra Prasad

   then we may write

   x = the first president of India

   and (*b*) the number x such that for all real numbers y, $xy = y$ is descriptive.

   1 = the real number x such that for all real numbers y, $xy = y$.

   The individual variables and constants either in the form of proper names or descriptions are classified as *terms*.

2. A *predicate* is a word or words, in a sentence which express the nature of the subject. For example, in the sentence

   (*a*) Every rational number is a real number

   Here '*is a real number*' is the predicate. But if this sentence is written as

   (*b*) For every x, if x is a rational number, then x is a real number.

   Here '*x is a rational number*' is new predicate.

Other examples of predicates which express the nature of the subject in a sentence are: *is black*, *is less than*, *divides*, *is the sum of*, etc. In mathematical logic, the word *predicate* is supplemented by including a variable as a place holder for the intended subject. For example, in the above sentence 'x is a rational number' becomes a *statement function* (also called *proposition function*) in the sense that for each value of x, a statement gets generated. Also, the statement such as *Ram loves* is not a predicate in grammar, but if x is added as a place holder for the object (of Ram's affections), which yields 'Ram loves x', then the result is a statement function.

12.15 QUANTIFIERS

A declarative sentence which contains one or more variables, and is not a statement, but becomes a statement when the variables in it are replaced by certain allowable choices is called an *open statement*. Here 'certain allowable choices' constitute the *universe* or *universe of discourse* for the variable values in the open statement. For example, the open statement

$p(x)$: The number $x + 4 \leq 10$ is an even integer. Then negative of $p(x)$, i.e. $\sim p(x)$ may be read as:

$\sim p(x)$: The number x is an integer less than 10.

Similarly, if any open statement contains two variables, then such statements are denoted either by $p(x, y)$ or $r(x, y)$ or $q(x, y)$, etc. For example,

$p(x, y)$: The numbers $x + 2$, $x - y$ and $x + 2y$ are even integers.

When we make certain substitutions for x and y for the set of integers (i.e. universe in this case), we get the open statements

$p(7)$: The number 7 is less than 10, (a false statement).

$\sim p(15)$: The number 15 is not less than 10, (a true statement).

$p(6, 4)$: The numbers $6 + 2 = 8$, $6 - 4 = 2$ and $6 + 8 = 14$ are even integers (a true statement).

Thus some substitutions result in true statements and others in false statements. Hence, we make the following true statements:

(*i*) For some x, $p(x)$

(*ii*) For some $x, y, p(x, y)$

Consequently the statements 'For some x, $p(x)$' and 'For some x, $\sim p(x)$' are both true. However, the open statement $\sim p(x)$ is the negation of the open statement $p(x)$. Similar results are also true for open statements involving two or more variables.

The phrases (or words) of the type 'For at least' 'For every . . .', 'Such that . . .', 'There exists an . . .' etc. are said to quantify the open statements. Above results are due to two types of quantifiers: *Universal* and *Existential quantifiers.*

12.15.1 Types of Quantifiers

Universal quantifier The symbol $\forall$ (for all) is used to represent universal quantifier and is read in either of the following ways:

- $\forall x \in A : p(x)$
- $p(x), \forall x \in A$

Thus, the open statement 'For some x $p(x)$' become $\forall x\, p(x)$ in symbolic form.

Similarly, the phrase '*For every x, y*' or '*For any x, y*' is denoted by $\forall x \forall y$ or $\forall x, y$. Open statements such as

$p(x)$: The number x is an even integer.

$g(x)$: The number $2x$ is an even integer.

can be changed into the quantified statements as:

- $\forall x\, p(x)$, a false statement
- $\forall x\, q(x)$, a true statement

The variable x in an open statement is called its *free variable* because as x varies over the universe for an open statement the truth or false value of the statement may vary.

Existential quantifier The symbol $\exists$ (there exist) is used to represent existential quantifier and is read in either of the following ways:

- $\exists x \in A : p(x)$
- $\exists x \in A \;\; p(x)$
- $\exists x\, p(x)$
- $p(x)$ is true for every $x \in A$

Thus for an open statement '*For some x, p (x)* becomes $\exists x\, p(x)$ symbolic form. Similarly, in symbolic form the phrase 'For some x and y $p(x, y)$' is expressed as $\exists x\, \exists y\, p(x, y)$ or $\exists x, y\, p(x, y)$.

In general, the following logical equivalence holds.

1. (*a*) $\forall x \forall y\, p(x, y) \Leftrightarrow \forall y \forall x\, p(x, y)$ (Commutative property)

   (*b*) $\forall xy\, p(x, y) \Leftrightarrow \forall yx\, p(x, y)$

2. $\forall x, y, z\, p(x, y, z) \Leftrightarrow \forall y, x, z\, p(x, y, z) \Leftrightarrow \forall x, z, y\, p(x, y, z)$ (Associative property)
3. $\exists x, y\, p(x, y), \Leftrightarrow \exists y, x\, p(x, y)$
4. $\forall x \exists y\, p(x, y) \neq \exists y \forall x\, p(x, y)$.

For example, the statement $p(x) : x + y = 15$ is read as 'There exist an integer y so that for all integer x, $x + y = 15$'. This statement is false because for any integer y, the only value that x can have is $17 - y$ to satisfy the equation $x + y = 17$.

Example 38 Given the following open statements

$p(x) : x > 0$ $\quad$ $q(x)$: x is odd $\quad$ $r(x)$: x is a perfect square

$s(x)$: x is divisible by 3 $\quad$ $t(x)$: x is divisible by 2.

Write the following statements in symbolic form

(*i*) At least one integer is odd.

(*ii*) There exists a positive integer that is odd.

(*iii*) If x is odd, then x is not divisible by 2.

(*iv*) No odd integer is divisible by 2.

(*v*) There exists an odd integer divisible by 2.

(*vi*) If x is odd and x is perfect square, then x is divisible by 3.

Solution (*i*) $\exists x\, q(x)$ $\quad$ (*ii*) $\exists x\, [p(x) \wedge q(x)]$

(*iii*) $\forall x\, [q(x) \Rightarrow \sim t(x)]$ $\quad$ (*iv*) $\forall x\, [q(x) \Rightarrow \sim t(x)]$

(*v*) $\exists x\, [q(x) \wedge t(x)]$ $\quad$ (*vi*) $\exists x\, [\{q(x) \wedge r(x)\} \Rightarrow s(x)]$.

12.15.2 Negation of Quantified Statements

Negation of a quantified statement change the quantifier and negate the original statement as mentioned below:

(*i*) $\sim \forall x : p(x) \equiv \exists x : \sim p(x)$ $\quad$ (*ii*) $\sim \exists x : p(x) \equiv \forall\ x : \sim p(x)$

(*iii*) $\sim [\forall x \sim p(x)] \equiv \exists x \sim \sim p(x) \equiv \exists x\, p(x)$ $\quad$ (*iv*) $\sim [\exists x \sim p(x) \equiv \forall x \sim \sim p(x) \equiv \forall x\, p(x)$

12.15.3 Properties of Quantifiers

(*i*) $\exists x\, (p(x) \Rightarrow q(x)) \equiv \forall x\, p(x) \Rightarrow \exists x\, q(x)$

(*ii*) $\exists x\, p(x) \Rightarrow \forall x\, q(x) \equiv \forall x\, (p(x) \Rightarrow q(x))$

(*iii*) $\exists x\, (p(x) \vee q(x)) \equiv \exists x\, p(x) \vee \exists x\, q(x)$

$\forall x\, [p(x) \wedge q(x)] \equiv [\forall x\, p(x) \wedge \forall x\, q(x)]$

$[\forall x\, p(x) \vee \forall x\, q(x)] \Rightarrow \forall x\, [p(x) \vee q(x)]$

(*iv*) $\sim (\exists x \sim p(x)) \equiv \forall x\, p(x)$

In general, the negation of a quantified predicate is logically equivalent to the propcsition obtained by replacing each $\forall$ by $\exists$ and vice versa, as well as replacing the predicate itself by its negative as shown in Table below 12.67

Table 12.67

Statement	*Negation*
all true $\forall x\, p(x)$	$\exists x\, [\sim p(x)]$ at least one false
at least one false $\exists x[\sim p(r)]$	$\forall x\, p(x)$ all true
all false $\forall x\, [\sim p(x)]$	$\exists x\, p(r)$ at least one true
at least one true $\forall x\, p(x)$	$\forall x\, [\sim p(x)]$ all false

Example 39 Over the universe of four-wheelers, let A (x) : x is a four wheeler, B (x) x is a car and C (x) : x is manufactured by Maruti Udyog Ltd. Express the following statements statements using quantifiers

(*a*) Every car is a four-wheeler manufactured by MUL.

(*b*) There are cars that are not manufactured by MUL.

(*c*) Every four-wheeler is a car.

Solution (*a*) $(\forall x)\, (A(x) \wedge \sim B(x)) \Rightarrow C(x)$

(*b*) $(\exists x)\, (A(x) \wedge C(x))$

(*c*) $(\forall x)\, (A(x) \Rightarrow B(x))$.

Example 40 Over the universe of animals, let

A (x) : x is a whale; B (x) : x is a fish; C (x) : x lives in water

Translate the following in your own words:

(*i*) $\exists x\, (\sim C(x))$ (*ii*) $(\exists x)\, (B(x) \wedge \sim A(x))$

(*iii*) $(\forall x)\, (A(x) \wedge C(x)) \Rightarrow B(x)$.

Solution (*i*)There exists an animal which does not live in water.

(*ii*) There exists a fish that is not a whale.

(*iii*) Every whale that lives in the water is a fish.

Example 41 Find the negation of the following statement $\forall x \exists y\, [\{p(x, y) \wedge q(x, y)\} \Rightarrow r(x, y)]$

Solution $\sim [\forall x \exists y\, [\{p(x, y) \wedge q(x, y)\} \Rightarrow r(x, y)]]$

$\Leftrightarrow \exists x\, [\sim \exists y\, [\{p(x, y) \wedge q(x, y)\} \Rightarrow r(x, y)]]$

$\Leftrightarrow \exists x \forall y \sim [\{p(x, y) \wedge q(x, y)\} \Rightarrow r(x, y)]$

$\Leftrightarrow \exists x \forall y \sim [\sim \{p(x, y) \wedge q(x, y)] \vee r(x, y)]$

$\Leftrightarrow \exists x \forall y\, [\sim \sim \{p(x, y) \wedge q(x, y)\} \wedge \sim r(x, y)]$

Example 42 Use quantifiers to say that $\sqrt{3}$ is not a rational number. [*UPTU, BTech, 2007-08*]

Solution Let $p(x)$ be a prime number and $q(x)$: $\sqrt{x}$, where x is prime number. Then $r(x)$: is rational for all x, and $q(x) \Rightarrow \sim r(x)$, i.e. square root of every prime number is not rational.

Example 43 Let $p(x)$ be 'x is mammal'. Let $q(x)$ be 'x is an animal'. Let $r(x)$ be 'x is warm blooded'. Translate into a formula: 'Every mammal is warm, blooded.'
(*b*) Translate into English $(\exists x)\ (q(x) \wedge (\sim(p(x)))$ [*UPTU, BTech, 2007-08*]

Solution Formula for 'every mammal is warm blooded' is

$$\forall x(\ p(x) \rightarrow r(x))$$

$(\exists x)\ (q(x) \wedge (\sim p(x))) \equiv$ there are some animals that are not mammals.

Example 44 Negate the proposition and show it by quantifiers. 'All integers are greater than θ'. [*UPTU, BTech, 2007-2008*]

Solution Let $p(x)$ denote 'x is integer' and $q(x)$ denote 'x is greater than q'. Then the given proposition can be written as $(\exists x \in p(x))\ (\sim q(x))$

Review Questions

1. Define a statement and its truth value.
2. Define truth table. Explain its uses.
3. What are logical connectives? Explain the following terms:
   (*a*) Negation (*b*) Conjunction (*c*) Disjunction.
4. Define conditional and biconditional statements.
5. Explain the following terms by giving suitable example:
   (*a*) Converse (*b*) Inverse (*c*) Contrapositve.
6. Explain the term tautologies and fallacies in the context of truth tables.
7. Define various axioms and theorems in the context of various connectives used in Boolean algebra.
8. What do you understand by an argument?
9. Explain the concept of compound statement. Give an example.

Self-practice Problems

1. Which of the following sentences are not logical statements?
   (*i*) What are you doing? (*ii*) Mohan is a brave boy.
   (*iii*) What is your opinion about students unrest at present?
   (*iv*) What is your portfolio? (*v*) Why did you buy XYZ company stock?
   (*vi*) Rate of inflation exceeded 5.5 in January 2002.
2. Express the following compound statements in words taking p for popular and q for qualified.
   (*i*) $p \wedge q$ (*ii*) $\sim p \vee \sim q$ (*iii*) $p \vee (\sim q \wedge q)$ (*iv*) $\sim q \vee p$ (*v*) $\sim (p \vee q)$

3. Express the following statements through appropriate symbols:

(*i*) He is practical but not quiet. (*ii*) He is neither quiet nor practical.

(*iii*) It is false that he is practical and quiet. (*iv*) It is not true that he is an idealist but not quiet.

4. Construct the truth table for the following:

(*i*) $(p \wedge q) \Rightarrow p$ (*ii*) $\sim (p \wedge q) \Leftrightarrow \sim p \vee \sim q$ (*iii*) $\sim (p \vee q)$

(*iv*) $(p \vee \sim q) \wedge p$ (*v*) $\sim (p \vee q) \vee (\sim p \wedge \sim q)$ (*vi*) $(p \vee q) \vee (\sim q)$.

5. Write which of these is a tautology or a fallacy.

(*i*) $p \Rightarrow q \wedge q \Rightarrow p$ (*ii*) $p \wedge q \Rightarrow p$ (*iii*) $p \vee \sim (p \wedge q)$

(*iv*) $((q \rightarrow q) \wedge (q \rightarrow r)) \rightarrow (p \rightarrow r)$ (*v*) $p \vee \sim (p \wedge q)$

(*vi*) $(p \wedge q) \wedge (p \vee q)$ (*vii*) $\sim p \Rightarrow (p \Rightarrow q)$ (*viii*) $(\sim q \wedge (p \Rightarrow q)) \Rightarrow \sim p$

(*ix*) $\sim (p \wedge q) \Leftrightarrow (\sim p) \vee (\sim q)$ (*x*) $\sim (p \Rightarrow q) \Rightarrow q$.

6. Using truth table prove that the following identities are true:

(*i*) $p \wedge (p \wedge r) \equiv (p \wedge q)$ (*ii*) $(p \wedge q) \equiv (p \wedge r) \wedge (p \wedge r)$ (*iii*) $\sim (p \wedge q) \equiv \sim p \vee \sim q$

(*iv*) $\sim p \wedge \sim q \equiv \sim (p \vee q)$ (*v*) $p \vee (q \vee r) \equiv (p \vee q) \wedge (p \vee r)$.

7. Indicate the relevant law operating the following propositions:

(*i*) $[(p \Rightarrow q) \wedge (r \Rightarrow q)] \Leftrightarrow [(p \vee r) \Rightarrow p]$ (*ii*) $[(p \wedge \sim q) \Rightarrow (r \wedge \sim r)] \Leftrightarrow p \Rightarrow q$.

(*iii*) $\sim (p \vee q) = \sim p \wedge \sim q$ (*iv*) $\sim (p \Leftrightarrow q) = q \Leftrightarrow \sim p$

8. Which is the logical opposite of the following words?

(*i*) good (*ii*) less (*iii*) equal (*iv*) braves.

9. Let *p*, *q*, *r* and *s* represent simple statements. Assume *p* is false, *q* is true, *r* is false and *s* is true. Determine the truth value of each of the statement expressed below:

(*i*) $q \wedge r$ (*ii*) $r \vee p$ (*iii*) $p \wedge p$ (*iv*) $p \vee s$ (*v*) $\sim q$ (*vi*) $q \vee s$

10. Write each sentence in the 'if . . ., then' form

(*i*) All ducks are birds. (*ii*) Vertical angles are equal.

(*iii*) Any two parallel lines are coplanar.

11. Consider the statements:

p : You will work hard. q : You will become wealthy.

Translate each of the symbolic statements into one sentence.

12. Compare the following statements:

(*i*) $p \Rightarrow q$. (*ii*) If *p*, then *q*.

(*iii*) *p* is sufficient condition for *q*. (*iv*) *p* only if *q*.

(*v*) *q* is sufficient condition for *p*.

13. By means of truth table, prove that

(*i*) $(p \wedge q) \vee r \equiv (p \vee r) \wedge (q \vee r)$ (*ii*) $(p \vee q) \wedge r \equiv (p \wedge r) \vee (q \wedge r)$

14. By means of truth table, show that $\sim (p \Leftrightarrow q) = \sim p \Leftrightarrow q = p \Leftrightarrow \sim q$

15. Write the converse, inverse and contrapositive of each statement

(*i*) If I move fast, then I may get the grade. (*ii*) If I run fast, then I can win the race.

16. Show by means of truth tables that each of the following statements has the same truth values as $p \Rightarrow q$ (*i*) $(\sim p) \vee q$, and (*ii*) $p \wedge (\sim p)$

17. Give the reasons for each step required to show that the following argument is valid

(*i*) $(p \vee q) \wedge \sim(\sim p \wedge q) \Leftrightarrow p$

(*ii*) $\sim[\sim \{(p \vee q) \wedge r\} \vee \sim q] \Leftrightarrow q \wedge r$

(*iii*) $[(\sim p \vee q) \wedge (p \wedge (p \wedge q))] \Leftrightarrow p \wedge q$

(*iv*) $(p \vee q \vee r) \wedge (p \wedge t \wedge \sim q) \wedge (p \vee \sim t \vee r) \Leftrightarrow p \vee [r \wedge (t \vee \sim q)]$

(*v*) $[(p \wedge \sim q) \vee (\sim p \wedge q)]$

(*vi*) $(p \Rightarrow q) \wedge [\sim q \wedge (r \vee \sim q)] \Leftrightarrow \sim (p \vee q)$

(*vi*) $[(\sim p \vee \sim q) \Rightarrow (p \wedge q \wedge r)] \Leftrightarrow p \wedge q$

18. For any statements *p* and *q*, prove that

(*i*) $\sim (p \downarrow q) \Leftrightarrow (\sim p \uparrow \sim q)$

(*ii*) $\sim (p \uparrow q) \Leftrightarrow (\sim p \downarrow \sim q)$

19. Establish the validity of the arguments

(*i*) p
$p \Rightarrow \sim q$
$\sim q \Rightarrow \sim r$
$\overline{\therefore \sim r}$

(*ii*) $p \Rightarrow r$
$r \Rightarrow s$
$r \vee \sim s$
$\sim t \vee u$
$\sim u$
$\overline{\therefore \sim p}$

(*iii*) $p \Rightarrow r$
$\sim p \Rightarrow q$
$q \Rightarrow s$
$\overline{\therefore \sim r \Rightarrow s}$

(*iv*) $(\sim p \vee \sim q) \Rightarrow (r \wedge s)$
$r \Rightarrow t$
$\sim r$
$\overline{\therefore p}$

20. Let *p* denotes 'It rains' and *q* denotes. 'The atmospheric humidity increases'. Write the following statements in symbolic form:

(*i*) Atmospheric humidity increases only if it rains.

(*ii*) Sufficient condition for it to rain is that atmospheric humidity increases.

(*iii*) Necessary condition for it to rain is that atmospheric humidity increases.

(*iv*) Whenever atmospheric humidity increases it rains.

21. Write the negation of following statements:

(*i*) If he studies, he will pass the examination.

(*ii*) Ram is tall, therefore, he is slim.

(*iii*) If it rains, then Ram does not drive the car.

(*iv*) If *x* is less than zero, then *x* is not positive.

22. Write down the following statement in the symbolic form and find its negation: 'If he is successful' then he will join 'either M.B.A. or C.A.'

23. Write the negation of the following statements:

(*i*) He swims if and only if the water is warm.

(*ii*) India will be prosperous if and only if its citizens are industrious.

(*iii*) If it is cold, then he wears coat but no sweater.

24. Test the validity of the following arguments:

(*i*)
$$\begin{array}{l} p \vee q \\ \sim q \\ \hline \therefore p \wedge \sim q \end{array}$$

(*ii*)
$$\begin{array}{l} p \vee q \\ \sim p \\ \hline \therefore q \end{array}$$

(*iii*)
$$\begin{array}{l} q \vee (p \vee r) \\ \sim r \\ \hline \therefore p \vee q \end{array}$$

(*iv*)
$$\begin{array}{l} p \to q \\ p \to r \\ \hline \therefore p \to (q \wedge r) \end{array}$$

(*v*)
$$\begin{array}{l} p \to r \\ \sim p \\ \hline \therefore \sim q \end{array}$$

(*vi*)
$$\begin{array}{l} p \to (q \vee (\sim r)) \\ q \to p \wedge r \end{array}$$

25. Test the validity of the following arguments:

(*i*) If I like mathematics, then I will study. Either I study or I fail. Therefore, if I fail, then I do not like mathematics.

(*ii*) If you work hard, you pass. You did not pass. Therefore, you did not work hard.

(*iii*) If it rains, Ram will be sick. Ram was not sick. Therefore, it did not rain.

(*iv*) If 7 is less than 4, then 7 is not a prime number. 7 is not less than 4. Therefore, 7 is a prime number.

(*v*) If two sides of a triangle are equal, then the opposite angles are equal. Two sides of a triangle are not equal. Therefore, the opposite angles are not equal.

(*vi*) If the rainfall is good, the yield of crops is good. The yield of crops was not good. Therefore, the rainfall was not good.

(*vii*) If my brother stands first in the class, I give him a T-shirt. Either he stood first or I was out of station. I did not give my brother a T-shirt. Therefore, I was out station.

26. Write the converse, inverse and contrapositive for the statement: $(\sim p) \Rightarrow q$.

27. Show by a truth table that the inverse of $p \Rightarrow q$ is equivalent to converse of $p \Rightarrow q$.

28. Test the validity of the following statement: 'If a boy gets high marks in examination, he is either deligent or intelligent. The boy has got high marks in examination. He is either deligent or intelligent.'

29. Over the set of integers, let A (x) : $x + y$ is even; B (x) : x is a prime number; and C (x) : $x + y$ is even. Write following into simple English sentences.

(*i*) $\forall x$ A (x), $\exists x$ A (x)

(*ii*) $(\forall x)$ $((\exists y)$ (C $(x, y)))$; $(\$ x)$ $((\forall y)$ C $(x, y))$

(*iii*) $\forall x$ $(\sim$ B (x) ; $(\exists y)$ $(\sim$ A $(x))$

(*iv*) $\sim$ $(\exists x$ A$(x))$; $\sim$ $(\forall x$ B $(x))$

30. Use the propositions

p : Ram is drug edict; q : Ram is hospitalised

to give examples using sentences of each of the following laws:

(*i*) Idempotent law (*ii*) Commutative law (*iii*) DeMorgan's law

31. Write the converse, contrapositive and inverse of the following sentences:

(*i*) If it is raining, the grass is wet. (*ii*) It is raining if it is cloudy.

(*iii*) Rain is sufficient for it to be cloudy. (*iv*) Rain is necessary for it to be cloudy.

32. Construct truth tables to determine whether each of the following is a tautology, a contingency or contradiction (absurdity):

(*i*) $p \wedge \sim p$ (*ii*) $p \Rightarrow (q \Rightarrow p)$ (*iii*) $(p \Rightarrow q) \wedge (p \vee q)$

(*iv*) $q \vee (\sim q \wedge p)$ (*v*) $(q \wedge p) \vee (q \wedge \sim p)$ (*v*) $(p \wedge q) \Rightarrow p$

33. Let p : The examination is hard, and q : The grades are low. Write each of the following symbolically.

(*i*) The examination is not hard nor are the grades low.

(*ii*) The examination is not hard and the grades are low.

(*iii*) The grades are low only if the examination is hard.

(*iv*) The grades are low if the examination is hard.

34. Prove the following using a direct proof and then an indirect proof.

(*i*) When it rains, Amit does not go to school. Amit is going to school, show that it is not raining.

(*ii*) If I do not go to work, I will go to cinema. I will not go to cinema, show that I will go to work.

(*iii*) Either Manisha is a good girl or Amit is a good boy. If Anjali cries, then Manisha is not a good girl. Amit is not a good boy. Does Anjali cry?

35. Determine whether following arguments are valid.

(*i*) Premise : When students study, they get good marks. These students do not study.
Conclusion : These students do not receive good marks.

(*ii*) Premise : If Anjali studies, she will not fail in the paper. If she does not play tennis too often, she will study. Anjali failed in the paper.
Conclusion : She played tennis too often.

Hints and Answers

1. (*i*) and (*ii*) are not logical statements.

2. (*i*) He is popular and qualified. (*ii*) He is neither popular nor qualified.
(*iii*) He is popular as he is unpopular and qualified.

3. (*i*) $p \wedge \sim q$ (*ii*) $\sim q \wedge \sim p$ (*iii*) $\sim (p \wedge q)$ (*iv*) $\sim (\sim p \vee \sim q)$.

5. Tautology: (*ii*), (*iii*), (*iv*), (*v*); fallacy: (*vi*)

8. (*i*) Not good (*ii*) not less (*iii*) unequal (*iv*) not brave

9. (*i*) F (*ii*) F (*iii*) F (*iv*) T (*v*) F (*vi*) T

10. (*i*) If the creature is a duck, then it is a bird.
(*ii*) If two angles are vertical angles, then they are equal.
(*iii*) If two lines are parallel, then they are coplaner.

11. (*i*) If you all work hard, then you will become wealthy.
(*ii*) If you will become wealthy, then will work hard.
(*iii*) If you will not work hard, then you will not become wealthy.

12. All statements are equivalent.

14.

p	q	$p \Leftrightarrow q$	$\sim(p \Leftrightarrow q)$	$\sim p$	$q \Leftrightarrow \sim p$
T	T	T	F	F	F
T	F	F	T	F	T
F	T	F	T	T	T
F	F	T	F	T	F

18.

p	q	$\sim(p \downarrow q)$	$(\sim p \uparrow \sim q)$	$\sim(p \uparrow q)$	$(\sim p \downarrow \sim q)$
T	T	T	T	T	T
T	F	F	F	T	T
F	T	F	F	T	T
F	F	F	F	F	F

19. (*i*)

1. $p \Rightarrow \sim q$	Premise
2. $\sim q \Rightarrow \sim r$	Premise
3. $p \Rightarrow \sim r$	From (1) and (2), Law of syllogism
4. p	Premise
5. $\therefore \sim r$	From (3) and (4), Law of Detachment

(*ii*)

1. $p \Rightarrow r,\ r \Rightarrow s$	Premises
2. $p \Rightarrow s$	From (1), Law of syllogism
3. $t \vee \sim s$	Premise
4. $\sim s \vee t$	From (3)
5. $s \Rightarrow t$	From (4) because $\sim s \vee t \Leftrightarrow s \Rightarrow t$
6. $p \Rightarrow t$	From (2) and (5), Law of syllogism
7. $\sim t \vee u$	Premise
8. $t \Rightarrow u$	From (7) because $\sim t \vee u \Leftrightarrow t \Rightarrow u$
9. $p \Rightarrow u$	From (6) and (8), Law of syllogism
10. $\sim u$	Premise
11. $\therefore \sim p$	From (9) and (10), Modus Tollens

(*iii*)

1. $p \Rightarrow r$	Premise
2. $\sim r \Rightarrow \sim p$	From (1) because $p \Rightarrow r \Leftrightarrow \sim p \Rightarrow \sim r$
3. $\sim p \Rightarrow q$	Premise
4. $\sim r \Rightarrow q$	From (2) and (3), Law of syllogism
5. $q \Rightarrow s$	Premise
6. $\therefore \sim r \Rightarrow s$	From (4) and (5), Law of syllogism

(*iv*)

1. $r \Rightarrow t$	Premise
2. $\sim t$	Premise

3. $\sim r$	From (1) and (2), Modus Tollens
4. $\sim r \vee \sim s$	From (3), Rule of Disjunctive Amplification
5. $\sim(r \wedge s)$	From (4), DeMorgan's laws
6. $(\sim p \vee \sim q) \Rightarrow (r \wedge s)$	Premise
7. $\sim(\sim p \vee \sim q)$	From (5) and (6), Modus Tollens
8. $p \wedge q$	From (7) DeMorgan's law and Law of double negation
9. $\therefore p$	From (8), Rule of conjunctive simplification.

20. (*i*) Statement tells 'q only if p'. In symbolic form it can be written as: $q \rightarrow p$.

(*ii*) It shows that q is sufficient condition for p. Therefore, statement in symbolic form is $q \rightarrow p$.

(*iii*) It shows that q is the necessary condition for p. Therefore, statement in symbolic form is $p \rightarrow q$.

(*iv*) Whenever atmospheric humidity increases it rains. If atmospheric humidity increases, then it rains.

In symbolic form, statement is $q \rightarrow p$.

21. (*i*) Let p be the statement 'He studies' and q be the statement 'He passes the examination'.

In symbolic form, statement is $p \rightarrow q$.

We know $\sim (p \rightarrow q) = p \wedge \sim q$, so negation of the statement is: He studies and he will not pass the examination.

(*ii*) Let p be the statement 'Ram is tall' and q be the statement 'Ram is slim'.

Symbolically it can be written as: $p \rightarrow q$.

Now $\sim (p \rightarrow q) = p \wedge \sim q$, so negation of the statement is: Ram is tall and Ram is not slim.

(*iii*) Let p be the statement 'It rains' and q be the statement 'Ram does not drive the car'.

Symbolically it can be written as $p \rightarrow q$. Now negation of $p \rightarrow q = p \wedge \sim q$, so negation of the statement is: 'It rains and Ram drives the car'.

(*iv*) Let p be the statement 'x is less than zero' and q be statement 'x is not positive'.

Symbolically it can be written as: $p \rightarrow q$.

Since $\sim (p \rightarrow q) = p \wedge \sim q$, negation of the statement $p \rightarrow q$ is: 'x is less than zero and x is positive'.

22. Let p : He is successful, q : He joins M.B.A., r : He joins C.A.

Symbolically we can write these statements as: $p \rightarrow (q \vee r)$.

Its negation is $\sim [p \rightarrow (q \vee r)] = p \wedge \sim (q \vee r) = p \wedge (\sim q \wedge \sim r)$.

The negation of the given statement is: 'He is successful but joins neither M.B.A. nor C.A.'

23. (*i*) Let p be the statement 'He swims' and q be the statement 'the water is warm'.

Symbolically it is represented as: $p \Leftrightarrow q$.

Since $\sim (p \leftrightarrow \sim q \equiv q \leftrightarrow \sim p \equiv q \Leftrightarrow \sim p$, negation of the given statement is either of the following two statements.

'He swims if and only if the water is not warm'.

'He does not swim if and only in the water is warm'.

(*ii*) Let p be the statement 'India is prosperous', and q be the statement 'citizens are industrious'.

The negation of $p \leftrightarrow q$ is equivalent to either of the following statements:

'India is prosperous if and only if its citizens are not industrious'.

'India will not be prosperous if and only if its citizens are industrious'.

(*iii*) Let p be the statement 'It is cold' and q be the statement 'He wears a coat' and r be the statement 'He wear a sweater'.

Symbolically we may write as: $p \rightarrow (q \wedge \sim r)$

Negation of the above statement is:

$$\sim [p \rightarrow (q \wedge \sim r)] = p \wedge [\sim (q \wedge \sim r)] = p \wedge (\sim q \vee r)$$

Now $\sim q \vee r$ is the statement: 'He wears a sweater or no coat' and p is the statement 'It is cold'.

$p \wedge (q \vee \sim r)$ is the statement: 'It is cold and he wears a sweater or no coat'.

24. (*i*) Valid (*ii*) Valid (*iii*) Valid
(*iv*) Valid (*v*) Invalid (*vi*) Invalid

25. (*i*) Invalid (*ii*) Valid (*iii*) Valid
(*iv*) Invalid (*v*) Invalid (*vi*) Valid
(*vii*) Valid

26. Converse $q \Rightarrow (\sim p)$; inverse $p \Rightarrow (\sim q)$; Contrapositive $(\sim q) \Rightarrow p$.

28. Let p = boy has got high marks in examination.

The conclusion to be tested is: He is deligent or intelligent. Hence, the argument is:

$(p \Rightarrow q) \wedge p \Rightarrow q$.

p	q	$p \Rightarrow q$	$p \Rightarrow q \wedge p$	$(p \Rightarrow q) \wedge p \Rightarrow q$
T	T	T	T	T
T	F	F	F	T
F	T	T	F	T
F	F	T	F	T

Since $(p \Rightarrow q) \wedge p \Rightarrow q$ is tautology, therefore, the statements are valid.

Chapter 13

Boolean Algebra

PREVIEW

Boolean algebra is a mathematical model of both propositional logic and set theory. For example, if zero (0) represents that a sentence is false and one (1) represents that this sentence is true, then we may use a truth table to show how the truth or falsity of any sentence depends on the truth or falsity of its component sentences. The concepts of Boolean algebra are used in electronic circuits of computer, telephone systems, industrial control devices, etc.

LEARNING OBJECTIVES

After studying this chapter, you should be able to

- appreciate areas of application of Boolean algebra
- prove basic theorems (results) on Boolean algebra
- develop the truth table for a Boolean expression
- use OR, AND and NOT gates as well as NAND and NOR gates to build truth function
- write a Boolean expression in sum-of-products form and product-of sum form for a given truth function.
- simplify Boolean expression using Karnaugh maps.

13.1 INTRODUCTION

The concept of Boolean algebra was first developed by an English mathematician, George Boole, in 1947, to simplify logical statements and solve logical problems. Claude Shannon, later on, in his paper 'A Symbolic Analysis of Relay and Switching Circuits' pointed out the use of Boolean algebra in the analysis and design of logical circuits. It is, therefore, used in designing of computers based on binary numbers 1 and 0. Numbers 1 and 0, correspond to two possible states of an electric circuit – *current flowing or not, switch on or off.*

As discussed earlier that there are three basic operations: $\wedge$ (*and*), $\vee$ (*or*) and ~ (*negation*) in case of logical statements, and $\cap$ (*intersection*), $\cup$ (*union*) and $\{\ \}^c$ (*complement*) in case of set theory. But in case of Boolean algebra all these operations are symbolised by AND, OR and NOT, respectively. These basic operations are called *logic operations*. All other operations can be expressed in terms of these basic operations. In this chapter, common mathematical symbols such as: plus (+), dot (.) and prime (′ or ‾) would be used for all above said operations, respectively.

The purpose of this chapter is to generalise the basic concepts of the algebra of sets, logic and binary arithmetic, and apply it to the development of electronic digital computers. Table 13.1 shows the list of notations widely used in Boolean algebra and their relationship with notations used in set theory and mathematical logic.

13.2 BOOLEAN ALGEBRA

Definition 1 A non-empty set B containing at least two distinct elements, 0 and 1, with two binary operations '+' (OR) and '·' (AND) as well as a unary operation '–' (NOT) is called a *Boolean algebra,* denoted by (B, +, ·, –, 0, 1) if and only if following axioms are satisfied.

(*a*) *Commutative laws*

$$a + b = b + a$$

and $\quad a \cdot b = b \cdot a,\quad$ for all $a, b \in$ B.

(*b*) *Distributive laws*

$$a \cdot (b + c) = a \cdot b + a \cdot c$$

and $\quad a + (b \cdot c) = (a + b) \cdot (a + c),\quad$ for all $a, b, c \in$ B

(*c*) *Identity laws*

$$a + 0 = 0 + a = a$$

and $\quad a \cdot 1 = 1 \cdot a = a,\quad$ for all $a \in$ B.

(*d*) *Complement laws*

$$a + \bar{a} = \bar{a} + a = 1$$

The element $\bar{a}$ is said to be the complement of the element $a \in$ A with respect to the operation (+) and if

$$a \cdot \bar{a} = \bar{a} \cdot a = 0,$$

then $\bar{a}$ is said to be complement of the element $a \in$ A with respect to the operation (·).

Table 13.1 Summary of Notations

Operation	*Set Theory*	*Logical Statements*	*Boolean Algebra*
Elements	$\{x, y, \ldots\}$	$[p, q, r, s, \ldots]$	$\{x, y, \ldots\}$
Operator AND	$A \cap B$	$p \wedge q$	$a \cdot b$
Operator OR	$A \cup B$	$p \vee q$	$a + b$
Operator NOT	$\overline{A}$	$- p$	$\bar{a}$
Tautology	U	T	1
Fallacy	Φ	F	0
Implication	$\rightarrow$ mapping	$\Rightarrow$	$\bar{a} + b$
Equivalence	one-one mapping $A \subseteq B$ or $\overline{A} \cup B = U$ or $A \cap \overline{B} = \Phi$	$\Rightarrow$ $p \Rightarrow q$ $\sim p \vee q$ is true $p \wedge \sim q$ is false	$\bar{a} + \bar{b} + ba$ $- a + b = 1$ $a \cdot (- b) = 0$
Idempotent Law	$A \cup A = A$ $A \cap A = A$ $A \cup \overline{A} = U$ $A \cap \overline{A} = \Phi$	$p \vee p = p$ $p \wedge p = p$ $p \vee \sim p$ is true $p \wedge \sim p$ is false	$a + a = a$ $a \cdot a = a$ $a + (- a) = 1$ $a \cdot (a - a) = 0$

Definition 2 Let A = $\{p, q, r, \ldots\}$ be any set of statements and let $\vee$ (OR), $\wedge$ (AND) and $\sim$ (Negation) be binary operations defined on A. Then the mathematical structure $[A, \vee, \wedge, \sim]$ is called a *Boolean algebra* if and only if the following axioms are satisfied.

(*a*) *Commutative laws*

$$p \vee q \Leftrightarrow q \vee p$$

and $\quad p \wedge q \Leftrightarrow q \wedge p,\quad$ for all $p, q \in A$.

(*b*) *Distributive laws*

$$p \vee (q \wedge r) \Leftrightarrow (p \vee q) \wedge (p \vee r)$$

and $p \wedge (q \vee r) \Leftrightarrow (p \wedge q) \vee (p \wedge r),\quad$ for all $p, q, r \in A$

(*c*) *Identity laws*

$$p \vee 0 = 0 \vee p = p$$

and $\quad p \wedge 1 = 1 \wedge p = p,\quad$ for all $p \in A$

(*d*) *Complement laws*

$$p \vee \sim p = 1$$

and $\quad p \wedge \sim p = 0$, for all $p \in A$.

The element $\sim p$ is called a complement of the element p.

(*e*) *DeMorgan's laws*

(*i*) $\sim (p \wedge q) \Leftrightarrow \sim p \vee \sim q$

(*ii*) $\sim (p \vee q) \Leftrightarrow \sim p \wedge \sim q$

Remarks

1. Since $A \cup A = A$ and $A \cap A = A$, in Boolean algebra we have $a + a = a$ and $a \cdot a = a$.
2. Each variable in Boolean algebra has either of two values: true (T) or false (F) which corresponds to 1 and 0, respectively.
3. Disjunction, $p \vee q$ Conjunction, $p \wedge q$

Disjunction, $p \vee q$ (p rows, q columns):

$\vee$	F	T
F	F	T
T	T	T

Conjunction, $p \wedge q$ (p rows, q columns):

$\wedge$	F	T
F	F	F
T	F	T

It may be noted that $p \vee q$ is false only when both p and q are false, otherwise $p \vee q$ is true. Similarly, $p \wedge q$ is false when either of them or both are false, otherwise true.

Now let us replace $\vee$ by (+) and $\wedge$ by $(\cdot)$, F by 0 and T by 1, we get following tables:

+	0	1
0	0	1
1	1	1

$\cdot$	0	1
0	0	0
1	0	1

If we compare it with ordinary arithmetic, we find that in logic, $1 + 1 = 1$, but in ordinary arithmetic, $1 + 1 = 2$.

4. Boolean algebra does not have operations for subtraction and division.
5. The distribution law: $a + (b \cdot c) = (a + b) \cdot (a + c)$ does not hold good in ordinary algebra.

13.3 BASIC THEOREMS ON BOOLEAN ALGEBRA

Theorem 13.1 The zero and unit elements are unique.

Proof (*a*) Let there be two zero elements 0 and 0_1 in the Boolean algebra A. Then

$$0_1 + 0 = 0_1 \text{ and } 0 + 0_1 = 0$$

Since $0_1 + 0 = 0 + 0_1$, therefore, $0_1 = 0$.

(*b*) Let there be two unit elements 1 and 1_1 in the Boolean algebra B. Then

$$1.1_1 = 1 \text{ and } 1_1.1 = 1_1$$

Since $1.1_1 = 1_1.1$, therefore, $1_1 = 1$.

Theorem 13.2 The complement (or inverse) of an element is unique.

Proof Let $\bar{a}_1$ and $\bar{a}_2$ be two complements of an element a in the Boolean algebra. Then

$$a + \bar{a}_1 = 1 \text{ and } a \cdot \bar{a}_1 = 0$$

Also $$a + \bar{a}_2 = 1 \text{ and } a \cdot \bar{a}_2 = 0$$

Now
$$\begin{aligned} \bar{a}_1 &= \bar{a}_1 \cdot 1 = \bar{a}_1 \cdot (a + \bar{a}_2) \\ &= \bar{a}_1 \cdot a + \bar{a}_1 \cdot \bar{a}_2 = 0 + \bar{a}_1 \cdot \bar{a}_2 \\ &= a \cdot \bar{a}_2 + \bar{a}_1 \cdot \bar{a}_2 = \bar{a}_2 (a + \bar{a}_1) = \bar{a}_2 \cdot 1 = \bar{c}_2 \end{aligned}$$

It follows that $\bar{a}_1 = \bar{a}_2$, which contradicts the assumption that element a has two different complements. Hence, each element in Boolean algebra has a unique complement.

Theorem (Idempoten Law) 13.3 (*i*) $a + a = a$ (*ii*) $a \cdot a = a$

Proof (*i*) We can write

$$\begin{aligned} \text{RHS} &= a = a + 0 && \text{[Identity]} \\ &= a + (a \cdot \bar{a}) && \text{[Complement law]} \\ &= (a + a) \cdot (a + \bar{a}) && \text{[Distributive law]} \\ &= (a + a) \cdot 1 = (a + a) = \text{LHS} && \text{[Complement law]} \end{aligned}$$

(*ii*)

$$\begin{aligned} \text{RHS} &= a = a \cdot 1 \\ &= a \cdot (a + \bar{a}) && \text{[Complement law]} \\ &= (a \cdot a) + (a \cdot \bar{a}) && \text{[Distributive law]} \\ &= (a \cdot a + 0) = (a \cdot a) = \text{LHS} && \text{[Complement law]} \end{aligned}$$

Theorem (Identity Law) 13.4 (*i*) $a + 1 = 1$ (*ii*) $a \cdot 0 = 0$

Proof (*i*) Since

$$\begin{aligned} \text{RHS} &= 1 = a + \bar{a} && \text{[Complement law]} \\ &= a + (\bar{a} \cdot 1) \\ &= (a + \bar{a}) \cdot (a + 1) && \text{[Distributive law]} \\ &= 1 \cdot (a + 1) = (a + 1) = \text{RHS} && \text{[Complement law]} \end{aligned}$$

(*ii*)

$$\begin{aligned} \text{LHS} &= (a \cdot 0) = 0 + (a \cdot 0) \\ &= a \cdot \bar{a} + a \cdot 0 && [\text{Since } a \cdot \bar{a} = 0] \\ &= a \cdot (\bar{a} + 0) && \text{[Distributive law]} \\ &= a \cdot \bar{a} = 0 = \text{RHS} \end{aligned}$$

Theorem (Involution Law) 13.5 (*i*) $\overline{(\bar{a})} = a$ (*ii*) $\bar{1} = \bar{0}$ (*iii*) $\bar{0} = 1$

Proof (*i*) Since $a + \bar{a} = 1$, $a \cdot \bar{a} = 0$, therefore, a is the complement of $\bar{a}$ and hence $a = \overline{(\bar{a})}$;

(*ii*) Since $1 + 0 = 1$, $1.0 = 0$, therefore, 0 is the complement of 1 and hence $0 = \bar{1}$.

(*iii*) As above, 0 is complement of 1, i.e. $0 = \bar{1} \Rightarrow \bar{0} = \overline{(\bar{1})} = 1$.

Theorem (DeMorgan's Law) 13.6 (*a*) $\overline{(a + b)} = \bar{a} \cdot \bar{b}$ or $(a \vee b)' = a' \wedge b'$ and (*b*) $\overline{(a \cdot b)} = \bar{a} + \bar{b}$ or $(a \wedge b)' = a' \vee b'$.

Proof (*a*) By definition, an element $\bar{x}$ is called the complement of x in Boolean algebra provided

$$x + \bar{x} = 1 \text{ and } x \cdot \bar{x} = 0$$

To prove these results we have to show that $(\bar{a} \cdot \bar{b})$ is the complement of $(a + b)$.

$$\begin{aligned} (a + b) + (\bar{a} \cdot \bar{b}) &= [(a + b) + \bar{a}] \cdot [(a + b) + \bar{b}] && \text{[Distributive law]} \\ &= [(a + \bar{a}) + b] \cdot [a + (b + \bar{b})] && \text{[Associative law]} \\ &= [1 + b] \cdot [a + 1] && \text{[Complement law]} \\ &= 1 \cdot 1 = 1 && \text{[Identity law]} \end{aligned}$$

Also, $(a + b) \cdot (\bar{a} \cdot \bar{b}) = (\bar{a} \cdot \bar{b}) \cdot (a + b)$ [Commutative law]

$= (\bar{a} \cdot \bar{b}) \cdot a + (\bar{a} \cdot \bar{b}) \cdot b$ [Distributive law]

$= (\bar{a} \cdot a) \cdot \bar{b} + \bar{a} \cdot (\bar{b} \cdot b)$ [Associative law]

$= 0 \cdot \bar{b} + \bar{a} \cdot 0$ [Complement law]

$= 0 + 0 = 0$ [Since $a + a = a$]

Since, $(\bar{a} \cdot \bar{b})$ is the complement of $(a + b)$, therefore, we can write $(\bar{a} \cdot \bar{b}) = \overline{(a+b)}$. To prove this result, we have to show that

$$(a \cdot b) + (\bar{a}+\bar{b}) = 1 \text{ and } (a \cdot b) \cdot (\bar{a}+\bar{b}) = 0.$$

Consider $(a \cdot b) + (\bar{a}+\bar{b}) = [a+(\bar{a}+\bar{b})] \cdot [b+(\bar{a}+\bar{b})]$ (Distributive law)

$= [(a + \bar{a}) + \bar{b}] \cdot [(b + \bar{a}) + \bar{b}]$ (Associative law)

$= [1 + \bar{b}] \cdot [\bar{a} + (b + \bar{b})]$

$= [1 + \bar{b}] \cdot [\bar{a} + 1] = 1 \cdot 1 = 1$ (Identity law)

and $(a \cdot b) \cdot (\bar{a} + \bar{b}) = (a \cdot b) \cdot \bar{a} + (a \cdot b) \cdot \bar{b}$ (Distributive)

$= (b \cdot a) \cdot \bar{a} + a \cdot (b \cdot \bar{b})$ (Associative law)

$= b \cdot (a \cdot \bar{a}) + a \cdot 0 = b \cdot 0 + 0$ (Complement law)

$= 0 + 0 = 0$ (Identity)

Theorem (Absorption Law) 13.7 (*i*) $a + (a \cdot b) = a$ and (*ii*) $a \cdot (a + b) = a$

Proof (*i*) LHS $= a + (a \cdot b) = a \cdot 1 + (a \cdot b)$ (Identity)

$= a \cdot (1 + b)$ (Distributive law)

$= a \cdot 1 = a =$ RHS (Identity law)

(*ii*) LHS $= a \cdot (a + b) = a \cdot a + a \cdot b$ (Distributive law)

$= a + (a \cdot b)$ (Idempotent law)

$= a \cdot (1 + b) = a \cdot 1 =$ RHS (Distributive law)

Theorem (Cancellation Law) 13.8 If (*i*) If $x + z = y + z$ and $x \cdot z = y \cdot z$, then $x = y$, where, $x, y, z \in$ B.

(*ii*) If $y \cdot x = z \cdot x$ and $y \cdot \bar{x} = z \cdot \bar{x}$, then $y = z$ where $x, y, z \in$ B.

Proof (*i*) Since, $x = x \cdot (x + z)$ [Absorption law]

$= x \cdot (y + z)$ [Given $x + z = y + z$]

$= x \cdot y + x \cdot z$ (Distributive law)

$= x \cdot y + y \cdot z$ [Given $x \cdot z = y \cdot z$]

$= y \cdot x + y \cdot z$ (Commutative law)

$= y \cdot (x + z)$ (Distributive law)

$= y \cdot (y + z)$ (Given $x + z = y + z$)

$= y$ (Absorption law)

(*ii*) Since, $y = y \cdot 1$

$= y \cdot (x + \bar{x})$ (Complement law)

$= y \cdot x + y \cdot \bar{x}$ (Distributive law)

$= z \cdot x + z \cdot \bar{x}$ (Given $y \cdot x = z \cdot x$ and $y \cdot \bar{x} = z \cdot \bar{x}$)

$= z \cdot (x + \bar{x})$ (Distributive law)

$= z \cdot 1 = z$ (Complement law)

Theorem (Associative Law) 13.9 (*i*) $x + (y + z) = (x + y) + z$ and (*ii*) $x \cdot (y \cdot z) = (x \cdot y) \cdot z$; for all $x, y, z \in B$.

Proof (*i*) Let $L = x + (y + z)$ and $M = (x + y) + z$. Then

$L \cdot x = [x + (y + z)] \cdot x$

$= x \cdot [x + (y + z)]$ (Commutative law)

$= x$ (Absorption law)

and $M \cdot x = [(x + y) + z] \cdot x$

$= (x + y) \cdot x + z \cdot x$ (Distributive law)

$= x \cdot (x + y) + z \cdot x$ (Commutative law)

$= x + z \cdot x$ (Absorption law)

$= x \cdot (1 + z)$ (Distributive law)

$= x \cdot 1 = x$ (Identity law)

Hence, $L \cdot x = M \cdot x$. Also

$L \cdot \bar{x} = [x + (y + z)] \cdot \bar{x}$

$= x \cdot \bar{x} + (y + z) \cdot \bar{x}$ (Distributive law)

$= 0 + (y + z) \cdot \bar{x} = (y + z) \cdot \bar{x}$ (Complement law)

$M \cdot \bar{x} = [(x + y) + z)] \cdot \bar{x}$

$= (x + y) \cdot \bar{x} + z \cdot \bar{x}$ (Distributive law)

$= [x \cdot \bar{x} + y \cdot \bar{x}] + z \cdot \bar{x}$

$= [0 + y \cdot \bar{x}] + z \cdot \bar{x}$ (Complement law)

$= y \cdot \bar{x} + z \cdot \bar{x} = (y + z) \cdot \bar{x}$ (Distributive law)

Hence, $L \cdot \bar{x} = M \cdot \bar{x}$.

Since $L \cdot x = M \cdot x$ and $L \cdot \bar{x} = M \cdot \bar{x}$, therefore, $L = M$ or $x + (y + z) = (x + y) + z$.

Example 1 Prove that:

(*a*) $a + (\bar{a} \cdot b) = a + b$ and $a \cdot (\bar{a} + b) = a \cdot b$

(*b*) $(a + b) \cdot (\bar{b} + c) + b \cdot (\bar{a} + \bar{c}) = a \cdot \bar{b} + a \cdot c + b$

Solution (*a*) $a + (\bar{a} \cdot b) = (a + \bar{a}) \cdot (a + b)$ (Distributive law)

$= 1 \cdot (a + b) = a + b$ (Complement, Identity)

Similarly, $a \cdot (\bar{a} + b) = (a \cdot \bar{a}) + (a \cdot b)$ (Distributive law)

$= 0 + (a \cdot b) = a \cdot b$ (Complement, Identity)

(*b*) $a \cdot b + c \cdot (\bar{a} + \bar{b}) = a \cdot b + c \cdot \overline{(a \cdot b)}$ [Since $\overline{(a \cdot b)} = \bar{a} + \bar{b}$]

$= (a \cdot b + c) \cdot ((a \cdot b) \cdot \overline{(a \cdot b)})$

$= (a \cdot b + c) \cdot 1 = a \cdot b + c.$ [Since $a + \bar{a} = 1$, for all a]

Also $(a + b) \cdot (\bar{b} + c) + b \cdot (\bar{a} + \bar{c})$

$$
\begin{aligned}
&= a \cdot (\bar{b} + c) + b \cdot (\bar{b} + c) + b \cdot (\bar{a} + c) \\
&= a \cdot (\bar{b} + c) + b \cdot (\bar{b} + c + \bar{a} + \bar{c}) \\
&= a \cdot (\bar{b} + c) + b \cdot (\bar{b} + \bar{a} + c + \bar{c}) \\
&= a \cdot (\bar{b} + c) + b \cdot (\bar{b} + \bar{a} + 1) \\
&= a \cdot (\bar{b} + c) + b \cdot (\bar{b} + 1) \quad [\text{Since } a + 1 = 1 \text{ and } \bar{a} + 1 = 1 \text{ for all } a \in B] \\
&= a \cdot (\bar{b} + c) + b \cdot 1 = a \cdot \bar{b} + a \cdot c + b.
\end{aligned}
$$

Example 2 Prove that

(a) $a + (\bar{a} \cdot c + b) = a + b + c$

(b) $a \cdot b + c \cdot (\bar{a} + \bar{b}) = a \cdot b + c$

(c) $(a + b) \cdot \bar{a} \cdot \bar{b} = 0$

(d) $\overline{(a+b)} + \overline{(a+\bar{b})} = \bar{a}$

(e) $a \cdot b + \left[\overline{(a+\bar{b}) \cdot b}\right] = 1$

(f) $[a + \bar{a} \cdot \bar{b}] \cdot [\bar{a} + a \cdot b] = \bar{a} \cdot \bar{b} + a \cdot b$

(g) $a \cdot b \cdot c + a \cdot b \cdot \bar{c} + a \cdot \bar{b} \cdot c + \bar{a} \cdot b \cdot c = a \cdot b + b \cdot c + c \cdot a$

Solution (a) LHS $= a + (\bar{a} \cdot c + b) = (a + \bar{a} \cdot c) + b = [(a + \bar{a}) \cdot (a + c)] + b$

$$
\begin{aligned}
&= (1 \cdot (a + c)) + b = (a + c) + b = a + (c + b) \\
&= a + (b + c) = a + b + c = \text{RHS}
\end{aligned}
$$

(b) LHS $= a \cdot b + c \cdot (\bar{a} + \bar{b}) = a \cdot b + c \cdot \overline{(a \cdot b)}$

$$
\begin{aligned}
&= (a \cdot b + c) \cdot (a \cdot b + \overline{(a \cdot b)}) \\
&= (a \cdot b + c) \cdot 1 = a \cdot b + c = \text{RHS}
\end{aligned}
$$

(c) LHS $= (a + b) \cdot \bar{a} \cdot \bar{b} = (a + b) \cdot \overline{(a+b)} = 0 = \text{RHS}$

(d) LHS $= \overline{(a+b)} + \overline{(a+\bar{b})} = \bar{a} \cdot \bar{b} + \bar{a} \cdot \bar{\bar{b}} = \bar{a} \cdot (\bar{b} + \bar{\bar{b}})$

$$
= \bar{a} \cdot (\bar{b} + b) = \bar{a} \cdot 1 = \text{RHS}
$$

(e) LHS $= a \cdot b + \left[\overline{(a+\bar{b}) \cdot b}\right] = a \cdot b + \left[\overline{b \cdot (a+\bar{b})}\right]$

$$
= a \cdot b + \overline{(a \cdot b + 0)} = a \cdot b + \overline{(a \cdot b)} = 1 = \text{RHS}
$$

(f) LHS $= [a + \bar{b} \cdot \bar{b}] \cdot [\bar{a} + a \cdot b] = (a + \bar{a}) \cdot (a + \bar{b}) \cdot (\bar{a} + a) \cdot (\bar{a} + b)$

$$
\begin{aligned}
&= 1 \cdot (a + \bar{b}) \cdot 1 \cdot (\bar{a} + b) = (a + \bar{b}) \cdot (\bar{a} + b) \\
&= (a + \bar{b}) \cdot \bar{a} + (a + \bar{b}) \cdot b = \bar{a} \cdot (a + \bar{b}) + b \cdot (a + \bar{b}) \\
&= \bar{a} \cdot a + \bar{a} \cdot \bar{b} + b \cdot a + b \cdot \bar{b} = 0 + \bar{a} \cdot \bar{b} + a \cdot b + 0 \\
&= \bar{a} \cdot \bar{b} + a \cdot b = \text{RHS}
\end{aligned}
$$

(g) $a \cdot b \cdot c + a \cdot b \cdot \bar{c} + a \cdot \bar{b} \cdot c + \bar{a} \cdot b \cdot c = a \cdot b \cdot c + a \cdot b \cdot c + a \cdot b \cdot \bar{c} + a \cdot \bar{b} \cdot c + \bar{a} \cdot b \cdot c$

(Idempotent law)

$$
\begin{aligned}
&= a \cdot b \cdot c + a \cdot b \cdot c + a \cdot b \cdot c + a \cdot b \cdot \bar{c} + a \cdot \bar{b} \cdot c + \bar{a} \cdot b \cdot c \\
&= (a \cdot b \cdot c + a \cdot b \cdot \bar{c}) + (a \cdot b \cdot c + a \cdot \bar{b} \cdot c) + (a \cdot b \cdot c + \bar{a} \cdot b \cdot c) \\
&= a \cdot b \cdot 1 + a \cdot c \cdot (b + \bar{b}) + b \cdot c \cdot (a + \bar{a}) \\
&= a \cdot b + a \cdot c \cdot 1 + b \cdot c \cdot 1 \\
&= a \cdot b + a \cdot c + b \cdot c = a \cdot b + b \cdot c + c \cdot a
\end{aligned}
$$

Example 3 For each element a in a Boolean algebra prove that

$$\left.\begin{array}{l} a+b=1 \\ a\cdot b=0 \end{array}\right\} \Rightarrow b=\bar{a}$$

Solution
$$\begin{aligned} \bar{a} &= a+0=\bar{a}+(a\cdot b) \\ &= (\bar{a}+a)\cdot(\bar{a}+b)=1\cdot(\bar{a}+b) \\ &= (a+b)\cdot(\bar{a}+b)=(b+a)\cdot(b+\bar{a}) \\ &= b+(a\cdot\bar{a})=b+0=b=\text{RHS} \end{aligned}$$

Remarks

1. The symbols 0 and 1 need not necessarily represent the number zero and one.
2. When parentheses are not used, the expression (′ or –) has precedence over (·), and operation (·) has precedence over (+). For example, $a+b\cdot c=a+(b\cdot c)$ and not $(a+b)\cdot c$. Also $a\cdot\bar{b}$ means $a\cdot(\bar{b})$ and not $\overline{(a\cdot b)}$ or $(a\cdot b)'$.
3. Boolean algebra has no operation equivalent to subtraction and division.
4. In Boolean algebra, the concellation law does not hold. For example, $a+b=a+c$ does not mean $b=c$ and $a\cdot b=a\cdot c$ does not imply $b=c$.

13.4 DUALITY PRINCIPLE

The *dual* of any statement in a Boolean algebra B is the statement that is obtained by interchanging operations (+) and (·) as well as interchanging the identities 0 and 1 in the original statement. This property of interchanging operations and identity elements in Boolean algebra is called *duality principle*. This principle ensures that if a theorem is proved based on the axioms of Boolean algebra, then a dual theorem obtained by interchanging (+) with (·) and 0 with 1 also holds. For example, if

$$a\cdot(b+c)=(a\cdot b)+(a\cdot c),$$

then the dual of it is given by

$$a+(b\cdot c)=(a+b)\cdot(a+c),$$

and similarly, the dual of $(a\cdot 0)+(1\cdot a)=a$ becomes $(a+1)\cdot(0+a)=a$.

Example 4 State whether the set A = $\{a, b\}$ with the operation defined in the tables below is Boolean algebra.

+	a	b
a	a	b
b	b	b

·	a	b
a	a	a
b	a	b

Solution (*i*) *Commutative law* Both the operations are commutative, since from tables we observe that

$$a+b=b+a=b \text{ and } a\cdot b=b\cdot a=a$$

(*ii*) *Associative law* Both the operations are associative, since from tables we observe that

$$a+(a+b)=(a+a)+b=a+b$$

and

$$a\cdot(a\cdot b)=(a\cdot a)\cdot b=a\cdot b$$

(*iii*) *Distributive law*

(*i*) $a + (a \cdot b) = (a \cdot a) \cdot b = a \cdot b$

(*ii*) $a \cdot (a + b) = a \cdot b = a$ and $a \cdot a + a \cdot b = a \cdot a = a$

Again, $b \cdot (a + b) = b \cdot b = b$ and $b \cdot a + b \cdot b = a + b = b$

(*iv*) *Identity element* Since

$a + a = a$ and $b + a = a + b = b$

therefore, a is the identity element for operation (+). Similarly, b is the identity element for operation (·), since $a \cdot b = a$ and $b \cdot b = b$.

(*v*) *Inverse* Since $a + b = b$, inverse of a is b, which is the identity element for operation (+) . Similarly, the inverse of b is a, since $b + a = a + b$ and $b \cdot a = a \cdot b = a$.

Since the mathematical structure [A, +, .] satisfies all the requirements, it is a Boolean algebra.

Example 5 Prove that $(a + b) \cdot (\bar{b} + c) + b \cdot (\bar{a} + \bar{c}) = a \cdot \bar{b} + a \cdot c + b$

Solution
$$\begin{aligned} \text{LHS} &= (a + b) \cdot (\bar{b} + c) + b \cdot (\bar{a} + \bar{c}) \\ &= a \cdot (\bar{b} + c) + b \cdot (\bar{b} + c) + b \cdot (\bar{a} + \bar{c}) \\ &= a \cdot (\bar{b} + c) + b \cdot (\bar{b} + \bar{a} + 1) \\ &= a \cdot (\bar{b} + c) + b \cdot 1 = a \cdot \bar{b} + a \cdot c + b = \text{RHS} \end{aligned}$$

Example 6 Prove that

(*i*) $(a \cdot b) \cdot c = a \cdot (b \cdot c)$ (*ii*) $(a + b) + c = a + (b + c)$

Solution (*i*) Let $p = a \cdot (b \cdot c)$ and $q = (a \cdot b) \cdot c$. Then we have to show that $p = q$.

(*a*) Since $a + p = a + q$, therefore

$$\begin{aligned} a + p &= a + a \cdot (b \cdot c) = (a + a \cdot b) \cdot (a + c) \\ &= a \cdot (a + c) = a \cdot a + a \cdot c \\ &= a + a \cdot c = a \\ &= a \cdot (a + c) = (a + a \cdot b) \cdot (a + c) \\ &= a + (a \cdot b) \cdot c = a + q \end{aligned}$$

(*b*) To prove, $\bar{a} + p = \bar{a} + q$, let

$$\begin{aligned} \bar{a} + p &= \bar{a} + a \cdot (b \cdot c) \\ &= (a + \bar{a}) \cdot (a + b \cdot c) = 1 \cdot (\bar{a} + b \cdot c) \\ &= \bar{a} + b \cdot c = (\bar{a} + b) \cdot (\bar{a} + c) \\ &= 1\,(\bar{a} + b) \cdot (\bar{a} + c) \\ &= (\bar{a} + a) \cdot (\bar{a} + b)\,(\bar{a} + c) \\ &= (a + a \cdot b) \cdot (\bar{a} + c) = \bar{a} + (a \cdot b \cdot c) = \bar{a} + q \end{aligned}$$

Let $p = q$. Then from (*a*) and (*b*), we have

$$\begin{aligned} (a + p) \cdot (\bar{a} + p) &= (a + q) \cdot (\bar{a} + q) \\ a \cdot \bar{a} + p &= a \cdot \bar{a} + q \\ 0 + p &= 0 + q \\ p &= q \end{aligned}$$

Hence, $a \cdot (b \cdot c) = (a \cdot b) \cdot c$

(*ii*) It is the dual of (*i*).

Example 7 If $x \in B$, then prove that

(a) $x \cup x = x$ (b) $x \cap x = x$ (c) $x \cup 1 = 1$ (d) $x \cap 0 = 0$

Solution (a) $x \cup x = (x \cup x) \cap 1$ [since 1 is the identity element]

$= (x \cup x) \cap (x \cup \bar{x})$ [since $\bar{x}$ is the complement]

$= x \cup (x \cap \bar{x}) = x \cup 0 = x$

(b) $x \cap x = (x \cap x) \cup 0 = (x \cap x) \cup (x \cap x)$

$= x \cap (x \cup \bar{x}) = x \cap 1 = x$

(c) $x \cup 1 = (x \cup 1) \cap 1 = (x \cup 1) \cap (x \cup \bar{x})$

$= x \cup (1 \cap \bar{x}) = x \cup \bar{x} = 1$

(d) $x \cap 0 = (x \cap 0) \cup 0 = (x \cap 0) \cup (x \cap \bar{x})$

$= x \cap (0 \cup \bar{x}) = x \cap x = 0$

Example 8 If $x \in B$, then prove that (a) $x \cup (x \cap y) = x$ and (b) $x \cap (x \cup y) = x$.

Solution (a) $x \cup (x \cap y) = (x \cap 1) \cup (x \cap y)$

$= x \cap (1 \cup y) = x \cap 1 = x$

(b) Left as an exercise for the reader.

Theorem 13.10 If [B, +, ·] is a Boolean algebra, then prove that the complement $\bar{a}$ of any element $a \in A$ is unique.

Solution Let $a \in A$ and suppose that its complement is not unique. If possible, suppose that $\bar{a}$ and $\bar{\bar{a}}$ are two complements of a such that

$a + \bar{a} = 1$, and $a \cdot \bar{a} = 0$

$a + \bar{\bar{a}} = 1$, and $a \cdot \bar{\bar{a}} = 0$

Now $\bar{a} = \bar{a} \cdot 1 = \bar{a} \cdot (a + \bar{\bar{a}})$

$= \bar{a} \cdot a + \bar{a} \cdot \bar{\bar{a}} = 0 + \bar{a} \cdot \bar{\bar{a}}$

$= a \cdot \bar{\bar{a}} + \bar{a} \cdot \bar{\bar{a}}$ [Assumption]

$= \bar{\bar{a}} (a + \bar{a}) = \bar{\bar{a}} \cdot 1 = \bar{\bar{a}}$

It follows that $\bar{a} = \bar{\bar{a}}$, which contradicts the assumption that element a has two different complements. Hence, every element in Boolean algebra has a unique complement.

Example 9 In a certain town all cars marked P belong to the police. The cars are not fitted with the wireless unless they are marked P. The police has no blue cars. Prove by using the Boolean algebra that no blue cars are fitted with the wireless.

Solution Let us denote by letters p, q, r and s the following statements

p = cars belong to the police q = cars are fitted with the wireless

r = cars are marked P s = cars are blue.

Therefore, by given condition, we have $r \Rightarrow p$, $q \Rightarrow r$, $p \Rightarrow s$

But if $r \Rightarrow p$ and $p \Rightarrow \bar{s}$, then $r \Rightarrow \bar{s}$

Also, if $q \Rightarrow r$ and $r \Rightarrow \bar{s}$, then $q \Rightarrow \bar{s}$

or $s \cdot q \Rightarrow s \cdot \bar{s} = 0$ [since $s \cdot \bar{s} = 0$]

Hence, it follows that no blue cars are fitted with the wireless.

13.5 BOOLEAN FUNCTIONS

To simplify the definition of Boolean functions, we shall be using symbols for operations from chapters on Set Theory and Mathematical Logic. Boolean functions are very helpful in understanding circuits in switching theory.

Definition Let [B, +, ·, –] be a Boolean algebra. Let $x_1, x_2, \ldots, x_n$ be variables (statements) in B which can assume values 0 and 1. The equation generated by the variables $x_1, x_2, \ldots, x_n$ is a valid combination of x_is with the operations defined on A.

If $(x_1, x_2, \ldots, x_n)$ is a Boolean expression, then the function $f(x_1, x_2, \ldots, x_n)$ generated by its n variables is defined as:

$$f : A_n \to A.$$

The Boolean expression represents a compound statement having all combinations of 0s and 1s for variables x_is in the expression.

The values 0 and 1 for a statement in the function $f(x_1, x_2, \ldots, x_n)$ implies that the statement is false and true, respectively and hence truth or falsity of $f(x_1, x_2, \ldots, x_n)$ depends on the truth or falsity of its component statements x_i.

Boolean functions may be used to represent output requirements of a circuit for any possible input values as shown in Fig. 13.1. Each variable x_i carries two indicater voltages, i.e. one for 0 and other for 1. Thus, for all such cases, the function f indicates output response. Such requirements are used while designing combinatorial and sequential computer circuits.

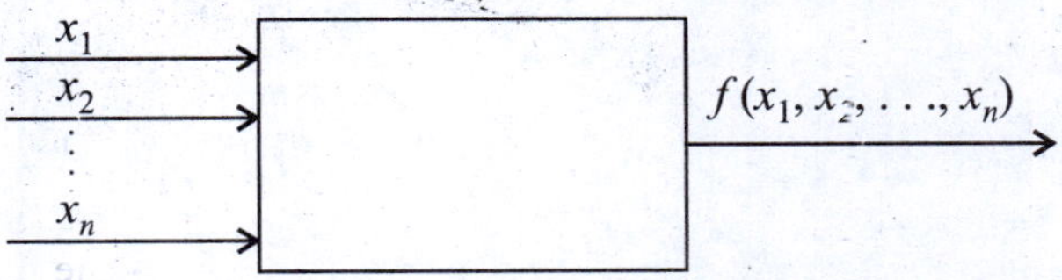

Fig. 13.1 Input-Output Diagram

Example 10 Show that (a) $\overline{[(x \cap y) \cup [(x \cup \bar{y}) \cap y]} = 1$ and (b) $\overline{[(x \cdot y)(\bar{z} + x\bar{y})]} = \bar{x} + \bar{y} + \bar{z}$

Solution (a) LHS $= \overline{[(x \cap y) \cup [(x \cup \bar{y}) \cap y]}$

$= (x \cap y) \cup [\overline{(x \cup \bar{y})} \cup \bar{y}] = (x \cap y) \cup [(\bar{x} \cap y) \cup \bar{y}]$

$= (x \cap y) \cup [(\bar{x} \cup \bar{y}) \cap (y \cup \bar{y})] = (x \cup y) \cup [(\bar{x} \cup \bar{y}) \cap 1]$

$= (x \cap y) \cup (\bar{x} \cap \bar{y}) = 1 = \text{RHS}$

(b) LHS $= \overline{[(x \cdot y)(\bar{z} + x\bar{y})]} = \overline{(x \cdot y)} + \overline{(\bar{z} \cdot x\bar{y})}$

$= \bar{x} + \bar{y} + [z \cdot (\bar{x} + y)] = \bar{x} + \bar{y} + z\bar{x} + zy$

$= \bar{x} + \bar{x} \cdot z + \bar{y} + yz = \bar{x} + \bar{y} + \bar{z} = \text{RHS}$ [Since $x + \bar{x}y = x + y$ and $x + xy = x$]

The results can be verified by the following truth table. Let $a = x \cap y$, $b = (x \cup \bar{y}) \cap y$.

x	y	$a = x \cap y$	$\bar{y}$	$x \cup \bar{y}$	$b = (x \cup \bar{y}) \cap y$	$\bar{b}$	$a \cup \bar{b}$
1	1	1	0	1	1	0	1
1	0	0	1	1	0	1	1
0	1	0	0	0	0	1	1
0	0	0	1	1	0	1	1

Example 11 Prove that $\overline{[\{(\bar{x} \cap \bar{y}) \cup z\} \cap (x \cup z)]} = \bar{x} \cap \bar{z}$

Solution

$$\text{LHS} = \overline{[\{(\overline{\bar{x} \cap \bar{y}}) \cup z\} \cap (x \cup z)]} = [\overline{\{(\overline{\bar{x} \cap \bar{y}}) \cup z\}} \cup \overline{(x \cup z)}]$$

$$= [\overline{\{(\overline{\bar{x} \cap \bar{y}}) \cup z)} \cup (\overline{\bar{x} \cap \bar{z}})]$$

$$= (\bar{x} \cap \bar{y} \cap \bar{z}) \cup (\bar{x} \cap \bar{z}) = \bar{x} \cap \bar{z} = \text{RHS} \qquad [\text{since } \bar{x} \cap \bar{y} \cap \bar{z} \subseteq \bar{x} \cap \bar{z}]$$

Example 12 Simplify, $\overline{[(x \cup \bar{y}) \cap (x \cap \bar{y} \cap \bar{z})]}$

Solution

$$\overline{[(x \cup \bar{y}) \cap (x \cap \bar{y} \cap \bar{z})]} = \overline{(x \cup \bar{y})} \cup (x \cap \bar{y} \cap z) = (\bar{x} \cap y) \cup (x \cap \bar{y} \cap z)$$

$$= [(x \cup y) \cap (y \cup z)] \cap [(\bar{x} \cup z) \cap (\bar{x} \cup \bar{y})]$$

Example 13 Simplify the Boolean expression:

$$x + \bar{x} \cdot (x + y) + y \cdot z$$

Solution We have,

$$x + \bar{x} \cdot (x + y) + y \cdot z = x + (\bar{x} \cdot x + \bar{x} \cdot y) + y \cdot z \qquad [\text{Distributive law}]$$

$$= x + (0 + \bar{x} \cdot y) + y \cdot z \qquad [\text{Since } \bar{a} \cdot a = 0]$$

$$= x + \bar{x} \cdot y + y \cdot z \qquad [\text{Associative law}]$$

$$= (x + \bar{x}) \cdot (x + y) + y \cdot z \qquad [\text{Distributive law}]$$

$$= 1 \cdot (x + y) + y \cdot z \qquad [\text{Since } x + \bar{x} = 1]$$

$$= (x + y) + y \cdot z = x + y \cdot (1 + z) \qquad [\text{Since } 1 \cdot a = a]$$

$$= x + y \cdot 1 = x + y \qquad [\text{Since } 1 + z = 1]$$

Example 14 Using properties of Boolean algebra, prove that for each x in a Boolean algebra, if $x + y = 1$ and $x \cdot y = 0$, then $y = \bar{x}$.

Solution We have,

$$\bar{x} = \bar{x} + 0 \qquad [\text{Since } x = x = 0]$$

$$= \bar{x} + x \cdot y \qquad [\text{Since } x \cdot x = 0 \text{ (given)}]$$

$$= (\bar{x} + x) \cdot (\bar{x} + y) \qquad [\text{Distributive law}]$$

$$= 1 \cdot (\bar{x} + y) \qquad [\text{Since } x + \bar{x} = 1]$$

$$= (x + y) \cdot (\bar{x} + y) \qquad [\text{Since } x + y = 1 \text{ (given)}]$$

$$= (y + x) \cdot (y + \bar{x}) \qquad [\text{Commutative law}]$$

$$= y + (x \cdot \bar{x}) \qquad [\text{Distributive law}]$$

$$= y + 0 = y. \qquad [\text{Since } x \cdot \bar{x} = 0]$$

Example 15 Simplify the Boolean expression:

$$(x + y) \cdot (x + z) \cdot (\overline{\bar{x} \cdot y})$$

Solution We have,

$$(x + y) \cdot (x + z) \cdot (\overline{\bar{x} \cdot y}) = (x + y) \cdot (x + z) \cdot [(\overline{\bar{x}}) \cdot \bar{y}] \qquad [\text{DeMogran's law}]$$

$$= (x + y) \cdot (x + z) \cdot (x + \bar{y}) \qquad [\text{Since } (\overline{\bar{x}}) = x]$$

$$= (x + y) \cdot (x + z) \cdot (x + \bar{y}) \qquad [\text{Since } (a + b) \cdot (a + c) = a + b \cdot c)]$$

$$= x + (y \cdot z) \cdot \bar{y} = x + \bar{y} \cdot (y \cdot z) = x + (\bar{y} \cdot y) \cdot z$$

$$= x + 0 \cdot z = x + 0 = x.$$

13.6 APPLICATIONS OF BOOLEAN ALGEBRA

One of the practical applications of Boolean algebra is to facilitate the logical design of certain electrical network (or circuit). An electrical switch has only two states: *closed* and *open*, i.e. one allowing the current to pass (closed state) and one not allowing current to pass (open state). A common household light switch is an example of a two-state switch.

The standard notations used for Boolean algebra operations in the logic design of switching circuits are: (+) for $\vee$ or $\cup$; ($\cdot$) for $\wedge$ or $\cap$, and (–) for ~ or ^{c} (complement).

In Boolean algebra digits 0 and 1 do not repres-ent actual numbers. Several terms that are used synonymously with 0 and 1 in the digital logic field are given in the following table.

Logic 0	*Logic 1*
False	True
Off	On
Open	Closed
No	Yes
F	T

13.6.1 Open and Closed Switches

Consider electrical networks as shown in Figs. 13.2 (*a*) and (*b*). In these diagrams *p* represents a switch, B represents a battery and L represents an electric lamp. *A switch is said to be open if L does not light otherwise closed*, i.e. an open switch does not allow the flow of current while a closed switch allows the flow of current. The value of a closed switch when L light is 1 (one) and value of an open switch when L does not light is 0 (zero). The states of the switch and lamp are shown in Table 13.2.

Table 13.2 States of a Switch

State of Switch *p*	*State of Lamp* *L*
0 (open)	0 (OFF)
1 (closed)	1 (ON)

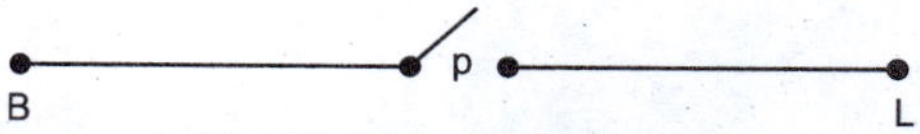

Fig. 13.2(*a*) Open Switch

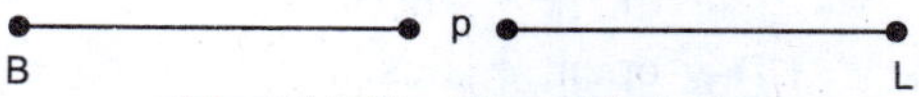

Fig. 13.2(*b*) Closed Switch

13.6.2 Boolean Multiplication (Switches in Series)

The logical multiplication in Boolean algebra is also referred to as AND operation $p \wedge q$ or pq circuit. The logical multiplication of two variables p and q is expressed as: $p \cdot q = r$ and is read as: p AND q equals r. Each of the variable p, q and r may have only the value 0 or 1. An application of multiplication operation is as follows:

If in an electric circuit two switches p and q are connected as shown in Fig. 13.3, then the switches are said to be in *series*.

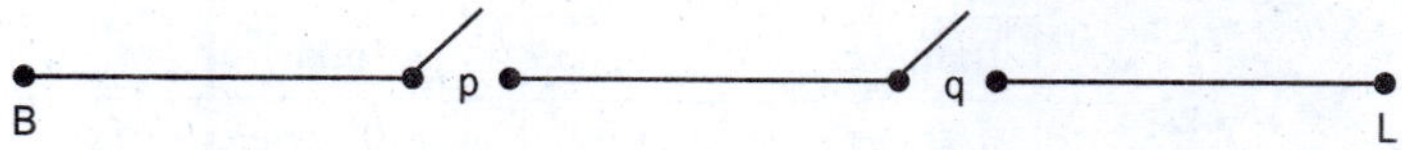

Fig. 13.3 Switches in Series

In general, four different states may exist in this types of circuits, namely,

(*i*) p open, q open (*ii*) p open, q closed

(*iii*) p closed, q closed (*iv*) p colsed, q open

Clearly the lamp L will light up in only one of the four cases, i.e. when both the switches are closed The states of these switches and lamp is similar to the truth table for $p \wedge q$ in mathematical logic. The different states of switches which are connected in series and position of lamp are shown in Table 13.3

Table 13.3 States of Two Switches in Series (Logical Multiplication or AND Operation)

Input, p	Input, q	Output, L($p.q$)
0 (open)	0 (open)	0 (OFF)
0 (open)	1 (closed)	0 (OFF)
1 (closed)	0 (open)	0 (OFF)
1 (closed)	1 (closed)	1 (ON)

There can be more than two switches in a series. If there are three, then there are $2^3 = 8$ possibilites.

13.6.3 Boolean Addition (Switches in Parallel)

The logical addition in Boolean algebra is also referred to as OR operation. The logical addition of two variables p and q is expressed as: $p + q = r$ and is read as: p OR q equals r. Each of the variable p, q and r may have only the value 0 or 1.

If two switches r and s in an electrical circuit are connected as shown in Fig. 13.4, then the switches are said to be in *parallel*.

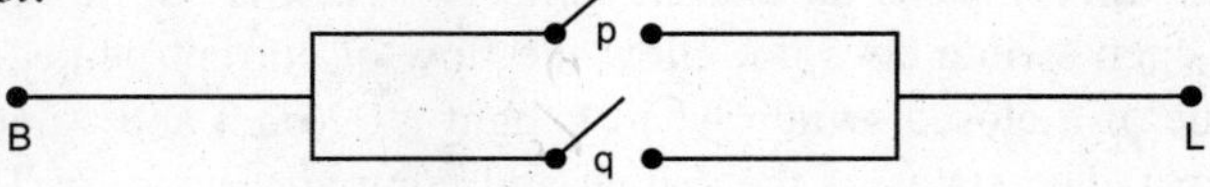

Fig. 13.4 Switches in Parallel

In general four different states may exist in this types of circuits such as:

(*i*) p open, q open

(*ii*) p open, q closed

(*iii*) p closed, q open

(*iv*) p closed, q closed

The current will not flow when both switches are in open state. In other word, lamp L will light up if and only if either or both the switches are closed. Table 13.4 shows the states of these switches and lamp is analogous to the table of $p \vee q$ in mathematical logic. The different states of switches which are connected in parallel and position of lamp are shown in Table 13.4.

Table 13.4 States of Switches in Parallel (Logical Addition OR Operation)

Input, p	Input, q	Output, L($p+q$)
0 (open)	0 (open)	0 (OFF)
0 (open)	1 (closed	1 (ON)
1 (closed)	0 (open)	1 (ON)
1 (closed)	1 (closed)	0 (OFF)

13.6.4 Boolean Complementation

As mentioned earlier, complementation operation represented by symbol '–' or ''' is defined as $\overline{0} = 1$ and $\overline{1} = 0$.

The complementation operation is also referred to as NOT operation. Thus the process of complementing is also called negation as shown below:

Input	*Output*
0	1
1	0

13.6.5 Complex Circuits

Consider few electric circuits which function based on *AND* and/or *OR* operations as shown in Fig. 13.5.

(*i*) The lower branch in the circuit shown in Fig. 13.5 is *AND* ciucuit, which is represented by $p \cdot q$ in Boolean algebra [$p \cap q$ in set theory; $p \wedge q$ in logic].

The two branches together comprise an *OR* circuit which is represented by $p + q \,.\, q$ in Boolean algebra [$p \cup (p \cap q)$ in set theory $p \vee (p \wedge q)$ in logic].

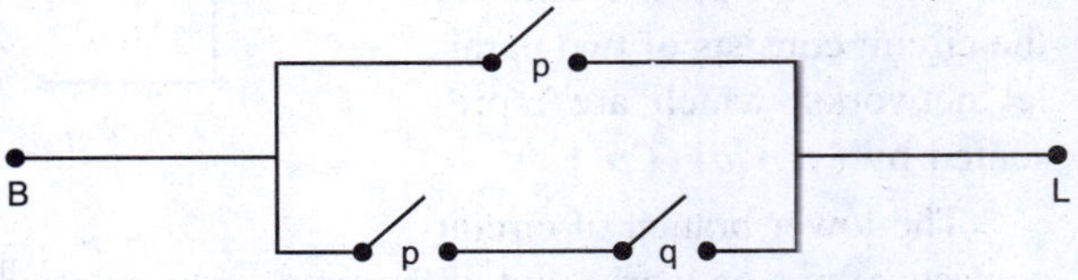

Fig. 13.5 Circuit, $p + p \,.\, q$

(*ii*) Branches q and r form an *OR* circuit represented by $q + r$ in Boolean algebra. When combine with p it forms an *AND* circuit, which is represented by $p \cdot (q + r)$ as shown in Fig. 13.6.

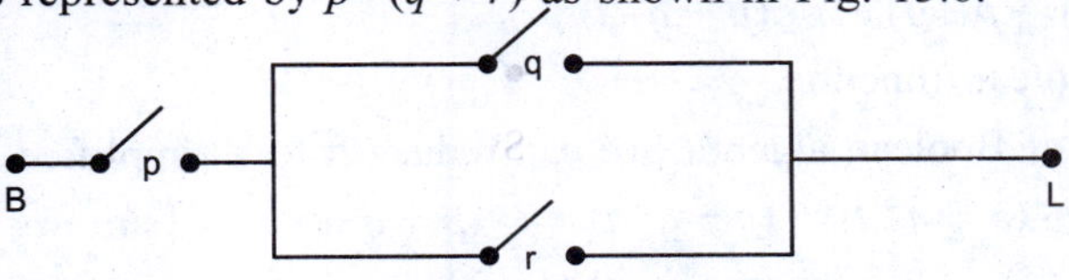

Fig. 13.6 Circuit, $p{\cdot}(q + r)$

(*iii*) Circuit showing $p \cdot q + p \cdot r$ is shown in Fig. 13.7.

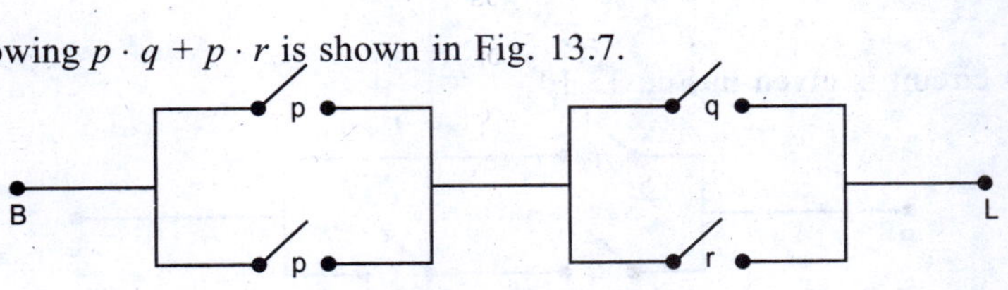

Fig. 13.7 Circuit, $p \cdot q + p \cdot r$

(*iv*) Circuit: $(p + q).\, r\, (u + v + w)$ is shown in Fig. 13.8.

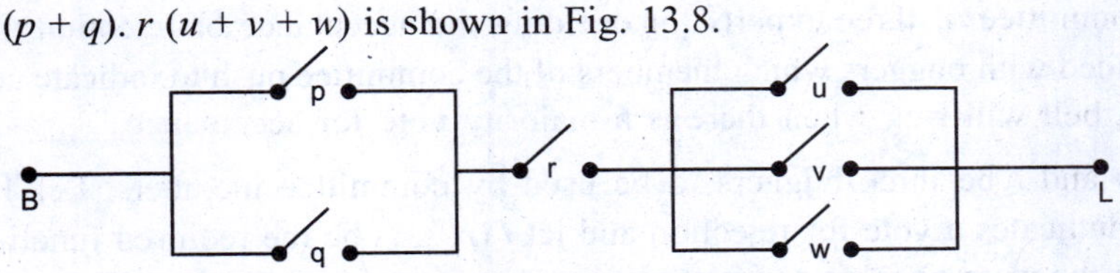

Fig. 13.8 Circuit, $(p + q)\,.\, r \cdot (u + v + w)$

A summary of symbols used in set theory, mathematical logic, and Boolean algebra is shown in Table 13.5.

Table 13.5 Summary of Symbols

Set Theory	*Logic*	*Boolean Algebra*	*Circuits*
A ∪ B	$p \vee q$	$a + b$	Switches in parallel
A ∩ B	$p \wedge q$	$a \cdot b$	Switches in series
∪	T	1	Current flows
Φ	F	0	No current

Example 16 Find the Bool-ean function of the switching circuit given in Fig. 13.9 and then its equivalent simplified circuit. Also draw the simplified circuit.

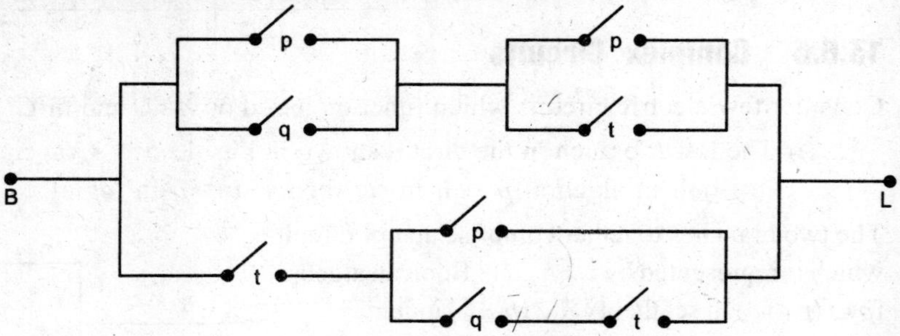

Fig. 13.9

Solution The upper branch of the circuit consists of two parallel networks, which are represented by $(p + q) \cdot (p + t)$.

The lower branch of circuit consists of two networks, out of those one is parallel and other in series which are equivalent to $t \cdot (p + q \cdot t)$.

The upper and lower branches together form a parallel network and is represented as:

$$(p + q) \cdot (p + t) + t \cdot (p + q \cdot t).$$

This is the required Boolean function.

Now with the help of Boolean algebra, we can reduce it to a simplified circuits as follows:

$$(p + q) \cdot (p + t) + t \cdot (p + q \cdot t) = (p + q \cdot t) + t \cdot (p + q \cdot t) \quad [\text{since } a + b \cdot c = (a + b) \cdot (a + c)]$$

$$= (p + q \cdot t)\,(1 + t) \quad [\text{since } a \cdot (b + c) = a \cdot b + a \cdot c]$$

$$= (p + q \cdot t) \cdot 1 \quad [\text{since } 1 + a = 1]$$

$$= p + q \cdot t$$

The equivalent circuit is given in Fig. 13.10.

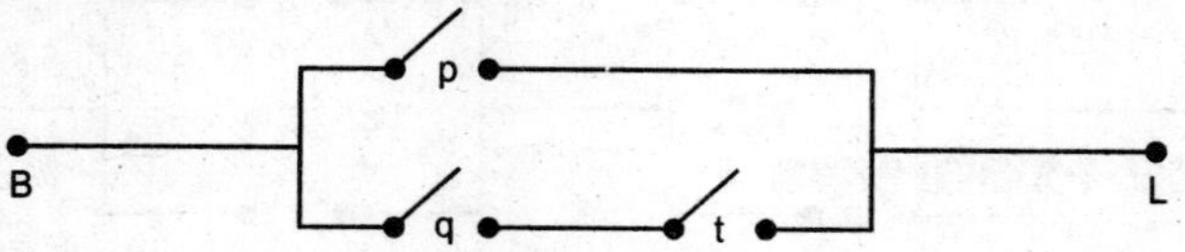

Fig. 13.10

Example 17 A committee of three experts for deciding the acceptance or rejection of photographs for exhibition is provided with buggers which members of the committee push to indicate acceptance. Design a circuit so that a bell will ring when there is a majority vote for acceptance.

Solution Let r, s and t be three buggers to be used by committee members. Let 1 indicates a vote for acceptance, 0 indicates a vote for rejection and let $f(r, s, t)$ be the required function whose truth or falsity depends on the truth or falsity of its components r, s and t. Table 13.6 shows the truth or falsity of $f(r, s, t)$.

Table 13.6

r	s	t	$f(r, s, t)$
1	1	1	0
1	1	0	1
1	0	1	0
1	0	0	1
0	1	1	0
0	1	0	1
0	0	1	1
0	0	0	0

From Table 13.6, function f (only when $f = 1$) if formed appears as:

$$f = r \cdot s \cdot \bar{t} + r \cdot \bar{s} \cdot \bar{t} + \bar{r} \cdot s \cdot \bar{t} + \bar{r} \cdot \bar{s} \cdot t$$

$= r \cdot s \cdot (t + \bar{t}) + r \cdot \bar{s} \cdot t + \bar{r} \cdot s \cdot t$

$= r \cdot s \cdot 1 + r \cdot \bar{s} \cdot t + \bar{r} \cdot s \cdot t = (r \cdot s + r \cdot \bar{s} \cdot t) + \bar{r} \cdot s \cdot t$

$= (r \cdot s + r \cdot \bar{s}) \cdot (r \cdot s + t) + \bar{r} \cdot s \cdot t$ [since $a + b \cdot c = (a + b) \cdot (a + c)$]

$= r \cdot (s + \bar{s}) \cdot (r \cdot s + t) + \bar{r} \cdot s \cdot t$

$= r \cdot 1\,(r \cdot s + t) + \bar{r} \cdot s \cdot t$ [since $a + \bar{a} = 1$]

$= r \cdot r \cdot s + r \cdot t + \bar{r} \cdot s \cdot t = r \cdot s + r \cdot t + \bar{r} \cdot s \cdot t$ [since $a \cdot a = a$]

$= r \cdot t + (r \cdot s + \bar{r} \cdot s \cdot t)$

$= r \cdot t + (r \cdot s + \bar{r} \cdot s) \cdot (r \cdot s + t)$ [since $a + b \cdot c = (a + b) \cdot (a + c)$]

$= r \cdot t + s \cdot (r + \bar{r}) \cdot (r \cdot s + t)$

$= r \cdot t + s \cdot 1 \cdot (r \cdot s + t)$ [since $a + \bar{a} = 1$]

$= r \cdot t + r \cdot s \cdot s + s \cdot t = r \cdot t + r \cdot s + s \cdot t$ [since $a \cdot a = a$]

The required circuit is shown in Fig. 13.11.

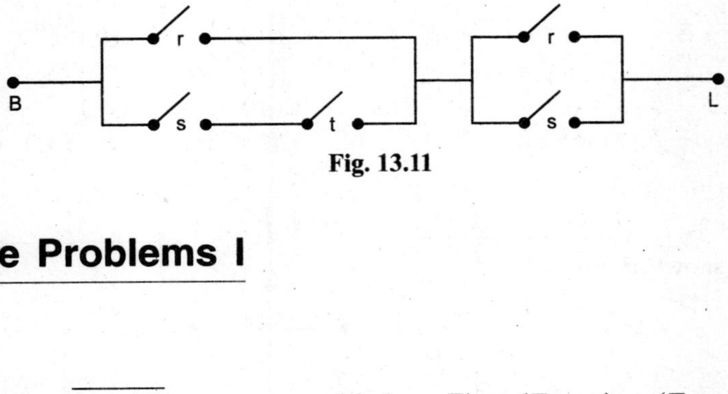

Fig. 13.11

Self-practice Problems I

1. Simplify:
   (*i*) $[(\bar{x} \cap \bar{y}) \cup z] \cap \overline{(x \cup \bar{y})}$ (*ii*) $(x \cup \bar{y}) \cap (\bar{x} \cup y) \cap (\bar{x} \cup \bar{y})$
   (*iii*) $[(x \cup y) \cap (z \cup \bar{y})] \cup [y \cap (\bar{x} \cup \bar{z})]$
2. Simplify: (*i*) $(x \cup y) \cap (\bar{x} + \bar{y})$ (*ii*) $(x \cap y \cap z) \cup (\bar{x} \cap \bar{y} \cap \bar{z})$
3. Simplify: (*i*) $(x \wedge \bar{y}) \vee [x \wedge \overline{(y \wedge z)}] \vee z$ (*ii*) $(x \wedge y) \vee (x \wedge \bar{y}) \vee (\bar{x} \vee y)$

4. Simplify:

(*i*) $\bar{a} + (a + \bar{b} + \bar{c})$ (*ii*) $(a + b) \cdot (\bar{a} + \bar{b})$

(*iii*) $a \cdot b \cdot (\bar{a} + \bar{b})$ (*iv*) $a \cdot b \cdot (a \cdot b + \bar{a} \cdot c)$

(*v*) $a \cdot (a + b) + \overline{[(\bar{b} + a) \cdot b]}$ (*vi*) $a + \bar{a} \cdot (a + b) + b \cdot c$

(*vii*) $(a + \bar{b}) \cdot (\bar{a} + b) \cdot (\bar{a} + \bar{b})$

5. Factorise and simplify:

(*i*) $(x \cdot \bar{y} + y \cdot z) \cdot (x \cdot z + y \cdot \bar{z})$ (*ii*) $\bar{y} \cdot (y + z) \cdot (y + \bar{z})$

(*iii*) $(x + y \cdot z) \cdot (\bar{y} + x)(\bar{y} + \bar{z})$

6. By using the distributive law of Boolean algebra factorize the following:

(*i*) $d + a \cdot b \cdot c$ (*ii*) $a \cdot b + c \cdot d$ (*iii*) $a \cdot \bar{b} + \bar{a} \cdot b$

7. Find the complement. (*i*) $p + q \cdot \bar{r}$

(*ii*) $(p + q) \cdot (q + \bar{r})$

8. Draw the circuits of $x \cdot (x \cdot y + \bar{x} + x \cdot \bar{y})$.

9. Write down the function which represent the circuit shown in Fig. 13.12 and simplify.

10. Prove: (*i*) $\overline{(a + b)} + \overline{(a + b)} = \bar{a}$

(*ii*) $a \cdot (\bar{a} + b) = \bar{a} \cdot b + a + a + b$

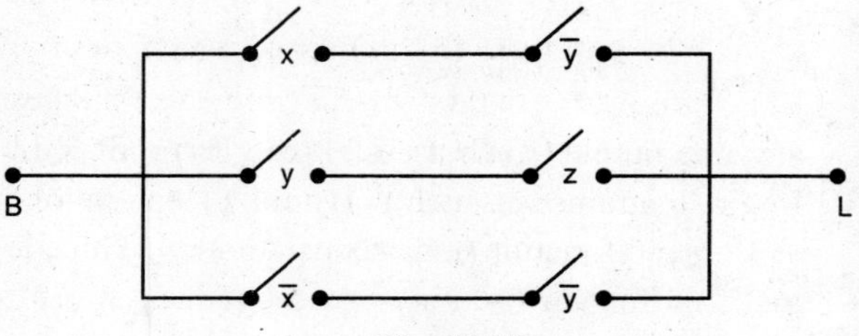

Fig. 13.12

Hints and Answers

1. (*i*) $\bar{x} \cap y$ (*ii*) $\bar{x} \cap \bar{y}$ (*iii*) $x \cup y$.

2. (*i*) 0 (*ii*) 1

3. (*i*) $x + z$ (*ii*) 1

4. (*i*) $\bar{a} \cdot \bar{b} + \bar{a} \cdot \bar{c}$ (*ii*) $a \cdot \bar{b} + \bar{a} \cdot b$ (*iii*) 0 (*iv*) $a \cdot b$

(*v*) 1, (*vi*) $a + b$ (*vii*) $\bar{a} \cdot \bar{b}$

5. (*i*) $x \cdot z$ (*ii*) 0 (*iii*) $x \cdot (\bar{y} + \bar{z})$

6. (*i*) $(d + a) \cdot (d + b) \cdot (d + c)$ (*ii*) $(a + c) \cdot (b + c) \cdot (a + d) \cdot (b + d)$

(*iii*) $(a + b) \cdot (\bar{a} + \bar{b})$

7. (*i*) $\bar{p} \cdot (\bar{q} + r)$ (*ii*) $\bar{p} \cdot \bar{q} + \bar{p} \cdot r = \bar{p} \cdot (\bar{q} + r)$

8. The circuit is shown in Fig. 13.13.

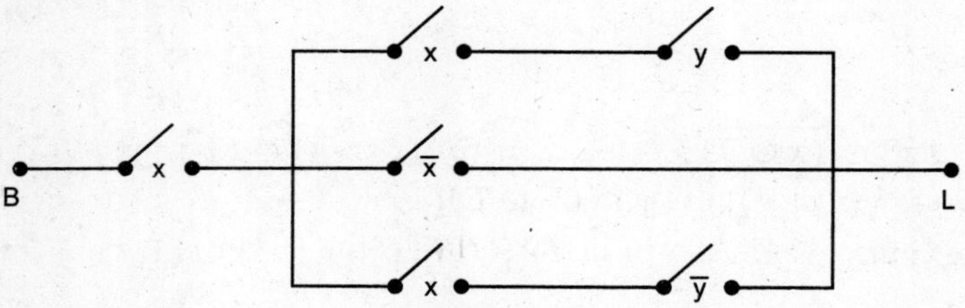

Fig. 13.13

9. The required function is

$$x \cdot \bar{y} + y \cdot z + \bar{x} \cdot \bar{y} = y \cdot z + \bar{y}\,(x + \bar{x}) = y \cdot z + \bar{y} \cdot 1 = \bar{y} + y$$

$$= (\bar{y} + y)(\bar{y} + z) = 1 \cdot (\bar{y} + z) = \bar{y} + z.$$

13.7 LOGIC GATES AND CIRCUITS

Logic circuits (or networks) are structures in a digital computer consist of several elementary circuits (also called switches) arranged in series and/or parallel. These circuits can be represented by logic diagrams as shown in Figs. 13.1 to 13.11. Each of these circuits can be viewed as a *logic gates* or *machines G,* consist of one or more input device and exactly one output device. Each input device sends a signal – a *bit* in the form of binary digit 0 or 1 to G and G processes the set of bits to yield an output bit. Hence, an *n*-bit sequence of 0 and 1 may be assigned to each input device and G processes the input sequences one bit at a time to produce an *n*-bit output sequence.

A Boolean function can be viewed as the description of particular logic circuit having several inputs and one output. Such a circuit can be produced by Boolean polynomials of the type: $r \cdot s$, $r + s$, $\bar{r}$, etc. Logic diagrams for such polynomials can be one way of constructing a circuit having same description as Boolean function under consideration. Thus, logic diagrams of these polynomials represent alternative methods for constructing the desired circuit in terms of logic gates.

In logic diagrams, gates are represented by symbols and inputs and output by arrowed lines labelled by letters. The following are gates symbols with their names which are used in the switching theory.

13.7.1 OR Gate

It is a circuit that has two or more inputs and its output is equal to the OR sum of the inputs. An input voltage indicator to gate has only two states: either 1 (high) or 0 (low). If x_1 and x_2 are input variables, then the output $f(x_1, x_2) = x_1 + x_2$ (or $x_1 \vee x_2$) is shown in Fig 13.14 and its truth table is shown in Table 13.7(*a*). In OR Gate, voltage indicator of the output is 1 (high) if at least one of the inputs is 1 (high). The output is 0 if all inputs are 0.

Table 13.7(*a*) Truth Table for OR Gate

Input		*Output*
x_1	x_2	$f(x_1, x_2) = x_1 + x_2$
1	1	1
1	0	1
0	1	1
0	0	0

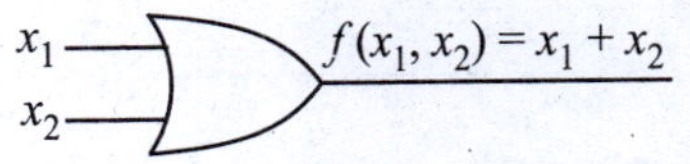

Fig. 13.14 Two-Input OR Gate Symbol

This concept can also be extended to more than two inputs. Figure 13.5 shows a three-input OR Gate symbol. The truth table is shown in Table 13.7(*b*).

Table 13.7(b) Truth Table for OR Gate and its Truth Table

Input			*Output*
x_1	x_2	x_3	$f(x_1, x_2, x_3) = x_1 + x_2 + x_3$
0	0	0	0
0	0	1	1
0	1	0	1
0	1	1	1
1	0	0	1
1	0	1	1
1	1	0	1
1	1	1	1

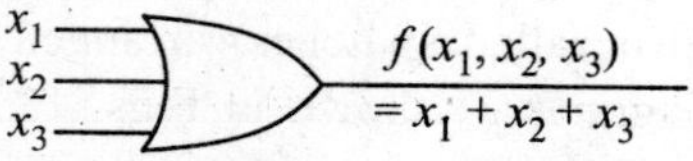

Fig. 13.15 Three-Input OR Gate Symbol

13.7.2 AND Gate

It is a circuit that has two or more inputs and its output is equal to the AND product of the inputs. For example, if x_1 and x_2 are input variables, then the output $f(x_1, x_2) = x_1 \cdot x_2$ (or $x_1 \wedge x_2$) is shown in Fig. 13.16 and its truth table is shown in Table 13.8(*a*). In two-input AND Gate, the voltage indicator of output is 1 (high) only if both inputs are 1 (high), otherwise the output is 0 (low).

Table 13.8(a) Truth Table for AND Gate

Input		*Output*
x_1	x_2	$f(x_1, x_2) = x_1 \cdot x_2$
1	1	1
1	0	0
0	1	0
0	0	0

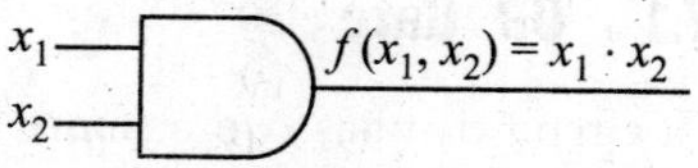

Fig. 13.16 Two-Input AND Gate Symbol

This concept can also be extended to more than two inputs. Figure 13.17 shows a three-inputs AND Gate symbol and its truth is shown in Table 13.8(*b*).

Table 13.8(b) Truth Table for AND Gate

Input			*Output*
x_1	x_2	x_3	$f(x_1, x_2, x_3) = x_1 \cdot x_2 \cdot x_3$
0	0	0	0
0	1	0	0
0	1	0	0
0	1	1	0
1	0	0	0
1	0	1	0
1	1	0	0
1	1	1	1

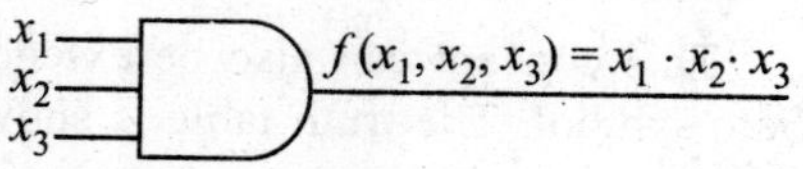

Fig. 13.17 Three-Input AND Gate Symbol

13.7.3 NOT (or Inverter) Gate

It is a circuit that has only one input and one output variable. Its output variable is the complement of the input variable. The output is 1 (high) if the input is 0 (low) and vice versa. If x_1 is the input variable, then the output is x'_1 or $\bar{x}_1$ as shown in Fig. 13.18. The truth table for NOT Gate is shown in Table 13.9.

Table 13.9 Truth Table for NOT Gate

Input x_1	*Output* $f(x_1) = \bar{x}_1$
0	1
1	0

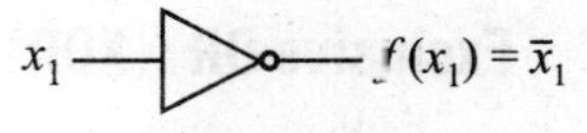

Fig. 13.18 NOT Gate Symbol

13.7.4 NOR and NAND Gates

Boolean algebra has three basic operations, namely, AND, OR and NOT, while other operations can be defined in terms of these operations. The NOR operation is the complement of OR operation and is also written as not-OR and uses an OR symbol followed by a small circle. Similarly, NAND operation is the complement of AND operation and is written as not-AND and uses an AND symbol followed by a small circle.

The truth table for NOR Gate is shown in Table 13.10. It shows that for every combination of input variables, the output of the NOR Gate is the complement of the output of the OR Gate.

Table 13.10 Truth Table for NOR Gate

Input		*Output*	
x_1	x_2	*OR* $f(x_1, x_2) = x_1 + x_2$	*NOR* $f^*(x_1, x_2) = \overline{(x_1 + x_2)}$
0	0	0	1
0	1	1	0
1	0	1	0
1	1	1	0

A combination of an OR Gate and NOT Gate is shown in Figs. 13.19(*a*) and (*b*), respectively.

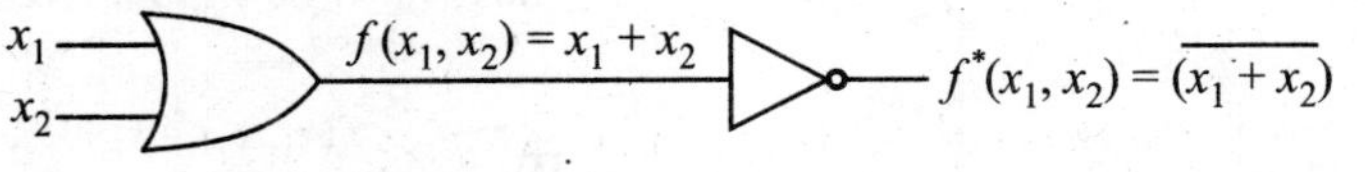

Fig 13.19(*a*) NOR Operation Logic Diagram

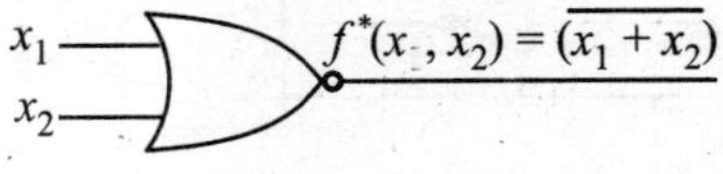

Fig 13.19(*b*) NOR Gate Symbol

The truth table for NAND Gate is shown in Table 13.11. It may be noted that for each combination of input variables, the output of NAND Gate is the complement of the output of AND.

Table 13.11 Truth Table for NAND Gate

Input		*Output*	
x_1	x_2	*AND* $f(x_1, x_2) = x_1 \cdot x_2$	*NAND* $f^*(x_1, x_2) = \overline{(x_1 \cdot x_2)}$
0	0	0	1
0	1	0.	1
1	0	0	1
1	1	1	0

The logic diagram of NOR operation and logic symbol for NOR Gate are shown in Figs. 13.20(*a*) and (*b*), respectively.

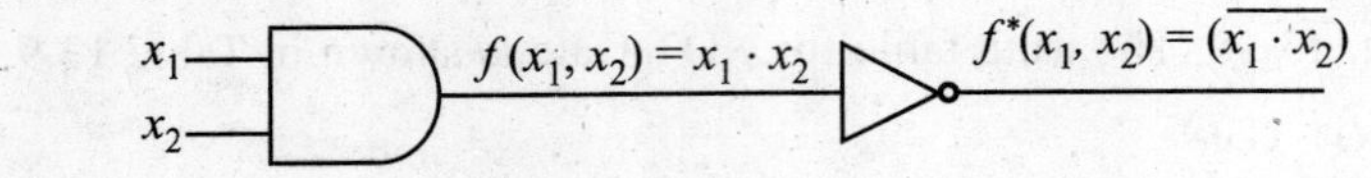

Fig. 13.20(*a*) NAND Operation Logic Diagram

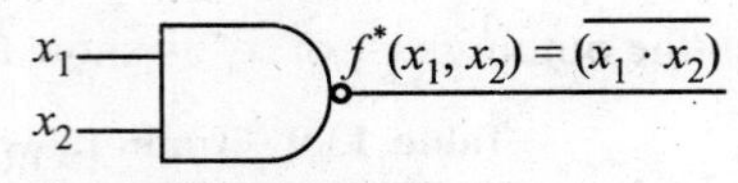

Fig. 13.20(*b*) NAND Gate Symbol

13.7.5 Exclusive-OR (XOR) Gate

The Exclusive-OR Gate is different than OR Gate because it includes only words that have an odd number of 1s. The truth table for a two-inputs XOR Gate is shown in Table 13.12.

Table 13.12 Truth Table for XOR Gate

Input		*Output*
x_1	x_2	$f(x_1, x_2) = \bar{x}_1 \cdot x_2 + x_1 \cdot \bar{x}_2$
0	0	0
0	1	1
1	0	1
1	1	0

In Boolean algebra, the sign ⊕ stands for XOR operation. Thus,

$$f(x_1, x_2) = \bar{x}_1 \cdot x_2 + x_1 \cdot \bar{x}_2 \quad \text{or} \quad f(x_1, x_2) = x_1 \oplus x_2$$

The summary of the output $f(x_1, x_2)$ for XOR operations is as follows:

$$0 \oplus 0 = 0; \qquad 0 \oplus 1 = 1$$

$$1 \oplus 0 = 1; \qquad 1 \oplus 1 = 0$$

Figures 13.21(*a*) and (*b*) show a two inputs XOR gate and its symbol, respectively.

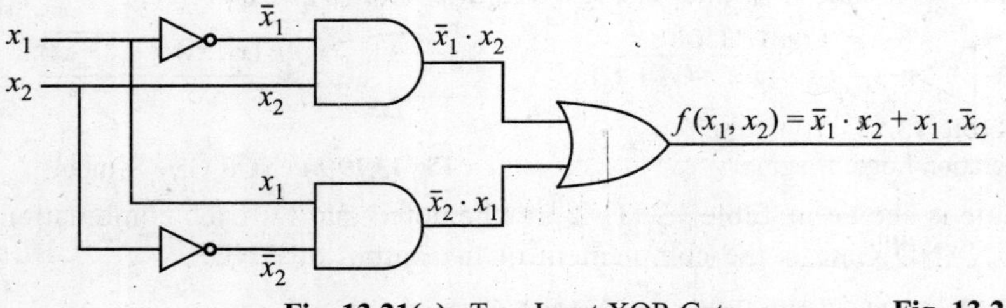

Fig. 13.21(*a*) Two-Input XOR Gate

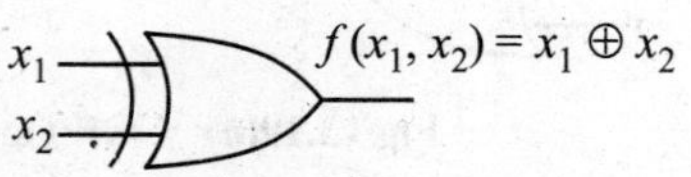

Fig. 13.21(*b*) Two-Input XOR Gate Symbol

From Table 13.12, it may be noted that the output is 1 only when number of binary 1 inputs are odd, otherwise the output is 0. An XOR Gate, irrespective of number of inputs, recognises only those words which have an odd number of binary 1s.

The logic diagram and logic symbol for XOR shown is Figs. 13.22(a) and (b), respectively.

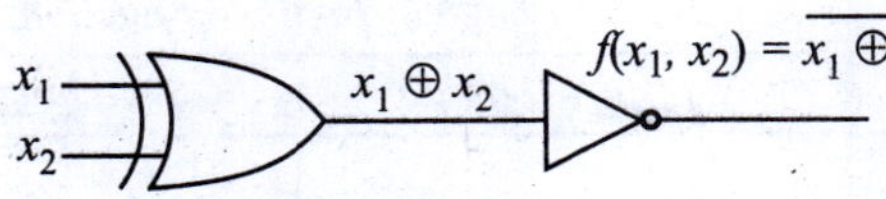

Fig. 13.22(*a*) XNOR Gate

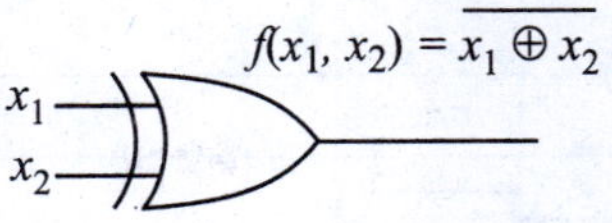

Fig. 13.22(*b*) XNOR Gate Symbcl

Remarks 1. More than two gates can be interconnected to form a logic network. A logic network with three inputs x_1, x_2 and x_3 and an output expression is shown in Figs. 13.23(a), (b) and (c).

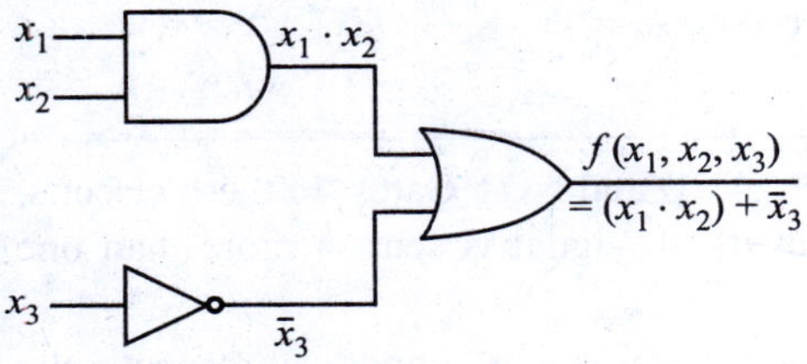

Fig. 13.23(*a*)

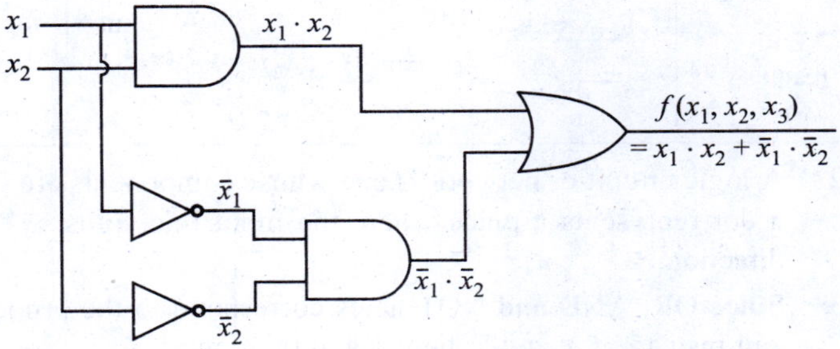

Fig. 13.23(*b*)

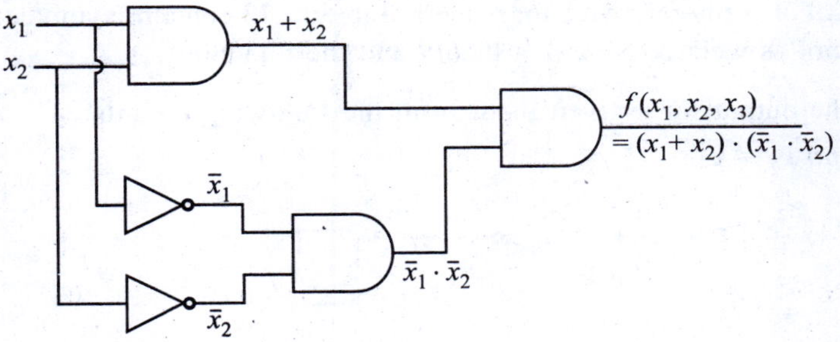

Fig. 13.23(*c*)

Table 13.13 Amalogy among Symbols

Operation	*Symbol*		*Logical/Boolean Function*	
	Input	*Output*	*Logic*	*Switching Theory*
AND (and)	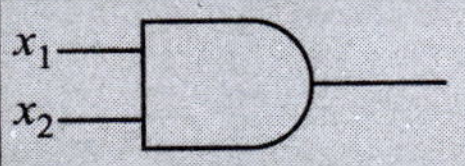	$f(x_1, x_2) = x_1 \cdot x_2$	$f(x_1, x_2) = x_1 \wedge x_2$	$f(x_1, x_2) = x_1 \cdot x_2$
OR (or)		$f(x_1\ x_2) = x_1 + x_2$	$f(x_1, x_2) = x_1 \vee x_2$	$f(x_1\ x_2) = x_1 + x_2$
NOT (not)	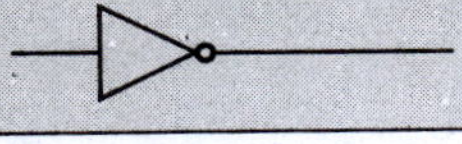	$f(x_1) = \bar{x}_1$	$f(x_1) = \bar{x}_1$	$f(x_1) = \bar{x}_1$

Contd...

Table 13.13 Amalogy among Symbols

Operation	*Symbol*		*Logical/Boolean Function*	
	Input	*Output*	*Logic*	*Switching Theory*
NAND (not-and)	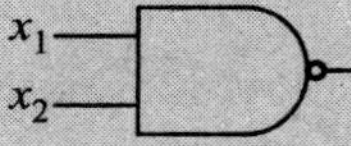	$f(x_1, x_2) = \overline{(x_1 \cdot x_2)}$	$f(x_1, x_2) = \overline{x_1 \wedge x_2}$	$f(x_1, x_2) = \overline{(x_1 \cdot x_2)}$
NOR (not-or)	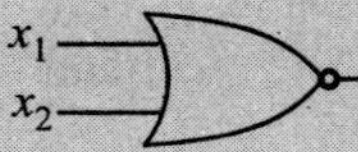	$f(x_1, x_2) = \overline{x_1 + x_2}$	$f(x_1, x_2) = \overline{x_1 \vee x_2}$	$f(x_1, x_2) = \overline{x_1 + x_2}$
XOR (exclusive -or)	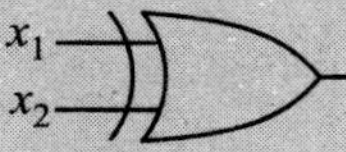	$f(x_1, x_2) = x_1 \oplus x_2$	$f(x_1, x_2) = x_1 \oplus x_2$	$f(x_1, x_2) = x_1 \oplus x_2$

2. A logic circuit or network is one whose components are OR, AND and NOT Gates. In these circuits, a dot represents a place where the input line splits so that its bit signal is sent in more than one direction.
3. Since OR, AND and NOT gates correspond to the propositions $p \vee q$, $p \wedge q$ and $\sim p$, respectively, and instead of T and F, here 1 and 0 are used, therefore, logic circuits also form a Boolean algebra.

From discussion on gates, it is clear that the circuit will allow the current to flow if and only if $f(x_1, x_2, x_3)$ is true. For ready reference for readers, Table 13.13 contains symbols used in mathematical logic, Boolean algebra as well as switching theory and their amalogy.

Example 18 Find the output for a given input from the following circuits:

(*a*) Inputs $x_1 = 0$ and $x_2 = 0$

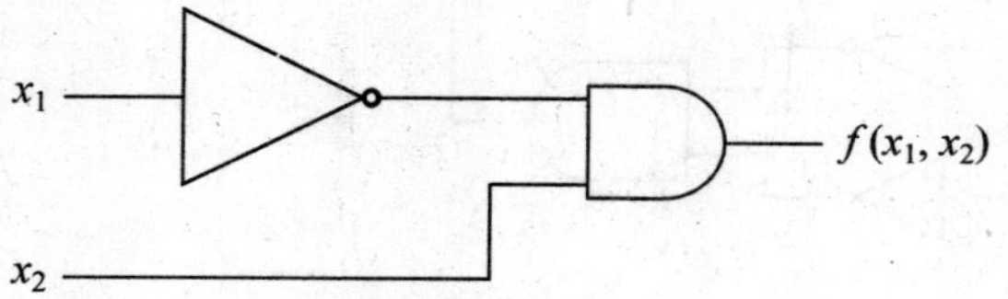

(*b*) Inputs $x_1 = 1$, $x_2 = 0$ and $x_3 = 1$.

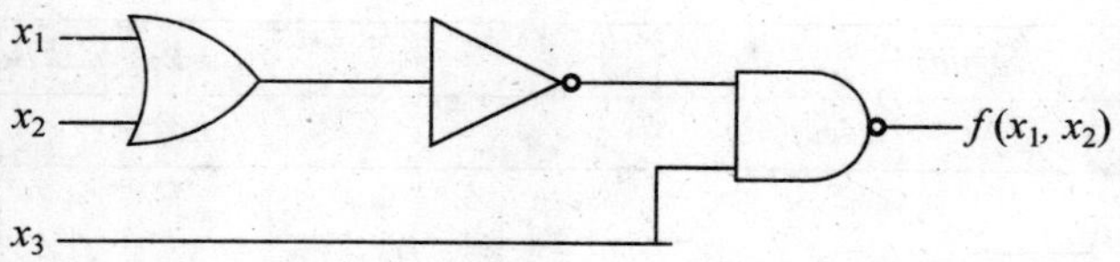

(*c*) Inputs $x_1 = 0$ and $x_2 = 1$.

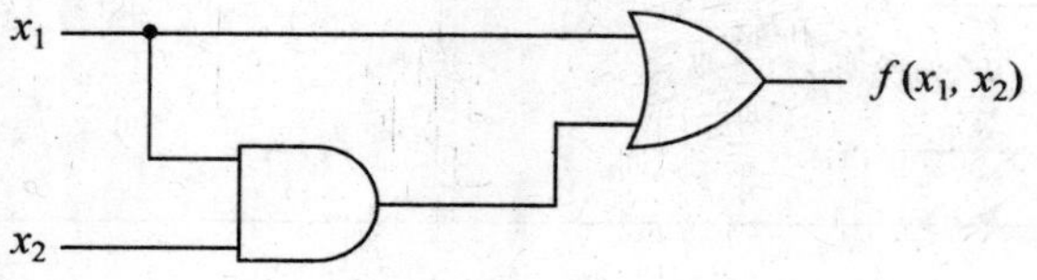

(*d*) Inputs $x_1 = 0$, $x_2 = 1$ and $x_3 = 1$.

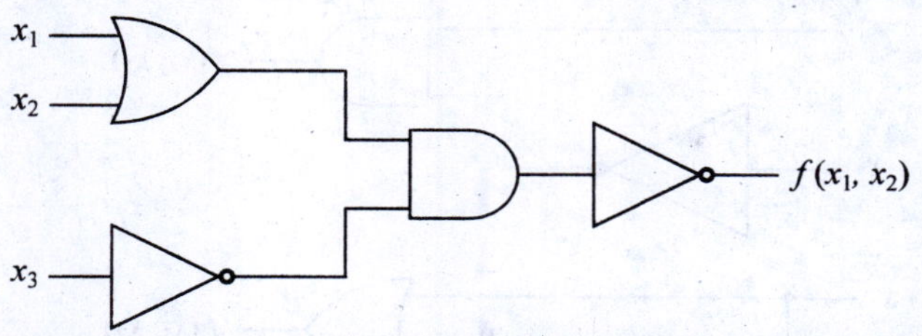

Solution (*a*) We move from left to right in the given circuit tracing the action of each gate on the input values.

The NOT gate changes the input $x_1 = 1$ to 0 and so both the inputs to the AND gate are 0. Consequently the output is: $s = 0$, as shown in Fig. 13.24(*a*).

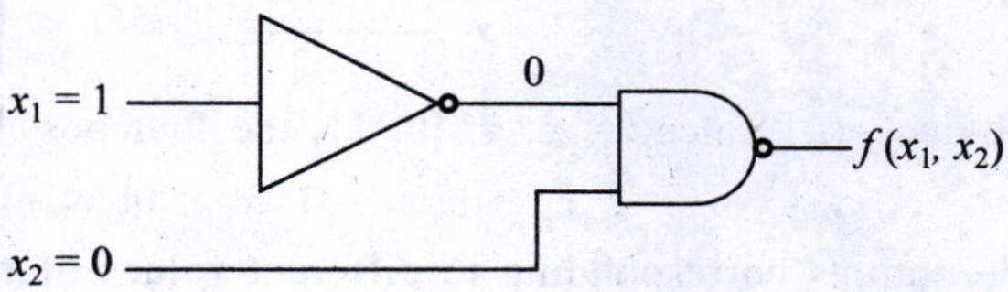

Fig. 13.24(*a*)

(*b*) The inputs to the OR gate are 1 and 0, therefore, its output is 1. NOT gate changes the input 1 to 0. Thus inputs to AND gate are 0 and 1, and hence, the output is $s = 0$, as shown in Fig. 13.24(*b*).

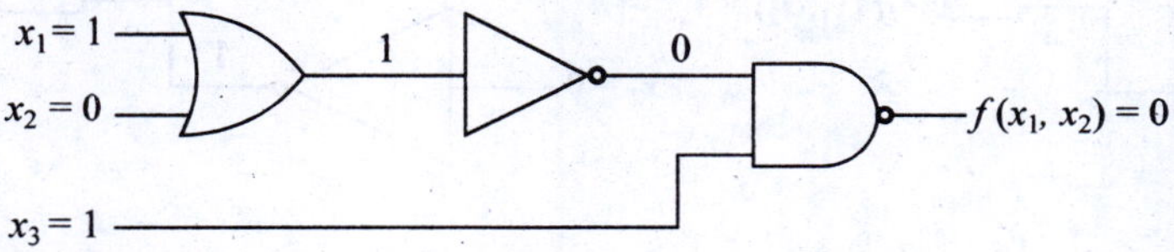

Fig. 13.24(*b*)

(*c*) The inputs to the AND gate are 0 and 1, therefore, its output is 0. Thus, the inputs to the OR gate are 0 and 0, and hence the output is $s = 0$, as shown in the Fig. 13.24(*c*).

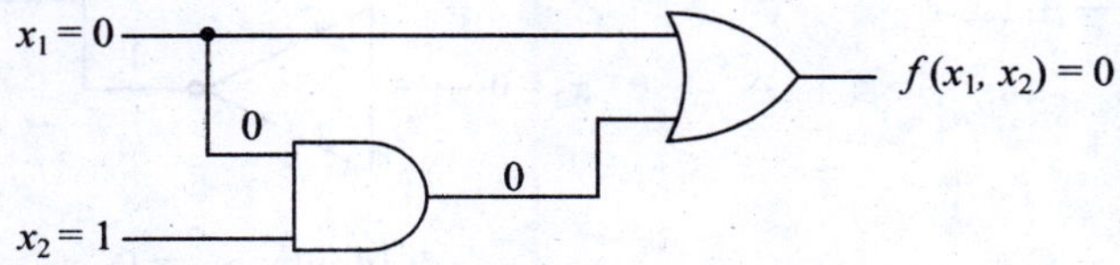

Fig. 13.24(*c*)

(*d*) The inputs to the OR gate are 0 and 1, therefore, its output is 1. NOT gate changes the input 1 to 0. Thus, the inputs to the AND gate are 1 and 0 and its output is 0. The NOT gate changes the input 0 to 1 and hence, the output is $s = 1$. This is shown in Fig. 13.24(*d*).

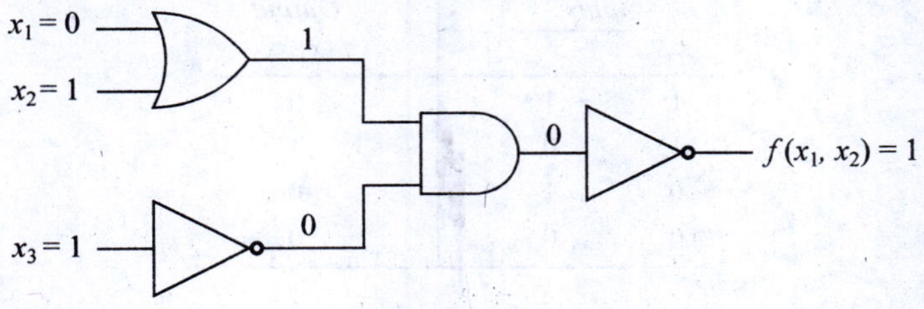

Fig. 13.24(*d*)

Example 19 Construct an input/output table of the following circuits:

(*a*)

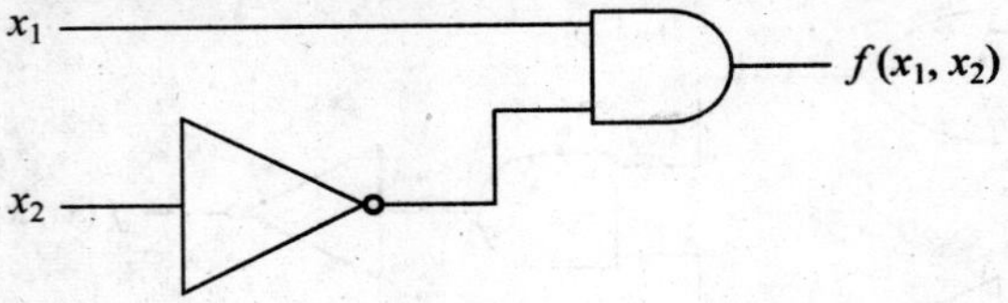

(*b*)

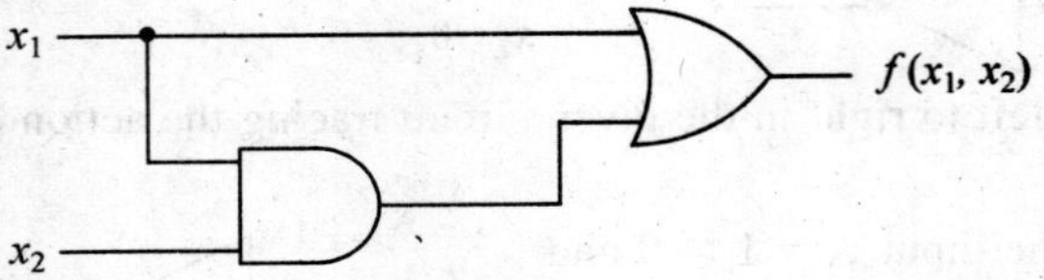

Solution (*a*) Since $x_1, x_2 \in \{0, 1\}$, the four possible values of x_1 and x_2 are

$$x_1 = 1, x_2 = 1; x_1 = 0; x_2 = 0; x_1 = 0, x_2 = 1; x_1 = 0, x_2 = 0.$$

The outputs corresponding to different values of x_1 and x_2 are shown in Figs. 13.25(*a*) to (*d*).

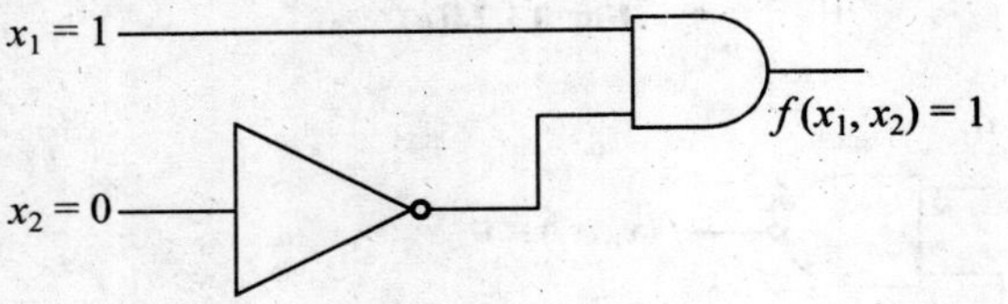

Inputs: $x_1 = 1, x_2 = 1$

(*a*)

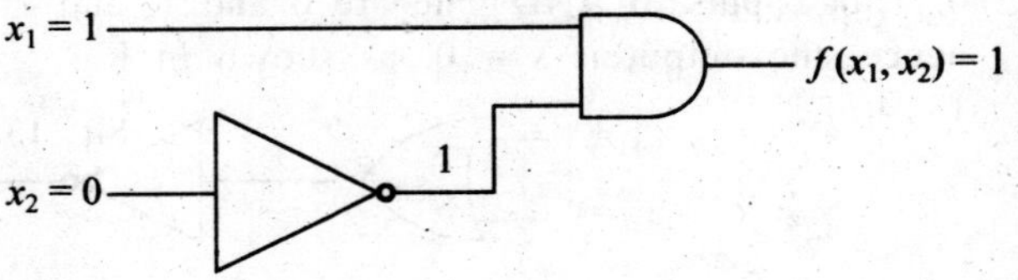

Inputs: $x_1 = 1, x_2 = 0$

(*b*)

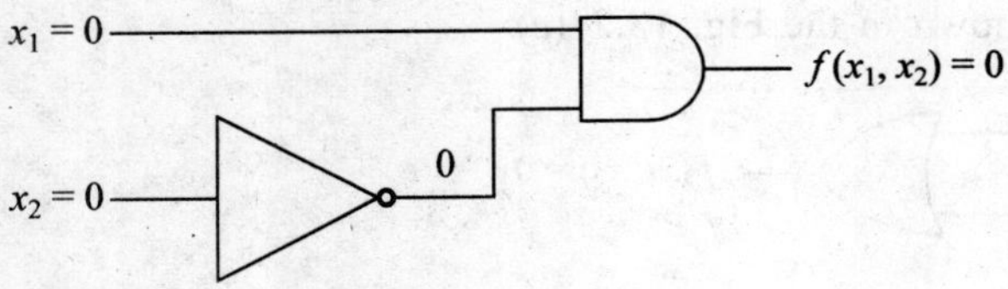

Inputs: $x_1 = 0, x_2 = 1$

(c)

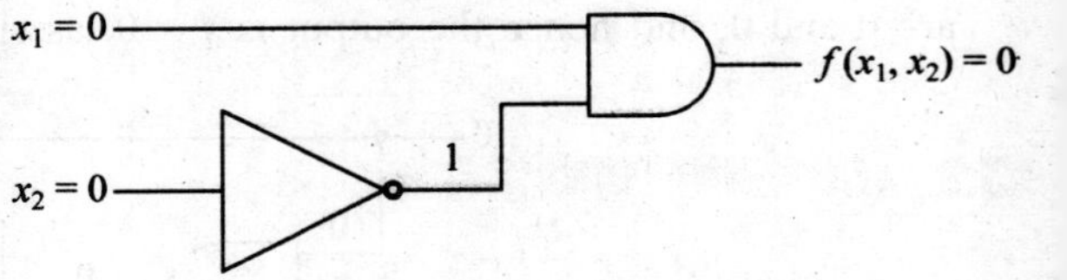

Inputs: $x_1 = 0, x_2 = 0$

(*d*)

Fig. 13.25

The required input/output table is shown below:

Inputs		*Output*
x_1	x_2	$f(x_1, x_2)$
1	1	0
1	0	1
0	1	0
0	0	0

(*b*) Since $x_1, x_2 \in \{0, 1\}$, the four possible values of x_1 and x_2 are

$$x_1 = 1, x_2 = 1; x_1 = 1, x_2 = 0; x_1 = 0, x_2 = 1; x_1 = 0, x_2 = 0.$$

The outputs corresponding to different values of x_1 and x_2 are shown in Figs. 13.26(*a*) to (*d*).

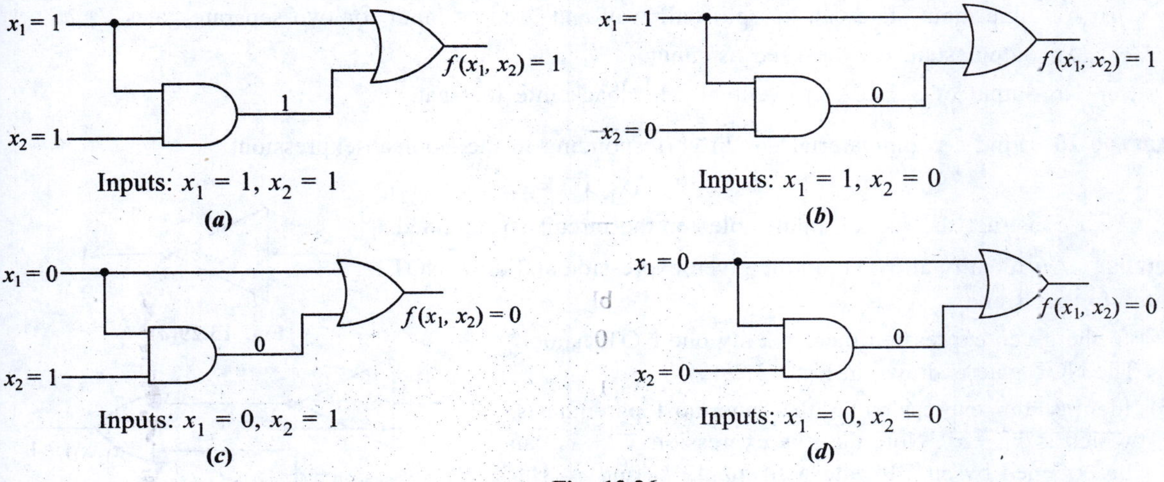

Inputs: $x_1 = 1, x_2 = 1$
(*a*)

Inputs: $x_1 = 1, x_2 = 0$
(*b*)

Inputs: $x_1 = 0, x_2 = 1$
(*c*)

Inputs: $x_1 = 0, x_2 = 0$
(*d*)

Fig. 13.26

The required input/output table is shown below:

Inputs x_1	x_2	*Output* $f(x_1, x_2)$
1	1	1
1	0	1
0	1	0
0	0	0

13.8 COMBINATORIAL CIRCUITS

A circuit in which the output is uniquely defined for each combination of inputs is called a *combinatorial circuit*.

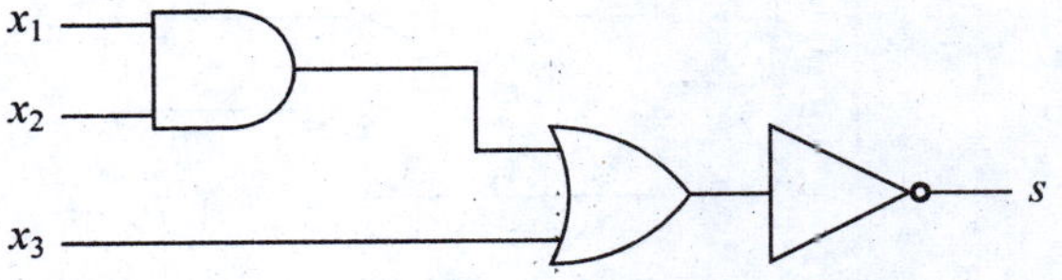

Fig. 13.27 Combinatorial Circuit

For example, in the circuit shown in Fig. 13.27, the output s is uniquely defined for each combination of inputs x_1, x_2 and x_3.

In the circuit if $x_1 = 1$ and $x_2 = 0$, then the inputs to the AND gate are 1 and 0 and so the output of the AND gate is 0 (minimum of 1 and 0) which is the input to the NOT gate which yields the output 1, i.e. $s = 1$. But the Fig. 13.28 states that $x_2 = s$, i.e. 0 = 1, which is a contradiction. Thus, the output s is not uniquely defined. Such a circuit is not a combinatorial circuit.

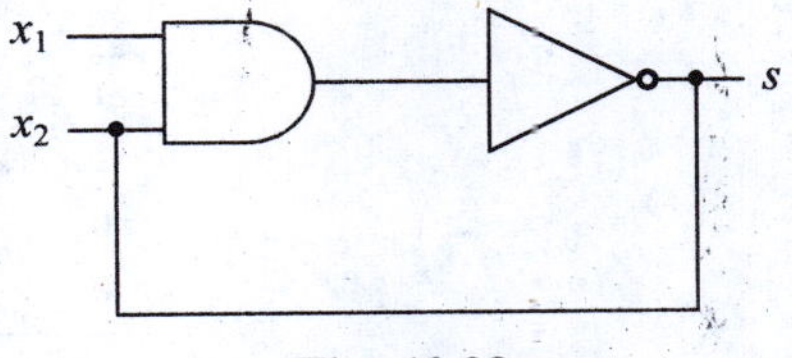

Fig. 13.28

While combining the three basic gates, viz., AND gate, OR gate and NOT gate, if the following four rules are followed, the resulting circuit is always a combinatorial circuit.

(*i*) Never combine two input states.

(*ii*) A single input state can be split half way and used as input for two separate gates.

(*iii*) An output state can be used as input.

(*iv*) No output of a gate can eventually feedback into that gate.

Example 20 Find a combinatorial circuit corresponding to the Boolean expression

$$f(x_1, x_2) = x_1 \cdot (\overline{x}_1 + x_2)$$

Also construct the input/output table for the circuit so obtained.

Solution (*a*) Identity all NOTs in the given expression and draw NOT gates for each one.

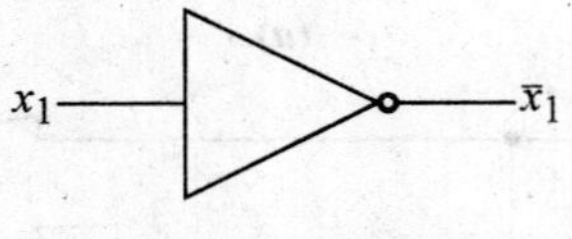

Fig. 13.29(*a*)

In the given expression there is only one NOT, namely $\overline{x}_1$. The NOT gate is drawn in Fig. 13.29(*a*).

(*b*) Identity the expression in the innermost parenthesis, which is $\overline{x}_1 + x_2$. Note that the expression $\overline{x}_1 + x_2$ can be obtained by an OR gate with inputs $\overline{x}_1$ and x_2. But the input $\overline{x}_1$ is the output of the NOT gate given in Fig. 13.29(*a*). So add an OR gate to this NOT gate with inputs $\overline{x}_1$ and x_2 as shown in Fig. 13.29(*b*).

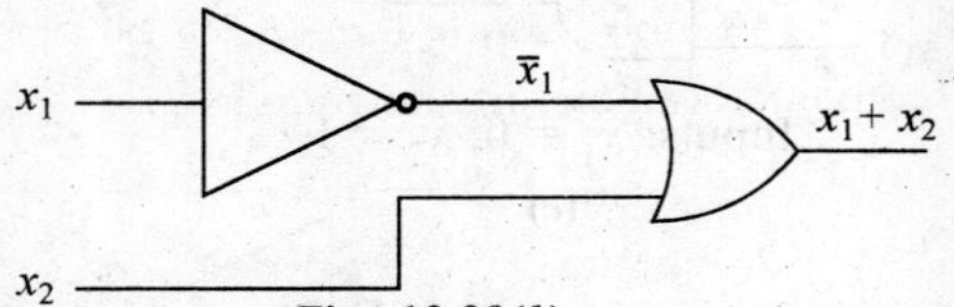

Fig. 13.29(*b*)

(*c*) Identity the expression in the next innermost parenthesis and draw the circuit using Steps (*a*) and (*b*). The expression $x_1 \cdot (\overline{x}_1 + x_2)$ is the next and the final one. This can be obtained by an AND of x_1 and $(\overline{x}_1 + x_2)$. So add an AND gate to the circuit in Fig. 13.29(*c*) with inputs x_1 and $(\overline{x}_1 + x_2)$ to obtain the desired combinatorial circuit as shown in Fig. 13.29(*c*).

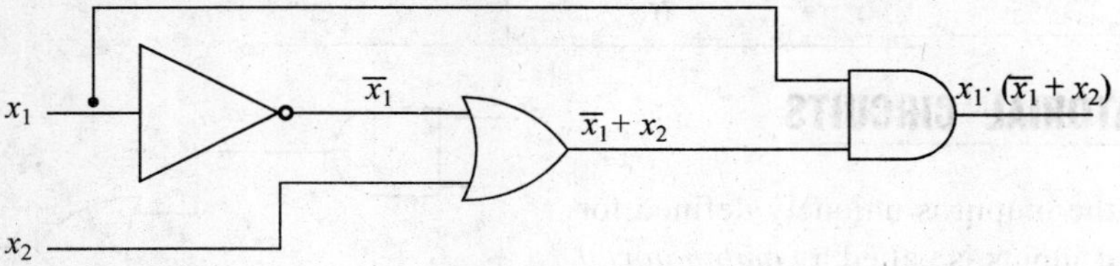

Fig. 13.29(*c*)

The input/output table for the combinatorial circuit in Fig. 13.29(*c*) is given below.

Inputs		*Output*
x_1	x_2	$f(x_1, x_2) = x_1 \cdot (\overline{x}_1 + x_2)$
1	1	1
1	0	0
0	1	0
0	0	0

Example 21 Find a combinatorial circuit corresponding to the Boolean expression

$$f(x_1, x_2, x_3) = [x_1 + (\overline{x}_2 \cdot x_3)] + x_3$$

Also, construct the input/output table for the circuit so obtained.

Solution In the given expression there is only one NOT, viz., $\overline{x}_2$. The NOT gate is drawn in Fig. 13.20(*a*).

The expression in the innermost parenthesis is $\overline{x}_2 \cdot x_3$. It can be obtained by an AND gate with inputs $\overline{x}_2$ and x_3. Since $\overline{x}_2$ is the output of the NOT gate given in Fig. 13.30(*a*), therefore, add an AND gate to this NOT gate to get circuit shown in Fig. 13.30(*b*).

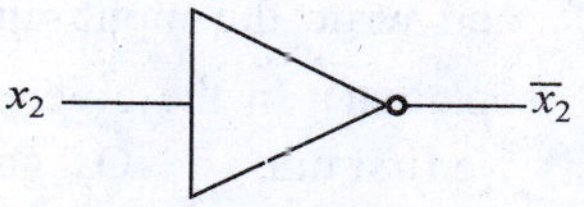

Fig. 13.30(*a*)

The expression in the next innermost parenthesis is $x_1 + (\overline{x}_2 \cdot x_3)$ which can be obtained by an OR gate with inputs x_1 and $\overline{x}_2 \cdot x_3$. So add an OR gate to circuit shown in Fig. 13.30(*a*) to get circuit shown in Fig. 13.30(*b*).

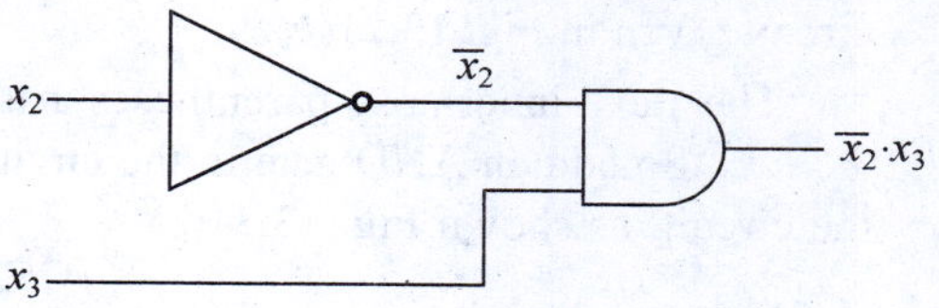

Fig. 13.30(*b*)

The final expression $[x_1 + (\overline{x}_2 \cdot x_3)] + x_3$ can be obtained by an OR gate with inputs $x_1 + (\overline{x}_2 \cdot x_3)$ and x_3. Add an OR gate to circuit shown in Fig. 13.30(*c*) to obtain the required combinatorial circuit shown in Fig. 13.30(*d*).

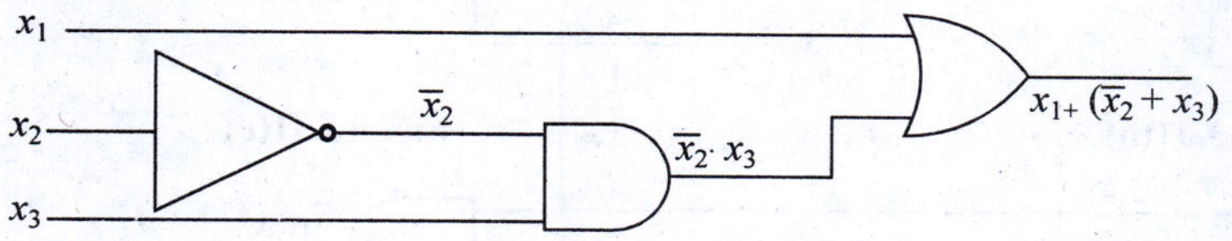

Fig. 13.30(c)

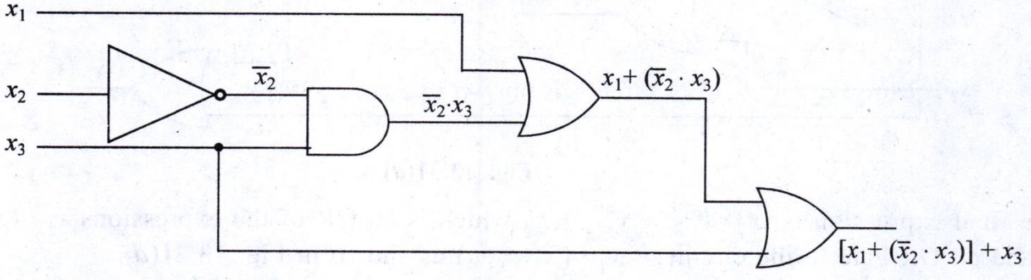

Fig. 13.30(*d*)

The input/output table for the combinatorial circuit in Fig. 13.30(*d*) is given below.

Inputs			*Output*
x_1	x_2	x_3	$f(x_1, x_2, x_3)$
1	1	1	1
1	1	0	1
1	0	1	1
1	0	0	1
0	1	1	1
0	1	0	0
0	0	1	1
0	0	0	0

Example 22 Find a combinatorial circuit corresponding to the Boolean expression

$$f(x_1, x_2, x_3) = [x_1 \cdot (\overline{x}_2 + x_3)] + x_2$$

and write the input/output table for the circuit so obtained.

Solution In the given expression there is one NOT gate namely, $\overline{x}_2$ and so we first draw a NOT gate with input x_2 as shown in Fig. 13.31(a). Identify the expression in the innermost parenthesis, i.e. $\overline{x}_2 + x_3$. This is an OR of $\overline{x}_2$ and x_3. So add an OR gate to the NOT gate with inputs $\overline{x}_2$ and x_3 to obtain the circuit given in Fig. 13.21(b).

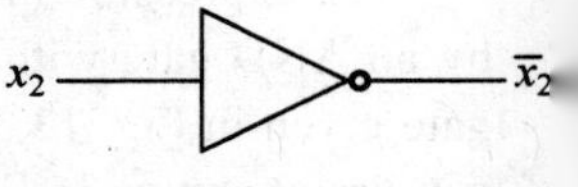

Fig. 13.31(a)

The next innermost parentheses has the expression $x_1 \cdot (\overline{x}_2 + x_3)$ which is an AND of x_1 and $\overline{x}_2 + x_3$. So add an AND gate to the circuit given in Fig. 13.31(b) with inputs x_1 and $(\overline{x}_2 + x_3)$ to obtain the circuit, as shown Fig. 13.31(c).

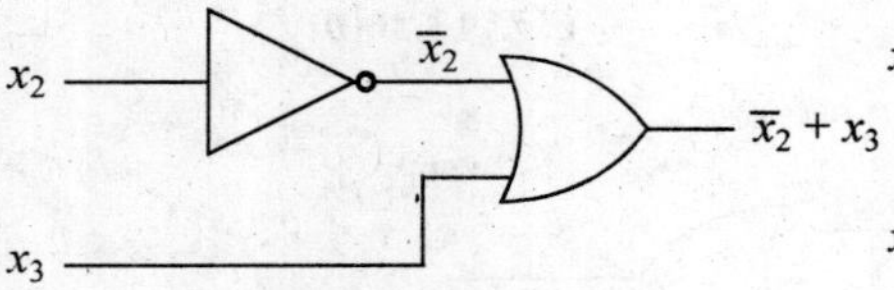

Fig. 13.31(b)

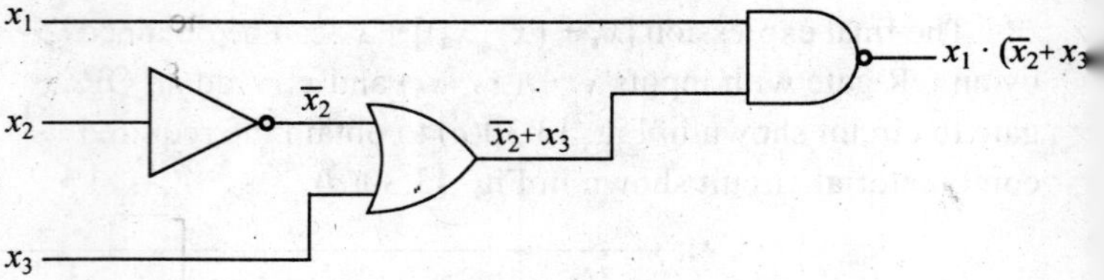

Fig. 13.31(c)

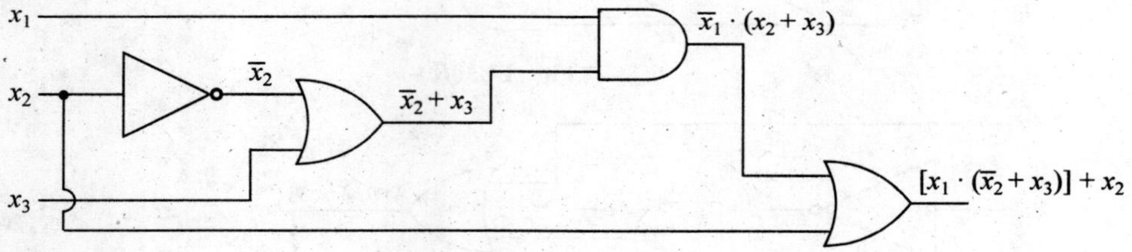

Fig. 13.31(d)

The final expression is $[x_1 \cdot (\overline{x}_2 + x_3)] + x_2$ which is an OR of the expressions $x_1 \cdot (\overline{x}_2 + x_3)$ and x_2. So add an OR gate to the circuit of $x_1 \cdot (\overline{x}_2 + x_3)$ as shown in Fig. 13.21(d).

The input/output table for the combinatorial circuit is given below.

Inputs			*Output*
x_1	x_2	x_3	$f(x_1, x_2, x_3) = [x_1 \cdot (\overline{x}_2 + x_3)] + x_2$
1	1	1	1
1	1	0	1
1	0	1	1
1	0	0	1
0	1	1	1
0	1	0	1
0	0	1	0
0	0	0	0

Example 23 Construct a combinatorial circuit from the following input/output table.

Inputs		Output
x_1	x_2	$f(x_1, x_2)$
1	1	1
1	0	0
0	1	0
0	0	1

Solution The Boolean expression corresponding to the given input/output table is given by

$$f(x_1, x_2) = x_1 \cdot x_2 + \bar{x}_1 \cdot \bar{x}_2$$

To construct the combinatorial circuit corresponding to the given Boolean expression, we first construct the NOT gates corresponding to $\bar{x}_1$ and $\bar{x}_2$ and then construct AND gates corresponding to $x_1 \cdot x_2$ and $\bar{x}_1 \cdot \bar{x}_2$. Finally, join these AND gates with an OR gate to obtain the required circuit.

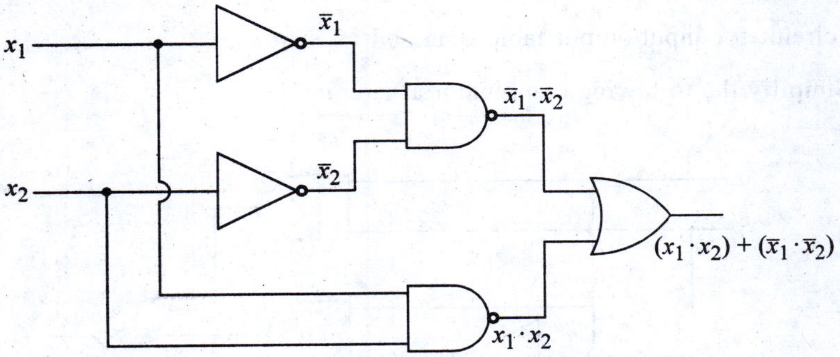

Fig. 13.32

13.8.1 Equivalent Combinatorial Circuits

Two combinatorial circuits are said to be *equivalent* if for same inputs, they produce the same outputs. The input/output tables of equivalent combinatorial circuits are identical.

Example 24 Show that the combinatorial circuits given below are equivalent.

Solution The input/output tables for the two circuits are given in the following tables are equivalent.

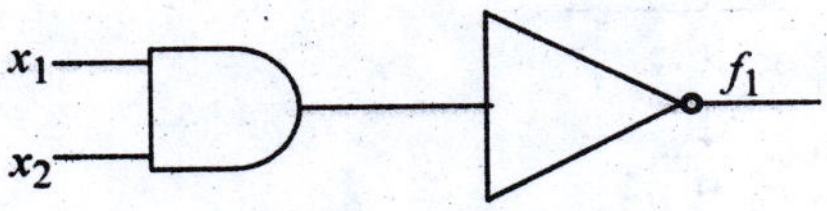

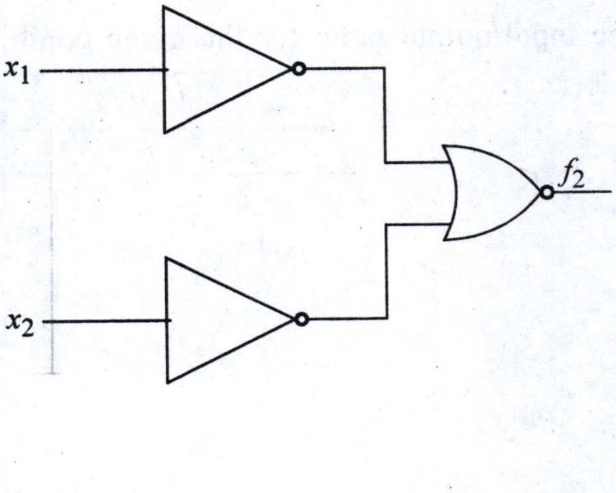

Table (*a*)

Inputs		*Output*
x_1	x_2	f_1
1	1	0
1	0	1
0	0	1
0	1	1

Table (*b*)

Inputs		*Output*
x_1	x_2	f_2
1	1	0
1	0	1
0	0	1
0	1	1

Since the two tables are identical, therefore the two circuits are equivalent.

Method to simplify a combinatorial circuit A combinatorial circuit can be simplified by following the steps:

1. Construct an input/output table.
2. Obtain the Boolean expression corresponding to input/output table, constructed in Step 1 and simplify it.
3. Draw the circuit for input/output table obtained in Step 2.

Example 25 Simplify the following combinatorial circuit.

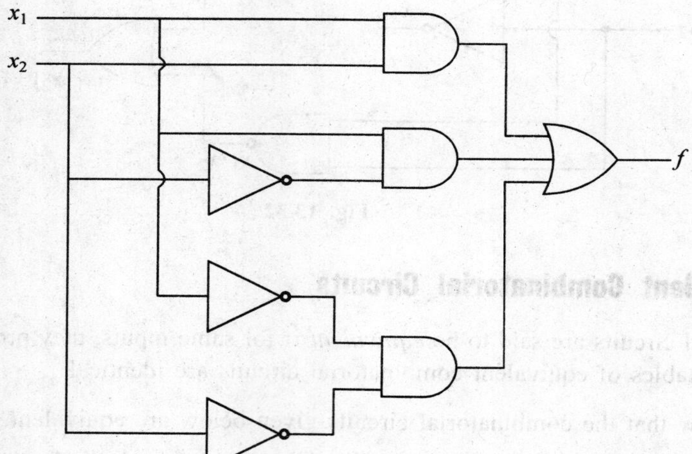

Solution The input/output table for the given combinatorial circuit is given below:

Inputs		*Output*
x_1	x_2	$f(x_1, x_2)$
1	1	1
1	0	1
0	1	0
0	0	1

The Boolean expression corresponding to the input/output table is

$$x_1 \cdot x_2 + x_1 \cdot \overline{x}_2 + \overline{x}_1 + \overline{x}_2 = x_1 \cdot (x_2 + \overline{x}_2) + \overline{x}_1 \cdot \overline{x}_2 = x_1 \cdot 1 + \overline{x}_1 \cdot \overline{x}_2 \qquad [\text{Since } x + \overline{x} = 1]$$

$$= x_1 + \overline{x}_1 \cdot \overline{x}_2 = (x_1 + \overline{x}_1) \cdot (x_1 + \overline{x}_2)$$

$$= 1 \cdot (x_1 + \overline{x}_2) = x_1 + \overline{x}_2$$

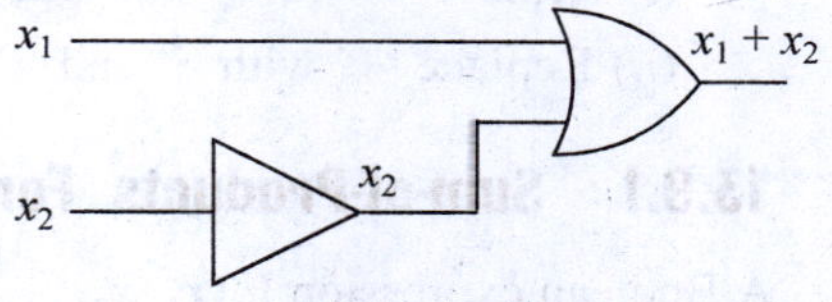

Fig. 13.33

The simplified circuit is shown in Fig. 13.33.

Example 26 Simplify the following combinatorial circuit.

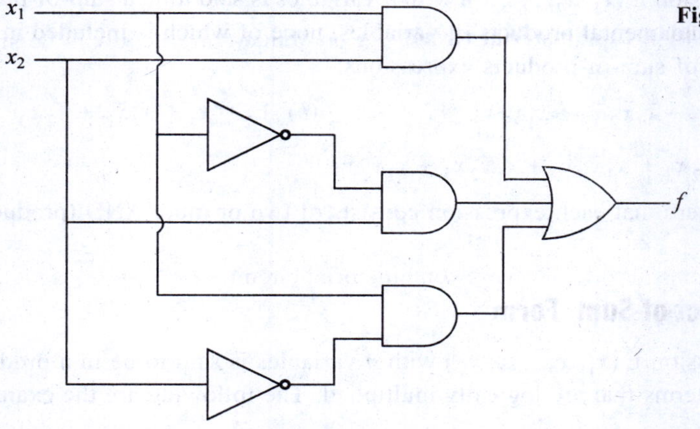

Solution The input/output table for the given combinatorial circuit is given below:

Inputs		*Output*
x_1	x_2	$f(x_1, x_2)$
1	1	1
1	0	1
0	1	1
0	0	0

The Boolean expression corresponding to this input/output table is

$$x_1 \cdot x_2 + \overline{x}_1 \cdot x_2 + x_1 \cdot \overline{x}_2 = x_2 \cdot x_1 + x_2 \cdot \overline{x}_1 + x_1 \cdot \overline{x}_2 \qquad [\text{Since } x \cdot y = y \cdot x]$$

$$= x_2 \cdot (x_1 + \overline{x}_1) + x_1 \cdot \overline{x}_2 \qquad [\text{Since } x \cdot (y + z) = x \cdot y + x \cdot z]$$

$$= x_2 \cdot 1 + x_1 \cdot \overline{x}_2 \qquad [\text{Since } x + \overline{x} = 1]$$

$$= x_2 + x_1 \cdot \overline{x}_2 = (x_2 + x_1) \cdot (x_2 + \overline{x}_2)$$

$$= (x_2 + x_1) \cdot 1 = x_2 + x_1$$

The simplified combinational circuit is shown in Fig. 13.34.

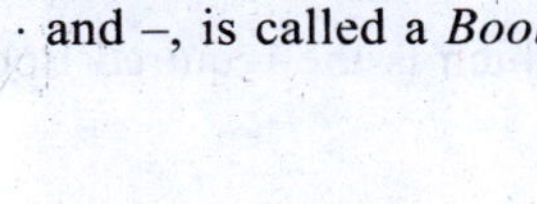

Fig. 13.34

13.9 BOOLEAN EXPRESSION

Let $x_1, x_2, \ldots, x_n$ be a set of n-input variables (letters or symbols). Then an expression, written as E $(x_1, x_2, \ldots, x_n)$ built up from these n variables using the Boolean operations +, · and –, is called a *Boolean expression*.

DeMorgan's laws: (*i*) $\overline{(a+b)} = \bar{a}\bar{b}$ and (*ii*) $\overline{(a \cdot b)} = \bar{a}+\bar{b}$ can be combined into a single law to complement a Boolean expression. Following rules are used to obtain the complement of the Boolean expression

(*i*) Replace 0 with 1 and vice versa

(*ii*) Replace '+' with '·' and '·' with '′' or '–' (sign of complement)

13.9.1 Sum-of-Products Form

A Boolean expression E ($x_1, x_2, \ldots, x_n$) with n variables is said to in a sum-of-product form if it is a sum of two or more fundamental products of variables, none of which is included in another. The following are the examples of sum-of-products expressions:

(*a*) $E_1 = x_1\bar{x}_3 + \bar{x}_1x_2\bar{x}_3 + x_1x_2x_3$ (*b*) $E_2 = x_1 + x_1\bar{x}_2 + \bar{x}_2x_3$

(*c*) $E_3 = x_1x_2x_3 + x_1\bar{x}_2x_3 + \bar{x}_1\bar{x}_2x_3 + \bar{x}_1x_2\bar{x}_3$

It may be noted here that each expression consists of two or more AND (product) terms that are ORed together.

13.9.2 Product-of-Sum Form

A Boolean expression E ($x_1, x_2, \ldots, x_n$) with n variables is said to be in a product-of-sum form if it is a sum of several terms that are logically multiplied. The following are the examples of product-of-sum form expressions:

(*a*) $E_1 = (x_1 + x_2)(\bar{x}_1 + \bar{x}_2)$ (*b*) $E_2 = (x_1 + x_2)(x_3 + x_4)$

(*c*) $E_3 = (x_1 + x_2 + x_3)(x_1 + \bar{x}_2 + x_3)(x_1 + \bar{x}_2 + \bar{x}_3)$

It may be noted here that each expression consists of two or more OR (sum) terms that are ANDed together.

Example 27 Obtain the sum of product in canonical form of the following expression

$(x_1 \vee x_2)' \vee (x_1' \wedge x_3)$ [*UPTU, MCA, 2002*]

Solution Given

$(x_1 \vee x_2)' = x_1' \wedge x_2'$ (De-morgain's law)

$= (x_1' \wedge x_2') \wedge 1$ (identity law)

$= (x_1' \wedge x_2') \wedge (x_3 \vee x_3')$ (complementation law)

$= (x_1' \wedge x_2' \wedge x_3) \vee (x_1' \wedge x_2' \wedge x_3')$ (distributive law) ...(*i*)

Again

$(x_1' \wedge x_3) = (x_1' \wedge x_3) \wedge 1$ (identity law)

$= (x_1' \wedge x_3) \wedge (x_2 \vee x_2')$ (complementation law)

$= (x_1' \wedge x_3 \wedge x_2) \vee (x_1' \wedge x_3 \wedge x_2')$ (distributive law)

$= (x_1' \wedge x_2 \wedge x_3) \vee (x_1' \wedge x_2' \wedge x_3)$ (commutative law) ...(*ii*)

Putting values from (*i*) and (*ii*) in the given expression, we get

$$(x_1' \wedge x_2' \wedge x_3) \vee (x_1' \wedge x_2' \wedge x_3') \vee (x_1' \wedge x_2 \wedge x_3) \vee (x_1' \wedge x_2' \wedge x_3)$$

which is the required Boolean expression.

Example 28 Construct an input/output table for the Boolean functions defined as follows:

(*a*) $f(x_1, x_2) = (x_1 \cdot \overline{x}_2) + x_1$

(*b*) $f(x_1, x_2, x_3) = (x_1 + x_2) \cdot \overline{x}_3$

(*c*) $f(x_1, x_2) = \overline{(x_1 \cdot x_2)} + x_2$

(*d*) $f(x_1, x_2, x_3) = \overline{(x_1 \cdot x_2)} + x_3$

Solution (*a*) The domain of the function is determined by the number of variables in the Boolean expression of the function. Since the number of variables is two, the domain is given by

$$(\{0, 1\})^2 = \{0, 1\} \times \{0, 1\} = \{(0, 0), (0, 1), (1, 0), (1, 1)\}$$

Now,
$$f(0, 0) = (0 \cdot \overline{0}) + 0 = (0 \cdot 1) + 0 = 0 + 0 = 0$$
$$f(0, 1) = (0 \cdot \overline{1}) + 0 = (0 \cdot 0) + 0 = 0 + 0 = 0$$
$$f(1, 0) = (1 \cdot \overline{0}) + 1 = (1 \cdot 1) + 1 = 1 + 1 = 1$$
$$f(1, 1) = (1 \cdot \overline{1}) + 1 = (1 \cdot 0) + 1 = 0 + 1 = 1.$$

Thus, the input/output table is given by

Inputs		*Output*
x_1	x_2	$(x_1 \cdot \overline{x}_2) + x_1$
0	0	0
0	1	0
1	0	1
1	1	1

(*b*) We have, $f(x_1, x_2, x_3) = (x_1 + x_2) \cdot \overline{x}_3$. Since the number of variables is three, the domain is given by

$$(\{0, 1\})^3 = (\{0, 1\})^2 \times \{0, 1\} = \{(0, 0), (0, 1), (1, 0), (1, 1)\} \times \{0, 1\}$$
$$= \{(0, 0, 0), (0, 1, 0), (1, 0, 0), (1, 1, 0), (0, 0, 1)\ (0, 1, 1), (1, 0, 1), (1, 1, 1, 1)\}$$

Now
$$f(0, 0, 0) = (0 + 0) \cdot \overline{0} = 0 \cdot 1 = 0$$
$$f(0, 1, 0) = (0 + 1) \cdot \overline{0} = 1 \cdot 1 = 1$$
$$f(1, 0, 0) = (1 + 0) \cdot \overline{0} = 1 \cdot 1 = 1$$
$$f(1, 1, 0) = (1 + 1) \cdot \overline{0} = 1 \cdot 1 = 1$$
$$f(0, 0, 1) = (0 + 0) \cdot \overline{1} = 0 \cdot 0 = 0$$
$$f(0, 1, 1) = (0 + 1) \cdot \overline{1} = 1 \cdot 0 = 0$$
$$f(1, 0, 1) = (1 + 0) \cdot \overline{1} = 1 \cdot 0 = 0$$
$$f(1, 1, 1) = (1 + 1) \cdot \overline{1} = 1 \cdot 0 = 0$$

Thus, the input/output table is given by

Inputs			Output
x_1	x_2	x_3	$(x_1 + x_2) \cdot \overline{x}_3$
0	0	0	0
0	1	0	1
1	0	0	1
1	1	0	1
0	0	1	0
0	1	1	0
1	0	1	0
1	1	1	0

(*c*) We have, $f(x_1, x_2) = \overline{(x_1 \cdot x_2)} + x_2$. Since the number of variables is two, the domain is given by

$$(\{0, 1\})^2 = \{0, 1\} \times \{0, 1\}$$
$$= \{(0, 0), (0, 1), (1, 0), (1, 1)\}$$

Now,
$$f(0, 0) = \overline{(0 \cdot 0)} + 0 = \overline{0} + 0 = 1 + 0 = 1$$
$$f(0, 1) = \overline{(0 \cdot 1)} + 1 = \overline{0} + 1 = 1 + 1 = 1$$
$$f(1, 0) = \overline{(1 \cdot 0)} + 0 = \overline{0} + 0 = 1 + 0 = 1$$
$$f(1, 1) = \overline{(1 \cdot 1)} + 1 = \overline{1} + 1 = 0 + 1 = 1$$

Thus, the input/output table is given by

Inputs		Output
x_1	x_2	$\overline{(x_1 \cdot x_2)} + x_2$
0	0	1
0	1	1
1	0	1
1	1	1

(*d*) We have, $f(x_1, x_2, x_3) = \overline{(x_1 \cdot x_2)} + \overline{x}_3$. Since number of variables is three, the domain is given by

$$(\{0, 1\})^3 = (\{0, 1\})^2 \times \{0, 1\}) = \{(0, 0), (0, 1), (1, 0), (1, 1)\} \times \{0, 1\}$$
$$= \{(0, 0, 0), (0, 1, 0), (1, 0, 0), (1, 1, 0), (0, 0, 1), (0, 1, 1), (1, 0, 1) (1, 1, 1)\}$$

Now,
$$f(0, 0, 0) = \overline{(0 \cdot 0)} + 0 = \overline{0} + 0 = 1 + 0 = 1$$
$$f(0, 1, 0) = \overline{(0 \cdot 1)} + 0 = \overline{0} + 0 = 1 + 0 = 1$$
$$f(1, 0, 0) = \overline{(1 \cdot 0)} + 0 = \overline{0} + 0 = 1 + 0 = 1$$
$$f(1, 1, 0) = \overline{(1 \cdot 1)} + 0 = \overline{1} + 0 = 0 + 0 = 0$$
$$f(0, 0, 1) = \overline{(0 \cdot 0)} + 1 = \overline{0} + 1 = 1 + 1 = 1$$
$$f(0, 1, 1) = \overline{(0 \cdot 1)} + 1 = \overline{0} + 1 = 1 + 1 = 1$$
$$f(1, 0, 1) = \overline{(1 \cdot 0)} + 1 = \overline{0} + 1 = 1 + 1 = 1$$
$$f(1, 1, 1) = \overline{(1 \cdot 1)} + 1 = \overline{1} + 1 = 0 + 1 = 1$$

Example 29 Construct the function whose input/output table is given below.

Inputs			Output
x_1	x_2	x_3	$f(x_1, x_2, x_3)$
1	1	1	1
1	1	0	0
1	0	1	0
1	0	0	1
0	1	1	0
0	1	0	1
0	0	1	0
0	0	0	0

[*UPTU, BTech, 2003*]

Solution We construct the required function as follows:

(*a*) Identity all rows of the table where the output is 1. Note that first, fourth and sixth rows have output 1.

(*b*) Form the combination $(x_1 \cdot x_2 \cdot x_3)$ (as there are only three inputs namely x_1, x_2, and x_3) for the rows identified in Step 1 and place a '–' over each $x_i(1 \le i \le 3)$ whose value is 0 in the respective row. We obtain the following combinations

$x \cdot x_2 \cdot x_3$ for the first row

$x_1 \cdot \overline{x}_2 \cdot \overline{x}_3$ for the fourth

and $\overline{x}_1 \cdot \overline{x}_2 \cdot \overline{x}_3$ for the sixth row

(*c*) Applying OR gate to all the combinations obtained in Step 2, we get the Boolean expression

$$(x_1 \cdot x_2 \cdot x_3) + (x_1 \cdot \overline{x}_2 \cdot \overline{x}_3) + (\overline{x}_1 \cdot x_2 \cdot \overline{x}_3)$$

(*d*) Write $f(x_1 \cdot x_2 \cdot x_3) = (x_1 \cdot x_2 \cdot x_3) + (x_1 \cdot \overline{x}_2 \cdot \overline{x}_3) + (\overline{x}_1 \cdot x_2 \cdot \overline{x}_3)$ which is the required function. Since the function is represented by a Boolean expression it is a Boolean function.

Example 30 Construct the function whose input/output table is given below

Inputs			Output
x_1	x_2	x_3	$f(x_1, x_2, x_3)$
1	1	1	1
1	1	0	1
1	0	1	1
1	0	0	0
0	1	1	0
0	1	0	0
0	0	1	0
0	0	0	1

Solution We construct the function as follows:

(*a*) 1st, 2nd, 3rd and 8th rows have output 1.

(*b*) Form the combination $x_1 \cdot x_2 \cdot x_3$ and place '–' over each variable whose value is 0 in the respective row.

$$\begin{aligned} &x_1 \cdot x_2 \cdot x_3 && \text{for first row} \\ &x_1 \cdot x_2 \cdot \bar{x}_3 && \text{for second row} \\ &x_1 \cdot \bar{x}_2 \cdot x_3 && \text{for third row} \\ &\bar{x}_1 \cdot \bar{x}_2 \cdot \bar{x}_3 && \text{for eighth row} \end{aligned}$$

(*c*) Apply OR gate to all combinations obtained in Step 2 to get the Boolean expression

$$(x_1 \cdot x_2 \cdot x_3) + (x_1 \cdot x_2 \cdot \bar{x}_3) + (x_1 \cdot \bar{x}_2 \cdot x_3) + (\bar{x}_1 \cdot \bar{x}_2 \cdot \bar{x}_3)$$

Thus the Boolean functions is: $f(x_1, x_2, x_3) = (x_1 \cdot x_2 \cdot x_3) + (x_1 \cdot x_2 \cdot \bar{x}_3) + (x_1 \cdot \bar{x}_2 \cdot x_3) + (\bar{x}_1 \cdot \bar{x}_2 \cdot \bar{x}_3)$.

13.10 MINTERM AND MAXTERM

When each term of a Boolean expression is either in normal or complemented form, it is said to be in *canonical form*. A Boolean expression E is said to be a *complete sum-of-product* expression if it is an expression where each product involves all *n* variables.

(*i*) When a sum-of-products form of Boolean expression is in canonical form, each product term is called a *minterm.* Each minterm is obtained from an AND (i.e. $\wedge$ or $\cdot$) term of variables which are being complemented if corresponding binary number is 0, otherwise 1.

(*ii*) When a product-of-sums form of Boolean expression is in canonical form, each sum term is called a *maxterm.* Each maxterm is obtained from an OR (i.e. $\vee$ or +) term of variables which are being normal if corresponding binary number is 0, otherwise 1.

If a Boolean expression contains *n* variables, then maximum possible number of minterms and maxterms is 2^n as shown in Table 13.14.

Table 13.14 Minterms and Maxterms

x_1	x_2	x_3	*Minterms*	*Maxterms*
0	0	0	$\bar{x}_1\bar{x}_2\bar{x}_3$	$x_1 + x_2 + x_3$
0	0	1	$\bar{x}_1\bar{x}_2x_3$	$x_1 + x_2 + \bar{x}_3$
0	1	0	$\bar{x}_1x_2\bar{x}_3$	$x_1 + \bar{x}_2 + x_3$
0	1	1	$x_1\bar{x}_2\bar{x}_3$	$x_1 + \bar{x}_2 + \bar{x}_3$
1	0	0	$\bar{x}_1x_2x_3$	$\bar{x}_1 + x_2 + x_3$
1	0	1	$x_1\bar{x}_2x_3$	$\bar{x}_1 + x_2 + \bar{x}_3$
1	1	0	$x_1x_2\bar{x}_3$	$\bar{x}_1 + \bar{x}_2 + x_3$
1	1	1	$x_1x_2x_3$	$\bar{x}_1 + \bar{x}_2 + \bar{x}_3$

Recall that a function $f: A_n \rightarrow A$, where A is the Boolean algebra {0, 1}, is given by some Boolean expression and is useful in constructing a logic diagram for the function. If E $(x_1, x_2, \ldots, x_n)$ is a Boolean expression with *n* variables, then the function that it produces is generated by substituting all combinations of 0s and 1s for x_is in the expression. Let $f: A_3 \rightarrow A$ be produced by the expression E $(x_1, x_2, x_3) = x_1 + x_2 x_3 = x_1 \vee (x_2 \wedge x_3)$. Then the truth table of f is shown in Table 13.15.

Table 13.15 Truth Table of $f(x_1, x_2, x_3) = x_1 + x_2x_3$

x_1	x_2	x_3	$f(x_1, x_2, x_3)$
0	0	0	0
0	0	1	0
0	1	0	0
0	1	1	1
1	0	0	1
1	0	1	1
1	1	0	1
1	1	1	1

As a reverse process truth table of Boolean function can also be used in writting corresponding Boolean expression in 'sum-of-products' and 'product-of-sums' forms as shown in Table 13.16.

Table 13.16 Boolean Sum-of-Products and Product-of-Sums Expressions

x_1	x_2	x_3	$f(x_1, x_2, x_3)$	*Sum Terms*	*Product Terms*
0	0	0	1	—	$\bar{x}_1\bar{x}_2\bar{x}_3$
0	0	1	1	—	$\bar{x}_1\bar{x}_2x_3$
0	1	0	1	—	$\bar{x}_1x_2\bar{x}_3$
0	1	1	0	$x_1 + \bar{x}_2 + \bar{x}_3$	—
1	0	0	1	—	$x_1\bar{x}_2\bar{x}_3$
1	0	1	1	—	$x_1\bar{x}_2x_3$
1	1	0	0	$\bar{x}_1 + \bar{x}_2 + x_3$	—
1	1	1	0	$\bar{x}_1 + \bar{x}_2 + \bar{x}_3$	—

The sum-of-products and product-of-sums expressions be obtained from sum terms and product terms given in Table 13.16 as follows by using definition of Boolean algebra and its properties.

$$
\begin{aligned}
\text{Sum-of-products expression} &= \bar{x}_1\bar{x}_2\bar{x}_3 + \bar{x}_1\bar{x}_2x_3 + \bar{x}_1x_2\bar{x}_3 + x_1\bar{x}_2\bar{x}_3 + x_1\bar{x}_2x_3 \\
&= \bar{x}_1\bar{x}_2\,(\bar{x}_3 + x_3) + x_1\bar{x}_2\,(\bar{x}_3x_3) + \bar{x}_1x_2\bar{x}_3 \\
&= \bar{x}_1\bar{x}_2 + x_1\bar{x}_2 + \bar{x}_1x_2\bar{x}_3 \\
&= \bar{x}_2\,(\bar{x}_1x_1) + \bar{x}_1x_2\bar{x}_3 \\
&= \bar{x}_2 + \bar{x}_1x_2\bar{x}_3 = \bar{x}_2 + \bar{x}_1\bar{x}_3
\end{aligned}
$$

$$
\begin{aligned}
\text{Product-of-sums expression} &= (x_1 + \bar{x}_2 + \bar{x}_3)(\bar{x}_1 + \bar{x}_2 + x_3)(\bar{x}_1 + \bar{x}_2 + \bar{x}_3) \\
&= (x_1 + \bar{x}_2 + \bar{x}_3)(\bar{x}_1 + \bar{x}_2) \\
&= \bar{x}_2\,(x_1 + \bar{x}_1) + \bar{x}_2\,(1 + \bar{x}_3) + \bar{x}_1\bar{x}_3 \\
&= \bar{x}_2 + \bar{x}_1\bar{x}_3
\end{aligned}
$$

Truth Table for a Logic Circuit

The steps of constructing truth task for a logical circuit when output is given by a Boolean sum of product expressions in the inputs:

Step 1 Write down the sequences for the input x_1, x_2, ... and their complements.

Step 2 Find each product appearing in the output is such that a product $x_1 \cdot x_2 \ldots = 1$ is in a position in which all the x_1, x_2, ... have value 1 in that position.

Step 3 Find the sum of the product such that $x_1 + x_2 + \ldots = 0$ in a position in which all the x_1, x_2, have value 0 in the position.

Example 31 Find out the output of the logic circuit represented by $Y = x_1 \cdot x_2 \cdot x_3 + x_1 \cdot \bar{x}_2 \cdot x_3 + \bar{x}_1 \cdot x_2$, where $x_1 = 00001111$, $x_2 = 00110011$, $x_3 = 01010101$ are the 8-bit special sequences.

Solution We have $x_1 = 00001111$, $x_2 = 00110011$, and $x_3 = 01010101$. Then

$\bar{x}_1 = 11110000$, $\bar{x}_2 = 11001100$, and $\bar{x}_3 = 10101010$, and

$$x_1 \cdot x_2 \cdot x_3 = 0\,0\,0\,0\,0\,0\,0\,1$$
$$x_1 \cdot \bar{x}_2 \cdot x_3 = 0\,0\,0\,0\,0\,1\,0\,0$$
$$\bar{x}_1 \cdot x_2 = 0\,0\,1\,1\,0\,0\,0\,0$$

Thus $Y = 0\,0\,1\,1\,0\,1\,0\,1$

13.11 KARNAUGH MAP

The process of writing a Boolean function in terms of minterms or maxterms of the Boolean function can be simplified by using a graphical method called *Karnaugh map* provided number of variables involved in an expression are not too large. Here we shall demonstrate the use of Karnaugh map procedure to simplify a Boolean function involving 2, 3 and 4 variables.

For a Boolean function of n variables, the Karnaugh map will have 2^n squares, representing a minterm (product term) of the Boolean expression. Each square contains either 0 or 1. The value 1 in a square represents a product term present in the expression, otherwise 0 is written in the square. Once a K-map has been filled with 0s and 1s, the sum-of-products expression for the output can be obtained by ORing together squares that contain 1.

Case 1 When *two variables* x_1 and x_2 and their negations can be combined in four ways by cross reference of row and cloumn. Table 13.17(a) shows the K-map corresponding to Boolean expression E (x_1, x_2) for two variables x_1 and x_2 and Table 13.17(b) shows the alternative way of representing two variables using binary digits 0 and 1. The digit 0 and 1 written above the top horizontal line represents $\bar{x}_1$ and x_1, respectively. Similarly, 0 and 1 written in the left side of vartical line represents $\bar{x}_2$ and x_2, respectively.

Table 13.17 Two Variable Karnaugh Map

	y	$\bar{y}$
x	$x \cdot y$	$x \cdot \bar{y}$
$\bar{x}$	$\bar{x} \cdot y$	$\bar{x} \cdot \bar{y}$

(a)

$x \backslash y$	0	1
0	00	01
1	10	11

(a)

Case 2 Same as above, K-map corresponding to Boolean expression E (x_1, x_2, x_3) for three variables x_1, x_2 and x_3 is shown in Table 13.18(a) and the alternative way of repersenting three variables using binary digits 0 and 1 is shown in Table 13.18(b).

Table 13.18 Three Variable Karnaugh Map

	yz	$y\bar{z}$	$\bar{y}\bar{z}$	$\bar{y}z$
x	xyz	$xy\bar{z}$	$x\bar{y}\bar{z}$	$x\bar{y}z$.
$\bar{x}$	$\bar{x}yz$	$\bar{x}y\bar{z}$	$\bar{x}\bar{y}\bar{z}$	xyz

(a)

x \ yz	00	01	11	10
0	000	010	011	010
1	100	101	111	110

(b)

Case 3 K-map corresponding to Boolean expression E (x_1, x_2, x_3, x_4) for four variables x_1, x_2, x_3 and x_4 is shown in Table 13.19(a) and the alternative way of representing four variables using binary digits 0 and 1 is shown in Table 13.19(b).

Table 13.19 Four-Variable Karnaugh Map

	yz	$y\bar{z}$	$\bar{y}\bar{z}$	$\bar{y}z$
wx	$wxyz$	$wxy\bar{z}$	$wx\bar{y}\bar{z}$	$wx\bar{y}z$
$w\bar{x}$	$w\bar{x}yz$	$w\bar{x}y\bar{z}$	$w\bar{x}\bar{y}\bar{z}$	$w\bar{x}\bar{y}z$
$\bar{w}\bar{x}$	$\bar{w}\bar{x}yz$	$\bar{w}\bar{x}y\bar{z}$	$\bar{w}\bar{x}\bar{y}\bar{z}$	$\bar{w}\bar{x}\bar{y}z$
$\bar{w}x$	$\bar{w}xyz$	$\bar{w}xy\bar{z}$	$\bar{w}x\bar{y}\bar{z}$	$\bar{w}x\bar{y}z$

(a)

wx \ yz	00	01	11	10
00	0000	0001	0011	0010
01	0100	0101	0111	0110
11	1100	1101	1111	1110
10	1000	1001	1011	1010

(b)

13.11.1 Looping

In the K-map corresponding to Boolean expression E (x_1, x_2) with two variables x_1, x_2, squares containing binary digit 1 can be represented by two rectangles. The horizontal square represents the output of the expression E (x_1, x_2) corresponding to $\bar{x}_1$ whereas vertical square represents the output corresponding to $\bar{x}_2$. Thus E $(x_1, x_2) = \bar{x}_1 + \bar{x}_2$ (or $\bar{x}_1 \vee \bar{x}_2$). The process of combining these digits is called *looping*.

Similarly, any two adjacent squares containing binary digit 1 can be looped to form a pair. Such a loop eliminates the variable that appears in complemented and uncomplemented forms. A square which is not adjacent to any other square of E (x_1, x_2) represents the AND function of variables. The output expression can be obtained by ORing all the reduced terms and terms containing single digit 1.

Illustrations

1. $E_1 = x_1x_2 + x_1x_2'$

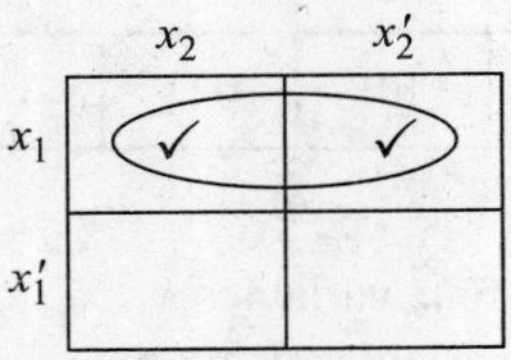

2. $E_2 = x_1x_2 + x_1'x_2' + x_1'x_2$

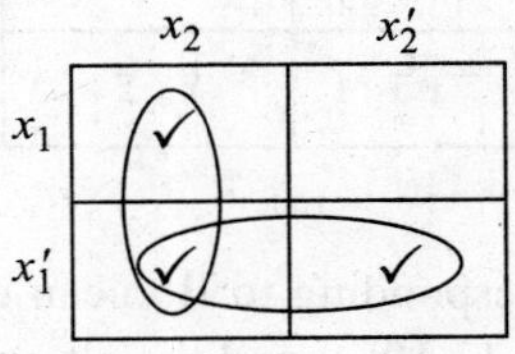

3. $E_3 = x_1x_2 + x_1'x_2'$

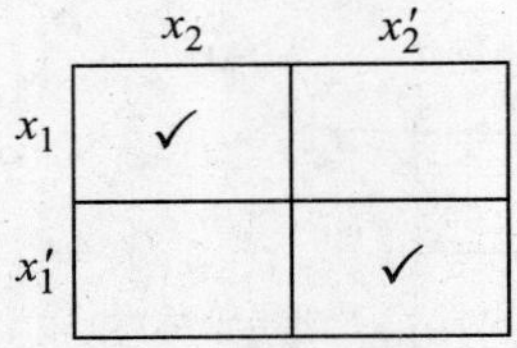

4. $E_4 = x_1x_2x_3 + x_1x_2x_3' + x_1'x_2x_3'$

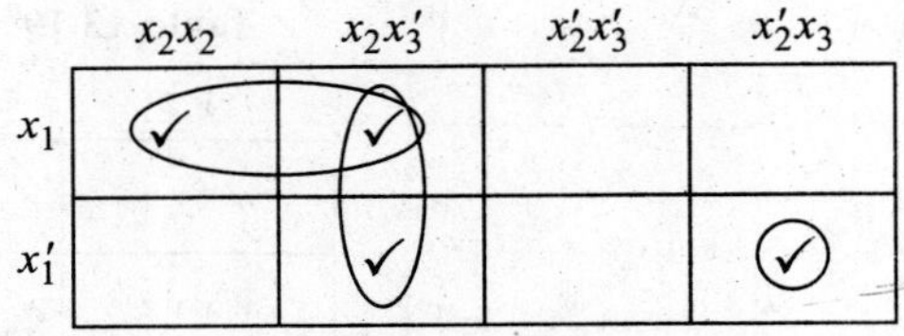

5. $E_5 = x_1x_2x_3 + x_1x_2x_3' + x_1x_2'x_3 + x_1'x_2x_3 + x_1'x_2'x_3$

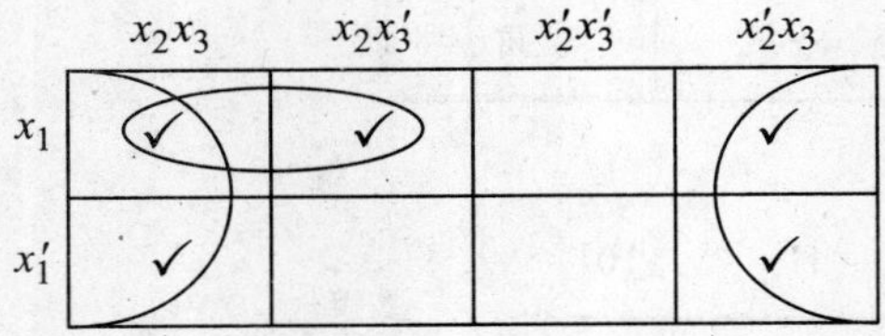

6. $E_6 = x_1x_2x_3 + x_1x_2x_3' + x_1'x_2x_3' + x_1'x_2'x_3' + x_1'x_2'x_3$

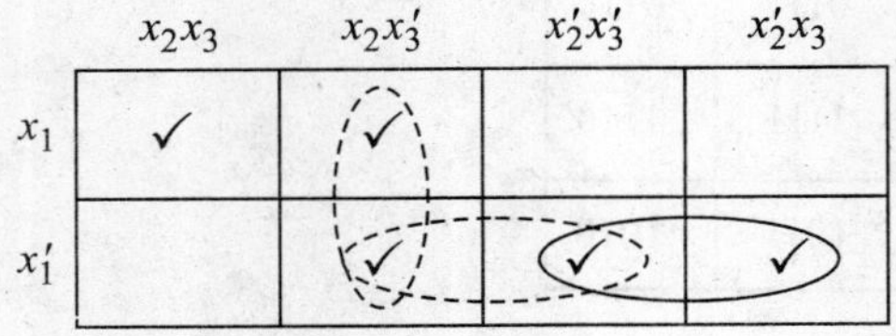

7. $E_7 = x_1'x_2'x_3' + x_1'x_2x_3' + x_1'x_2'x_3 + x_1'x_2'x_3 + x_1'x_2x_3$

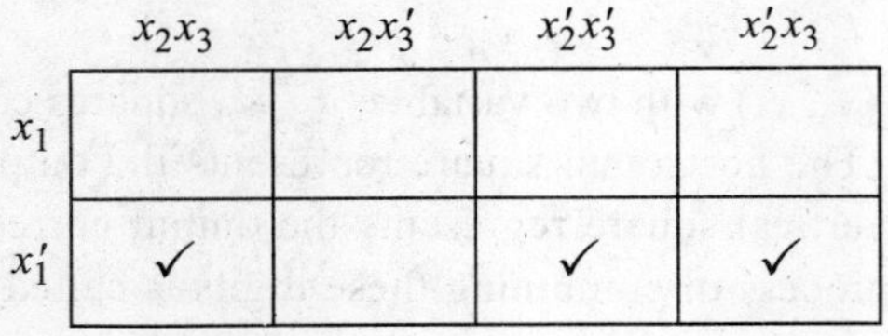

Example 32 Simplify, $E(x_1, x_2) = \bar{x}_1\bar{x}_2 + x_1\bar{x}_2$.

Solution Let $f : A_2 \to A$ be the Boolean function whose truth table is shown in Table 13.20(*a*). Then the corresponding Boolean expression $E(x_1, x_2)$ for f is shown by two-variable K-map by putting 1 in the squares which corresponds to $\bar{x}_1\bar{x}_2$ and $x_1\bar{x}_2$. The remaining squares are filled up with 0s as shown in Table 13.20(*b*).

Two adjacent squares $\bar{x}_1\bar{x}_2$ and $x_1\bar{x}_2$ containing 1 have been enclosed in a rectangle. These two terms can be looped to give a resultant that eliminates variable x_1 since it appears in both uncomplemented and complemented forms. This can also be verified algebraically as follows:

$$E(x_1, x_2) = \bar{x}_1\bar{x}_2 + x_1\bar{x}_2 = \bar{x}_2(\bar{x}_1 + x_1) = \bar{x}_2 \cdot 1 = \bar{x}_2$$

Table 13.20(*a*)

x_1	x_2	$f(x_1, x_2)$
0	0	1
0	1	1
1	0	0
1	1	0

Table 13.20(*b*)

	$\bar{x}_1$ 0	x_1 1
$\bar{x}_2$ 0	1	1
x_2 1	0	0

Example 33 Simplify $E(x_1, x_2) = \bar{x}_1\bar{x}_2 + \bar{x}_1x_2 + x_1\bar{x}_2$.

Solution The truth table of Boolean function $f : A_2 \to A$ is shown in Table 13.21(*a*). The two-variable K-map of the resulting Boolean expression $E(x_1, x_2)$ is shown in Table 13.21(*b*). There are two pairs of 1s and can be combined as shown in Table 13.21(*b*). The digit 1 appears both in first column and first row has been enclosed twice, therefore, decomposition into rectangles is not unique and that we should try to use largest possible rectangles. The group of horizontal 1-squares gives $\bar{x}_2$ and vartical 1-squares give $\bar{x}_1$. Hence, the simplified expression is $E(x_1, x_2) = \bar{x}_1 + \bar{x}_2$.

Table 13.21(*a*)

x_1	x_2	$f(x_1, x_2)$
0	0	1
0	1	1
1	0	1
1	1	0

Table 13.21(*b*)

	$\bar{x}_1$ 0	x_1 1
$\bar{x}_2$ 0	1	1
x_2 1	1	0

Example 34 Simplify the following Boolean expressions.

(*a*) $E(x_1, x_2, x_3) = x_1x_2\bar{x}_3 + x_1x_2x_3$

(*b*) $E(x_1, x_2, x_3) = \bar{x}_1\bar{x}_2\bar{x}_3 + x_1\bar{x}_2\bar{x}_3$

(*c*) $E(x_1, x_2, x_3) = \bar{x}_1\bar{x}_2\bar{x}_3 + \bar{x}_1x_2\bar{x}_3 + x_1x_2\bar{x}_3 + x_1\bar{x}_2x_3$

Solution (*a*) Three-variable K-map of the given expression is shown in Table 13.22.

Table 13.22 K-Map of $x_1x_2\bar{x}_3 + x_1x_2x_3$

	$\bar{x}_1\bar{x}_2$ 00	$\bar{x}_1x_2$ 01	x_1x_2 11	$x_1\bar{x}_2$ 10
$\bar{x}_3$ 0	0	0	1	0
x_3 1	0	0	1	0

The squares containing 1-values in Table 13.22 are grouped together. These squares represent terms $x_1x_2\bar{x}_3$ and $x_1x_2x_3$ leaving variable x_3 since it appears in both uncomplemented and complemented forms. The simplified Boolean expression becomes E $(x_1, x_2) = x_1x_2$.

(*b*) The Boolean expression E (x_1, x_2, x_3) is shown in Table 13.23. In a K-map, the top and bottom rows of squares are considered to be adjacent. Thus, the two 1s in this map can be looped to provide a simplified result $f(x_1, x_2) = \bar{x}_2\bar{x}_3$.

Table 13.23 K-Map of $\bar{x}_1\bar{x}_2\bar{x}_3 + x_1\bar{x}_2\bar{x}_3$

	$\bar{x}_1\bar{x}_2$ 00	$\bar{x}_1x_2$ 01	x_1x_2 11	$x_1\bar{x}_2$ 10
$\bar{x}_3$ 0	1	0	0	1
x_3 1	0	0	0	0

(*c*) Three variables K-map of the given Boolean function is shown in Table 13.24. These are two pairs of 1s which can be combined. Only 1 representing the term $x_1\bar{x}_2x_3$ is isolated. The group of first two horizontal 1 square gives $x_2\bar{x}_3$. Hence the simplified result is,

$$f(x_1, x_2, x_3) = \bar{x}_1\bar{x}_3 + x_2\bar{x}_3 + x_1\bar{x}_2x_3$$

Table 13.24 K-Map of $\bar{x}_1\bar{x}_2\bar{x}_3 + \bar{x}_1x_2\bar{x}_3 + x_1\bar{x}_2x_3$

	$\bar{x}_1\bar{x}_2$ 00	$\bar{x}_1x_2$ 01	x_1x_2 11	$x_1\bar{x}_2$ 10
$\bar{x}_3$ 0	1	1	1	0
x_3 1	0	0	0	1

Self-practice Problems II

11. Explain Boolean variables and Boolean functions.

12. Explain AND and OR operations with suitable examples of logic statements and electrical switches.

13. Show that the exclusive-OR function, $f(x_1, x_2, x_3, x_4) = x_1 \oplus x_2 \oplus x_3 \oplus x_4$ is an odd function.

14. Simplify the following Boolean expressions:

(*a*) $E = x_1 (x_1 + x_2) (x_1 + x_1x_2)$
(*b*) $E = x_1\bar{x}_2 + \bar{x}_1\bar{x}_2 + \bar{x}_1x_2 + x_1x_2$
(*c*) $E = (x_1\bar{x}_2 + x_2)(x_1x_2 + \bar{x}_2)$
(*d*) $E = x_1x_2x_3 + x_1\bar{x}_2x_3 + x_1x_2\bar{x}_3 + x_1\bar{x}_2\bar{x}_3$
(*e*) $E = \bar{x}_1x_2(x_2 + x_3) + x_2x_3(\bar{x}_2 + \bar{x}_1)$
(*f*) $E = \bar{x}_1x_2\bar{x}_3 + x_1x_2\bar{x}_3 + x_1\bar{x}_2\bar{x}_3 + \bar{x}_1\bar{x}_2\bar{x}_3$
(*g*) $E = \bar{x}_1x_2(\bar{x}_1 + x_3) + x_1\bar{x}_2(\bar{x}_2 + x_3)$

15. Show that

(*a*) $x_1 + \bar{x}_1x_2 = (x_1 + \bar{x}_1)(x_1 + x_2)$
(*b*) $\bar{x}_1(x_1 + \bar{x}_2)(x_1 + \bar{x}_3) + \bar{x}_3(x_3 + \bar{x}_1\bar{x}_2) = \bar{x}_1\bar{x}_2\bar{x}_3$
(*c*) $(x_1x_2 + x_3)(x_1x_2 + \bar{x}_3) = x_1x_2$

16. What are (*a*) sum-of-products form, and (*b*) product-of-sums form of Boolean expression? Explain with suitable examples.

17. What is the canonical form of Boolean expression? Explain minterms and maxterms.

18. Draw Karnaugh map and simplify the following Boolean expressions:

(*a*) $x_1x_2 + x_1\bar{x}_2$
(*b*) $x_1x_2 + \bar{x}_1x_2 + \bar{x}_1\bar{x}_2$
(*c*) $x_1x_2 + \bar{x}_1\bar{x}_2$
(*d*) $x_1x_2x_3 + x_1x_2\bar{x}_3 + \bar{x}_1x_2\bar{x}_3 + \bar{x}_1\bar{x}_2\bar{x}_3$
(*e*) $x_1x_2x_3 + x_1x_2\bar{x}_3 + \bar{x}_1x_2\bar{x}_3 + \bar{x}_1\bar{x}_2x_3 + \bar{x}_1\bar{x}_2\bar{x}_3$
(*f*) $x_1x_2x_3 + \bar{x}_1x_2\bar{x}_3 + x_1x_2\bar{x}_3 + \bar{x}_1x_2x_3$
(*g*) $\bar{x}_1\bar{x}_2x_3 + \bar{x}_1\bar{x}_2\bar{x}_3 + x_1x_2\bar{x}_3 + \bar{x}_1x_2\bar{x}_3$
(*h*) $x_1\bar{x}_2\bar{x}_3 + x_1x_2\bar{x}_3 + x_1x_2x_3 + x_1\bar{x}_2x_3 + \bar{x}_1\bar{x}_2x_3$

19. Consider the Boolean expression

$$E(x_1, x_2, x_3) = (\bar{x}_1x_3) + (x_2\bar{x}_3) + (x_2 + x_3)$$

(*a*) Simplify this expression using Boolean algebra laws.
(*b*) Write this expression in minterm form.

20. Simplify the following Boolean expression using Boolean algebra laws.

(*a*) $E(x_1, x_2, x_3) = x_1x_2 + \bar{x}_1x_3 + x_2x_3$
(*b*) $E(x_1, x_2, x_3) = x_3(x_2 + x_3)(x_1 + x_2 + x_3)$
(*c*) $E(x_1, x_2, x_3) = x_1 + x_2(x_1 + x_2) + x_1(\bar{x}_1 + x_2)$
(*d*) $E(x_1, x_2, x_3) = ((x_1 + x_2) + (x_1 + x_3))\, x_1\bar{x}_2$

21. Construct gate diagram corresponding to the following Boolean functions.

(*a*) $f(x_1, x_2, x_3) = ((x_1 + x_2) + (x_1 + x_3)) \cdot x_1 \cdot \bar{x}_2$

(*b*) $f(x_1, x_2, x_3) = x_1 \cdot \bar{x}_2 \cdot \bar{x}_3 + x_1 \cdot \bar{x}_2 \cdot x_3$

(*c*) $f(x_1, x_2, x_3) = (x_1 + x_2) \cdot \overline{((x_1 + x_2) \cdot x_3)}$.

22. Draw the on-off circuit diagram for the following Boolean functions:

(*a*) $f(x_1, x_2, x_3) = x_1 + ((x_2 \cdot (x_1 + \bar{x}_2))$ (*b*) $f(x_1, x_2, x_3) = ((x_1 + x_2) + (x_1 + x_3)) \cdot x_1 \cdot \bar{x}_3$

(*c*) $f(x_1, x_2, x_3) = (x_1 + x_2) \cdot \overline{((x_1 + x_2) \cdot x_3)}$.

23. Construct Karnaugh maps for functions whose truth tables are given below:

x_1	x_2	$f(x_1, x_2)$
0	0	1
0	1	0
1	0	0
1	1	1

x_1	x_2	x_3	$f(x_1, x_2, x_3)$
0	0	0	1
0	0	1	1
0	1	0	0
0	1	1	0
1	0	0	1
1	0	1	0
1	1	0	1
1	1	1	0

24. What is a NOR Gate and a NAND Gate? Under what input condition is the output of a NOR Gate equal to 1? For what input condition is the NAND Gate output equal to 0?

25. Write the Boolean expression corresponding to the following switching circuit network. Use law of Boolean algebra to simplify the circuit. Construct the network for the simplified circuit.

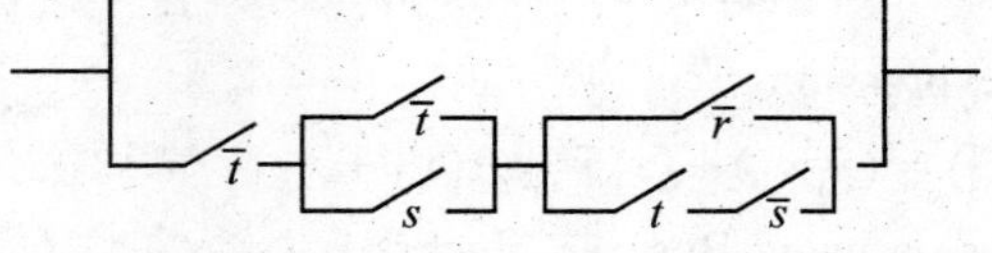

26. Find the boolean expression corresponding to each switching circuit:

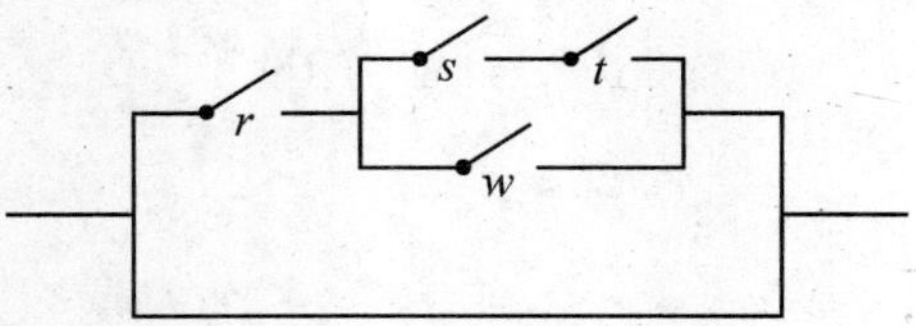

(*a*)

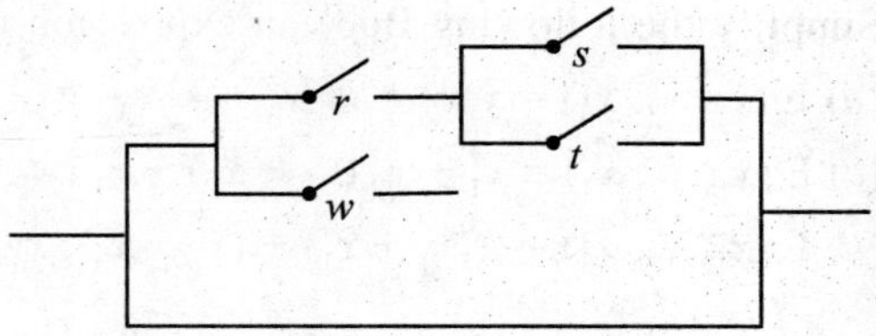

(*b*)

Hints and Answers

14. (*a*) $E = x_1$ (*b*) $E = 1$ (*c*) $E = x_1$ (*d*) $E = x_1$ (*e*) $E = \bar{x}_1 x_2$

(*f*) $E = \bar{x}_3$ (*g*) $E = \bar{x}_1 x_2 + x_1 \bar{x}_2$

18. (*a*)

	x_1	$\bar{x}_1$
x_2	1	0
$\bar{x}_2$	1	0

(*b*)

	x_1	$\bar{x}_1$
x_2	1	1
$\bar{x}_2$	0	1

(*c*)

	x_1	$\bar{x}_1$
x_2	1	0
$\bar{x}_2$	0	1

(*d*)

	x_1x_2	$\bar{x}_1x_2$	$\bar{x}_1x_2$	$\bar{x}_1\bar{x}_2$
x_2	1	0	1	1
$\bar{x}_2$	1	0	0	1

(*e*)

	x_1x_2	$x_1\bar{x}_2$	$\bar{x}_1\bar{x}_2$	$\bar{x}_1x_2$
x_2	1	1	1	1
$\bar{x}_2$	1	0	0	0

(*f*) $x_2\bar{x}_3 + \bar{x}_2 x_4 + x_1 \bar{x}_2 x_3$

19. (*a*) $E(x_1, x_2, x_3) = x_2 + (\bar{x}_1 x_3)$

(*b*) $E(x_1, x_2, x_3) = (x_1 x_2 x_3) + (x_1 x_2 \bar{x}_3) + (\bar{x}_1 x_2 x_3) + (\bar{x}_1 x_2 \bar{x}_3) + (\bar{x}_1 \bar{x}_2 x_3)$

20. (*a*) $E(x_1, x_2, x_3) = x_1 x_2 + \bar{x}_1 x_3$ (*b*) $E(x_1, x_2, x_3) = x_3$

(*c*) $E(x_1, x_2, x_3) = x_1 + x_2$ (*d*) $E(x_1, x_2, x_3) = x_1 \bar{x}_2$.

21. (*a*)

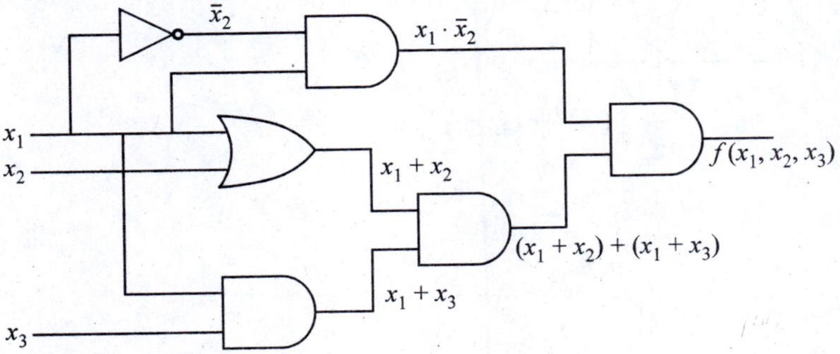

(*b*)

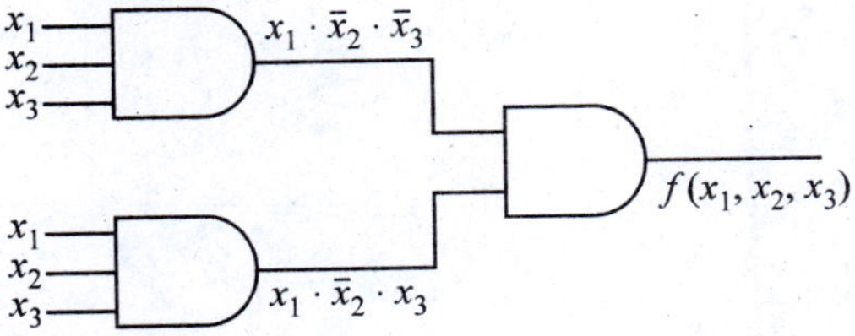

(*c*)

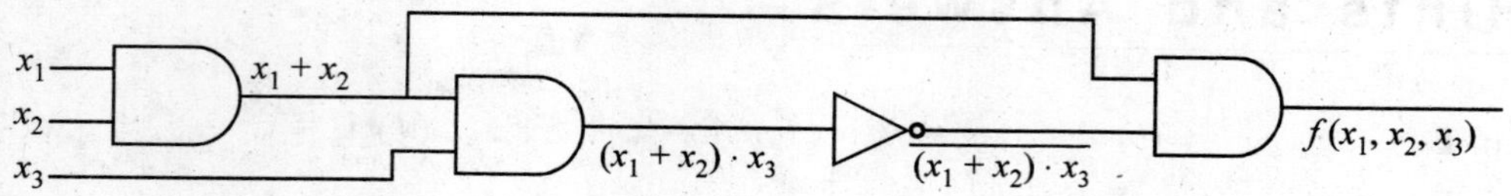

22. (*a*)

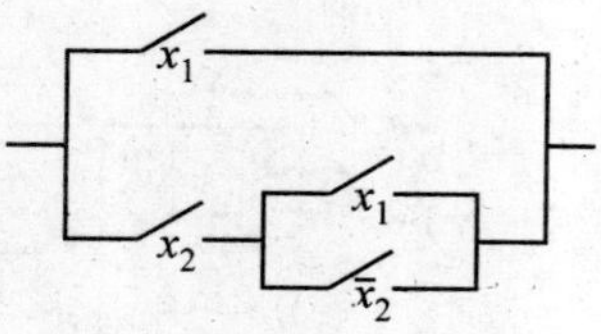

(*b*)

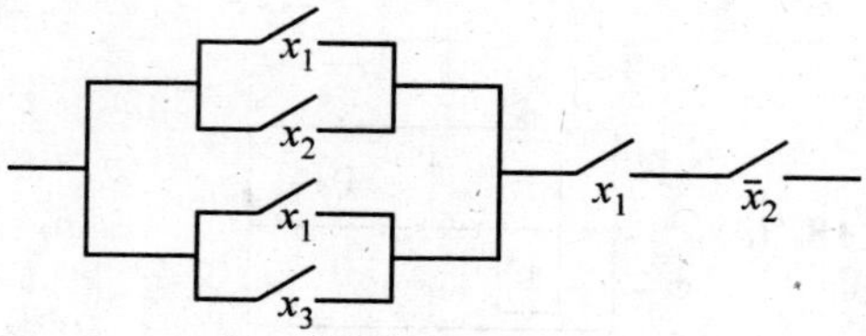

(*c*)

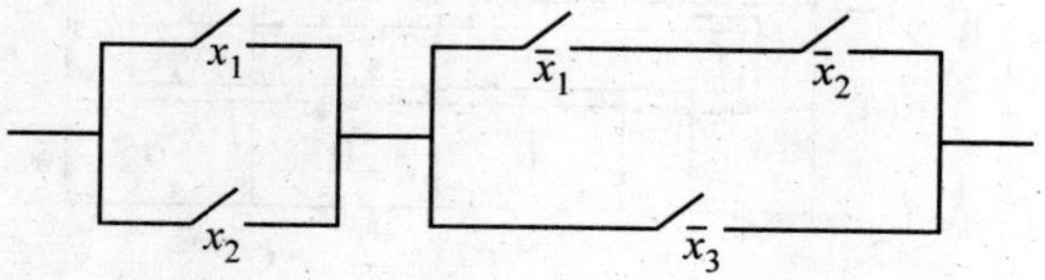

23.

	$\bar{x}_2$	x_2
x_1	1	0
$\bar{x}_1$	0	1

	$\bar{x}_2\bar{x}_3$ 00	$\bar{x}_2 x_3$ 01	$x_2 x_3$ 11	$x_2\bar{x}_3$ 10
$\bar{x}_3$ 0	1	0	0	1
x_3 1	0	0	0	0

25. $rs + st$

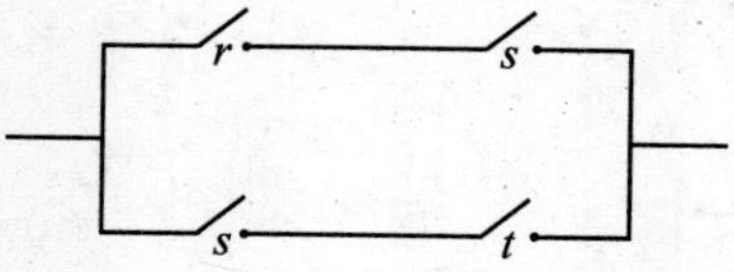

26. (*a*) $r(w + st)$ (*b*) $r(s + t) + w$

Chapter 14

Posets and Lattices

PREVIEW

A lattice is also a mathematical structure with two binary operations—join and meet, whereas a poset is a partially ordered set and is written as (A, R), where R represents a relation on a set A.

LEARNING OBJECTIVES

After studying this chapter, you should be able to

- determine whether a given mathematical structure is a lattice
- differentiate between variety of lattices and their properties.
- identify join, reducible elements.

14.1 INTRODUCTION

Recall from Chapter 1 that a non-empty set A on which a partial ordering relation, (generally denoted by $\leq$, a relation that is reflexive, anti-symmetric and transitive) is defined is called a *partially ordered set* or *poset* and is written as $(A, \leq)$. In this chapter, we will discuss those posets in which every subset $\{a, b\}$ consisting of only two elements – greatest lower bound (glb) and least upper bound (lub).

Lower bound Let A be a poset and $a, b \in A$ be any two arbitrary elements. An element $c \in A$ is the lower bound of a and b if $c \leq a$ and $c \leq b$, i.e. element c precedes every-element of A.

If a lower bound $c \in A$ and $u \leq c$ whenever u is a lower bond of A, i.e. c succeeds every other lower bound of A, then it is *infimum* of A and is denoted by $inf(a_1, a_2, \ldots, a_n)$ for all $a_i \in A$ $(i = 1, 2, \ldots, n)$.

An element $g \in A$ is called the *greatest lower bound* (glb) of a and b if and only if $g \leq a$ and $g \leq b$.

Upper bound Let A be a poset and $a, b \in$ A be any two arbitrary elements. An element $d \in$ A is the upper bound of a and b if $a \leq d$ and $b \leq d$, i.e. element d suceeds every element of A.

If an upper bound $d \in$ A and $d \leq e$ whenever e is an upper bound of A, i.e. d precedes every other upper bound of A, then it is *supremum* of A and is denoted by $sup\ (a_1, a_2, \ldots, a_n)$ for all $a_i \in$ A $(i = 1, 2, \ldots, n)$.

Further, an element $a \in$ A in a poset is called a *greatest* (*maximum*) *element* in A if $x \leq a$ for all $x \in$ A. Also an element $a \in$ A is called a *least* (*minimum*) *element* in A if $a \leq x$ for all $x \in$ A. The greatest and the least elements in a poset are frequently denoted by 1 and 0, respectively. An element in a poset (A, $\leq$) is a greatest (or least) element if and only if it is a least (or greatest) element of the dual poset (A, $\geq$).

The glb of the subset $\{a, b\}$ is denoted by $a \wedge b$ and is called the *meet* (or *product*) of a and b. Similarly, the lub of $\{a, b\}$ is denoted by $a \vee b$ and is called the *join* (or *sum*) of a and b. Since both lub and glb produces unique results on the subset $\{a, b\}$ consisting of two elements in a poset, therefore, meet ($\wedge$) and join ($\vee$) are both regarded as binary operations.

From above discussion, we conclude that $a \vee b = \text{sub}\{a, b\}$ and $a \wedge b = \inf\{a, b\}$.

Remarks

1. A subset of a poset may not have lower or upper bound.
2. Supremum or infimum of a set may or may not exist.
3. Supermum or infimum of set A may exist but may not belong to A.
4. A lower or upper bound may or may not belong of a subset of poset itself.
5. There may exists more than one lower or upper bounds of a set.
6. The greatest element is always the supremum but the converse is not true. That is a = sup (A) is the greatest element if and only if $a \in$ A.
7. The least element is always the infimum but converse is not true. That is, b = inf (A) is the least element if and only if $b \in$ A.

Illustrations

1. Suppose the partially ordered set S = {3, 4, 5, 6, 7, 8, 9, 10} and the partial order relation is defined as: '$\leq$' (less than or equal to). If A $\subseteq$ S and A = {3, 5, 6}, then upper bounds of A are 6, 7 and 8 and lower bounds of A are 3 and 5. Hence, *Sup* (*A*) = 6 and *Inf* (*A*) = 3 both belong to A.
2. Let S = {1, 2, 3, ... 10} and the partially ordered relation be 'divides'.

   If A = {5, 8} is the subset of *S*, then A has no upper bound because there is no element $a \in S$ such that every element of A, i.e. 5 and 8 precedes a. Its lower bound is 1 $\in$ S, that precedes every element of A. But, if A = {2, 5}, then 10 is its upper bound as every element of A divides 10 and 1 is its lower bound.

14.2 TOTALLY ORDERED SET OR CHAIN

If every pair of elements in a non-empty set A is comparable, then the set A is said to be *totally ordered* (*Linearly ordered set, Chain or Toset*) with respect to the given relation and the relation is said to be *simple (or linear) ordering* on the set A.

Theorem 14.1 Prove that the digraph of a partial order has no cycle of length greater than one,

Proof Let any digraph of the partial order relation, $\leq$ on the set A has a cycle of length $n \geq 2$. Then there will exist distinct elements a_1, a_2, ..., $a_n \in$ A such that

$$a_1 \leq a_2,\ a_2 \leq a_3,\ \ldots\ a_{n-1} \leq a_n,\ a_n \leq a_1$$

If the transitive relation $\leq$ is applied $(n - 1)$ times, then it may be proved that $a_1 \leq a_n$. But due to antisymmetric property of the relation $\leq$ and above two results we have $a_1 = a_n$. This contradicts the assumption that the elements a_1, a_2, ..., a_n are all district.

Hence, the assumption that the relation $\leq$ has a cycle of length greater than one is not true.

Illustrations

Partially Ordered sets

(*a*) Let A be a set of positive integers (= I) and the relation R be 'divides', i.e. a R b (a divides b), or $a \leq b$. Then (I, $\leq$) is a poset and the relation $\leq$ (divides) is a partial order relation because three conditions of binary relation are satisfied.

Since the relation of 'divisibility' is not a partial ordering on the set of integers because 5 divide (–5) and (–5) divide 5 but $5 \neq -5$. This implies that the relation is not antisymmetric.

(*b*) Let S be a super set, then the relation 'subset' $\subseteq$ is a partial ordering on S and (S, $\subseteq$) is a poset.

(*c*) The set of all subgroups of a given group form a poset under the relation 'subset'.

Totally Ordered set

(*a*) The set of positive integers, I and the relation, $\leq$ (less than or equal to) do not form a totally or linearly ordered set. However, the set I with relation 'divisibility' is not totally or linearly ordered set, because elements like 3 and 5 that are not comparable, i.e. neither 3 | 5 nor 5 | 3.

(*b*) Let P(A) be the power set of a set A = $\{a, b]$ having two or more elements. Although the set P(A) is not linearly ordered by relation, $\subseteq$ because $\{a\}$ and $\{b\}$ may be uncomparable, yet each of its subset $\{\phi\}$, $\{a\}$, {A} and $\{\phi\}$, $\{b\}$, {A} are linearly ordered.

14.3 PRODUCT SET AND PARTIAL ORDER RELATION

Theorem 14.2 If (A, $\leq$) and (B, $\leq$) are posets, then (A $\times$ B, $\leq$) is a poset with partial order defined by $(a, b) \leq (a', b')$ if $a \leq a'$ in A and $b \leq b'$ in B.

Proof Let $(a, b) \in$ A $\times$ B. Then $(a, b) \leq (a', b')$ because $a \leq a'$ in A and $b \leq b'$ in B (reflexive properly of posets A and B). Thus the relation, $\leq$ satisfies the reflexive property in A $\times$ B.

To prove that relation, $\leq$ also, satisfies antisymmetric property in A $\times$ B we need to show that $(a, b) \leq (a', b')$ and $(a', b') \leq (a, b)$ implies $(a, b) = (a', b')$, where $a, a' \in$ A and $b, b' \in$ B.

We know that $\quad a \leq a'$ and $a' \leq a$ implies that $a = a'$ (set A is antisymmetric)

and $\quad b \leq b'$ and $b' \leq b$ implies that $b = b'$ (set B is antisymmetric)

These results show that $(a, b) \leq (a', b')$ and $(a', b') \leq (a, b)$. This implies that $(a, b) = (a', b')$ in (A $\times$ B). Hence, the relation $\leq$ is antisymmetric in (A $\times$ B).

To prove that the relation, $\leq$ is transitive in $A \times B$, let

$$(a, b) \leq (a', b') \text{ and } (a', b') \leq (a'', b'') \text{ where } a, a', a'' \in \text{A and } b, b', b'' \in \text{B.}$$

We know that $a \le a'$ and $a' \le a''$ which implies that $a \le a''$ (as A is transitive)

Also $b \le b'$ and $b' \le b''$ which implies that $b \le b''$ (as B is transitive)

These results show that $(a, b) \le (a'', b'')$. Hence, that relation, $\le$ is transitive.

Since all three properties of binary set are satisfied by A × B, therefore (A × B, $\le$) is a poset.

14.4 HASSE DIAGRAMS OF PARTIALLY ORDERED SETS

Let A be a partially order set with respect to the relation $\le$. That is, (A, $\le$) is a poset. Then any two elements $a, b \in$ A may be related as:

- $a < b$ read as a strictly precedes b if $a \le b$ and $a \ne b$.
- $b > a$ read as b strictly succeeds a if $b > a$ and $b \ne a$.

Also, if a precedes b and b precedes c, then a precedes c ($\le$ being transitive relation).

Immediate Predecessor and Successor

Let (A, $\le$) be a poset and $a, b \in$ A. Then element a is an immediate predecessor of b or b is an immediate successor of a or b is a cover of a (written as $a << b$) provided $a < b$ and no element c of the set A lies between a and b.

Hasse Diagram

If A is a finite set and (A, $\le$) is a poset, then an element $a \in$ A either does not have a successor (there is no $b \in$ A such that $a < b$ in which case a is the maximal element) or element a must have immediate successor (otherwise element a shall have infinite chain of successors). This contradicts the finiteness of the set A. A similar argument may be given in the case for predecessors.

This concept helps in representing finite posets with the help of diagrams known as *Hasse diagrams.*

The *Hasse diagram* of a finite partially order set A is a diagram whose vertices represent the elements of A. If $a, b \in$ A and b is an immediate successor of a ($a << b$), then an edge (or line) is drawn directed from a to b. However, instead of drawing a line directed from a to b, the element b may be placed at a higher level than a and join a and b by a straight line.

But there can be directed cycles in the Hasse diagram because the partial order relation is antisymmetric. Thus Hasse diagram need not be connected.

To draw a Hasse diagram of a partially ordered relation, a diagraph of the given relation is drawn and then delete.

(*i*) all the loops, and

(*ii*) all the edges implied by transitivity properly.

This implies that, if $a \le b$ and $b \le c$, then a line between (a and b) and (b and c) may be drawn that but not between (a and c), although $a \le c$ by transitivity.

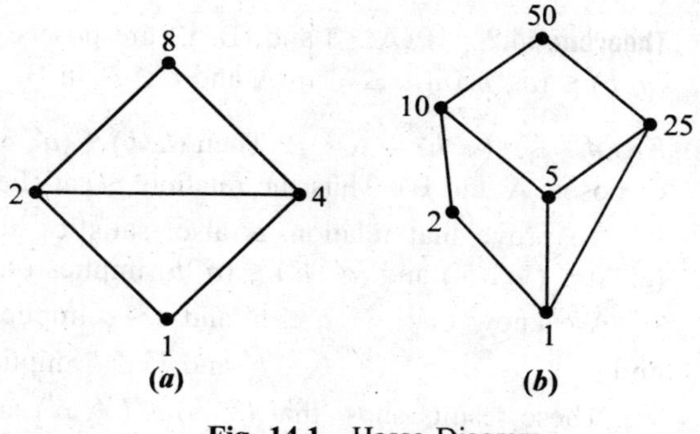

Fig. 14.1 Hasse Diagram

Let D_n be the set of all positive divisor of a positive integer, n. For example, if $n = 8$, then $D_8 = \{1, 2, 4, 8\}$. The Hasse diagram of D_8 is shown in Fig. 14.1(a). If $n = 50$, then $D_{50} = \{1, 2, 5, 10, 25, 50\}$. The Hasse diagram of D_{50} is shown in Fig. 14.1(b).

Example 1 Find the upper and lower bounds, supremum and infimum of a set A = {3, 6} which is a subset of a poset S = {1, 2, 3, 6, 8, 9, 12, 24, 36} with the relation R given by $x < y$, i.e. x divides y. Also draw its Hasse diagram.

Solution Given: S = {1, 2, 3, 6, 8, 9, 12, 24, 36} and the subset A = {3, 6}. The Hasse diagram of S is shown in Fig. 14.2.

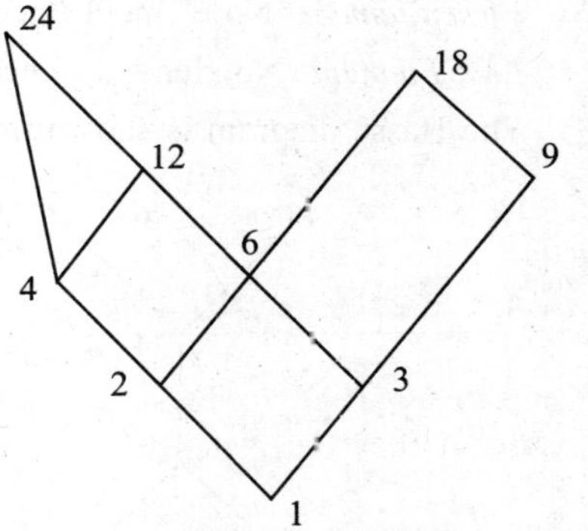

Fig. 14.2 Hasse Diagram

The upper bonds of A are 12 and 36 because each element of A divides 12 and 36. *Sup* (*A*): 9 because divides 36 also.

The Lower bounds of A are 1 and 2 as these divides every element of A. *Inf* (*A*): 2 as 1 also divides 2 or 2 succeeds the other lower bound.

Example 2 Draw Hasse diagram of the poset (S, ≤) where S = {1, 2, 4, 8, 24, 48} and $x \le y$ if $x \mid y$, i.e. x divides y. If A = {4, 8} is a subset of S, find upper and lower bounds, supremum and infimum of A.

Solution Given S = {1, 2, 4, 8, 24, 32} and the subset A = {4, 8}. The Hasse diagram of S is shown in Fig. 14.3.

The upper bounds of A are 24 and 48 as each element of A divides 24 and 48 *Sup*(*A*): 24 as it precedes (or divides) every other upper bound.

The lower bounds of A are 1, 2, 4 and 8 as each of these elements precedes (or divides) all the elements of A. *Inf* (*A*): 8 as it succeeds every other lower bound as shown in Fig. 14.3.

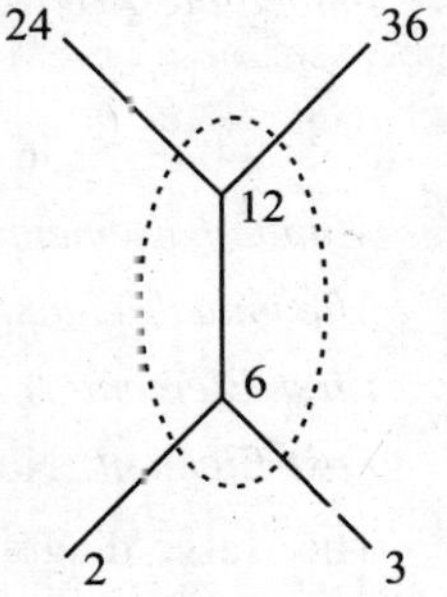

Fig. 14.3

Example 3 Hasse diagram of poset S = {1, 2, 3, 4, 5, 6, 7, 8) is given below. If A = {4, 5, 7} is a subset of S, find upper bounds, lower bounds, supremum and infimum of A.

Solution Upper bounds of A are 1, 2 and 3 as every elements of A precedes 1, 2 and 3. *Sup*(*A*): 3 as it precedes every other bound.

The lower bound of A is 8 as it precedes every element of A. *Inf* (*A*) : 8 as shown Fig. 14.4

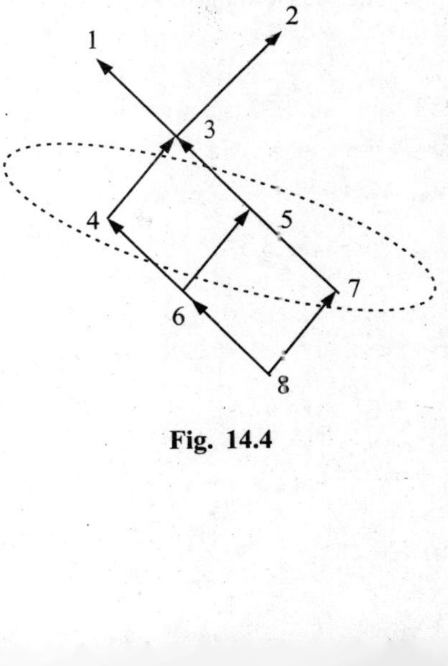

Fig. 14.4

Illustrations

1. Let A = {4, 6, 12, 24, 48, 72} along with the relation ≤ such that $a \le b$ when 'a devides b', then Hasse diagram is as shown Fig. 14.5.

   Immediate Predecessors: 4 << 12, 6 << 12, 12 << 24, 24 << 48, 24 << 72.

   Minimal Elements: 4 and 6 as 48 and 72 succeed all other elements, but no element succeed them.

Maximal Elements: 48 and 72 as there elements precedes other elements but no element precedes them.

First Element: No element (since neither $4 \leq 6$ nor $6 \leq 4$).

Last Element: No element (since neither $48 \leq 72$ nor $72 \leq 48$).

The Hasse diagram is shown in Fig. 14.5.

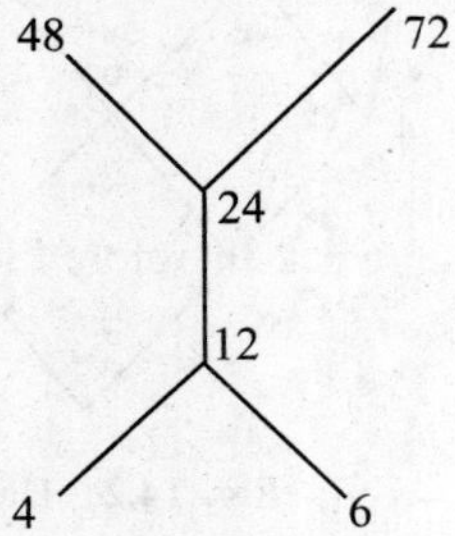

Fig. 14.5 Hasse Diagram

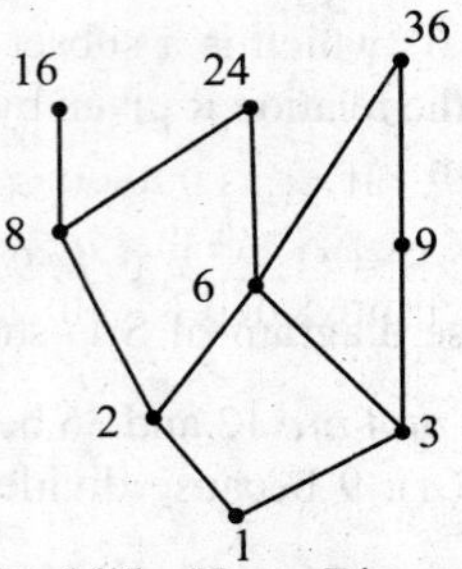

Fig. 14.6 Hasse Diagram

2. Let A = {1, 2, 3, 6, 8, 9, 16, 24, 36} along with the relation, ≤ such that $a \leq b$ when '*a* devides *b*'

Immediate Predecessors:

1 << 2, 1 << 3, 2 << 6, 2 << 8, 3 << 6, 3 << 9, 8 << 16, 8 << 24,

6 << 24, 6 << 36, 9 << 36,

Minimal Elements: 1

Maximal Elements: 24 and 36

First Element: 1 as it precedes every other element of A.

Last Element: No element.

The Hasse diagram is shown in Fig. 14.6.

3. Let S be the power set of A = {1, 2, 3} along with the relation ≤ such that $a < b$. Then

ϕ << {1}, ϕ << {2}, ϕ << {3}, {1} << {1, 2}, {1} << {1, 2}, {1, 3} << {1, 2, 3} when $a \leq b$

First Element: ϕ

Last Element: {1, 2, 3}

The Hasse diagram is shown in Fig. 14.7 [Since all the lines are not vertical, arrows have been used.]

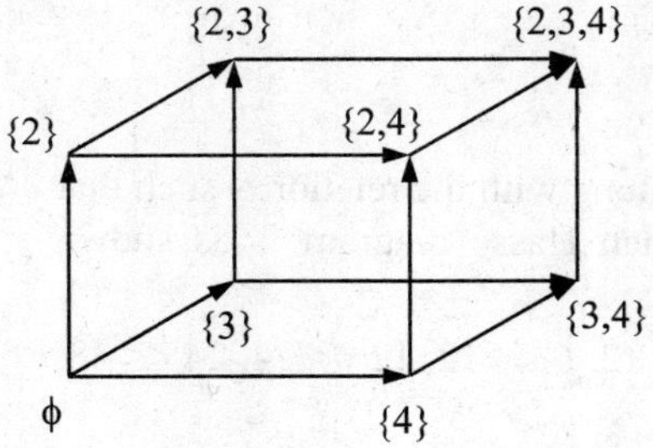

Fig. 14.7 Hasse Diagram

4. Let A = {1, 2, 3, 4, 5, 6} along with partial ordered relation < such that $a < b$ when a/b.

Since element 6 does not cover 1 or 1 is not the immediate predecessor of 6, there is no line joining 1 and 6. Similarly 2 covers 1, 6 covers 2 , 4 does not cover 1. The Hasse diagram is shown in Fig. 14.8.

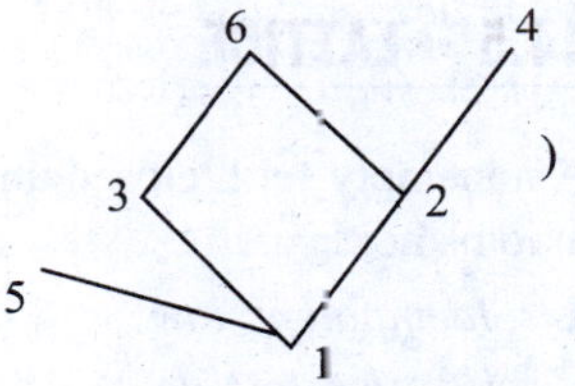

Fig. 14.8 Hasse Diagram

Theorem 14.3 Let A be a finite nonempty poset with partial order ≤. Then A has at least one minimal element and one maximal element.

Proof Suppose $a \in A$. If 'a' is not a maximal element of A, we need to find elements say a_1 and $a_2 \in A$ such that $a \le a_1$ and $a_1 \le a_2$. Since set A is a finite set, therefore there should be an element a_k which preceded all other elements and no other succeed it, i.e. $a < a_1 < a_2 \ldots < a_{k-1} < a_k$. Consequently $a_k \in A$ is the maximal element. Proceeding in the same way, it may also be proved that the dual poset (A , ≥) has a maximal element and hence the poset (A, ≤) has a minimal element.

Theorem 14.4 A poset has at most one greatest element and atmost one least element.

Proof Suppose, if possible, a and b are greatest elements of poset A.

Since both a and b are assumed to be the greatest elements, $a \le b$ and $b \le a$. This implies that $a = b$. Hence, greatest element in any poset is unique (antisymmetry property).

Similarly, let x and y be the least elements of A. since both x and y are least elements of A, $x \le y$ and $y \le x$. Since both x and y are least elements of A, $x \le y$ and $y \le x$. This implies that $x = y$ (antisymmetry property). Hence, least element in any poset is also unique.

Remark

A *maximal element* of a set A may strictly succeed other elements, although no other element strictly succeed maximal element. That is, some of the elements may precede it while some may not precede it.

A *greatest element* does not precede any other element of the set A, i.e. greatest element succeeds all the other elements of set A. In any set there may be more than one maximal elements while the greatest element is only one.

Similarly, the *minimal element* of set A may strictly precede other elements, although no other element of A strictly precede it.

A *least element* of set A does not succeed any other element of the set A, i.e. a least element precedes all the other elements of the set. In any set there may be more than one minimal elements, while the least element is only one.

Hasse diagram Fig. 14.9(a) shows d, f and g as maximal elements but there is no greatest element while in Fig. 14.9(b) elements a and b are minimal elements, but there is no least element.

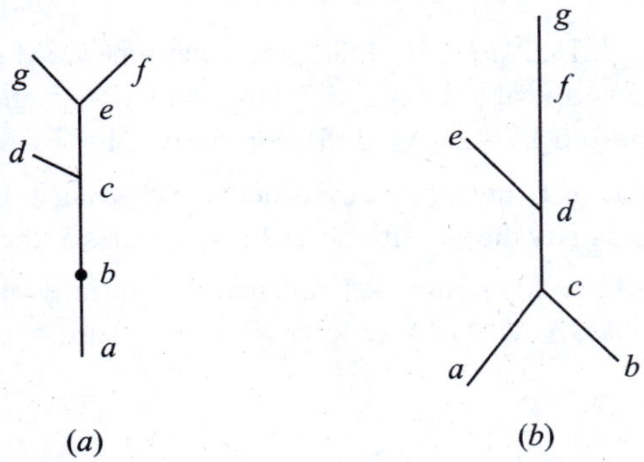

Fig. 14.9 Hasse Diagram

14.5 LATTICE

A nonempty set L closed under two binary operations, $\wedge$ and $\vee$, is called a lattice (L, $\wedge$, $\vee$) provided following axioms hold:

1. *Idempotent law*
   (*a*) $a \wedge a = a$ (*b*) $a \vee a = a$, for all $a \in L$
2. *Commutative law*
   (*a*) $a \wedge b = b \wedge a$ (*b*) $a \vee b = b \vee a$, for all $a, b \in L$
3. *Associative law*
   (*a*) $(a \wedge b) \wedge c = a \wedge (b \wedge c)$ (*b*) $(a \vee b) \vee c = a \vee (b \vee c)$, for all $a, b, c \in L$
4. *Absorption law*
   (*a*) $a \wedge (a \vee b) = a$ (*b*) $a \vee (a \wedge b) = a$, for all $a, b, c \in L$

Lattice as a Poset A lattice is a poset (L, $\leq$) in which every 2-element subset $\{a, b\}$ has a lub and glb. That is, poset (L, $\leq$) is a lattice if for every $a, b \in L$, lub (a, b) and glb (a, b) exist in L.

The lower and upper bounds in lattices are defined as follows:

1. If $a \leq a \vee b$ and $b \leq a \vee b$, then $a \vee b$ (read as *a join b*) is the *upper bound* of elements $a, b \in L$.

   Also, if $a \leq c$ and $b \leq c$, then $a \vee b \leq c$, then $a \vee b = \text{lub}\{a, b\}$ for all $a, b \in L$. It is also denoted as, $a \vee b = \sup(a, b)$.

2. If $a \wedge b \leq a$ and $a \wedge b \leq b$, then $a \wedge b$ (read as *a meet b*) is the *lower bound* of elements $a, b \in L$. Also, if $c \leq a$ and $c \leq b$, then $c \leq a \wedge b$, then $a \wedge b = \text{glb}\{\text{a, b}\}$ for all $a, b \in L$. It is also denoted as, $a \wedge b = \inf(a, b)$.

Illustrations

1. Let I_+ be the set of all positive integers and R be the relation 'divides' defined in I_+ such that for all $a, b \in I_+$, $a R b$ if and only if a divides b. Then the partially ordered set (I_+, $\div$) is a lattice in which 'join' and 'meet' of each pair of elements a and b is:

   $a \vee b$ = LCM (a, b), the least common multiple of a and b, i.e. the smallest integer in I_+ which is divisible by both a and b.

   $a \wedge b$ = GCD (a, b), the greatest common divisor of a and b, i.e. the largest integer which divides both a and b.

   For example, let $I_+ = \{1, 3, 5, 7, 15, 21, 35, 105\}$. Then to determine LCM of elements 3 and 5, we consider an element $l \in I_+$ such that l is divisible by 3 and 5. Here both $l = 15$ and $l = 105$ satisfy this condition. As 15 also divides 105, therefore 15 = LCM (3, 5). Similarly, 21 = LCM (3, 7).

   Further to determine GCD of 15 and 35, we consider an element $g \in L$ such that g divides both 15 and 35. Here both $g = 1$ and $g = 5$ satisfy the condition. As 1 also divides 5, therefore 5 = GCD (15, 35).

2. Let P (A) be the power set of any non-empty set A. Then the partially ordered set (P (A), $\leq$) is a lattice, because for any two subsets $A_1, A_2 \in P(A)$ we have $A_1 \vee A_2 = \sup\{A_1, A_2\} = A_1 \cup A_2$ and $A_1 \wedge A_2 = \inf\{A_1, A_2\} = A_1 \cap A_2$.

. The Hasse diagram shown in Fig. 14.10 does not represent a lattice because $6 \vee 7$, i.e. *sup*({6, 7}) does not exist.

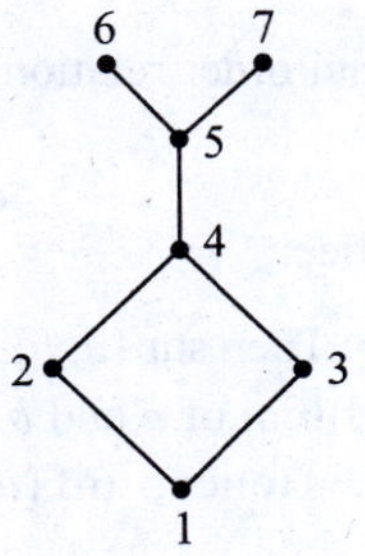

Fig. 14.10 Hasse Diagram

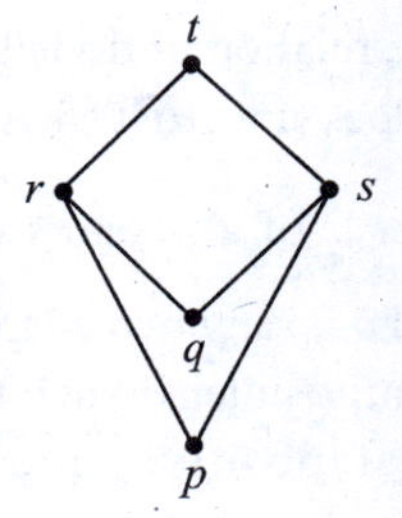

Fig. 14.11 Hasse Diagram

4. The Hasse diagram shown in Fig. 14.11 does not represent a lattice because $r \vee s$, i.e. inf({r, s)) does not exist.
5. The Hasse diagrams shown in Figs. 14.12(a to c) represent posets but not lattice.

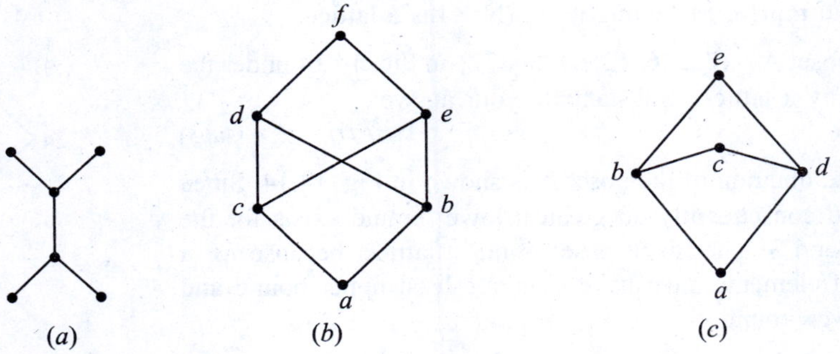

Fig. 14.12 Hasse Diagrams

6. The poset with the following Hasse diagram shown in Fig. 14.13 is also not a lattice.

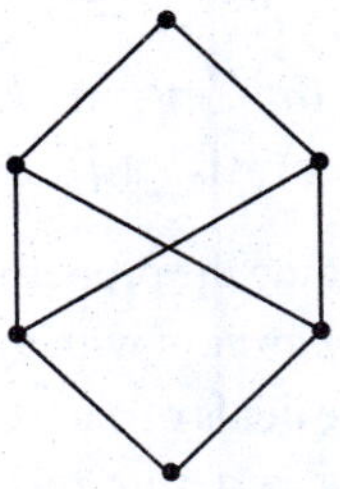

Fig. 14.13 Hasse Diagrams

The Hasse diagram of a lattice is always a combination of closed polygons because any two of its elements have a common predecessor and a common successor.

Example 4 Let A = {2, 3, 4, 6} and let divisibility be the relation on A. Then the (A, |) is a poset but not a lattice.

Solution Since the relation of divisibility on the set {2, 3, 4, 6} is a partial order relation, (A, |) is a poset. Since sup {4, 6} does not exist in A, therefore (A, |) is not a lattice.

Example 5 The set L of all factors of 12 under divisibility forms a lattice.

Solution Let a and b be any two elements in the set. L = {1, 2, 3, 4, 6, 12}. Then sup$\{a, b\}$ under divisibility relation is least positive integer such that $a \mid c$ and $b \mid c$. Thus, sup$\{a, b\}$ = lcm of a and b. Similarly, inf$\{a, b\}$ is the greatest positive integer d such that $d \mid a$ and $d \mid b$. Hence, inf$\{a, b\}$ = gcd of a and b.

Example 6 Let (N, ≤) be a partial ordered set, where N is the set of natural numbers and ≤ is the 'less than or equal to' relationship among natural numbers. Then (N, ≤) is a lattice.

Solution Let a and b be any two natural numbers. Then sup$\{a, b\}$ is the least natural numbers c such that $a \le c$ and $b \le c$. Clearly $c = \max\{a, b\}$. Thus, sup$\{a, b\} = \max\{a, b\}$. Similarly, inf$\{a, b\} = \min\{a, b\}$. Since max$\{a, b\}$ and min$\{a, b\}$ belong to N, (N, ≤) is a lattice.

Example 7 Is the poset A = (2, 3, 6, 12, 24, 36, 72} in Fig. 14.14 under the relation of divisibility a lattice? Substantiate your answer.

[*UPTU, MCA, 2005*]

Solution The Hasse diagram of the poset A is shown in Fig. 14.14. Since no lower bound and consequently no greatest lower bound exists for the pair of elements 2 and 3. The given poset is not a lattice, because for a lattice every pair of elements must have a unique least upper bound and unique greatest lower bound.

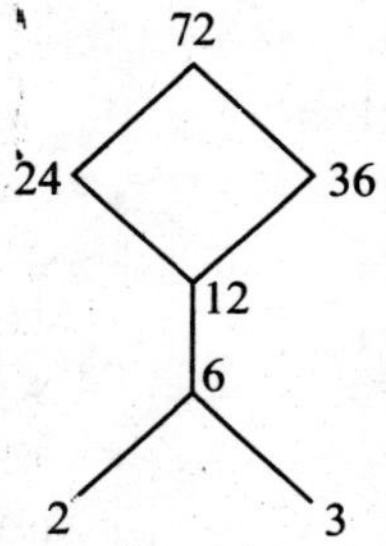

Fig. 14.14 Hasse Diagrams

14.6 DUALITY

The dual of any statement in a lattice (L, ∧, ∨) is defined to be the statement that is obtained by interchanging ∧ and ∨.

For example, dual of

$$a \vee (b \wedge a) = a \text{ is } a \wedge (b \vee a) = a$$

also dual of $(a \wedge b) \wedge c = a \wedge (b \wedge c)$ is $(a \vee b) \vee c = a \vee (b \vee c)$

Principle of duality Any valid statement concerning the properties of lattices, yields another valid statement by replacing the relation (*i*) ≤ with ≥, and (*ii*) *join operation* with the *meet operation* and vice versa.

The lattices (L, ≤) and (L, ≥) are called the duals of each other. That is, for any partial order relation ≤ on a set L, the converse relation ≥ is also a partial ordering relation on L because for any element $a, b \in$ L, lub(a, b) in (L, ≤) is equal to the glb (a, b) in (L, ≥) and vice versa.

Remarks

1. A lattice is said to be *complete* if every subset has glb and lub.
2. A lattice is said to be *modular* if for all elements $a, b, c \in$ L, $a \ge c \Rightarrow a \cap (b \cup c) = (a \cap b) \cup c$.
3. The relation $a \le b$ defined by $a \wedge b = a$ or $a \vee b = b$ is a partial order on a lattice L.

Theorem 14.5 Let (L, $\leq$) be a lattice. then prove that for every element $a, b \in$ L.

(*a*) $a \wedge b = a$ if and only if $a \leq b$.

(*b*) $a \vee b = b$ if and only if $a \leq b$.

(*c*) $a \wedge b = a$ if and only if $a \vee b = b$.

Proof (*a*) Suppose $a \leq b$, and we know that $b \leq b$. Then by definition of least upper bound we have $a \vee b \leq b$. Since $a \vee b$ is an upper bound of a and b, therefore $b \leq a \vee b$, and hence $a \vee b = b$.

Conversely, suppose $a \vee b = b$. Since $a \leq a \vee b = b$, therefore $a \leq b$.

(*b*) Leave as an exercise for the reader.

(*c*) Combining (*a*) and (*b*), we get (*c*).

Theorem 14.6 Let (L, $\leq$) be lattice and $a, b, c \in$ L. Then the following results hold:

(*a*) If $a \leq b$ and $a \leq c$, then $a \leq b \vee c$.

(*b*) If $a \leq b$ and $a \leq c$, then $a \leq b \wedge c$.

Proof (*a*) From the definition of join operation, we know that $b \vee c = \text{sub}\{b, c\}$, so $b \leq b \vee c$.

Hence, $a \leq b$ and $b \leq b \vee c \Rightarrow a \leq b \vee c$ (Transitivity law)

(*b*) Suppose $a \leq b$ and $a \leq c$. Then a is a lower bound of $\{b, c\}$. Thus, $a \leq b \wedge c$ is the greatest lower bound of $\{b, c\}$.

Corollary Let (L, $\leq$) be a lattice and (L, $\geq$) be its dual. Then for all $a, b, c \in$ L, we have

(*a*) $a \geq b$ and $a \geq c \Rightarrow a \geq b \wedge c$.

(*b*) $a \geq b$ and $a \geq c \Rightarrow a \geq b \vee c$.

Theorem 14.7 Let (L, $\leq$) be a lattice. Then prove that

(*a*) $b \leq c \Rightarrow \begin{cases} a \wedge b \leq a \wedge c \\ a \vee b \leq a \vee c \end{cases}$

(*b*) $a \leq b$ and $c \leq d \Rightarrow \begin{cases} a \vee c \leq b \vee d \\ a \wedge c \leq b \wedge d \end{cases}$

for every element $a, b, c, d \in$ L.

Proof (*a*) From Theorem 14.7(*a*) we know that $b \leq c$ if and only if $b \wedge c = b$. To show that $a \wedge b \leq a \wedge c$, it is enough to show that

$$(a \wedge b) \wedge (a \wedge c) = a \wedge b.$$

Now
$$\begin{aligned}(a \wedge b) \wedge (a \wedge c) &= (a \wedge a) \wedge (b \wedge c) \\ &= a \wedge (b \wedge c) = a \wedge b, \text{ since } b \wedge c = b\end{aligned}$$

This result implies that $a \wedge b \leq a \wedge c$.

Further
$$\begin{aligned}(a \vee b) \vee (a \vee c) &= (a \vee a) \vee (b \vee c) \\ &= a \vee (b \vee c) = a \vee b, \text{ from Theorem 14.7.}\end{aligned}$$

This result implies that $a \vee b \leq a \vee c$.

(*b*) Suppose that $a \leq b$ and $c \leq d$. Then by definition of join operation $\vee$, we have

$$b \leq b \vee d \text{ and } d \leq b \vee d$$

Also by transitivity, we have

$$a \leq b \text{ and } b \leq b \vee d \Rightarrow a \leq b \vee d, \text{ and}$$
$$c \leq d \text{ and } d \leq b \vee d \Rightarrow c \leq b \vee d$$

Thus $b \vee d$ is an upper bound of a and c. Since $a \vee c$ is the least upper bound of a and c, we have, $a \vee c \leq b \vee d$.

Further, suppose that $a \leq b$ and $c \leq d$. Then by definition of meet operation $\wedge$, we have

$$a \wedge c \leq a \text{ and } a \wedge c \leq c$$

Also by transitivity, we have

$$a \wedge c \leq a \text{ and } a \leq b \Rightarrow a \wedge c \leq b, \text{ and}$$
$$a \wedge c \leq c \text{ and } c \leq d \Rightarrow a \wedge c \leq d$$

Thus $a \wedge c$ is a lower bound of b and d, hence $a \wedge c \leq b \wedge d$ because $b \wedge d$ is the greatest lower bound of b and d.

Theorem 14.8 In any lattice (L, ≤), the distributive inequalities:

(*a*) $a \wedge (b \vee c) \geq (a \wedge b) \vee (a \wedge c)$

(*b*) $a \vee (b \wedge c) \leq (a \vee b) \wedge (a \vee c)$

holds for any $a, b, c \in$ L.

Proof (*a*) We know that

$a \wedge b$ and $a \wedge b \leq b \leq b \vee c \Rightarrow a \wedge b$ is a lower bound of $\{a, b \vee c\}$ and therefore $a \wedge b \leq a \wedge (b \vee c)$ because $a \wedge (b \vee c)$ is the greatest lower bound of $\{a, b \vee c\}$.

Again $\quad a \wedge c \leq a$ and $\quad a \wedge c \leq c \leq b \vee c \Rightarrow a \wedge c \leq a \wedge (b \vee c)$

From these two results, we conclude that $a \wedge (b \vee c)$ is an upper bound of $\{a \wedge b, a \wedge c\}$.

and hence $\quad (a \wedge b) \vee (a \wedge c) \leq a \wedge (b \vee c)$.

(*b*) Left as a exercise for the reader.

Theorem 14.9 In any lattice (L, ≤), the modular inequality

$$a \leq c \Leftrightarrow a \vee (b \wedge c) \leq (a \vee b) \wedge c.$$

holds for any $a, b, c \in$ L.

Proof We know that $a \leq c \Leftrightarrow a \vee c = c$. Substituting c for $a \vee c$ in the right hand side of the inequality

$$a \vee (b \wedge c) \leq (a \vee b) \wedge (a \vee c)$$

we get $a \leq c \Leftrightarrow a \vee (b \wedge c) \leq (a \vee b) \wedge c$.

Example 8 Show that dual of a lattice is a lattice. [*CCS Univ., MSc(Maths), 2005*]

Solution Let (L, R) be a given lattice and let (L, $\overline{\text{R}}$) be its dual, where $\overline{\text{R}}$ is defined as $x\overline{\text{R}}y$ if and only if $y\text{R}\,x$. Then we need to prove that (L, $\overline{\text{R}}$) is a poset.

If $x, y \in$ L, then sup (x, y) exists in R because (L, R) is a lattice. Let $x \vee y = \sup\{x, y\}$ in (L, R). Then $x\,\text{R}\,(x \vee y)$ and $y\,\text{R}\,(x \vee y) \Rightarrow z\,(x \vee y)\,\overline{\text{R}}\,x$ and $(x \vee y)\overline{\text{R}}\,y$.

$\Rightarrow x \vee y$ is a lower bound of $\{x, y\}$ in (L, $\overline{\text{R}}$)

To show that $x \vee y$ is the greatest lower bound of $\{x, y\}$ in (L, $\overline{\text{R}}$), let z be any lower bound of

(x, y) in $(L, \overline{R})$, then $z\, \overline{R}\, x$ and $z\, \overline{R} y$, so that

$$x\ R\ z \text{ and } x\ R\ z \Rightarrow x\ R\ z \text{ and } z\ R\ y$$
$$\Rightarrow z \text{ is a upper bound of } \{x, y\} \text{ in } (L, R)$$
$$\Rightarrow (x \vee y)\ R\ z \text{ since } x \vee y = \sup\{x, y\} \text{ in } (L, R)$$
$$\Rightarrow z\ \overline{R}\ (x \vee y)$$
$$\Rightarrow x \vee y \text{ is the greatest lower bound of } (x, y) \text{ in } (L, \overline{R}).$$

Similarly, it can be shown that $x \wedge y$ is the least upper bound in $(L, \overline{R})$. Hence, $(L, \overline{R})$ is a lattice.

Example 9 For any positive integers $a, b \in N$, show that

$$\max\{a, \min.\ (a, b)\} = a$$
$$\min\{a, \max.\ (a, b)\} = a$$

Solution Let $(N, \leq)$ be a lattice with respect to the set of natural numbers N. It is given that

$$a \vee b = \max\{a, b\} \quad \text{and} \quad a \wedge b = \min\{a, b\}$$

Applying absorption properties,

$$a \wedge (a \vee b) = a \text{ and } a \vee (a \wedge b) = a$$

we get $\min\{a, \max\{a, b\}\} = a$ and $\max\{a, \min\{a, b\}\} = a$ for any positive integers a and b.

Example 10 For any positive integers a and b, show that

$$\text{lcm}\{a, \gcd(a, b)\} = a \quad \text{and} \quad \gcd\{a, \text{lcm}(a, b)\} = a$$

Solution The relation $\leq$ is defined as $a \leq b$ if and only if a divides b in the lattice $(N, \leq)$. We know that

$$a \vee b = \text{lcm}(a, b) \quad \text{and} \quad a \wedge b = \gcd(a, b)$$

Applying absorption laws,

$$a \wedge (a \vee b) = a \quad \text{and} \quad a \vee (a \wedge b) = a$$

we get $\text{lcm}(a, \gcd\{a, b)\} = a$ and $\gcd\{a, \text{lcm}(a, b)\} = a$.

14.7 TYPES OF LATTICES

14.7.1 Sublattices

Let $(L, \wedge, \vee)$ be a lattice. Then a nonempty subset S of the lattice L is said to be sublattice if it is closed under the operations $\wedge$ and $\vee$ of L. That is, if $a \vee b \in S$ and $a \wedge b \in S$ for every $a, b \in S$.

For example, the lattice D_n of all divisors of n is a sublattice of the lattice of the set of positive integers I_+ under the relation of divisibility.

Example 11 Let $S_1 = \{a_1, a_2, a_4, a_6\}$; $S_2 = \{a_3, a_5, a_7, a_8\}$ and $S_3 = \{a_1, a_2, a_4, a_8\}$

be the subsets of the set $L = \{a_1, a_2, \ldots, a_n\}$.

If $(L, \leq)$ is the lattice, then prove that S_1, S_2 and S_3 are also lattice under the same relation.

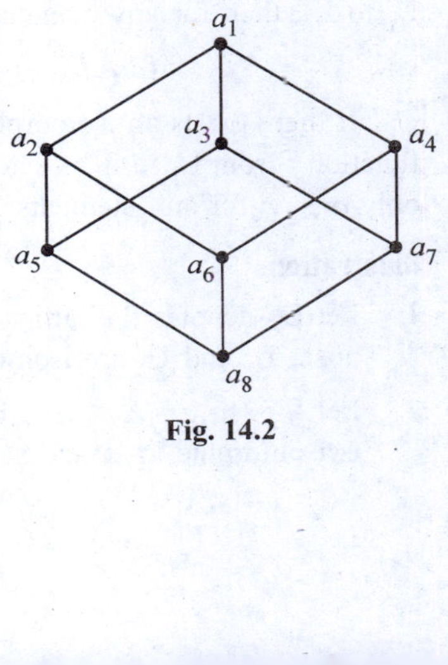

Fig. 14.2

Solution Consider the lattice $(L, \leq)$ shown in Fig. 14.2. Subsets S_1 and S_2 of $(L, \leq)$ are sublattices. But subset S_3 is not a sublattice of $(L, \leq)$ since $a_2 \wedge a_4 = a_6 \notin S_3$, for $a_2, a_4 \in S_3$.

Example 12 Let D_n be the set of all positive divisor of a positive integer n. Let I_+ be set of all positive integers. Prove that D_n is sublattice of I_+ under the relation of divisibility.

Solution Since $D_n \subseteq I_+$, and every pair of elements $a, b \in D_n$ has a greatest lower bound (glb) and a least upper bound (lub), therefore D_n is a sublattice of I_+.

Example 13 Consider the lattice L shown in Fig. 14.2. Find all sublattices with five elements.

Solution The sublattices of L are as under.

$S_1 = \{a_1, a_2, a_4, a_5, a_7\}$; $S_2 = \{a_1, a_2, a_3, a_4, a_7\}$;

$S_3 = \{a_1, a_3, a_5, a_6, a_7\}$; $S_4 = \{a_1, a_2, a_3, a_6, a_7\}$, and so on.

Example 14 Consider the lattice L shown in Fig. 14.3. Find partially ordered subsets which are not sublattice of L.

Solution The Hasse diagrams S_1 and S_2 shown in Fig. 14.3(*b*) and Fig. 14.3(*c*), respectively are not sublattice.

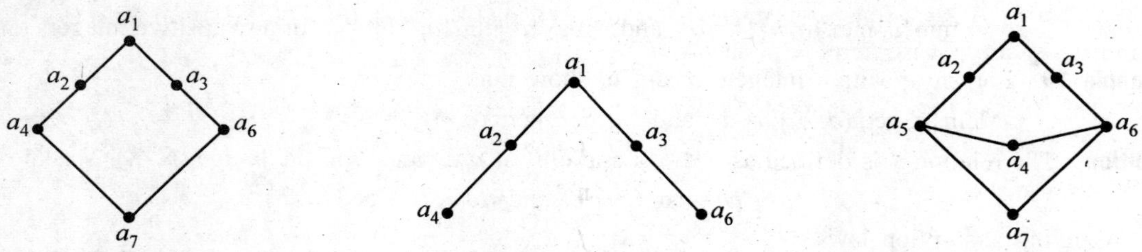

(*a*) (*b*) $a_5 \wedge a_6 \notin S_1$, $a_5 \vee a_6 \notin S_1$ (*c*) $a_5 \vee a_6 \notin S_2$

Fig. 14.3

14.7.2 Isomorphic Lattices

Let $(L_1, \leq_1)$ and $(L_2, \leq_2)$ be two posets. If the function (mapping) f defined from L_1 to L_2 as

$$f: L_1 \to L_2 \text{ such that } a \leq_1 b \Rightarrow f(a) \leq_2 f(b), \text{ for all } a, b \in L,$$

then f is a isomorphism,

We know that L_1 is a lattice if and only if L_2 is a lattice. However, if f is one-one onto function from L_1 to L_2, then for any elements $a, b \in L_1$ we have

$$f(a \wedge b) = f(a) \wedge f(b) \quad \text{and} \quad f(a \vee b) = f(a) \vee f(b).$$

If there exists an isomorphism between two lattices, then such lattices are said to be *isomorphic*. The function f from L_1 to L_2 is said to be *order preserving* relative to the ordering $\leq_1$ in L_1 and $\leq_2$ in L_2 if and only $a \leq_1 b$ for any elements $a, b \in L_1$ and $f(a) \leq_2 f(b)$ in L_2.

Illustrations

1. Let D_6 denotes the lattice L_1 and power set P (A) of the set A = $\{a, b\}$ denotes the lattice L_2. Since poset L_1 and L_2 are isomorphic, therefore both lattices L_1 and L_2 are isomorphic.
2. Let S = $\{a_1, a_2, \ldots, a_n\}$ be a set containing n elements and P (S) be its power set. The lattice P (S) is isomorphic to lattice S.

Example 15 Show that the lattice L = {1, 2, 3, 6} under divisibility relation and the lattice (P (S), $\subseteq$), where S = {a, b} are isomorphic.

Solution We define a mapping f: L $\to$ P (S) as follows:

$$f(1) = \phi, \quad f(2) = \{a\}, \quad f(3) = \{b\}, \quad f(6) = \{a, b\}$$

This implies that f is one-to-one and onto. Also, for all a, $b \in$ L

$$f(a \wedge b) = f(a) \wedge f(b) \quad \text{and} \quad f(a \vee b) = f(a) \vee f(b)$$

Thus f is an isomorphism and hence lattice L is isomorphic to the lattice (P (S), $\subseteq$).

Example 16 Let L = {1, 2, 3, 4, 6, 12} be a lattice. Consider lattices (L, |) and (L, $\leq$) where | is the divisibility relation on L and $\leq$ is the 'less than or equal to' relation on L. Show that lattice (L, |) and (L, $\leq$) are not isomorphic.

Solution We define a mapping f : (L, |) $\to$ (L, $\leq$), such that,

$$3 \wedge 4 = 1 \in (L, |) \Rightarrow f(3 \wedge 4) = f(1).$$

But $f(3) \wedge f(4) = f(3)$ or $f(4)$ depends upon values of $f(3)$ and $f(4)$. In any case, $f(3 \wedge 4) \neq f(3) \wedge f(4)$. Thus, (L, |) and (L, $\leq$) are not isomorphic to each other.

Theorem 14.9 If the lattices L and M are isomorphic and f denotes an isomorphism, then f preserves the ordering relation. That is, for any a, $b \in$ L $a \leq b \Rightarrow f(a) \leq f(b)$.

Proof Suppose f : L $\to$ M is an isomorphism. Then for all a, $b \in$ L and $a \leq b$, we have

$$a = a \wedge b \text{ or } f(a) = f(a \wedge b) = f(a) \wedge f(b) \Rightarrow f(a) \leq f(b).$$

Thus f preserves ordering relation.

Theorem 14.10 Let L and M be two lattices. If f: L $\to$ M is an isomorphism, then $f(a)$ is the least element of M, for any least elements $a \in$ L.

Proof Let $a \in$ L be any least element. Then

$$a \leq x \Rightarrow f(a) \leq f(x) \quad \text{for all } x \in \text{L and } f(a) \in \text{M}.$$

Since f is onto, any element $y \in$ M is of the form $f(x)$ for some $x \in$ L. Thus, $f(a)$ is the least element of M.

Theorem 14.11 Let f: L$\to$M is an isomorphism such that $x < y$ for any element x, $y \in$ M. Then there exist a, $b \in$ L such that $f(a) = x$, $f(b) = y$ and $a < b$.

Proof Since f is onto and x, $y \in$ M, therefore exist a and c in L such that $f(a) = x$ and $f(c) = y$. Now

$$f(a \vee c) = f(a) \vee f(c) \qquad \text{[Since } f \text{ is isomorphism]}$$

$$= x \vee y = y \qquad \text{[since } x < y\text{]}$$

If $a = a \vee c$, then $f(a) = f(a) \vee f(c) = y$. This implies that $x = y$, which is a contradiction because $a \leq a \vee c$. Hence, $a \neq a \vee c$, so $a < a \vee c$.

If we take $b = a \vee c$, then $a < b$ and $f(a) = x$, $f(b) = y$. This completes the proof of the theorem.

14.7.3 Bounded Lattices

A lattice L is said to be *bounded* if it has both a least element (lower bound) zero and a greatest elemer (upper bound) one.

If L is a bounded lattice, then for any element $a \in$ L, we have the following identities

(*i*) $0 \leq a \leq \mathrm{I}$

(*ii*) $a \vee 0 = a, \quad a \wedge 0 = 0$

(*iii*) $a \wedge \mathrm{I} = a, \quad a \vee \mathrm{I} = \mathrm{I}$

Illustrations

1. The set of non-negative integers with usual ordering $0 < 1 < 2 < 3 \ldots$ has a least element (lower bound zero but there is no greatest element (upper bound).
2. The power set P (U) of all subsets of a universal set U is a bounded lattice since it has lower bound Φ (empty set) and the upper bound U (universal set itself).
3. The Hasse diagram is shown in Fig. 14.4 represents a bounded lattice because 1 is the least elemen and 6 is the greatest element.
4. The Hasse diagram shown in Fig. 14.5 represent a bounded lattice because 1 is the least element an 8 is the greatest element.

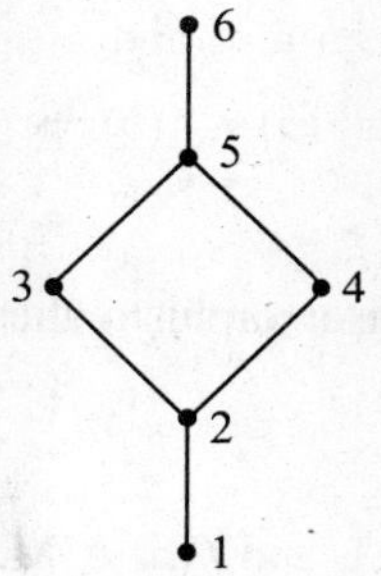

Fig. 14.4 Hasse Diagram

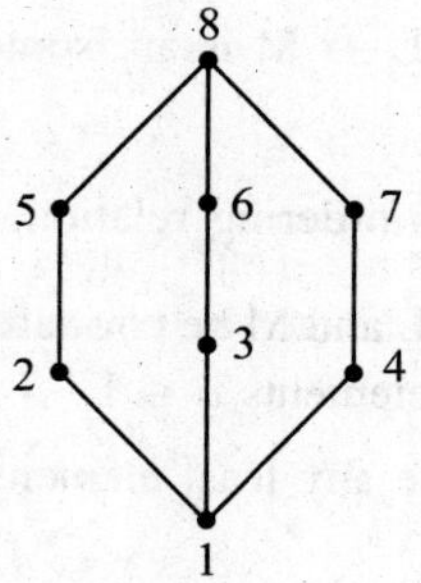

Fig. 14.5 Hasse Diagram

Theorem 14.12 Prove that every finite lattice L is bounded. That is, if L = $\{a_1, a_2, \ldots, a_n\}$ is a finite lattice, then L is bounded.

Proof To show that the finit lattice L = $\{a_1, a_2, a_3, \ldots, a_n\}$ contains both least element and greatest element, put

$$\begin{aligned} b_1 &= a_1 \\ b_2 &= a_2 \wedge b_1 \\ b_3 &= a_3 \wedge b_2 \\ &\vdots \\ b_n &= a_n \wedge b_{n-1} \end{aligned}$$

This shows that $b_1, b_2, \ldots, b_n \in$ L and $b_n \leq a_i$ for all $i = 1, 2, \ldots, n$.

Thus b_n is the least element of L. By associativity of the operation $\wedge$, we get $b_n = a_1 \wedge a_2 \wedge a_3 \ldots \wedge a_n$ as the least element (lower bound)

Similarly, we can show that $a_1 \vee a_2 \vee \ldots \vee a_n$ is the greatest element (upper bound) of L. Hence, L is bounded.

Theorem 14.13 If (L, $\leq$) is a lattice with least element 0 and greatest element 1. Then for any $a \in$ L, show that

(*a*) $a \vee 1 =$ and $a \wedge 1 = a$

(*b*) $a \vee 0 = a$ and $a \wedge 0 = 0$.

Proof (*a*) Let a be any element of lattice L. Since 1 is the greatest of L, therefore $a \vee 1 \leq 1$. Also $a \vee 1$ is the supremum of a and 1, we have $1 \leq a \vee 1$. These two inequalities implies that $a \vee 1 = 1$.

Further since $a \wedge 1$ is the infimum of a and 1, therefore $a \wedge 1 \leq a$. Also $a \leq a$ and $a \leq 1$, we have $a \leq a \vee 1$. These two inequalities implies that $a \wedge 1 = a$. Since 0 is the least element of L and a is any element of L, therefore $0 \leq a$ and $a \leq a$ or $a \vee 0 \leq a$ Also $a \leq a \vee 0$. Thus, we have $a \vee 0 = a$.

(*b*) As $a \wedge 0$ is the minimum of elements a and 0, so $a \wedge 0 \leq 0$. Also $0 \leq a$ so we get $0 \leq a \wedge 0$. This implies that $a \wedge 0 = 0$.

14.7.4 Complete Lattice

A lattice L is said to be complete if every non-empty subset of L has a least upper bound and a greatest lower bound.

Theorem 14.14 Every finite lattice is complete. [*CCS Univ., MSc(Maths), 2004*]

Proof Every non-empty subset of a finite set L shall always be finite. Since every finite subset of a lattice is a lattice and hence shall have least upper bound and a greatest lower bound. Hence L shall be a complete lattice.

Illustrations

1. The lattice (N, $\leq$), (Z, $\leq$), (R, $\leq$) etc. are not complete because each subset of these lattices does not posses a unique greatest lower bound (N, divides) is also not complete lattice.
2. Let A be any infinite set and the lattice be denoted by L = (P(A), *n*, *u*) where P(A) is the power set of A. Then for any non-empty subset B of P(A), the lattice is written $L_1 = (B, \cap, \cup)$.

   If every non-empty subset of L has unique least upper bond and greatest upper bond, then L is complete.
3. If A is a set, then lattice of its power set P(A) under the partial order $\subseteq$ is distributive. The supremum of a subset of two elements belongs to P(S) is their union and infimum of a subset of two elements belongs to P(S) is their intersection, i.e. $A \cap B \in P(S)$

$$A \cup B = A \vee B \text{ and } A \cap B = A \wedge B$$

   Also these two operations of union and intersection distributes over each other. Therefore the lattice {P(S), $\cup$, $\cap$} is distributive under $\subseteq$. [*UPTU., MCA, 2002; B Tech, 2008*]

14.7.5 Distributive Lattices

A lattice L is said to be *distributive* if for any elements a, b, $c \in$ L, the following distributive laws hold.

(*i*) $a \wedge (b \vee c) = (a \wedge b) \vee (a \wedge c)$

(*ii*) $a \vee (b \wedge c) = (a \vee b) \wedge (a \vee c)$

If above laws do not hold, then L is said to be *non-distributive.*

Illustrations

1. The lattice of power set P (S) of finite set S is distributive under the operations $\cup$ and $\cap$ (the join and meet, respectively).
2. The lattice L_1 shown in Fig. 14.6(*a*) is non-distributive lattice, whereas lattice L_2 shown in Fig. 14.6(*b*) is distributive, as can be seen by verifying the distributive properties for all ordered triples chosen from the elements of lattices L_1 and L_2.

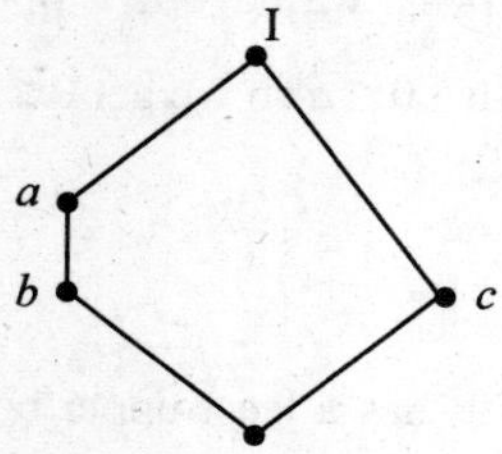

Fig. 14.6(*a*) Non-distributive Lattice L_1

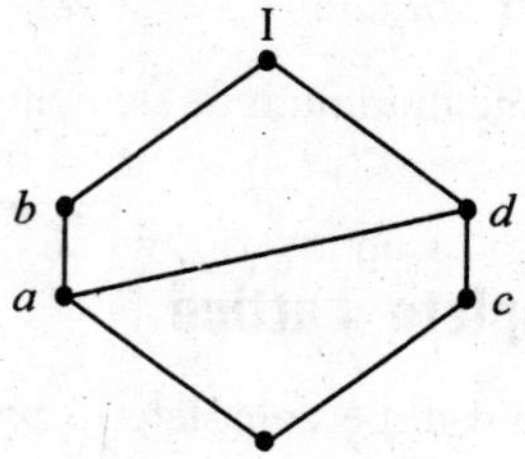

Fig. 14.6(*b*) Non-distributive Lattice L_2

Example 17 Show that the lattice shown below is not distributive.

Solution Consider $a_1 \wedge (a_2 \vee a_3) = a_1 \wedge \text{I} = a$, and $(a_1 \wedge a_2) \vee (a_1 \wedge a_3) = 0 \vee 0 = 0$

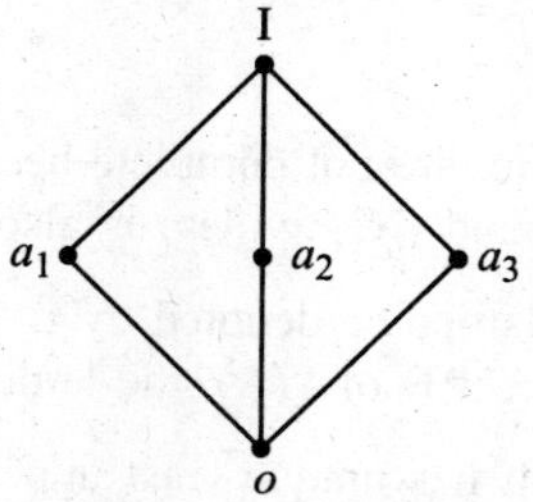

Fig. 14.7 Lettice

As $a_1 \neq 0$, the lattice shown in Fig. 14.7 is not distributive.

Example 18

(*a*) Let S = $\{a, b, c\}$ and A = P(S). Draw the Hasse diagram of the poset A with the partial order c.

[*UPTU., BTech, 2007*]

(*b*) Prove that if a and b are elements in bounded, distributive lattice and if a has a complement a', then $a \vee (a' \wedge b) = a \vee b$ and $a \wedge (a' \vee b) = a \wedge b$ [*UPTU, MCA, 2007*]

Solution: (*a*) As $\phi << \{a\}, \phi << \{b\}, \phi << \{c\}, \{a\} << \{a, b\}, \{a\} << \{a, c\},$

$\{a, b\} << \{a, b, c\}, (a, c\} << \{a, b, c\}, \{b, c\} << (a, b, c\}.$

The required Hasse diagram is shown in Fig. 14.8.

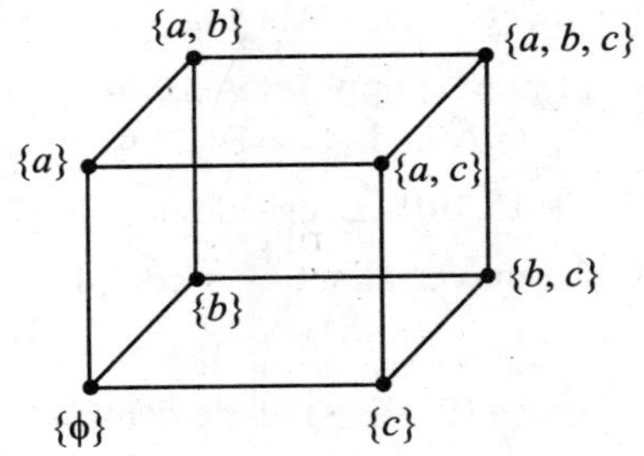

Fig. 14.8 Hasse Diagram

(*b*) L.H.S. $= a \vee (a' \wedge b) = (a \vee a') \wedge (a \vee b)$ [property of distributive lattice]

$= I \wedge (a \vee b)$ (property of complement)

$= a \vee b$ (property of bounded lattice)

Again $a \wedge (a' \vee b) = (a \wedge a') \vee (a \wedge b)$ (distributive property)

$= o \vee (a \wedge b)$ (complement property)

$= a \wedge b$ (boundedness property)

Theorem 14.15 Prove that a lattice L is distributive if and only if following inequalities hold.

(*a*) $a \vee (b \wedge c) \leq (a \vee b) \wedge (a \vee c)$

(*b*) $a \wedge (b \vee c) \geq (a \wedge b) \vee (a \wedge c)$

Proof (*a*) For any elements $a, b, c \in$ L, we know that $a \leq (a \vee b)$ and $a \leq (a \vee c)$. This implies that

$$a \leq (a \vee b) \wedge (a \vee c)$$

and $$b \wedge c \leq b \leq a \vee b; \quad b \wedge c \leq c \leq a \vee c$$

Combining above inequalities, we have

$$b \wedge c \leq (a \vee b) \wedge (a \vee c)$$

and hence $$a \vee (b \wedge c) \leq (a \vee b) \wedge (a \vee c)$$

(*b*) Similarly, using the principle of duality, the inequality (*b*) $a \wedge (b \vee c) \geq (a \wedge b) \vee (a \wedge c)$ holds if and only if inequality (*a*) holds.

14.7.6 Complemented Lattices

Let L be a bounded lattice with lower bound 0 and upper bound 1. An element x in a bounded lattice L is said to be the *complement* of its another element $y \in$ L provided

(*i*) $x \wedge y = 0$, and (*ii*) $x \vee y = 1$

The complement x of an element $y \in$ L can also be denoted by $\bar{x}$. Obviously $\bar{0} = 1$ and $\bar{1} = 0$.

A lattice L is said to be *complemented lattice* if it is bounded and every element in L has at least one complement.

Illustrations

1. In the lattice shown in Fig. 14.6(*a*), every element has a complement. The element *c* has two complements *a* and *b*. However, the elements *a* and *d* in Fig. 14.6(*b*) have no complements.
2. The lattice P (A) of all subsets of a nonempty set A is complemented and each element $A_1 \in P(A)$ has a complement so that $A_1 \vee A_2 = A$ and $A_1 \wedge A_2 = \Phi$.
3. For $n = 30$, $D_{30} = \{1,2, 3, 5, 6, 10, 15, 30\}$. Every element in D_{30} has a complement.
4. The lattice P (U) of all subset of a universal set U is complemented, and each subset $A \in U$ has the unique complement.

Theorem 14.16 Prove that the complement (if exists) of elements in a bounded distributive lattice is unique.

Proof Let x_1 and x_2 be the complements of an element x in a bounded distributive lattice L. Then we have

$$x \vee x_1 = I \qquad x \vee x_2 = I$$
$$x \wedge x_1 = 0 \qquad x \wedge x_2 = 0$$

By using distributive law, we get

$$x_1 = x_1 \vee 0 = x_1 \vee (x \wedge x_2) = (x_1 \vee x) \wedge (x_1 \vee x_2)$$
$$= I \wedge (x_1 \vee x_2) = x_1 \vee x_2 \quad ...(i)$$

and
$$x_2 = x_2 \vee 0 = x_2 \vee (x \vee x_1) = (x_2 \vee x) \wedge (x_2 \vee x_1)$$
$$= I \wedge (x_2 \vee x_1) = x_2 \vee x_1 \quad ...(ii)$$

Hence, from Eqs (*i*) and (*ii*), we have $x_1 = x_1 \vee x_2 = x_2 \vee x_1 = x_2$.

Theorem 14.17 Let L be a complemented distributive lattice. Then prove that DeMorgan's laws

(*a*) $(a \vee b)' = a' \wedge b'$

(*b*) $(a \wedge b)' = a' \vee b'$

hold in L for all $a, b \in L$, where a' denotes the complement of a.

Solution (*a*) To prove this result, we need to show that

$$(a' \wedge b') \vee (a \vee b) = 1 \text{ and } (a' \wedge b') \wedge (a \vee b) = 0$$

Now
$$(a' \wedge b') \vee (a \vee b) = [(a' \vee a) \vee b] \wedge [b' \vee (a \vee b)]$$
$$= (1 \vee b) \wedge [a \vee (b \vee b')]$$
$$= (1 \vee b) \wedge (a \vee 1) = 1 \wedge 1 = 1$$

and
$$(a' \wedge b') \wedge (a \vee b) = [(a' \wedge b') \wedge a] \vee [(a' \wedge b') \wedge b]$$
$$= [(a' \wedge a) \wedge b'] \vee [a' \wedge (b' \wedge b)]$$
$$= (0 \wedge b') \vee (a' \wedge 0) = 0 \vee 0 = 0$$

This shows that $a' \wedge b'$ is the complement of $a \vee b$, i.e. $(a \vee b)' = a' \wedge b'$.

(*b*) To prove this result we need to show that

$$(a' \vee b') \vee (a \wedge b) = 1 \text{ and } (a' \vee b') \wedge (a \wedge b) = 0$$

Now $(a' \vee b') \vee (a \wedge b) = [(a' \vee b') \vee a] \wedge [(a' \vee b') \vee b]$

$$= [(a' \vee a) \vee b'] \wedge [a' \vee (b' \vee b)]$$

$$= (1 \vee b') \wedge (a' \vee 1) = 1 \wedge 1 = 1$$

Similarly $(a' \vee b') \wedge (a \wedge b) = 0$. This shows that $a' \vee b'$ is the complement of $a \wedge b$, i.e. $(a \wedge b)' = a' \vee b'$.

Theorem 14.18 If L is a distributive lattice and let b' be the complement of an element b in L and $a \wedge b' = 0$, then show that $a \leq b$.

Proof Since $a \wedge b' = 0$, we have

$$(a \wedge b') \vee b = 0 \vee b = b$$

But according to distributive law, we have

$$(a \vee b) \wedge (b' \vee b) = b$$

or $$a \vee b = b \Rightarrow a \leq b$$

Theorem 14.19 Every sub-lattice of a distributive lattice is a sublattice.

Proof Let S be a sub-lattice of a distributive lattice L. Let $a, b, c \in$ S, then $a, b, c \in$ L. Thus

$$a \wedge (b \vee c) = (a \wedge b) \vee (a \wedge c) \in \text{L}$$

Since S is closed in $\wedge$ and $\vee$, we have

$$a \wedge (b \vee c) = (a \wedge b) \vee (a \wedge c) \in \text{S}.$$

Hence, S is distributive.

Theorem 14.20 The dual of a distributive lattice is a distributive lattice.

Proof Let (L, $\wedge$, $\vee$) be a distributive lattice. Then to show that (L, $\overline{\wedge}$, $\overline{\vee}$) is also distributive, where $\overline{\wedge} = \vee$ ad $\overline{\vee} = \wedge$. For all $a, b, c \in$ L, we have

$$a \mathbin{\overline{\wedge}} (b \mathbin{\overline{\vee}} c) = (a \mathbin{\overline{\wedge}} b) \mathbin{\overline{\vee}} (a \mathbin{\overline{\wedge}} c)$$

or $$a \vee (b \wedge c) = (a \vee b) \wedge (a \vee c).$$

This result holds because (L, $\wedge$, $\vee$) is a distributive lattice. Hence, (L, $\overline{\wedge}$, $\overline{\vee}$) is a distributive lattice.

Theorem 14.21 In a lattice L with least element 0 and greatest element 1, show that 0 is the unique complement of 1 and 1 is the unique complement of 0.

Proof We know that

$$0 \wedge 1 = 0 \text{ and } 0 \vee 1 = 1$$

This also implies that 0 and 1 are complements of each other. To show that each of these two complements are unique, we choose another complement a (say) of 0, so that

$$0 \wedge a = 0 \text{ and } 0 \vee a = 1$$

But $0 \vee a = a$, so $a = 1$.

Thus the complement of 0 is unique. Similarly, we can show that 0 is the only complement of 1.

Theorem 14.22 Two bounded lattices L_1 and L_2 are complemented if and only if $L_1 \times L_2$ is complemented.

Proof Let L_1 and L_2 be two complemented lattices and let 0, 1 and 0′, 1′ are least and greatest elements of L_1 and L_2, respectively. Then (0, 0′) and (1, 1′) will be the least and greatest elements of $L_1 \times L_2$. If $(a_1, a_2) \in L_1 \times L_2$, then $a_1 \in L_1$ and $a_2 \in L_2$.

Since L_1 and L_2 are complemented, there exists $a'_1 \in L_1$ and $a'_2 \in L_2$ such that $a_1 \wedge a'_1 = 0$, $a \vee a' = 1$ and $a_2 \wedge a'_2 = 0'$ and $a_2 \vee a'_2 = 1'$.

We shall now show that (a'_1, a'_2) is the complement of (a_1, a_2) in $L_1 \times L_2$.

$$(a_1, a_2) \vee (a'_1, a'_2) = (a_1 \wedge a'_1, a_2 \wedge a'_2) = (0, 0')$$

and $$(a_1, a_2) \vee (a'_1, a'_2) = (a_1 \vee a'_1, a_2 \vee a'_2) = (1, 1')$$

This shows that (a'_1, a'_2) is the complement of (a_1, a_2) in $L_1 \times L_2$. Hence, $L_1 \times L_2$ is complemented.

Conversely, let $L_1 \times L_2$ be complemented. Then we need to show that L_1 and L_2 are complemented. If $a_1 \in L_1$ and $a_2 \in L_2$, then $(a_1, a_2) \in L_1 \times L_2$.

Since $L_1 \times L_2$ is complemented, there exists $(a_1, a_2) \in L_1 \times L_2$ such that

$$(a_1, a_2) \wedge (a'_1, a'_2) = (0, 0') \text{ and } (a_1, a_2) \vee (a'_1, a'_2) = (1, 1')$$

In other words

$$(a_1 \wedge a'_1, a_2 \wedge a'_2) = (0, 0') \text{ and } (a_1 \vee a'_1, a_2 \vee a'_2) = (1, 1')$$

$$a_1 \wedge a'_1 = 0,\ a_2 \wedge a'_2 = 0' \text{ and } a_1 \vee a'_1 = 1,\ a_2 \vee a'_2 = 1'$$

$$a_1 \wedge a'_1 = 0,\ a_1 \vee a'_1 = 1 \text{ and } a_2 \wedge a'_2 = 0,\ a_2 \vee a'_2 = 1'.$$

These equations show that a'_1 and a'_2 are complements of a_1 and a_2, respectively. Hence, L_1 and L_2 are complemented.

Theorem 14.23 The complement of an element a in a bounded distributive lattice, if it exits, is unique.

[*UPTU, BTech, 2008*]

Proof Let a_1 and a_2 be the complement of $a \in L$. Then,

$$a \vee a_1 = 1, \quad a \vee a_2 = I \qquad (i)$$

$$a \wedge a_2 = o, \quad a \wedge a_2 = o \qquad (ii)$$

Now $a_1 = (a_1 \vee o) = a_1 \vee (a \wedge a_2)$, from Eqn (*ii*)

$= (a_1 \vee a) \wedge (a_1 \vee a_2)$, by distributive law

$= (a \vee a_1) \wedge (a_1 \vee a_2)$, by commutative law

$= I \wedge (a_1 \vee a_2) = a_1 \vee a_2$, by Eqn (*i*)

Similarly, $a_2 = a_2 \vee o = a_2 \vee (a \wedge a_1) = (a_2 \vee a) \wedge (a_2 \vee a_1)$

$= (a \vee a_2) \wedge (a_1 \vee a_2)$

$= I \wedge (a_1 \vee a_2) = a_1 \vee a_2$

Hence, $a_1 = a_2$

Theorem 14.24 Two bounded lattices A and B are complemented if and only if A × B is complemented.

[*CCS Univ., MSc(Maths), 2004*]

Proof Let A and B be two complemented lattices and suppose 0 and 1 and 0′ and 1′ are the universal bounds of A and B respectively. Then (0, 0′) and (1, 1′) will be least and greatest elements of A × B.

Let $(a, b) \in A \times B$ be any element. Then $a \in A$ and $b \in B$. As A and B are complemented lattices, therefore there exists elements $a' \in A$ and $b' \in B$ such that

$$a \wedge a' = 0 \text{ and } a \vee a' = 1$$

$$b \wedge b' = 0' \text{ and } b \vee b' = 1'$$

Now, $(a, b) \wedge (a', b') = (a \wedge a', b \wedge b') = (0, 0')$

$(a, b) \vee (a', b') = (a \vee a', b \vee b') = (1, 1')$

This implies that (a', b') is complement of (a, b) in $A \times B$. As (a, b) is arbitrary in $A \times B$, therefore $A \times B$ is complemented.

Conversely: Let $A \times B$ is complemented lattice. If

$(a, b) \in A \times B$ then there exist $(a', b') \in A \times B$ such that

$$(a, b) \wedge (a', b') = (0, 0') \text{ and } (a, b) \vee (a', b') = (1, 1')$$

This implies that $(a \wedge a', b \wedge b') = (0, 0')$ and $(a \vee a', b \vee b') = (1, 1')$

or $a \wedge a' = 0$ and $b \wedge b' = 0'$

$a \vee a' = 1$ and $b \vee b' = 1$

or a' is complement of a in A and b' is complement of b in B

Since $(a, b) \in A \times B$ is an arbitrary element, A and B are complemented.

Theorem 14.25 Dual of a complemented lattice is complemented. [*CCS Univ.,MSc(Maths), 2004*]

Proof Let (L, R) be a complemented lattice with 0 and 1 as least and greatest elements respectively. If (L, R′) be the dual of (L, R), then 1 and 0 are its least and greatest elements respectively.

Let $a \in L$ be any element. Since (L, R) is complemented, therefore there exists $a' \in L$ such that

$a \wedge a' = 0$ and $a \vee a' = 1$ in L, i.e. $0 =$ glb of $\{a, a\}$ in L

Also, $0 \, R \, a$ and $0 \, R \, a'$ in (L, R) and $a \, R' \, o$ and $a' \, R'$ in (L, R′)

This implies that 0 is the upper bound of $\{a, a'\}$ in (L, R′).

Let $y \in$ (L, R′) be another upper bound of $(a, a'\}$. Then

$a \, R' \, y$ and $a' \, R' \, y$ in (L, R′) and $y \, R \, a$ and $y \, R \, a'$ in (L R)

This implies that y is the lower bound of $\{a, a'\}$ in (L, R).

As 0 is the glb of $\{a, a'\}$ in (L, R). Thus $y \, R \, 0$ in (L, R) and $0 \, R' \, y$ in (L, R′). Also y is the least upper bound of $\{a, a'\}$ in (L, R′).

Similarly, it can also be proved that 1 is the g.l.b of $\{a, a'\}$ in (L, R′). Hence, (L, R′) is complemented.

Example 19 Show that the lattice P (S), where P (S) is the power set of a finite set S is complemented.

Solution We know that the complement of any subset A of S is given by $A' = S - A$. Now

$$A \vee (S - A) = 1 \text{ and } A \wedge (S - A) = 0$$

as $A \cup (S - A) = S$ and $A \cap (S - A) = \phi$.

This shows that A has its complement in P (S). Hence, the given lattice is complemented.

14.7.7 Modular Lattices

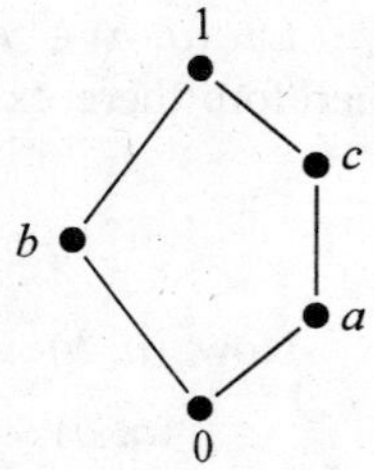

Fig. 14.9

A lattice L is said to be modular lattice if

$$a \vee (b \wedge c) = (a \vee b) \wedge c \text{ and } a \leq c$$

for all $a, b, c \in$ L.

Illustrations

1. The pentagonal lattice shown in Fig. 14.9 is not modular because
$$a \leq c \text{ but } a \vee (b \wedge c) \ (= a) \neq (a \vee b) \wedge c \ (= c).$$
Therefore, the pentagonal lattice is not modular.
2. Every chain is a modular lattice, because we cannot find any triplet a, b, c in a lattice such that $a < c$ and b is not comparable with a or c, therefore every chain is modular.

Theorem 14.26 Show that sublattice of a modular lattice is modular.

Proof Let S be a sublattice of a modular lattice L. If $a, b, c \in$ S with $a \leq c$, then $a, b, c \in$ L and hence,

$$a \vee (b \wedge c) = (a \vee b) \wedge c$$

Since S is closed with respect to $\wedge$ and $\vee$, the above result holds in S, and hence S is modular.

Theorem 14.27 The dual of a modular lattice is modular.

Proof Let (L, $\leq$) be a modular lattice. To show that the dual (L, $\geq$) of L is also modular, we need to prove result

$$a \wedge (b \vee c) = (a \wedge b) \vee c \text{ for all } a, b, c \in \text{L with } c \leq a$$

Since (L, $\leq$) is modular and $c \leq a$, we have

$$c \vee (b \wedge a) = (c \vee b) \wedge a$$

or

$$c \vee (a \wedge b) = (b \vee c) \wedge a$$

or

$$a \wedge (b \vee c) = (a \wedge b) \vee c$$

This equation is same as above. Hence, (L, $\leq$) is modular.

Theorem 14.28 If two elements b and c in a modular lattice are comparable and $a \wedge b = a \wedge c$, $a \vee b = a \vee c$ for any element $a \in$ L, then $b = c$.

Proof Suppose $b \geq c$. Then

$$b = b \wedge (a \vee b) = b \wedge (a \vee c) = (b \wedge a) \vee c$$
$$= (a \wedge c) \vee c = c, \text{ by given condition}$$

Hence, $b = c$.

Theorem 14.29 Every distributive lattice is modular. [*CCS Univ., Msc(Maths), 2004*]

Proof Let L be a distributive lattice. Then we need to show that

$$a \leq c \Rightarrow a \vee (b \wedge c) = (a \vee b) \wedge c \text{ for all } a, b, c \in \text{L}$$

Since L is distributive, we have

$$a \vee (b \wedge c) = (a \vee b) \wedge (a \vee c)$$

Also, $a \leq c \Rightarrow a \vee c = c$

From these equations, we have $a \leq c \Rightarrow a \vee (b \wedge c) = (a \vee b) \wedge c$. Thus, L is modular lattice.

Example 14.30 Show that N × N is a modular lattice, where N is a lattice under $\leq$ relation.

Solution Let (a_1, a_2), (b_1, b_2) and (c_1, c_2) be any three elements of N × N. Suppose

$$(a_1, a_2) \wedge (b_1, b_2) = (a_1, a_2) \wedge (c_1, c_2) \quad \text{...}(i)$$

and $$(a_1, a_2) \vee (b_1, b_2) = (a_1, a_2) \vee (c_1, c_2) \quad \text{...}(ii)$$

We have to show that $(b_1, b_2) = (c_1, c_2)$

From Eq. (*i*), we have $a_1 \wedge b_1 = a_1 \wedge c_1$ and $a_2 \wedge b_2 = a_2 \wedge c_2$ ...(*iii*)

From Eq. (*ii*), we have $a_1 \vee b_1 = a_1 \vee c_1$ and $a_2 \vee b_2 = a_2 \vee c_2$ ...(*iv*)

Combining Eqs (*iii*) and (*iv*), we have

$$a_1 \wedge b_1 = a_1 \wedge c_1 \text{ and } a_1 \vee b_1 = a_1 \vee c_1$$

and $$a_2 \wedge b_2 = a_2 \wedge c_2 \text{ and } a_2 \vee b_2 = a_2 \vee c_2$$

Since (N, $\leq$) is a chain, therefore it is modular, and hence $b_1 = c_1$ and $b_2 = c_2$ as well as $(b_1, b_2) = (c_1, c_2)$
Thus N × N is modular.

14.8 JOIN IRREDUCIBLE ELEMENTS

Let L be a lattice with least element (lower bound) zero. An element a ($\neq 0$) in L is called *join irreducible* provided $a = b_1 \vee b_2$, where b_1 and b_2 are two immediate predecessors of element a as shown in Fig. 14.10(*c*). This relationship implies $a = b_1$ or $a = b_2$. Obviously, 0 is join irreducible.

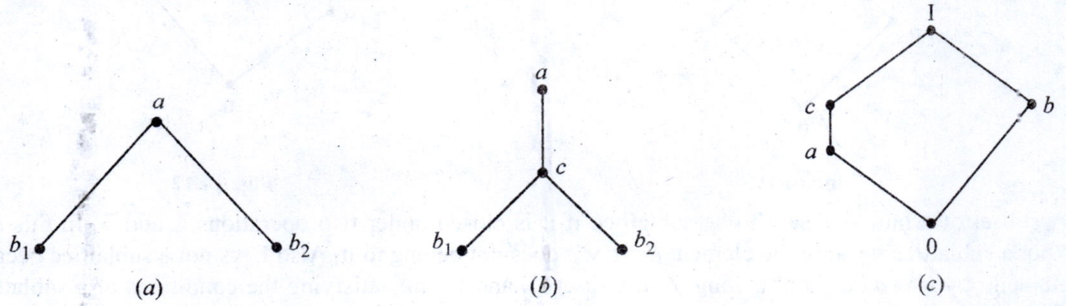

Fig. 14.10 Example of Join Irreducible Element

Also, if as shown in Fig. 14.10(*b*), the element c is the immediate predecessor of a, then

$$a \neq b_1 \vee b_2 = \sup (b_1, b_2).$$

This shows that an element $a \neq 0$ in L is join irreducible if and only if it has unique immediate predecessor. In other words, elements in L which immediately succeed 0 (also called *atoms*), are join irreducible. For example, the element c in Fig. 14.10(*c*) is not an atom but is join irreducible because element, a is its only immediate predecessor.

Example 14.31 Consider the lattice L shown in Fig. 14.11.

(*a*) Find the non-zero join irreducible elements and atoms of L.

(*b*) Is L distributive lattice?

(*c*) Is L complemented lattice?

(*d*) Find complements (if exists) for elements *c*, *d* and *e*.

(*e*) Which of the following subsets of L are sublattices?

$$L_1 = \{0, c, e, I\};\ L_2 = \{0, c, b, I\};\ L_3 = \{c, d, a, I\};\ L_4 = \{0, d, a, I\}$$

Solution (*a*) The non-zero elements in Fig. 14.11 with a unique immediate predecessor are *a*, *b*, *c* and *e*. Hence, these are join irreducible elements. Among these elements only *c* and *e* immediately succeed 0, and hence *c* and *e* are the atoms in L.

(*b*) The lattice L is not distributive because one of its sublattice $L_1 = \{0, c, a, b, I\}$ is isomorphic to the non-distributive lattice shown in Fig. 14.12.

(*c*) The lattice L is not complemented because element *d* has no complement.

(*d*) From Fig. 14.11, it may be noted that $c \wedge b = 0$ and $c \vee b = I$ and, therefore, *c* and *b* are complements, but *d* has no complement.

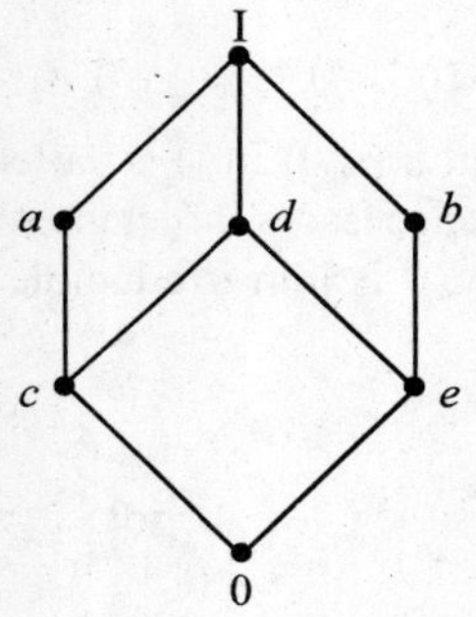

Fig. 14.11

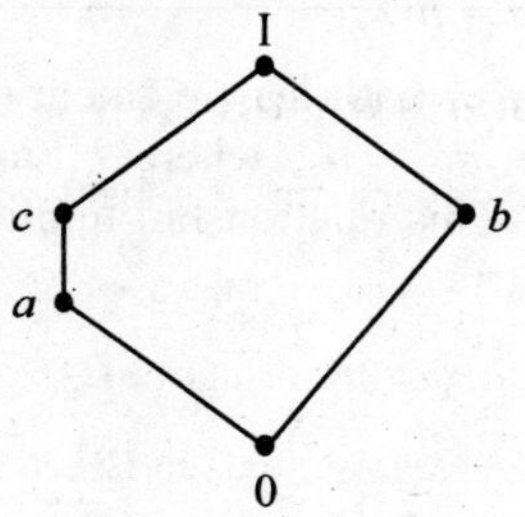

Fig. 14.12

(*e*) A subset of a lattice is said to be sublattice if it is closed under two operations $\wedge$ and $\vee$. In this case L_1 is not a sublattice because the element $d = c \vee e$ does not belong to it. Also L_4 is not a sublattice because the element $c = d \wedge a$ does not belong to it. Subset L_2 and L_3 are satisfying the conditions of a sublattice, therefore these are sublattices.

Example 14.32 Consider the lattice $D_{30} = \{1, 2, 3, 5, 6, 10, 15, 30\}$, the divisor of 30 ordered by divisibility.

(*a*) Draw the Hasse diagram of D_{30}.

(*b*) Which elements are join irreducible and atoms?

(*c*) Find the complements of 2 and 10, if exists.

Solution (*a*) The Hasse diagram of D_{30} is shown in Fig. 14.13.

(*b*) Since an element $a \neq 0$, is join irreducible if and only if it has a unique immediate predecessor. Thus elements 2, 3 and 5 are join irreducible.

Since atoms are the elements which immediately succeed 0 (a lower bound), therefore atoms are 2, 3, and 5.

(*c*) Elements 2 and 10 in D_{30} have no complements.

Self-practice Problems

1. Find the dual of the following statements:

   (a) $(a \wedge b) \vee c = (b \vee c) \wedge (c \vee a)$ (b) $(a \wedge b) \vee a = a \wedge (b \vee a)$

2. Show that the following distributive laws hold for any lattice:

   (a) $a \vee (b \wedge c) \leq (a \vee b) \wedge (a \vee c)$ (b) $a \wedge (b \vee c) \geq (a \wedge b) \vee (a \wedge c)$

   (c) $(a \wedge b) \vee (c \wedge d) \leq (a \vee c) \wedge (b \vee d)$

   (d) $(a \wedge b) \vee (b \wedge c) \vee (c \wedge a) \leq (a \vee b) \wedge (b \vee c) \wedge (c \vee a)$

3. If $a \leq b$ and $c \leq d$ in a lattice L, then prove that $a \wedge c \leq b \wedge d$.

4. If $a \leq b \leq c$ in a lattice L, then prove that

   (a) $a \vee b = b \wedge c$ (b) $(a \wedge b) \vee (b \wedge c) = (a \vee b) \wedge (a \vee c) = b$

5. Show that $a \wedge b = a \Leftrightarrow a \vee b = b$ in a lattice L.

6. Prove that every finite subset of a lattice has lub. and glb.

7. Show that a subset of an ordered poset is a sublattice.

8. If a bounded lattice has two or more elements, then show that $0 \neq I$.

9. If L is a complemented lattice with unique complements, then prove that the join irreducible elements in it, other than 0, are its atoms.

10. Let D_n denote the positive divisors of n ordered by divisibility.

    (a) Draw the Hasse diagrams of $D_{12} = \{1, 2, 3, 4, 6, 12\}$; $D_{15} = \{1, 3, 5, 15\}$ and $D_{16} = \{1, 2, 4, 8, 16\}$.

    (b) Find nonzero irreducible elements and atoms in D_{12} and D_{15}.

    (c) Is D_{12} and D_{15} distributive and complemented?

11. Let L be a bounded lattice. If $a, b \in$ L and $\bar{a}$ is the complement of a, then prove that

    (a) $a \wedge (\bar{a} \wedge b) = a \vee b$ (b) $a \wedge (\bar{a} \vee b) = a \wedge b$

12. Let L be distributive lattice. If $a \wedge x = a \wedge y$ and $a \vee x = a \vee y$ for any element $a \in$ L, then prove that $x = y$.

13. Prove that poset, D = {2, 3, 6, 12, 24, 36, 72} under the relation 'divides' is a lattice.

14. Show that the lattice D_{60} is distributive. Find the complement of each its elements, as well as elements which are join irreducible and atoms.

15. If A = $\{a, b, c\}$ is a nonempty set and the power set L = P (A) of A is the lattice, then prove that (L, ≤) is isomorphic to D_{42}.

16. Let N be the lattice of positive integers ordered by divisibility. Find the elements which are join irreducible and atoms.

17. (*a*) Show that the lattices shown in Fig. 14.14(*a*) and (*b*) are not distributive.

18. A lattice L is distributive if and only if the equations $a \wedge c = b \wedge c$ and $a \vee c = b \vee c$ imply $a = b$ for all $a, b, c \in$ L.

19. Prove that the non-empty intersection of two sublattices is again a sublattice of a given lattice.

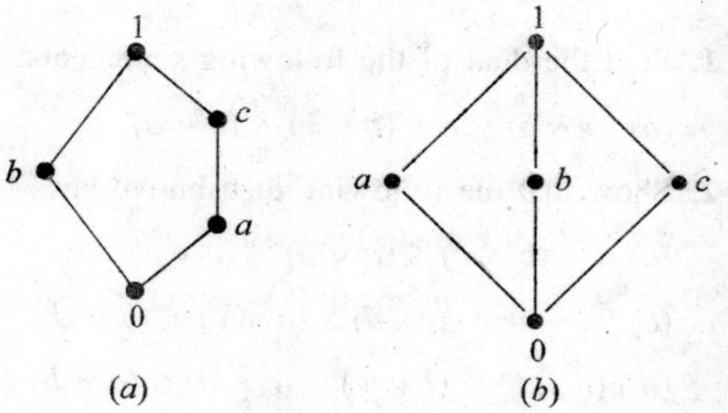

Fig. 14.14

20. If L is a lattice and $a, b \in$ L, then show that the subset $\{a, b\}$ of L is sublattice if and only if a and b are comparable.

21. Let (L, $\vee$, $\wedge$) be an algebraic system, where $\vee$ and $\wedge$ are binary operations satisfying the absorption law. Show that $\vee$ and $\wedge$ also satisfy the idempotent law.

22. Let f : L $\rightarrow$ M be an isomorphic from a lattice L onto a lattice M. If a is the greatest element of L, then $f(a)$ is the greatest element of M.

23. Let f : L $\rightarrow$ M be an isomorphism from a lattice L onto a lattice M. If $f(a) \leq f(b)$, then show that $a \leq b$.

24. Show that the inequality

$$a \leq c \Leftrightarrow a \vee (b \wedge c) \leq (a \vee b) \wedge c$$

in a lattice is self dual.

25. Prove that a poset (L, $\leq$) is a lattice if and only if every non-empty finite subset of L has supremum and infimum.

26. Show that a lattice L is modular if and only if for any elements a, b, $c \in$ L,

$$(a \wedge b) \vee (a \wedge c) = a \wedge ((a \wedge b) \vee c)$$

27. Show that in a bounded distributive lattice, the elements which have complements form a sublattice.

28. Prove that product L × M of two distributive lattices L and M is distributive.

29. Show that a modular lattice is non-distributive if and only if it contains a sublattice isomorphic with M_5.

30. Consider the lattice L shown in Fig. 14.15.
   (*a*) Which non-zero elements are join-irreducible?
   (*b*) Which elements are meet-irreducible?
   (*c*) Which elements are atoms?
   (*d*) Find complements (if exist) for the elements *a*, *b*, *c*.
   (*e*) Is L a complemented lattice?
   (*f*) Is L a complete lattice?

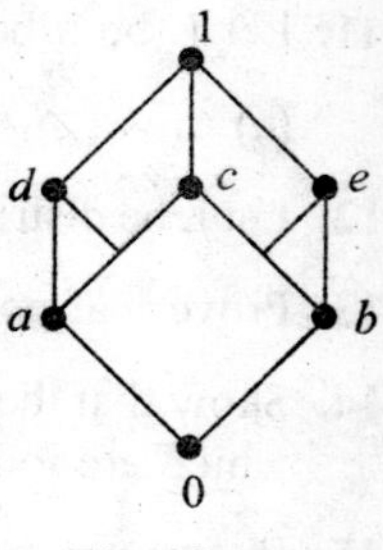

Fig. 14.15

31. Consider the lattice L = {1, 2, 3, 5, 6, 10, 15, 30} under divisibility relation.
   (*a*) Which elements are join-irreducible?
   (*b*) Which elements are atoms?
   (*c*) Which element is the greatest element?

32. Consider the lattice L = {1, 2, 3, 6, 9, 18} under divisibility relation.

(*a*) Which element is zero element of L and which element is 1?

(*b*) Which non-zero elements are join-irreducible?

(*c*) Which elements are atoms?

(*d*) Which elements are cover of 3?

(*e*) Find complements (if exist) for the elements 2, 3, 6.

(*f*) Is L a complemented lattice?

33. Find the complements of every element of the lattice of factors of 75 under divisibility relation.

34. Consider the lattice L = {0, 1, 2, 3, 6, 9, 18} under divisibility relation. What is the greatest element of this lattice? Is L complete? Is L complemented?

35. Show that the posets given below are lattices. Obtan the Hasse diagrams.

(*i*) $(S_6, |)$ (*ii*) $(S_8, |)$

(*iii*) $(S_{24}, |)$ (*iv*) $(S_{30}, |)$

where S_n is the set of all divisors of *n*.

36. Show that the posets given below are lattices:

(*i*) $(S_9, 1)$ (*ii*) $(S_{12}, 1)$ (*iii*) $(S_{15}, 1)$

Also obtain their Hasse diagram.

37. Determine whether the posets represented by the Hasse diagrams shown in Fig. 14.16 are lattices:

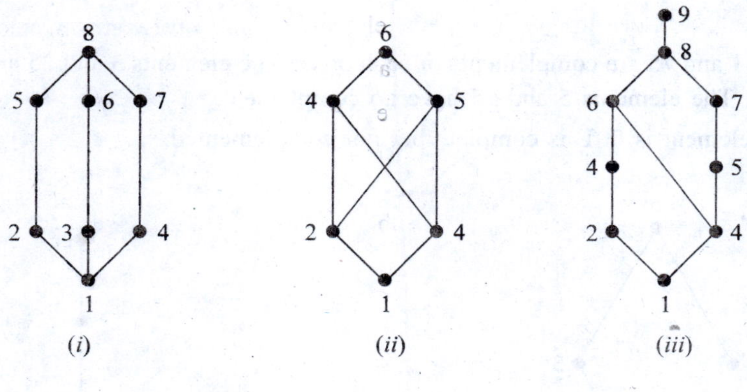

Fig. 14.16

Hints and Answers

17. (*a*) It is not distributive lattice, because

$a \wedge (b \vee c) = a$, whereas $(a \wedge b) \vee (a \wedge c) = 0$

(*b*) Not distributive lattice because

$c \wedge (a \vee b) = c$, whereas $(c \wedge a) \vee (c \wedge b) = a \vee 0 = a$

30. (*a*) The non-zero elements *a, b, d* and *e* are join-irreducible. Since $c = a \vee b$, *c* is not a join-irreducible element.

(*b*) The elements *c, d, e* and 1 are meet-irreducible elements.

(*c*) Only *a* and *b* are the atoms.

(*d*) Since $a \wedge e = 0$ and $a \vee e = 1$, *a* and *e* are complements of each other. Also, *b* and *d* are complements of each other. But the element *c* has no complement.

(*e*) L is not complemented since the element *c* has no complement.

(*f*) Since L is finite, it is complete.

31. (*a*) 2, 3 and 5 (*b*) 2, 3 and 5 (c) 30

32. (*a*) The least element zero is 1 and the greatest element is 18.

(*b*) The non-zero elements 2, 3 and 9 are join-irreducible.

(*c*) The elements 2 and 3 are atoms.

(*d*) The elements 6 and 9 are covers of 3.

(*e*) Complement of 2 is 9 while 3 and 6 have no complement.

(*f*) No.

33. The elements 1 and 75 are complements of each other. The elements 3 and 25 are also complements of each other. The elements 5 and 15 have no complement.

34. The greatest element is 0. L is complete but not complemented.

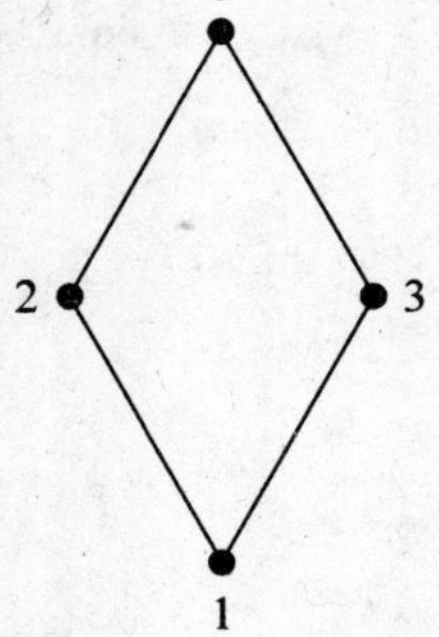

(*i*) $S_6 = \{1, 2, 3, 6\}$

(*ii*) $S_8 = \{1, 2, 4, 8\}$

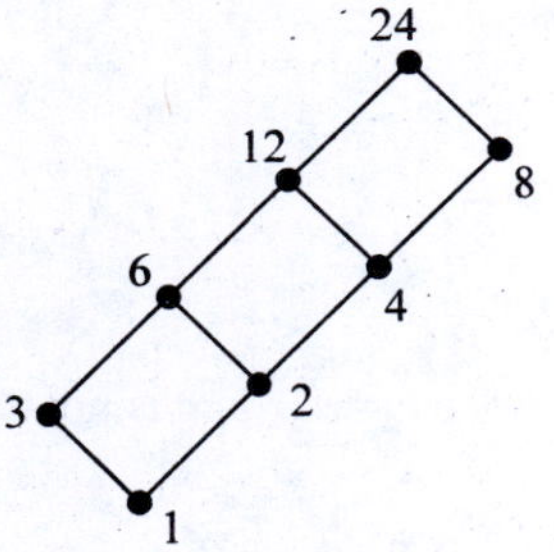

(*iii*) $S_{24} = \{1, 2, 3, 4, 6, 8, 12, 24\}$

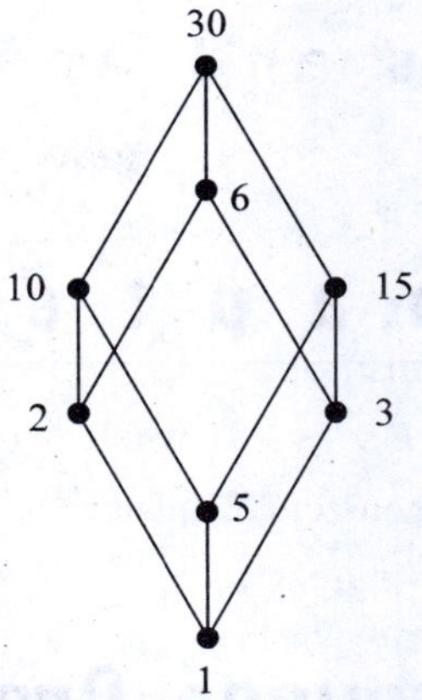

(*iv*) $S_{30} = \{1, 2, 3, 5, 6, 10, 15, 30\}$

36.

(*i*) $(S_9, 1)$

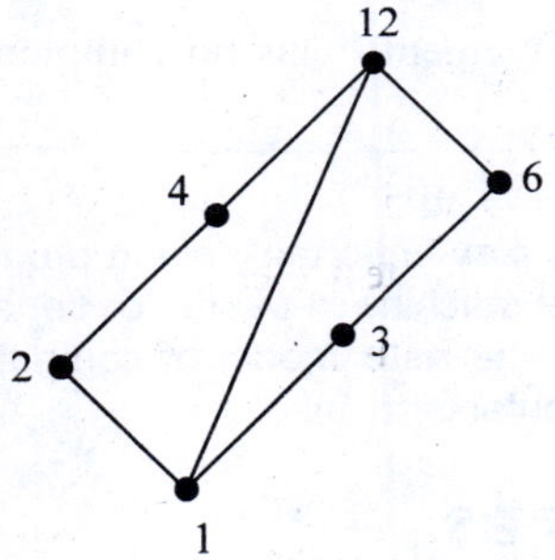

(*ii*) $(S_{12}, 1)$

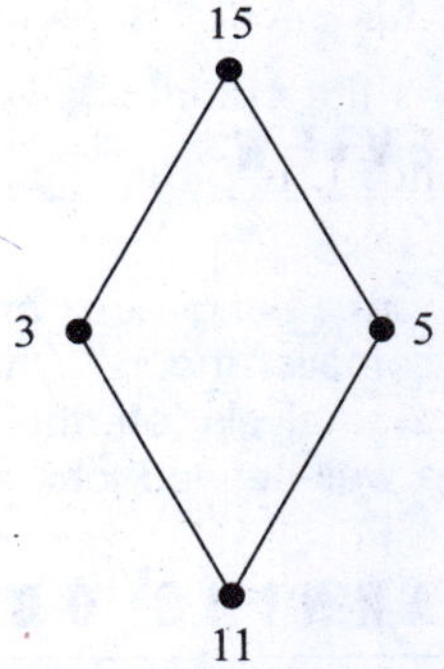

(*iii*) $(S_{15}, 1)$

37. (*i*) The poset represented by the given Hasse diagram is a lattice as every pair of elements of it has the least upper bound as well as the greatest lower bound.

(*ii*) The poset represented by the given Hasse diagram is not a lattice because the elements 2 and 3 have no least upper bound. The elements 4, 5 and 6 are the upper bounds but none of these three elements precedes the other two with respect to the ordering of the given poset.

(*iii*) The poset represented by the given Hasse diagram is a lattice as every pair of elements of it has the least upper bound as well as the greatest lower bound.

Chapter 15

Language, Grammar and Automata

PREVIEW

The formal languages and grammars play important role in programming languages and compiler theory. While finite-state machine is useful for certain jobs such as the lexical analysis job, the turning machine as a model of computation indicates that some well-defined jobs are not computable at all.

LEARNING OBJECTIVES

After studying this chapter, you should be able to

- understand the relationship between variety of formal languages and computational devices.
- construct the derivation of strings for a given grammar
- construct moves and transition table of a turning machine to perform certain computational task.
- develop finite-state lemma to recognize a given language
- understand equivalence of Moore and Mealy machines as well as DFA and NFA.

15.1 INTRODUCTION

In this chapter we shall discuss three important concepts useful in the study of computer science: *Language, Grammar* and *Finite Automata.* To communicate in any language of the world such as Hindi, English, German, French, etc. (also called *natural languages*), we need to know the letters of alphabet and then know how the letters are combined in a meaningful sentence. The set of rules of syntax that lead to the automatic translation of one natural language to another is referred as *formal language.* One simple method of developing a formal language is a *phase structure grammar* or simple *grammar*.

A *grammar* is an algebraic system which describes the process by which instances of a language can be constructed. It gives a mathematical structural descriptions of sentences for accepting or rejecting strings (words) in a language and hence determines membership criterion for a string in a language.

A *finite automata* (also called *finite state machine* or *sequential circuit*) is one of two basic types of central systems found in digital computers. Other type is a *combination circuit* or *gating network* without memory. However, sequential circuit in a digit computer helps to recongnise strings in a grammar found in compilers and interpretors for various programming languages. Obviously, this type of circuits have memory in which output depends not only on the input but also on the finite number of internal states where the machine *internally remembers* certain information when it is in a particular state. The control mechanism found in vending machines, elevators and traffic lights, etc. can be viewed as a finite state machines where at each step in time the machine is in a certain internal state, reads some input, yields some output depending only its internal state and input and then changes its internal state.

In this chapter, we shall discuss some set theoretical concepts in order to define a computable function in terms of a special machine which yields a non-negative integer output for any given non-negative integer input.

15.2 LANGUAGE: THE SET THEORY OF STRINGS

An *alphabet* is a non-empty finite set Σ of symbols. The word *symbol* used here is an abstract entity frequently expressed by using English letters or non-negative integers.

A *string* or *word* on the set Σ is a finite sequence of its elements (symbols). For example, if $\Sigma = \{a, b\}$, then the sequences $u = abba$ and $v = aabbb$ of letters are strings of Σ. If letters in a string are repeated as in string u and v, then we may also write these as: $u = a^2b^2$ and $v = a^2b^3$.

In any alphabet Σ elements cannot be formed from other elements of Σ by juxtaposition. For example, if $\Sigma = \{a, b\}$, then string ab is the juxtaposition of the letters a and b. Thus by convention, alphabet such as $\Sigma = \{0, 1, 2, 11, 12\}$ and $\Sigma = \{a, b, ab, bb\}$ are not considered.

Length of String The length of a string u, denoted by $|u|$ is the number of symbols in the string.

For example, if $\Sigma = \{0, 1, 2\}$, then $u = 01$, $v = 212$ and $w = 01212$ are possible strings on A. The length of these strings is: $|u| = 2$, $|v| = 3$ and $|w| = 5$.

Empty String The string consisting of no symbols or letters taken from an alphabet Σ is called an empty string. For an alphabet Σ, we define $\Sigma^o = \{\lambda\}$. Thus $|\lambda| = 0$. The set of all strings including empty string on the alphabet Σ is denoted by Σ^*. So the difference between sets Σ^+ and Σ^* is the presence of the element λ because $\lambda \in \Sigma^*$ only when $n = 0$. Thus $E^* = \Sigma^+ \cup \Sigma^o$ or $\Sigma^+ = \Sigma^* - \{\lambda\}$.

The symbol λ is not an element of any alphabet Σ and it should not be considered as blank (space) found in any alphabet. Thus, we should note that (*i*) $\{\lambda\} \not\subseteq \Sigma$ because $\lambda \notin \Sigma$, and (*ii*) $\{\lambda\} \neq \phi$ because $|\{\lambda\}| = 1 \neq 0 = |\phi|$.

15.2.1 Operations on Strings

The basic operation for strings is *concatenation operations*. The concatenation of strings is a string formed by writing the letters (symbols) of first string first followed by letters of second string. For example, if u = cat and v = house, are two strings of length 3 and 5, respectively, then concatenation of u and v written as uv = cathouse is the string of length 3 + 5 = 8. Similarly, if $u = abb$ and $v = aabb$ are two strings on $\Sigma = \{a, b\}$, then concatenation of u and v is written as: $uv = abbaabb = a^3b^4$.

Properties of Concatenation

1. If Σ is an alphabet and n is a non-negative integer, then the *power of* Σ is defined recursively as
(*a*) $\Sigma^1 = \Sigma$, and
(*b*) $\Sigma^{n+1} = \{ab : a \in \text{A and } b \in \Sigma^n\}$, where ab is the juxtaposition of letters a and b.
For example, if $\Sigma = \{0, 1\}$, then $\Sigma^2 = \{00, 01, 10, 11\}$. Also, for $n = 3$, we write $\Sigma^3 = \{ab : a \in \Sigma \text{ and } b \in \Sigma^2\}$. But as we know the form of the elements of Σ^2, we may consider the strings in Σ^3 as sequences of the form abc, for all $a, b, c \in \Sigma$. Thus when $\Sigma = \{a, b, c, d\}$, it would contain $4^3 = 64$ three-letter strings such as *aaa, acd, cdd, dca.*
In general, for any non-negative integer x, we find that $|\Sigma^n| = |\Sigma|^n$, because our concern is with arrangements (of size n) where we may allow to repeat any of the element of $|\Sigma|$.
2. If Σ is an alphabet, then
(*a*) $\Sigma^+ = \Sigma^1 \cup \Sigma^2 \cup \Sigma^3 \cup \ldots = \bigcup_{n=1}^{\infty} \Sigma^n = \bigcup_{n \in I_+} \Sigma^n$
(*b*) $\Sigma^* = \Sigma^o \cup \Sigma^1 \cup \Sigma^2 \ldots = \bigcup_{n=0}^{\infty} \Sigma^n$.
For example, if $\Sigma = \{0, 1, 2\}$, then the strings 0, 01, 120, 1122 belongs to both Σ^+ and Σ^*. The elements of Σ^+ and Σ^* are also termed as *words* or *sentences* in addition to the term string. The sets Σ^+ and Σ^* are infinite, but the elements of these sets are *finite strings* of symbols or letters. If $\Sigma = \{a, b\}$, then
$$\Sigma^* = \{\lambda, a, b, aa, ab, ba, bb, aaa, aab, \ldots\},$$
and $$\Sigma^+ = \Sigma^* - \{\lambda\}.$$
3. The concatenation operation on an alphabet Σ is *associative,* since for every $u, v, w \in \Sigma^*$, we have $u\,(vw) = (uv)\,w$.
4. The concatenation operation is not *commutative* (unless $|\Sigma| = 1$) on an alphabet Σ, since $uv \neq vu$ for every $u, v \in \Sigma^*$.
5. The empty string λ is an *identity element* for concatenation operation on an alphabet Σ. That is
$$\lambda u = u\lambda = u \text{ for every } u \in \Sigma^*$$
6. If $u, v \in \Sigma^+$, then we may write $u = u_1\, u_2 \ldots u_m$ and $v = v_1 v_2 \ldots v_n$, for $m, n \in I_+$ and $u_1, u_2, \ldots, u_m; v_1, v_2; \ldots v_n \in \Sigma$. Two strings u and v are said to be equal, i.e. $u = v$ if $m = n$ and $u_i = v_i$ for all $1 \le i \le m$.
7. For every $u, v, w \in \Sigma^*$, we have
$$wu = wv \Rightarrow u = v \text{ (left cancellation)}$$
$$uw = vw \Rightarrow u = v \text{ (right cancellation)}$$
8. For $u, v \in \Sigma^*$, we have
$$|uv| = |u| + |v|$$
For example, if $u = 01$, $v = 212$ are strings in Σ^* for $\Sigma = \{0, 1, 2\}$, then we find that
$$|uv| = |01212| = 5 = 2 + 3$$
$$= |01| + |212| = |u| + |v|$$
9. For each $u \in \Sigma^*$, the power of u is defined as:
$$u^0 = \lambda,\ u^1 = u,\ u^2 = uu,\ u^3 = uuu, \ldots,\ u^{n+1} = uu^n, \text{ for } n \in N.$$

For example, if $\Sigma = \{0, 1\}$, and $u = 0, 1$, then $u^0 = \lambda$, $u^1 = 01$, $u^2 = 0101$, $u^3 = 010101$ and so on. For all $n > 0$, u^n consists of a string of 0's and 1's. Thus $|u^2| = 4 = 2|u|$; $|u^3| = 6 = 3|u|$ and in general $|u^n| = n|u|$.

Prefix and Suffix of a String For a string $w = uv$ formed by concatenation of two strings $u, v \in \Sigma^*$ the string u is called a *prefix* of w and we write $u < w$. But if $v \neq \lambda$, then u is said to be a *proper prefix*. For example, the string $w = 125$ has four prefix λ, 1, 12, 125. It may be noted that λ is always a prefix of any string w including itself and the string w is always a prefix of itself. Similarly, the string v is called a *suffix* of w, it is a *proper suffix* when $v \neq \lambda$.

In general, for an alphabet Σ and $u_i \in \Sigma$ for all $1 \leq i \leq n$, each of $\lambda, u_1, u_1u_2, u_1u_2u_3, \ldots$ and $v_1v_2 \ldots v_n$ is a prefix of the string $u = u_1u_2 \ldots u_n$ and $\lambda, u_n, u_{n-1}u_n, u_{n-2}u_{n-1}u_n, \ldots$ and $u_1u_2u_3 \ldots u_n$ are all suffixes of u.

Illustrations

1. Let $\Sigma = \{0, 1, 2\}$. Consider the string $w = 01122$. Then each of the strings λ, 0, 01, 011, 0112 and 01122 is a prefix of w. Except for 01122 itself, each string is a proper prefix.

   Similarly, each of the strings λ, 2, 22, 122, 1122 and 01122 is a suffix of w, where the first five strings are proper suffixes.
2. If $|u| = |7|$, $|v| = |5|$ and $w = uv$, then w has u as a proper prefix and v as a proper suffix. There are in total 12 proper prefixes and 12 proper suffixes since λ is both a proper prefix as well as proper suffix for every string in Σ^+.

Substring and Initial Segments Let $u = u_1u_2u_3 \ldots u_n$ be any string on an alphabet Σ. Any sequence $v = a_j a_{j+1} \ldots a_k$ is called a *substring* (or *subword*) of u. In particular, the substring $v = a_1a_2 \ldots a_k$ beginning with the first letter of v is called an *initial segment* of u.

For example, if $u = xyzx$, then we have

(*i*) Substrings: λ, x, y, z, xy, yz, zx, xyz, yzx, $xyzx$

(*ii*) Initial segments: λ, x, xy, xyz, $xyzx$.

Note that the word xz is not a substring of v because $u = \lambda u$.

15.3 LANGUAGES

The elements of the set Σ^* which represents sentences may not be meaningful or sensibly constructed. The specification of the meaningful sentence is called *syntax* of a language. Thus to write or generate a syntactically correct program in some programming language, it is important to understand the syntax of that language to solve a problem. Formal languages are used to communicate with computers and are helpful in the design of programming languages, compilers, operating systems and software.

Definition For a given alphabet Σ, any subset of Σ^* is called a language over Σ and is denoted by L. This includes the subset ϕ also called *empty language*. Recall that Σ^* denotes the set of all words on Σ.

A string in a language L is called a *sentence* of Σ.

Illustrations If $\Sigma = \{a, b\}$, then $\Sigma^* = \{\lambda, a, b, aa, ab, ba, bb, aaa, aab, \ldots\}$. Thus the following subsets of Σ^* are languages over Σ:

(*a*) $L_1 = \{a, aa, ab, bb\}$ (*b*) $L_2 = \{\lambda, a, b, aa, abb, ba, \ldots\}$

(c) $L_3 = \{a^m b^m : m > 0\}$ (d) $L_4 = \{b^m ab^n : m, n \geq 0\}$
(e) $L_5 = \{a, aa, aab, \ldots\}$ (f) $L_6 = \{a, ab, ab^2, \ldots\}$

If a language has finite number of sentences, then it is called a *finite language.* For example, languages L_2, L_5 and L_6 are infinite. The complement of a language with respect to Σ^* is defined as: $L' = \Sigma^* - L$.

15.3.1 Operations on Languages

Let L_1 and L_2 be two languages over an alphabet Σ. Then the *concatenation* of L_1 and L_2, denoted by L_1L_2 is the language defined as:

$$L_1L_2 = \{ab : a \in L_1, b \in L_2\}.$$

This definition implies that the concatenation of two languages L_1 and L_2 is the set of all words which come from the concatenation of a word from L_1 with a word from L_2. For example, if L_1 = {hope, meaning} and L_2 = {less, ful}, then

L_1L_2 = {hopeless, meaningless, hopeful, meaningful}

1. Let $L_1 = \{a, ab, c\}$ and $L_2 = \{\lambda, b\}$. Then

$$L_1L_2 = \{a, ab, c, abb, cb\} \text{ and } L_2L_1 = \{a, ab, c, ba, bab, bc\}$$

2. Let $L_1 = \{a, b\}$ and $L_2 = \{a^2, ab, b^2\}$. Then

$$L_1L_2 = \{a^3, a^2b, ab^2, ba^2, ab^2, b^3\}$$

3. Let $L_1 = \{a, b^2\}$ and $L_2 = \{a, a^2, a^3\}$. Then

$$L_1L_2 = \{a^2, a^3, a^4, b^2a, b^2a^2, b^2a^3\} \text{ and } L_1L_1 = \{a^2, ab^2, b^2a, b^4\}$$

Remark Concatenation may also be compared with the cross product. Since in general $A \times B \neq B \times A$ and $AB \neq BA$. If languages L_1 and L_2 are finite, then $|L_1 \times L_2| = |L_2 \times L_1|$ but $|L_1L_2| \neq |L_2L_1|$. For example, in Illustration (1) above, we have

(i) $|L_1L_2| = 5 \neq 6 = |L_2L_1|$

(ii) $|L_1L_2| = 5 \neq 6 = 3.2 = |L_1|\,|L_2|$

Such cases arise due to two ways to represent ab : (i) ab for $a \in L_1$ and $b \in L_2$, and (ii) $ab\lambda$ for $ab \in L_1$ and $\lambda \in L_2$.

However, in general, for finite languages L_1 and L_2 the result $|L_1\,L_2| \leq |L_1|\,|L_2|$ is true.

15.3.2 Power of a Language

For a given language $L \in \Sigma^*$, we define other languages also called *power of the language* as follows:

(i) $L^0 = \{\lambda\}$, $L^1 = L$, $L^2 = L\,L$ for all $n > 1$

$L^{n+1} = L^n L = \{ab : a \in L \text{ and } b \in L^n\}$

(ii) The unary operation L^* of a language L, called the *Kleene closure* of L in the honour of the American logician Stephen Cole Kleene (1909–1994). This operation is defined as follows:

$$L^* = L^0 \cup L^1 \cup L^2 \ldots = \bigcup_{n=0}^{\infty} L^n.$$

The set L^* contains all words formed by catenating a finite number of words, possibly zero from L. The plus or positive closure of L denoted by L^+ is defined as:

$$L^+ = L^1 \cup L^2 \cup L^3 \ldots = \bigcup_{n=1}^{\infty} L^n$$

Illustrations

1. Let $L = \{a\}$ over an alphabet $\Sigma = \{a, b, c\}$. Then
 $L^0 = \{\lambda\}$; $L^n = \{a^n\}$ for all $n \in N$.
 $L^+ = \{a^n : n \geq 1\}$ and $L^* = \{a^n : n \geq 0\}$
2. Let $L = \{ab, c\}$ over an alphabet $\Sigma = \{a, b, c\}$. Then
 $L^0 = \{\lambda\}$ and $L^2 = \{abab, abc, cab, c^2\}$

Important Results

1. $L^+ \subseteq L^*$ for all languages L
2. $L^* = \{\lambda\} \cup L^+$ but $L^+ \neq L^* - \{\lambda\}$. For example, if
 $L = \{\lambda, a\}$, then $L^+ = \{a^n: n > 0\}$ which implies that $L^* = L^+$ and $L^+ \neq L^* - \{\lambda\}$. However, $L^+ = L^* - \{\lambda\}$ if and only if $\lambda \notin L$.

Example 1 Let $L_1 = \{a, ab, a^2\}$ and $L_2 = \{b^2, aba\}$ be two languages over the alphabet $\Sigma = \{a, b\}$. Then find out $L_1 L_2, L_2$.

Solution
$$L_1L_2 = \{a, ab, a^2\}\ \{b^2, aba\}$$
$$= \{ab^2, ab^3, a^2b^2, a^2ba, a^2b^2a, a^3ba\}$$
$$L_2^2 = L_2L_2 = \{b^2, aba\}\ \{b^2, aba\}$$
$$= \{b^4, b^2aba, abab^2, aba^2ba\}$$

Example 2 If $\Sigma = \{a, b, c\}$; and $L = \{b^2\}$, $L^2 = \{b^4\}$, $L^3 = \{b^6\}$, $L^4 = \{b^8\}, \ldots$, then find out L^* and L^+.

Solution
$$L^* = L^0 \cup L^1 \cup L^2 \cup L^3 \cup \ldots$$
$$= \{\lambda\} \cup \{b^2\} \cup \{b^4\} \cup \{b^6\} \cup \ldots = \{\lambda, b^2, b^4, b^6, \ldots\}$$
$$L^+ = L^1 \cup L^2 \cup L^3 \cup \ldots$$
$$= \{b^2\} \cup \{b^4\} \cup \{b^6\} \cup \ldots = \{b^2, b^4, b^6, \ldots\}$$

Remark In algebra we know that $a^2 = b^2 \Rightarrow a = b$ for all $a, b \in R$. But if $\Sigma = \{x, y\}$ and $L_1 = \{\lambda, x, x^3, x^4, \ldots\} = \{x^n : n \geq 0\} - \{x^2\}$ and $L_2 = \{x^n : n \geq 0\}$, then
$L_1^2 = L_2^2$, but $L_1 \neq L_2$. It is because of the reason that $\lambda \in \Sigma$ but we may have $\lambda \in L_1 \subseteq \Sigma^*$.

Important Results For an alphabet Σ and languages L_1 and L_2, which are subsets of Σ^*, we have

(i) $L_1 \subseteq L_1L_2^*$

(ii) $L_1 \subseteq L_2^* L_1$

(iii) $L_1 \subseteq L_2 \Rightarrow L_1^+ \subseteq L_2^+$

(iv) $L_1 \subseteq L_2 \Rightarrow L_1^* \subseteq L_2^*$

(v) $L_1L_1^* = L_1^* L_1 = L_1^+$

(vi) $L_1^* L_1 = L_1^* = (L_1^*)^* = (L_1^*)^+ = (L_1^+)^*$

(vii) $(L_1 \cup L_2)^* = (L_1^* \cup L_2^*)^* = (L_1^* L_2^*)^*$

15.4 REGULAR EXPRESSIONS AND REGULAR LANGUAGES

Regular operations such as ( ), *, ∪, λ that are used to develop expressions for describing languages are called *regular expressions*. Thus the regular expression r over a non-empty alphabet Σ is a word (string) which attaches the letters of Σ with the help of operations ( ), *, ∪ and λ. For example, $(0 \cup 1)\ 0^*$ is the language consisting of all the strings beginning with 0's or 1's followed by any number of 0's. Here $(0 \cup 1)$ means $\{0\} \cup \{1\}$ and the value of this expression is the language {0, 1}. Also 0^* means $\{0\}^*$ and its value is the language consisting of all strings combining any number of 0's.

If Σ = {0, 1}, then it can be written for the regular expression as $(0 \cup 1)$. Generally, if Σ is any alphabet, then the regular expression describes the language consisting of all strings of length 1 over Σ.

Following are the examples of regular expression over an alphabet Σ:

(*i*) Each letter $a \in \Sigma$ is a regular expression.

(*ii*) Empty word λ and empty expression ϕ (language that does not contain any string) are regular expressions.

(*iii*) If r is a regular expression, then its closure expression r^* is a regular expression.

(*iv*) If r_1 and r_2 are regular expressions, then $r_1 + r_2$ (union of two expressions) is a regular expression.

(*v*) If r_1 and r_2 are regular expressions, then their concatenation $r_1 r_2$ is a regular expression.

(*vi*) If r is a regular expression, then (r) is also a regular expression.

Illustrations For an alphabet $\Sigma = \{a, b\}$

1. $r = (a + b)^*\ (a + bb)$ is the regular expression and $L(r) = \{a, bb, aa, abb, ba, bbb, \ldots\}$ is the language. Here $(a + b)^*$ denotes any string of a's and b's while $(a + bb)$ denotes either a or double b. So L (r) is the set of all strings on $\{a, b\}$ terminated by either a or bb.
2. $r = (aa)^*\ (bb)^*b$ is the regular expression of all strings with even number of a's followed by an odd number of b's. Thus $L(r) = \{a^{2n}\ b^{2m+1} : n \geq 0, m \geq 0\}$.

Remarks

1. For an alphabet Σ = {0, 1}, we get a regular expression r such that

$$L(r) = \{w \in \Sigma^* : w \text{ has at least one pair of consecutive zeros}\}.$$

Thus L (r) is a language of strings in which each string must have 00, however, we have no idea of preceding and succeeding strings. For example:

(*i*) any string $(0 + 1)^*$ denotes all strings of 0's and 1's. So we may get a regular expression of the form: $r = (0 + 1)^*\ 00\ (0 + 1)^*$, that denotes all strings of 0's and 1's with at least two consecutive 0's.

(*ii*) The regular expression $(1 + 10)^*$ denotes all strings of 0's and 1's beginning with 1 and not having two consecutive 0's.

(*iii*) The regular expression $(0 + 1)^*\ 011$ denotes all strings of 0's and 1's ending in 011.

(*iv*) $0^*\ 1^*\ 2^*$ denotes any number of 0's followed by any number 1's followed by any number of 2's.

2. A regular expression for the language $L(r) = \{w \in \{0, 1\}^* : w$ has no pair of consecutive zeros$\}$ is $r = (1 + 01)^*\ (0 + \phi)$.
3. In writing regular expressions, many parentheses may be omitted provided we assume that * has more precedence than concatenation (or +), and that concatenation has higher precedence than +. For example, $((0(1^*) + 0) = 01^* + 0$ and we abbreviate rr^* by r^+.

Example 3 Let $\Sigma = \{a, b\}$ be an alphabet, show that

(*a*) $a^*(a + b)^*$ (*b*) $aa^*(a + b)^*a$ (*c*) $(a + b)^* (a + \phi)$

are regular expressions over Σ.

Solution (*a*) Since $a, b \in \Sigma$, therefore a^* and b^* are regular expressions over Σ. Since $a + b$ is the regular expression, so $(a + b)^*$ is also a regular expression over Σ. Hence, by concatenation of two regular expressions the $a^* (a + b)^*$ is a regular expression.

(*b*) Since a and a^* are regular expressions, by concatenation aa^* is a regular expression. As $(a + b)^*$ is a regular expression, by concatenation $aa^*(a + b)^*$ is the regular expression. When regular expression $aa^* (a + b)^*$ concatenation with regular expression a, we get the required regular expression $aa^* (a + b)^* a$.

(*c*) Since a, b and ϕ are regular expressions, therefore $a + \phi$ and $(a + \phi)^*$ are also regular expressions. By concatenation of two regular expressions $a + \phi$ and $(a + \phi)^*$ we get a regular expression $(a + \phi)$ ($(a + \phi)^*$ over Σ.

Example 4 Describe the following sets by regular expressions:

(*a*) L_1 = set of all strings of 0's and 1's ending in 00

(*b*) L_2 = set of all strings of 0's and 1's beginning with 0's and ending with 1.

(*c*) $L_3 = \{\lambda, 11, 1111, \ldots\}$

Solution (*a*) Any string in L_1 is obtained by concatenating any string over the set $\{0, 1\}$ and string 00. Also $\{0, 1\}$ is represented as $0 + 1$. Thus L_1 can be described as $(0 + 1)^*\ 00$.

(*b*) Any string in L_2 is obtained by concatenating 0 with any string over the set $\{0, 1\}$ and 1. Thus, L_2 can be described as $0(0 + 1)^*\ 1$.

(*c*) Any string in L_3 is either λ or a string of even number of 1's, i.e. a string of the form $(11)^n$, $n \geq 0$. Thus L_3 can be described as $(11)^*$.

Regular Set A subset of Σ^* is called a regular set corresponding to regular expression r provided following conditions are true

1. The expression ϕ corresponds to the empty set ϕ.
2. The expression λ corresponds to the set $\{\lambda\}$.
3. If $x \in \Sigma$, then regular expression x corresponds to the set $\{x\}$.
4. If the regular expressions r_1 and r_2 corresponds to the set r_1 and r_2, then
   (*a*) regular expression $r_1 \cup r_2$ or $r_1 + r_2$ corresponds to the set $r_1 \cup r_2$.
   (*b*) regular expression r_1r_2 corresponds to the set $r_1r_2 = \{x_1x_2 : x_1 \in r_1 \text{ and } x_2 \in r_2\}$
5. If regular expression r corresponds to a subset R of Σ^*, then the regular expression r^* corresponds to the set R^*.

15.4.1 Identities of Regular Expressions

Two regular expressions r_1 and r_2 are equivalent if these represent the same set of strings

1. $\phi + r = r$
2. $\phi r = r\phi = \phi$
3. $\lambda r = r\lambda = r$
4. $\lambda^* = \lambda$ and $\phi^* = \lambda$
5. $r + r = r$
6. $r^*\ r^* = r^*$

7. $rr^* = r^*r$
8. $(r^*)^* = r^*$
9. $\lambda + rr^* = r^* = \lambda\, r^*r$
10. $(r_1r_2)^*r_1 = r_1(r_2\, r_1)^*$
11. $(r_1 + r_2)^* = r_1^* + r_2^*$
12. $(r_1 + r_2)\, r_3 = r_1r_3 + r_2r_3$ and $r_3(r_1 + r_2) = r_3r_1 + r_3r_2$

15.4.2 Regular Language

The language L (r) represented by the regular expression r over an alphabet Σ, is as follows:

(*i*) If $r = \phi$, then L $(r) = \phi$

(*ii*) If $r = \lambda$, then L $(\lambda) = \{\lambda\}$ and L $((\,)) = \phi$ empty set

(*iii*) If $r = a$, for some $a \in \Sigma$, then L $(a) = \{a\}$

(*iv*) L (r^*) = (L (r))*, the Kleene closure of L (r)

(*v*) L $(r_1 \vee r_2)$ = L $(r_1) \cup$ L (r_2), union of two languages

(*vi*) L $(r_1\, r_2)$ = L (r_1) L (r_2), concatenation of two languages.

The language L is called *regular language* over an alphabet Σ if there exists a regular expression r over Σ such that L = L (r).

Example 5 Let $\Sigma = \{a, b, c\}$. Describe the language L (r) for each regular expression:

(*a*) $r = ab^*\, c^*$ (*b*) $a^* \cup b^* \cup c^*$

Solution (*a*) Since L (r) = L (a) L $((b))^*$ L $((c))^*$ consists of all words in a, b, c with a as first letter, therefore it contains exactly one a which is the first letter.

(*b*) L (r) = (L (a))* $\cup$ (L (b))* $\cup$ (L (c))*. This language contains λ and all words only in a or only in b or only in c. That is L $(r) = \{\lambda, a, a^2, \ldots, b, b^2, \ldots, c, c^2, \ldots\}$.

Example 6 Let $\Sigma = \{a, b, c\}$ and $w = ac$. State whether or not w belongs to L (r) where

(*a*) $r = a\, b^*\, c^*$ (*b*) $r = (a^*\, b \cup c)^*$

Solution (*a*) $w = ac$ belongs to L (r) because $w = a\, \lambda\, c$ and $\lambda \in$ L (b^*) and $c \in$ L (c^*).

(*b*) It does not belong to L (r) because L $(a^*\, b)$ consists of words a^nb. Thus, if a appears in a word in L(r), then a can only be followed by an a or b not by c.

Example 7 Let $\Sigma = \{0, 1\}$. Show that the following expressions are regular expressions over Σ

(*a*) 0* (0 + 1) (*b*) 00* (0 + 1)* (*c*) (01)* (01 + 1)*

Also find regular sets corresponding to these regular expressions.

Solution (*a*) Since $0, 1 \in \Sigma$, these are regular expressions.

As 0 + 1 is a regular expression, so 0* and (0 + 1)* are also regular expressions. By concatenation of 0* and (0 + 1)* we have 0* (0 + 1)*.

(*b*) Since 0, 1 and 0* (0 + 1)* are regular expressions, therefore 00* (0 + 1)* 1 is also a regular expression.

(*c*) Since 0, 1, 1* and 01 are regular expressions, therefore 01 is regular and hence (01)* and (01 + 1)* are regular expressions. By concatenation of (01)* and (01 + 1*) we have the regular expression (01)* (01 + 1*).

The regular set corresponding to (*a*) consists of all sequences of 0's and 1's. Thus the regular set is Σ^*.

The regular set corresponding to (*b*) contains all sequences of 0's and 1's which begins with at least one 0 and end with at least one 1.

The regular set corresponding to (*c*) is the set of all sequences of the form xy where $x = (01)^n$, for $n \geq 0$ and $y = 1^m$ for $m \geq 0$. Thus, this set contains the strings of the form: 1111, 01, 010101, 010101011111 and 011.

Example 8 Let $r_1 = aa^*$, $r_2 = a + b^*$ and $r_3 = (a + b)^*$ be regular expressions over the alphabet $\Sigma = \{a, b\}$. Find the corresponding languages $L(r_i)$ for $i = 1, 2, 3$.

Solution Since $r_1 = aa^*$, it contains all strings a^* for $n \geq 1$. Thus $L(r_1)$ consists of all positive powers of a.

Again since $r_2 = a + b^*$, $L(r_2)$ consists of a and all strings b^n for $n \geq 0$. That is $L(r_2) = \{a, \lambda, b, b^2, \ldots\}$.

Let $r_3 = (a + b)^*$. Since $L(a + b) = L(a) \cup L(b)$

$$= \{a\} \cup \{b\} = \{a, b\} = \Sigma, \text{ therefore } L(r_3) = L((a + b)^*)$$

Also $(L(a + b))^* = \Sigma^*$. Hence, $L(r_3)$ contains all strings over Σ.

Example 9 Describe the following languages by regular expression r over the alphabet $\Sigma = \{a, b\}$.

(*a*) $L_1 = \{a^m b^n : m, n > 0\}$

(*b*) $L_2 = \{b^m a b^n : m, n > 0\}$

Solution (*a*) L_1 consists of all strings (words) beginning with one or more a's followed by one or more b's. Hence we may take $r = aa^* \, bb^*$ or $r = a^* \, abb^*$.

(*b*) L_2 consists of all strings with exactly one a. Hence, we may take $r = bb^* \, abb^*$ or $r = b^*b \, ab^*b$.

Example 10 Prove the following equation

$$(1 + 00^*1) + (1 + 00^*1)\,(0 + 10^*1)^*\,(0 + 10^*1) = 0^*1(0 + 10^*1)^*$$

Solution LHS $= (1 + 00^*1) + (1 + 00^*1)\,(0 + 10^*\,1)^*\,(0 + 10^*1)$

$= (1 + 00^*1)\,[\lambda + (0 + 10^*1^*(0 + 10^*1)$, since $(r_1 + r_2)\,r = r_1 r + r_2 r$ and $r(r_1 + r_2) = rr_1 + rr_2$

$= (1 + 00^*1)\,(0 + 10^*1)^*$, since $\lambda + rr^* = r^* = \lambda + r^*r$

$= (\lambda + 00^*)1\,(0 + 10^*1)^* = 0^*1(0 + 10^*1)^* =$ RHS

Self-practice Problems I

1. Let $\Sigma = \{a, b, c, d, e\}$. Find (*a*) $|\Sigma^2|$ and $|\Sigma^3|$ and (*b*) How many strings in Σ^* have length at most 5.
2. Given the strings $u = a^2bab^2$ and $v = bab^2$, find the string (*a*) uv, (*b*) vu, (*c*) λu and (d) $u\lambda v^2$. Also find their length.
3. Given the strings $u = a^2ba^3b^2$ and $v = bab^2$, find (*a*) $|u|$ (*b*) $|v|$, (*c*) $|uv|$, and $|v^2|$.
4. Given the strings $u = a^2b^2$ and $v = b^3ab$, find (*a*) uvu and (*b*) $u\lambda v$.
5. For any strings u and v, show that (*a*) $|uv| = |u| + |v|$ and (*b*) $|uv| = |vu|$.

6. For the alphabet $\Sigma = \{0, 1\}$, the languages L_1, L_2, L_3 are defined as:

$$L_1 = \{0, 1, 00, 11, 000, 111, 0000, 1111\}$$
$$L_2 = \{x : 2 \leq |x|, x \in \Sigma^*\}$$
$$L_3 = \{x : 2 \geq |x|, x \in \Sigma^*\}$$

Determine the following languages (subsets) of Σ^*

(*a*) $L_1 \cap L_2$, (*b*) $L_1 \cap L_3$, (*c*) $L_2 \cup L_3$, and (*d*) $L_1 - L_2$.

7. For the alphabet $\Sigma = \{0, 01\}$, discribe the elements of Σ^*.

8. Let $L_1 = \{a, b\}$ and $L_2 = \{b, c\}$ be languages over $\Sigma = \{a, b, c\}$. Find

(*a*) $(L_1 \cup L_2)^*$ (*b*) $(L_1 \cap L_2)^*$ (*c*) $L_1^* \cup L_2^*$

(*d*) $L_1^* \cap L_2^*$ (*e*) $(L_1 - L_2)^*$ (*f*) $(L_2 - L_1^*)^*$

9. Let $\Sigma = \{a, b\}$. Describe the following languages over Σ (subsets of Σ^*)

(*a*) $L_1 = \{(ab)^m : m > 0\}$ (*b*) $L_2 = \{a^2 b^m a^3 : m > 0\}$

(*c*) $L_3 = \{b^m a b^n : m \geq 0, n \geq 0\}$ (*d*) $L_4 = \{a, ab, ab^2, \ldots\}$

10. For the alphabet $\Sigma = \{0, 1\}$, describe the strings in Σ^* for each of the following languages

(*a*) $\{01\}$ (*b*) $\{000\}$ (*c*) $\{0, 010\}$

11. For the alphabet $\Sigma = \{0, 1\}$, the language L be defined as follows:

(*i*) The symbols 0, 1 are both in L.

(*ii*) For each string $x \in \Sigma$, the string $01 \in \Sigma$ (recursive process)

(*a*) Find four different strings two of length 3 and two of length 5 in Σ.

(*b*) Use the given recursive process to show that $0001111 \in \Sigma$.

12. Let $L_1 = \{a, ab, b^2\}$ and $L_2 = \{b^2, aba\}$ be languages over $\Sigma = \{a, b\}$. Find

(*a*) L_1L_2, (*b*) L_2L_1 (c) L_1L_1

13. (*a*) If $L = \{a^2, ba\}$, find L^0, L^2 and L^3.

(*b*) If $L = \{ab, c\}$ over $\Sigma = \{a, b, c\}$, find L^0, L^3, L^{-3}.

14. Given $\Sigma = \{a, b, c\}$, find L^* for (*a*) $L = \{b^2\}$, (*b*) $L = \{a, b\}$ and (c) $L = \{a, b, c^3\}$.

15. Does $(L^2)^* = (L^*)^2$. If not, how are they related?

16. Given $u = abcd$. (*a*) Find all subwords of u, and (*b*) which of them are initial segments.

17. Given $A = \{ab, bc, ca\}$, verify whether following strings belongs to Σ^*.

(*a*) *abc* (*b*) *ababab* (*c*) *abba* (*d*) *bcabbab*

18. Let $\Sigma = \{a, b, c\}$. Describe the language for each regular expression

(*a*) $r = ab^*c^*$ (*b*) $r = a^* + b^* + c^*$

19. Let $\Sigma = \{a, b, c\}$ and $w = abc$. State whether or not u belong to L (r), where

(*a*) $r = ab^*c^*$ (*b*) $r = (a^*b + c)^*$ (*c*) $r = a^*(b + c)^*$

(*d*) $r = a^* + (b + c)^*$ (*e*) $r = ab^*(bc)^*$ (*f*) $r = a^* + (b + c)^*$

20. Represent the following sets by regular expression:

(*a*) {0, 1, 2} (*b*) $\{|2n + 1| : n > 0\}$

(*c*) $\{a^2, a^5, a^8, \ldots\}$ (*d*) $\{a^n : n$ is divisible by 2 or 3 or 5$\}$

(*e*) $\{\lambda, ab\}$

21. Given $\Sigma = \{a, b\}$, describe the language for (*a*) $r = abb^*a$, (*b*) $r = b^*ab^*ab^*$ and (*c*) $r = ab^* + a^*$

22. Describe the following sets by regular expression

(*a*) {01, 10} (*b*) {101} (*c*) $\{\lambda, 1, 11, 111, \ldots\}$ (*d*) $\{1, 11, 111, \ldots\}$

23. Express each of the following sets using a regular expression.

(*a*) The set of strings of one or more 0's followed by 1.

(*b*) The set of all strings of 0's and 1's end with 00.

24. Describe in words the strings in each of the following regular sets

(*a*) $1(0)^*$ (*b*) $1^*0(0)^*$ (*c*) (0^*1)

25. Let $\Sigma = \{a, b\}$. Find regular expression *r* such that words in L (*r*)

(*a*) begins with a^2 and ends with b^2.

(*b*) are all even numbers of *a*'s.

26. Describe the sets represented by the following regular expressions

(*a*) $a(b + c)$ (*b*) $ab + c^*$ (*c*) $ab(bc)^*$

Hints and Answers

1. (*a*) $|\Sigma^2| = \Sigma\Sigma = 5 \times 5 = 25$; $|\Sigma^3| = 125$ (*b*) 3906

2. Write letters of first string followed by the letters of the second string

(*a*) $uv = (a^2bab^2)(bab^2) = a^2bab^3ab^2$; $|uv| = 10$

(*b*) $vu = (bab^2)(a^2bab^2) = bab^2a^2bab^2$; $|vu| = 10$

(*c*) $\lambda u = u\lambda = u = a^2bab^2$, since λ is the empty string

(*d*) $u\lambda v^2 = uv^2 = (a^2bab^2)(bab^2)(bab^2) = (a^2bab^3ab^2)(bab^2)$
$= a^2bab^3ab^3ab^2$; $|u\lambda v^2| = 14$

3. Count the number of letters in each string

(*a*) $|u| = 8$;

(*b*) $|v| = 4$,

(*c*) $uv = (a^2ba^3b^2)(bab^2) = a^2ba^3b^3ab^2$; $|uv| = 12$

(*d*) $v^2 = vv = (bab^2)(bab^2) = bab^3ab^2$; $|v^2| = 8$

4. (*a*) $uvu = (a^2b^2)(b^3ab)(a^2b^2) = (a^2b^5ab)(a^2b^2) = a^2b^5aba^2b^2$

(*b*) Since λ is the empty string, $u\lambda v = uv = (a^2b^2)(b^3ab) = a^2b^5ab$.

5. Let (*a*) $|u| = r$ and $|v| = s$. Then *uv* consists of $r + s$ letters. Thus $|uv| = r + s = |u| + |v|$

(*b*) $|uv| = |u| + |v| = |v| + |u| = |vu|$

6. (*a*) $L_1 \cap L_2 = (00, 11, 000, 111, 0000, 1111\}$ (*b*) $L_1 \cap L_3 = \{0, 1, 00, 11\}$
(*c*) $L_2 \cup L_3 = \Sigma^*$ (*d*) $L_1 - L_2 = \{0, 1\}$
7. The set of all strings which begin and end with 01 and have at least two 1's between every pair of 0's.
8. (*a*) $(L_1 \cup L_2)^*$ = set of all strings in $\Sigma^* = \{a, b, c\}^*$
(*b*) $L_1 \cap L_2 = \{b\}$, so $(L_1 \cap L_2)^* = \{b^n : n \geq 0\}$
(*c*) $L_1^* \cup L_2^*$ = set of all strings in Σ^* which are in L_1^* or in L_2^*
(*d*) $L_1^* \cap L_2^* = \{b^n : n \geq 0\}$
(*e*) $L_1 - L_2 = \{a\}$, so $\{L_1 - L_2)^* = \{a^n : n \geq 0\}$
(*f*) $L_2 - L_1^* = \{c\}$, so $\left(L_2 - L_1^*\right)^* = \{c^n : n \geq 0\}$
9. (*a*) L_1 consists of all strings that begin with *a*, alternating with *b* and ending with *b*.
(*b*) L_2 consist of all strings that begin with a^2, end with a^3 with one or more *b*'s between them.
(*c*) L_3 consists of all strings with exactly one *a*.
(*d*) L_4 consists of all strings that begin with and followed by 0's or more *b*'s.
10. (*a*) Σ^* consists of all strings *x* of even length. If $x \neq \lambda$, then *x* begins with 0 and ends with 1 and the symbols (0 and 1) alternate.
(*b*) Σ^* consists of those strings that are made up of $3n$ 0's for $n \in$ N.
(*c*) Any string $x \in \Sigma^*$ provided (*i*) *x* is a string of *n* 0's for $n \in$ N, or (*ii*) *x* is a string that starts and ends with 0 and has at least one 1 and atleast two 0's between any two 1's.
11. (*a*) The strings 001 and 011 have length 3 and belong to Σ. The words 00011 and 00111 have length 5 and also belong to Σ.
(*b*) Since $1 \in \Sigma$, therefore repeating it three times, we get
(*i*) $1 \in \Sigma \Rightarrow 011 \in \Sigma$ (*ii*) $011 \in \Sigma \Rightarrow 00111 \in \Sigma$, and
(*iii*) $00111 \in \Sigma \Rightarrow 0001111 \in \Sigma$.
12. (*a*) Concatenate words in L_1 with words in L_2 to obtain

$$L_1L_2 = \{a, ab, b^2\}\{b^2, aba\} = \{ab^2, a^2ba, ab^3, ababa, b^4, b^2aba\}$$
$$L_2L_1 = \{b^2, aba\}\ \{a, ab, b^2\} = \{b^2a, a^2ba, b^2ab, a^2b^2a\ b^4, abab^2\}$$
$$L_1L_1 = \{b^2, aba\}\ \{b^2, aba\} = \{b^4, b^2aba, abab^2, aba^2ba\}$$

13. (*a*) (*i*) $L^0 = \{\lambda\}$ by definition
(*ii*) All two-word sequences from L
$L^2 = \{a^2, ba\}\{a^2, ba\} = \{a^4, a^2ba, ba^3, baba\}$
(*iii*) L^3 = All three-word sequence from L
$= \{a^6, a^4ba, a^2ba^3, a^2baba, ba^5, ba^2ba, baba^3, bababa\}$
(*b*) (*i*) $L^0 = \{\lambda\}$
(*ii*) $L^3 = \{ababab, ababc, abcab, abc^2, cabab, cabc, c^2ab, c^3\}$
(*iii*) The negative power of a language is not defined.

14. (*a*) $L^* = \{b^n : n$ is even number including $\lambda\}$

(*b*) $L^* = \{a, b\}$

(*c*) L^* consists of all words from Σ and the length of each maximal subwords are composed of c's divisible by 3.

15. Not true, because $(L^2)^* \subseteq (L^*)^2$.

16. (*a*) Subwords of u are : λ, a, b, c, d, ab, bc, cd, abc, bcd, $abcd$. Here the word acd is not the subword of u even all letters belong to u.

(*b*) The initial segments are: λ, a, ab, abc, $abcd$.

17. (*a*) No (*b*) yes (*c*) No (*d*) No

18. L (r) consists of all strings (words) w in a, b, c such that

(*a*) w contains only one a that is the first letter

(*b*) w is described only in a or b or c. That is

$$L(r) = \{\lambda, a, a^2, \ldots, b, b^2, \ldots, c, c^2, \ldots\}$$

19. (*a*) Yes, because $w = a\lambda c$ and $\lambda \in L(b^*)$ and $c \in L(c^*)$.

(*b*) No, because $L(a^*b) = \{a^n b : n \in N\}$. Thus, if a appears in a word in L (r), then it can only be followed by either a or b not by c.

(*c*) Yes (*d*) No (*e*) Yes (*f*) No

20. (*a*) $0 + 1 + 2$ (*b*) $1(11)^*$ (*c*) $aa\,(aaa)^*$ (*d*) $(aa)^* + (aaa)^* + aaaaa$ (*e*) $\lambda + ab$

21. (*a*) L (r) consists of words that begins and ends in a enclosing one or more b's.

(*b*) L (r) consists of all words with exactly two a's.

(*c*) $r = ab^* + c^*$ is not a regular expression because symbol + is not one of the symbols adjoined to Σ for forming a regular expression.

22. (*a*) Since {01, 10} is the union of sets {01} and {10}, therefore {01, 10} is represented by $01 \cup 10$.

(*b*) Since 101 is obtain by concatenation of 1, 0 and 1, the set {101} is represented by 101.

(*c*) The set $\{\lambda, 1, 11, 111\}$ is represented by $(1)^*$.

(*d*) Any element in the set {1, 11, 1111, . . .} can be obtained by concatenating 1 and any element of $(1)^*$. Thus the set {1, 11, 1111, . . .} is represented by $1(1)^*$.

23. (*a*) Since the set of one or more 0's is represented by $0\{0\}^*$, therefore the set of one or more 0's followed by 1 is obtained by concatenation of $0\,(0)^*$ and 1 and hence is represented by $0\,(0)^*1$.

(*b*) The set {0, 1} is represented by $0 \cup 1$. The set of strings of 0's and 1's is represented by $(0 \cup 1)^*$. Hence, the set of all strings of 0's and 1's ending in 00 is obtained by concatenation of $(0 \cup 1)^*$ and 00. So it is represented as: $(0 \cup 1)^*\,00$.

24. (*a*) Any number of 1's followed by any number of 0's including no zero.

(*b*) Any number of 1's followed by one or more 0's.

(*c*) Any string not ending with 0.

25. (*a*) $r = a^2(a + b)^*b^2$ (*b*) $r = (b^*\,ab^*\,ab^*)^* = s^*$

26. (*a*) $\{ab, ac\}$ (*b*) $ab + \{c^n : n \geq 0\}$ (*c*) $\{ab, abbcbc, abcbcbcbc, \ldots\}$

15.5 GRAMMAR

A grammar consists of a finite set of rules that specify the syntax of the language and hence helps in structuring the sentences of a language. In formal languages, the grammar helps in the following two types of problems:

(*a*) Generation of the valid strings in the language.

(*b*) Identify whether a given string is in the language.

In order to understand the technical definition of grammar that is used in the application of programming languages, it is desirable to understand typical rule of English grammar. Any sentence in English is made up of a *noun phrase* followed by a *verb phrase,* i.e.

< Sentence > → < Noun phrase > < Verb phrase >

The noun phrase is made up of an *article* and the verb phrase is made up of a *verb* and *object phrase* or *adverb.* That is

<Verb phrase> → <Verb> <Object phrase or Adverb>

Thus a sentence is made up of an *article* followed by a *noun,* followed by a *verb,* and then an *adverb,* i.e.

<Sentence> → <Article> <Noun> <Verb> <Adverb>.

For example, in a sentence like *'The sun rises in the east'*, we associate the actual words as:

<article> → the, <noun> → sun, <verb> → rises, and <adverb> → east.

This example should help us to define grammar and the language it generates.

Definition A phrase structure grammar or simply a grammar G *consists* of four parts: V, Σ, S and P, where

(*i*) V is a finite set, also called vocabulary, whose elements are called variables.

(*ii*) Σ is a finite subset of V whose elements are called terminals.

(*iii*) $V \cap \Sigma = \phi$

(*iv*) S is a special variable (non-terminal symbol) in the set V, called the *start symbol* that begins the generation of any sentence in the language.

(*v*) P is a finite set whose elements are ordered pair (α, β) usually written as $\alpha \to \beta$ where α and β are strings (words) on $V \cup \Sigma$ and has at least one non-terminal on its left side. Elements of P are *called productions* or *production rules*. These rules specify how sentences in the language can be made up. A grammar is denoted by G = (V, Σ, S, P).

Remark The set V consisting of terminal and non-terminal symbols is called *vocabulary* of the language. The elements of V (non-terminals) are intermediate symbols that are used to describe the structure of the sentences. As mentioned earlier, the set V is written as:

V = {< sentence >, < article >, < noun >, < verb > < adverb >}

Similarly, the elements of the set Σ (terminals) are used to make up sentences in the language. For example,

Σ = {the, sun, rises, east}

The sentences are obtained (*i*) starting with S, (*ii*) replacing words using productions and (*iii*) terminating when a string of terminals is obtained.

In the above example, production P consists of the following productions

< sentence > → < article > < noun > < verb > < adverb >

↓ ↓ ↓ ↓

the sun rise east

Illustrations

1. Let V = {A, B, S, a, b}, Σ = {a, b}, P = {$S \xrightarrow{1} AB, A \xrightarrow{2} Aa, B \xrightarrow{3} Bb, A \xrightarrow{4} a, B \xrightarrow{5} b$} with S as the start symbol. The productions may be written abbreviated as follows:

$$S \to AB,\ A \to (Aa, a),\ B \to (Bb, b).$$

2. Let G = (V, Σ, S, P) be a phase-structure grammar, and let w_1 and w_2 be two strings over N. If there are strings u and v over V and a production $\alpha \to \beta$ such that $w_1 = u\alpha v$ and $w_2 = u\beta v$, then w_1 *derives* w_2 and this is written as $w_1 \Rightarrow w_2$.

   Successive strings are derived by applying the productions of the grammar in any order. If there are finite number of strings such that

$$w_1 \Rightarrow w_2,\ w_2 \Rightarrow w_3, \ldots, \Rightarrow w_n,\ n \geq 0$$

   then we say that w_1 derives w_n and is written as $w_1 \Rightarrow w_n$.

3. Let Σ = {0, 1}, V = {S} and P = {$S \to 0S1, S \to \lambda$}. Then G = {V, Σ, S, P} is a grammar, where λ denotes the null string.

15.5.1 Language of a Grammar G

When production rules are applied in different order, a given expression can generate many strings. The set of all such strings is the language generated or defined by the grammar.

Definition Let G = (V, Σ, S, P) be a grammar. Then the language L (G) generated by G is the set of all strings of terminals which are derivable from the starting state S, i.e.

$$L\,(G) = \{w \in \Sigma^*,\ S^* \Rightarrow w\}.$$

The elements of L (G) are usually called sentences.

If $w \in L\,(G)$, then the sequence $S \Rightarrow w_1 \Rightarrow w_2 \ldots \Rightarrow w_n \Rightarrow w$ is the derivation of the sentence w which does not contain any non-terminal symbol. The strings S, $w_1, w_2, \ldots, w_n$ which contain non-terminal symbols as well as terminals are called *sentential* forms of the derivation.

Example 11 Find the language L (G) generated by the grammar with V = {S, A, B}, Σ = {a, b} and productions P = {$S \to aB, B \to b, B \to bA, A \to aB$}.

Solution $S \to aB \to ab$ which is obtained by using first production. Applying productions in succession, we get

$$L\,(G) = \{(ab)^n = ababab \ldots ab : n \in N\}$$

Example 12 Find the language L(G) generated by the grammar with V = {S, A}, Σ = {a, b, c} and productions P = {$S \to aSb, aS \to Aa, Aab \to c$}.

Solution Applying first production $(n - 1)$ times, we get $S \Rightarrow a^nSb^n$, $n > 0$. Now

$$S \Rightarrow a^nSb^n \Rightarrow a^mAabb^m, \text{ where } m = n - 1 \geq 0$$

$$\Rightarrow a^mcb^m \text{ since } Aab \to c;\ m \geq 0.$$

Hence, $L\,(G) = \{a^mcb^m : m \geq 0\}$

Example 13 Find the language L (G) generated by the grammar with V = {S}, S = {0, 1} and productions P = {S → 0S1, S → λ}.

Solution Since S → λ is a production, S* → λ. This implies that λ ∈ L (G), for all $n \geq 1$.

Also
$$S \Rightarrow 0S1 \Rightarrow 00S11 = 0^2S1^2$$
$$\Rightarrow 00^2S1^21 = 0^3S1^3$$
$$\Rightarrow 0^nS1^n = 0^n1^n \text{ because } S \rightarrow \lambda.$$

Hence, $0^nS1^n \in$ L (G) for all $n \geq 0$.

To show that L (G) ⊆ $\{0^n1^n : n \geq 0\}$, we may start with $w \in$ L (G). When production S → λ is applied we get empty string λ, which is of the form $\lambda = 0^n1^n$ for $n = 0$. If production S → 0S1 is applied first, then it produces strings by concatenating 0 at the beginning and 1 at the end in succession. Thus, we get the derivation of w of the form $S \Rightarrow 0^nS1^n \Rightarrow 0^n1^n$, because S → λ. So L (G) = $\{0^n1^n : n \geq 0\}$.

Example 14 Let V = {S, C}, Σ = {a, b}, P = {S → aCa, C → aCa, C → b}. Find L (G).

Solution Here S ⇒ aCa ⇒ aba, since C → b. Again
$$S \Rightarrow aCa \Rightarrow aaCaa = a^2Ca^2, \text{ since } C \rightarrow aCa$$

Now applying C → aCa, $(n - 1)$ times, we get
$$S \Rightarrow a^nCa^n \Rightarrow a^nba^n, \text{ since } C \rightarrow b$$

Hence, $a^nba^n \in$ L (G) for all $n \geq 1$, so $\{a^nba^n : n \geq 1\} \subseteq$ L (G)

Let w be any string in L (G). Since S → aCa is the only production having S on the left side. S → aCa is applied to derive string in L (G). Applying C → b, we get S ⇒ aba. Thus $aba \in$ L (G). But when S → aCa is applied repeatedly, we get $S \Rightarrow a^nba^n$. Hence, any string $w \in$ L (G) is of the form $S \Rightarrow a^nba^n$ with $n \geq 1$. So L (G) = $\{a^nba^n : n \geq 1\}$.

Example 15 Find the phase-structure grammar that generate the set L = $\{a^nb^{2n} : n \geq 1\}$.

Solution Let V = {S}, Σ = {a, b}. The set of productions can be obtained by considering first $abb \in$ L. Then for any $x \in$ L we have $axbb \in$ L, so we have the production S → abb and S → aSbb. Hence L (G) = (V, Σ, S, P) where V = {S}, Σ = {a, b} and P = {S → abb, S → aSbb}.

Example 16 Let V = {S, A, B}, Σ = {a, b} starting symbol S and production P = {S → aABa, A → baABb, B → Aab, aA → baa, bBb → $abab$}. Find the language generated by G = (V, Σ, S, P).

Solution Since L (G) = {$w \in$ T* : S* ⇒ w), therefore we need to start with a production where S is in the left side. So we start with

$$S \Rightarrow aABa \Rightarrow baaBa, \text{ since } aA \rightarrow baa$$
$$\Rightarrow baaAaba, \text{ since } B \rightarrow Aab$$
$$\Rightarrow babaaaba, \text{ since } aA \rightarrow baa$$

Hence, L (G) = {$babaaaba$}, as S ⇒ $babaaaba$.

Example 17 Consider the grammar G with V ={S, A, B}, Σ = {a, b} and P = {S → aABa, A → baABb B → Aab, aA → baa, bBb → $abab$}. Verify whether the string w = $baabbabaaabbaba \in$ L (G).

Solution Here
$$S \Rightarrow aAB\,a \Rightarrow baaBa \Rightarrow baaAaba$$
$$\Rightarrow baaba\ ABbaba$$
$$\Rightarrow baabbaa\ Bbaba$$
$$\Rightarrow baabbaa\ Aabbaba$$
$$\Rightarrow baabbabaaabbaba = w$$

Hence, $w \in$ L (G).

Example 18 Consider the grammar G with V = {S}, $\Sigma = \{a, b\}$ and P = $\{S \to aSa, S \to bSb, S \to aa, S \to bb, S \to \lambda\}$. Show that

(*a*) L (G) has no strings of odd length

(*b*) any string in L (G) is of length $2n$, $n \geq 0$

(*c*) the number of strings of length $2n$, is 2^n.

Solution (*a*) and (*b*) By applying given productions except $S \to \lambda$, a variable is replaced by two terminals and at most one variable. This process increases the number of terminals by 2 in each derivation, except $S \to \lambda$. Hence, the results (*a*) and (*b*) are proved.

(*c*) Let w be any string of even length $2n$. Then it is written as $w = a_1 a_2 \ldots a_n a_n \ldots a_1$ involving n parameters $a_1, a_2, \ldots, a_n$, where each a_i can either be a or b. Hence, the number of such strings will be 2^n.

Example 19 Consider the grammar G with V = {S, A}, $\Sigma = \{0, 1, 2\}$ and P = $\{S \to 0SA2, S \to 012, 2A \to A2, 1A \to 11\}$. Test whether strings (*a*) $w = 00112$ and (*b*) $w = 001122$ are in L (G).

Solution (*a*) Length of string $|w| = 5$ and $S \Rightarrow 0SA2 \Rightarrow 0012A2 \Rightarrow 001A22$. Since length of any string $|w|$ in this case is more than 5, therefore $00112 \notin L(G)$.

(*b*) Length of string $|w| = 6$. Now

$$S \Rightarrow 0SA2 \Rightarrow 0012A2 \Rightarrow 001A22 \Rightarrow 001122 \in L(G).$$

15.5.2 Types of Grammars

Grammars are classified according to the types of productions that are allowed to define them. The following classifications is due to Noam Chomsky.

1. A grammar G is said to be of *Type 1* if every production is of the form $\alpha \to \beta$ where $|\alpha| \leq |\beta|$ or of the form $\alpha \to \lambda$. Here λ is the null string and α, β are arbitrary strings on V.
2. A grammar G is said to be of *Type 2* if every production is of the form $A \to \beta$, i.e. where the left hand side string is always a single non-terminal symbol.
3. A grammar G is said to be of *Type 3* if every production is of the form $A \to \alpha$ or $A \to aB$, i.e. where the left hand side string is a single non-terminal symbol and the right hand side string is either terminal symbol or a terminal symbol followed by a non-terminal symbol or of the form $S \to \lambda$.
4. A grammar G with no restriction on its productions is called a *Type* 0 grammar.

Grammars are also classified in the following forms:

1. A grammar G is said to be a *context sensitive grammar* if the production in G have the specialised form

$$\alpha_1 A \alpha_2 \to \alpha_1 \beta \alpha_2$$

where A denotes non-terminal symbol and α_1, α_2 and β are words (strings) in V.

The name 'context-sensitive' comes from the fact that the non-terminal symbol A can be replaced by β in a word only when A lies between α_1 and α_2. However, more often a grammar is said to be context sensitive if productions merely are of the form $\alpha \to \beta$, where $|\alpha| = |\beta|$ or of the form $\alpha \to \lambda$.

2. A grammar G is said to be *context free* if productions are of the form $A \to \alpha$. That is, the left side is a single non-terminal symbol and the right side is a word in one or more symbols.

The name 'context free' comes from the fact that the non-terminal symbol A on the left of a production can be replaced in a string β wherever A appears.

3. A grammar G is said to be *regular* if productions are of the form $A \to a$ or $A \to aB$, or $S \to \lambda$, where A and B are non-terminal symbols and '*a*' is a terminal symbol. That is, in this type of grammar the left side is a single non-terminal symbol and the right side is either a single terminal symbol or a terminal symbol followed by a non-terminal symbol.

It may be observed that a phase-structure grammar, context-sensitive grammar, context free grammar and regular grammar, is also known as type 0, type 1, type 2 and type 3 grammar, respectively.

These grammars form a grammatical hierarchy, called the *Chomsky hierarchy.* Observe that every Type 3 grammar is Type 2 grammar, every Type 2 grammar is a Type 1 grammar and every Type 1 grammar is a Type 0 grammar.

15.5.3 Backus-Naur Form (BNF)

The BNF form is another way of describing the productions of a context-free (Type 2) grammar. This form is used to specify the syntacture rules of computer languages such as Java, Pascal, Algol. The specific features of BNF form are

(*i*) The symbol $::=$ is used instead of the symbol $\to$.

(*ii*) Every non-terminal symbol is enclosed in brackets $<\,>$.

(*iii*) All productions with the same non-terminal symbol on the left side are combined into one statement with all the right side symbols listed on the right of $::=$ separated by vertical bar.

Illustrations

1. The productions $S \to aA$, $S \to B$, $S \to aBb$ are combined into one statement as:

$$<S> ::= a<A> \mid <B> \mid a<B>b$$

2. The productions $S \to aB$, $B \to bA$, $B \to b$, $B \to a$, $A \to aB$, $A \to a$ are combined into one statement as:

$$<S> ::= a<B>,$$

$$<B> ::= b<A> \mid b \mid a,$$

$$<A> ::= a<B> \mid a$$

3. The productions $S \to aA$, $A \to aAB$, $B \to b$, $A \to a$ are written as:

$$<S> ::= a<A>;\ <A> ::= a<A><B> \mid a;\ <B> ::= b$$

Example 20 Determine the type of the grammar G which consists of the productions

(*a*) $P = \{S \to aA, A \to aAB, B \to b, A \to a\}$ with starting symbol S

(*b*) $P = \{S \to b, S \to aA, A \to b, A \to aA, A \to bB, B \to a, B \to aB\}$

(*c*) $P = \{S \to aAB, AB \to a, A \to b, B \to AB\}$

(*d*) $P = \{S \to b, S \to aA, A \to b, A \to aA, A \to bB, B \to a, B \to aB\}$

Solution (*a*) G is a context free or Type 2 grammar since each production is of the form $A \to a$, i.e. a single non-terminal symbol on the left and right side is a word in one or more symbols.

(*b*) G is a regular or Type 3 grammar since each production is of the form $A \to a$ or $A \to aB$, i.e. a single non-terminal symbol on the left and the right side is either a single terminal symbol or a terminal symbol followed by a non-terminal symbol.

(*c*) G is a grammar of Type 0, since one of the production is $AB \to a$.

(*d*) Length of left side of each production does not exceed the length of the right side, so G is a context-sensitive or Type 1 grammar.

Definition A language is said to be of Type *k* language for $k = 0, 1, 2, 3$ provided it is specified by a Type *k* grammar only.

For example, the language $L = \{a^n b^n : n \geq 1\}$ is a Type 2 language because it can only be specified by Type 2 grammar with productions of the form: $A \to aAb$, and $A \to ab$.

Each type of grammar corresponds to a specific type of machine which accepts or produces a language of given type. Table 15.1 shows Chomsky hierarchy between type of grammars, their characteristics and machines.

Table 15.1 Chomsky Hierarchy

Grammar Type	*Characteristics of Grammar*	*Machine Type*
Type 0	Phase-structure grammar or Unrestricted	Turning machine
Type 1	Context-sensitive	Linear bounded automata
Type 2	Context-free	Pushdown automata
Type 3	Regular or Right linear	Finite state machine (automata)

We shall discuss turning machines and finite state automata (FSA) latter in this chapter.

Example 21 Consider the language $L = \{a^n b^m : n \neq m\}$. Find context-free grammar G which generates L.

Solution When $n = m$, the grammar G with productions $S \to ab$, $S \to aSb$ will generate L.

When $n > m$, we first generate a string with equal number of *a*'s and *b*'s and then add these extra *a*'s on the left. This gives $S \to AS_1$, $S_1 \to aS_1b$, $S_1 \to \lambda$, $A \to aA$, $A \to a$.

When $n < m$, the set of productions with the same reasoning is: $S \to AS_1$, $S \to S_1B$, $S_1 \to aSb$, $S_1 \to \lambda$, $A \to aA$, $A \to a$, $B \to bB$, $B \to b$.

In all these cases G is context-free language since each left side is a single non-terminal symbol.

15.5.4 Derivation Trees of Context-free Grammar

For a context-free (Type 2) grammar G, any derivation helps to interpret a string (word) in L (G). A derivation is represented graphically by means of an ordered rooted tree T, called a *derivation tree* or *parse tree*. The root of the tree represents the non-terminal starting symbol, whereas vertices of the tree are labelled with the left sides of a production. The interior vertices correspond to non-terminal symbols that arise in the derivation. The leaves of the tree represent the terminal symbols.

For example, if the production $A \to w$ arise in the derivation, then the vertex that represents A has children vertices that represent symbols in the word, in order from left to right. For the production $A \to \lambda$, vertex labelled A has only one child (vertex) λ, i.e. a vertex labelled λ has no siblings (children).

The strings of symbols obtained by reading the leaves of the tree from left to right, ignoring λ's, is said to be the *yield* of the tree.

Remark Derivation trees exist only for context-tree (Type 2) and regular (Type 3) grammars.

Example 22 Consider the grammar G with

$V = \{S, A, B\}$, $\Sigma = \{a, b\}$ and $P = \{S \to aAS, S \to a, A \to SbA, A \to ba\}$

Construct the derivation tree of the word (*i*) *abba*, (*ii*) *aabbaa*

Solution (*a*) The word *abba* can be derived as follows:

$$A \Rightarrow SbA \Rightarrow abA \Rightarrow abba$$

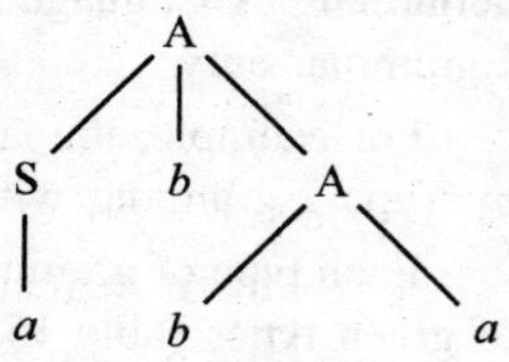

Fig. 15.1

Figure 15.1 shows this derivation.

(*b*) The word *aabbaa* can be derived as follows

$$S \Rightarrow aAS \Rightarrow ASbAS \Rightarrow aabAS \Rightarrow aabAa \Rightarrow aabbaa$$

Example 23 Figure 15.2 is the derivation tree of a string *w* in a language L of a context-tree grammar G. Find (*a*) the yield, and (*b*) which terminals, non-terminals and production must belong to G?

Solution (*a*) The yield of the tree is the string of symbols obtained by reading leaves of the tree from left to right, i.e. $w = ababbb$

(*b*) Terminal symbol $\Sigma = \{a, b\}$, non-terminal symbols $V = \{S, A, B\}$ and production $P = \{S \to aAB, A \to Bab, B \to bB, B \to b\}$.

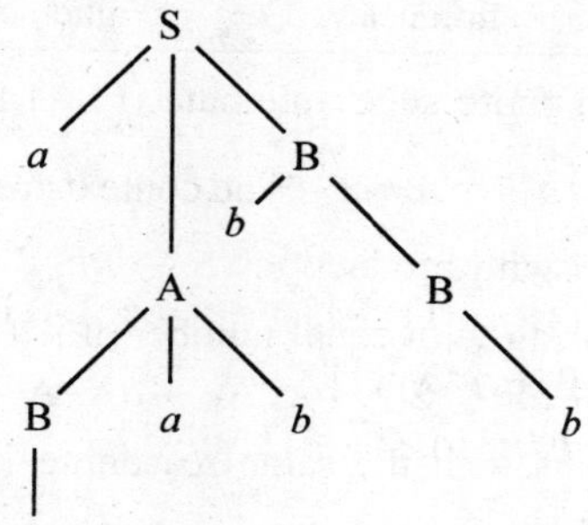

Fig. 15.2

Self-practice Problems II

27. Consider the grammar G with $V = \{S, A, B\}$, $\Sigma = \{a, b\}$, and $P = \{S \to AB, S \to bA, A \to a, B \to ba\}$. Find L(G).

28. Consider the grammar G with $V = \{S, 0, 1\}$, $\Sigma = \{0, 1\}$ and $P = \{S \to 11S, S \to 0\}$. Find L (G).

29. Determine the type of grammar

(*a*) $S \to bA, A \to aA, B \to aB, B \to b, A \to a$

(*b*) $S \to aA, A \to bB, A \to b, A \to a, B \to aA, B \to a$

(*c*) $S \to bA, A \to BA, bB \to a, B \to b, A \to a$

(*d*) $S \to bAB, A \to aB, B \to a, A \to bB$

30. Find the grammar G that generates the language L consisting of all words in a and b such that no two a's appear together.

31. Find a grammar G that generates the language L consisting of all words of the form $b^n a b^n$ for $n \geq 0$.

32. Write each grammar in Problem 3 in Backus-Naur form.

33. Consider the context-free grammar with productions $S \to bAS$, $S \to b$, $A \to aS$.

Find the derivation tree of the word $w = babbabb$.

34. Let $V = \{S, A, B\}$, $\Sigma = \{a, b, c\}$ consists of all words in a, b and c such that a's followed by b's followed by c's.

(*a*) Find a regular expression r such that $L = L(r)$.

(*b*) Find a regular grammar that generates the language L.

35. Construct context-tree grammar to generate the following languages

(*a*) $\{0^n 1^n : m \neq n, n \geq 1\}$ (*b*) $\{0^m 1^n : 1 \leq m \leq n\}$

36. Construct regular grammar to generate the following languages

(*a*) $\{a^{2n} : n \geq 1\}$ (*b*) $\{a^l b^m c^n : l, m, n \geq 1\}$ (*c*) $\{(ab)^n : n \geq 1\}$

37. Show that the grammar $G_1 = \{S\}, \{a, b\}, P_1, S\}$ is equivalent to the grammar $G_2 = (\{S, A, B, C\}, \{a, b\}, \{P_2, S\}$ where $P_1 = \{S \to aSb, S \to ab\}$, $P_2 = \{S \to AC, C \to SB, S \to AB, A \to a, B \to b\}$

38. Construct a grammar to generate the language

$$L(G) = \{(ab)^n : n \geq 1\} \cup \{(ba)^n : n \geq 1\}$$

Hints and Answers

27. $S \Rightarrow AB \Rightarrow aB \Rightarrow aba$. Also $S \Rightarrow bA \Rightarrow ba$. Thus $L(G) = \{abc, ba\}$

28. $S \Rightarrow 11S \Rightarrow 110$; $S \Rightarrow 11S \Rightarrow 1111S \Rightarrow 11110$, and so on. $L(G) = \{0, 110, 11110, \ldots\}$

29. (*a*) Type 2, (*b*) Type 3, (*c*) Type 0, (*d*) Type 1.

30. $S \to (a, b, aB, bA)$, $A \to (bA, ab, a, b)$, $B \to (b, bA)$

31. $S \to aSa$, $S \to b$

32. (*a*) <S> : : = *a* <A>, <B> : : = *a* <B> | *b*, <A> : : = *a* <A> | *a*

(*b*) <S> : : *a* <A>, <A> : : = *b* <B> |*b*| *a*, <B> : : = *a* <A>|*a*

(*c*) <S> : : = *b* <A> <B>, <A> : : = *a* <B> | *b* <B>, <B> : : = *a*

33.

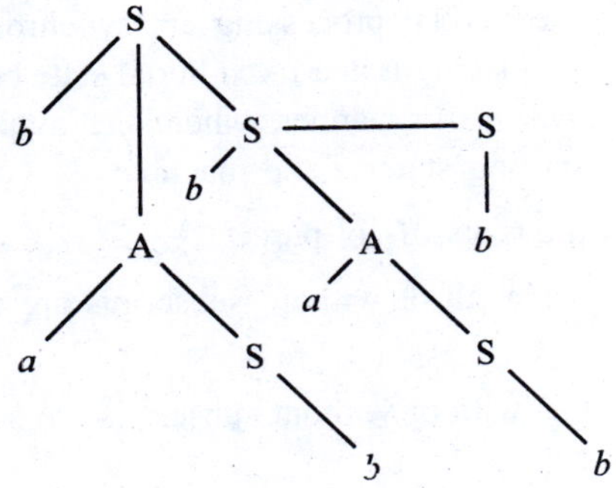

34. (*a*) Let $r = aa^*\ bb^*\ cc^*$. Then $L = L(r)$

(*b*) The regular grammar with the following productions which generates language L

$$S \to aA,\ A \to aA,\ A \to bB,\ B \to bB,\ B \to c,\ B \to cC,\ C \to c,\ C \to cC.$$

35. (*a*) $G = (\{S, A, B\}), \{0, 1\}, P, S\})$, where $P = \{S \to 0S1,\ S \to 0A,\ S \to 1B,\ S \to 0,\ S \to 1,\ A \to 0A,\ A \to 0,\ B \to 1B,\ B \to 1\}$. Using $S \to 0S1$, $S \to 0$, and $S \to 1$ repeatedly, we can get 0^m1^n where m and n differ by 1. To get more 0's than 1's in the string, we should apply productions $A \to 0A$, $A \to 0$, $S \to 0A$, repeatedly. Similarly, to get more 1's, apply $S \to 1B$, $B \to 1B$ and $B \to 1$ repeatedly.

36. (*a*) $S \to aA,\ A \to aS,\ S \to aB,\ B \to a$

(*b*) $S \to aA,\ A \to aA,\ A \to bB,\ B \to bB,\ C \to cC,\ C \to c$

(*c*) $S \to aA,\ A \to bS,\ S \to aB,\ B \to b$

37. $L(G_1) = \{a^nb^n : n \geq 1\}$. In grammar G_2, the productions are: $S \Rightarrow AC \Rightarrow ASB \Rightarrow A^{n-1} \mid SB^{n-1}$. Also $S \Rightarrow AB \Rightarrow ab$. Hence, $S \Rightarrow a^nb^n$ for all $n \geq 1$. Thus $L(G_2) = \{a^nb^n : n \geq 1\} = L(G_1)$.

38. To get the language L (G), apply the productions: $S \to A,\ A \to abA,\ A \to ab,\ S \to B,\ B \to baB,\ B \to ba$.

15.6 FINITE-STATE MACHINE

A finite-state machine (FSM) is an abstract model of a machine which accept discrete inputs, produces discrete output and has the internal memory to keep track of certain information about previous inputs. At any instant of time the machine can be in any of the finite number of states (also called configurations that summarises the past memory of inputs and react to subsequent inputs). The next state of the machine at any point in time is determined by the present state and subsequent input and produces an output.

The major features of a finite-state machine are as follows:

1. At any given time, the machine can be in only one of *finite number of states*. These states are called the *internal states* of the machine, and at a given time the total memory available to the machine is the knowledge of internal state it is in at that moment.
2. The machine accepts as input only a finite number of symbols. The finite set of all possible inputs constitute the *input alphabet* for the machine.
3. An *output* and a *next state* are determined by each combination of inputs and internal states. The finite set of all possible outputs constitute the *output alphabet* for the machine.
4. We assume that the machine has an internal clock, and the sequential processing are synchronised by separate and distinct clock pulses. The machine is initiated by placing it in a fixed initial state before the clock is started and that the machine operates in a *deterministic* manner, where the output is completely determined by the total input provided and the starting state of the machine.

Thus a finite-state machine (or complete sequential machine) M consists of six parts:

(*i*) Σ, finite set of input alphabet. Symbols a_i ($i = 1, 2, \ldots, n$) with or without subscripts are input symbols, i.e. $\Sigma = \{a_1, a_2, \ldots, a_n\}$.

(*ii*) S, a finite set of *internal states*. Symbols s_i ($i = 1, 2, \ldots, n$) with or without subscripts are states. The states s_0 is the initial state, i.e. $S = \{s_0, s_1, s_2, \ldots, s_n\}$.

(*iii*) O a finite set of *output alphabet,* i.e. $O = \{x, y, z\}$.

(*iv*) f, a *transition (or next-state) function* or *mapping* from $S \times \Sigma$ into S, i.e. $f: S \times \Sigma \to S$. This function describes the effect of input on the states during transition.

(*v*) g a *output function* from $S \times \Sigma$ into O, i.e. $g : S \times \Sigma \to O$.

Such a machine is written as: $M = \{\Sigma, S, s_0, O, f, g\}$.

For example, if the machine is in state s_1 at time t_i and given the input a at this state, then the output at time t_i is $g(s_1, a) = x$. This output is followed by a transition of the machine at time t_{i+1} to the next internal state given by $f(s_1, a) = s_2$.

As mentioned earlier, we assume that the machine receives its first input at time t_0 and the machine is in its starting state s_0.

Illustration For the sake of simplicity, we define a finite-state machine M with two input symbols, three internal states and three output symbols:

(*i*) $\Sigma = \{a, b\}$ (*ii*) $S = \{s_0, s_1, s_2\}$ (*iii*) $O = \{x, y\}$

(*iv*) Transition (next-state) function $f : S \times \Sigma \to S$ defined by

$f(s_0, a) = s_0$ $f(s_1, a) = s_2,$ $f(s_2, a) = s_0$

$f(s_0, b) = s_1$ $f(s_1, b) = s_1$ $f(s_2, b) = s_1$

(*v*) Output function $g : S \times \Sigma \to O$ defined by

$g(s_0, a) = x$ $g(s_1, a) = x$ $g(s_2, a) = x$

$g(s_0, b) = x$ $g(s_1, b) = x$ $g(s_2, b) = y$

15.6.1 State Table and State Diagram of a Finite State Machine

A finite-state machine M can be represented in compact form by means of

(*i*) a *state-table* also called *transition table of M,* and

(*ii*) a labelled directed graph called the *state diagram* or *transition diagram of M.*

The state table combines the state function f and the output function g for all pairs of states and inputs. The first column of the table list the (present) states for the machine. Entries in the second row are the elements of the input alphabet Σ listed once under f and then again under g. The six numbers in the last two columns (and last three rows) are elements of the output alphabet O.

Table 15.2 State Transition Table

State	*f (Input)*		*g (Output)*	
	a	*b*	*a*	*b*
s_0	s_0	s_1	x	x
s_1	s_2	s_1	x	x
s_2	s_0	s_1	x	y

A state diagram helps us to know the output string directly without actually listing sequence of transition states. In this diagram each internal state s_i $(i = 1, 2, \ldots, n)$ is represented by a circle (also called vertex of a labelled directed graph) with s_i inside the circle. For states s_i and s_j of $f(s_i, a) = s_j$ for $a \in \Sigma$

and $g(s_i, a) = x$ for $x \in O$, then such input and output functions are represented in the state diagram by drawing a *directed edge* (or *arc*) from circle s_i to the circle s_j and labelling the edge with the input a and output x as shown in Fig. 15.4.

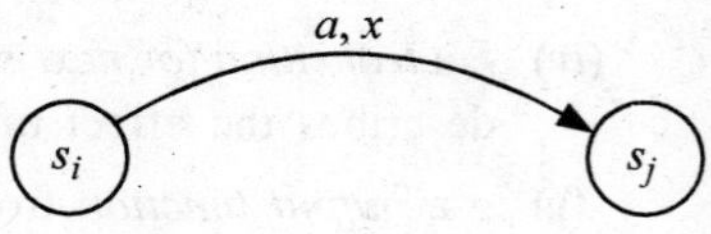

Fig. 15.4

With these conventions, the state diagram for the machine M of Table 15.2 is shown in Fig. 15.5.

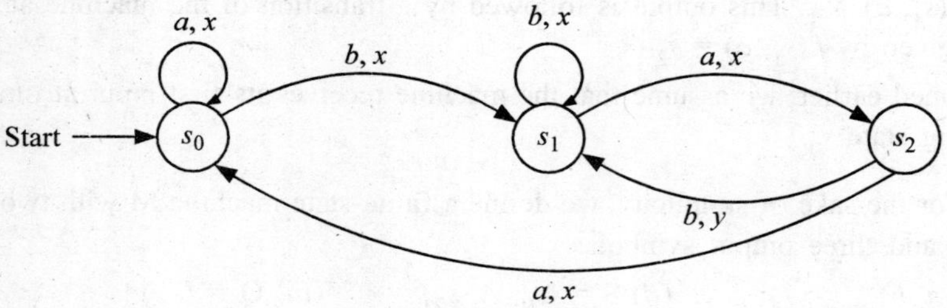

Fig. 15.5 State Diagram

15.6.2 Input and Output Tapes

To show the dynamic quality of a finite-state machine M, suppose M is given by a word (string) of input symbols: $u = a_1 a_2 \ldots a_m$.

If we call these symbols as 'input tape', the machine M reads these inputs symbols one by one and or simultaneously, move through a sequence of states: $v = s_0\, s_1\, s_2, \ldots, s_m$, while printing a string of output symbols : $w = x_1\, x_2\, x_3 \ldots$ on an output tape. Hence, the initial state s_0 and the input string u determine the strings v and w by

$$s_i = f(s_{i-1}, a_i) \text{ and } x_i = g(s_{i-1}, a_i) \text{ for all } i.$$

Example 23 Let M be the finite-state machine with state table given in the following page.

State	*f*		*g*	
	0	1	0	1
s_0	s_1	s_2	0	0
s_1	s_0	s_1	0	1
s_2	s_1	s_2	1	0
s_3	s_1	s_2	1	1

(*a*) Find the input set Σ, the state set S, the output set O and the initial state of M.

(*b*) Draw the state diagram of M.

(*c*) Find the output string of the input string 01001.

Solution (*a*) The input set $\Sigma = \{0, 1\}$, state set $S = \{s_0, s_1, s_2, s_3\}$, output set $O = \{0, 1\}$ and s_0 is the initial state.

(*b*) The state diagram of M is shown in Fig. 15.6.

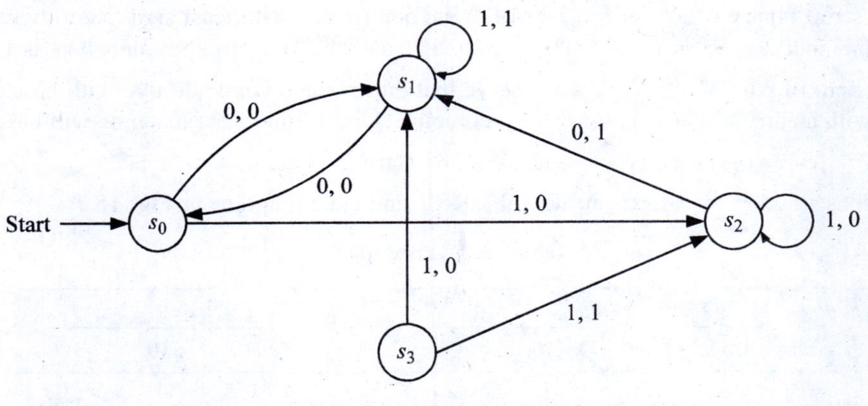

Fig. 15.6 State Diagram

(*c*) Beginning at the initial state s_0, the input string 01001 from state to state can be represented by arranging arrows as:

$$\longrightarrow s_0 \xrightarrow{0} s_0 \xrightarrow{1} s_2 \xrightarrow{0} s_1 \xrightarrow{0} s_0 \xrightarrow{1} s_2$$

This yields the following sequence of states and output words

$$v = s_0\, s_0\, s_2\, s_1\, s_0\, s_2 \text{ and } w = 01001.$$

15.6.3 Binary Addition

A finite-state machine M can also be used for adding binary numbers. Since writing 0's to the left of given number do not change the value of the number, we can assume that binary numbers to be added have the same number of digits. The rules of binary addition are:

$$0 + 0 = 0, \quad 1 + 0 = 0, \qquad 0 + 1 = 1$$
$$1 + 1 = 10 = 0 + \text{carry of 1 into next column}$$
$$1 + 1 + 1 = 11 = 1 + \text{carry of 1 into next column}$$

The last case occurs when bits in a certain position are 1 and there is a carry from the previous position.

Suppose 00111 and 01101 be two binary numbers, which are given as input to machine M Then

$$\begin{array}{r} 0\,0\,1\,1\,1 \\ +\,0\,1\,1\,0\,1 \\ \hline 1\,0\,1\,0\,0 \end{array}$$

Here when binary numbers 0 and 1 are added we carry 1 to the next column. So the sum is 10. If we assume that the digits of the two numbers to be added appear one by one with least significant digits coming in first, then the input is the string of pair of digits: 00, 01, 10, 11, depending upon whether we are seeking 0 + 0, 0 + 1, 1 + 0 or 1 + 1. The two digits in each pair are digits in the numbers being added. Thus, the set of input symbols (or alphabet) is Σ = {00, 01, 10, 11} and the set of output symbols (alphabet) O = {0, 1}.

Example 24 Design a finite-state machine that performs serial addition.

Solution The serial binary adder accepts pair of digits one by one with least significant digits come first. Thus, the set of input alphabets is Σ = {00, 01, 10, 11} and the set of output alphabets is O = {0, 1}.

Given an input of pair of digits *xy*, we observe that either *x* and *y* are additive with no carry or *x* and *y* are additive with a carry bit 0 (or 1) to the next column. Thus, a finite-state machine will have two states:

S = {carry(*c*), no carry(*n*)} with *n* as the starting state.

The functions *f* and *g* are shown in the Table 15.3 and state diagram in Fig. 15.7

Table 15.3 State Table

State	*f (Input)*				*g (Output)*			
	00	01	10	11	00	01	10	11
n	*n*	*n*	*n*	*c*	0	1	1	0
c	*n*	*c*	*c*	*c*	1	0	0	1

In Fig. 15.7 we observe that if input is 00 to *n*, then the output is 0 and remains in state *n*; if input is 11 to *n*, then there is a carry 1 and output is 0 and therefore there is a change in the state from n to *c*. When 11 is input to *c*, then 1 + 1 = 11 = 1, and remains in state *c*.

From Table 15.3, we observe, for example, *f* (*n*, 01) = *n* and *g* (*n*, 01) = 0 because *n* indicates a carry of 1 from the addition of the previous bits. The input 01 indicates that we are adding 0 and 1 (and carrying 1). Thus the sum is 10 and *g* (*n*, 01) = 0 for the 0 in 10. The carry is remembered in *n* = *f* (*n*, 01).

Example 25 Design a finite-state machine M where the input set is {0, 1, 2} and output set contains digits which are modulo 3 sum of the digits in the input sequence.

Solution The machine M that we design will have three states: S = {s_0, s_1, s_2}, where

s_0 = output is 0, i.e. modulo 3 sum of all input digits is 0

s_1 = output is 1, i.e. modulo 3 sum of all input digits is 1

s_2 = output is 2, i.e. modulo 3 sum of all input digits is 2.

The input and output functions are shown in Table 15.4 and state diagram in Fig. 15.8.

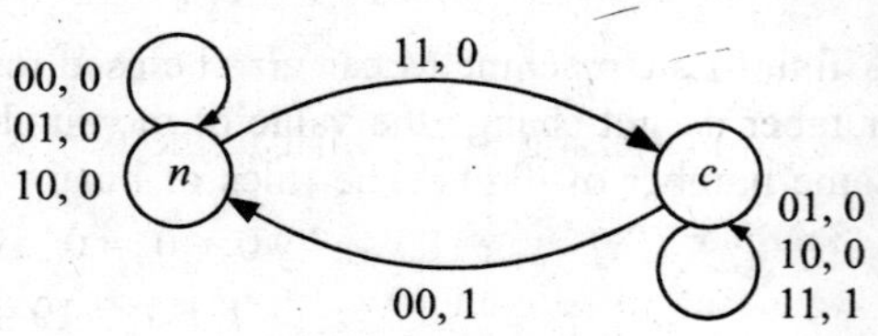

Fig. 15.7 State Diagram

Table 15.4 State Table

State	*f (Input)*			*g (Output)*		
	0	1	2	0	1	2
s_0	s_0	s_1	s_2	0	1	2
s_1	s_1	s_2	s_0	0	1	2
s_2	s_2	s_0	s_1	0	1	2

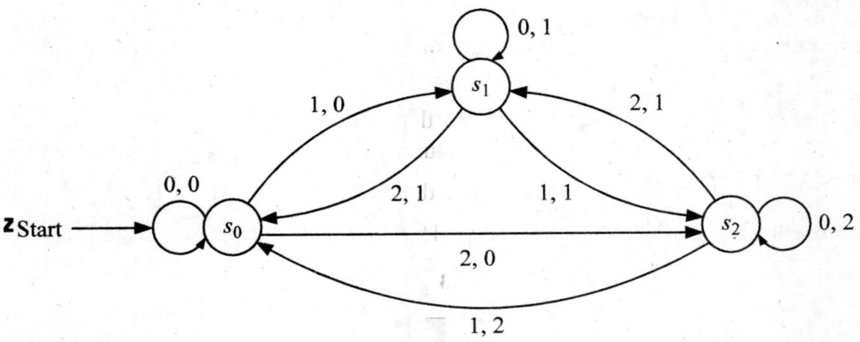

Fig. 15.8 State Diagram

Example 26 Let M be a finite-state machine with state table given below:

State	*f*			*g*		
	a	*b*	*c*	*a*	*b*	*c*
s_0	s_0	s_3	s_2	0	1	1
s_1	s_1	s_1	s_3	0	0	1
s_2	s_1	s_2	s_3	1	1	0
s_3	s_2	s_3	s_0	1	0	1

(*a*) Find the input set Σ, the state set S, the output set O and initial state of M.

(*b*) Draw the state diagram of M.

(*c*) Starting at s_0, what is the output for the input string *abbccc*?

Solution (*a*) The input set $\Sigma = \{a, b, c\}$, the state set $S = \{s_0, s_1, s_2, s_3\}$, the output set $O = \{0, 1\}$ and s_0 is the initial state.

(*b*) The state diagram of M is shown in Fig. 15.9.

(*c*) Starting with initial state s_0, the input string *abbccc* from state to state can be represented by labelled input symbols as follows:

$$\xrightarrow{\text{start}} s_0 \xrightarrow{a,\,0} s_0 \xrightarrow{b,\,1} s_3 \xrightarrow{b,\,0} s_3 \xrightarrow{c,\,1} s_0 \xrightarrow{c,\,1} s_2 \xrightarrow{c,\,0} s_3$$

This yields the following sequence *u* of states and output string *w* as:

$$u = s_0\, s_0\, s_3\, s_3\, s_0\, s_2\, s_3; \quad w = 010110.$$

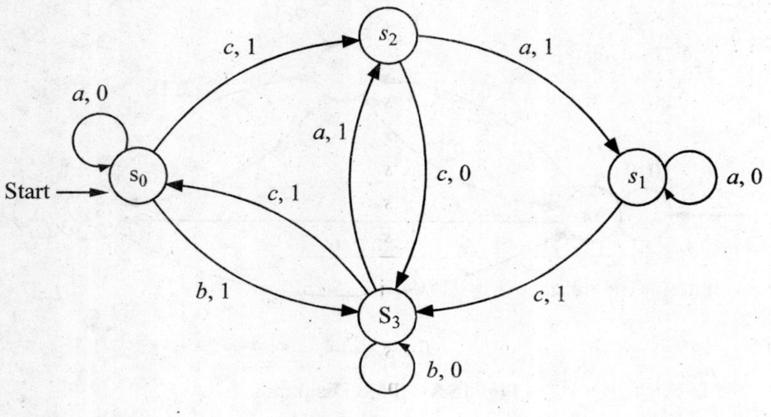

Fig. 15.9 State Diagram

Example 27 Given the state diagram shown in Fig. 15.10 of a finite-state machine M, find

(*a*) output string for the input string 110111, starting with initial state s_0.

(*b*) input set Σ, output set O and state set S

(*c*) state table for machine M

(*d*) states so that the input string 10010 produces the output 10000.

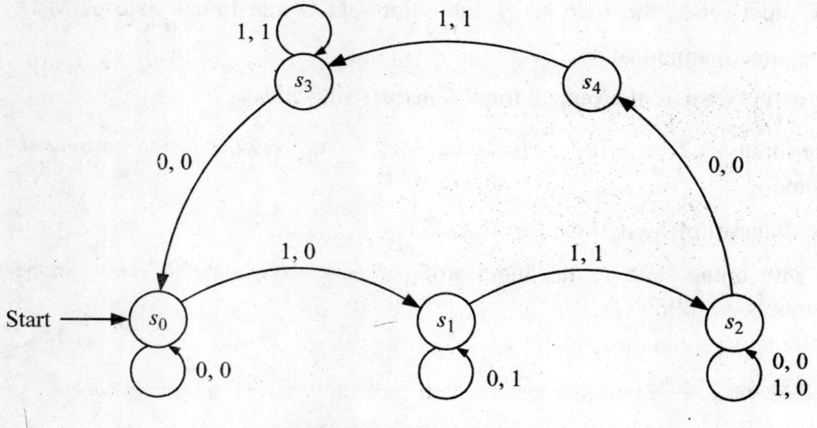

Fig. 15.10 State Diagram

Solution (*a*) Beginning at the initial state s_0, arrows which are labelled by input symbol is as follows:

$$\longrightarrow s_0 \xrightarrow{1,\,0} s_1 \xrightarrow{1,\,1} s_2 \xrightarrow{0,\,0} s_2 \xrightarrow{1,\,0} s_2 \xrightarrow{0,\,0} s_4 \xrightarrow{1,\,1} s_3$$

This yields the sequence w = 010001 of output string

(*b*) Input set $\Sigma = \{0, 1\}$, output set O = $\{0, 1\}$, state set S = $\{s_0, s_1, s_2, s_3, s_4\}$ with s_0 as initial state.

(*c*) The state table for machine M is shown in Table 15.5.

Table 15.5 State Table

State	*f*		*g*	
	0	1	0	1
s_0	s_0	s_1	0	0
s_1	s_1	s_2	1	1
s_2	s_2	s_2	0	0
s_3	s_0	s_3	0	1
s_4	s_2	s_3	0	1

(*d*) The sequence of states with input string 10010 is as follows:

$$\longrightarrow s_1 \xrightarrow{1,\,1} s_2 \xrightarrow{0,\,0} s_2 \xrightarrow{0,\,0} s_4 \xrightarrow{1,\,0} s_3 \xrightarrow{0,\,0} s_2$$

This yields the sequence w = 10000 of output string.

15.7 FINITE-STATE AUTOMATA

A finite-state automation is a special type of finite-state machine, in which there is no output, and some of its states are distinguished as accepting states. There are two types of automaton.

(*a*) Deterministic finite-state automaton

(*b*) Non-deterministic finite-state automaton

15.7.1 Deterministic Finite-state Automaton (DFA)

A deterministic finite-state automaton or simply an automaton M consists of five parts.

1. A finite non-empty set Σ of inputs, called alphabet.
2. A finite non-empty set S of (internal) states.
3. A next-state function (also called *transition function*) *f* from S × Σ into S, i.e. $f(s, a)$ is a state for each state s and input symbol a. This function or mapping describes the change of states during the transition. It is usually represented by a state transition table or a state transition diagram.
4. A subset F of set S of accepting (final) states.
5. An initial state $s_0 \in$ S.

Such a finite automaton is denoted by M = (Σ, S, F, s_0, f).

State Transition Diagram A typical state transition diagram (or system) is shown in Fig. 15.11. This diagram is a finite directed labelled graph so that

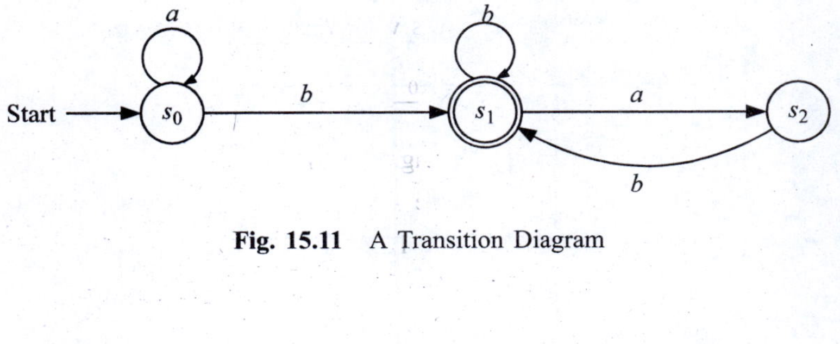

Fig. 15.11 A Transition Diagram

(*i*) Each vertex (node) represents a state,

(*ii*) The directed edges are represented by arrows and indicate the transition from state s_i to state s_j and labelled with an input a if $f(s_i, a) = s_j$.

(*iii*) The initial state s_0 is represented by a circle with an arrow pointing towards it. There is one final state indicated by double circles, and the other states are represented by just a circle.

Definition A string w is said to be accepted by a finite automaton M provided.

(*a*) there exists a path which originates from some initial state, goes along the arrows and terminates at some final state, and

(*b*) the path value obtained by concatenation of all edge-labels of the path is equal to w.

Properties of Transition Function Let $f(s_i, w) = s_j$ be a unique state such that there is a path in the transition diagram from s_i to s_j. Thus, we define

(*a*) $f(s, \lambda) = s$, where λ denotes no input. That is, without reading an input symbol, the finite automaton cannot change state.

(*b*) For all strings w and input symbols a,

$$f(s, wa) = f(f(s, w), a)$$

That is, to find the state after reading a non-empty input string wa, find the state $f(s,w)$ after reading w.

Trap State A state from which a *dfa* can never come out is called a trap state.

Example 28 Consider the transition diagram shown in Fig. 15.12.

(*a*) Find its states

(*b*) Find its input symbols

(*c*) Find its initial state

(*d*) Find its accepting states

(*e*) Find $f(s_2, 1)$

(*f*) Write its next-state table.

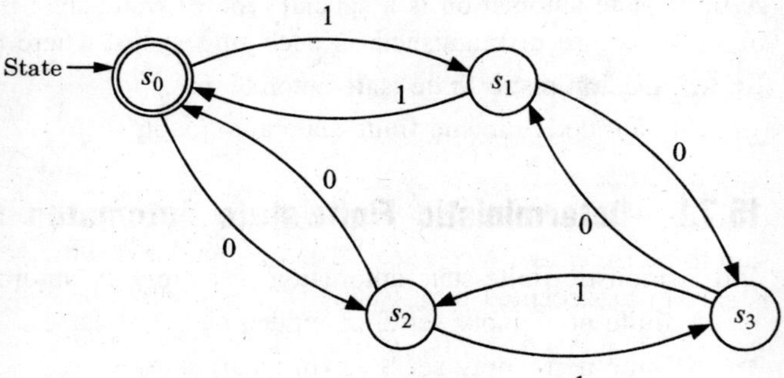

Fig. 15.12

Solution (*a*) The set of states S = (s_0, s_1, s_2, s_3) because these are the labels of the circles.

(*b*) The set of input alphabet $\Sigma = \{0, 1\}$, because symbols 0 and 1 are the labels of arrows.

(*c*) The initial state is s_0, since unlabelled arrow points to s_0.

(*d*) The accepting state is s_0, since it is marked with double circle.

(*e*) $f(s_2, 1) = s_3$, since there is arrow from s_2 to s_3 labelled 1.

(*f*) The next-state table is shown in Table 15.6.

Table 15.6 Next-state Table

State	*f (Inputs)*	
	0	1
s_0	s_2	s_1
s_1	s_3	s_0
s_2	s_0	s_3
s_3	s_1	s_2

Example 29 Consider the finite automaton whose transition function f is given in Table 15.6 where $S = \{s_0, s_1, s_2, s_3\}$, $\Sigma = \{0, 1\}$, $F = \{s_0\}$. Give the entire sequence of states for the input string 110101.

Solution From transition Table 15.6 we note that $f(s_0, 1) = s_1$ and $f(s_1, 1) = s_0$. Thus

$$f(s_0, 11) = f(s_0, 1), 1) = f(s_1, 1) = s_0$$

Similarly $f(s_0, 110) = f(f(s_0, 11), 0) = f(s_0, 0) = s_2$

State	f (Inputs)	
	0	1
$\rightarrow$ (s_0)	s_2	s_1
s_1	s_3	s_0
s_2	s_0	s_3
s_3	s_1	s_2

Continuing in this manner, we find that $f(s_0, 1101) = s_3$, $f(s_0, 11010) = s_1$ and finally $f(s_0, 110101) = s_0$.

The entire sequence of states is: $s_0 \xrightarrow{1} s_1 \xrightarrow{1} s_0 \xrightarrow{0} s_2 \xrightarrow{1} s_3 \xrightarrow{0} s_1 \xrightarrow{1} s_0$.

Remark Finite automaton accepts all strings of 0's and 1's in which numbers of 0's and the number of 1's are even.

15.7.2 Language Recognised by an Automaton

When a string of input symbols is fed into a finite-state automaton in sequence, the states of automaton changed after every successive input symbols. The final state of automaton determines whether the string is accepted by a finite-state automaton or not. A string that sends the automaton to an accepting state is said to be *accepted* (*recognised*) by the automaton, otherwise said to be rejected. Mathematically, a string is said to be accepted by a finite-state automaton $M = (\Sigma, S, F, s_0, f)$ when

$$f(s_0, a) = s_m \text{ for some } s_m \in F.$$

That is, M recognise the string if the final state s_m is an accepting state in F.

The *language accepted* by M, denoted by L(M) is the set of all strings that are accepted by M, i.e.

$$L(M) = \{x : f(s_0, x) \in F\}.$$

A language L is said to be a *finite state language* when a finite-state machine accepts all its sentences. Thus a language L is the finite state language if there exists a finite-state machine M that accepts a sentence w if and only if $w \in L$.

(Kleene) Theorem 15.1 A language L over an alphabet is said to be regular, denoted by L (M) if and only if there exists a finite-state automaton M such that L = L (M).

Illustration The transition diagram shown in Fig. 15.13 shows that the language $L = \{awa : w \in S\}$, where input set $\Sigma = \{a, b\}$ is regular or create a *dfa*. That is, the string that starts with a and ends with a is accepted.

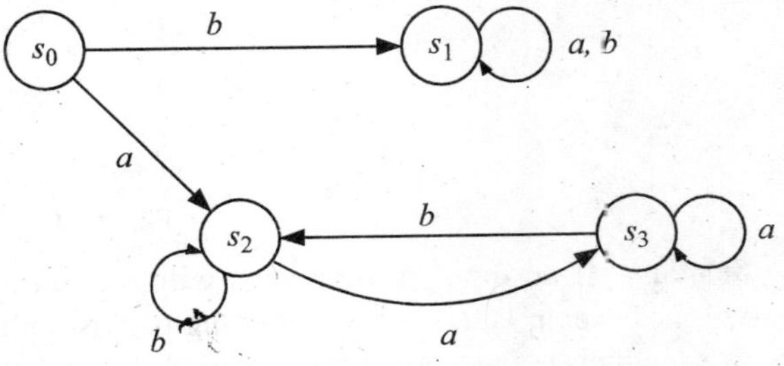

Fig. 15.13 Transition Diagram

Example 30 Consider the transition diagram shown in Fig. 15.14.

(*a*) To what state does the automaton reach when the symbols of the following strings are input to it in sequence starting from the initial state?

(*i*) $w_1 = 1001$ (*ii*) $w_2 = 010110$

(*b*) Which of the strings in part (*a*) are accepted by M?

(*c*) What is the language accepted by M?

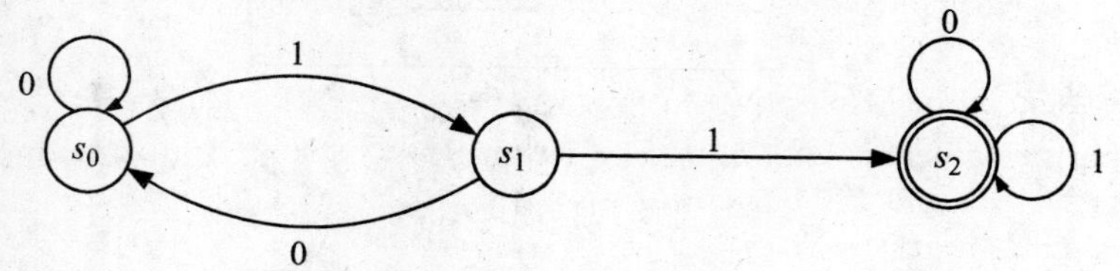

Fig. 15.14 Transition Diagram

Solution (*a*) (*i*) The string $w_1 = 1001$ is determined by the path

$$s_0 \xrightarrow{1} s_1 \xrightarrow{0} s_0 \xrightarrow{0} s_0 \xrightarrow{1} s_1$$

(*ii*) The string $w_2 = 010110$ is determined by the path

$$s_0 \xrightarrow{0} s_0 \xrightarrow{1} s_1 \xrightarrow{0} s_0 \xrightarrow{1} s_1 \xrightarrow{1} s_2 \xrightarrow{0} s_2$$

Since the final state s_2 is an accepting state, therefore string w_1 is not accepted, while w_2 is accepted.

(*b*) The language L (M) will consists of all words *w* from Σ which do not have two successive 00's. This is clear from the fact that the states s_1 and s_2 are rejecting states.

Example 31 Describe the language L = L (M) accepted by *dfa* whose transition diagram is shown in Fig. 15.15.

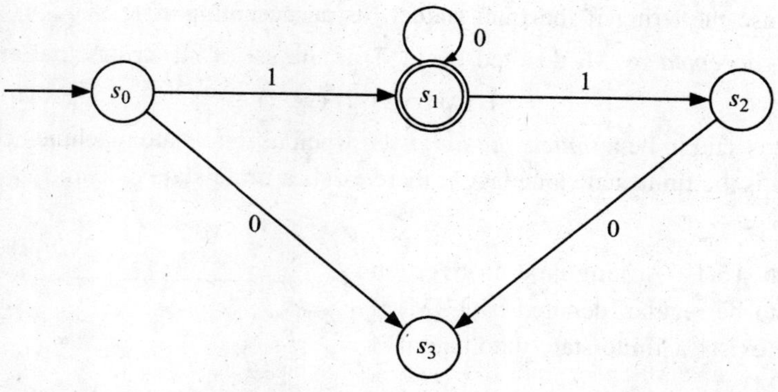

Fig. 15.15 Transition Diagram

Solution If an input string begins with 0, then machine move to state s_3, a non-terminating state and it will never exit. But, if the input string begins with symbol 1, then it terminates at s_3. Otherwise terminates in s_1, which is the only accepting state. Hence, strings of 0 and 1 symbols that begin with 1 is accepted.

Example 32 Draw the transition diagram of a finite-state automaton M that accepts those strings from $\Sigma = \{a, b\}$ which have even number of *a*'s, and (*b*) exactly one *a*.

Solution The finite-state automaton M needs to have two states.

s_0 : start state when the input string contains even number of *a*'s (a accepting state)

s_1 : next state when the input string contains odd number of *a*'s.

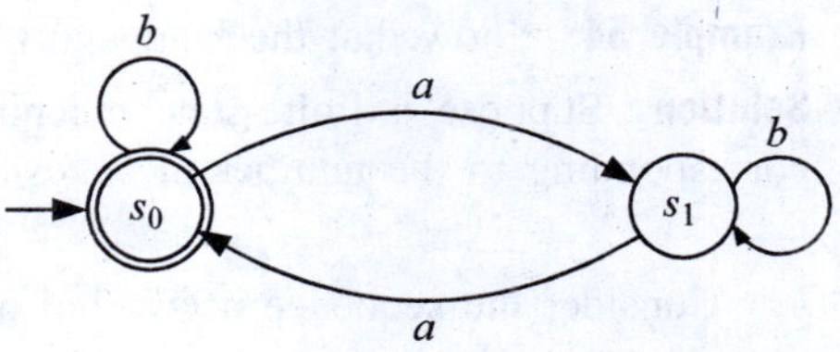

Fig. 15.16 Transition Diagram

The transition diagram of M is shown in Fig. 15.16.

If the input string begins with *b*, then M terminates at s_0 only. But if the input string begins with *a*, then M will move to s_1 and will come back to s_0 when s_1 has an input a. Hence, the string containing even number of *a*'s is accepted by M.

(*b*) The desired automaton M needs to have three states s_0, s_1 and s_2 as shown in Fig. 15.17.

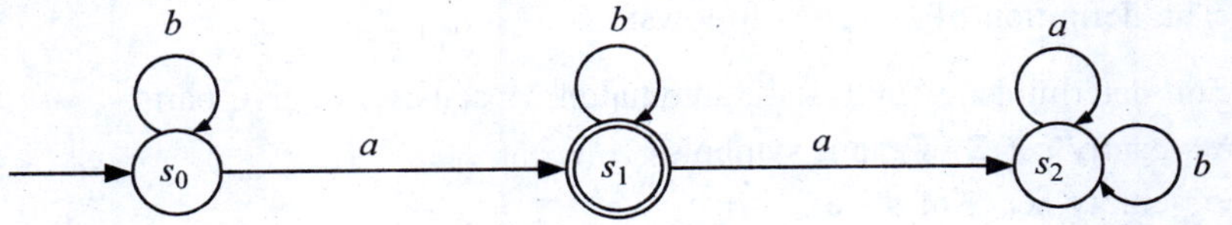

Fig. 15.17 Transition Diagram

If the input string begins with *b* at s_0, then it will terminate at s_0. But, if the input string begins with *bab*, then M will move to s_1 and will terminate at s_1, which is the accepting state. However, if the input string is *baba*, then M will move to s_2. Since M cannot return to the accepting state s_1 once reaching to s_2, therefore input string contains exactly one *a*.

Example 33 Draw a transition diagram of a finite-state automaton M that accepts binary sequences that ends with the digits 011 over the set $\Sigma = \{0, 1\}$.

Solution The automaton M must have three states as shown in the Fig. 15.18.

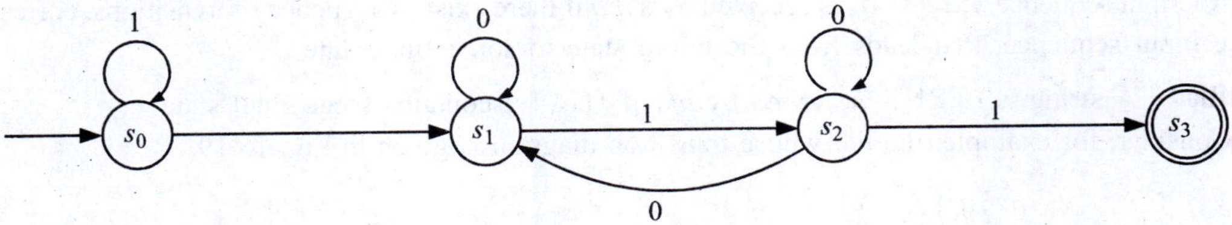

Fig. 15.18 Transition Diagram

Let any sequence of digits in Σ that ends with 011 be $w = x_1x_2 \ldots x_i\ 011$.

If M is in state s_0 and the input string is *x*, then it will move to states s_1, s_2 and s_3 as shown in Fig. 15.18. Suppose M is in state s_1 after receiving the input $x_1x_2 \ldots x_i$. If the next input symbol is 0, then M will remain in s_1 as shown in Fig. 15.18 but as soon as next input symbol is 1, M moves to s_2 and then to s_3 on receiving another input symbol 1. Since state s_3 is the accepting state, therefore string *w* is accepted by M.

Example 34 Show that the language $L = \{a^m : m = n^2, n \geq 1\}$ is not a finite-state language.

Solution Suppose a finite-state machine M consists of N states accepts the given language L. Now corresponding to the number of states N, we choose a sufficiently large integer number n such that

$$(n + 1)^2 - n^2 > N$$

Consider the sequence of symbol of $(n + 1)^2$ a's as follows:

$$aaa \ldots aaa \ldots aaa.$$

Since number of a's between $(n + 1)^2$ and n^2 are more than N, the machine M will come to a state s_i more than once. Thus, it is required to remove certain number of a's between $(n + 1)^2$ and n^2 so that a sequence so obtain will be accepted by M. But this sequence of symbols does not belong to L, as it contains a's such that $(n + 1)^2 \geq aaa \ldots \geq n^2$. Hence L is not a finite-state language.

15.7.3 Non-deterministic Finite-state Automaton

A non-deterministic finite-state automaton (*nfa*) allows zero, one or more transitions from a state on the same input symbol. The definition of *nfa* is as follows:

Definition A non-deterministic finite-state automaton M consists of five parts:

1. A finite non-empty set Σ of input symbols.
2. A finite non-empty set S of states.
3. A next-state function (or mapping) f from $S \times \Sigma$ into P (S), which is the power set of S, i.e. the set of all subsets of S which are equal to 2^S.
4. A subset F of set the S of accepting (final) states.
5. An initial state $s_0 \in S$.

Observe that the difference between the *dfa* and *nfa* is only in terms of next-state function f. For *dfa*, the outcome is a state, i.e. an element of S whereas for *nfa* the outcome is a subset of S. In *dfa*, for a given input string w and state s_i, there will be exactly one path labelled w starting at s_i. To determine whether a string is accepted by a *dfa*, it suffices to check this one path. For an *nfa* there may be many paths labelled w and all must be checked to see whether one or more terminate at a final state.

An input sequence $a_1a_2 \ldots a_n$ is accepted by a *nfa* if there exists a sequence of transitions, corresponding to the input sequence, that leads from the initial state to some final state.

Definition A string $w \in \Sigma^*$ is *accepted* by *nfa* if $f(s_0, w)$ contains some final state.

Consider, for example, the *nfa* whose transition diagram is given in Fig. 15.19.

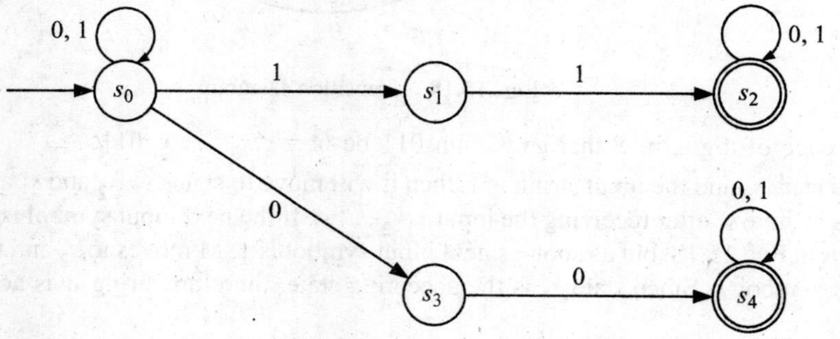

Fig. 15.19 Transition Diagram

The sequence of states for the input string w = 01001 is accepted by *nfa* because there exists a sequence of transitions through the states: $s_0 \xrightarrow{0} s_0 \xrightarrow{1} s_0 \xrightarrow{0} s_3 \xrightarrow{0} s_4 \xrightarrow{1} s_4$. This *nfa* accepts all strings with either two consecutive 0's or two consecutive 1's.

The state transition table defining the next-state function for *nfa* is shown in Fig. 15.20 is shown in Table 15.7.

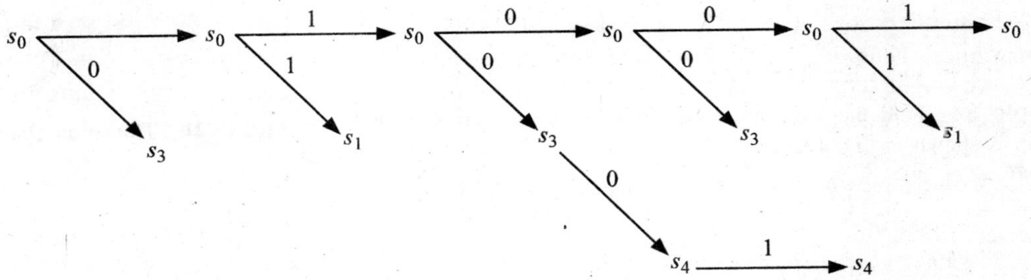

Fig. 15.20 States of NFA while Processing w = 01001

Table 15.7 Function for NFA

State	f: 0	f: 1
s_0	$\{s_0, s_3\}$	$\{s_0, s_1\}$
s_1	ϕ	$\{s_2\}$
s_2	$\{s_2\}$	$\{s_2\}$
s_3	$\{s_4\}$	ϕ
s_4	$\{s_4\}$	$\{s_4\}$

For the given initial state s_0, state set S $=\{s_0, s_1, s_2, s_3, s_4\}$ and input set $\Sigma = \{a, b\}$, the entries in the state transition Table 15.7 can be verified as:

$$f(s_0, 0) = \{s_0, s_3\}$$
$$f(s_0, 01) = f(f(s_0, 0), 1\} = f(\{s_0, s_3\}, 1) = f(s_0, 1) \cup f(s_3, 1) = \{s_0, s_1\}$$

Similarly, we compute

$$f(s_0, 010) = (s_0, s_3); f(s_0, 0100) = \{s_0, s_3, s_4\}, \text{ and}$$
$$f(s_0, 01001) = \{s_0, s_1, s_4\}.$$

Example 35 For a non-deterministic finite-state automaton shown in Fig. 15.21, find (*a*) initial state, (*b*) state set, (*c*) input set, and (*d*) state table defining the next-state function.

Solution (*a*) Initial state is s_0,

(*b*) State set S $= \{s_0, s_1, s_2\}$

(*c*) Input set $\Sigma = \{0, 1\}$

(*d*) The state table showing the next-state function is given in Table 15.8.

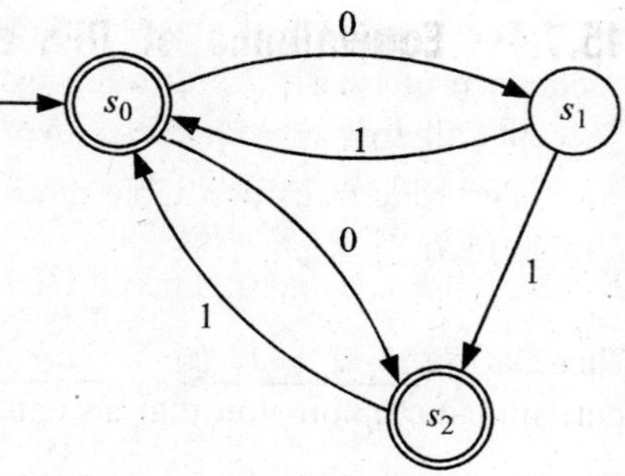

Fig. 15.21 Transition Diagram

Table 15.8

	f	
State	*0*	*1*
s_0	$\{s_0, s_1, s_2\}$	$\{s_1, s_2\}$
s_1	ϕ	$\{s_1, s_2\}$
s_2	ϕ	ϕ

Example 36 Find the language recognised by *nfa* whose transition diagram is shown in Fig. 15.22.

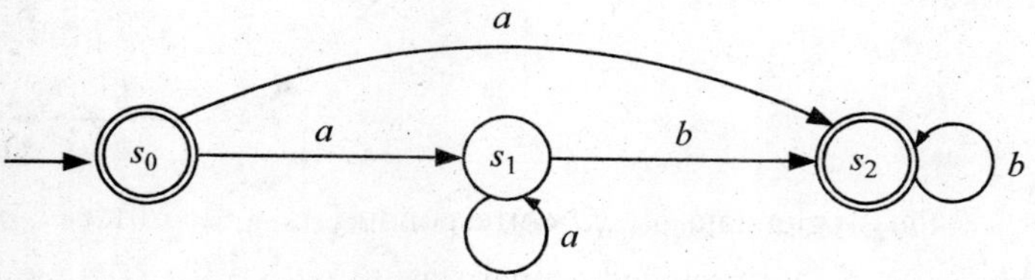

Fig. 15.22 Transition Diagram

Solution Since the initial state is also an accepting state, the language accepted by *nfa* is λ. Another accepting state is s_0. To reach up to accepting state s_2 we may follow two paths:

(*i*) $s_0 \xrightarrow{a} s_2$ and (*ii*) $s_0 \xrightarrow{a} s_1 \xrightarrow{b} s_2$.

Thus the language accepted by M is

$$L(M) = \{\lambda, 0\} \cup \{a^m\, b^n : m, n \geq 1\}$$

Example 37 Draw a transitive diagram of the *nfa* whose transition function table is given below:

	f	
State	*0*	*1*
s_0	$\{s_0, s_1\}$	$\{s_2\}$
s_1	ϕ	$\{s_1\}$
s_2	$\{s_1, s_2\}$	ϕ

Solution The transition diagram of *nfa* is shown in Fig. 15.23.

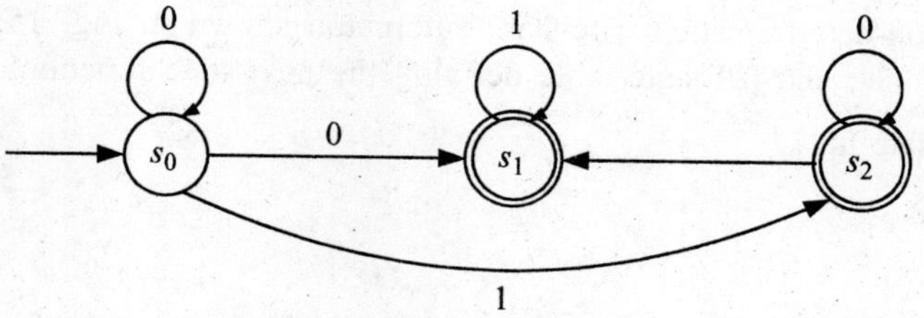

Fig. 15.23 Transition Diagram

15.7.4 Equivalence of DFA and NFA

The equivalence between *dfa* and *nfa* is as follows:

1. A *dfa* can simulate the behaviour of *nfa*, i.e. for each *nfa* we can construct an equivalent *dfa* (one which accepts the same language) by increasing the states of *dfa* to correspond to set of states of *nfa*.
2. Any *nfa* is a more general machine without being more powerful.

Theorem 15.1 If L is the set of states accepted by non-deterministic automaton, then there exists a deterministic automaton that accepts L.

Proof Let $M = \{\Sigma, S, f, s_0, F\}$ be a *nfa* accepting L. Then we construct a *dfa* $M' = \{\Sigma, S', f', S', s_0', F'\}$ such that

(*i*) The states of M′ are all subsets of the set of states of M, i.e. $S' = 2^s$. The elements of the (states) in S′ are denoted by $S' = [s_1, s_2, \ldots, s_i]$ where $s_1, s_2, \ldots, s_i$ belong to S.

(*ii*) $s_0' = [s_0]$

(*iii*) F′ is the set of all subsets of S containing an element of F.

(*iv*) $[s_1, s_2, \ldots, s_i]$ is the single state of *dfa* corresponding to a set of states of *nfa*.

Observe that a string $w \in L(M)$ provided a final state is one of the possible states that reaches on processing the string *w*. Thus, a final state in *nfa* (i.e. an element of F′) is any subset of S containing some final state of *nfa*.

Let us now define f'

$$f'([s_1, s_2, \ldots, s_i], a) = f(s_1, a) \cup f(s_2, a) \ldots \cup f(s_i, a) = \{t_1, t_2, \ldots, t_j\}$$

if and only if

$$f(\{s_1, s_2, \ldots, s_i\}, a) = \{t_1, t_2, \ldots, t_j\}$$

i.e. f' applied to an element $[s_1, s_2, \ldots, s_i]$ of S′ is computed by applying f to each state of S represented by $[s_1, s_2, \ldots, s_i]$. By applying f to each of $s_1, s_2, \ldots, s_i$ and taking their union, we get a new set of states $t_1, t_2, \ldots, t_j$. The new set so obtained has an element $\{t_1, t_2, \ldots, t_j\} \in S'$. This element represent the value of $f'([s_1, s_2, \ldots, s_i\}, a)$.

It is easy to prove by induction, the length $|w|$ of the input string w that

$$f'(s_0', w) = \{s_1, s_2, \ldots, s_i\}$$

if and only $f(s_0, w) = \{s_1, s_2, \ldots, s_i\}$

If $|w| = 0$, then it implies that $f(s_0, \lambda) = \{s_0\}$ and by definition of f', we get $f'(s_0, \lambda) = s_0' = [s_0]$. So, w must be λ.

Suppose the result is true for all input strings u of length $|u| \le m$. Let $w = ua$ be a string of length $m + 1$, where $|u| = m$ and $a \in \Sigma$. Suppose $f(s_0, u) = \{t_1, t_2, \ldots, t_j\}$ and $f(s_0, ua) = \{r_1, r_2, \ldots, r_k\}$. Since $|u| \le m$, therefore by induction hypothesis we have

$$f'(s_0', u) = [t_1, t_2, \ldots, t_j]$$

Also $f(s_0, ua) = f(f(s_0, u), a) = f(\{t_1, t_2, \ldots, t_j\}, a) = \{r_1, r_2, \ldots, r_c\}$. Hence, $f'(s_0', ua) = f'(f'(s_0', u), a) = f'((t_1, t_2, \ldots, t_j), a) = \{r_1, r_2, \ldots, r_k\}$.

This proves that $f'(s_0', ua) = \{r_1, r_2, \ldots, r_k\}$ if and only if $f(s_0, ua) = \{r_1, r_2, \ldots, r_k\}$. Thus by induction $f'(s_0', ua)$ is true for all input strings w.

To complete the proof, let $x \in L(M)$ provided $f(s_0, w) \in F$. But it is true only if $f'(s_0, w) \in F'$. So, $x \in L(M) \Leftrightarrow x \in L(M')$ and hence $L(M) = L(M')$ and *dfa* accept L.

The Algorithm to Convert NFA to DFA

1. Construct a transition graph for *dfa* with vertex $\{s_0\}$. Call this vertex as initial vertex.
2. (*a*) Take any vertex $\{s_0, s_1, \ldots, s_i\}$ of the graph that has no outgoing edge for some $a \in \Sigma$.
   (*b*) Compute functions $f(s_1, a), f(s_2, a), \ldots, f(s_i, a)$ and form the union of these functions to get the set $\{s_1, s_2, \ldots, s_j\}$.
   (*c*) If such a set exists, then create a vertex in the transition graph labelled $\{s_1, s_2, \ldots, s_j\}$
   (*d*) Add to the graph an edge from $\{s_1, s_2, \ldots, s_i\}$ to $\{s_1, s_2, \ldots, s_j\}$ and label it with $a \in \Sigma$.
3. Every state in the graph whose label contains any $s_j \in F$ is identified as final vertex.
4. If *nfa* accept λ, the vertex $\{s_0\}$ in *dfa* is also identified as final vertex.

Example 38 Construct a *dfa* equivalent to $M = \{\Sigma, S, f, s_0, F\}$ where $\Sigma = \{0, 1\}$, $S = \{s_0, s_1\}$, $F = \{s_0\}$. The transition table is shown below:

State	*f*	
	0	*1*
s_0	s_0	s_1
s_1	s_1	s_0, s_1

Solution For a *dfa* M′ equivalent to M, the following informations are available

(*i*) states of M′ are subsets of $\{s_0, s_1\}$, i.e. ϕ, $\{s_0\}$, $\{s_1\}$, $\{s_0, s_1\}$.
(*ii*) $\{s_0\}$ is the initial state.
(*iii*) $\{s_0\}$, and $\{s_0, s_1\}$ are the final states as these states are the only states containing s_0, and
(*iv*) f is defined by the transition table given below, where states s_0 and s_1 appear in rows corresponding to s_0 and s_1 and column corresponding to 0. So $f(\{s_0, s_1\}, 0) = \{s_0, s_1\}$.

State	*f*	
	0	*1*
ϕ	ϕ	ϕ
$\{s_0\}$	$\{s_0\}$	$\{s_1\}$
$\{s_1\}$	$\{s_1\}$	$\{s_0, s_1\}$
$\{s_0, s_1\}$	$\{s_0, s_1\}$	$\{s_0, s_1\}$

Example 39 Construct *dfa* equivalent to *nfa* whose state table is shown below:

State	*f*	
	a	*b*
s_0	$\{s_0, s_1\}$	$\{s_2\}$
s_1	$\{s_0\}$	$\{s_1\}$
s_2	ϕ	$\{s_0, s_1\}$

Solution The transition diagram based on state stable of *nfa* is shown in Fig. 15.24.

(*i*) Starting with s_0 and all the edges from s_0 with input symbol *a*, we get a set $\{s_0, s_1\}$. Similarly with edges with input symbol *b*, we get $\{s_2\}$, as shown in Fig. 15.25.

(*ii*) Consider the set $\{s_0, s_1\}$ of vertices, s_0 and s_1 with input symbol *a*, we find s_0 on a transit to $\{s_0, s_1\}$ and s_1 on a transit to $\{s_0\}$. Thus, taking union of these sets we get

$$\{s_0\} \cup \{s_0, s_1\} = \{s_0, s_1\} \text{ at } s_0 \xrightarrow{b} \{s_2\}$$

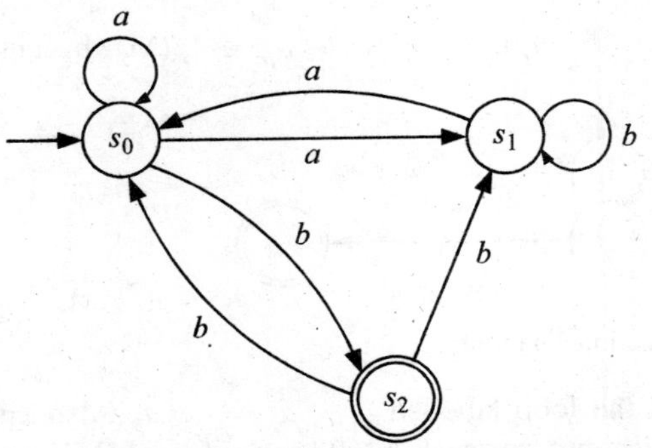

Fig. 15.24 Transition Diagram

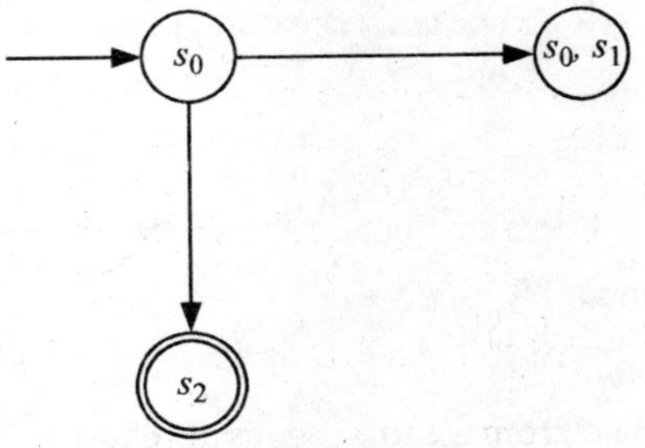

Fig. 15.25

Also $s_1 \xrightarrow{b} \{s_1\}$. Thus with input symbol *b*, we have $\{s_0, s_1\} \rightarrow \{s_1, s_2\}$ as shown in Fig. 15.26. Continuing in the same manner, we construct a state diagram of *dfa* as shown in Fig. 15.26.

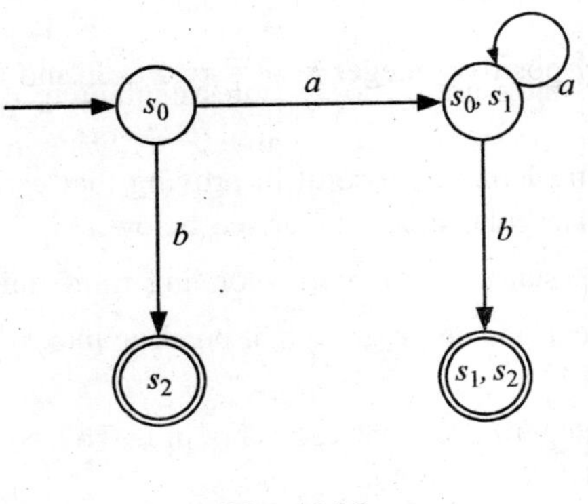

Fig. 15.26

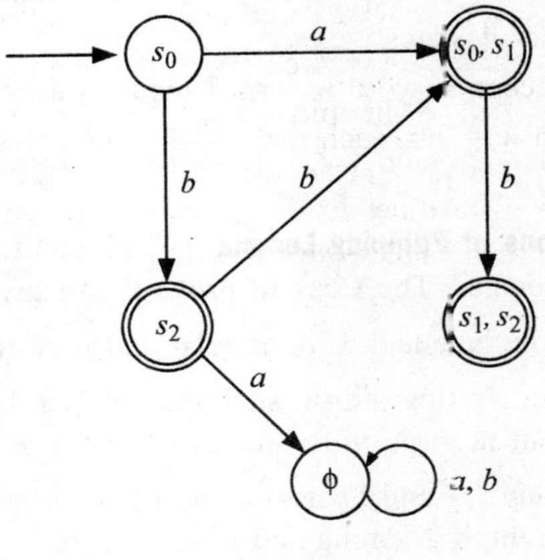

Fig. 15.27

15.8 PUMPING LEMMA FOR REGULAR SETS

The necessary condition for an input string to belong to a regular set is called *pumping lemma* because it provides a method of pumping (or generating) many input string from a given string. This lemma is also used as a tool to prove certain languages as non-regular. Thus, it is very useful in the development of algorithms to prove whether the language accepted by a given finite-automaton is finite or infinite.

Suppose an automaton M over an input set Σ has n states. Consider an input string $w = a_1 a_2 \ldots a_m$ over Σ accepted by M such that $|w| = m \geq n$. Let $(s_0, s_1, \ldots, s_n)$ be the corresponding sequence of states determined by the word w. Since $m > n$, it is not possible for each of the $n + 1$ states $s_0, s_1, s_2, \ldots, s_n$ to be distinct, since there are only n different states. Thus there are two states i and j $0 \leq i < j \leq n$, such that $s_i = s_j$. Let

$$x = a_1 a_2 \ldots, a_i, y = a_{i+1} a_{i+2} \ldots a_j \text{ and } z = a_{j+1} a_{j+2} \ldots a_m$$

Since $i < j$, the string y is of length at least 1 and hence $j \leq n$, its length is no more than as shown in Fig. 15.28.

If $s_m \in$ F, i.e. $a_1 a_2 \ldots a_m \in$ L (M), then $xz = a_1 a_2 \ldots a_i\, a_{j+1}\, a_{j+2} \ldots a_m \in$ L (M), because the

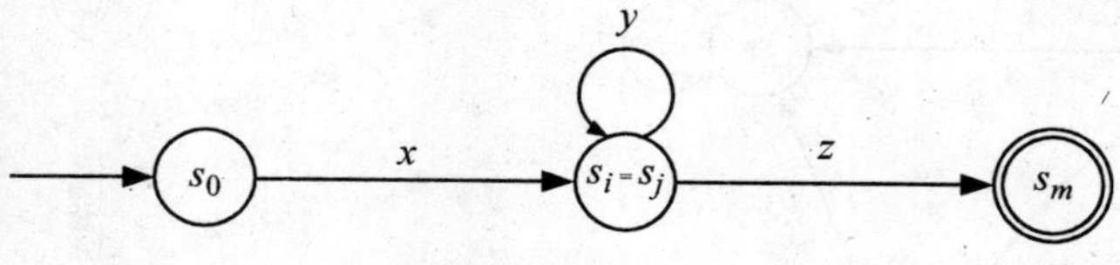

Fig. 15.28 Transition Diagrams

sequence from s_0 to s_m goes through s_j but not around the loop labelled $a_{i+1}\, a_{i+2} \ldots a_j$. Also xy ends in $s_i = s_j$ and therefore xy^k also ends in $s_i = s_j$. This implies that, for each $k \geq 0$ $w_k = xy^k z \in$ L(M) and ends in s_m, which is an accepting state. Hence, we conclude that for any long string accepted by finite automaton it is possible to find a substring near the beginning of the string that may be pumped (repeated as many times as we like) and the resulting string will also be accepted by finite automaton.

Pumping Lemma Let M be a finite automation over a regular set, L such that

(a) M has n states

(b) M accepts a word w form L where $| w | \geq n$.

Then $w = xyz$, such that $|xy| \leq n$, $|y| \geq 1$ and for every positive integer k, $w = xy^k z \in$ L and accepted by M.

Applications of Pumping Lemma The statement of pumping lemma is useful in proving that certain sets are not regular. The steps to prove that a given set is not regular are summarised below:

1. Let the language L be regular and n be the number of states in the corresponding finite automaton.
2. Select a string w in L such that $|w| \geq n$. Using statement of pumping lemma, break w into x, y and z, so that $w = xyz$ with constraints $|xy| \leq n$ and $|y| \geq 1$.
3. For any x, y and z parts determined in Step 2, find an integer i such that $xy^i z$ is not in L. This contradicts the pumping lemma and hence we may conclude that L is not regular.

Remark The selection of a suitable integer i such that $xy^i z \notin$ L depends on n, x, y and z. In few cases we may have to prove $xy^i z \notin$ L by considering $|xy^i z|$ or we may have to use the structure of strings in L.

Example 40 Show that the set $L = \{0^{i^2} : i \geq 1\}$ is not regular. That is, the set L which consists of all strings of 0's whose length is a perfect square is not regular.

Solution We prove the result by the method of contradiction.

(*a*) Let L be a regular language, and let n be the integer that represents number of states in finite automaton accepting each sentence in L.

(*b*) Let $w = 0^n1^n$. Then $|w| \geq 2n > n$. By pumping lemma we may write $w = xyz$, where $|xy| \leq n$ and $1 \leq |y| \leq n$, and $xy^iz \in$ L.

(*c*) By contradiction, we want to prove that $xy^iz \notin$ L for some integer *i*. The string *y* may be in any of the following three forms:

(*i*) $y = 0^k$ for some $k \geq 1$, i.e. *y* consists of only 0's

(*ii*) $y = 1^k$ for some $k \geq 1$, i.e. *y* consists of only 1's

(*iii*) $y = 0^k1^j$ for some $k, j \geq 1$.

In case (*i*) if we take $i = 0$, then we get $xy^oz = 0^o1^n$ or $xz = 0^{n-k}1^n$. Since $k \geq 1$, and $n - k \neq n$, therefore $xz \notin$ L, which is a contradiction.

In case (*ii*), if we again take $i = 0$, then $xy^oz = xz = 0^n1^{n-k}$. Since $n \neq n - k$, $xz \notin$ L, which is also a contradiction.

In case (*iii*), if we take $i = 2$, then $xy^2z = 0^{n-k}(0^k\ 1^j)^z\ 1^{n-j} = 0^n\ 1^j\ 0^k\ 1^n \notin$ L. As xy^2z is not of the form 0^i1^i, $xy^2z \notin$ L, which is a contradiction.

Since in all three cases above we arrive at a contradiction, we conclude that L is not regular.

Example 41 Show that the set L = $\{ww : w \in \{a, b\}^*\}$ is not regular.

Solution We prove the result by the method of contradiction.

(*a*) Let L be a regular language and let *n* be an integer that represents number of states in a finite automaton accepting each sentence in L.

(*b*) Let $ww = a^n\ b\ a^nb$ be a string in L, so that $|ww| = 2(n + 1) > n$. By pumping lemma, we may write L = $ww = xyz$ where $|xy| \leq n$ and $|y| \neq 0$.

(*c*) By contradiction, we now want to prove that $xy^iz \notin$ L. The string *y* may be in any of the following two states:

(*i*) $y = a^k$ for some $k \geq 1$, i.e. *y* consists of only *a*'s and has no *b*'s.

(*ii*) y^{no} consists of only one *b*.

It may be noted here that if y consists of two *b*'s, then $|y| \geq n + 2$, and hence $|y| \leq |xy| < n$. If we take $i = 0$ in case (*i*), then $xy^0z = xz$ will be of the form $a^mb\ a^nb$, where $m = n - k < n$. Since *xz* cannot be written as *uu* for $u \in \{a, b\}$, this implies that $xz \notin$ L, which is a contradiction.

But, if we take $i = 0$ in case (*ii*), then $xy^0z = xz$ has only one *b* because *y* consists of one *b* and that is removed from *xyz*. This implies that $xz \notin$ L, which is again a contradiction, because *xz* should have even number of *a*'s and *b*'s.

Since in both the cases we arrive at a contradiction, we conclude that L is not regular.

Example 42 Show that the set L = $\{a^p : p$ is a prime$\}$ is not regular.

Solution Let L be a regular language and *n* be number of states in the finite automaton accepting L.

If *p* is a prime number greater than *n*, i.e $w = a^p$, where $|w| = p$, then by pumping lemma $w = xyz$ such that $|xy| \leq n$ and $|y| \neq 0$ and hence *x*, *y*, and *z* are simply strings of *a*'s. So $y = a^m$ for some $1 \leq m \leq n$.

Suppose $i = p + 1$, then $|xy^iz| = |xyz| + |y^{i-1}| = p + (i - 1)\ m = p + pm = p\ (1 + m)$ which is not a prime number. Hence, xy'z Ï L, which is a contradiction, so L is not regular.

15.9 FINITE AUTOMATON WITH OUTPUT

The finite automaton discussed before has its output limited to a binary digit, i.e. it *accept* the string or *do not accept* the string. In this section, we consider an automaton in which output can be chosen from some other alphabet. If value of the output function Z (t) in general, is a function of the present state q (t) and the present input x (t), then

$$Z(t) = \lambda [q(t), x(t)]$$

where λ is called the output function. Such a machine is called *Mealy machine*. If the output function $z(t)$ depends only on the present state q (t) and is independent of the present input x (t), i.e. $Z(t) = \lambda(q(t))$, then such a machine is called a *Moore machine*.

15.9.1 Moore Machine

A Moore machine consists of six components $(\Sigma, S, \Delta, f, \lambda, s_0)$ where S, Σ, f and s_0 are the same as in deterministic automaton (*dfa*); Δ is the *output alphabet* and λ is the output function (or mapping) from S to Δ, i.e. $f : s \to \Delta$ gives the output associated with each state in S. The output of a Moore machine in response to input string $a_1\ a_2\ \ldots\ a_n$ for $n \geq 0$ is

$$\lambda(s_0)\,\lambda(s_1)\ldots\lambda(s_n)$$

where $s_0\ s_1\ \ldots\ s_n$ is the sequence of states such that $f(s_{i-1}, a_i) = s_i$ for $1 \leq i \leq n$.

A deterministic finite-state automaton (*dfa*) may be viewed as a special case of a Moore machine where the output alphabet is {0, 1} and state s is 'acceting state' provided $\lambda(s) = 1$. The output sequence has length $n + 1$ and the first output is $\lambda(s_0)$ for all output strings. For example, the state transition table that describes a Moore machine is shown in Table 15.9, where initial state s_0 is marked with an arrow.

Table 15.9 State Transition in Moore Machine

Present State	*Next State f*		*Output*
	a = 0	*a = 1*	λ
→ (s_0)	s_1	s_2	1
s_1	s_3	s_2	0
s_2	s_2	s_1	1
s_3	s_0	s_3	1

For the input string 01010, the sequence of states $s_0 \xrightarrow{0}$ $s_0 \xrightarrow{1} s_2 \xrightarrow{1} s_2 \xrightarrow{0} s_1$ gives output sequence 001001.

15.9.2 Mealy Machines

A Mealy machine consists of six components $(\Sigma, S, \Delta, f, \lambda, s_0)$ where all symbols are as in the Moore machine, except that λ is the function (mapping) from S × Σ to Δ. That is, $\lambda(s, a)$ gives the output associated with the transition from state s to input a. The output in response to input string $a_1a_2\ldots a_n$ for $n \geq 0$ is $\lambda(s_0, a_1)\,\lambda(s_1, a_2)\ldots\lambda(s_{n-1}, a_n)$, where $s_0, s_1, \ldots, s_n$ is the sequence of states such that $f(s_{i-1}, a_i) = s_i$ for $1 \leq i \leq n$.

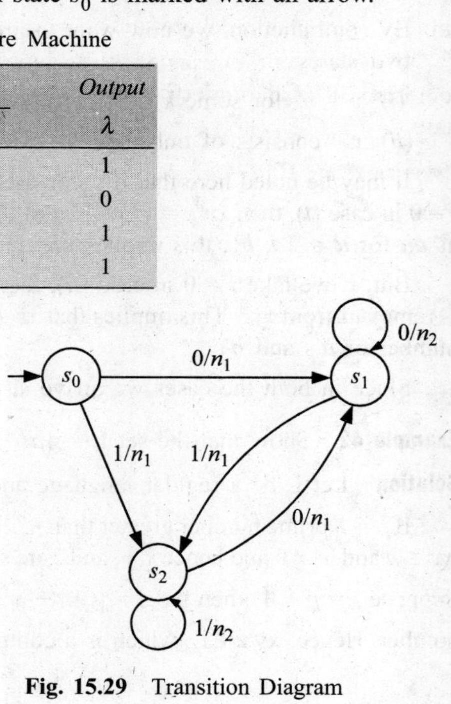

Fig. 15.29 Transition Diagram

If the input string is of length n, then the output string is also of length n rather than $n + 1$ as in the case of Moore machine. Also for the input string $\in$, the output is also $\in$.

The state transition table, for example, that describes a Mealy machine is shown in Table 15.10.

Table 15.10 State Transition in Mealy Machine

Present State	*Next State*			
	a = 0		*a = 1*	
	State	*Output*	*State*	*Output*
→ (s_1)	s_1	1	s_2	0
s_2	s_4	1	s_4	1
s_3	s_2	1	s_3	1
s_4	s_3	0	s_1	1

For example, let M = $(\Sigma, S, \Delta, f, \lambda, s_0)$ be a Mealy machine, where $\Sigma = \{0, 1\}$, $S = \{s_0, s_1, s_2\}$, $\Delta = \{n_1, n_2\}$ as shown in Fig. 15.29. In this figure a label a/b on an arc from a state s_i to another state s_j has been used to indicate $f(s_i, a) = s_j$ and $\lambda(s_i, a) = b$. For the input string 01100, the transition states is given by $s_0 \xrightarrow{0} s_1 \xrightarrow{1} s_2 \xrightarrow{1} s_2 \xrightarrow{0} s_1 \xrightarrow{0} s_1$ and the output string is $n_1 n_1 n_2 n_1 n_2$.

15.9.3 Equivalence of Moore and Mealy Machines

Let M represents a Moore or Mealy machine, and let $Z_M(w)$ be output corresponding to an input string w. If M_1 is a Mealy machine and M_2 be a Moore machine, then the output $Z_{M_1}(w)$ and $Z_{M_2}(w)$ cannot be equal, because length of output string $|Z_{M_2}(w)|$ for a Mealy machine is one less than the length of output string $|Z_{M_2}(w)|$ for a Moore machine for each input string of length $|w|$.

If we neglect the response of a Moore machine to input string $\in$, then M_1 and M_2 can be considered equivalent if for all input strings w, $bZ_{M_1}(w) = Z_{M_2}(w)$ where b is the output of Moore machine for its initial state.

Theorem 15.3 If $M_1 = (\Sigma, S, \Delta, f, \lambda, s_0)$ is a Moore machine, then there is an equivalent Mealy machine M_2.

Proof Let $M_2 = (\Sigma, S, \Delta, f, \lambda', s_0)$ be a Mealy machine and we define its output function λ' as a function of present state and input symbol as:

$$\lambda'(s_i, a_i) = \lambda(f(s_i, a_i) \text{ for all states } s_i \text{ and input symbols } a_i.$$

Then both M_1 and M_2 enter the same sequence of states on the same input. The transition function in M_2 gives the same output as that of M_1 with the states entered.

Example 43 Construct a Mealy machine which is equivalent to the Moore machine given below:

Transition Table of Moore Machine

Present State	*Next State*		*Output*
	a = 0	*a = 1*	
→ (s_0)	s_1	s_2	1
s_1	s_3	s_2	0
s_2	s_2	s_1	1
s_3	s_0	s_3	1

Solution For the given Moore machine, form a pair of next state and the corresponding output for every input symbol. For example, the states s_1 and s_3 in the next state column with associated output 0 and 1, respectively. The transition table for Mealy machine in given is Table 15.11.

Table 15.11 Mealy Machine

Present	Next State			
	a = 0		*a = 1*	
State	*State*	*Output*	*State*	*Output*
→ (s_0)	s_1	0	s_2	1
s_1	s_3	1	s_2	1
s_2	s_2	1	s_1	0
s_3	s_0	1	s_3	1

Remark If two states have identical output, then such states can be deleted from the transition table.

Theorem 15.4 If $M_1 = (\Sigma, S, \Delta, f, \lambda, s_0)$ be a Mealy machine, then there is an equivalent Moore machine M_2.

Proof Make pairs $[s_i, s_j]$ of states of Moore machine M_2, having a state and an output symbol of Mealy machine M_1. Define a transition function and out function as:

$$f'([s_i, s_j], a) = [f(s_i, a), \lambda(s_i, a)]$$

and

$$\lambda'([s_i, s_j]) = s_j$$

This definition inplies that the second component s_j of a pair of states $[s_i, s_j]$ of M_2 is the output made by M_1 on some transition into state s_i.

Hence, in general, we conclude that when M_1 enters states $s_0, s_1, \ldots, s_n$ on input $a_1a_2 \ldots a_n$ and produces output $x_1, x_2, \ldots, x_n$, then M_2 enters states $[s_0, x_0], [s_1, x_1] \ldots, [s_n, x_n]$ and also produces output $x_0, x_1, x_2, \ldots, x_n$.

Example 44 Construct a Moore machine which is equivalent to the Mealy machine given below:

Transition Table of Mealy Machine

Present State	Next State			
	a = 0		*a = 1*	
	State	*Output*	*State*	*Output*
(s_1)	s_1	1	s_2	0
s_2	s_4	1	s_4	1
s_3	s_2	1	s_3	1
s_4	s_3	0	s_1	1

Solution Choose any state, say s_i in the next state column for $a = 0$ and $a = 1$ of the table and note down different outputs associated with this state in those columns. Split s_i into several different states, the number of such states being equal to the number of different outputs associated with s_i.

For example, in the given table, the state s_1 is associated with output 1, s_2 is associated with outputs. 1 and 0; s_3 is associated with outputs 0 and 1, and state s_4 is associated with single output 1.

We choose the state s_2 and split it into two different states s_{20} and s_{21}. Similarly, state s_3 is splited into two different states s_{30} and s_{31}. The new transition table is shown in Table 15.12.

Table 15.12 New Mealy Machine

Present State	Next State			
	$a = 0$		$a = 1$	
	State	Output	State	Output
→ (s_1)	s_1	0	s_{20}	0
s_{20}	s_4	1	s_4	1
s_{21}	s_4	1	s_4	1
s_{30}	s_{21}	1	s_{31}	1
s_{31}	s_{21}	1	s_{31}	1
s_4	s_{30}	0	s_1	1

Rearrange pairs of states and outputs appear in the next column as shown in Table 15.13.

Table 15.13 Revised State Table

Present State	Next State		Output
	$a = 0$	$a = 1$	
→ (s_1)	s_1	s_{20}	1
s_{20}	s_4	s_4	0
s_{21}	s_4	s_4	1
s_{30}	s_{21}	s_{31}	0
s_{31}	s_{21}	s_{31}	1
s_4	s_{30}	s_1	1

In Table 15.13 it may be observed that the initial state s_1 is associated with output 1. This implies that for a null imput $\in$, an output 1 is produced when start from state s_1. Thus, the Moore machine accepts a sequence of input of zero length, which is not accepted by a Mealy machine. This problem is solved by either neglecting the response of a Moore machine to null input $\in$ or adding a new starting state s_0, whose output is zero but state transition is same as that of s_1. The new transition table so obtained is shown in Table 15.14.

Table 15.14 Transition Table of Moore Machine

Present State	Next State		Output
	$a = 0$	$a = 1$	
→ (s_0)	s_1	s_{20}	0
s_1	s_1	s_{20}	1
s_{20}	s_4	s_4	0
s_{21}	s_4	s_4	1
s_{30}	s_{21}	s_{31}	0
s_{31}	s_{21}	s_{31}	1
s_4	s_{30}	s_1	1

Example 45 Let a Mealy machine be represented by transition diagram shown in Fig. 15.30. Construct an equivalent Moore machine.

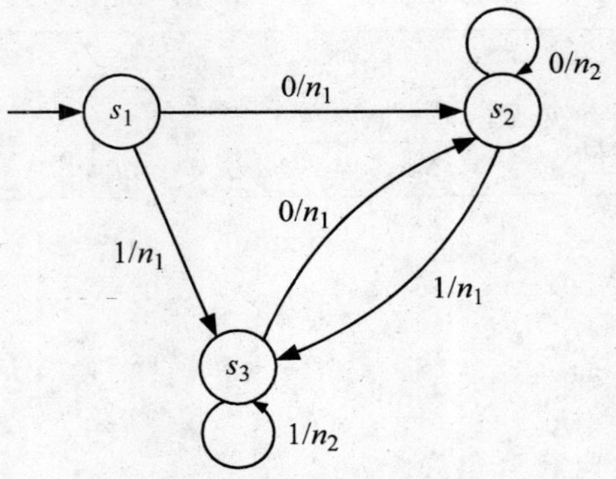

Fig. 15.30 Transition Diagram

Solution The transition table corresponding to the given transition diagram is shown in Table 15.15.

Table 15.15 Transition Table Corresponding to Given Transition Diagram

Present State	Next State			
	a = 0		*a = 1*	
	State	*Output*	*State*	*Output*
→ (s_1)	s_2	n_1	s_3	n_1
s_2	s_2	n_2	s_3	n_1
s_3	s_2	n_1	s_3	n_2

The transition Table 15.15 is reconstructed as the Mealy machine transition table. The new entries are shown in Table 15.16.

Table 15.16 Transition Table for Moore Machine

Present State	*Next State*		*Output*
	a = 0	*a = 1*	
→ (s_1)	s_{21}	s_{31}	$\in$
s_{21}	s_{22}	s_{31}	n_1
s_{22}	s_{22}	s_{31}	n_2
s_{31}	s_{21}	s_{32}	n_1
s_{32}	s_{21}	s_{32}	n_2

The transitions and outputs of Moore machine are shown in Fig. 15.31.

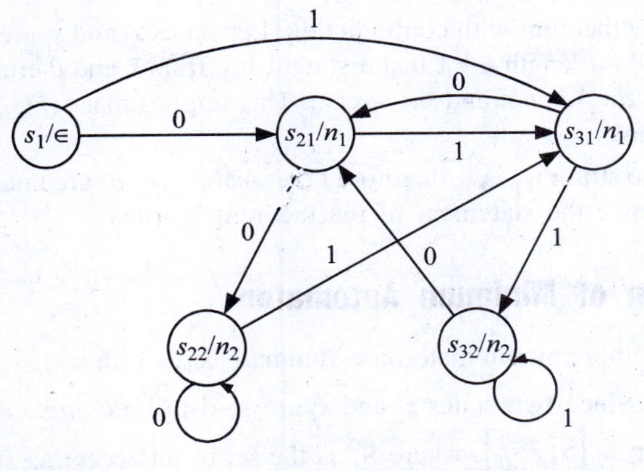

Fig. 15.31 Moore Machine Constructed from Mealy Machine

15.10 MINIMIZATION OF FINITE AUTOMATA

Minimisation of a finite automaton M_1 is the process of constructing another automaton M_2 with the following properties:

1. Both M_1 and M_2 emit same output given the same input when they start from their respective initial (start) state.
2. Automaton M_2 has minimum number of states equivalent to a given automaton M_1.

Definition Two states s_i and s_j of a finite automaton are said to be *equivalent* (denoted by $s_i \equiv s_j$) if and only if for each input string. $a \in \Sigma^*$ $f(s_i, a)$ and $f(s_f, a)$ are accepting states or both of them are non-accepting states.

Since there are infinite number of strings in Σ^*, it is difficult to construct $f(s_i, a)$ and $f(s_f, a)$ for every $x \in \Sigma^*$. So the following definition on equivalent states will make it easy to understand this concept.

Definition (Equivalent States) Two states s_i and s_j of a finite automaton M are said to be *k-equivalent* if both $f(s_i, a)$ and $f(s_j, a)$ are accepting states or non-accepting states for all strings a of length k or less.

Properties of Equivalent States

1. The relation '*k*-equivalence' on the set of states S in M is an equivalence relation. That is, this relation is reflexive, symmetric and transitive.
2. The equivalence relation partition the set S into equivalence classes denoted by π and π_k. The elements of π_k are *k-equivalence classes*. That is, π_0, π_1 and π_2 represent partition corresponding to 0-equivalence, 1-equivalence and 2-equivalence, respectively.
3. If two states s_i and s_j are k-equivalent for all $k \geq 0$, then they are equivalent.
4. If states s_i and s_j are $(k + 1)$-equivalent, then they are k-equivalent.
5. $\pi_k = \pi_{k+1}$ for some k.

Theorem 15.5 Two states s_i and s_j are (k + 1)-equivalent if they (*i*) *k*-equivalent and (*ii*) $f(s_i, a)$ and $f(s_j, a)$ are also *k*-equivalent for every $a \in \Sigma$.

Proof We shall prove the theorem with contradiction. Let states s_i and s_j are not $(k + 1)$-equivalent. Then there exists a string $w = aw_1$ of length $k + 1$ that distinguish s_i from s_j and therefore $f(s_i, aw_1)$ is an accepting (final) state, whereas $f(s_j, aw_1)$ is not and vice-versa. This implies that $f(f(s_i, a), w_1)$ is an accepting state whereas $f(f(s_j, a), w_1)$ is not.

Since length $|w_1|$ of the string w_1 is k, therefore $f(s_i, a)$ and $f(s_j, a)$ are not k-equivalent. Thus, we arrive at a contradiction, and hence the statement of the theorem is true.

15.10.1 Construction of Minimum Automaton

Steps of construction of minimum automaton are summarised as follows:

1. **Construction of π_0** Since two states s_i and s_j are said to be *0-equivalent* (π_0) if these states have the same output, i.e. $\pi_0 = [S_1^0, S_0^2]$, where S_1^0 is the set of all accepting (final) states and $S_2^0 = S - S_1^0$.
2. **Construction of π_{k+1} from π_k** Let S_i^k be any subset of π_k and $s_i, s_j \in S_i^k$. Then states s_i and s_j are $(k + 1)$-equivalent if and only if $f(s_i, a)$ and $f(s_j, a)$ are k-equivalent for every $a \in \Sigma$. Hence to prove that s_i and s_j are $(k + 1)$-equivalent, we need to verify whether $f(s_i, a)$ and $f(s_j, a)$ for every $a \in \Sigma$, are in the same equivalence class in π_k. If not, then S_i^k is further divided into $(k + 1)$-equivalence classes and repeat the above procedure for every S_i^k in π_k to get all the elements of π_{k+1}.
3. Construct π_n for $n = 1, 2, \ldots$ until $\pi_n = \pi_{n+1}$.
4. **Construction of minimum state automaton** For the required minimum state automaton, the states are the elements of π_n, i.e. they are equivalence classes obtained in Step 3. Prepare state table by replacing a state by the corresponding equivalence class $[s]$.

Example 46 Construct π_0, π_1 and π_2 from the following transition table of an automaton.

State	*f (Input)* 0	1	*Output*
$\rightarrow$ (s_0)	s_1	s_3	0
s_1	s_0	s_3	0
s_2	s_5	s_0	0
s_3	s_6	s_1	0
s_4	s_0	s_6	1
s_5	s_6	s_2	1
s_6	s_0	s_2	1

Solution Let $S = \{s_0, s_1, s_2, \ldots, s_6\}$ be a finite set of states and let S_i^k denotes the equivalence class of state s_i under the relation k-equivalence. Thus

$$S_0^0 = \{s_i : s_0 \text{ is 0-equivalent } s_i\}$$

$$= \{s_i : s_0 \text{ and } s_i \text{ have the same output}\} = \{s_0, s_1, s_2, s_3\}$$

As $S_0^0 \neq S$, we take the next element of S not in $[S_0]_0$, we get

$$S_4^0 = \{s_i : s_4 \text{ is 0-equivalent to } s_i\} = \{s_4, s_5, s_6\}$$

Hence, $\pi_0 = \{\{s_0, s_1, s_2, s_3\}, \{s_4, s_5, s_6\}\}$

Similarly, we have 1-equivalent and 2-equivalent as follows:

$$\pi_1 = \{\{s_0, s_1\}, \{s_2, s_3\}, \{s_4\}, \{s_5\}, \{s_6\}\}$$

$$\pi_2 = \{\{s_0, s_1\}, \{s_2, s_3\}, \{s_4\}, \{s_5\}, \{s_6\}\}$$

Example 47 Construct a minimum state automaton equivalent to a given automaton whose transition table is given below:

Transition Table

State	*Input*	
	a	*b*
→ (s_0)	s_0	s_3
s_1	s_2	s_5
s_2	s_3	s_4
s_3	s_0	s_5
s_4	s_0	s_6
s_5	s_1	s_4
(s_6)	xs_1	s_3

Solution Since there is only one accepting state s_6, therefore

$$S_1^0 = \{s_6\} \text{ and } S_2^0 = S - S_1^0$$

Thus $\pi_0 = \{\{s_6\}, \{s_0, s_1, s_2, s_3, s_4, s_5\}\}$.

As $\{s_6\}$ cannot be partitioned further, this gives $S_1' = \{s_6\}$. Consider s_0 and $s_1 \in S_2^0$. The entries under *a*-column in the transition table corresponding to s_0 and s_1 are s_0 and s_1; these lie in S_2^0. The entries under *b*-column corresponding to s_0 and s_1 are s_3 and s_5; s_3 and $s_5 \in S_2^0$. Thus s_0 and s_1 not 1-equivalent. Similarly, s_0 is 1-equivalent to the remaining states. s_2, and s_6, but s_0 and s_4 are not 1-equivalent. Thus $\{s_0, s_1, s_2, s_3, s_5\}$ is a subset in π_1 and $S_2' = \{s_0, s_1, s_2, s_3, s_5\}$. Hence,

$$\pi_1 = \{\{s_6\}, \{s_0, s_1, s_2, s_3, s_5\}, \{s_4\}\}$$

Again, as $\{s_6\}$ cannot be partitioned further, therefore $\{s_6\} \in \pi_2$. Consider s_0 and $s_1 \in S_2^0$. The entries under *a*-column are s_0 and s_2; these lie in the same equivalence class in π_1. The entries in *b*-column are s_3 and s_5, these also lie in the same equivalence class in π_1, s_0 and s_1 are not 2-equivalent. Similarly, s_0 is not 2-equivalent with s_2, s_3 and s_5. Also s_1 is not 2-equivalent with s_2, s_3 and s_5.

Consider s_2 and any of the states s_3 and s_5. States s_2 and s_3 are not 2-equivalent. However, for s_2 and s_5, entries under *b*-column are s_4 and s_4. So s_2 and s_5 are 2-equivalent. Hence $\{s_0, s_1, s_2, s_3, s_5\}$ is partitioned into $\{s_2, s_5\}$ and $\{s_0, s_1, s_3\}$. Thus

$$\pi_2 = \{\{s_6\}, \{s_4\}, \{s_0, s_1, s_3\}, \{s_2, s_5\}\}.$$

Similarly, we get

$$\pi_3 = \{\{s_6\}, \{s_4\}, \{s_0\}, \{s_1\}, \{s_3\}, \{s_2, s_5\}\}$$
$$\pi_4 = \{\{s_6\}, \{s_4\}, \{s_0\}, \{s_1\}, \{s_3\}, \{s_2\}, \{s_5\}\}$$

Since π_4 = S, π_4 gives the equivalence classes and the minimum state automaton is M itself.

15.11 TURNING MACHINES

The Turning machine is equivalent in computing power to the digital computer and also to all the most general mathematical notations of computation. That is, it is capable of performing any calculations which can be performed by any computing machine.

15.11.1 Turning Machine Model

The Turning machine can be viewed as a finite-state automaton connected to a Read/Write(R/W) head which moves back and forth along an infinite tape. As shown in Fig. 15.32, it has a finite control, an input tape that is divided into a number of squares (cells), and the tape head that scans one cell at a time. Each cell of the tape may be blank or hold exactly one of a finite number of tape symbols. At each step in time, the machine M is in a certain internal state s_i scanning one of the tape symbols a_j on the tape. Initially, the n ($n \geq 0$) left most cells hold the input, which is a string of symbols chosen from a subset of the tape symbols called the *input symbols.* The remaining infinite number of cells hold the blank, which is a special tape symbol that is not an input symbol.

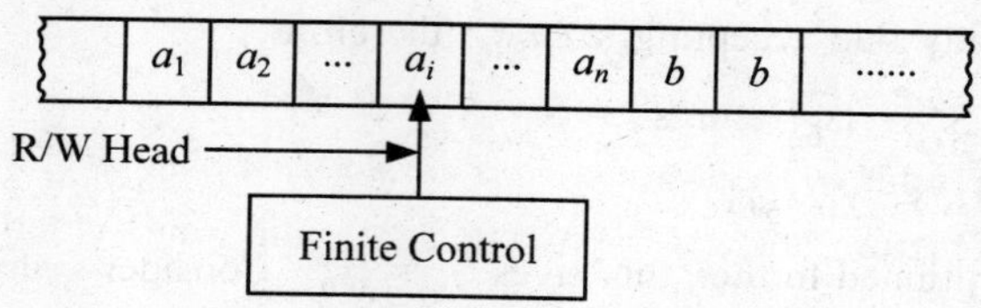

Fig. 15.32 Basic Turning Machine

The input end the output from the finite state automaton are effected by the R/W head which scan one cell at a time. In a single move depending upon the symbol scanned by the R/W head and the state of the 'finite control', the Turning machine

(*i*) changes its state.

(*ii*) prints a symbol on the tape in the cell under R/W head, replacing what was written there before

(*iii*) moves its R/W head along the tape one cell to left or right.

Remark In the above discussion, we consider the Turning machine that uses an finite two-way tape. However, other definitions with finite two-way tape are equivalent.

Definition A Turning machine consists of seven components, which is M = {Σ, S, Γ, f, s_0, b, F} where

Γ is the non-empty set of allowable *tape symbols.*

S is the *finite state set.*

b, is the set of *input symbols* of Σ is the subset of Γ not including b.

Γ, that is the *blank.*

f is the *next-move (transition) function* or mapping that maps the states of finite automaton and tape symbols to states, tape symbols and movement of the head, i.e.

$$f : S \times \Gamma \rightarrow S \times \Gamma \times \{L, R, N\}$$

Here L denotes 'left', R denotes 'right' and N denotes 'no movement or stay'. F a set of final states is the subset of S, i.e. F ≤ S.

In Figs. 15.33 (*a*) to (*c*). The Turning machine is shown in three informal stages and their corresponding expression. In Fig. 15.33 (*a*), M is in state s_3 scanning the third symbol where $a_1a_2a_3ba_2a_2$ is printed on the tape, where b is a blank symbol. This picture is expressed as $\alpha = a_1a_2s_3ba_2a_2$, where state s_3 of M is written before the tape symbol a_3 which is being scanned. It may be noted that the expression α_1 is using only the tape alphabet Σ except for the state symbol s_3 which is not at the end of the expression as it appears before the tape symbol a_3 that machine is scanning.

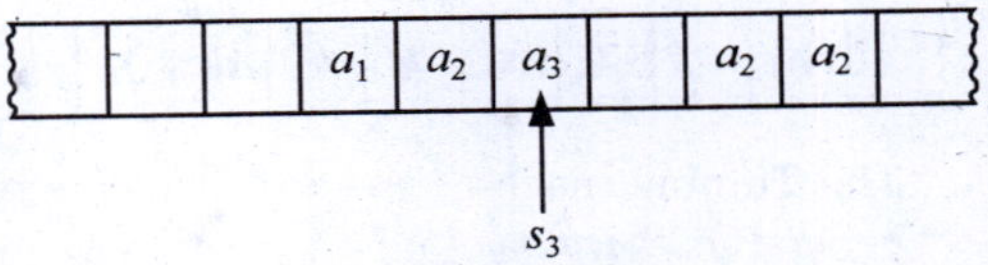

Fig. 15.33(*a*)

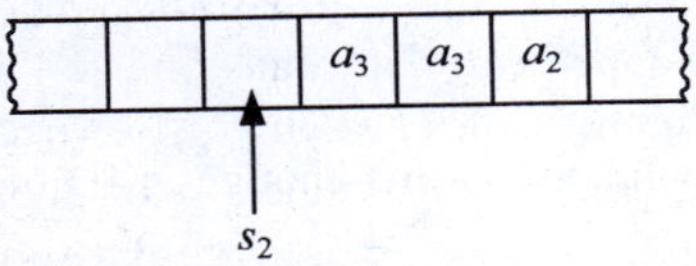

Fig. 15.33(*b*)

$\alpha_2 = s_2ba_3a_3a_2$

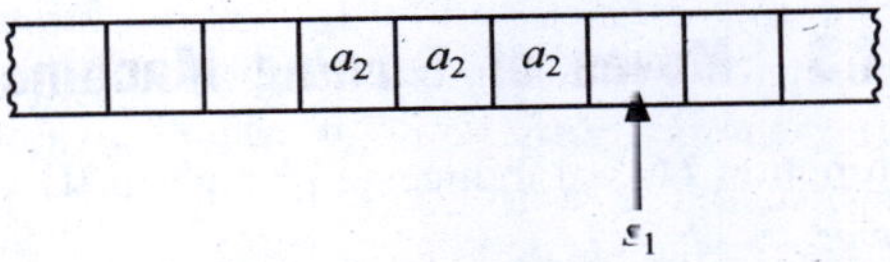

Fig. 15.33(*c*)

$\alpha_3 = a_2a_2a_2s_1b$

Remarks

1. A *picture* is an expression of the form: $\alpha = p\, s_i\, a_k\, q$, where p and q are tape expressions possibly blank.
2. If $\alpha = ps_ia_kq$, is a picture, then M is in state s_i scanning the symbol a_k, and the *expression on the tape* is pa_kq, i.e. a picture without its state symbol s_i.

   Since at each step in machine M is in state s_i scanning the symbol a_k, it can do three things simultaneously.

   (*a*) It Erases the scanned symbol a_k and replace it with another tape symbol a_t, where we permit $a_t = a_k$.

   (*b*) It changes its internal states s_i to a state s_j, where we permit $s_j = s_i$.

   (*c*) It moves one square to the left, moves to the right or does not move at all.

15.11.2 Instantaneous Description of Turning Machine

'Snapshots' of a Turning machine in action give its 'instantaneous descriptions'. The ID of a Turning machine is defined in terms of the entire input string and the current state.

Definition An ID of the Turning machine M is expressed by a string $\alpha_1 p \alpha_2$, where p is the current state of M and $\alpha_1\alpha_2$ is the string in Γ^*, i.e. it represents the contents of the tape up to the rightmost non-blank symbol or the symbol to the left of R/W head whichever is rightmost. A blank may also occur in $\alpha_1\alpha_2$. The R/W head is assumed to be scanning the leftmost symbol of α_2 or if $\alpha_2 = \in$, the R/W head scanned a blank.

Illustration A snapshot of Turning machine is shown in Fig. 15.34(*a*). The Turning machine is in state s_2 scanning the third symbol a_3. This picture may be represented by the expression $\alpha = b\, a_1\, a_1\, s_3\, a_2\, a_2\, a_1\, bb$. The symbol a_3 is written to the right of state s_3 of M. The non-blank symbols to the left of symbol a_3 form the string $a_1 a_1$ and to the right of a_3 form the string $a_2\, a_2\, a_1$. Thus, the ID of M is shown in Fig. 15.34(*b*).

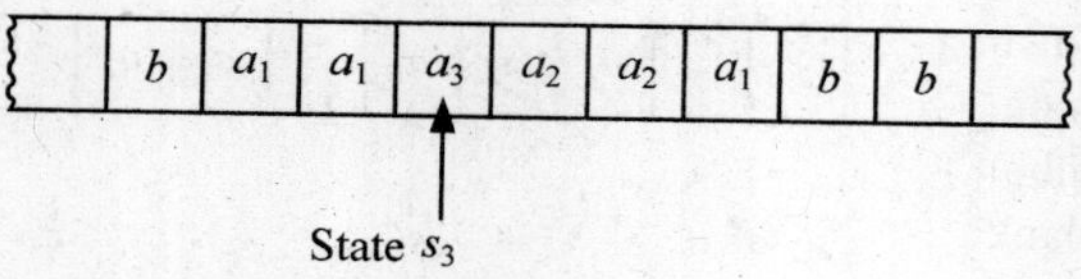

Fig. 15.34 (*a*) Snapshot of Turning Machine

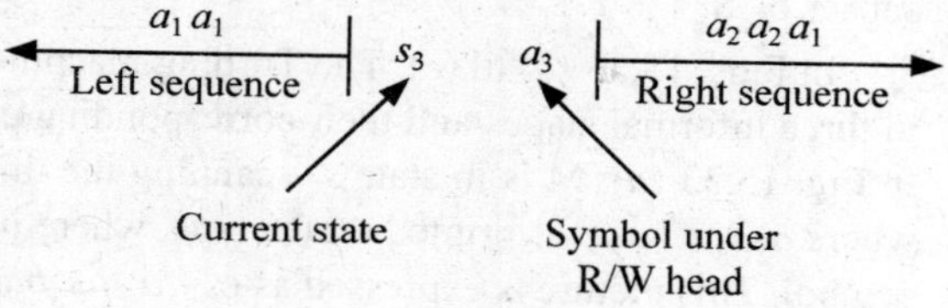

Fig. 15.34 (*b*) Representation of ID

Observe that (*i*) for constructing ID, we insert the current state in the input string to the left of the symbol under R/W head, and (*ii*) blank symbols may also occur as part of the left or right substring.

15.11.3 Moves of Turning Machine

The function $f(s, a)$ induces a change in ID of the Turning machine, and this change in ID of M is called *a move*.

Let $f(s, a_i) = (t, y, \text{L})$, and let $a_1 a_2 \ldots a_n$ be an input string to be processed, where R/W head is at symbol a_i. If $i - 1 = n$, then a_i is considered to be blank. Thus for $i = 1$, there is no next ID because R/W head is not allowed to fall off the left end of the tape. If $i > 1$, then ID before processing a_i is: $a_1\, a_2 \ldots a_i\, s,\ a_{i+1} \ldots a_n$. Hence after processing of a_i, we write the resulting ID as:

$$a_1\, a_2 \ldots a_{i-1}\; s\; a_i\; a_{i+1} \ldots a_n \underset{\text{M}}{\vdash} a_1\, a_2 \ldots a_{i-2}\; s\; a_{i-1}\; y\; a_{i+1} \ldots a_n$$

If any suffix of $a_{i-1}\; y\; a_{i+1} \ldots a_n$ is blank, then that is deleted from above ID. However, if $f(s, a_i) = (t, y, \text{R})$, then the change in ID of M is written as:

$$a_1 a_2 \ldots a_{i-1}\; s\; a_i\; a_{i+1} \ldots a_n \underset{\text{M}}{\vdash} a_1 a_2 \ldots a_{i-1}\; y\; t\; a_{i+1} \ldots a_n$$

In case $i - 1 = n$, the string $a_1 a_2 \ldots a_n$ is empty and right side of this ID is longer than the left side.

Two ID's I_j and I_k related as $\text{I}_j \underset{\text{M}}{\vdash} I_k$ defines a relation among them. If one ID results from another by some finite number of moves including zero move, then they are related by the symbol $\overset{*}{\underset{\text{M}}{\vdash}}$. This symbol denotes the reflexive transitive closure of the relation $\underset{\text{M}}{\vdash}$. In particular, the relation $\text{I}_1 \overset{*}{\underset{\text{M}}{\vdash}} \text{I}_n$ can be written as $\text{I}_1 \underset{\text{M}}{\vdash} \text{I}_2 \underset{\text{M}}{\vdash} \text{I}_3 \ldots \underset{\text{M}}{\vdash} \text{I}_n$ for some IDs $\text{I}_2, \ldots, \text{I}_{n-1}$.

15.11.4 Transition Table of Turning Machines

Let $\text{S} = \{s_0, s_1, s_2, s_3, s_4\}$, $\Sigma = \{0, 1\}$, $\Gamma = \{0, 1, x, y, b\}$ and $\text{F} = \{s_4\}$, where s_0 is the initial state and s_4 is final (accepting) state. If we define the transition function as: $f(s, a) = (0, 1, x, y, b)$, then s_0 is entered initially and immediately prior to each replacement of a leftmost 0 by an x. State s_1 is used to search right skipping over 0's and y's until it finds the leftmost 1. If R/W head finds 1, then it changes it to y and enters into state s_2. Now s_2 searches left of an x and enters state s_0 upon finding it, moving right, to the leftmost 0, as it changes state. As R/W head moves right up to state s_1 and find there b or x before 1, then the input is rejected due to the reason that either there are too many 0's or the input is not in 0* 1*. The transition table is given in Table 15.17 describes f.

Table 15.17 Transition Table of a Turning Machine

Present State	*Tape Symbols*				
	0	1	x	y	b
Initial State → s_0	x R s_1	–	–	y R s_3	–
s_1	0 R s_1	y L s_2	–	y R s_1	–
s_2	0 L s_2	–	x R s_0	y L s_2	–
s_3	–	–	–	y R s_3	b R s_4
(s_4)	–	–	–	–	–

Example 48 Consider a Turning machine M described in Table 15.17. Draw the computation sequence of the input string 001.

Solution The computation of M is carried out in terms of the elements of the tape and the current state. As explained earlier, when the string in the tape is $a_1a_2 \ldots a_j\ a_{j+1} \ldots a_n$, and machine M in state s, is to read a_{j+1}, and we write it as: $a_1\ a_2 \ldots a_j\ s\ a_{j+1} \ldots a_n$.

For the input string 001b, we get the following sequence

$$s_0 001b \vdash xs_1 01b \vdash x_0 s_1 1b \vdash xs_2 0y_1 b$$
$$\vdash s_2 x0y1b \vdash xs_0 0y1b \vdash xxs_1 y1b \vdash xxys_1 1$$
$$\vdash xxs_2 yy \vdash xs_2 xyy \vdash xxs_0 yy \vdash xxys_3 y$$
$$\vdash xxyys_3 \vdash xxyybs_4$$

15.11.5 Grammars and Turning Machines

A language that is recognised (or accepted) by a Turning machine M is said to be *recursively enumerable* language whose strings can be enumerated or listed by TM). If L is the set of strings in the tape set such that any Turning machine halts on input w in the accepting state, then we write L = L (M) and we say that M recognises the language L. Thus, an input $w \notin$ L (M) when M fail to halt on input w or when M halts on w but not in the accepting state.

Remark A language L is recognisable by a Turning machine if and only if L is a Type 0 language.

Example 49 Let $\Sigma = \{0, 1\}$, and language L consists of all strings beginning with one or more a's and followed by one or more b's, i.e. $L = \{a^m\ b^n : m, n > 0\}$. Find a Turning machine which recognises L.

Solution The following are three cases to visualise M:

(*i*) It moves right over all the a's. Thus, we may have the sequences $\alpha_1 = s_0 s_1$R and $\alpha_2 = s_1 s_1$R.

(*ii*) It moves right over all the b's. Thus, we may have the sequences $\alpha_3 = s_1 1 1 s_2$R and $\alpha_4 = s_2 1 1 s_2$R.

(*iii*) It halts in the final state s_y when moves to blank symbol b. Thus, we have the sequence $\alpha_5 = s_2 00 s y_R$.

Since M does not accept an input string (word) which does not belong to L, therefore we need following sequences

$\alpha_6 = s_0bbs_nR$, $\alpha_7 = s_011s_nR$, $\alpha_8 = s_1bbs_nR$ and $\alpha_9 = s_200s_nR$. Observe that α_6 is used when $w = \lambda = b$(empty word), α_7 is used when w is an expression beginning with 1; α_8 is used when w contains only 0's and α_9 is used when w contains a symbol 0 following a symbol 1.

Self-practice Problems III

39. Construct the state transition diagram for the finite state machine based on the following state tables:

(*a*)

State	*f*		*g*	
	0	1	0	1
s_0	s_1	s_0	1	0
s_1	s_2	s_1	0	1
s_2	s_3	s_1	1	1
s_3	s_2	s_1	0	0

(*b*)

State	*f*		*g*	
	0	1	0	1
s_0	s_0	s_4	1	1
s_1	s_0	s_3	0	1
s_2	s_0	s_2	0	0
s_3	s_1	s_2	1	1
s_4	s_1	s_0	1	0

40. Let M be a finite-state machine with the following state table:

State	*f*			*g*		
	a	*b*	*c*	*a*	*b*	*c*
s_0	s_0	s_1	s_2	0	1	0
s_1	s_1	s_1	s_0	1	1	1
s_2	s_2	s_1	s_0	1	0	0

(*a*) Draw the state-diagram.

(*b*) Find the output string for the input string *aabbcc*.

41. Construct a state table corresponding to the state-diagram shown below.

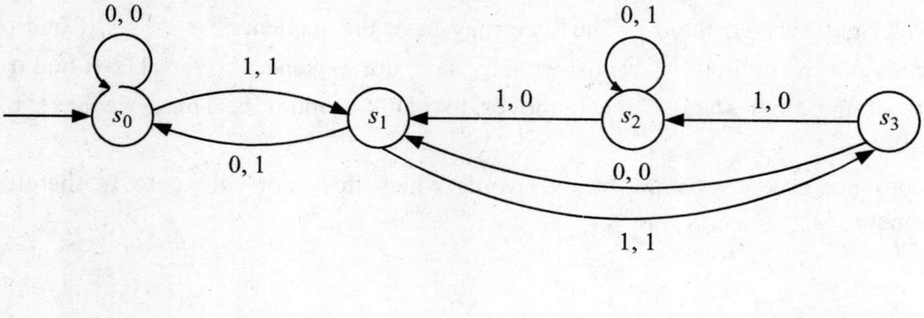

42. Design a finite-state machine having an output of 1 exactly when the input sequence ends with the digits 101.

43. Construct a finite-state machine which will accept all words over the input set $\Sigma = \{a, b\}$, where number of b's are divisible by 3.

44. Construct a finite-state machine which will accept all words over the input set $\Sigma = \{a, b\}$ which ends in two b's.

45. Construct an automaton that excepts the language
(a) $L(M) = \{a^n b^m : m, n \geq 0\}$ and (b) $L(M) = \{b^m a b^n : m, n \geq 0\}$

46. Given *dfa* $M = (\{a, b\}, \{s_0, s_1, s_2\}, f, s_0, \{s_2\})$ where

$f(s_0, a) = \{s_0, s_1\};\ \ f(s_0, b) = s_2$

$f(s_1, a) = s_0;\ \ f(s_1, b) = s_1$

$f(s_2, b) = \{s_0, s_1\}$

Construct *dfa* M′ equivalent to M.

47. Given *dfa* $M = (\{0, 1\}, \{s_0, s_1, s_2\}, f, s_0, \{s_1\})$, where

$f(s_0, 0) = s_0,\ \ f(s_0, 1) = s_1$

$f_1(s_1, 0) = s_0,\ \ f(s_1, 1) = s_2$

$f(s_2, 0) = s_2,\ \ f(s_2, 1) = s_1$

Construct the transition diagram of M and check whether it accepts string 01 or not.

48. Construct a deterministic finite-state automaton with the following conditions:
(a) Accepts all strings on the input set $\Sigma = \{0, 1\}$ except those containing the substring 001.
(b) Accpets all strings on the input set $\Sigma = \{0, 1\}$, start and end with the same symbol.

Hints and Answers

39. (a)

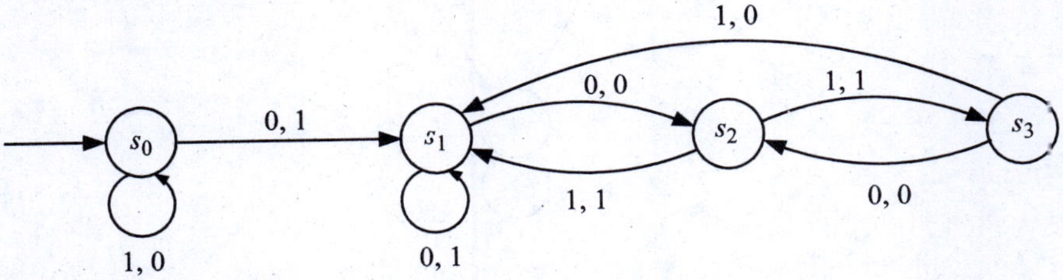

(b)

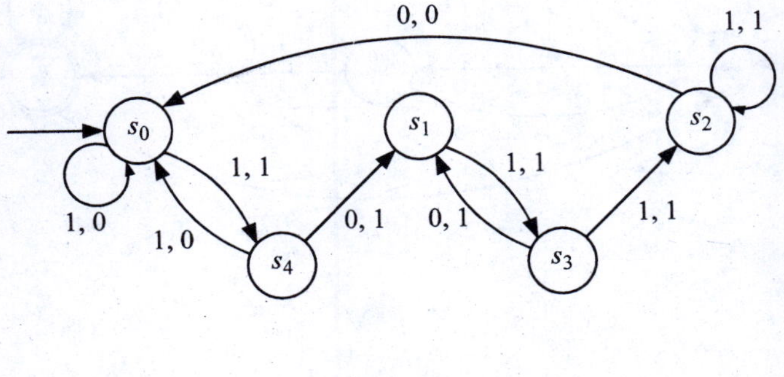

40. (*a*)

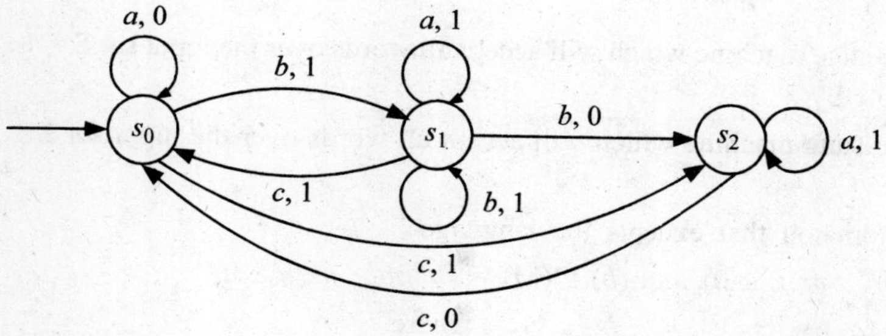

(*b*) $\longrightarrow s_0 \xrightarrow{0} s_0 \xrightarrow{0} s_0 \xrightarrow{1} s_1 \xrightarrow{1} s_1 \xrightarrow{1} s_1 \xrightarrow{0} s_2$

This yields the output string: 001110.

41.

State	f		g	
	0	1	0	1
s_0	s_1	s_0	1	0
s_1	s_3	s_0	1	1
s_2	s_1	s_2	0	1
s_3	s_2	s_1	0	0

43.

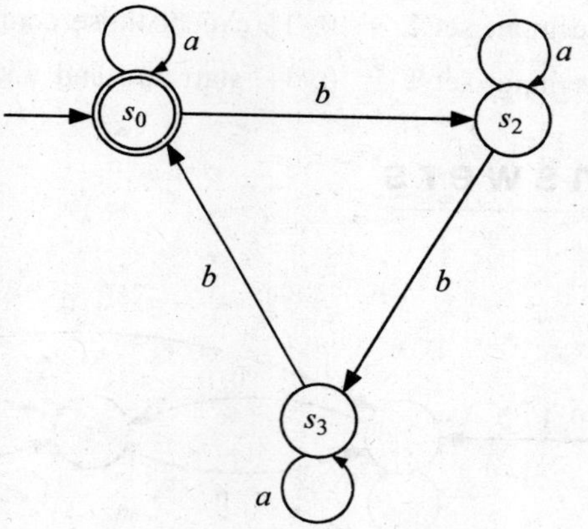

44.

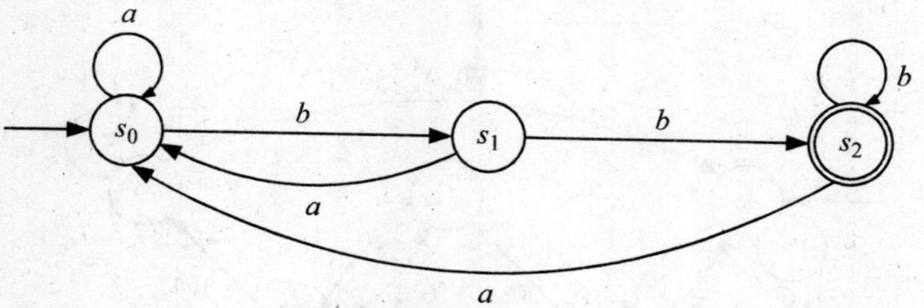

45. (*a*)

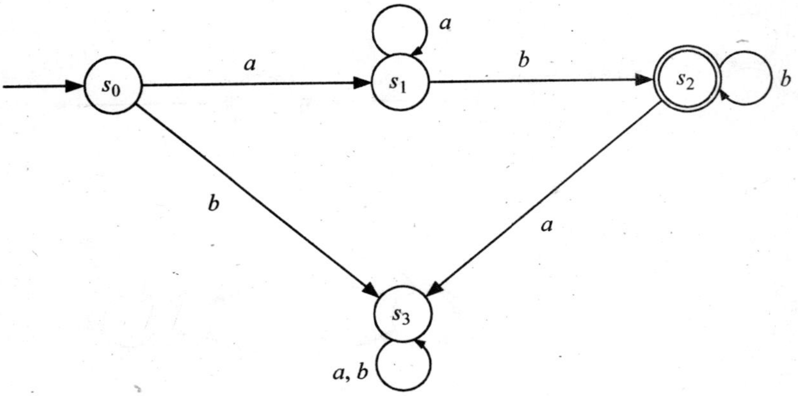

(*b*)

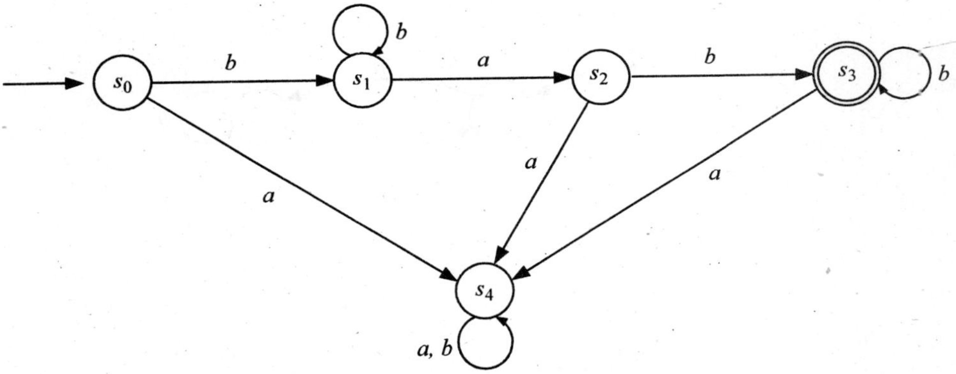

46. The *dfa* M′ equivalent to M is defined as: M′ = ({a, b}, 2^S, f, F′, [s_0])

where F = {[s_2], [s_0, s_2], [s_1, s_2], [s_0, s_1, s_2]). The state table of M′ is given below:

States	Input	
	a	*b*
[s_0]	[s_0, s_1]	[s_2]
[s_2]	ϕ	[s_0, s_1]
[s_0, s_1]	[s_0, s_1]	[s_1, s_2]
[s_1, s_2]	[s_0]	[s_0, s_1]

47.

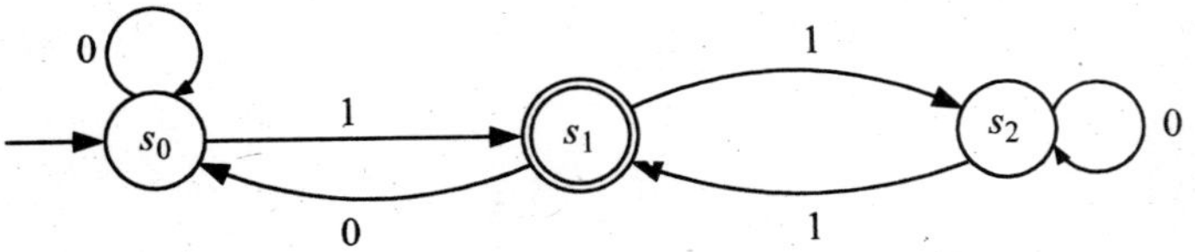

48. (*a*)

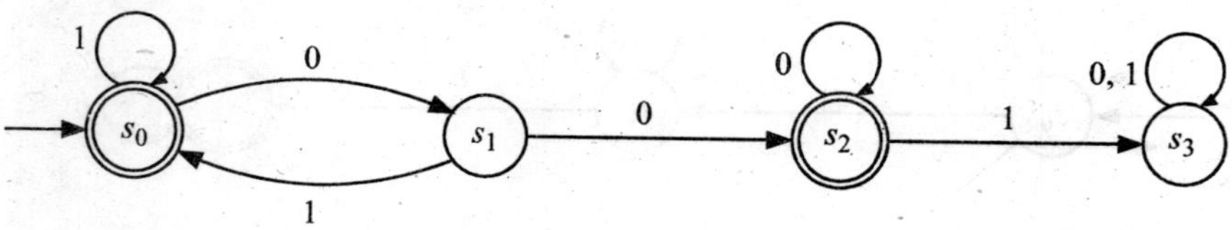

(*b*)

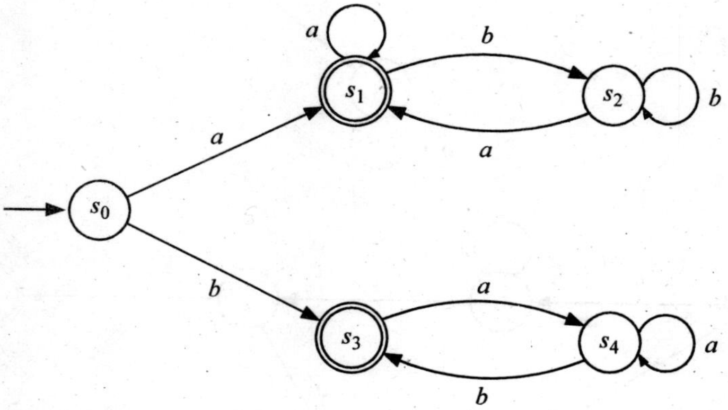

References

Alan Doerr and Kenneth Levasseur, *Applied Discrete Structures For Computer Science*, Galgotia Publications, New Delhi, 2001.

Aho, Alfred V., and Jeffrey D. Vllman, *Principle of Compiler Design,* Addison-Wesley, Reading Mass., 1977.

Bala Chandran, V.K., *Theory and Problems of Combinatories,* Schaum's Outline Series, McGraw-Hill, Singapore, 1995.

Ben-Ari, M., *Principles of Concurrent Programming* , Prentice-Hall, Englewood Cliffs, NJ, 1982.

Berge, C., *The Theory of Graphs and Its Applications*, New York, Wiley, 1962.

Deo, N., *Graph Theory with Application to Engineering and Computer Science,* Prentice-Hall of India, New Delhi, 2000.

Denning, P.J., J.B. Dennis and J.L. Quatitz, *Machines, Languages and Computation*, Prentice-Hall, Englewood Cliffs NJ, 1978.

Dornhoff, L.L. and F.E. Horn, *Applied Modern Algebra*, Macmillan, New York, 1978.

Ford, R.L. and D.R. Fulkerson, *Flows in Networks*, Princeton University Press, NJ, 1962.

Fraleigh, John B., *A First Course in Abstract Algebra (3rd Edn.)*, Addison-Wesley, Reading, Mass., 1982.

Gersting, J.L., *Mathematical Structures for Computer Science,* W.H. Freeman, San Francisco, 1982.

Gourdin, A. and M. Bounahrat, *Applied Numerical Methods*, Prentice-Hall of India, New Delhi, 1996.

Gupta, A.K. and S.K. Sarkar, *Mathematics of Computing*, Wheeler Publishing, New Delhi, 1999.

Hill, F.J. and G.R. Peterson, *Switching Theory and Logical Design* (*2nd Edn.*), Wiley, NY, 1974.

Hohn, F.E., *Applied Boolean Algebra* (*2nd Edn.*), Macmillan, NY, 1966.

Hopcroft, J.E. and J.D. Ullman, *Formal Languages and Their Relation to Automata,* Addition-Wesley, Reading, MA, 1969.

Hopcroft, J.E. and J.D. Ullman, *Introduction to Automata Theory, Languages and Computation*, Narosa Publishing House, New Delhi, 1998.

Kolman, B., R.C. Busby, and S. Ross, *Discrete Mathematical Structures* (*3rd Edn.*), Prentice-Hall of India, New Delhi, 2002.

Lipschutz, S., *Discrete Mathematics (2nd Edn.), Schaum's Easy Outlines*, Tata McGraw-Hill, New Delhi, 1997.

Lipson, J.D. *Elements of Algebra and Algebraic Computing*, Addition-Wesley, Reading, MA, 1981.

Liu, C.L., *Elements of Discrete Mathematics*, Tata McGraw-Hill, New Delhi, 2001.

Mathews, J.H., *Numerical Methods for Mathematics, Science and Engineering* (*2nd Edn.*), Prentice-Hall of India, New Delhi, 1994.

Mondelson, E., *Boolean Algebra and Switching Circuits,* Schaum's Outline Series, McGraw-Hill, NY, 1973.

Mishra, K.L.P. and N. Chandrasekaran, *Theory of Computer Science* (*2nd Edn.*), Prentice-Hall of India, New Delhi, 2002.

Rajaraman, V., *Computer-Oriented Numerical Methods*, Prentice-Hall of India, New Delhi, 1997.

Prather, R.E., *Discrete Mathematics of Structures for Computer Science,* Houghton Mifflin, 1976.

Rogers, D.F. and J.A., Adams, *Mathematics for Computer Graphics,* McGraw-Hill, Singapore, 1989.

Sahni, S., *Concepts in Discrete Mathematics,* Frindley, MN, Camelor, 1981.

Tucker, A., *Applied Combinatories*, John Wiley & Sons, NY, 1995.

Verendra Kumar, *Digital Technology: Principles and Practice,* New Age International, New Delhi, 1997.

Wand, M., *Induction, Recursion, and Programming,* North-Halland, NY, 1980.

Williams, G., *Computational Linear Algebra with Models* (*2nd Edn.*), Allyn and Bacon, Boston, 1978.

Index